Climate Change Management

Series Editor

Walter Leal Filho, International Climate Change Information and Research Programme, Hamburg University of Applied Sciences, Hamburg, Germany

The aim of this book series is to provide an authoritative source of information on climate change management, with an emphasis on projects, case studies and practical initiatives – all of which may help to address a problem with a global scope, but the impacts of which are mostly local. As the world actively seeks ways to cope with the effects of climate change and global warming, such as floods, droughts, rising sea levels and landscape changes, there is a vital need for reliable information and data to support the efforts pursued by local governments, NGOs and other organizations to address the problems associated with climate change. This series welcomes monographs and contributed volumes written for an academic and professional audience, as well as peer-reviewed conference proceedings. Relevant topics include but are not limited to water conservation, disaster prevention and management, and agriculture, as well as regional studies and documentation of trends. Thanks to its interdisciplinary focus, the series aims to concretely contribute to a better understanding of the state-of-the-art of climate change adaptation, and of the tools with which it can be implemented on the ground.

Notes on the quality assurance and peer review of this publication

Prior to publication, the quality of the works published in this series is double blind reviewed by external referees appointed by the editor. The referees are not aware of the author's name when performing the review; the referees' names are not disclosed.

Uttam Mukhopadhyay · Subhasis Bhattacharya ·
Pradip Chouhan · Suman Paul ·
Indrajit Roy Chowdhury · Uday Chatterjee
Editors

Climate Crisis, Social Responses and Sustainability

Socio-ecological Study on Global Perspectives

Springer

Editors
Uttam Mukhopadhyay
Department of Geography
Vidyasagar College
Kolkata, West Bengal, India

Subhasis Bhattacharya
Department of Economics
Sidho-Kanho-Birsha University
Purulia, West Bengal, India

Pradip Chouhan
Department of Geography
University of Gour Banga
Malda, West Bengal, India

Suman Paul
Department of Geography
Sidho-Kanho-Birsha University
Purulia, West Bengal, India

Indrajit Roy Chowdhury
Department of Geography and Applied Geography
University of North Bengal
Darjeeling, West Bengal, India

Uday Chatterjee
Department of Geography
Bhatter College
Dantan, West Bengal, India

ISSN 1610-2002 ISSN 1610-2010 (electronic)
Climate Change Management
ISBN 978-3-031-58260-8 ISBN 978-3-031-58261-5 (eBook)
https://doi.org/10.1007/978-3-031-58261-5

This Springer imprint is published by the registered company Springer Nature Switzerland AG
The registered company address is: Gewerbestrasse 11, 6330 Cham, Switzerland

Paper in this product is recyclable.

Dedicated to
Young Scholars in the Field of Atmospheric Science, Climatology, Environmental Science, Hazard and Disaster Management, Geography, Social Science and Sustainability Science and Policy Makers

Foreword

The book *Climate Crisis, Social Responses and Sustainability: Socio-ecological Study on Global Perspectives*, edited by Uttam Mukhopadhyay, Subhasis Bhattacharya, Pradip Chouhan, Suman Paul, Indrajit Roy Chowdhury, and Uday Chatterjee, is an important source of knowledge and insight in light of the challenges posed by climate change, inequality, and environmental harm. This book aims to enhance our comprehension of the intricate relationship between human societies and the environment by working collaboratively.

The book is divided into four parts, each offering valuable insights into different aspects of the climate crisis. In Part I, readers are introduced to the primary themes and goals of the book, providing a thorough understanding of the social and ecological issues we face. Part II examines ecological perspectives on the climate crisis, including the impact of rapid urbanization on microclimates, the vulnerability to natural disasters, and the response of agricultural systems to climate change. By conducting thorough research and analyzing data, these chapters reveal the complex connections between human actions and the environment, emphasizing the importance of sustainable practices in environmental management and urban development. In Part III, the book delves into sociological perspectives, exploring how the climate crisis intersects with social justice, health, and migration. These sections explore the psychological impact of climate anxiety and the socioeconomic effects of disease outbreaks. They highlight the significance of implementing comprehensive approaches that prioritize human welfare and fairness in combating climate change. Part IV of the volume scrutinizes sustainable initiatives that aim to address the climate crisis by considering the sustainable development goals (SDGs). These chapters provide practical tips for creating climate-resilient communities and supporting sustainable livelihoods by utilizing indigenous knowledge, community resilience, and new adaptation strategies. As I contemplate the extensive research presented in this volume, I am conscious of the significant obstacles we face in our shared endeavor to address the climate crisis. However, despite these challenges, there is cause for hope. The diversity of perspectives and innovative solutions showcased in this book demonstrate how people can remain resilient, imaginative, and united in the face of adversity.

I commend the editors and contributors for their committed efforts to enhance understanding and encourage discourse on this pressing issue. I sincerely hope that this updated compilation will motivate individuals to continue prioritizing sustainability, equity, and the well-being of our planet, as we strive to construct a fairer and more adaptable world for future generations.

March 2024

Prof. Sudip Mitra
Visiting Professor at Keio University, Japan, Founding Head, Centre for Disaster Management and Research, Professor, School of Agro and Rural Technology
Indian Institute of Technology Guwahati (IITG)
Guwahati, India

Preface

Climate change is apprising deeply since the intervention of human involvement, and it creates several challenges and issues seeking the global crisis which have been assumed the reciprocal impact on society as well as on the ecology in the global perspective. Issue of climate change turns over the communities into more vulnerable and they are facing several disproportionate challenges in terms of extreme weather events such as extremity of flood, unprecedented impact on ecosystem and human well-being too, incidence of natural as well as anthropogenic disaster and several mountable effect such as climate crisis and agricultural responses, interlinking relation between climate change and urban environmental issues, most importantly several discussion seeking to incorporate in this proposed book about the transformation of ecological adaptation to climate crisis for sustainable development in connection with SDGs. Moreover, society and community are some of the underlying platforms which might be responded positively due to this climate crisis issue, it brings the impact on human health and well-being, several impacts must be highlighted on the mental as well as physical health, and it leads to food security issue due to inadaptability case of cropping pattern and diversification. Several studies highlighted the burning issues like social stress which leads to migration in the global perspective. Therefore, issues of climate crisis outbreak the poverty and drought as the human adaptability are not in a position to balance to resolve the environmental crisis, which leads to search alternative livelihood options to sustain in the society. Climate change also creates extreme phenomenon like outbreak of diseases which leads to occurrence of epidemic and pandemic diseases globally, crisis of climate also decline the patches of green space and it creates a reciprocal relationship with the changing phenomenon of climate change and ecosystem services. The report of Intergovernmental Panel on Climate Change (IPCC) has noted the fact that resolving the climate crisis issue that brings the principles of procedural and meaningful justice which leads to sustainable development outcomes, in this connection measures and policies should be expanded for the beneficiaries of the society as well as communities too. In addition to changeable phenomenon of climate crisis, addressing the climate resilience action plan and disaster risk reduction (DRR) statue must be implemented in connection with community participation, collaborative programs with the administration, meaningful innovations, and technical assistance to resolve the climatic crisis must be judgmental discussion in the present-day context. In connection to implement the mitigating plan, long-term impact assessment with sustainable solutions touching the SDGs is more relevant to discussion which would create the backbone of the proposed theme of the book. Therefore, this kind of comprehensive theme has not been discussed yet, which is more meaningful to bring the policy measures with community outlook.

Climate crisis leads to several socio-ecological issues which need to be discussed with some empirical case studies from the contextual global evidences. Climatic crisis generates several social responses which are associated with mitigating issues in addition to sustainable development goals. Under these circumstances, several loopholes interlinked with climatic crisis need to be exposed in the present-day context. The books cover climate change holistically, including global and regional scales, ecosystems, agriculture, energy, and sustainability. This book also addresses the variety of challenges associated with climate change, along with possible solutions. Climate Change, Community Response and Resilience focuses on recent climate change research to help researchers and graduate students in natural resource management, ecological vulnerability, landscape ecology, social engineering, environmental science and management, and sustainability get up to speed on the science of climate change so far and establish a baseline for how to effectively move into the future. This book explores the effects of climate extremes as well as disaster risk reduction measures. This book also assesses climate change policies and initiatives, as well as provides insights from independent assessments of the major international organizations that promote climate action in developing nations. It will appeal to students and instructors at both undergraduate and graduate levels, especially in programs that promote multi-disciplinary and trans-disciplinary study of contemporary complex issues

Additionally, this book aims to provide an interdisciplinary understanding of the issues mainly from the lenses of geography, economics, and sociology as well as environmental studies too. Given the focus of this study, it is believed that an approach that harmonizes the cognitive domain from different discipline is appropriated. A combination of chapters using qualitative as well as quantitative methods also made this book exclusive from others. We believe that this edited volume will surely contribute the knowledge domain with some relevant chapters' discussion in the contemporary time and lead to reduce the gap of knowledge.

Kolkata, West Bengal, India — Uttam Mukhopadhyay
Purulia, West Bengal, India — Subhasis Bhattacharya
Malda, West Bengal, India — Pradip Chouhan
Purulia, West Bengal, India — Suman Paul
Darjeeling, West Bengal, India — Indrajit Roy Chowdhury
Dantan, West Bengal, India — Uday Chatterjee

Acknowledgments

We extend our heartfelt gratitude to the individuals who played a crucial role in the conception and publication of *Climate Crisis, Social Responses and Sustainability: Socio-ecological Study on Global Perspectives* book. Our appreciation goes out to the contributors, colleagues, reviewers, and researchers whose expertise, support, guidance, and dedication significantly contributed to the development of this comprehensive book, addressing the current state, challenges, and sustainable solutions of climate change. Their expertise, unwavering support, guidance, and dedication, groundbreaking research, sharing relevant examples, and motivating positive change have left an indelible mark on the book. We also express our gratitude to our families for their unwavering support and to the readers for their commitment to sustainability. Together, we can make a meaningful impact in addressing the climate crisis and shaping a more sustainable future. A special acknowledgment goes to the publishing team at Springer for their guidance, expertise, and steadfast commitment to excellence. Their meticulous editing, design, and production efforts have transformed the book into a thorough and easily accessible reader resource.

Kolkata, West Bengal, India — Uttam Mukhopadhyay
Purulia, West Bengal, India — Subhasis Bhattacharya
Malda, West Bengal, India — Pradip Chouhan
Purulia, West Bengal, India — Suman Paul
Darjeeling, West Bengal, India — Indrajit Roy Chowdhury
Dantan, West Bengal, India — Uday Chatterjee

Contents

Editors and Contributors

About the Editors

Dr. Uttam Mukhopadhyay is Associate Professor and Head of the Department of Geography (undergraduate and postgraduate), at Vidyasagar College, University of Calcutta. He is also Director of the Research Centre of Vidyasagar College, Kolkata. He has been associated with college teaching for the last 37 years. He has taught at City College of Commerce, Gokhale Memorial Girls' College, and Presidency College, and at present, he is teaching at Vidyasagar College where he has completed 26 successful years. Before venturing into college teaching, he initially started his career as Planner in the Planning Department of KMDA. Dr. Mukhopadhyay has simultaneously contributed largely to the social field and is associated with a great deal of charity work. His writing and views as Social Expert on various aspects of life, society, and education are highly valued and published in different newspapers, leading TV channels, articles, and journals from time to time. He is also associated with a lot of research work where his research papers get accepted in various national and international journals. He has successfully completed several prestigious government-funded projects from various social and economic perspectives. The majority of his work involves tribal society, environment, urban planning, and resource potentiality. Apart from all this, he has authored numerous textbooks at the school and college levels.

Dr. Subhasis Bhattacharya completed his Masters in Economics from Calcutta University and Doctoral degree from University of North Bengal. He has 23 years of academic experience comprised of teaching learning and research activities mainly in Cooch Behar College, University of Gour Banga, and presently in Sidho-Kanho-Birsha University. The research work of Dr. Bhattacharya mainly deals with health economics, environmental economics, development economics, resource economics, and financial economics. Specially multivariate analysis, multinomial distribution, data envelopment analysis, and stochastic frontier analysis are the preferred fields of econometrics of Dr. Bhattacharya.

Dr. Pradip Chouhan is Professor at the Department of Geography at University of Gour Banga, Malda, West Bengal, India. Earlier he was Assistant Professor at the Department of Geography, Jadavpur University, Kolkata. His areas of research interest are fertility behavior, public health, maternal health, and child health. Actively engaged in teaching and research in population geography for nearly two decades, Prof. Chouhan has published more than 65 research papers in Scopus and Web of Science-indexed journals, edited 3 books, and authored 1 book. He is a reviewer of Scopus and Web of Science-indexed journals. He has completed a major research project funded by the Indian Council of Social Science Research, Ministry of Human Resource, Government of India. Dr. Chouhan has also completed one minor research project funded by the University Grants Commission. He has successfully supervised 4 Ph.D. and 8 M.Phil. scholars.

Dr. Suman Paul is Professor of Geography at Sidho-Kanho-Birsha University, Purulia, West Bengal, India. He received his B.Sc. and M.Sc. degrees from the University of Calcutta, India, in 2000 and 2002, respectively. He achieved National Scholarship and stood 1st position (1st Class FIRST) at undergraduate level at Calcutta University. His areas of specialization are urban geography, environmental hazards and disaster analysis, GIS, and remote sensing. He completed his Ph.D. on Urban Geography from the Department of Geography, University of Calcutta, West Bengal, India, entitled Emerging Urban Centres—Status and Possibilities: A Case Study of the District of North 24 Parganas, West Bengal. He was associated as Assistant Professor at Post Graduate Department of Geography with more than nine years of experience at Krishnagar Government College (Nadia, WB) and Acharya B. N. Seal Government College (Cooch Behar, WB). He has presented several research papers in international and national, and state-level seminars and conferences. He has published a number of articles in reputed peer-reviewed international, national journals and in edited books. He had completed two Major Research Projects funded by Department of Planning and Statistics (WB) and Department of Science and Technology and Biotechnology (WB) and one ongoing Research Program funded by ICSSR, India.

Dr. Indrajit Roy Chowdhury started his academic work in the field of urban and regional planning, transport geography, GIS, and remote sensing. He has more than 50 research publications in high-quality peer-reviewed journals in the topics covering from urban environmental and infrastructural issues, regional planning, women empowerment, and gender-related issues on in the various reputed international and national journals. His research has been funded by the Department of Science and Technology, Government of India and Government of West Bengal. Dr. Roy Chowdhury has recently completed a project on the 'Problem and Prospect and Opportunities of Homestay Tourism in Sub-Himalayan North Bengal' funded by the University of North Bengal. Dr. Roy Chowdhury has been funded by ICSSR, New Delhi, for Major Research Project, entitled 'Understanding Quality of Life (QoL) of the Tribe in the hilly region of Darjiling and Kalimong District, West Bengal, India: A Future Road Map Towards Sustainable Development.' He has served as a reviewer for many International journals. Dr. Roy Chowdhury did his graduation in Geography Honours from Asutosh College, Kolkata, in 2005 and did his Masters in Geography from University of Calcutta in 2007. He was awarded Ph.D. degree by the University of Calcutta in 2015. His area of research in Ph.D. was 'Urban Transport and Vehicular Emissions in Kolkata city' (http://hdl.handle.net/10603/163994). He is Assistant Professor in the Department of Geography and Applied Geography, University of North Bengal, West Bengal, India. A number of research scholars are also carrying out M.Phil. and Ph.D. research work under his supervision.

Dr. Uday Chatterjee is Assistant Professor at the Department of Geography, Bhatter College, Dantan, Paschim Medinipur, West Bengal, India, and Applied Geographer with a Doctoral Degree in Applied Geography at Ravenshaw University, Cuttack, Odisha, India. His areas of research interest cover urban planning, social and human geography, applied geomorphology, hazards and disasters, environmental issues, disaster governance, community-based disaster risk management, climate change adaptation, urban risk management, and disaster. He has delivered ten invited lectures in University Grants Commissions (UGC) sponsored national seminars and various academic departments of different colleges in India. In addition, he presented 20 papers in national and international seminars/conferences held in India as well as chaired and co-chaired more than five technical sessions. He has successfully guided project dissertations to undergraduate students. He has also conducted (Convener) one Faculty Development Program on 'Modern methods of teaching and advanced research methods' sponsored by the Indian Council of Social Science Research (ICSSR), Government of India. Currently, Dr. Uday Chatterjee has completed the Special Issue (S.I.) of Urbanism, Smart Cities and Modelling, *Geojournal*, Springer as Lead Editor, book series editor Development in Environmental Science, Elsevier (https://www.elsevier.com/books-and-journals/book-series/developments-in-environmental-science). His research work has been funded by the West Bengal Pollution Control Board (WBPCB) Government of West Bengal, India. He has served as a reviewer for many International journals. Currently, Dr. Chatterjee doing the international project, in collaboration with Indonesia, Malaysia, and Japan funded by Asia Pacific-Global Change Research (APN), and he is the guest editor of the special issue Social Ecology, Human-Well-being and Sustainability, Global Social Welfare journal, Springer. He has published 30 research papers, 10 edited books, 2 author books, 15 book chapters, and 2 conference proceedings.

Contributors

Ansar Abbas Faculty of Economics and Business, Surabaya, Indonesia; Hamdard University, Islamabad, Pakistan

S. M. Akram Ullah Department of Political Science, University of Rajshahi, Rajshahi, Bangladesh

Dodi Al Vayed Master of Government and Administrative Affairs, Jusuf Kalla School of Government, Muhammadiyah University Yogyakarta, Yogyakarta, Indonesia

Iwan Setiajie Anugrah National Research and Innovation Agency (BRIN), Jakarta, Indonesia

Shoukat Ara Division of Environmental Sciences, SKUAST-K, Srinagar, India

Iqra Binti Ayoub Division of Environmental Sciences, SKUAST-K, Srinagar, India

Dipanjan Bagchi Department of Psychology, CHRIST (Deemed to be University) Delhi NCR, Ghaziabad, India

Emon Bagchi Department of Geography, Heramba Chandra College, Kolkata, West Bengal, India

Debangshu Banerjee Department of Zoology, Tarakeswar Degree College, Hooghly, WB, India

Jhoney Barui Department of Geography, Vidyasagar College, University of Calcutta, Kolkata, India

Tirthankar Basu Research Scholar, Independent, Malda, West Bengal, India

Subhasis Bhattacharya Department of Economics, Sidho-Kanho-Birsha University, Purulia, WB, India

Arghadeep Bose Department of Geography and Applied Geography, University of North Bengal, Siliguri, West Bengal, India

Debabrata Chanda Department of Geography, Asutosh College, Kolkata, India

Ram Chandra Centre for Rural Development and Technology, Indian Institute of Technology Delhi, New Delhi, India

Pradip Chouhan Department of Geography, University of Gour Banga, Malda, West Bengal, India

Indrajit Roy Chowdhury Department of Geography and Applied Geography, University of North Bengal, Siliguri, West Bengal, India

Arijit Das Department of Geography, University of Gour Banga, Malda, West Bengal, India

Chandan Surabhi Das Department of Geography, Barasat Government College, Barasat, West Bengal, India

Semanti Das Department of Geography, Chandrakona Vidyasagar Mahavidyalaya, Paschim Medinipur, West Bengal, India

Premjeet Das Gupta Department of Urban and Regional Planning, School of Planning and Architecture, Bhopal, Bhopal, Madhya Pradesh, India

Sarthak Dash School of Humanities, KIIT Deemed to be University, Bhubaneshwar, India

Aju David Raj Department of Environmental Science, School of Earth Science Systems, Central University of Kerala, Kasaragod, India

Anu David Raj Forest Research Institute, Deemed to be University, Dehradun, India

Saptashish Deb Centre for Rural Development and Technology, Indian Institute of Technology Delhi, New Delhi, India

Tinku Casper D'Silva Kerala State Council for Science, Technology and Environment—National Transportation Planning and Research Centre (KSCSTE—NATPAC), Government of Kerala, Thiruvananthapuram, Kerala, India; Centre for Rural Development and Technology, Indian Institute of Technology Delhi, New Delhi, India

Yogia Dutta Department of Geography, Haldia Government College, East Midnapur, India

Munyaradzi A. Dzvimbo Department of Geography, University of Free State, Bloemfontein, South Africa

Dian Ekowati Faculty of Economics and Business, and Head of Planning, Surabaya, Indonesia

Rudy G. Erwinsyah Research Center for Social Welfare Village and Connectivity, National Research and Innovation Agency, Jakarta, Indonesia

Siti Fatimah Research Center for Social Welfare, Village and Connectivity. National Research and Innovation Agency (BRIN), Jakarta, Indonesia

Rakotoarisoa Maminirina Fenitra Faculty of Economics and Business, Universitas Indonesia, Depok, Indonesia

Rutwik Gandhe The Bhopal School of Social Sciences, Habibganj, Bhopal, Madhya Pradesh, India

Sanatan Ghosh Department of Geography, Krishnath College, Berhampore (University of Kalyani), Murshidabad, West Bengal, India

Habibullah Research Center for Social Welfare, Village and Connectivity, National Research and Innovation Agency (BRIN), Jakarta, Indonesia

Fatwa Nurul Hakim Research Center for Social Welfare, Village and Connectivity. National Research and Innovation Agency (BRIN), Jakarta, Indonesia

Imran Hossain Department of Political Science, Varendra University, Rajshahi, Bangladesh;
Department of Political Science, University of Rajshahi, Rajshahi, Bangladesh

M. Alie Humaedi Research Center for Social Welfare Village and Connectivity, National Research and Innovation Agency, Jakarta, Indonesia

Dian Karinawati Imron Research Center for Social Welfare Village and Connectivity, National Research and Innovation Agency, Jakarta, Indonesia

Awani Irewati PRP-BRIN (National Research and Innovation Agency), Jakarta City, Indonesia

Sheeba Joseph Department of Social Work, The Bhopal School of Social Sciences, Habibganj, Bhopal, Madhya Pradesh, India

Ajinkya Kanitkar RV College of Architecture, Bengaluru, Karnataka, India

Nanigopal Kapasia Department of Geography, Malda College, Malda, West Bengal, India

Jhikmik Kar Department of Geography, University of Calcutta, Kolkata, West Bengal, India

Prosenjit Kayal Department of Geography and Applied Geography, University of North Bengal, Siliguri, West Bengal, India

Tanuja Khan The Bhopal School of Social Sciences, Habibganj, Bhopal, Madhya Pradesh, India

Suhail A. Lone Department of Earth Sciences, University of Kashmir, Srinagar, India

A. K. M. Mahmudul Haque Department of Political Science, University of Rajshahi, Rajshahi, Bangladesh

Suranjan Majumder Department of Geography and Applied Geography, University of North Bengal, Siliguri, West Bengal, India

Pintu Mandal Department of Geography, Sidho-Kanho-Birsha University, Purulia, WB, India

Bernadin Dwi Mardiatmi Faculty of Economics and Business, Universitas Pembangunan Nasional Veteran Jakarta, South Jakarta, Indonesia

Admire Mare Department of Communication and Media, University of Johannesburg, Johannesburg, Gauteng, South Africa

Biplob Kumar Modak Department of Zoology, Sidho-Kanho-Birsha University, Purulia, WB, India;
Department of Zoology, Sidho-Kanho-Birsha University, Purulia, West Bengal, India

Chumki Mondal Department of Geography, Khandra College, Khandra (Kazi Nazrul University), Paschim Bardhaman, West Bengal, India

Subhajit Mondal Department of Geography, University of Calcutta, Kolkata, West Bengal, India

Doreen Doreen Mrimi Department of Biology, College of Natural and Mathematical Sciences, University of Dodoma, Dodoma, Tanzania

Adrika Mukhopadhyay Department of Geography, Calcutta University, Kolkatta, WB, India

Uttam Mukhopadhyay Department of Geography, Vidyasagar College, University of Calcutta, Kolkata, West Bengal, India

Susanta Nath Department of Zoology, Government General Degree College Singur, Hooghly, WB, India

Joram Ndlovu School of Social Sciences, Howard College, University of KwaZulu Natal, Durban, South Africa

Mduduzi Ndlovu Department of Development Studies, Lupane State University, Lupane, Zimbabwe

Kelvin Ngongolo Department of Biology, College of Natural and Mathematical Sciences, University of Dodoma, Dodoma, Tanzania

Douglas Nyathi School of Public Management, Governance and Public Policy, University of Johannesburg, Johannesburg, Gauteng, South Africa

R. Padmapriya Department of Environmental Science, School of Earth Sciences, Central University of Rajasthan, Ajmer, India

Lumgailu Panmei Department of Forestry and Biodiversity, Tripura University, Suryamaninagar, Agartala, India

Upasana Patgiri School of Planning and Architecture, Bhopal, Bhopal, Madhya Pradesh, India

Suman Paul Department of Geography, Sidho-Kanho-Birsha University, Purulia, WB, India

Reza Amarta Prayoga Research Center for Social Welfare, Village and Connectivity, National Research and Innovation Agency (BRIN), Jakarta, Indonesia

Sugyanta Priyadarshini School of Humanities, KIIT Deemed to be University, Bhubaneshwar, India

Lis Purbandini Research Center for Social Welfare, Village and Connectivity. National Research and Innovation Agency (BRIN), Jakarta, Indonesia

Subrata Purkayastha Department of Geography, North Eastern Hill University, Shillong, India

Margubur Rahaman Department of Migration and Urban Studies, International Institute for Population Sciences (IIPS), Mumbai, India

Hiranmay Rishi Department of Geography, North Eastern Hill University, Shillong, India

Risyanto Research Center for Social Welfare, Village and Connectivity, National Research and Innovation Agency (BRIN), Jakarta, Indonesia

Avijit Roy Department of Geography, Malda College, Malda, West Bengal, India

Subham Roy Department of Geography and Applied Geography, University of North Bengal, Siliguri, West Bengal, India

Subham Kumar Roy Department of Geography, Tilka Manjhi Bhagalpur University, Bhagalpur, Bihar, India

Jayanta Saha Department of Geography, Sidho-Kanho-Birsha University, Purulia, WB, India

Bhaskar Samanta Department of Geography, Krishnagar Government College, Krishnanagar, WB, India

Shyamal Santra Department of Geography, Bankura Zilla Saradamani Mahila Mahavidyapith, Bankura, WB, India

Rachmini Saparita Research Center for Social Welfare, Village and Connectivity. National Research and Innovation Agency (BRIN), Jakarta, Indonesia

Mainak Sarkar Department of Zoology, Bankura Christian College, Bankura, West Bengal, India;
Department of Zoology, Sidho-Kanho-Birsha University, Purulia, West Bengal, India

Thiru Selvan Department of Forestry and Biodiversity, Tripura University, Suryamaninagar, Agartala, India

Hari Harjanto Setiawan National Research and Innovation Agency (BRIN), Jakarta, Indonesia

Nuzul Solekhah Research Center for Social Welfare, Village and Connectivity. National Research and Innovation Agency (BRIN), Jakarta, Indonesia

Akancha Srivastava Jindal School of Psychology and Counseling, O.P. Jindal Global University, Sonipat, Haryana, India

Fendy Suhariadi Centre for Postgraduate Program, Surabaya, Indonesia

Sukanta Chandra Swain School of Humanities, KIIT Deemed to be University, Bhubaneshwar, India

Imam Syafi'i PRP-BRIN (National Research and Innovation Agency), Jakarta City, Indonesia

Mijak Tampung Makekal United Group, Jambi, Indonesia

Bhawna Tushir Department of Psychology, CHRIST (Deemed to be University) Delhi NCR, Ghaziabad, India

Sunitha Vijayan Kerala State Council for Science, Technology and Environment—National Transportation Planning and Research Centre (KSCSTE—NATPAC), Government of Kerala, Thiruvananthapuram, Kerala, India

Sharaniya Vijitharan Department of Bio-science, Faculty of Applied Science, University of Vavuniya, Vavuniya, Sri Lanka

Eko Wahyono Research Center for Social Welfare, Village and Connectivity (BRIN), National Research and Innovation Agency, Jakarta, Indonesia

Retnaningdyah Weningtyastuti Research Center for Social Welfare, Village and Connectivity, National Research and Innovation Agency (BRIN), Jakarta, Indonesia

Djoko Puguh Wibowo Research Center for Social Welfare, Villages, and Connectivity, National Research and Innovation Agency (BRIN), Jakarta, Indonesia

Sjarief Widjaja Faculty of Marine Affairs and Fisheries Technology, Institut Teknologi Sepuluh Nopember (ITS), Surabaya City, Indonesia

Part I
Introduction

Chapter 1
Climate Crisis Impact on Ecosystem Services and Human Well-Being

Aju David Raj, R. Padmapriya, and Anu David Raj

Abstract Ecosystem services encompass a broad spectrum of benefits to human society, making a substantial contribution to human well-being. Any disturbances in the ecosystem directly affect the services they offer, thus impacting human welfare and endangering the essential requirements for human survival. Therefore, comprehending how climate change affects these services and human society is of paramount importance in addressing the issue. The potential negative outcomes of climate change can lead to substantial global socio-economic and environmental issues. This chapter, therefore, explores the global implications of climate change on ecosystem services and human welfare. A thorough review of existing research will improve our comprehension of the complex relationship between ecosystem services, climate change, and their influence on human well-being. This knowledge will equip decision-makers with essential insights for devising critical measures to mitigate the potential threats.

Introduction

Climate change constitutes an omnipresent and burgeoning worldwide peril to the integrity of biodiversity and the stability of ecosystems (Weiskopf et al. 2020). Human actions, largely due to the emission of greenhouse gases, have unquestionably led to global warming, raising global surface temperatures by 1.1 °C above pre-industrial

A. David Raj (✉)
Department of Environmental Science, School of Earth Science Systems, Central University of Kerala, Kasaragod, India
e-mail: ajudraj77@gmail.com

R. Padmapriya
Department of Environmental Science, School of Earth Sciences, Central University of Rajasthan, Ajmer, India

A. David Raj
Forest Research Institute, Deemed to be University, Dehradun, India

U. Mukhopadhyay et al. (eds.), *Climate Crisis, Social Responses and Sustainability*, Climate Change Management, https://doi.org/10.1007/978-3-031-58261-5_1

levels between 2011 and 2020. The global emission of greenhouse gases has consistently increased from 2010 to 2019, with variations in past and present contributions originating from unsustainable energy practices, changes in land use, lifestyle decisions, consumption, and production patterns across different regions, both internationally and within nations, and at the individual scale (IPCC 2023). Ecosystem Services (ESs) refer to the direct and indirect advantages provided by ecosystems for human well-being and survival. Ecosystem valuation is an approach used to ascribe a financial value to an ecosystem and its vital goods and services, often referred to as Ecosystem Service Value (ESV) (Sannigrahi et al. 2019). Costanza et al. (1997) assessed the present economic worth of 17 ecosystem services across 16 biomes, drawing upon existing research findings and a limited number of original calculations. The estimated total value of the entire biosphere, much of which exists beyond conventional market transactions, falls in a range of US$16–54 trillion annually, with an approximate average of US$33 trillion per year. It's important to note that this estimate is a conservative baseline, given the inherent uncertainties involved. To put this in perspective, the global gross national product is approximately US$18 trillion per year. The cumulative impacts of human-induced climate change, along with other human-related pressures like changes in land use, the spread of invasive species, and the growing global population, will lead to unprecedented consequences for human well-being (Whitmee et al. 2015).

Ecosystem management endeavours aimed at optimizing the yield of a single ecosystem service frequently lead to significant reductions in the delivery of other ecosystem services (Bennett et al. 2009). An inadvertent outcome of human cultivation and alteration of ecosystems has been unanticipated or adverse decreases in various other services provided by ecosystems. Globally, this phenomenon has led to an increase in specific services like food and timber production while simultaneously observing a reduction in the majority of other services, including flood control, preservation of genetic resources, and pollination (Millennium Ecosystem Assessment 2005). A comprehensive global assessment of these transformations has disclosed that more than 60% of the services or societal advantages furnished by biotic systems have been eroded due to human actions, with the most substantial depletion occurring within the last half century (Millennium Ecosystem Assessment 2005). Hua et al. (2021) conducted a research study in the Tibetan Plateau of China, wherein they examined the responsiveness of ecosystem services to historical climate variations and projected the possible impact of a global crisis on future ecosystem service vulnerability. Their specific objectives included the assessment of spatial patterns and long-term tendencies in primary ecosystem services, the quantification of how susceptible these services are to climate fluctuations, and an investigation into how ecosystem services may be affected across various prospective climate scenarios. Ecosystem services have the capacity to support both mitigation and adaptation initiatives by extending their contribution (Locatelli 2016). Each year, terrestrial ecosystems capture about 3 billion metric tons of atmospheric carbon through net growth, constituting 30% of human-induced CO_2 emissions (Canadell and Raupach 2008). Effectively governed ecosystems can assist societies in adjusting to present climate risks and forthcoming climate shifts by furnishing a variety of ecosystem services

(Pramova et al. 2012), local climate regulation in agriculture and cities, the preservation of coastal regions through land stabilization and the absorption and dissipation of wave energy, the protection of watersheds through groundwater recharge, and preventing soil erosion (Locatelli 2016).

Climate change is amplifying the intensity of global disasters and extreme climate conditions, exerting particularly disastrous impacts on developing nations and communities with limited resources. Immediate consequences, such as gradual environmental shifts, elevated temperatures, and natural catastrophes, are expected to give rise to subsequent effects, including social and economic pressures, population displacement, and conflicts (Weissbecker 2011). Climate change primarily affects human well-being by disturbing the ecosystem services furnished by nature. The migration of mosquitoes due to global warming poses a health hazard in numerous nations. Malaria, the most widespread mosquito-borne illness, has been a long-standing threat to nearly 50% of the global population, resulting in over 200 million documented cases in 2014 (WHO 2016). Norman Myers, a scholar from Oxford University, provided estimates for the long-term displacement of people directly attributable to climate change. According to him, with the onset of global warming, up to 200 million individuals may experience the adverse effects of disruptions in monsoon systems and other rainfall patterns, prolonged and severe droughts, as well as rising sea levels leading to coastal inundation (Myers 2005). The economic consequences of the climate crisis are multifaceted and wide-ranging. When the global average temperature increases by 2.5 °C, the economic well-being of an average individual would be akin to a decrease in income of about 1.3%, as suggested by the average of 11 impact assessments for this 2.5 °C warming scenario (Tol 2018). These challenges will result in the disturbance of physical, biological, economic, and environmental aspects of global society, consequently affecting human well-being. Existing research demonstrates a significant gap in the literature regarding the comprehensive integration of ecosystem services and climate change, and the multifaceted connections between these vital systems.

This chapter addresses the impact of the climate crisis in a more comprehensive and integrated way. The primary objective of this chapter to deepen understanding of the complex interplay between ecosystem services and climate change, specifically analysing the consequences of this interaction for human well-being. It recognizes the significance of the interconnectedness of ecosystem services and societal welfare. In order to achieve effective solutions, it is imperative to possess a comprehensive understanding of the interconnections and interdependencies among various elements. By considering the association between the ecosystem and human beings, the chapter presents an innovative and collaborative strategy to address the complexities presented by the global climate crisis. The chapter aims to study the distinct ramifications of the climate crisis on human wellbeing. Overall, the chapter's novelty stems from its all-encompassing, interconnected, and multi-faceted approach, which acknowledges the interdependencies among ecosystems, human welfare, and the consequences of climate change and promotes sustainable development. It also delivers a case study of ecosystem services assessment in a Himalayan watershed using the InVEST-SDR model.

Climate Change

Climate change refers to enduring transformations in temperature regimes and meteorological patterns. These alterations may manifest as inherent responses to variations in solar irradiance or consequential volcanic outbursts. Nevertheless, since the 1800s, anthropogenic activities have assumed the paramount role in instigating climate change, mainly originating from the burning of fossil fuels, specifically coal, oil, and natural gas (United Nations 2023). It is commonly recognized that human-induced climate change is poised to inflict deleterious consequences upon numerous individuals, particularly those who are socioeconomically marginalized. Specifically, it is anticipated to culminate in inundations, heightened thermal discomfort, nutritional instability, arid conditions, and augmented susceptibility to maladies transmitted through water and vectors (Caney 2015). Human-induced emissions of greenhouse gases (GHGs) and other radiative forcing agents have resulted in a global average temperature increase of approximately 1.2 °C (Fig. 1.1) by the year 2020 (World Meteorological Organization 2020). The Arctic region is presently experiencing warming in its oceanic and terrestrial aspects, along with an increased frequency of extreme heat events. Similarly, the Antarctic Peninsula and specific regions of West Antarctica have demonstrated significant warming patterns. Forecasts suggest that the Arctic, Antarctic Peninsula, West Antarctica, and certain parts of East Antarctica are anticipated to undergo continued warming at a rate surpassing the global average (IPCC 2021). To mitigate the repercussions stemming from additional warming (IPCC 2019), the signatories of the Paris Agreement reached a consensus to constrain the rise in global temperatures to a level significantly lower than 2 °C, with a dedicated commitment to endeavour towards limiting it to 1.5 °C. To facilitate this goal, a periodic global stocktake process has been established to evaluate the requisite reductions in emissions at intervals of five years (Terhaar et al. 2022).

Climate change represents the quintessential exemplar of externalities on a global scale (Tol 2009). The quantifiable and monetizable consequences of climate change encompass its effects on agriculture, forestry, water resources, coastal areas, energy utilization, air quality, and human well-being (Tol 2008). Air quality exhibits a significant dependence on meteorological conditions, rendering it sensitive to the influence of climate change. Studies employing General Circulation Models (GCMs) in conjunction with Chemical Transport Models (CTMs) indicate that climate change is poised to impact particulate matter (PM) concentrations in regions characterized by pollution by an estimated range of 0.1–1 $\mu g\ m^{-3}$ in the forthcoming decades. Furthermore, the escalating prevalence of wildfires, attributed to climate change, has the capability to emerge as a progressively important source of PM pollution (Jacob and Winner 2009). The study conducted by Marshall and Randhir (2008) reveals that the watershed's hydrological dynamics, including water patterns and fluxes, were observed to undergo notable shifts due to the effects of snowmelt and evaporation. Climate scenarios have brought about alterations in watershed processes that have substantial implications for both the volume and characteristics of water over the study duration. These alterations manifest as reduced annual stream flow, heightened

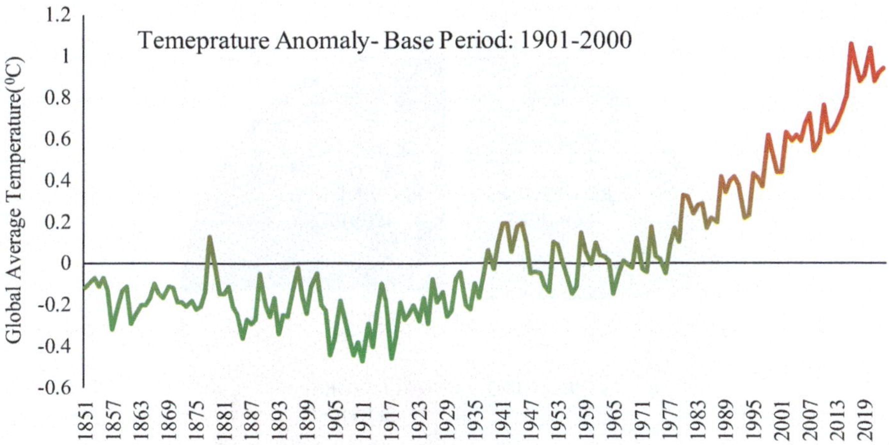

Fig. 1.1 Temporal change in global average temperature. *Source* Temperature anomaly data https://www.ncei.noaa.gov/access/monitoring/climate-at-a-glance/global/time-series

runoff during the winter and spring seasons, and amplified yearly sediment loading. In general, the subtle yet pervasive impact of climate change on our daily existence introduces heightened uncertainties across various dimensions of individuals' lives.

Ecosystem Services

Ecosystem services encompass the array of conditions and mechanisms by which natural ecosystems, in conjunction with the constituent species therein, support and enhance human existence. These services encompass the preservation of biodiversity and the generation of essential ecosystem commodities, including seafood, fodder, lumber, biomass energy sources, organic fibers, numerous pharmaceuticals, industrial materials, and their respective precursors (Daily 1997). One of the most widely employed categorizations for ecosystem services emanates from the seminal Millennium Ecosystem Assessment (MA), a comprehensive endeavour that engaged a consortium of more than 1300 scientists. This classification system delineates ecosystem services into four primary classes: supporting, regulating, provisioning, and cultural services (Millennium Ecosystem Assessment 2005). Provisioning services, encompassing the provision of sustenance and freshwater; regulating services, entailing the management of climatic conditions and disease control; supporting services, relating to the facilitation of nutrient cycling and oxygen generation; and cultural services, encapsulating benefits for spiritual and leisure purposes (Fig. 1.2). While provisioning services, like utilizing resources such as fishing for sustenance and timber for construction, are commonly recognized by the majority, there is notably less awareness regarding regulating and supporting services, such as biological control and habitat for species. Consequently, these critical services are

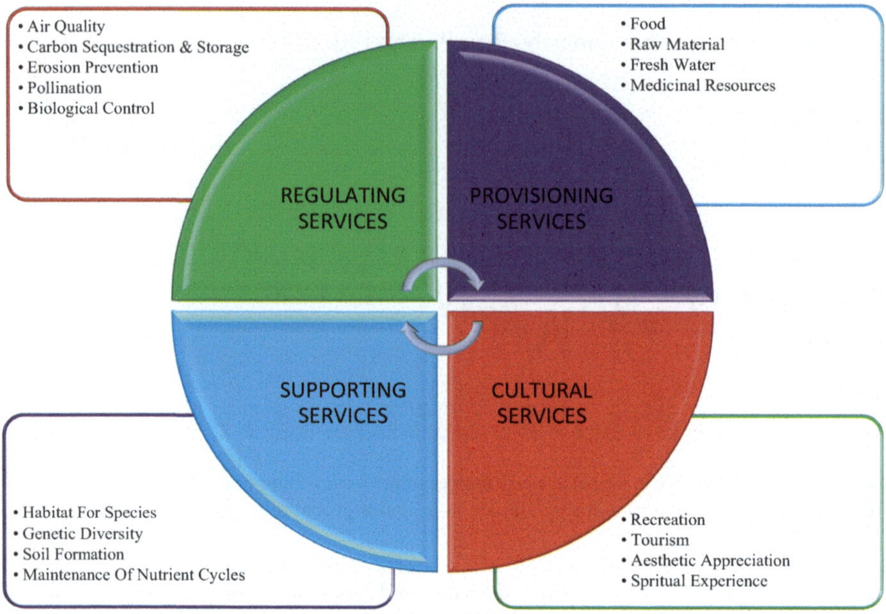

Fig. 1.2 Illustration showing classification of ecosystem services according to millennium assessment. *Source* Millennium Ecosystem Assessment (2005)

often underappreciated or overlooked (Burkhard and Maew 2017). The consequences of climate change on species, populations, and ecosystems have implications for the accessibility and provision of ecosystem services.

Climate Change Impact on Ecosystem Services

Provisioning Services

A provisioning service refers to any form of advantage to human beings that can be derived from the natural environment (The National Wildlife Federation 2023). Ecosystems assume a pivotal function in furnishing resources via provisioning services, constituting the most overt nexus between ecosystem services and human welfare (Edwards-Jones et al. 2011). The delivery of freshwater ecosystem services holds a position of utmost significance and is intricately shaped by the simultaneous impacts of climate change and human interventions, with both these factors exerting profound and interrelated influences on this crucial aspect of our environment (Fu et al. 2017). In ecologically sensitive areas like the Yellow River Basin (YRB) in China, freshwater provisioning services assume a critical role in the management of water resources. These services are geared towards ensuring the reliable and sustainable supply of water to meet the demands of agriculture, while also serving as a protective measure for the preservation of other vital ecosystem functions (Zhang et al. 2022). Malmstrom et al. (2009) employed remote sensing methodologies to assess the alterations in forage provisioning services resulting from restoration initiatives. Their study yielded essential insights for adaptive management, highlighting the considerable resistance to transformation exhibited by invaded grasslands. Furthermore, their findings imply that enhancing the functional diversity of restoration seed mixes could potentially bolster forage availability while mitigating the risk of weed reinvasion.

Agriculture stands as a predominant mode of terrestrial land administration on a global scale, with agroecosystems encompassing approximately 40% of the Earth's terrestrial surface (Power 2010). Nowadays, the agroecosystem confronts significant challenges within the framework of both global warming and extensive human-induced land use change (Deng et al. 2013). The swift urbanization process and climate change have already exerted direct or indirect impacts on the food provisioning services provided by agroecosystems (Chen 2007). Brecka et al. (2018) propose that climate change exerts major negative effects on the provision of timber from boreal forests, yet it may also instigate adjustments in operational practices. Furthermore, climate change seems to benefit pioneer species such as Pinus spp. and Populus spp. while potentially disadvantaging late successional species like Picea spp. and Abies spp. Nevertheless, our understanding of the economic implications of climate change on timber production is still in a nascent stage of development. Biofuels assume a crucial role in the process of reducing carbon emissions in the transportation sector. They offer a low-carbon alternative for current technologies, particularly for light-duty vehicles in the short term, and present valuable options for heavy-duty trucks, ships, and aircraft in the long term, where few cost-effective alternatives are available. Notably, in 2022, the demand for biofuels reached a historic peak of 4.3 exajoules (equivalent to 170,000 million liters), surpassing pre-pandemic levels observed in 2019 (International Energy Agency 2023). Biofuels produced from crops grown through traditional agricultural practices are improbable to mitigate the effects of climate change, and in fact, they may exacerbate the pre existing challenges associated with the management of water resources, the quality of water, and the utilization of land, especially when contrasted with petroleum-based fuels (Delucchi 2010).

Medicinal plants offer numerous advantages to human health, particularly in regions lacking access to Western medical treatments. However, like other species, these plants face potential threats due to alterations in temperature and precipitation patterns, disturbances in mutualistic interactions, heightened occurrences of pests and pathogens, and the adverse effects of human-induced habitat fragmentation, which hinder natural migration patterns (Applequist et al. 2020). Climate change has a profound influence on the vital secondary compounds present in pharmaceutical plants. Increased levels of carbon dioxide, ozone exposure, drought, and cold conditions represent significant abiotic factors responsible for changing environmental conditions, which in turn have direct or indirect impacts on the development, yield, and formation of these secondary compounds in plants. These substances are progressively finding applications in medical therapies as a safer option to address worries regarding the adverse effects and resistance to multiple drugs linked to conventional allopathic treatments (Gupta et al. 2019). Available evidence indicates that these occurrences are affecting medicinal and aromatic plants, raising noteworthy concerns warranting further discussion (Das et al. 2016).

Regulating Services

Ecosystems offer 'regulating services,' such as preserving air and soil quality, managing floods and diseases, and facilitating crop pollination. These services often operate imperceptibly and consequently tend to be undervalued. When these services are compromised, the ensuing consequences can be significant and challenging to remediate (FAO 2023). The state of air quality is intricately linked to meteorological conditions and, as a result, is highly susceptible to the impacts of the global climate crisis. Recent research has sought to quantify this climate-induced effect by examining the associations between air quality and various meteorological factors (Jacob and Winner 2009). A pronounced association between heightened ozone levels and temperature is a consistently observed characteristic in regions with high pollution levels, including areas with predominantly warm climates like the southwestern United States (Wise and Comrie 2005). This correlation typically pertains to situations characterized by pollution, specifically when ozone levels exceed approximately 60 parts per billion (ppb). The observed associations between concentrations of particulate matter (PM) and meteorological factors are less robust compared to those observed for O_3 (Wise and Comrie 2005). Koch et al. (2003) observe an inverse relationship between sulphate levels and cloud cover in Europe when assessed across synoptic timeframes. Their interpretation attributes this association, in part, to the connection between clouds and precipitation and, in part, to a decline in the photochemical oxidation of SO_2. These factors collectively outweigh the influence of clouds in enhancing the aqueous-phase generation of sulphate. so, Actions aimed at mitigating climate change through decreased reliance on fossil fuel combustion frequently yield additional advantages by reducing the immediate health-related consequences of air pollution (Kinney 2008).

The prospect of promoting forest expansion as an approach for the capture and storage of carbon dioxide has garnered significant attention, driven by apprehensions regarding the potential impact of the greenhouse effect on global climate change (Newell and Stavins 2000). Forests, as a dominant land-based ecosystem, encompass roughly 30–43% of the Earth's total land surface, serving as a crucial reservoir for fiber, fuel, and resources essential for human consumption (Dixon et al. 1994). Furthermore, forests offer a range of services, encompassing the provision of wildlife habitats, the purification of water resources, carbon sequestration, and contributions to climate crisis mitigation (Lindquist et al. 2012). The cumulative carbon stored within forest vegetation approximately amounts to 359 billion metric tons (Allen et al. 2010). Gaining insights into the cumulative consequences of these climate alterations, including global warming, increased CO_2 levels, shifts in precipitation patterns, and nitrogen deposition, plays a critical role in predicting carbon behaviour in forest ecosystems (Hui et al. 2017).

Climatic extremes, such as uncommonly intense heatwaves, prolonged droughts, and abnormal or exceptionally intense precipitation, whether through direct or indirect mechanisms, have been associated with the rapid increase in global atmospheric carbon dioxide levels and the subsequent rise in temperatures. These events are anticipated to occur with greater frequency as the climate continues to warm (Drijfhout et al. 2015). Climate change has induced alterations in the occurrence, strength, and geographical spread of extreme events and will persist as a catalyst for further transformations in the future (Bell et al. 2018). Climate change has already yielded significant consequences for marine, freshwater, and terrestrial ecosystems, with a pronounced impact observed in the Arctic region (Constable 2022). Certain planktonic species in polar regions seem to exhibit heightened susceptibility to the consequences resulting from the acidification of the ocean's waters (Thor et al. 2018). Hence, climate change and extreme events are anticipated to have adverse repercussions on the ecological equilibrium of the ecosystem.

Food security, the diversity of available food sources, human nutrition, and food pricing are all significantly dependent on animal pollinators. However, animal pollination is facing significant challenges due to factors such as habitat degradation, unsustainable agricultural methods, and various consequences of climate change (FAO 2023). The capacity of global climate change to disrupt existing conservation and biological control efforts is evident, as it has the potential to unsettle established initiatives aimed at preserving and managing ecosystems and their inhabitants. Climate change possesses the capacity to bring about alterations in abiotic elements such as temperature, levels of precipitation, humidity, and the patterns of wind, which could subsequently influence the timing of life cycles for both predator and prey species, as well as the character and intensity of their interactions. Foreseeing the influence of climate change on predator and prey communities constitutes a significant research endeavour (Schmitz and Barton 2014). A comprehensive comprehension of how climate change influences soil erosion is imperative for devising efficacious strategies for soil preservation and sustainable management (Raj et al. 2022a, b). According to Borrelli et al. (2020), the 2015 baseline model forecasts a global potential soil erosion rate of Pg yr^{-1}, with existing conservation agriculture (CA) operations expected to lower this figure by approximately 5%. Their prospective scenarios indicate that alterations in socio-economic factors affecting land utilization will result in either a reduction (SSP1-RCP2.6–10%) or an escalation (SSP2-RCP4.5 + 2%, SSP5-RCP8.5 + 10) in water erosion by the year 2070. A study conducted by Raj et al. (2022a, b) in the Shivalik region of the Himalayas, employing a calibrated APEX model, revealed that the average annual soil loss may experience a growth of up to 46.94%, 38.80%, 35.34%, and 25.93% under the A2, B2, RCP 4.5, and RCP 8.5 scenarios, respectively, compared to the base period (1985–2014). The vulnerability assessment suggests that a majority of the areas are at varying levels of risk, ranging from slight to highly vulnerable. Furthermore, the study found that forested land cover can mitigate this sensitivity to a certain extent.

Cultural Services

The Millennium Assessment (Millennium Ecosystem Assessment 2005) characterizes cultural services as the intangible advantages that individuals derive from ecosystems. These encompass a wide array of elements such as cultural diversity, spiritual and religious significance, knowledge systems, educational value, inspiration, aesthetic appreciation, social bonds, connection to place, cultural heritage significance, recreational opportunities, and ecotourism experiences. While the significance of cultural services has been consistently acknowledged, in rare cases where additional attention is given, they are frequently described as 'intangible,' 'subjective,' and challenging to measure using biophysical or monetary metrics. The cultural services offered by ecosystems encompass recreational opportunities, both mental and physical well-being, aesthetic appreciation, inspiration for artistic and cultural expressions, tourism, spiritual experiences, and a sense of belonging. Tourism is acknowledged as a sector highly susceptible to the impacts of climate variability. It is also significantly affected by changes in the environment and socioeconomic conditions influenced by climate change. Additionally, it is a sector that is expanding and, in turn, contributing to human-induced climate change (Scott et al. 2012). Frequent natural disasters can lead to a decline in the tourist influx to a particular area. Psychological unease regarding safety levels and available services contributes to a decrease in the flow of tourists (Bateman and Department of Transport Großbritannien 2002). Tourism is intricately connected to the provision of recreational services by ecosystems, meaning that any impact on tourism has a corresponding effect on recreational services. Consequently, this interconnectedness influences the mental and physical well-being of individuals.

The repercussions of climate change on tourists have corresponding effects on the well-being of local communities. Tourism in mountainous areas has developed into a substantial means of sustaining the livelihoods of communities situated in these regions. This has resulted in economic benefits, creating a wide range of employment opportunities, both directly within the tourism sector and indirectly in related industries, such as offering accommodations, animal rentals, handicraft sales, and undertaking roles such as guiding mountain treks, culinary preparation, and transportation services. As an illustration, in 2014, Nepal hosted around 100,000 tourists, primarily drawn to trekking and mountaineering activities, contributing approximately 2% to the nation's GDP (MCTCA 2015). Glacier tourism serves as a significant catalyst for economic growth in various regions. In Norway, for instance, it draws as many as 30,000 tourists annually (Furunes and Mykletun 2012). Numerous glaciers hold sacred or possess profound symbolic significance within their respective local communities (Gagné et al. 2014). As an illustration, the African tribes living in the vicinity of the base of Mount Kilimanjaro hold the belief that the glacier on the mountain serves as a sacred dwelling place (Mölg et al. 2008). Within China, indigenous communities hold the perspective that climate change has interfered with their spiritual domain and longstanding ceremonial practices (Shijin and Dahe 2015).

Supporting Services

Ecosystems inherently rely on the continuity of fundamental natural processes; this encompasses critical processes like photosynthesis, the cycling of nutrients, the creation of soil, and the functioning of the hydrological cycle. These processes are indispensable not only for the sustenance of fundamental life forms but also for the overall health and balance of ecosystems and the well-being of human populations. Devoid of these foundational services, the existence of provisioning, regulating, and cultural services would be untenable (National Wildlife Federation 2023). Climate change exerts significant impacts on river systems, leading to consequential alterations in the suitability of their habitats for aquatic fauna (Inoue and Berg 2017). To date, human-induced climate change has resulted in diminished productivity in ocean ecosystems, disrupted dynamics within food webs, decreased populations of species that shape habitats, changes in the spread of species, and an increased prevalence of diseases. Although specific spatial and temporal nuances remain uncertain, it is evident that climate change is causing fundamental and substantial transformations within ocean ecosystems (Hoegh-Guldberg and Bruno 2010). Modest temperature elevations lead to heightened metabolic rates, which in turn govern various aspects of life history characteristics, population dynamics, and ecosystem functions (O'Connor et al. 2007). However, outside of this specific range, acclimatization becomes ineffective, mortality risks rise, fitness decreases, and populations either decline significantly or face the risk of local extinction (Hochachka and Somero 2002).

Elevated temperatures and shifts in precipitation patterns, both in terms of quantity and frequency, linked to climate change, will modify biogeochemical and hydrological cycles. Climate change projections suggest an increased release of methane from ecosystems worldwide due to rising levels of carbon dioxide (CO_2) and temperature. Nevertheless, the magnitude of this amplifying effect remains uncertain and hinges on variables that influence microbial activity within methane cycling processes across diverse ecosystems. One such variable is the introduction of nitrogen through atmospheric deposition or fertilizer application, This phenomenon can elicit a wide spectrum of impacts on the processes of microbial methane production and its subsequent consumption (Bodelier and Steenbergh 2014). Soil genesis is affected by a multitude of factors, including climatic elements such as temperature and precipitation. These weather-related parameters have a direct effect on soil formation by facilitating the presence of biomass and creating conditions conducive to weathering processes. Sustained and gradual climate warming results in irreversible alterations to the mineral composition of soils. Variations in external components affecting soil formation, such as temperature and precipitation, will trigger modifications in internal factors, encompassing energy, hydrological processes, and biological elements (Pareek 2017). Prolonged and gradual shifts in abiotic environmental factors bring about adjustments in natural soil-forming processes. These alterations impact the soil's hydrological patterns, mineral composition evolution, and the pace of organic matter creation and decomposition. The resultant physical and chemical attributes of the soil play a pivotal role in the productivity and environmental integrity of cultivated land. Hence, it is imperative to assess the prospective consequences of the interplay between climate change and soil dynamics (Gelybó et al. 2018).

Human Well-Being

Well-being refers to a favourable state that individuals and communities encounter. Comparable to health, it represents a valuable resource for our day-to-day existence, a resource profoundly shaped by socio-economic and environmental conditions. Well-being encompasses not only the quality of an individual's life but also the ability of both individuals and societies to actively engage in meaningful contributions to the betterment of the world (WMO 2023). Human well-being has been recognized as a qualitative framework that serves to uphold and enhance the quality of life (Costanza et al. 2007). Comprehending human well-being is a fundamental endeavour for both scholars and policymakers. Assessing whether there has been an enhancement in human well-being, especially given that over 1 billion individuals exist on incomes of less than $1 per day, is of paramount significance (McGillivray 2007). Society faces a twofold task: to maintain and enhance ecosystem services at an adequate level while simultaneously promoting human well-being and poverty reduction (Leemans and De Groot 2003).

According to the Millennium Ecosystem Assessment, human well-being is comprised of security, access to basic necessities for a quality life, good health, positive communal relationships, and the freedom to make choices (Millennium Ecosystem Assessment 2005). In this chapter, we define human well-being as a combination of factors including food security, water resources, economic stability, human health (both mental and physical), environmental cleanliness, and issues related to displacement and migration (Fig. 1.3). We have structured it in this manner to facilitate the identification and assessment of the influence of climate change on human well-being.

Fig. 1.3 Illustration showing various components of human well-being (authors structured it in this manner from existing literature)

Climate Change Impact on Human Well-Being

Human Health (Mental and Physical)

Numerous research studies have delved into the assessment of the effects of climate change on human health, providing a comprehensive understanding of the various impacts involved (Hasegawa et al. 2016). The key effects of climate change include a These encompass the increase in diseases transmitted by vectors, waterborne diseases, and foodborne illnesses, as well as the injuries and fatalities resulting from extreme hydro-geological events. Additionally, there are heat-related illnesses and fatalities induced by heat waves, characterized by thermal stressors, along with adverse consequences contributing to malnutrition (Markandya and Chiabai 2009). As a result of both short-term and long-term climate and environmental changes, a variety of psychological health outcomes have already been identified. This includes the elevated occurrence of conditions such as anxiety disorders, mood disorders, and post-traumatic stress disorders, increased violence and conflict, and strong emotional responses like despair, fear, helplessness, and suicidal ideation. Additionally, a decreased sense of self and identity from isolation has been identified as a climate-related mental health outcome (Bourque and Cunsolo Willox 2014). Temperature extremes at both the maximum and lowest are also anticipated to rise. Due to the anticipated rise in both the frequency and severity of heat waves, it is expected that there will be a corresponding uptick in the mortality rate. Risks from cold waves and other extreme cold conditions include hypothermia (Majra and Gur 2009). The influence of climate change is substantial when it comes to the proliferation of vector-borne diseases, including well-known conditions such as dengue and malaria. The simplest linkages are made by temperature, which affects the vectors' ability to bite, their ability to survive and reproduce, and the ability of the infections they carry to survive and develop (Campbell-Lendrum et al. 2015). Ozone, wildfire smoke, and various types of pollen are instances of pollutants, whether caused by human activities or natural processes, that are anticipated to worsen due to climate change. The chances of adverse health impacts are higher in areas where a large population is exposed, with ozone and pollen having a greater likelihood of affecting many people compared to wildfire smoke. Individuals with pre-existing conditions like asthma, allergies, or other respiratory ailments may be especially susceptible to respiratory problems (Kinney 2008).

West African nations like Ghana and Nigeria are among those countries in Africa that have been impacted by extreme heatwaves. Heatwaves have had a serious negative impact on workers' health, resulting in sunburns, insomnia, irritation, and weariness and lowering productivity (Opoku et al. 2021). The health of Sri Lanka's population of approximately 20 million people is experiencing detrimental consequences due to the impact of climate change. The nation is especially susceptible to diseases spread by vectors, including malaria, filariasis, chikungunya, dengue, and Japanese encephalitis. Another issue is diarrhoea brought on by droughts and floods. Warmer weather is predicted to make it more difficult to acquire clean water, which will aggravate the spread of infectious diseases like typhoid fever, hepatitis A and E, diarrhoea, and skin conditions (World Health Organization 2008). In Latin America, the influence of climate change on human wellness is significant. The growing occurrence of illnesses including heat exhaustion, malaria, dengue, cholera, and other water-borne diseases in Latin America is indicative of the influence of climate change on human health. By 2030, it is predicted that the largest relative hazards in the area will be from coastal floods, which cause diarrhoea and vector-borne illnesses (Rodríguez-Morales et al. 2010).

The analysis provided in this study sheds light on the numerous difficulties that climate change presents to human health. It is clear that action is necessary; these range from the evident threats presented by vector-borne diseases and extreme weather events to the nuanced yet substantial realm of mental health consequences. The need for adaptive methods becomes more obvious when temperature extremes intensify and the frequency of health-endangering incidents rises. Increasing resilience against hydro-geological disasters, reducing the spread of diseases like dengue and malaria, and treating the complex web of mental health consequences are all essential elements of a comprehensive strategy. The best course of action, though, is a coordinated, worldwide effort. We must adopt sustainable practices, cut emissions, and put the health of our environment and communities first. We can work to create a future where the influence of climate change on human health is minimized and the wellness of people and society is maintained by understanding the seriousness of the situation and actively participating in solutions.

Food Security

According to the Food and Agriculture Organization (FAO), food security is characterized as a state in which every individual, without exception, possesses the means, both in physical, social, and economic terms, to obtain an adequate, safe, and nutritious food supply that aligns with their dietary requirements and personal food preferences, thereby enabling them to lead a vigorous and wholesome life (World Food Summit 1996). Food systems are under growing environmental pressure as well as constant consumer growth. The global climate crisis will have an impact on the quantity, quality, and reputation of the food we produce (Myers et al. 2017). Climate change, such as less precipitation, can affect crop yields, which will decrease food supply and have an impact on crop prices, which will subsequently affect food accessibility (Ziervogel and Ericksen 2010). Droughts followed by heavy rainfall can make it more probable for flooding and decrease soil water absorption; this situation may give rise to fungal infestations affecting leafy, root, and tuber crops within regions characterized by water runoff (Rosenzweig et al. 2001). Between 15 and 35% of the fluctuation in wheat, oilseeds, and coarse grain yields worldwide can be associated with the El Niño-Southern Oscillation (ENSO) phenomenon and the recurring patterns of droughts and periods of flooding. Due to this sensitivity, agriculture will be affected by a changing climate (Schmidhuber and Tubiello 2007). The anticipated effect on seashores is that rising sea levels will increase groundwater salinity, causing desertification and decreasing land productivity and becoming a threat to coastal agriculture (Neupane 2011).

All aspects of the global, national, and local food systems will be impacted by climate change. Nearly 1 billion individuals in developing nations struggle to feed themselves. In times of crisis, another 5–10% are susceptible to "acute" food insecurity (Saina et al. 2013). Natural disasters like droughts, floods, landslides, and hailstorms with huge stones are becoming more frequent in Nepal. This increased frequency is correlated with climatic variance and has had a negative influence on agriculture by reducing soil fertility and agricultural output. The probability that the majority of people will experience food insecurity has increased due to numerous crop failures (Shrestha and Nepal 2016). When people started burning fossil fuels, the rice crop in India is responsive to a rise in minimum temperature, but the wheat crop is more susceptible to a rise in maximum temperature. The production of both rice and wheat suffers in north-west India due to acute water constraints and temperature stress. Recent research from the Indian Agricultural Research Institute indicates that there may be a loss of 4–5 million metric tons of wheat yield for every 1 °C rise in temperature throughout the growing season (Ahmad et al. 2011). The growth and production of rice crops have been influenced by the rise in temperature and modifications to the patterns of precipitation. A drop in crop duration and yield may result from the temperature increase, which is predicted to be between 1.5 and 2.0 °C. A 7-day reduction in crop duration for every 0.5 °C increase in temperature might lead to a 0.45 tonne per hectare decline in output (Sinha and Swaminathan 1991).

To effectively tackle the intricate network of issues spawned by climate change in the realm of food systems, a comprehensive and future-oriented approach is imperative. The protection of global food security necessitates both adaptation and mitigation measures. Implementing resilient farming techniques, creating crops resistant to drought, and improving water management systems are all part of agriculture's adaptation to climate change. In addition, addressing the underlying causes of these issues requires a focus on climate crisis mitigation. The key elements of this mitigation endeavour include reducing greenhouse gas emissions, encouraging sustainable land use, and encouraging international cooperation. The dynamic interaction between adaptation and mitigation emerges as a key mechanism for navigating the complexity of climate change's influence on food security. It organizes a resilient future where the assurance of sustenance resonates globally.

Water Resources

In recent times, there has been a growing global awareness and concern about the substantial environmental and economic consequences of climate change on water resources, attracting heightened attention from communities across the world (Luo et al. 2019). Climate change is poised to exert its influence on water resources by modifying precipitation patterns, altering runoff dynamics, impacting sea levels, and inducing shifts in population distribution. Elevated temperatures are anticipated to expedite the hydrologic cycle, thereby influencing the occurrence of floods, droughts, and evapotranspiration processes (Frederick and Major 1997). It is anticipated that climate change will have repercussions on water resources and agricultural activities in the southern regions of Brazil, namely Santa Catarina and Rio Grande do Sul. Currently, the region boasts abundant water supplies for large-scale rainfed cultivation of maize and soybeans. Projections for the next two decades indicate continued water availability, but there's a potential threat of increased flooding, emphasizing the need for proactive investment in drainage infrastructure to secure long-term agricultural resilience (Rosenzweig et al. 2004).

The El Niño/Southern Oscillation (ENSO) phenomena exert influence on global weather patterns, thereby impinging upon the availability of freshwater resources. ENSO events have repercussions on diverse regions and their respective natural resource sectors worldwide. A case in point is the formidable El Niño occurrence of 1997 and 1998. Empirical evidence showcases a substantial elevation in ocean temperatures, contributing to widespread flooding and setting records for snowfall along the western coast of the United States (Thomson et al. 2003). Extensive historical data gathered from rivers and lakes across North America showed a similar pattern of rising water temperatures and changing precipitation patterns. The study showed that these climate-related changes had a major impact on water quality, with increased temperatures and extreme weather events causing heightened nitrogen levels, hazardous algal blooms, and a reduction in overall water quality. The results support the concerns expressed in the statement supplied by demonstrating the real influence of climate change on the quality of surface water (Haj-Amor and Bouri 2020).

The instances observed in Brazil and the broader consequences of El Niño events stand as stark reminders of how climate change significantly affects water resources worldwide. Mounting evidence underscores the urgency of addressing changing temperatures, shifting precipitation patterns, and the overall dynamics of the hydrological cycle. To reduce hazards like increasing flooding, proactive measures are essential, especially in sensitive areas like Brazil's southern states. Effective water resource planning also requires an understanding of the complex interactions between stressors brought on by the climate and changes in land use brought on by humans. In areas prone to flooding, adaptation strategies should include the construction of a solid drainage system. To lessen the effects on water availability and quality, it is crucial to put water conservation measures into place, increase irrigation efficiency, and promote sustainable agriculture practices. Additionally, incorporating regional ecological thresholds into water resource management plans is essential to boosting resilience in the face of climate change. Communities can endeavour to ensure that water supplies are sustainable and adaptable in the context of continuous climate change by embracing these adaptive solutions.

Displacement and Migration

Climate change is anticipated to exert a progressively adverse impact on both environmental degradation and socioeconomic frameworks reliant on the environment, possibly culminating in substantial population displacement (Warner et al. 2010). Climatic factors, including temperature patterns and the occurrence of extreme events, are currently being examined in conjunction with the expanding dataset on human migration, both at macro and micro levels (Beine and Jeusette 2021). Bangladesh exemplifies heightened susceptibility, with around 26 million people, or 16% of the population, expected to migrate due to climate-induced hardships, emphasizing the urgent need for comprehensive policies (Kartiki 2011).

In Burkina Faso, decreased rainfall and environmental stress are driving migration trends, as indicated by a study involving 8644 individuals. Short- and long-term movements are more prevalent, especially within rural areas, reflecting adaptation strategies to climate-related challenges (Fussell et al. 2014). Similarly, in Northern Australia, indigenous groups temporally migrate to urban centers during the wet season to evade flooding impacts on remote locations (Schwerdtle et al. 2018). The Mekong Delta of Vietnam experiences increasing migration propelled by worsening environmental stressors, notably flooding. People look for other coping mechanisms as floods worsen and affect livelihoods depending on farming and fishing, which include urbanization and seasonal labor migration, with governmental relocation programs seeking to address the impacts while potentially disrupting established social networks (Warner 2010). As demonstrated by instances like the Koshi River basin incident in 2008 and the Mahakali River accidents in 2013, flash flooding poses a serious threat to agricultural production in Nepal. Over 65,000 families were forced to leave their livelihoods due to the Koshi River flood in 2008, while 2600 people were forced to leave their homes due to the Mahakali River flash flood in 2013 (Chapagain and Gentle 2015).

These diverse cases underscore the urgency of comprehending the potential ramifications of environmental changes on human mobility (Gemenne 2011). Effectively addressing this gap necessitates an integrative approach encompassing climate adaptation policies, international collaboration, robust early warning systems, and community empowerment initiatives. These measures strive to foster resilience and long-term sustainability amid continuous environmental alterations.

Pollution

The spatial and temporal dispersion of pollution levels is significantly shaped by climatic and weather conditions (Kinney 2018). Human-induced contamination of ecosystems, including soil, water, and air, disrupts the natural functioning of the soil, water, and air cycles. As the significance of soil, water, and air quality increases, public consciousness of the environment also grows, leading to greater support for environmental conservation (Celik 2020). Multiple factors come into play, including shifts in photochemical reaction rates, emissions originating from natural sources, the deposition of pollutants, and the patterns of atmospheric circulation, collectively influencing the interplay between climate and air quality (Silva et al. 2013). The incineration of biomass and fossil fuels releases harmful substances that contribute to both air pollution and climate change. Given that air pollution is connected to early global mortality, this dual challenge poses existential hazards to humans and wildlife. It is essential to take immediate action to address these interconnected issues, including measures for reducing both air pollution and climate change (Al-Delaimy et al. 2020). Interactions between air pollution and climate change degrade air quality, which has an effect on children's health. There are 850 million children worldwide who are affected by overlapping environmental and climate shocks (Perera and Nadeau 2022). Climate change factors like heavy rain and flooding cause dampness and the growth of mold in impacted homes, affecting the quality of indoor air (Eguiluz-Gracia et al. 2020).

Water pollution emerges as a pivotal concern, instigating modifications in essential water quality parameters. This perturbation extends to the introduction of toxic by-products, exacerbating the intricate challenges associated with global warming and broader environmental threats. Supporting this concern is the fact that around 20% of the global population lacks adequate access to safe and clean drinking water, and 40% grapple with unsatisfactory sanitary conditions. These repercussions span from contaminating drinking water to emitting greenhouse gases, thereby intensifying challenges linked to climate change (Inyinbor Adejumoke et al. 2018). Increased storminess brought on by climate change may increase litter input into the marine ecosystem, which in turn diminishes the capacity of ecosystems and species to endure the challenges posed by climate change. Marine debris has a serious negative influence on coastal populations in India, particularly in the states of Tamil Nadu and Kerala. The population of Kerala is 35 million, and in Tamil Nadu, it is 83 million, including 11 million in Chennai. The majority of the population resides in coastal and urban areas, placing them in direct exposure to the impacts of marine pollution, along with the associated risks to both livelihoods and ecosystems (Lincoln et al. 2023).

The primary origins of soil contamination are intricately tied to human actions, specifically the release of industrial wastes, runoff from agricultural activities, and the discharge of untreated sewage into the environment. It is worth noting that the influence of climate change on soil pollution is relatively indirect and constrained in its scope. It can lead to issues such as the thawing of permafrost in regions like North Western Siberia, resulting in the release of polycyclic aromatic hydrocarbons and petroleum hydrocarbons (Opekunova et al. 2022), thereby causing soil pollution. A study conducted by Eswar et al. (2021) indicates that salt-affected soils, covering approximately 1060.1 million hectares worldwide, are gradually expanding due to the influence of climate change. Soil salinity tends to grow in response to increasing sea levels and temperatures.

The complex network of interdependencies between pollution and the dynamics of climate change presents a grave and substantial threat to the well-being of our planet, harming ecosystems, human health, and the overall well-being of our environment. Implementing policies that lessen air pollution and restrict climate change is essential to overcoming these obstacles. Enhancing waste management procedures, encouraging eco-friendly projects, and switching to sustainable energy sources are crucial first steps. Public support for environmental protection can be greatly influenced through education and awareness efforts. Action must be taken right away. Inaction has serious repercussions that not only harm the present generation but also future ones. In order to maintain the delicate balance of our ecosystems and ensure the welfare of both people and wildlife, we may work towards a more sustainable and resilient future by putting into practice effective mitigation and adaptation methods. It is impossible to stress how urgent it is to address both pollution and climate change, and to meet the problems that lie ahead, a worldwide commitment that is steadfast is needed.

Economy

The phenomenon of climate change is on the verge of wielding a significant economic impact that will reverberate across a multitude of nations, with a specific emphasis on the heightened vulnerability of many lower-income countries. The emissions stemming from a solitary nation have far-reaching consequences that extend to all nations, as they play a role in the build-up of heat-retaining gases within the Earth's atmosphere, which subsequently leads to global warming (International Monetary Fund

2023). While there is a general agreement on the imperative need to decrease greenhouse gas emissions, the discourse among economists regarding the issue of climate change has proven to be notably contentious, potentially reflecting the broader divisions within climate research and policy. Notably, the disparities in the assessments of the incremental impacts of climate change are so wide-ranging that it seems the initial determination of carbon pricing may be more driven by political considerations than by strictly economic factors (Tol 2018). An estimated $44 trillion in economic value creation, which accounts for more than half of the global Gross Domestic Product (GDP), exhibits a significant level of reliance on nature and the services it provides. Consequently, these sectors face exposure to risks associated with the decline of natural resources. Among these industries, construction holds the highest dependency at $4 trillion, followed by agriculture at $2.5 trillion, and food and beverages at $1.4 trillion (Herweijer et al. 2020).

Research conducted by Fankhauser and Tol (2005) suggests that in the long term, under significant direct impacts, climate change could potentially lead to a decline in economic growth, even within the optimistic Solow-Swan model, resulting in a decrease in per capita income. However, it should be noted that achieving such an outcome would necessitate substantial direct damages, amounting to at least 15% of GDP in the case of a 3 °C warming scenario. Such a scenario is considered unlikely for most regions and is not typically associated with the ongoing trajectory of climate change. The insurance sector acknowledges that it is expected to shoulder a significant portion of the burden related to global warming. In 2011, it was reported as the costliest year on record for natural disasters, resulting in insured losses exceeding $126 billion for the industry. Mark Carney, the Governor of the Bank of England, emphasized the findings of the Bank's recent research, highlighting that climate change stands among the foremost risks confronting the insurance industry (Wade and Jennings 2016).

If global warming reaches a 2 °C increase, it is projected that 99% of coral reefs worldwide will be lost. This loss poses economic risks for the tourism sector as well. Coral reefs contribute significantly to the global economy, generating $36 billion annually through tourism. Of this amount, $19 billion comes from activities directly associated with coral reefs, such as diving and wildlife observation. The balance of this revenue is derived from tourism activities in regions neighbouring the reefs. These activities encompass experiences such as oceanfront vistas, coastal beaches, and the consumption of locally sourced seafood (Spalding et al. 2017). The consequences of deforestation extend beyond just the agriculture sector. According to the World Bank, the 2015 crisis in Indonesia, driven by deforestation-related fires, resulted in economic losses of $16 billion due to disruptions in economic activities and a decline in GDP growth (World Bank 2016). The pharmaceutical sector heavily relies on biodiverse tropical rainforests for innovations, as a quarter of the drugs employed in contemporary medicine are derived from plants found in these rainforests (Shah 2019). With tropical forests under threat from deforestation and wildfires, pharmaceutical companies are at risk of losing a substantial reservoir of untapped genetic resources that could potentially lead to significant advancements in both medical and commercial domains (Palhares et al. 2015). The global economy is intricately connected to the services rendered by nature. Therefore, any alteration in the planet's usual state can have repercussions on the world's economy, potentially resulting in various economic crises that could disproportionately impact low-income countries and marginalized populations worldwide. To address this, macroeconomic strategies within these nations must be adjusted to accommodate a higher frequency of weather-related shocks. This adjustment includes creating policy flexibility to effectively respond to such shocks (International Monetary Fund 2023).

The Nexus Between Ecosystem Services and Human Well-Being

Ecosystem services (ES) and human well-being (HWB) serve as a potent linkage between the planet and human community and have emerged as pivotal topics within the realm of sustainability science (Wu 2013). Ecosystem services are inextricably linked to various facets of human well-being, with a complex web of interconnections and mutual dependencies between them, whether through direct or indirect relationships (Fig. 1.4). The discernible and rapidly escalating impact of anthropogenic climate change is acknowledged as a significant catalyst for biodiversity decline and alterations in landscapes and ecosystems (Bellard et al. 2012). The multitude of ecosystem services and the diverse ways in which they are provided constitute the foundational support for human well-being, forming an intricate and vital underpinning of our quality of life (Kosanic and Petzold 2020). The current indicators employed for assessing and monitoring well-being primarily focus on immediate measures of material prosperity, such as per capita Gross National Product (GNP) and, to a significant extent, the Human Development Index (HDI). Consequently, it was not unexpected when the Millennium Ecosystem Assessment (MA) reported that nearly two-thirds of evaluated ecosystem services were in decline, despite an increase in global well-being over the past five decades. This outcome might appear counterintuitive, as one would anticipate that declines in ecosystem services should correspondingly reduce human well being (Duraiappah 2011).

The case of mining activities in Papua New Guinea serves as a vivid illustration of the intricate interplay between ecosystem services and the welfare of the human population. The extraction and mining operations lead to the direct and targeted depletion of arable land, agricultural plantations, and the delicate coral reefs located in the immediate surroundings. However, mining companies provide monetary compensation in the form of cash payments to offset these losses. Additionally, this monetary compensation can open avenues to access services that were hitherto beyond reach, such as enhanced housing, the installation of piped water supply systems, a more nutritious high-protein diet, access to Western medicines, and a range of healthcare choices. These expanded options contribute to substantial enhancements in the overall quality of life and well-being (McMichael et al. 2005). Transformations in ecosystems and their consequent effects on human well-being can manifest at various

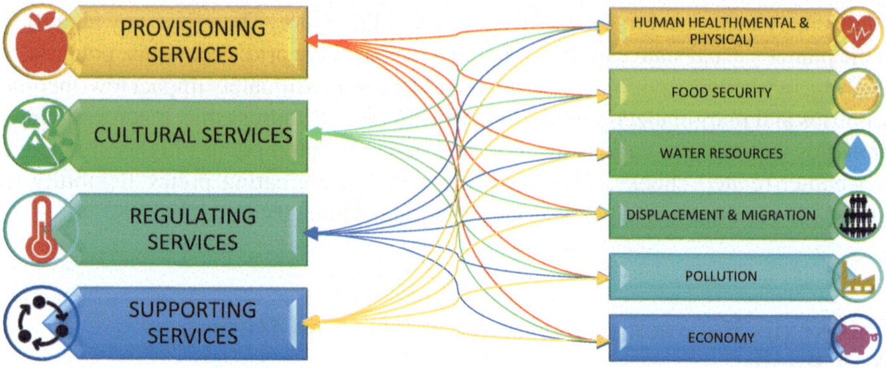

Fig. 1.4 Illustration showing the interconnectedness between ecosystem services and human well-being. *Source* Millennium Ecosystem Assessment (2005)

levels, including the global, regional, sub-regional, national, and local scales, often concurrently. To illustrate, let's consider fishing operations in the Caribbean Sea. These activities play an important role in the region's economic output, generate employment opportunities, and serve as a vital protein source for numerous individuals. The impacts of such fishing practices can be assessed across multiple tiers of analysis, ranging from the scale of local fishing communities to the national, sub-regional, and regional levels, extending even to the broader context of the entire North Atlantic Ocean (Berkes 2004). The presence of a greater number of ecosystem services does not inherently equate to increased human well-being, and the reverse is also not necessarily true. Enhancing overall human well-being necessitates an increase in total capital, which can still occur even if a few components experience a decline, as long as the others experience substantial growth to offset the decrease. If a specific ecosystem service diminishes, it can potentially be substituted with alternative sources. This approach is underpinned by the income generated through the utilization of both natural and human capital, as well as the growth of manufactured capital. This perspective emphasizes the way in which development broadens the spectrum of options that become accessible to individuals and communities, thereby enriching the scope of opportunities (Sen 1999).

The connection between the well-being of humanity and the condition of ecosystems is characterized by a mutual, two-way relationship where each influence and impacts the other in a continuous and interconnected manner. On one hand, it is influenced by the services that ecosystems offer to people. On the other hand, it is affected by the often-unintentional consequences of human actions on ecosystem functioning (McMichael et al. 2005). The intricacy of this relationship hinders our ability to fully understand the consequences of disruptions in ecosystem services on the deterioration of human well-being. Present-day assessments of human well-being have several limitations, and comprehending this connection is pivotal for evaluating the impacts stemming from anthropogenic climate change.

Ecosystem Service Assessment InVEST-Sediment Delivery Ratio Model (Soil Erosion)—Case Study

Study Area

The study region, encompassing a land area of 112.8 km^2, is situated within the Dhanaulti tehsil of the Tehri Garhwal district in Uttarakhand, India. The location of the study area extends from 78°7′51.164″ E to 78°18′24.636″ E and 30°36′38.468″ N to 30°20′44.727″ N (Fig. 1.5). It belongs to the Yamuna River basin. The climate in the basin is temperate, with dry winters and warm summers and a rainfall of 830.8 mm (Khan et al. 2018). It is a typical agricultural basin with forest cover in the upper part. The main field crops are rice, maize, pigeon pea, pearl millet, black gram wheat, and chickpea.

InVEST Model

The InVEST model, a specialized tool for evaluating ecosystem services, was created by the Natural Capital Project. Its purpose is to analyse and map ecosystem services, providing valuable support for decision-making related to natural resource management (Sharp et al. 2015). In essence, InVEST is tailored to offer insights for decisions about the management of natural resources. Its primary function is to furnish data on how alterations in ecosystems are expected to impact the distribution of advantages to human populations. These models are straightforward and require less data, making them suitable for use in various locations by individuals who may not be experts in the field. For this particular study, the sediment delivery ratio (SDR) model was used. The InVEST sediment delivery ratio (SDR) model is primarily utilized for mapping the transport of overland sediment and its delivery to streams. The model's outcomes reveal the significance of sediment retention and soil conservation as part of hydrological ecosystem services, particularly concerning reservoir management and the enhancement of water quality (Cong et al. 2020). The computation process hinges on the adaptation of the universal soil loss equation, known as RUSLE. To begin with, the model computes the yearly soil loss for every individual pixel and establishes a connectivity index, referred to as IC0 (Borselli et al. 2008). Following this, the SDR is determined for each pixel. The sediment load is then evaluated by considering the SDR and the quantity of annual soil loss. The input data for the SDR model include the digital elevation model (DEM), rainfall erosivity index (R), soil erodibility (K), and land use and land cover (LULC) data (Fig. 1.6).

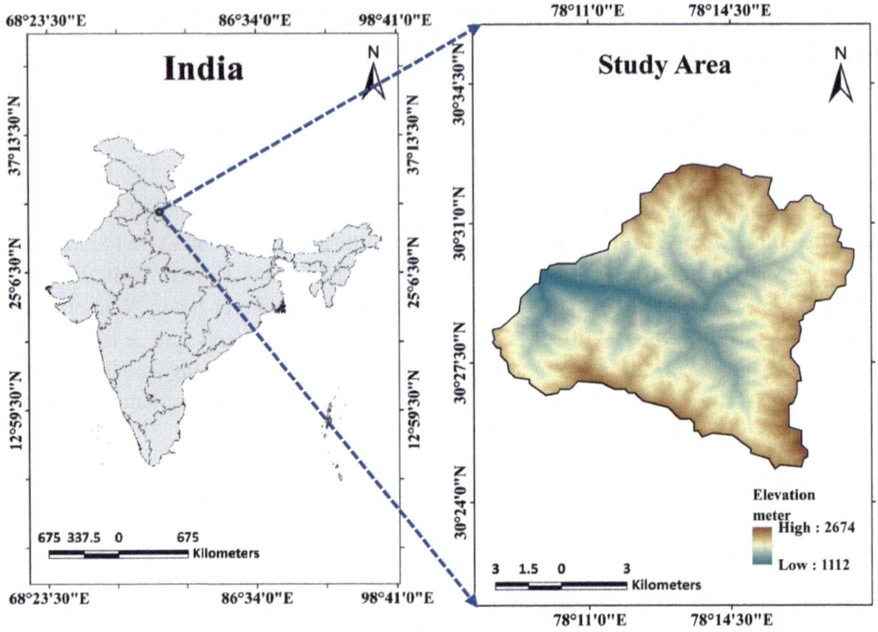

Fig. 1.5 Location of the study area

Salient Findings

The R factor was obtained from the Indian rainfall erosivity dataset (Raj et al. 2022a, b), and a single rainfall erosivity value of 1724 MJ mm h^{-1} ha^{-1} $year^{-1}$ has been used. The soil erodibility (K) factor was acquired from the Indian soil erodibility dataset (Raj et al. 2023) and ranges from 0.0007 to 0.0307 t ha ha^{-1} MJ^{-1} mm^{-1} (Fig. 1.6). The land use and land cover map of the study area were downloaded from the ESRI land use/land cover map (Karra et al. 2021). The cover factor (C) and practice factor (P) values were assigned based on Morgan (2009). In the soil retention model, we made adjustments to the values of Borselli's "k" parameter and IC0, as they play a crucial role in defining the association among hydrologic connectivity and the sediment delivery ratio, as described (Sharp et al. 2015). Borselli's k parameter to 2, IC_0 to 0.5, maximum SDR value to 0.8, and maximum L value to 122 were kept as per the default configuration because no data were available for calibration. These adjustments led to simulation results that were regarded as indicative of various outputs such as soil loss, retention, export, and deposition. These results would be useful for a qualitative assessment rather than a quantitative one.

The total potential soil loss (t ha^{-1} yr^{-1}), as determined by the original land cover using the RKLS equation (as depicted in Fig. 1.7a), is equivalent to the soil loss for bare soil and does not account for the C or P factors. Notably, areas with a higher potential for soil erosion are primarily situated near the valley regions, which

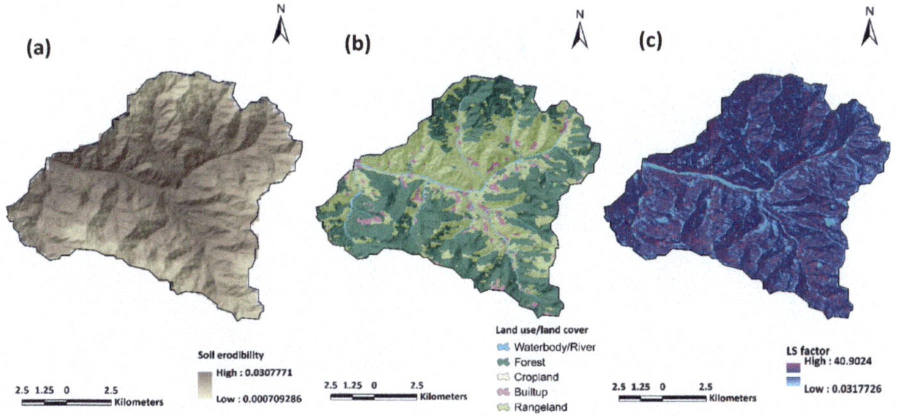

Fig. 1.6 **a** Soil erodibility (t ha ha^{-1} MJ^{-1} mm^{-1}), **b** land use/land cover (C & P), and LS factor map of the study area

are characterized by steep slopes. Additionally, most of these areas are covered by barren or rangeland, particularly with a south-facing aspect. On the other hand, the total potential soil loss per pixel in the original land cover, calculated using the RUSLE equation (as shown in Fig. 1.7b), reveals that forest cover has significantly reduced soil erosion, regardless of slope and aspect considerations, as evident from Figs. 1.6b and 1.7b.

Sediment export is the measure of the total sediment leaving each pixel and reaching the stream. In the context of soil conservation, the sediment export indicator exhibits a negative correlation with soil conservation efforts (Cong et al. 2020). Figure 1.8a highlights the highly vulnerable areas to soil erosion, characterized by steep slopes, bare soil, and a south-facing aspect. The impact of ecosystem services provided by vegetation cover is demonstrated in Fig. 1.7b. Specifically, "avoided erosion" refers to the contribution of vegetation to preventing soil erosion from entering a stream. This involves a combination of sediment retention on the local pixel and the trapping of erosion from upslope areas. It is evident that regions with forest cover are particularly effective in retaining soil in place, while this effect is less pronounced in rangeland or barren areas with minimal vegetation and steep slopes (Fig. 1.8c).

Summary and Conclusion

Human-induced climate change is exacerbating the severity of global disasters, posing a pervasive and escalating global threat to both the integrity of biodiversity and the stability of ecosystems. It impacts all facets of planetary existence, disrupting the natural rhythms of the globe. This disruption can impede the delivery of ecosystem

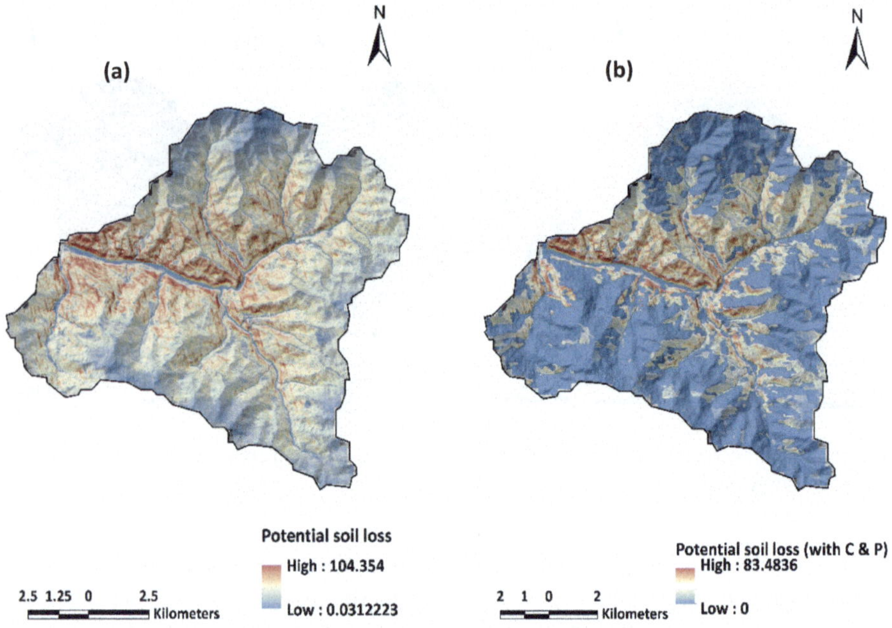

Fig. 1.7 **a** Potential soil loss without C & P factor; **b** with C & P factor

services, consequently influencing human well-being. Ecosystem services play a pivotal role in facilitating the harmonious functioning of human society. Disruptions to provisioning services, which encompass fundamental necessities like food, water, and medical resources, can have a direct impact on human well-being. Similarly, any hindrance in regulating services may disrupt daily life patterns and even lead to health issues. Interference with supporting services, including soil formation, nutrient cycling, and genetic diversity, can have a cascading effect on provisioning services and, consequently, human welfare. Furthermore, disturbances to cultural services may result in challenges related to mental health and ecosystem degradation.

The anticipated consequences of climate change encompass a wide spectrum of impacts on diverse dimensions of human well-being. These encompass physiological and psychological risks, challenges in food security, issues related to water resources, economic instability, pollution, migration, and other interconnected concerns. These adverse effects manifest as a complex interplay of short-term, long-term, and socially dynamic influences on individuals' lives and overall welfare. Particularly, low-income, under-resourced, and geographically vulnerable communities face a disproportionately heightened vulnerability to bearing the brunt of unfavourable outcomes arising from climate change. This chapter underscores the notion that disadvantaged communities, being particularly sensitive to even subtle shifts in these variables, highlight the paramount significance of addressing climate change and ameliorating its repercussions in safeguarding human welfare.

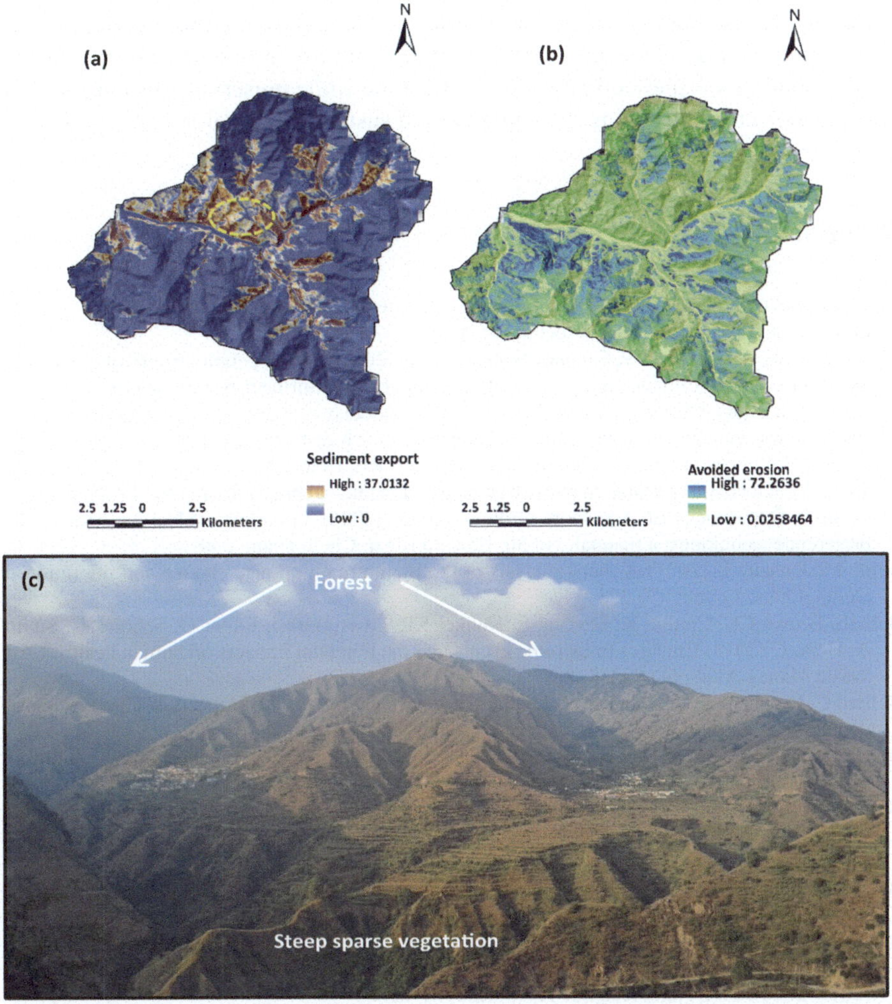

Fig. 1.8 **a** Total sediment exported from one pixel, **b** erosion avoided by vegetation cover and **c** field validation of the watershed (upper forest and lower side steep with bare soils)

This chapter offers a more holistic examination of the ramifications of the climate crisis. It acknowledges the crucial interplay between ecosystem services and human well-being, emphasizing that an all-encompassing comprehension of the intricate interconnections and dependencies among different facets is indispensable for effective solutions. The case study, which utilizes the InVEST-SDR model, illustrates the role of the forest ecosystem in retaining soils in their original position and preventing soil erosion. By delving into the nexus between ecosystems and humanity, the chapter introduces an innovative and collaborative approach to grappling with the intricacies posed by climate change. Climate change recognizes no geographical confines;

it exhibits a transboundary character. Consequently, addressing this global predicament from a worldwide standpoint is of great significance in discovering comprehensive solutions that effectively address the issue while preserving the equilibrium of ecosystems and advancing the attainment of sustainable development goals.

References

Ahmad J, Alam D, Haseen MS (2011) Impact of climate change on agriculture and food security in India. Int J Agric Environ Biotechnol 4(2):129–137

Al-Delaimy W, Ramanathan V, Sánchez Sorondo M (2020) Health of people, health of planet and our responsibility: climate change, air pollution and health. Springer Nature, p 419

Allen CD, Macalady AK, Chenchouni H et al (2010) A global overview of drought and heatinduced tree mortality reveals emerging climate change risks for forests. For Ecol Manag 259:660–684

Applequist WL, Brinckmann JA, Cunningham AB, Hart RE, Heinrich M, Katerere DR, Van Andel T (2020) Scientists' warning on climate change and medicinal plants. Planta Med 86(01):10–18

Bateman I, Department of Transport Großbritannien (2002) Economic valuation with stated preference techniques: a manual, vol 50. Edward Elgar, Cheltenham, p 480

Beine M, Jeusette L (2021) A meta-analysis of the literature on climate change and migration. J Demographic Econ 87(3):293–344

Bell JE, Brown CL, Conlon K, Herring S, Kunkel KE, Lawrimore J, Luber G, Schreck C, Smith A, Uejio C (2018) Changes in extreme events and the potential impacts on human health. J Air Waste Manag Assoc 68(4):265–287

Bellard C, Bertelsmeier C, Leadley P, Thuiller W, Courchamp F (2012) Impacts of climate change on the future of biodiversity. Ecol Lett 15(4):365–377

Bennett EM, Peterson GD, Gordon LJ (2009) Understanding relationships among multiple ecosystem services. Ecol Lett 12(12):1394–1404

Berkes F (2004) Fisheries institutions. Paper presented at Bridging scales and epistemologies: linking local knowledge and global science in multi-scale assessments, March, Alexandria, Egypt

Bodelier PL, Steenbergh AK (2014) Interactions between methane and the nitrogen cycle in light of climate change. Curr Opin Environ Sustain 9:26–36

Borrelli P, Robinson DA, Panagos P, Lugato E, Yang JE, Alewell C, Wuepper D, Montanarella L, Ballabio C (2020) Land use and climate change impacts on global soil erosion by water (2015–2070). Proc Natl Acad Sci 117(36):21994–22001

Borselli L, Cassi P, Torri D (2008) Prolegomena to sediment and flow connectivity in the landscape: a GIS and field numerical assessment. CATENA 75:268–277

Bourque F, Cunsolo Willox A (2014) Climate change: the next challenge for public mental health? Int Rev Psychiatry 26(4):415–422

Brecka AF, Shahi C, Chen HY (2018) Climate change impacts on boreal forest timber supply. Forest Policy Econ 92:11–21

Burkhard B, Maew J (2017) Mapping ecosystem services. Pensoft Publisher, Sofia, Bulgaria

Campbell-Lendrum D, Manga L, Bagayoko M, Sommerfeld J (2015) Climate change and vector-borne diseases: what are the implications for public health research and policy? Philos Trans R Soc B Biol Sci 370(1665):20130552

Canadell JG, Raupach MR (2008) Managing forests for climate change mitigation. Science 320(5882):1456–1457

Caney S (2015) Climate change. In: The Routledge handbook of global ethics. Routledge, pp 384–398

Celik S (2020) The effects of climate change on human behaviors. In: Environment, climate, plant and vegetation growth, pp 577–589

Chapagain B, Gentle P (2015) Withdrawing from agrarian livelihoods: environmental migration in Nepal. J Mt Sci 12:1–13
Chen J (2007) Rapid urbanization in China: a real challenge to soil protection and food security. CATENA 69(1):1–15
Cong W, Sun X, Guo H, Shan R (2020) Comparison of the SWAT and InVEST models to determine hydrological ecosystem service spatial patterns, priorities and trade-offs in a complex basin. Ecol Ind 112:106089
Constable AJ (2022) Imperatives for integrated science and policy in managing greenhouse gas risks to the Southern Polar Region. Glob Change Biol 28(15):4489–4492. https://doi.org/10.1111/gcb.16219
Costanza R, Fisher B, Ali S, Beer C, Bond L, Boumans R, Danigelis NL, Dickinson J, Elliott C, Farley J, Gayer DE, Glenn LM, Hudspeth T, Mahoney D, McCahill L, McIntosh B, Reed B, Rizvi SAT, Rizzo DM, Simpatico T, Snapp R (2007) Quality of life: an approach integrating opportunities, human needs, and subjective well-being. Ecol Econ 61:267–276
Costanza R, d'Arge R, De Groot R, Farber S, Grasso M, Hannon B, Limburg K, Naeem S, O'Neill RV, Paruelo J, Raskin RG (1997) The value of the world's ecosystem services and natural capital. Nature 387(6630):253–260
Daily GC (1997) Introduction: what are ecosystem services. In: Nature's services: societal dependence on natural ecosystems, vol 1, no 1
Das M, Jain V, Malhotra S (2016) Impact of climate change on medicinal and aromatic plants. Indian J Agric Sci 86(11):1375–1382
Delucchi MA (2010) Impacts of biofuels on climate change, water use, and land use. Ann N Y Acad Sci 1195(1):28–45
Deng X, Li Z, Huang J, Shi Q, Li Y (2013) A revisit to the impacts of land use changes on the human wellbeing via altering the ecosystem provisioning services. Adv Meteorol
Dixon RK, Solomon AM, Brown S et al (1994) Carbon pools and flux of global forest ecosystems. Science 263:185–191
Drijfhout S et al (2015) Proc Natl Acad Sci U S A 102:E5777
Duraiappah AK (2011) Ecosystem services and human well-being: do global findings make any sense? Bioscience 61(1):7–8
Edwards-Jones G, Cross PA, Harris I, Le Vay L, Rayment M, Scowen M, Waller P (2011) Provisioning services
Eguiluz-Gracia I, Mathioudakis AG, Bartel S, Vijverberg SJ, Fuertes E, Comberiati P, Cai YS, Tomazic PV, Diamant Z, Vestbo J, Galan C (2020) The need for clean air: the way air pollution and climate change affect allergic rhinitis and asthma. Allergy 75(9):2170–2184
Eswar D, Karuppusamy R, Chellamuthu S (2021) Drivers of soil salinity and their correlation with climate change. Curr Opin Environ Sustain 50:310–318
Fankhauser S, Tol RS (2005) On climate change and economic growth. Resour Energy Econ 27(1):1–17
FAO, Ecosystem Services & Biodiversity (ESB) (2023) https://www.fao.org/ecosystem-services-biodiversity/background/regulating-services/en/#:~:text=Maintaining%20the%20quality%20of%20air,regulating%20services'%20provided%20by%20ecosystems
The National Wildlife Federation, Ecosystem Services (2023) https://www.nwf.org/Educational-Resources/Wildlife-Guide/Understanding-Conservation/Ecosystem-Services
Frederick KD, Major DC (1997) Climate change and water resources. Clim Change 37(1):7–33
Fu B, Wang S, Liu Y, Liu J, Liang W, Miao C (2017) Hydrogeomorphic ecosystem responses to natural and anthropogenic changes in the Loess Plateau of China. Annu Rev Earth Planet Sci 45:223–243
Furunes T, Mykletun RJ (2012) Frozen adventure at risk? A 7-year follow-up study of Norwegian glacier tourism. Scand J Hosp Tour 12(4):324–348
Fussell E, Hunter LM, Gray CL (2014) Measuring the environmental dimensions of human migration: the demographer's toolkit. Glob Environ Chang 28:182–191

Gagné K, Rasmussen MB, Orlove B (2014) Glaciers and society: attributions, perceptions, and valuations. Wiley Interdiscip Rev Clim Change 5(6):793–808
Gelybó G, Tóth E, Farkas C, Horel Á, Kása I, Bakacsi Z (2018) Potential impacts of climate change on soil properties. Agrokém Talajt 67(1):121–141
Gemenne F (2011) Why the numbers don't add up: a review of estimates and predictions of people displaced by environmental changes. Glob Environ Chang 21:S41–S49
Gupta A, Singh PP, Singh P, Singh K, Singh AV, Singh SK, Kumar A (2019) Medicinal plants under climate change: impacts on pharmaceutical properties of plants. In: Climate change and agricultural ecosystems. Woodhead Publishing, pp 181–209
Haj-Amor Z, Bouri S (2020) Climate change impacts on coastal soil and water management. CRC Press
Hasegawa T, Fujimori S, Takahashi K, Yokohata T, Masui T (2016) Economic implications of climate change impacts on human health through undernourishment. Clim Change 136:189–202
Herweijer C, Evison W, Mariam S, Khatri A, Albani M, Semov A, Long E (2020) Nature risk rising: why the crisis engulfing nature matters for business and the economy. World Economic Forum and PwC. http://www3.weforum.org/docs/WEF_New_Nature_Economy_Report_2020.pdf
Hochachka PW, Somero GN (2002) Biochemical adaptation: mechanism and process in physiological evolution. Oxford University Press
Hoegh-Guldberg O, Bruno JF (2010) The impact of climate change on the world's marine ecosystems. Science 328(5985):1523–1528
Hui D, Deng Q, Tian H, Luo Y (2017) Climate change and carbon sequestration in forest ecosystems. In: Handbook of climate change mitigation and adaptation, vol 555, p 594
Inoue K, Berg DJ (2017) Predicting the effects of climate change on population connectivity and genetic diversity of an imperiled freshwater mussel, *Cumberlandia monodonta* (Bivalvia: Margaritiferidae), in riverine systems. Glob Change Biol 23(1):94–107
International Energy Agency, Tracking Biofuels Supply (2023) https://www.iea.org/energy-system/low-emission-fuels/biofuels
International Monetary Fund (2023) Climate and the economy. https://www.imf.org/en/Topics/climate-change/climate-and-the-economy
Inyinbor Adejumoke A, Adebesin Babatunde O, Oluyori Abimbola P, Adelani Akande Tabitha A, Dada Adewumi O, Oreofe Toyin A (2018) Water pollution: effects, prevention, and climatic impact. In: Water challenges of an urbanizing world, vol 33, pp 33–47
IPCC (2019) Special report on the ocean and cryosphere in a changing climate
IPCC (2021) Regional fact sheet—polar regions. In: Masson Delmotte V, Zhai P, Pirani A et al (eds) Climate change 2021: the physical science basis. Contribution of working group I to the sixth assessment report of the Intergovernmental Panel on Climate Change. https://www.ipcc.ch/report/ar6/wg1/resources/factsheets
IPCC, Lee H, Calvin K, Dasgupta D, Krinner G, Mukherji A, Thorne P (2023) Synthesis report of the IPCC sixth assessment report (AR6). Intergovernmental Panel on Climate Change, Geneva, Switzerland
Jacob DJ, Winner DA (2009) Effect of climate change on air quality. Atmos Environ 43(1):51–63
Karra Kontgis et al (2021) Global land use/land cover with Sentinel-2 and deep learning. In: IGARSS 2021–2021 IEEE international geoscience and remote sensing symposium. IEEE
Kartiki K (2011) Climate change and migration: a case study from rural Bangladesh. Gend Dev 19(1):23–38
Khan S, Kumar V, Chaurasia S (2018) A classed approach towards rainfall forecasting: machine learning method. Int J Swarm Intell 3(4):276–289
Kinney PL (2008) Climate change, air quality, and human health. Am J Prev Med 35(5):459–467
Kinney PL (2018) Interactions of climate change, air pollution, and human health. Curr Environ Health Rep 5:179–186
Koch D, Park J, Del Genio A (2003) Clouds and sulfate are anticorrelated: a new diagnostic for global sulfur models. J Geophys Res 108:4781

Kosanic A, Petzold J (2020) A systematic review of cultural ecosystem services and human wellbeing. Ecosyst Serv 45:101168

Leemans R, De Groot RS (2003) Millennium ecosystem assessment: ecosystems and human well-being: a framework for assessment

Lincoln S, Chowdhury P, Posen PE, Robin RS, Ramachandran P, Ajith N, Harrod O, Hoehn D, Harrod R, Townhill BL (2023) Interaction of climate change and marine pollution in Southern India: implications for coastal zone management practices and policies. Sci Total Environ, 166061

Lindquist EJ, D'Annunzio R, Gerrand A et al (2012) Global forest land-use change 1990–2005. FAO forestry paper no. 169. In: Food and Agriculture Organization of the United Nations and European Commission Joint Research Centre. FAO, Rome

Locatelli B (2016) Ecosystem services and climate change

Luo M, Liu T, Meng F, Duan Y, Bao A, Xing W, Feng X, De Maeyer P, Frankl A (2019) Identifying climate change impacts on water resources in Xinjiang, China. Sci Total Environ 676:613–626

Majra JP, Gur A (2009) Climate change and health: why should India be concerned? Indian J Occup Environ Med 13(1):11–16

Malmstrom CM, Butterfield HS, Barber C, Dieter B, Harrison R, Qi J, Riano D, Schrotenboer A, Stone S, Stoner CJ, Wirka J (2009) Using remote sensing to evaluate the influence of grassland restoration activities on ecosystem forage provisioning services. Restor Ecol 17(4):526–538

Markandya A, Chiabai A (2009) Valuing climate change impacts on human health: empirical evidence from the literature. Int J Environ Res Public Health 6(2):759–786

Marshall E, Randhir T (2008) Effect of climate change on watershed system: a regional analysis. Clim Change 89(3–4):263–280

McGillivray M (2007) Human well-being: issues, concepts and measures. Human well-being: concept and measurement. Palgrave Macmillan UK, London, pp 1–22

McMichael A, Scholes R, Hefny M, Pereira E, Palm C, Foale S (2005) Linking ecosystem services and human well-being. Island Press

MCTCA (2015) Nepal tourism statistics 2014. Government of Nepal Ministry of Culture, Tourism & Civil Aviation, Planning & Evaluation Division, Statistical Section. Singha Durbar, Kathmandu, Nepal

Millennium Ecosystem Assessment (2005) Washington, DC, Island Press

Mölg T, Hardy DR, Cullen NJ, Kaser G (2008) Tropical glaciers, climate change, and society. Darkening peaks: glacier retreat, science, and society. University of California Press, Berkeley, CA, pp 168–182

Myers N (2005) Environmental refugees: an emergent security issue. 13th Economic Forum, Prague

Myers SS, Smith MR, Guth S, Golden CD, Vaitla B, Mueller ND, Dangour AD, Huybers P (2017) Climate change and global food systems: potential impacts on food security and undernutrition. Annu Rev Public Health 38:259–277

Neupane D (2011) Impact of climate change on global food production system

Newell RG, Stavins RN (2000) Climate change and forest sinks: factors affecting the costs of carbon sequestration. J Environ Econ Manag 40(3):211–235

O'Connor MI, Bruno JF, Gaines SD, Halpern BS, Lester SE, Kinlan BP, Weiss JM (2007) Temperature control of larval dispersal and the implications for marine ecology, evolution, and conservation. Proc Natl Acad Sci 104(4):1266–1271

Opekunova MG, Opekunov AY, Kukushkin SY, Lisenkov SA, Vlasov SV, Somov VV (2022) Soil pollution with polycyclic aromatic hydrocarbons and petroleum hydrocarbons in the north of Western Siberia: spatial pattern and ecological risk assessment. Eurasian Soil Sci 55(11):1647–1664

Opoku SK, Filho WL, Hubert F, Adejumo O (2021) Climate change and health preparedness in Africa: analysing trends in six African countries. Int J Environ Res Public Health 18(9):4672

Palhares R et al (2015) Medicinal plants recommended by the World Health Organization: DNA barcode identification associated with chemical analyses guarantees their quality. PLoS ONE 10(5):e0127866

Pareek N (2017) Climate change impact on soils: adaptation and mitigation. MOJ Ecol. Environ. Sci 2:136–139

Perera F, Nadeau K (2022) Climate change, fossil-fuel pollution, and children's health. N Engl J Med 386(24):2303–2314

Power AG (2010) Ecosystem services and agriculture: tradeoffs and synergies. Philos Trans R Soc B Biol Sci 365(1554):2959–2971

Pramova E, Locatelli B, Djoudi H, Somorin OA (2012) Forests and trees for social adaptation to climate variability and change. Wiley Interdiscip Rev Clim Change 3:581–596

Raj AD, Kumar S, Sooryamol KR (2022a) Modelling climate change impact on soil loss and erosion vulnerability in a watershed of Shiwalik Himalayas. CATENA 214:106279

Raj R, Saharia M, Chakma S, Rafieinasab A (2022b) Mapping rainfall erosivity over India using multiple precipitation datasets. CATENA 214:106256

Raj R, Saharia M, Chakma S (2023) Mapping soil erodibility over India. CATENA 230:107271

Rodríguez-Morales AJ, Risquez A, Echezuria L (2010) Impact of climate change on health and disease in Latin America. In: Climate change and variability. Sciyo, Croatia, pp 463–486

Rosenzweig C, Iglesius A, Yang XB, Epstein PR, Chivian E (2001) Climate change and extreme weather events-Implications for food production, plant diseases, and pests

Rosenzweig C, Strzepek KM, Major DC, Iglesias A, Yates DN, McCluskey A, Hillel D (2004) Water resources for agriculture in a changing climate: international case studies. Glob Environ Chang 14(4):345–360

Saina CK, Murgor DK, Murgor FA (2013) Climate change and food security. Environ Change Sustain 10:55206

Sannigrahi S, Chakraborti S, Joshi PK, Keesstra S, Sen S, Paul SK, Kreuter U, Sutton PC, Jha S, Dang KB (2019) Ecosystem service value assessment of a natural reserve region for strengthening protection and conservation. J Environ Manage 244:208–227

Schmidhuber J, Tubiello FN (2007) Global food security under climate change. Proc Natl Acad Sci 104(50):19703–19708

Schmitz OJ, Barton BT (2014) Climate change effects on behavioral and physiological ecology of predator–prey interactions: implications for conservation biological control. Biol Control 75:87–96

Schwerdtle P, Bowen K, McMichael C (2018) The health impacts of climate-related migration. BMC Med 16(1):1–7

Scott D, Gössling S, Hall CM (2012) International tourism and climate change. Wiley Interdiscip Rev Clim Change 3(3):213–232

Sen AK (1999) Development as freedom. Oxford University Press, Oxford, p 336

Shah S (2019) Ethnomedicinal knowledge of indigenous communities and pharmaceutical potential of rainforest ecosystems in Fiji Islands. J Integr Med 17:244–249

Sharp R, Ricketts T, Guerry AD et al (2015) InVEST 3.2.0 beta user's guide. The Natural Capital Project, Stanford, pp 114–131

Shijin W, Dahe Q (2015) Mountain inhabitants' perspectives on climate change, and its impacts and adaptation based on temporal and spatial characteristics analysis: a case study of Mt. Yulong Snow, Southeastern Tibetan Plateau. Environ Hazards 14(2):122–136

Shrestha RP, Nepal N (2016) An assessment by subsistence farmers of the risks to food security attributable to climate change in Makwanpur, Nepal. Food Secur 8:415–425

Silva RA, West JJ, Zhang Y, Anenberg SC, Lamarque JF, Shindell DT, Collins WJ, Dalsoren S, Faluvegi G, Folberth G, Horowitz LW (2013) Global premature mortality due to anthropogenic outdoor air pollution and the contribution of past climate change. Environ Res Lett 8(3):034005

Sinha SK, Swaminathan MS (1991) Deforestation, climate change and sustainable nutrition security: a case study of India. Clim Change 19(1–2):201–209

Spalding M et al (2017) Mapping the global value and distribution of coral reef tourism. Marine Policy 82:104–113. https://thought-leadership-production.s3.amazonaws.com/2017/05/18/19/52/44/c655fbeeb0a5-4e48-a34f-2806ff724061/paper_coralreeftourism_spalding_2017.pdf (link as of 16 Dec 2019)

Terhaar J, Frölicher TL, Aschwanden MT, Friedlingstein P, Joos F (2022) Adaptive emission reduction approach to reach any global warming target. Nat Clim Chang 12(12):1136–1142
Thomson AM, Brown RA, Rosenberg NJ, Izaurralde RC, Legler DM, Srinivasan R (2003) Simulated impacts of el nino/southern oscillation on united states water resources 1. Jawra J Am Water Resour Assoc 39(1):137–148
Thor P, Bailey A, Dupont S, Calosi P, Søreide JE, De Wit P, Guscelli E, Loubet-Sartrou L, Deichmann IM, Candee MM, Svensen C, King AL, Bellerby RGJ (2018) Contrasting physiological responses to future ocean acidification among Arctic copepod populations. Global Change
Tol RSJ (2008) Why worry about climate change? A research agenda. Environ Values 17(4):437–470
Tol RSJ (2009) The economic effects of climate change. J Econ Perspect 23(2):29–51
Tol RS (2018) The economic impacts of climate change. Review of environmental economics and policy
United Nations (2023) What is climate change? https://www.un.org/en/climatechange/what-is-climate-change
Wade K, Jennings M (2016) The impact of climate change on the global economy. Schroders Talking Point
Warner K (2010) Global environmental change and migration: governance challenges. Glob Environ Chang 20(3):402–413
Warner K, Hamza M, Oliver-Smith A, Renaud F, Julca A (2010) Climate change, environmental degradation and migration. Nat Hazards 55:689–715
Weiskopf SR, Rubenstein MA, Crozier LG, Gaichas S, Griffis R, Halofsky JE, Hyde KJ, Morelli TL, Morisette JT, Muñoz RC, Pershing AJ (2020) Climate change effects on biodiversity, ecosystems, ecosystem services, and natural resource management in the United States. Sci Total Environ 733:137782
Weissbecker I (ed) (2011) Climate change and human well-being: global challenges and opportunities. Springer Science & Business Media
Whitmee S, Haines A, Beyrer C, Boltz F, Capon AG, de Souza Dias BF, Ezeh A, Frumkin H, Gong P, Head P, Horton R (2015) Safeguarding human health in the Anthropocene epoch: report of the Rockefeller Foundation-Lancet Commission on planetary health. The Lancet 386(10007):1973–2028
WHO (2016) Malaria fact sheet. http://www.who.int/mediacentre/factsheets/fs094/en/
Wise EK, Comrie AC (2005) Meteorologically adjusted urban air quality trends in the Southwestern United States. Atmos Environ 39:2969–2980
WMO (2023) Promoting wellbeing. https://www.who.int/activities/promoting-well-being#:~:text=Well%2Dbeing%20is%20a%20positive,social%2C%20economic%20and%20environmental%20conditions
World Bank (2016) The cost of fire: an economic analysis of Indonesia's 2015 fire crisis. http://pubdocs.worldbank.org/en/643781465442350600/Indonesia-forest-fire-notes.pdf (link as of 16 Dec 2019)
World Food Summit (1996) Food and Agricultural Organization. https://www.fao.org/3/w3548e/w3548e00.htm
World Health Organization (2008) Climate change and human health in Asia and the Pacific: from evidence to action: report of the regional workshop, Bali, Indonesia, 10–12 Dec 2007 (No. SEA-EH-555). WHO Regional Office for South-East Asia
World Meteorological Organization (2021) State of the global climate, 2020. https://library.wmo.int/doc_num.php?explnum_id=10618
Wu J (2013) Landscape sustainability science: ecosystem services and human well-being in changing landscapes. Landscape Ecol 28:999–1023
Zhang B, Tian L, Yang Y, He X (2022) Revegetation does not decrease water yield in the Loess Plateau of China. Geophys Res Lett 49(9):e2022GL098025
Ziervogel G, Ericksen PJ (2010) Adapting to climate change to sustain food security. Wiley Interdiscip Rev Clim Change 1(4):525–540

Hua T, Zhao W, Cherubini F, Hu X, Pereira P (2021) Sensitivity and future exposure of ecosystem services to climate change on the Tibetan Plateau of China. Landscape Ecol 36(12):3451–3471
Morgan RPC (2009) Soil erosion and conservation. John Wiley & Sons

Part II
Climate Crisis, Social Responses and Sustainability on Ecological Perspectives

Chapter 2
Assessment of Flood Vulnerability of the People Living in Tal and Diara Geomorphic Regions of Malda District, West Bengal

Hiranmay Rishi and Subrata Purkayastha

Abstract Malda is located at the interfluve of the Ganga and Brahmaputra river systems. This district is highly flood-prone, with both geomorphology and anthropogenic activity triggering floods and river bank erosion, leaving the population extremely vulnerable to flood hazards. Malda is divided into three geomorphic units: Tal, Diara, and Barind of which Tal and Diara are particularly prone to flooding. The present chapter attempts to quantify and compare the vulnerability to flood hazards of the local population belonging to different socioeconomic groups residing in the Tal and Diara geomorphic units. Additionally, the chapter investigates whether proximity from the two flood-producing rivers of Tal and Diara, Fulhar and Ganga, influences the degree of vulnerability of the Tal and Diara populations. The study is based on an examination of empirical data obtained through field investigations and focus group discussions in a pair of villages: Maniknagar and Ratua in Tal and Gopalpur and Nurpur in Diara. The study uses a weighted average approach to assess flood vulnerability by applying the Livelihood Vulnerability Index (LVI). The findings suggest the population's vulnerability varies by location within Tal and Diara, with settlements near major flood-causing rivers being more vulnerable. This chapter is significant as it examines and assesses the vulnerability of people impacted by floods at the grassroots level. This may be used by lawmakers and social scientists to assess vulnerability and lessen the flood hazard of the population living in similar ecological conditions.

Introduction

Catastrophic events such as floods are getting more frequent and severe as an outcome of climate change and global warming that the world is now experiencing. Flooding is one of the most common tragedies, affecting about 250 million people worldwide

H. Rishi (✉) · S. Purkayastha
Department of Geography, North Eastern Hill University, Shillong, India
e-mail: rishihiran@gmail.com

U. Mukhopadhyay et al. (eds.), *Climate Crisis, Social Responses and Sustainability*, Climate Change Management, https://doi.org/10.1007/978-3-031-58261-5_2

(Allaire 2018). In contrast to developed countries, where the population is more resilient to flood hazards owing to proper risk reduction, preparedness, and response systems at both the national and local levels, the people in emerging nations like India are greater vulnerable to flood risk due to inadequate infrastructure, as they lack early warning systems and organisational preparedness. Hence marginalised populations, especially from developing countries are disproportionately exposed and vulnerable to such hazards (World Bank 2018). It is important to note that, over the last 20 years, the traditional global approach to mitigating the impact of disasters such as floods has shifted from a prescriptive (top-down) to a more proactive approach in which the voices of the local community or affected households/individuals are considered (Wallemacq 2018). The Hyogo Framework for Action emphasises the importance of developing an indicator-based approach to measure disaster vulnerability and risk at the national and local levels to help decision-makers mitigate disaster consequences and foster proactive disaster risk reduction in order to develop resilient societies in disaster-affected countries (Heinkel et al. 2022; Nam et al. 2023; United Nations 2005). An analogous approach has been adopted in the present chapter, where indicators have been picked at the household level to quantify the vulnerability to floods of the villagers living in Tal and Diara, so as to build a more resilient society.

Vulnerability includes two elements, namely external and internal (Bohle 2016; Naheed 2021). The external element involves an individual's/household's/community's exposure and sensitivity to catastrophes as outlined by the Political Economy Frameworks (for instance, social inequities, unequal sharing of resources), the Human Ecology Perspectives (demographic features, and environmental management capacity), alongside the Entitlement Theory (which relates to the population's susceptibility or failure to access or handle resources through proper financial means). The internal component relates to the population's coping and adaptation strategies at the individual, household, and community levels. This includes the ability to foresee, deal with, resist, and recover from the effects of a given threat. The three notions viz. The crisis and conflict situation, i.e. control over assets and resources, as well as the ability to deal with crises and settle disagreements. The action of the affected individuals, how people behave in such crisis situations, and their reactions, which in turn are regulated by social, economic, or government limits. Finally, the access to assets, which effects the minimising of vulnerability through obtaining assets. As a result, the better preparedness, coping, and adaptation approaches, the less vulnerable the household is to climatic shocks. Similarly, the present chapter attempts to incorporate both the internal and external components of the vulnerability concerning the villagers at the household level residing in the Tal and Diara regions of Malda district by developing a Livelihood Vulnerability Index (LVI) constructed by Hahn et al. (2009).

Several scholarly efforts are referring to the assessment of vulnerability and adapting capacity of populations at the household level through the creation of indices (Ahsan and Warner 2014; Gbetibouo et al. 2010; Padhan and Madheswaran 2023; Pandey and Jha 2012). Nevertheless, most of these researches have difficulties in conceptualising the issue or data-related challenges in selecting and integrating relevant variables in respective indices, resulting in a lack of accurate and trustworthy evaluation. However, the IPCC Vulnerability Framework, which incorporates exposure, sensitivity, and adaptive capacity, is highly recommended for constructing a vulnerability index, particularly for riverine areas (Mitra and Kumar Mandal 2023; Patwardhan et al. 2007). This index combines/aggregates individual scores across multiple susceptibility factors in a geographical context, and recommends optimal coping and adapting measures for riverine communities at the household level (Smit and Wandel 2006).

Emphasising Chambers and Conway's (1992) Sustainable Livelihood Approach (SLA), which incorporates five types of assets, viz., natural, social, physical, human, and financial, Hahn et al. (2009) proposed the Livelihood Vulnerability Index (LVI), where household-level data are accumulated to develop community-level strategies, which are widely used by many scholars in various contexts (Al Mamun et al. 2023; Alam 2017; Panthi et al. 2016; Shahzad et al. 2021). Assessments of livelihood risk can help in decision-making at two levels, i.e. adaptation, and planning process (Singh 2023). The SLA approach combines climate exposure, sensitivity, and household adapting capacity that in turn serves to provide insight into the inequalities at the community level to climate-related jeopardy. Applying the Hahn et al. (2009) approach, the current chapter investigates to comprehend both the livelihood and climatic vulnerability of the villagers residing in two geomorphic divisions of Malda district, Tal and Diara, which are both riverine areas. To construct the LVI, information on selected indicators representing the five variables of the Sustainable Livelihood Approach (SLA), viz. social, physical, natural, human, and financial assets/capitals of the villagers at household level, were collected through field investigation and analysed to suggest the degree of vulnerability to floods of the locals dwelling in Tal and Diara. A brief review of the literature suggests that the bulk of climate-related vulnerability studies undertaken so far, particularly for the population living in riverine areas, rely on secondary data sources (Ahmad et al. 2023; Panthi et al. 2016; Rehman et al. 2021). Therefore, this chapter is unique in that it investigates the vulnerability of the selected villagers residing in Tal and Diara using primary information gathered from field investigations of indigenous households conducted in September–December 2018 following the severe flood event of August 2017. This delivers a clear picture of the degree of vulnerability during flood events at grassroots level, how local people perceive flood risk, their readiness, awareness, and mitigation measures meted out through their coping and adaptive strategies, which can be used by administrators and stakeholders alike to build a more flood resilient society across the Tal and Diara geomorphic units of Malda district.

The location of Malda district along the outlets of the Ganga–Brahmaputra drainage system renders this region highly flood-prone as there is massive discharge in the volume of water especially during the rainy season by the tributaries having their source in the snowclad ranges of the Himalayas in the north that feed the Ganga–Brahmaputra river system (Chakraborty et al. 2023; Rudra 2018). Malda possesses para-deltaic features resulting from deltaic hydrological activities in the past, as well as a massive basin area impacted by frequent river shifts, resulting in the formation of paleochannels, cutoff channels, and an active flood plain, making this area extremely susceptible to flood hazard (District Human Development Report 2007; Mehebub et al. 2015; Mukherjee and Pal 2018). Diara is an active alluvial floodplain, with floods enriching the soil almost every year. Tal is marked by marshy and muddy regions including vestiges of old meanders that are inundated consistently, especially during the rainy season (District Human Development Report 2007; Ghosh et al. 2022). As previously stated, significant rivers passing through Malda and its surroundings originate in the Himalayan or Sub-Himalayan region, making them susceptible to sudden inflows of massive amounts of water, particularly during heavy rainfall in the upper and middle catchment areas, and the majority of these rivers flow from north-west to south-east towards the neighbouring country of Bangladesh (Ghosh and Kar 2018). Since proximity to a river is an important cause of flooding, the western part of the district is characterised by a significant number of rivers flowing from north to south. As a consequence, low-lying active floodplains around the Tal and Diara rivers experience seasonal flooding (Ghosh et al. 2022). In addition, this region is marked by several river confluences, which trigger floods owing to the formation of massive flow accumulation at convergence points (Roca et al. 2009). Flooding in the Tal and Diara regions is primarily caused by a rapid rise in water level over a short period of time, along with a poor drainage system.

Aside from physical issues, anthropogenic causes have exacerbated flood risk in both Tal and Diara (Biswas and Anwaruzzaman 2019). The outstanding fertility of the soil, aided by regular floods, draws agrarians who make a decent living from different farming activities and fishing. This has resulted in a high concentration of population here. Tal and Diara both record higher population growth and density than the state average. High population density and land use change near riverine active floodplains and levees have increased the impact of flood hazards in both Tal and Diara. Furthermore, the construction of the controversial Farakka barrage across the Ganga in the 1970s sparked not only a water politics dispute between India and Bangladesh but also altered the Ganga's morphometry and geohydraulic regime, making Malda, which is located upstream of Farakka, more vulnerable to floods and river bank erosion (Ghosh et al. 2022; Rudra 2018). Thus, climate change, along with natural, demographic, and technological factors, has increased the frequency of flood threats in and around Malda, rendering the people here more exposed and vulnerable to such risks. This calls for proper assessment of disaster risk reduction, flood monitoring, and management in order to develop a resilient society. It is in this context, that this chapter focuses on evaluating the level of livelihood vulnerability of villagers in Tal and Diara in order to increase their resistance capability, as vulnerability and resilience are inseparable (Lin et al. 2017). The primary objective of this chapter is to evaluate and compare the local population's vulnerability, responses, resilience, and mitigation strategies to flood hazards. Furthermore, to evaluate whether the distance between Tal and Diara's major flood-causing rivers, i.e. Fulhar and Ganga, has a significant impact on the villagers' susceptibility. The assessment was carried out by performing case studies utilising empirical data from two sets of villages, one nearby the Fulhar River in Tal and the Ganga River in Diara, and the other set farther away from the two rivers in Tal and Diara.

Study Area

Malda district, as already mentioned, is mostly rural, with three distinct geomorphic units (Fig. 2.1): (a) Barind, comprised of historical alluvial deposits, are relics of a previous flood plain, now covering comparably higher land less prone to flooding (Majumdar and Mandal 2021). (b) Tal, a swampy region dominated by bog lands across the flood plain surrounding declining inland drainages that is very susceptible to floods (Mehebub et al. 2015). (c) Diara is flanked by rivers on three sides; it is a well-drained plain comprised of fresh alluvial deposits that are enhanced annually by floods; and the area is exceptionally productive, making it ideal for agriculture and related livelihood pursuits (Mukherjee and Pal 2018). Administratively, Diara is divided into five C.D. blocks. Three blocks, Manikchak, Kaliachak-II, and Kaliachak-III, are located along the River Ganga, Diara's main flood-causing river, while the other two blocks, i.e., English Bazar and Kaliachak-I, are located at a distance from the River Ganga. Tal presently consists of six C.D. blocks, two of which, Ratua-I and Harischandrapur-II, are located close to River Fulhar, Tal's major flood-causing river, and the other four blocks, Harischandrapur-I, Chanchal-I, Chanchal-II, and Ratua-II, are located more distant from River Fulhar (Fig. 2.1). Both Tal and Diara are extensively prone to alluviation and diluviation, as well as inundation, and are distinguished by shifting river courses coupled with narrow, meandering, and sluggish channels that tend to overflow during rainy season (Chakraborty et al. 2023). The porous soil profile here, combined with unconsolidated loosened soil material, results in the collapse of river banks in chunks, engulfing the river banks and exposing the rural population to loss of land, livelihood, property, and even life. Since villages are engulfed by rivers, population displacement is quite common here (Iqbal 2010).

Rationale of the Study

Flooding is unavoidable owing to global warming and climate change. This chapter seeks to explore the human reaction to floods, specifically how the vulnerable population responds to such conditions, where an understanding of the degree of vulnerability becomes essential. The degree of vulnerability of the people living in Tal and Diara is based on scientific analysis of relevant empirical data collected at the household level, where the five assets of the villagers, namely social, human, physical, natural, and financial, were measured using appropriate indicators as per the sustainable livelihood approach. This sort of micro-level investigation on the nature and dimension of vulnerability at the household level in two contrasting physiographic environments (Tal and Diara) has significant academic and practical implications. The investigation of the present situations of individuals living on flood plains and at a distance from main flood-causing rivers, namely Fulhar in Tal and Ganga in Diara, with varying levels of vulnerability. Consequently, reacting differently to the flood threat, as evidenced in their diverse immediate coping and subsequent adaptation

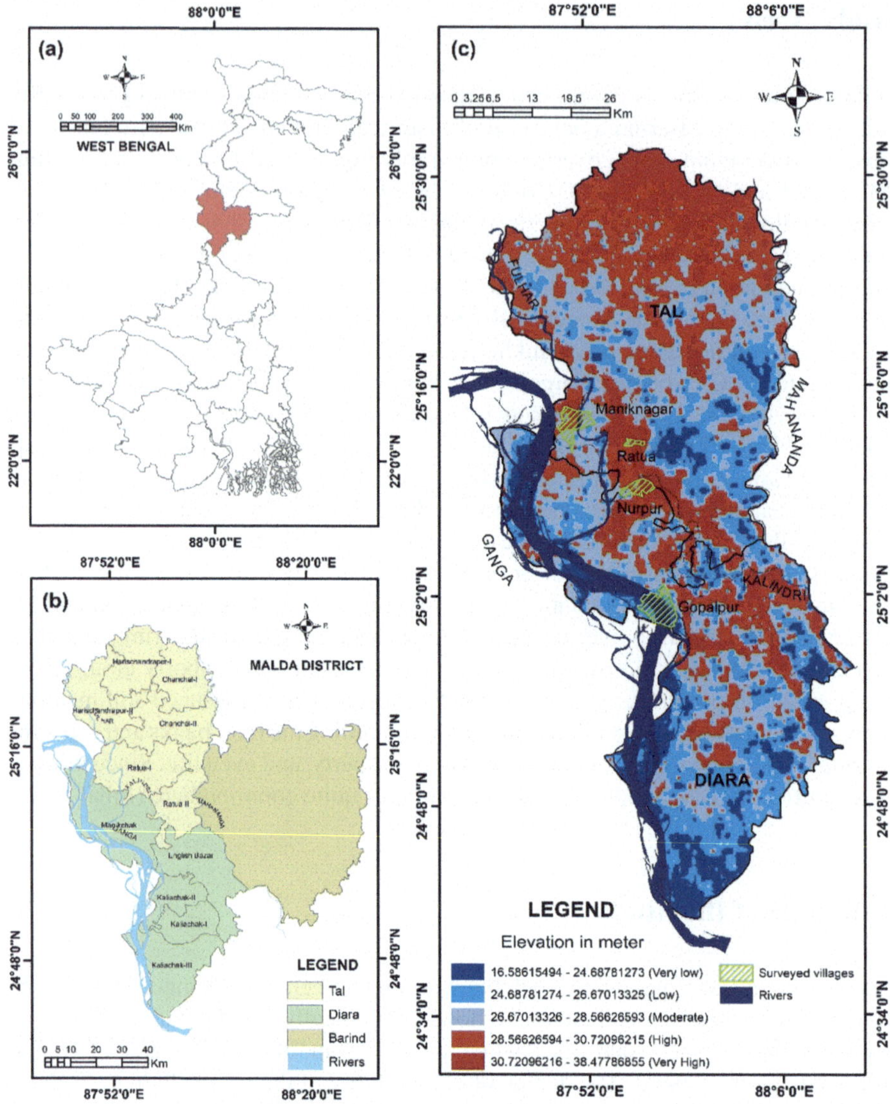

Fig. 2.1 Location of the study area: **a** Location of Malda district within West Bengal; **b** Three broad physiographic regions and C.D. block-wise division of Tal and Diara within the district; **c** Potential area subjected to inundation situation within Tal and Diara

techniques employed by the people in carrying out sustainable economic activities for their existence and living, may be replicated elsewhere in the world. This chapter focuses on the degree of vulnerability to flood hazards at the grassroots level, as well as the coping and adaptation mechanisms developed by impoverished rural populations living in rural Malda, which represents the marginalised population of developing economies with limited institutional interventions but the potential to benefit from flooding. As previously stated, the empirical findings may be replicated in parallel ecological situations around the globe to mitigate flood hazards, allowing people to adapt to floods and reap the benefits of floods.

Materials and Methods

The chapter is mainly based on empirical data collected during a field investigation from September to December of 2018, shortly after the major flood incident of August 2017. The two sets of villages were surveyed to obtain the necessary information on the vulnerability of the villagers living on the banks of the major flood-causing rivers, Ganga in Diara and Fulhar in Tal, as well as the population of the interior villages located a distance away from the two mentioned rivers. Data was gathered at two levels: focus group discussions among village elders and semi-structured questionnaires from chosen sample households in Tal and Diara.

To identify each pair of villages to be surveyed, a flood inundation map was first constructed by applying DEM (Digital Elevation Model) classification of Tal and Diara, obtained from the USGS (Fig. 2.1). Based on this map, two villages for field investigation were chosen: Maniknagar of Ratua-I block within Tal, alongside Gopalpur of Manikchak block within Diara, both of which are located at low elevations (18–24 m) and near the rivers Fulhar in Tal and Ganga in Diara, subsequently both are extremely prone to flooding. Similarly, two interior villages located at relatively higher elevations (28–35 m) within the same C.D. blocks of Tal and Diara, namely Ratua in Tal and Nurpur in Diara, have been chosen for fieldwork. Since they are at a higher elevation and placed distant from the principal flood-causing rivers of Tal and Diara, they are less impacted by floods (Fig. 2.1).

The Livelihood Vulnerability Index (LVI), proposed by Hahn et al. (2009), was used in this study to examine the differential effect of flood threats among the two settlements of Tal and Diara. Two processes were used in this technique, i.e. first, LVI was computed, which is a weighted mean of nine major components (health, knowledge and skills, livelihood, socio-demographic, social network, house type, and finance/income) under five capitals/assets such as human, natural, social, physical, and financial, with 30 indicators to represent all the five capitals/assets (Table 2.1). In the second stage, the nine major components of LVI were averaged independently to fit within the IPCC's three main contributing factors determining susceptibility to risk in this case floods, viz. exposure, sensitivity, and adaptive capacity of the local population (Hahn et al. 2009; Madhuri et al. 2014). The LVI score is scaled from 0 to 1, with 0 indicating the least vulnerability and 1 indicating the maximum vulnerability to flood hazards.

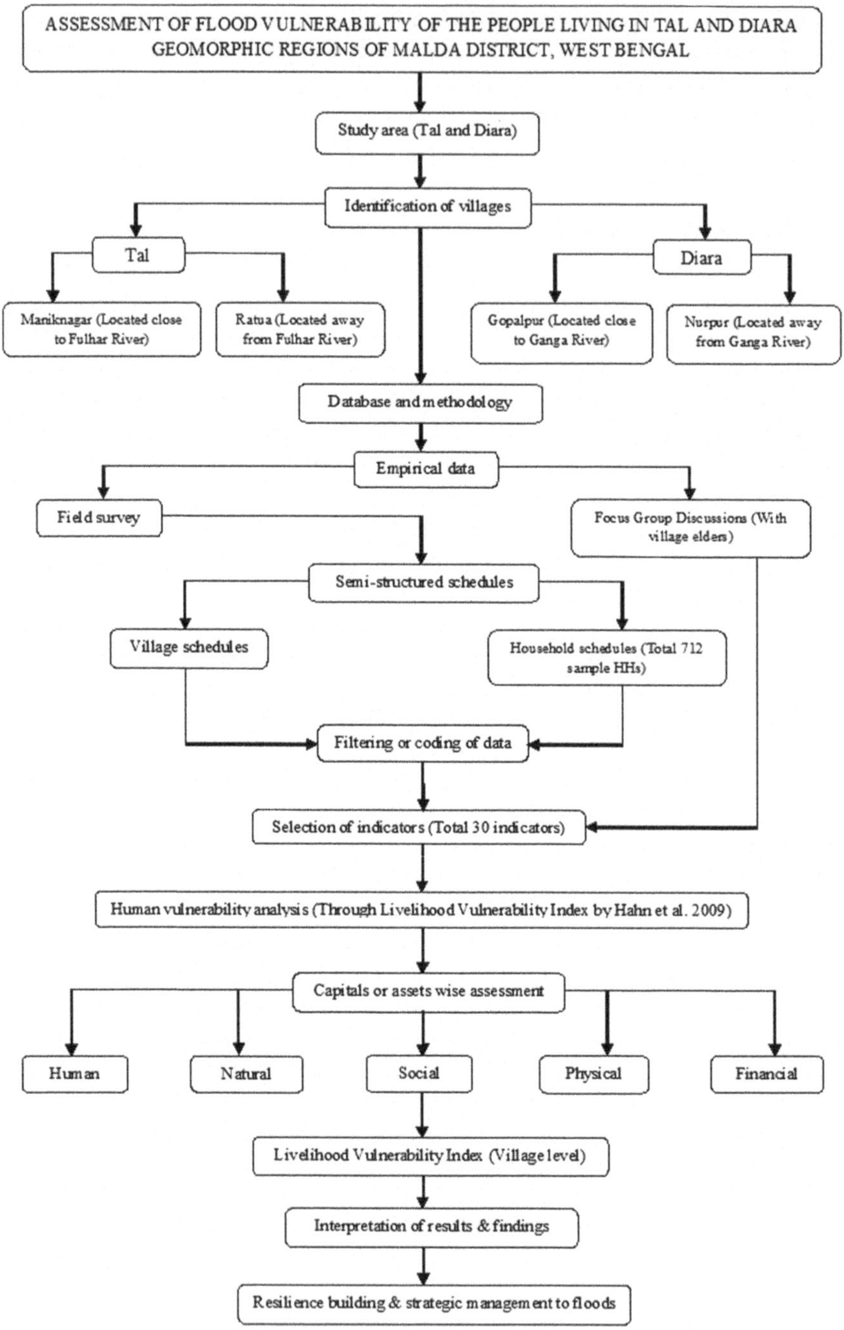

Fig. 2.2 Methodological flowchart of the study

Table 2.1 Capitals/assets, major components, and indicators used to calculate the Livelihood Vulnerability Index (LVI)

Capitals/assets	Major components	Indicators
Human	Health	Percentage of HHs reported sickness among family members during the recent flood incident
		Percentage of HHs reported inadequate access to treatment facilities at government hospitals during the most recent flood
		Percentage of HHs without proper sanitation facilities
		Percentage of HHs reporting no access to piped water supply
	Knowledge and skills	Percentage of HHs wherein the head of the HHs had not gone to school
		Percentage of HHs wherein the head of the HH has completed primary schooling
		Percentage of HHs in which the head of the HH has completed upper primary schooling
	Livelihood	Percentage of HHs with family members who work in another village, district, or state
		Percentage of HHs that rely solely on agriculture for their livelihoods
		Percentage of HHs claiming non-firming activities as their principal source of earnings
		Percentage of HHs who reported altering their livelihood practise
Natural	Land/water	Percentage of landless HHs
		Percentage of HHs with marginal land (< 1.0 ha)
		Percentage of HHs who rely on fishing during flood
	Natural disasters, and impact	Percentage of HHs reported loss of physical assets, i.e., house, machinery due to flooding in the past six years
		Percentage of HHs reported injuries or death of at least one family member caused by flood over the previous six years
		Percentage of HHs who reported moving to a safer location more than three times
Social	Socio-demographic	Percentage of dependent population (< 15 yrs and > 60 yrs)
		Proportion of HHs with female heads
		Average size of family
	Social networks	Percentage of HHs who received assistance from neighbours or relatives during floods

(continued)

Table 2.1 (continued)

Capitals/ assets	Major components	Indicators
		Percentage of HHs who do not have any family members involved in local organisations such as masjids, temples, flood committees, and so on
Physical	House type	Percentage of HHs who own *kutcha* dwellings
		Percentage of HHs stated that their dwellings were affected by floodwaters (partially or entirely submerged)
		Percentage of HHs with stilted houses
Financial	Finance/income	Percentage of BPL families
		Percentage of HHs without work during flood season
		Percentage of HHs who save extra money
		Percentage of HHs who reported borrowing money to deal with the floods
		Percentage of HHs without insurance (life and crop insurance)

Source Adopted from Hahn et al. (2009), indicators are slightly modified as per field situation

Computation of the Livelihood Vulnerability Index (LVI): A Combined Score Method

Despite the fact that every major component includes several sub-components, the LVI employs a balanced weighted average method (Alam 2017; Mitra and Kumar Mandal 2023), whereby a single sub-component counts equally to the total index. The LVI formulation takes the straightforward technique of assigning equal weights to each component.

Since each indicator is scored on a distinct scale, they are standardized first. For this, the UNDP procedure used in the Human Development Index (UNDP 2000) was utilised, with each indicator's original values scaled from 0 to 1. The calculation is illustrated in Eq. (2.1):

$$index_{S_v} = \frac{S_v - S_{\min}}{S_{\max} - S_{\min}} \tag{2.1}$$

where, S_v denotes the original sub-component for corresponding village v, and $S_{\min}$ and $S_{\max}$ represent the minimum and maximum values for each indicator, accordingly.

After standardising each indicator, the sub-components were combined by applying Eq. (2.2) to obtain the value associated with every sub-component:

$$M_v = \frac{\sum_{i=1}^{n} index_{s_v i}}{n} \tag{2.2}$$

where, M_v represents one of the nine major components for village v; $index_{S_v i}$ represents the sub-components, indexed by i, which comprise every single component; and n is the number of indicators in each sub-component.

To calculate the score for each of the village's nine major components, each livelihood asset or capital was averaged using Eq. (2.3) to get a vulnerability score. The values for each capital in a village were summed up using the same Eq. (2.3) to produce the village-level LVI:

$$LVI_v = \frac{\sum_{i=1}^{5} w_{M_i} M_{vi}}{\sum_{i=1}^{5} w_{M_i}} \tag{2.3}$$

This can be stated as

$$LVI_V = \frac{w_H H_v + w_N N_v + w_S S_v + w_P P_v + w_F F_v}{w_H + w_N + w_S + w_P + w_F} \tag{2.4}$$

where, LVI_v denotes the vulnerability score for one of the five livelihood capitals of village v, equivalent to the weighted mean of the nine key components, which constitute particular livelihood capital; w_{M_i} represents the weightings of every major component, which are computed from the number of sub-components that comprise each major component or capital, and contribute evenly to the cumulative LVI (Panthi et al. 2016).

LVI-IPCC

The IPCC framework defines livelihood vulnerability as the exposure, sensitivity, and adaptability of a system (Madhuri et al. 2014; Shah et al. 2013). The LVI-IPCC is assigned a score ranging from − 1 (least vulnerable) to 1 (most vulnerable). The IPCC Livelihood Vulnerability Index (LVI-IPCC) for both sets of villages was calculated using the following approach (Table 2.2).

Table 2.2 Classification of the major components as contributing factors (stated by the IPCC)

Contributing factors to vulnerability outlined by IPCC	Components
Exposure	Natural disasters, and impact
Adaptive capacity	Demographic, socio-economic, livelihood, social network
Sensitivity	Health, food, water, natural capital

Source Adapted from Hahn et al. (2009)

Computation of LVI-IPCC: An IPCC Framework Approach

Rather than incorporating the components into the LVI all at once, each of the sub-components was first integrated using the following Eq. (2.5) according to the major component categorization:

$$CF_v = \frac{\sum_{i=1}^{n} w_{M_i} M_{vi}}{\sum_{i=1}^{n} w_{M_i}} \tag{2.5}$$

where, CF_v is an IPCC-defined contributing factor (exposure, sensitivity, or adaptive capacity) represents village v, M_{vi} represent components for the village v ordered by i, w_{M_i} is the weighting for each component, while n denotes the number of components associated with each contributing factor.

The three contributing components were integrated using the following Eq. (2.6) after exposure, sensitivity, and adaptive capacity were calculated:

$$LVI\text{-}IPCC_v = (e_v - a_v) * s_v \tag{2.6}$$

where, $LVI\text{-}IPCC_v$ represents the LVI of the village v, employing the IPCC vulnerability system, e represents the estimated exposure value of the village v (equal to the natural disaster component), a is the estimated adaptive capacity value of the village v (comprised of a weighted sum of socio-demographic, livelihood strategies, social networks, and natural capital components), while s is the computed sensitivity value of the village v (weighted mean of health, food, and water components).

Sample Designing

The household survey was carried out in two pairs of villages within Tal and Diara using a stratified random sample technique. Considering the villages' large quantity of HHs, a total of 712 HHs were interviewed in the four selected villages (Table 2.3). The sample HHs were estimated utilising web-based assistance, that is, a sample size calculator by Creative Research Systems (https://www.surveysystem.com/sscalc.htm) that adopts the formulae proposed by Cochran (1963). The amount of HHs that were surveyed in each village was determined using the equation below (Eq. 2.9):

$$SS = \frac{Z^2 \times (p) \times (1 - p)}{C^2} \tag{2.9}$$

where, SS stands for sample size; Z denotes the standard score equivalent to a specific confidence level (for example, 1.96 for 95% confidence level); p represents the proportion of choosing an option, stated as a decimal (for example, 0.5 for required

Table 2.3 Sampling structure

Regions	Name of the C.D. blocks	Name of the villages	Total households	Sample households	Confidence level (%)	Confidence interval[a]
Tal	Ratua-I	Maniknagar	1880	178	95	7
		Ratua	1224	170	95	7
Diara	Manikchak	Gopalpur	2010	180	95	7
		Nurpur	2933	184	95	7
Total			8047	712		

[a] ± 7 denotes the margin of error. N.B: The sample HHs were picked using a web-based resource, namely the Creative Research Systems sample size calculator (https://www.surveysystem.com/sscalc.htm), which uses the equations proposed by Cochran (1963)

sample size); while C denotes the confidence interval in decimal form (for example, $0.04 = \pm 4$).

Adjustment for the Finite Population

Since acquiring the sample size from Eq. (2.9), it has become essential to alter this corresponding to the limited population to deliver appropriately greater details for the given population size. The equation used for rectification is as follows (Eq. 2.10):

$$SS = \frac{SS}{1 + \frac{SS-1}{pop}} \tag{2.10}$$

where, *pop* represents population

Results and Discussion

Approaches to Flood Hazard and Mitigation

It is worth noting that traditionally, hazards that led to disasters were seen as discrete environmental phenomena, with little connections made to the demographics and socioeconomic situations of the impacted population (Clar 2019). The previous approach to restoring normalcy in any hazard-affected population or region was more of a technical solution, relief, and rehabilitation efforts were taken up by the concerned authorities as they were of key importance. However, there has been a paradigm shift towards hazard mitigation since the United Nations' International Decade for Natural Disaster Reduction (IDNDR) in the 1990s and the UN International Strategy for Disaster Reduction (ISDR), with the current approach focusing on assessing people's vulnerability and resilience at the individual and community levels to mitigate flood hazards (Adedeji et al. 2012).

Floods and Vulnerability: Livelihood Vulnerability Index (LVI) of Tal and Diara

Flood is a natural occurrence that cannot always be considered a threat. However, certain floods might be viewed as hazards leading to disasters. Such flood threats are especially widespread in and around the Malda district, and they strike literally every year. A variety of strategies can be used to assess people's vulnerability (Padhan and Madheswaran 2023; Sai et al. 2021). As already mentioned, a modified version of the Livelihood Vulnerability Index (LVI) devised by Hahn et al. (2009) was utilised in this chapter to assess the degree of vulnerability of persons living in Tal and Diara (based on river proximity). The LVI has been computed with the help of five assets/ capitals related to livelihood, which have been addressed as follows:

Vulnerability Associated with Human Capital

Human capital refers to a community's or household's skills, education level, and good health, which are necessary for pursuing numerous livelihood options (Kucharčíková 2011). This study quantifies human capital using a weighted average score of three key components: health (four indicators), knowledge and skills (three indicators), and livelihood (four indicators). The vulnerability score of the riverfront settlements located in Tal and Diara, i.e., Gopalpur (0.54) in Diara and Maniknagar (0.39) in Tal, suggests that in terms of human capital, the population of the riverside villages reports a higher level of vulnerability in comparison to the interior villages of Nurpur (0.22) in Diara and Ratua (0.20) in Tal (Fig. 2.4; Tables 2.4 and 2.5). It is owing to the fact that the two mentioned riverbank villages have a greater proportion of less-educated HHs (80.50% in Gopalpur and 63% in Maniknagar) who lack formal education and skill development compared to the inner villages of Nurpur (40%) and Ratua (35%). The health status of these riverine communities is poor, with a greater percentage of HHs reporting a lack of sufficient medical services due to the absence of functioning public or community health centres in these areas. The lack of basic infrastructure, such as good sanitation (40% in Gopalpur and 18% in Maniknagar) and drinking piped water (97% in Gopalpur and 92% in Maniknagar), leaves riverine villages particularly vulnerable during and after flood events. Furthermore, life is tough in the riverfront areas of Gopalpur and Maniknagar since floods and bank erosion are prevalent here, resulting in the loss of productive agricultural land, property, livestock, livelihood, and sometimes even life. In comparison to the inner villages of Tal and Diara, a higher proportion (87.3% in Gopalpur and 63% in Maniknagar) of working-age people move out or migrate to other states in search of better employment opportunities. This migration is primarily seasonal, as migrant workers return to reap good agricultural produce after the flood. According to field data, a larger percentage of respondents (83% in Gopalpur and 36% in Maniknagar) had modified their major livelihood practises owing to flood and bank erosion in riverfront villages than in interior villages. Moreover, riverine communities rely more heavily on agriculture or fishing for earnings, leaving them more vulnerable. All of these conditions have led to the riverbank settlements' greater vulnerability in terms of human capital when compared to the inland villages.

Table 2.4 Inter-village variability of the Livelihood Vulnerability Index (LVI) in Tal

Major components	Maniknagar	Ratua	Tal	Capitals	Maniknagar	Ratua	Tal
Health	0.41	0.21	0.31	Human	0.39	0.20	0.29
Knowledge and skills	0.34	0.16	0.25				
Livelihood	0.40	0.22	0.31				
Land/water	0.58	0.37	0.47	Natural	0.43	0.19	0.31
Natural disasters, and impact	0.28	0.01	0.15				
Socio-demographic	0.27	0.16	0.21	Social	0.41	0.31	0.36
Social networks	0.62	0.55	0.58				
House type	0.68	0.19	0.43	Physical	0.68	0.19	0.43
Finance/income	0.79	0.31	0.55	Financial	0.79	0.31	0.55
Livelihood Vulnerability Index (LVI) score					0.50	0.23	0.36

Source Computed by the researchers

Table 2.5 Inter-village variability of the Livelihood Vulnerability Index (LVI) in Diara

Major components	Gopalpur	Nurpur	Diara	Capitals	Gopalpur	Nurpur	Diara
Health	0.54	0.25	0.39	Human	0.54	0.22	0.38
Knowledge and skills	0.45	0.20	0.32				
Livelihood	0.62	0.21	0.41				
Land/water	0.78	0.36	0.57	Natural	0.66	0.18	0.42
Natural disasters, and impact	0.55	0.01	0.28				
Socio-demographic	0.31	0.09	0.20	Social	0.44	0.28	0.36
Social networks	0.63	0.56	0.59				
House type	0.77	0.16	0.46	Physical	0.77	0.16	0.46
Finance/income	0.85	0.32	0.58	Financial	0.85	0.32	0.58
Livelihood Vulnerability Index (LVI) score					0.62	0.23	0.43

Source Computed by the researchers

Vulnerability Associated with Natural Capital

Natural capital is defined as access to natural resources that are necessary for livelihoods (Missemer 2018). Tal and Diara are both populated localities with a high population density; forest resources are few; as a result, access to land and water resources is crucial for livelihood practises, since agriculture and fishing are the main economic activity in both locations. As a result, the weighted mean of the two key components, namely land/water (with three indicators) and natural disaster and its consequences (that is actually flood and its impact) on the villagers over the previous six years from the date of the survey (comprising of three indicators), was measured and discussed. The overall score of natural capital is greater in the riverfront settlements of Tal and Diara, suggesting higher vulnerability, viz., Gopalpur (0.66) within Diara and Maniknagar (0.43) within Tal, compared to the inner villages of Nurpur (0.18) within Diara and Ratua (0.19) within Tal (Fig. 2.4).

Although the major components of natural capital are considered, cross-village heterogeneity is also evident in Tal and Diara (Tables 2.4 and 2.5). In regards to access to the extent of land holdings and significant sources of income from water resources, Gopalpur (0.78) and Maniknagar (0.58) show more susceptibility than Nurpur (0.36) and Ratua (0.37). This may be attributed to the fact that Gopalpur and Maniknagar have larger percentages of landless (89.31% in Gopalpur and 47.25% in Maniknagar) and marginal farmers (67% in Gopalpur and 52% in Maniknagar) in comparison to Ratua, and Nurpur. Even the proportion of HHs who rely on fishing as their primary source of income during floods is substantially higher in Maniknagar (75%) and Gopalpur (77%) than in Ratua and Nurpur. The riverfront rural communities of Tal and Diara, viz. Gopalpur and Maniknagar, are especially vulnerable to floods since they are located on the banks of the mighty Ganga in Diara and Fulhar in Tal, resulting in the loss of physical possessions (98.25% in Gopalpur and 76.56% in Maniknagar) and life, as well as injuries to HHs' family members (5.5% in Gopalpur and 3.82% in Maniknagar). Moreover, Gopalpur in Diara suffers from crucial river bank erosion of the River Ganga, making the villagers more vulnerable to the unavoidable effects of climatic shocks, as field data show that more than 60% of the HHs stated that they had moved from the river bank to a safer location at least three times since the date of the present survey. During floods, inhabitants typically fly out to a safer area, i.e., adjacent embankments, main roads, school buildings, or open spaces, and after flood water recedes, they return to their habitats and reap the advantages of flood.

Vulnerability Associated with Social Capital

The standing of households in society, in addition to its access to larger families along with social networks that act as safeguards during crisis situations, is referred to as social capital (Engbers et al. 2017). Tables 2.5 and 2.6 demonstrate that Gopalpur (0.44) of Diara and Maniknagar (0.41) of Tal are more vulnerable to social capital than Nurpur (0.28) of Diara and Ratua (0.31) of Tal. In terms of the other major components of social capital, viz., the socio-demographic component, Gopalpur and

Maniknagar are more vulnerable because the villages are comprised of a slightly larger proportion of female-headed HHs (6% in Gopalpur and 4% in Maniknagar), bigger family sizes (7 in Gopalpur and 6 in Maniknagar), and a higher share of dependent population (30.3% in Gopalpur and 26.4% in Maniknagar), contributing to higher vulnerability there (Fig. 2.3). This is due to the fact that many husbands (household heads) who migrate to big cities in search of better livelihoods do not return to their families for an extended period of time, resulting in a greater percentage of dependent people and female-headed HHs in riverbank villages. However, they send back remittances to their family, which is one of the major proactive coping strategies embraced by the villagers. As witnessed by the scores of the social networks major component, preparedness is stronger in the riverbank villages of Maniknagar and Gopalpur, indicating better support and backup from neighbours during flood-caused emergencies (Fig. 2.3). However, locals have a lower level of social connection and networking since riverside residents' addresses constantly change owing to changes in homestead locations caused by erosion and floods, leaving them vulnerable. At the same time, the absence of the Head of the HHs or any family member in any local groups, viz. masjid, temple, flood committee, prompted vulnerability in the riverfront villages as well as indicating a lack of function capacity of such entities, which needs proper attention from policymakers.

Table 2.6 Inter-village variance of LVI-IPCC within Tal and Diara

Contributing factors	Maniknagar	Ratua	Tal	Gopalpur	Nurpur	Diara
Exposure	0.28	0.01	0.15	0.55	0.01	0.28
Sensitivity	0.58	0.25	0.41	0.69	0.26	0.48
Adaptive capacity	0.40	0.27	0.34	0.52	0.25	0.38
LVI-IPCC	0.07	0.06	0.07	0.02	0.06	0.02

Source Computed by the researchers

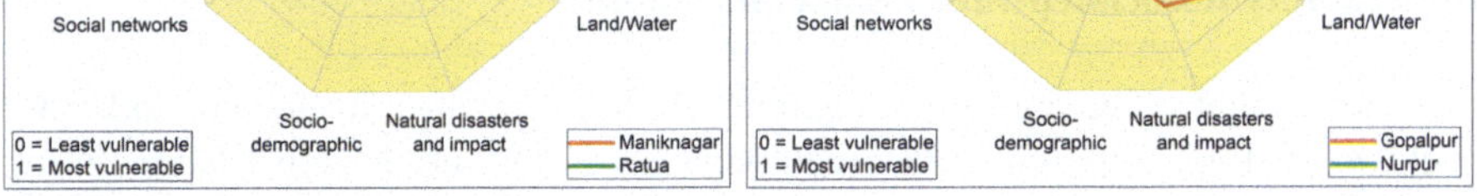

Fig. 2.3 Diagrammatic illustration of the LVI for selected Tal and Diara villages (major components wise). *Source* Field survey

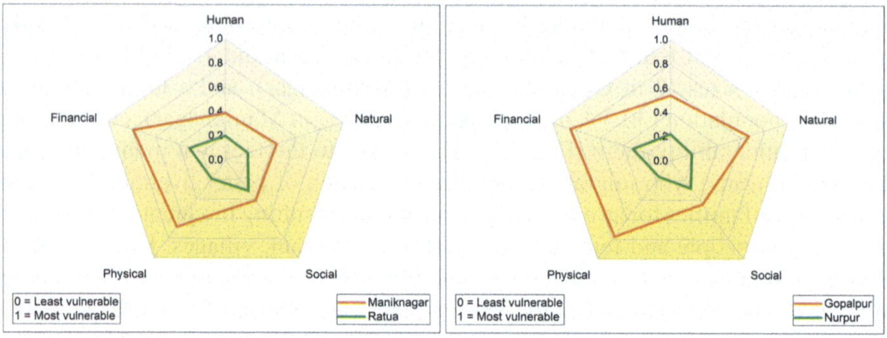

Fig. 2.4 Diagrammatic illustration of the LVI for selected Tal and Diara villages (capitals/assets wise). *Source* Field survey

Vulnerability Associated with Physical Capital

Availability of infrastructures like housing, tools, equipment, physical assets, and communication networks that are essential for long-term sustainable solutions is referred to as physical capital (Mayorga et al. 2004). Interestingly, riverside villages like Gopalpur (0.77) and Maniknagar (0.68) are more vulnerable to physical capital than inland villages like Ratua (0.19) and Nurpur (0.16) (Fig. 2.4; Tables 2.4 and 2.5).

This may be linked to the fact that the percentage share of HHs living in *kutcha* houses, which are easily impacted by floodwaters, is higher in riverfront villages (80.6% in Gopalpur and 61.2% in Maniknagar) than in the rural areas of Ratua and Nurpur, which have a negligible share of *kutcha* dwellings. This also reveals socioeconomic disparities and destitution in riverbank communities, which have an impact on the social well-being of people. Additionally, more respondents in Maniknagar (79.5%) and Gopalpur (81.3%) reported damage to their dwellings owing to recurrent flooding, particularly during the severe flood event of 2017, which was nearly nonexistent in the inner villages of Ratua and Nurpur. It should be emphasised that inhabitants of riverfront communities are the most vulnerable to flooding since they built their houses on stilts using less durable materials, suggesting a lower level of preparedness for flood threats.

Vulnerability Associated with Financial Capital

Financial capital is the ability to acquire cash, kind, savings, insurance, and other resources that are required to face emergencies during any hazard (Pandey et al. 2017). According to field statistics, the riverfront villages within Tal and Diara, namely Maniknagar (80%) and Gopalpur (92%), have a higher proportion of BPL HHs than the inland settlements of Ratua and Nurpur. Since HHs in Maniknagar (85%) and Gopalpur (97.52%) are mostly disadvantaged and live on limited funds, they are more inclined to borrow cash or in kind to cope with the financial situation, making them more financially susceptible compared to HHs in Ratua (9.2%) and Nurpur (4.25%). Field investigations reveal that destitute villagers borrow money mostly from village money lenders to fulfil their immediate needs, resulting in increased interest rates and becoming further entrenched in the vicious cycle of poverty. Unexpectedly, Gopalpur (97.37%) and Maniknagar (93.42%) have a greater proportion of HHs who do not have any type of insurance (mostly life and crop insurance) to protect against the negative effects of flood danger. Resultantly, the overall vulnerability score of financial capital is higher in both the riverbank settlements of Maniknagar (0.79) and Gopalpur (0.85) than in the interior villages of Ratua (0.31) and Nurpur (0.32), indicating a relatively higher degree of disparity among the riverside and the inland villages of Tal and Diara (Fig. 2.4; Tables 2.4 and 2.5).

In Tal and Diara, the overall LVI of selected villages shows inter-village disparity (Fig. 2.3). It is worth noting that waterfront villages such as Gopalpur (0.62) in Diara and Maniknagar (0.50) in Tal exhibit greater LVI scores, suggesting greater vulnerability since these settlements are heavily impacted by frequent flooding and river bank erosion caused by the Ganga in Diara and Fulhar in Tal. In contrast, the rural villages of Ratua (0.23) in Tal and Nurpur (0.23) in Diara have lower LVI scores because they are less impacted by floods and hence less susceptible (Tables 2.4 and 2.5). It should be noted that both in Tal and Diara, maximum vulnerability is associated with financial capital exhibited in the riverfront villages of Gopalpur and Maniknagar, followed by physical and natural capital vulnerability as compared to the inland villages (Fig. 2.4).

Assessment of LVI-IPCC in Tal and Diara

It is interesting to note that the flood exposure score in the riverfront settlement of Gopalpur in Diara is significantly higher (0.55) than in the other riverside village of Maniknagar in Tal (0.28). Likewise, flood sensitivity is higher in Gopalpur (0.69), which is located on the banks of the Ganga, than in Maniknagar (0.58), which is located on the banks of the Fulhar. Furthermore, Gopalpur has a higher adaptive capacity score (0.52) than Maniknagar (0.40) in Tal. As a result, the overall LVI-IPCC of Diara's Gopalpur (0.02) is much higher, suggesting more vulnerability than Tal's Maniknagar (− 0.07) (Fig. 2.5; Table 2.6). On the other hand, the inland villages, such as Ratua in Tal and Nurpur in Diara, indicated lower levels of vulnerability as contributing factors under LVI-IPCC, i.e., exposure, sensitivity, and adaptive capacity scores are minimal compared to the riverbank villages.

As an outcome, the riverbank village of Gopalpur in Diara reports a considerably higher degree of vulnerability and sensitivity to floods than the riverside settlement of Maniknagar. At the same time, the inhabitants of Gopalpur have developed better adaptive skills and resistance capacity as a result of their increased vulnerability to floods. On the other hand, the inland villages of both Tal and Diara, namely Ratua and Nurpur, have negligible scores of overall LVI-IPCC regardless of the other

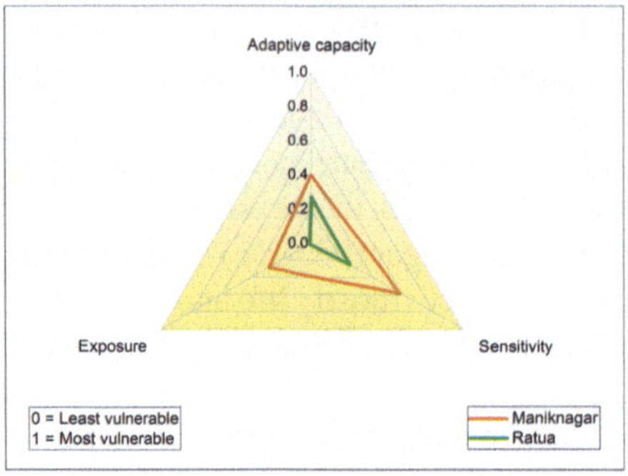

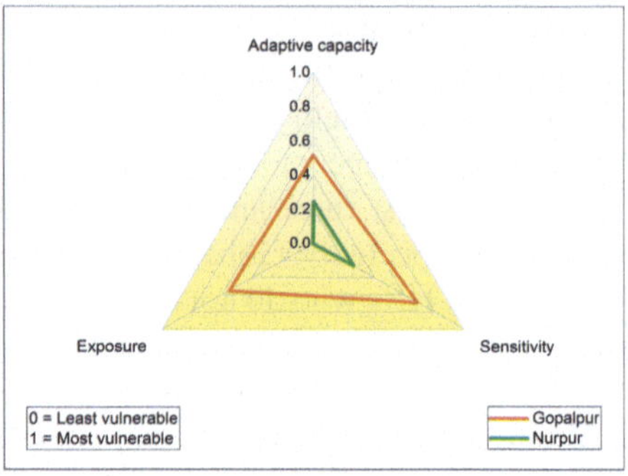

Fig. 2.5 Diagrammatic illustration of the LVI-IPCC for selected Tal and Diara villages. *Source* Field survey

contributing factors under LVI-IPCC, namely exposure, sensitivity, and adaptive capacity. Their scores suggest considerably lesser vulnerability to floods since these villages are located at higher elevation and are less experienced by the wrath of flood risks and river bank erosion than waterfront communities in both physiographic units.

Conclusion

Malda's location, coupled with its geomorphology, renders it exceedingly prone to floods, particularly in the shallow Tal and Diara geomorphic units. Amidst repeated flooding and river bank erosion, which intensified Tal and Diara people's vulnerability, disrupting their lives, livelihoods, and well-being, both geomorphic units of Malda report a very dense population. It means that the inhabitants residing here have acquired the ability to live with floods by adopting better preparation, coping, and adapting skills that render them more resistant to the negative consequences of flood hazards and reap the benefits of floods. The present chapter realised that people living near flood-prone rivers, in this case, Ganga represented by the village of Gopalpur in Diara, and Fulhar represented by the village of Maniknagar in Tal, are more vulnerable to flooding compared to the inland settlements of Nurpur in Diara and Ratua in Tal, have over time developed stronger flood-coping and adapting strategies, rendering them more resistant to flood hazard. The riverfront villagers of Maniknagar and Gopalpur's unique adaptive measures are reflected in their diversification of livelihood practices, switching from farming to non-farm occupations, and selling assets such as livestock and, in some cases, land to sustain themselves from flood vulnerability. In addition, there is a greater proportion of family members migrating to other states from these villages in quest of greater earning possibilities, with migrants regularly remitting cash/kind to support their families living in the riverfront villages. Field data reveal that villages nearby river banks are more vulnerable to flooding as they are not only adjacent to flood-producing rivers such as Ganga in Diara and Fulhar in Tal, but they also have a higher proportion of impoverished families living on handouts than interior villages. Conversely, the villagers of Ratua and Nurpur have comparatively better socioeconomic conditions than the residents of Maniknagar and Gopalpur. As opposed to the internal group of villages, the riverfront villages of both Tal and Diara report higher levels of vulnerability; however, the villagers here have developed not only a better level of diversification of livelihood activities but also a more diverse agricultural land usage with a wide range of crops grown. Therefore, the higher the vulnerability, the greater the resilience to mitigate the negative impact of floods through improved coping and adaptation practices portrayed in their settlements, housing, and economic activities. Field statistics suggest that riverine communities grow more cash crops with shorter production times than interior villages. *Makhana*, often known as foxnut, is widely grown in Tal's low and marshy areas. Mango orchards, sericulture, as well as food processing based on local agricultural resources, mainly mangoes, are emerging as significant livelihood alternatives in Diara. Cash crops such as watermelon, cucumber, and other vegetables are intensively produced on the river sandbanks of riverine villages. These are some of the few examples to indicate the divergent coping and adaptive techniques employed by the rural folks of Tal and Diara instead of increasing vulnerability and making the location overcrowded. Hence, locals choose to reside in Tal and Diara amidst flood havoc, resulting in loss of assets and livelihoods; the fertility of the soil here is enhanced with each flood episode, culminating in bumper agricultural production. In a wider context, it is necessary to examine coping and adaptation mechanisms at the household or community level when assessing vulnerability to hazards such as floods. This comprehensive policy for preventing catastrophes may not only enhance the management of risks, yet reduce the susceptibility of a household or group. The villagers here have learned to live with flood vulnerability, which can be replicated elsewhere in the world in similar ecological conditions, as the frequency of floods is bound to increase with climate change and global warming, and the riverine population must learn to live with extreme weather events such as floods. Furthermore, such micro-level research on vulnerability assessment assists stakeholders as well as policymakers in working together to broaden the population's access to basic assets/capital so that they may live a dignified life within a flood resilient society.

References

Adedeji OH, Odufuwa BO, Adebayo OH (2012) Building capabilities for flood disaster and hazard preparedness and risk reduction in Nigeria: need for spatial planning and land management. J Sustain Dev Afr 14(1):45–58

Ahmad D, Khurshid S, Afzal M (2023) Climate change vulnerability and multidimensional poverty in flood prone rural areas of Punjab, Pakistan: an application of multidimensional poverty index and livelihood vulnerability index. Environ Dev Sustain. https://doi.org/10.1007/s10668-023-04207-8

Ahsan MN, Warner J (2014) The socioeconomic vulnerability index: a pragmatic approach for assessing climate change led risks—a case study in the south-western coastal Bangladesh. Int J Disaster Risk Reduct 8:32–49. https://doi.org/10.1016/j.ijdrr.2013.12.009

Alam GMM (2017) Livelihood cycle and vulnerability of rural households to climate change and hazards in Bangladesh. Environ Manage 59(5):777–791. https://doi.org/10.1007/s00267-017-0826-3

Al Mamun A, Islam ARMT, Alam GMM, Sarker MNI, Erdiaw-Kwasie MO, Bhandari H, Mallick J (2023) Livelihood vulnerability of char land communities to climate change and natural hazards in Bangladesh: an application of livelihood vulnerability index. Nat Hazards 115(2):1411–1437. https://doi.org/10.1007/s11069-022-05599-y

Allaire M (2018) Socio-economic impacts of flooding: a review of the empirical literature. Water Secur 3:18–26. https://doi.org/10.1016/j.wasec.2018.09.002

Biswas R, Anwaruzzaman A (2019) Measuring hazard vulnerability by bank erosion of the Ganga river in Malda district using PAR model. J Geogr Environ Earth Sci Int 22(1):1–15

Bohle H-G (2016) Social vulnerability and livelihood security: towards an integrated framework for market risk assessment. In: Economic spaces of pastoral production and commodity systems. Routledge, pp 43–51

Chakraborty K, Das M, Saha S (2023) Application of multi-criteria decision-making approach for assessing flood susceptibility of the Tal-Diara and Barind Region in Malda District, India. In Islam A, Shit PK, Datta DK, Islam MS, Roy S, Ghosh S, Das BC (eds) Floods in the Ganga–Brahmaputra–Meghna Delta. Springer International Publishing, pp 203–230. https://doi.org/10.1007/978-3-031-21086-0_9

Chambers R, Conway G (1992) Sustainable rural livelihoods: practical concepts for the 21st century. Institute of Development Studies, UK

Clar C (2019) How demographic developments determine the management of hydrometeorological hazard risks in rural communities: the linkages between demographic and natural hazards research. Wiley Interdiscip Rev Water 6(6):e1378

Cochran WG (1963) Sampling techniques, 2nd edn. Wiley

Development and Planning Department, G. o. W. B. (2007) District human development report: Malda. HDRCC

Engbers TA, Thompson MF, Slaper TF (2017) Theory and measurement in social capital research. Soc Indic Res 132:537–558

Gbetibouo GA, Ringler C, Hassan R (2010) Vulnerability of the South African farming sector to climate change and variability: an indicator approach. Nat Resour Forum 34(3):175–187. https://doi.org/10.1111/j.1477-8947.2010.01302.x

Ghosh A, Kar SK (2018) Application of analytical hierarchy process (AHP) for flood risk assessment: a case study in Malda district of West Bengal, India. Nat Hazards 94(1):349–368. https://doi.org/10.1007/s11069-018-3392-y

Ghosh A, Dey P, Ghosh T (2022) Integration of RS-GIS with frequency ratio, fuzzy logic, logistic regression and decision tree models for flood susceptibility prediction in Lower Gangetic Plain: a study on Malda District of West Bengal, India. Journal of the Indian Society of Remote Sensing 50(9):1725–1745. https://doi.org/10.1007/s12524-022-01560-5

Hahn MB, Riederer AM, Foster SO (2009) The Livelihood Vulnerability Index: a pragmatic approach to assessing risks from climate variability and change—a case study in Mozambique. Global Environ Change 19(1):74–88. https://doi.org/10.1016/j.gloenvcha.2008.11.002

Heinkel S-B, Thiebes B, Zin Mar T, Toe A, Tin Tin K, Win Lei M, Saw Sandar O, Miller C, Willkomm M, Win M, Zin Nwe M, Khin Khin S, Spohner R, Kraas F (2022) Disaster preparedness and resilience at household level in Yangon, Myanmar. Nat Hazards 112(2):1273–1294. https://doi.org/10.1007/s11069-022-05226-w

Iqbal S (2010) Flood and erosion induced population displacements: a socio-economic case study in the Gangetic riverine tract at Malda district, West Bengal, India. J Hum Ecol 30(3):201–211

Kucharčíková A (2011) Human capital–definitions and approaches. Hum Resour Manage Ergon 5(2):60–70

Lin K-HE, Lee H-C, Lin T-H (2017) How does resilience matter? An empirical verification of the relationships between resilience and vulnerability. Nat Hazards 88(2):1229–1250. https://doi.org/10.1007/s11069-017-2916-1

Madhuri M, Tewari HR, Bhowmick PK (2014) Livelihood vulnerability index analysis: an approach to study vulnerability in the context of Bihar. Vulnerability, Livelihood, Sensitivity, Adaptive capacity, Exposure 6(1). https://doi.org/10.4102/jamba.v6i1.127

Majumdar S, Mandal S (2021) Acceptance of BANCS model for predicting stream bank erosion potential and rate in the left bank of Ganga river of Diara region in Malda district, North East India. Spat Inf Res 29(1):43–54. https://doi.org/10.1007/s41324-020-00334-w

Mayorga FDdO, Khan AS, Mayorga RD, Lima PVP (2004) Social capital, physical capital and vulnerability of rural workers: the case of the communities of Lustal and Sitio Lagoa in the county of Taua, Ceará. Revista de Economia e Sociologia Rural 42(1):111–132

Mehebub S, Raihan A, Nuhul H, Haroon S (2015) Assessing flood inundation extent and landscape vulnerability to flood using geospatial technology: a study of Malda district of West Bengal, India. Forum Geografic XIV(2):156–163. https://doi.org/10.5775/fg.2067-4635.2015.144.d

Missemer A (2018) Natural capital as an economic concept, history and contemporary issues. Ecol Econ 143:90–96. https://doi.org/10.1016/j.ecolecon.2017.07.011

Mitra R, Kumar Mandal D (2023) Assessment of livelihood vulnerability in the riparian region of the Tista River, West Bengal, India. GeoJournal 88(1):811–839. https://doi.org/10.1007/s10708-022-10645-0

Mukherjee K, Pal S (2018) Channel migration zone mapping of the River Ganga in the Diara surrounding region of Eastern India. Environ Dev Sustain 20(5):2181–2203. https://doi.org/10.1007/s10668-017-9984-y

Naheed S (2021) Understanding disaster risk reductionrisk reduction and resilience: a conceptual framework. In: Eslamian S, Eslamian F (eds) Handbook of disaster risk reduction for resilience: new frameworks for building resilience to disasters. Springer International Publishing, pp 1–25. https://doi.org/10.1007/978-3-030-61278-8_1

Nam BH, Choi S, Copeland T, Kim YJ (2023) Social vulnerability and geohazards: review and implications. In D'Amico S, De Pascale F (eds) Geohazards and disaster risk reduction: multidisciplinary and integrated approaches. Springer International Publishing, pp 3–37. https://doi.org/10.1007/978-3-031-24541-1_1

Padhan N, Madheswaran S (2023) An integrated assessment of vulnerability to floods in coastal Odisha: a district-level analysis. Nat Hazards 115(3):2351–2382. https://doi.org/10.1007/s11069-022-05641-z

Pandey R, Jha SK, Alatalo JM, Archie KM, Gupta AK (2017) Sustainable livelihood framework-based indicators for assessing climate change vulnerability and adaptation for Himalayan communities. Ecol Indicators 79:338–346. https://doi.org/10.1016/j.ecolind.2017.03.047

Pandey R, Jha S (2012) Climate vulnerability index-measure of climate change vulnerability to communities: a case of rural Lower Himalaya, India. Mitig Adapt Strat Glob Change 17(5):487–506. https://doi.org/10.1007/s11027-011-9338-2

Panthi J, Aryal S, Dahal P, Bhandari P, Krakauer NY, Pandey VP (2016) Livelihood vulnerability approach to assessing climate change impacts on mixed agro-livestock smallholders around the Gandaki River Basin in Nepal. Reg Environ Change 16(4):1121–1132. https://doi.org/10.1007/s10113-015-0833-y

Patwardhan A, Semenov S, Schnieder S, Burton I, Magadza C, Oppenheimer M, Pittock B, Rahman A, Smith J, Suarez A (2007) Assessing key vulnerabilities and the risk from climate change. Clim Change, 779–810

Rehman S, Sahana M, Kumar P, Ahmed R, Sajjad H (2021) Assessing hazards induced vulnerability in coastal districts of India using site-specific indicators: an integrated approach. GeoJournal 86(5):2245–2266. https://doi.org/10.1007/s10708-020-10187-3

Roca M, Martín-Vide JP, Moreta PJM (2009) Modelling a torrential event in a river confluence. J Hydrol 364(3):207–215. https://doi.org/10.1016/j.jhydrol.2008.10.020

Rudra K (2018) Flood in the GBM Delta. In: Rivers of the Ganga-Brahmaputra-Meghna Delta: a fluvial account of Bengal. Springer International Publishing, pp 125–136. https://doi.org/10.1007/978-3-319-76544-0_9

Sai KVRS, Manjunath A, Krishnaiah S (2021) Assessment of flood vulnerability index for Krishna river basin in Krishna, Guntur and Prakasam Districts of Andhra Pradesh. Int J Syst Assur Eng Manage. https://doi.org/10.1007/s13198-021-01222-8

Shah KU, Dulal HB, Johnson C, Baptiste A (2013) Understanding livelihood vulnerability to climate change: applying the livelihood vulnerability index in Trinidad and Tobago. Geoforum 47:125–137. https://doi.org/10.1016/j.geoforum.2013.04.004

Shahzad L, Shah M, Saleem M, Mansoor A, Sharif F, Tahir A, Hayyat U, Farhan M, Ghafoor G (2021) Livelihood vulnerability index: a pragmatic assessment of climatic changes in flood affected community of Jhok Reserve Forest, Punjab, Pakistan. Environ Earth Sci 80(7):252. https://doi.org/10.1007/s12665-021-09562-1

Singh A (2023) Introduction to the international handbook of disaster research. In Singh A (ed) International handbook of disaster research. Springer Nature Singapore, pp 3–11. https://doi.org/10.1007/978-981-19-8388-7_223

Smit B, Wandel J (2006) Adaptation, adaptive capacity and vulnerability. Global Environ Change 16(3):282–292. https://doi.org/10.1016/j.gloenvcha.2006.03.008

UNDP (2000) Human development report. Oxford University Press, New York

United Nations (2005) Hyogo framework for action 2005–2015: building the resilience of nations and communities to disasters. H. Kobe

Wallemacq P (2018) Economic losses, poverty & disasters: 1998–2017. Centre for Research on the Epidemiology of Disasters, CRED

World Bank (2018) Learning: to realize education's promise. W. B. Group

Chapter 3
Assessing the Rural Livelihood Vulnerability and Risk to River Flood in Lower Gangetic West Bengal Applying IPCC-AR5 Methodology

Pintu Mandal, Adrika Mukhopadhyay, Jayanta Saha, Shyamal Santra, Bhaskar Samanta, Subhasis Bhattacharya, and Suman Paul

Abstract The flood management Master plan for Ghatal in West Bengal, India, has gained national attention due to ongoing floods and aftermath, exacerbated by heavy rains in Chhattisgarh and the May 2021 *Yaas* storm. Floodwaters in areas like Amta-I and II, Udanarayanpur, Arambag, Khanakul-I and II, Pursurah, Daspur-I and II, and Ghatal caused severe damage to lives and livelihoods, highlighting climate change's impact on lower elevations. The study retrospectively examines the vulnerability and risk outlined in IPCC-AR5, evaluating micro-level livelihood vulnerability and risk across the 124 Panchayats in the three districts. This study examines the risk of catastrophic floods in the Shilabati, Dwarkeshwar, and Mundeshwari rivers due to DVC dams, using hybrid techniques to quantify livelihood sensitivity and risk. The overall hazards derived from IPCC-AR5 components, with an index of 0.418, indicate that households in the study region relying on nature are highly vulnerable to climate change. Farming households in riverine areas exhibit greater vulnerability, with an index of 0.689, compared to those in non-riverine areas, with an index of 0.372, based on the IPCC-AR5 risk assessment. The study extends its examination to IPCC-AR5-based risk assessments at the household level across various

P. Mandal · J. Saha · S. Paul (✉)
Department of Geography, Sidho-Kanho-Birsha University, Purulia, WB 723104, India
e-mail: suman.krish.2007@gmail.com

A. Mukhopadhyay
Department of Geography, Calcutta University, Kolkatta, WB, India

S. Santra
Department of Geography, Bankura Zilla Saradamani Mahila Mahavidyapith, Bankura, WB 722101, India

B. Samanta
Department of Geography, Krishnagar Government College, Krishnanagar, WB 741101, India

S. Bhattacharya
Department of Economics, Sidho-Kanho-Birsha University, Purulia, WB 723104, India

U. Mukhopadhyay et al. (eds.), *Climate Crisis, Social Responses and Sustainability*, Climate Change Management, https://doi.org/10.1007/978-3-031-58261-5_3

settings. Utilizing a violin plot to illustrate quartiles and risk levels among residents in different administrative districts, the study underscores the benefit of the IPCC-AR5 approach. It identifies adaptation methods and climate-resilient paths within the current boundaries of the system to mitigate the effects of climate change. Despite the increasing prevalence of climate change-related uncertainties, the study reveals that local residents have adapted to living in a stressful environment.

Introduction

In the contemporary era, the predominant global focus revolves around the shifting climate and its ramifications, marked by more frequent, intense, and extreme weather events. These climate-related occurrences encompass escalating temperatures, rising sea levels, heightened rainfall, increased humidity, and floods. A substantial portion of the world's population resides in flood-prone areas, directly or indirectly relying on the fertility of these lands. Human encroachment, alterations to river systems and ecosystem degradation in these regions invariably culminate in adverse consequences, foremost among them being floods (WMO 2018). Generally, floods arise as a consequence of a confluence of unfavourable meteorological, hydrological, and physical conditions (Gashaw and Legesse 2011). But damages to property and infrastructure, adverse effects on human health, agriculture, forestry and fisheries production and tourism are the common consequences of the flood (Rakib et al. 2017; Hoq et al. 2021). Occurrences of flooding have impacted over 196 million individuals across more than 90 countries worldwide (UNDP 2004). In recent decades from 2006 to 2015, the world has experienced annual average death of 5709 people by flood (Guha-Sapir et al. 2012). An estimated 2156 floods affected around 2.6 billion people during the course of the last 30 years, killing 206,303 people and causing 386 billion USD in losses, according to estimates from the Centre for Research on the Epidemiology of Disasters (Guha-Sapir et al. 2004). Floods are becoming more often as a natural phenomenon, and climate change is predicted to make them more intense and larger in scope. Around the world, many places encounter different kinds of flooding. For example, riverine flooding frequently causes havoc in Asia and Africa, whereas flash floods occur on the American continent (Jonkman and Kelman 2005; Jamshed et al. 2020).

Flooding is a frequent occurrence in various regions of India, driven by diverse climates and precipitation patterns. Annually, specific areas in India confront destructive floods with significant adverse effects, as noted by Mishra and Mishra (2010). More than 80% of the annual precipitation in Asia's monsoon-dominated region happens between June and September. Numerous variables, such as excessive rainfall, tidal surges, cyclonic occurrences, substantial flows from dams and barrages, and irresponsible human actions, are responsible for India's ongoing floods. India's floodplains, deltas, and coastal regions have all been significantly and extensively impacted by these elements (Sanyal and Lu 2005). Based on estimates from the Central Water Commission, in 2012, about 49.82 million hectares, or 15% of India's total land area, were classified as flood-prone zones. From 1900 to 2015 nearly 847 million people were affected and 72,039 people were killed in India due to floods. Between 1953 and 2016, floods and heavy rainfall wreaked havoc in India, affecting a staggering 460 million hectares of land, claiming the lives of 105,471 people, and causing damage totalling 347,581 crore to crops, houses, and public utilities (Flood Damage Statistics 2018). Despite Uttar Pradesh, Bihar, Odisha, Assam, and West Bengal being identified as the five flood-prone regions, their severity has witnessed a significant surge over the years. This alarming trend puts millions of people at risk in the foreseeable future. West Bengal, encompassing 42.30% of the flood-prone area, stands out as one of the most vulnerable states to flood risks nationwide (Sanyal and Lu 2005). The escalating intensity of floods in these areas highlights the urgent need for comprehensive measures to mitigate the potential devastation and protect the lives and livelihoods of the affected populations.

Vulnerability, as defined by the Intergovernmental Panel on Climate Change, pertains to the extent to which a system is at risk of being unable to withstand the negative impacts of climate change and extremes such as droughts, floods, and cyclones. The susceptibility to floods varies temporally, spatially, and among different groups and individuals. Notably, there exists a disparity in flood vulnerability among developing nations, largely attributed to their insufficient coping mechanisms and less robust institutional frameworks in the face of natural hazards (Yohe and Tol 2002). Research conducted by various scholars has demonstrated that the severity of floods is influenced by climate change, impacting rural, impoverished, disabled, elderly, and marginalized populations. This exacerbates pre-existing social vulnerabilities (Kasperson and Kasperson 2012; Vincent 2004; Dasgupta et al. 2014; Szabo et al. 2016). Additionally, local factors like population pressure, changes in land use, intensified agricultural practices, and urbanization contribute to increased vulnerability at the household level during floods (Hahn et al. 2009; Szabo et al. 2016). These factors force vulnerable people to displace and increase trapped populations (Adger et al. 2014; Das et al. 2020).

The evaluation of vulnerabilities related to hazards, disasters, and climate change is now a crucial aspect of risk assessment, as highlighted by Birkmann et al. (2013). There is a growing necessity to predict and evaluate the vulnerability and associated risks of severe hazards such as floods due to their escalating adverse impacts on social, economic, and environmental aspects (Wanders et al. 2014). Various theoretical frameworks exist for vulnerability assessment across different scales (local, regional, global), encompassing diverse components (exposure, susceptibility/sensitivity, capacity), and dimensions (social, economic, physical, environmental, institutional), depending on the specific focus of research (IPCC 2014; Birkmann et al. 2013; Cutter 1996; Turner et al. 2003; Wisner et al. 2004; Bohle 2007; Birkmann 2006; Füssel and Klein 2006). Many scholars around the world have already studied the risk and vulnerability related to flood hazards and most of the studies are from developing and less developed countries. Different models, methods, and variables have been used by them with different objectives. Brouwer et al. (2007) delved into the intricate interplay among environmental risk, poverty, and vulnerability in Bangladesh, a country known for its extreme poverty and susceptibility to floods. The study demonstrated that households with low income and limited access to productive natural assets are more exposed to the risk of flooding. Additionally, community-level disparities in income and asset distribution tend to intensify at higher levels of risk exposure. The findings underscored that the least prepared individuals face the greatest risk of flooding. In a related context, Mansur et al. (2016) highlighted the vulnerability of urban sectors in the Amazon Delta and Estuary (ADE) to various risks. Factors such as flood hazards, poverty, and a lack of basic structural facilities, including insufficient drinking water and inadequate wastewater collection, contribute to the potential risks faced by these areas. Unplanned rapid urbanization has created the most vulnerable situation urban sector of the ADE. Tanoue et al. (2016) assessed the long-term global flood vulnerability trend utilizing a comprehensive river and inundation model. This model relied on the ratio of reported flood loss or damage to flood exposure. The impact of historical shifts in population distribution, coupled with population growth scenarios, was considered in evaluating the population exposed to floods. The findings indicated that economic growth, along with various historical and socioeconomic advancements, could mitigate the adverse effects on both people and resources. In a parallel study, Imran et al. (2019) employed the partial least squares structural equation model (PLS-SEM) to examine flood vulnerability. The model incorporated latent variables representing geographical characteristics of houses and socioeconomic dimensions of households. The research demonstrated that social indicators were primarily influenced by the direct effects of spatial neighbourhoods and socioeconomic conditions, as well as the indirect effects of land tenure and demographics. Fekete endeavoured to map flood vulnerability in Germany using Social Vulnerability Indices. The findings underscored that specific social demographics, such as the elderly, financially disadvantaged individuals, and urbanites, faced elevated risks. To comprehend the factors contributing to vulnerability and identify vulnerable communities in Phu Huu and Ta Danh villages in the Giang province of the Mekong Delta in Vietnam, the Livelihood Vulnerability Index (LVI) developed by Can et al. (2013) was employed. The outcomes indicated that those in poverty were most susceptible to flooding and climate variability. Additionally, a higher percentage of female household heads or landless/small-farm households increased flood vulnerability. In a similar vein, Hoq et al. (2021) investigated livelihood vulnerability to flooding hazards in the flood-prone Haor ecosystem of Bangladesh. Employing the Livelihood Vulnerability Index (LVI) and the IPCC framework of vulnerability, their study revealed that Haor households were more vulnerable to flood hazards and natural disasters, particularly concerning food, water, and health. The results also highlighted that higher vulnerability stemmed from the combination of low adaptive capacity and heightened sensitivity and exposure. In their study, Das et al. (2020) endeavoured to establish a connection between the methodologies of the IPCC's AR4 and AR5. Their goal was to assess the vulnerability and risk of climate change in the CD blocks of the Indian Bengal Delta. This assessment relied on bio-physical and socio-economic variables, and the researchers applied Principal Component Analysis (PCA) in their analysis. The result has proved that in a similar geographical condition the vulnerability and risk vary with the difference in socio-economic variables and hence validated the IPCC's framework of assessment.

In 2007, the Intergovernmental Panel on Climate Change's (IPCC) fourth assessment report (AR4) established vulnerability as a construct determined by exposure, sensitivity, and adaptive capacity (Hahn et al. 2009; Olsson et al. 2014; Das et al. 2020). Subsequently, the IPCC's fifth assessment report (AR5) in 2014 expanded the vulnerability assessment framework by introducing the concept of risk (Das et al. 2020). This chapter employs the methodology derived from IPCC's AR5 to evaluate the vulnerability and risk of river flood impact on rural livelihoods in lower Gangetic West Bengal.

Background of the Study

With a high population density in the low-lying alluvial plain and a largely agrarian landscape, West Bengal is grappling with a severe flood threat. The region has a long-standing history of recurrent flooding. In pre-independence era, the undivided Bengal (Including Bangladesh) experienced worst floods in the years 1787, 1877, 1889, 1892, 1895, 1900, 1902, 1905, 1906, 1913, 1942, and 1946. After post-independence, the worst floods occurred in West Bengal in the subsequent years of 1954, 1959, 1968, 1971, 1978, 1993, 1995, 1999, and 2000 and this scenario is continuing. Almost every year, people from many places became homeless and lost their crops due to flooding (Mukhopadhyaya and Dasgupta 2014). Flood hazards in West Bengal have had a profound impact, affecting four million people, resulting in the tragic loss of 187 lives, and causing an average damage of 708 crore INR (Flood Damage Statistics 2018). The expansive 37,660 km^2 area of West Bengal is inherently flood-prone, as highlighted by Das and Bandyopadhyay (2015). Reports from the Irrigation and Waterways Department of the Government of West Bengal (IWD) indicate that 42% of the state is vulnerable to flooding. However, contrasting studies, such as those by Kale (2003a, b) and Nath et al. (2008), argue that a staggering 55.8% of the state, approximately twenty million people, is susceptible to floods. The factors contributing to the occurrence of floods in West Bengal vary across regions. Hydro-geomorphic and hydro-meteorological events, coupled with human-induced changes such as flat riverine topography, concentrated rainfall, cyclonic activities, discharge from dams and barrages, and drainage congestion, collectively exacerbate the flood situation in the riverine plain and deltaic areas of West Bengal. These phenomena have led to irreparable losses in terms of both property and human life (Sanyal and Lu 2006; Jha and Bairagya 2013; Mollah 2013; Das and Bandyopadhyay 2015; Mujibur-rehman 2015; Dandapat and Panda 2017). The northern districts of West Bengal face floods due to high intense rainfall, the landslide in the hilly areas, river bank erosion, siltation in the river beds, and changes in river dynamics. The occurrence of floods in the central region is attributed to a deficient drainage system, bank erosion along the Ganga-Bhagirathi River, substantial and intense rainfall, and the overflow of downstream rivers. Meanwhile, in the southern part of the state, frequent flooding is a result of intense rainfall, embankment destruction, cyclones, inadequacies in drainage systems, as well as the controlled release of water from dams and barrages (Dasgupta and Mukhopadhyay 2014). Consequently, a comprehensive assessment of flood risk and effective flood management are imperative for identifying vulnerable areas and implementing appropriate measures (Hagen and Lu 2011).

Methods

Selection of Study Area

The present study area is geographically located in the lower courses of the Silabati, Dwarkeswar, and Mundeshari river basins (between 22°20′5″ N and 23°05′37″ N and 87°35′22″ E to 88°12′8″ E) and administratively falls under the Hugli, Howrah and Paschim Medinipur district of the state of West Bengal (Fig. 3.1). It is evident from the historical flood record of Bengal to till the date that the Gram Panchayats under these blocks have been seriously affected due to flooding (Fig. 3.2). The upper courses of these rivers lie in the Western plateau and Rarh region of Bengal (eastern corridor of The Chotanagpur plateau). The entire areas of these watersheds belong to tropical monsoon climate condition which receives annual rainfall ranging from 250 to 300 mm mostly between June to September. Heavy rainfall in the upper courses, as well as entire watersheds, causes flooding in these areas. The discharge of water from the dams and barrages of the DVC (Damodar Valley Corporation) is another problem that flooded these areas most of the time. Therefore a total of 202 Gram Panchayats under these blocks has been selected for the study.

Conceptual Framework

The Fifth Assessment Report (AR5) from the Intergovernmental Panel on Climate Change (IPCC) introduces a novel methodology and terminology, aligning with the concept of disaster risk, distinct from the vulnerability perspective presented in the previous IPCC-AR4. According to IPCC-AR5, risk is characterized as "the potential for adverse effects when something valuable is at stake, and the outcome is uncertain, considering diverse values." It is commonly expressed as the probability of hazardous events or trends multiplied by the consequences if they occur, with a predominant focus on the hazards posed by climate change. In the IPCC-AR5 framework, the terms 'exposure' and 'vulnerability' are employed differently compared to the IPCC-AR4. Exposure is defined as "the presence of people, livelihoods, species, ecosystems, environmental functions, services, resources, infrastructure, or economic, social, or cultural assets in locations that may be adversely affected." Vulnerability, on the other hand, is described as "the inclination or predisposition to be adversely affected,"

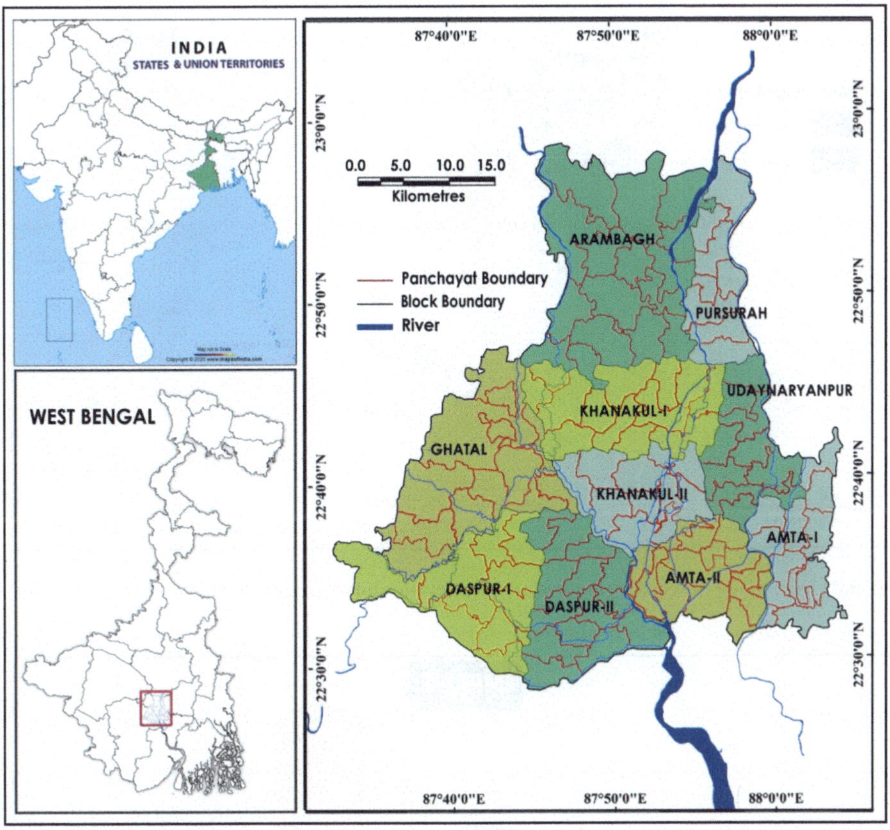

Fig. 3.1 Location of the study area

encompassing the concepts of sensitivity (susceptibility to damage) and adaptive capacity. According to AR5, risk extends beyond the potential occurrence of natural or human induced events, encompassing physical impacts that may lead to loss of life, injury, health effects, as well as damage and loss to property, infrastructure, livelihoods, service provision, ecosystems, and environmental resources (Fig. 3.3).

Database and Materials

For mapping, the extent of flooding across the whole area under research, the Sentinel-1 SAR C band Interferometric Wide (IW) swath mode GRD datasets from the Copernicus Cloud computing plate have been used. The satellite has a viewing swath of 250 km and collects data in dual-polarization with a spatial resolution of 5–20 m.

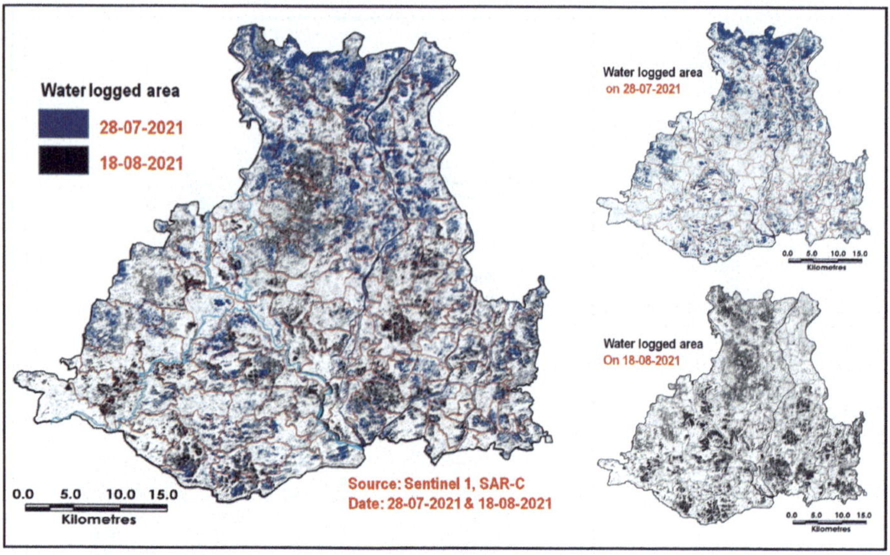

Fig. 3.2 Water logged scenario in the study area during July–August, 2021

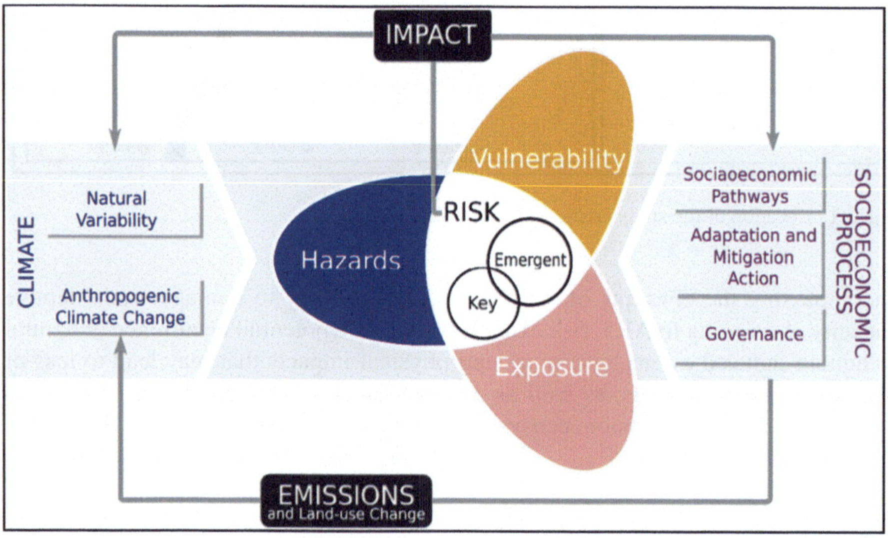

Fig. 3.3 The contributing factors of risk (adopted from IPCC AR5, 2014, p. 1046)

The vulnerability and risk methodological frameworks for this study were developed using the IPCC-AR5 working definitions. In streamlining the risk methodology framework, AR5 has disaggregated the term 'vulnerability' into 'sensitivity' and 'adaptive capability.' In simpler terms, the assessment of risk now considers four key elements: hazard (H), exposure (E), sensitivity (S), and adaptive capacity (AC). A total of twenty biophysical and socio-economic variables, deemed theoretically significant and pertinent to policy, have been identified across four principal components—natural hazards, exposure to hazards, sensitivity, and adaptive capacity. This selection process is grounded in a thorough examination of literature and available secondary data sets (Table 3.1). The socio-economic data sets used in this analysis may be found in the Census of India (2001 and 2011), the Bureau of Applied Economics and Statistics (2011), and the United Nations Development Programme—India (2009 and 2010).

Table 3.1 Components and indicators for assessing flood risk based on hazard, exposure index, sensitivity and adaptivity capacity

Components	Indicators		Unit of measurement
Hazard	Flood frequency (veryhigh—4, high—3, moderate—2, low—1)	H_1	Number
	Distance from active river channel (within 1 km of buffer—4, within 2 km of buffer—3, within 3 km of buffer—2 outside of 3 km of buffer—1)	H_2	km
	Flood affected area to total area	H_3	%
Exposure	Female headed HHs	E_1	%
	Share of HHs Working in Cultivation	E_2	%
	% of HHs with 3 or more rooms with pucca walls and pucca roof	E_3	%
Sensitivity	Dependency ratio	S_1	Ratio
	Female population	S_2	%
	Deprived HHs	S_3	%
	SC and ST HHs	S_4	%
	Share of HHs casual worker	S_5	%
	Share of HHs income less than Rs. 5000	S_6	%
	HHs owning land	S_7	%
	Working force to total population	S_8	%
Adaptive capacity	Illiterate population	AC_1	%
	HHs having safe drinking water facility	AC_2	%
	HHs having toilet/sanitation facility	AC_3	%
	HHs having electricity	AC_4	%
	HHs having mobile connectivity	AC_5	%
	HHs having banking facility	AC_6	%

Source Developed by the authors

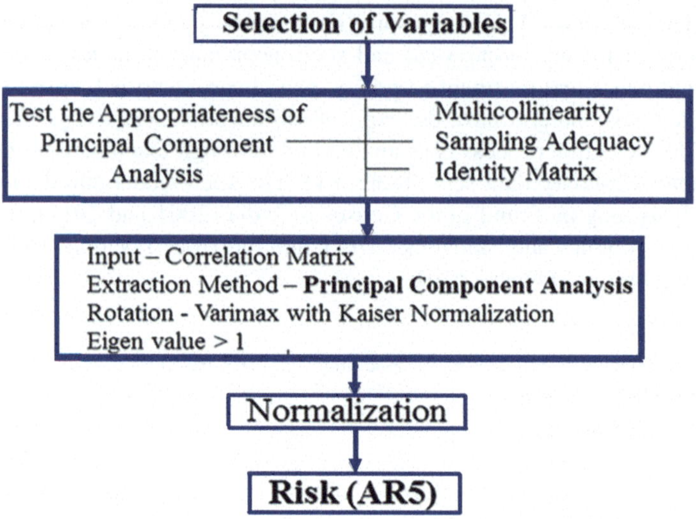

Fig. 3.4 Methodological framework for risk assessment based on IPCC AR-5

Analytical Techniques

Since the data lacks normalization, PCA (Principal Component Analysis) utilized the correlation matrix as an input for extracting principal components. Following Kaiser's 'eigenvalue greater than one' criterion, only components with eigenvalues (representing component variances) exceeding 1.0 were retained. To enhance interpretability, varimax (orthogonal) rotation was applied (refer to Fig. 3.4). Composite index calculation involved estimating component score coefficients. These coefficients, multiplied by the proportion of each component's variance in SPSS software, yielded values for contributing factors aligned with the IPCC, applicable to all Panchayats.

Hotspot Analysis

Hotspot analysis is conducted to detect clusters within a specific area, signifying hot and cold spots corresponding to high and low values of a given variable. The Getis-Ord G* statistics are employed to pinpoint these hot and cold spots. The Z value of the Getis-Ord G* statistics, serving as the dispersal parameter, is utilized to distinguish between hot spots and cold spots at confidence levels of 90%, 95%, and 99%. The analysis defines neighbors as 'contiguity edges corner' (refer to Fig. 3.5). The calculation of Getis-Ord G* statistics follows the formula provided below (Mondal 2020).

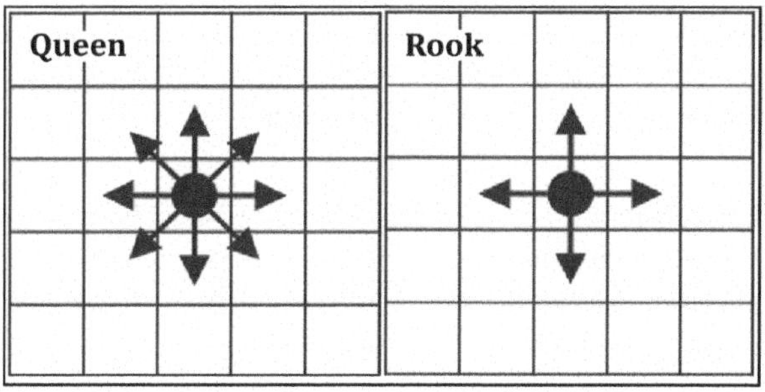

Fig. 3.5 'Queen' and 'rook' definition of neighbourhood

$$G_i^* = \frac{\sum_{j=1}^{n} w_{ij} w_j - \overline{x} \sum_{j=1}^{n} w_{ij}}{S\sqrt{\frac{n \sum_{j=1}^{n} w_{ij}^2 - \left(\sum_{ij}^{n} w_{ij}\right)^2}{n-1}}} \tag{3.1}$$

where x_j the attribute is the value for feature j; $w_{i,j}$ is the spatial weight between feature I and j; n is equal to the total number of features and

$$\overline{x} = \frac{\sum_{j}^{n} x_j}{n} \tag{3.2}$$

$$S = \sqrt{\frac{\sum_{j=1}^{n} x_j^2}{n} - (\overline{x})^2} \tag{3.3}$$

For the PCA analysis in this study, SPSS version 21 was employed. Additionally, ArcGIS version 10.3 was utilized for conducting the hotspot analysis.

Results

Assessing and quantifying flood vulnerability and risk is essential due to variations across different locations. Following normalization, PCA results were employed to analyze and chart vulnerability and risk for all 124 Panchayats. In risk analysis based on IPCC-AR5, seven components explained 76.05% of the total data variation in hazard, exposure, and adaptive capacity. The first four components contributed to 57.01% of the overall variance, with the initial component representing the highest volatility at 23.17% (see Table 3.2). Subsequent to PCA, and standardizing and aggregating hazard (H), exposure (E), sensitivity (S), and adaptive capacity (A), the risk for the seven components of Panchayats in India's lower Gangetic Plain was computed.

Table 3.2 Results of components and indicators using principal component analysis for assessing flood risk based on hazard, exposure index, sensitivity and adaptivity capacity

Indicators (i)	Component						
	1	2	3	4	5	6	7
Flood frequency	0.058	−0.023	−0.096	0.812	−0.111	0.223	0.197
Distance from channel	−0.182	0.589	−0.089	0.239	−0.188	0.337	−0.326
Flood affected area in %	0.208	0.039	0.174	0.808	−0.159	0.047	−0.102
Female headed HHs	0.487	0.737	−0.021	0.132	−0.122	−0.181	−0.016
% share of HHs working in cultivation	−0.162	0.523	−0.446	−0.099	−0.012	0.553	−0.174
% of HHs with 3 or more rooms with pucca walls and pucca roof	−0.75	−0.122	0.054	0.211	−0.123	−0.306	−0.002
Dependency ratio	0.457	0.428	−0.074	0.052	−0.161	−0.45	0.075
% of female population	−0.087	0.102	0.012	0.136	0.856	0.027	0.105
% of deprived HHs	0.923	−0.169	0.125	−0.062	0.111	0.033	−0.023
% of SC & ST HHs	0.708	0.214	−0.237	−0.18	0.173	0.272	0.09
% of share of HHs casual worker	0.701	−0.265	0.319	0.137	0.046	−0.078	−0.06
% share of HHs income less than Rs. 5000	0.796	0.021	0.06	−0.025	0.046	0.354	−0.246
% HHs owning land	−0.519	0.521	−0.434	0.119	−0.095	0.231	−0.011
% share of working force to total population	−0.253	0.168	−0.016	0.366	0.704	−0.115	−0.061
% of illiterate population	0.487	0.737	−0.021	0.132	−0.123	−0.181	−0.016
% of HHs having safe drinking water facility	0.195	−0.206	0.079	0.02	−0.102	0.329	0.805
% of HHs having toilet/ sanitation facility	−0.154	0.262	0.785	−0.063	0.104	0.233	−0.076
% of HHs having electricity	−0.296	0.352	0.723	−0.072	−0.069	0.18	−0.165
% of HHs having mobile connectivity	−0.291	0.347	0.663	−0.027	−0.105	0.135	0.175
% of HHs having banking facility	−0.476	0.493	−0.085	−0.106	0.004	0.025	0.174
Initial Eigen values	4.63	2.85	2.21	1.7	1.46	1.32	1.020
Weightage of the components	1.00	0.38	0.23	0.15	0.11	0.09	0.07
Percentage of variance	23.16	14.26	11.06	8.52	7.32	6.61	5.11
Cumulative percentage of variance	23.16	37.43	48.49	57.02	64.34	70.94	76.05

Source Computed by the authors

Status of the Components Associated with Flood Hazard

Flood hazard has been evaluated based on flood frequency (during 2000–2016), distance from the active channel, and flooded area of the Panchayats (Fig. 3.6a and b). The estimated values of flood hazard vary from 1.0 to 0.07. The highest value of hazard i.e. 1.0 has been found in Harali Udaynarayanpur and Kurchi Shibpur GP of Udaynarayanpur CD Block of Howrah district. All the GPs under the study area have been groped into five classes. A very high hazard value (> 0.75) has been observed in Binla Krishnabati GP (0.831) of Amta-II block, Bhabanipur Bidhanchandrapur (0.831), Harali Udaynarayanpur (1.000) Kurchi Shibpur (1.000) Rampur-Bihibhusutasand (0.813), Singiti (0.813) GPs of Udaynarayanpur block, Pursurah-II (0.927) of Pursurah block, Goura (0.754) of Daspur-II block and Mohanpur-I (0.813) and Mohoapur-II (0.927) of Ghatal block. The highest numbers of GPs having high hazard values (0.51–0.075) have been found in Khanakul-II, Pursurah, and Ghatal blocks. In Khanakul-II block, 9 out of 11 GPs, in Pursurah, 4 out of 8 GPs, in Ghatal block 6 out of 13 GPs, and in Amta-I block 8 out of 12 GPs have been subjected to high flood hazards. The flood frequency and the percentage of the affected area have been found the most important contributors to flood hazards (Fig. 3.7a).

The exposure has been measured using three components i.e. female head HHs, percentage share of HHs working in cultivation, and percentage of HHs with less than 1 room with kaccha walls and roof. The exposure value ranges from 0.830 to 0.081. The Khepuladakshinibarh GP in the Daspur-II block of Howrah district has a very high exposure value of 0.830. The Daspur-I, Daspur-II, Basudebpur GPs in Daspur-I block, Chaipat, Gochhati, Kamalpur, Sahalak GPs of Daspur-II block, Ajabnagar-II, Mohanpur-I GPs in Ghatal block, Chiladangi, Pursurah-II GPs in Pursurah block,

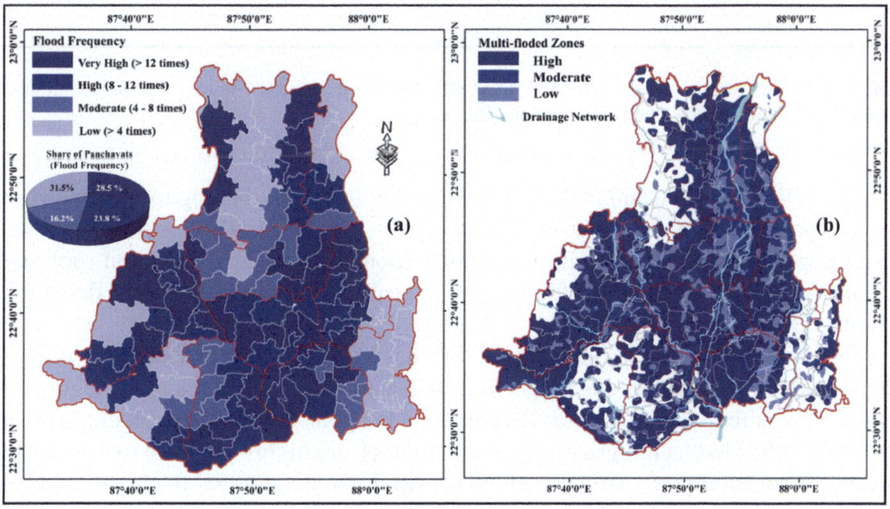

Fig. 3.6 Panchayat level: **a** Flood frequency and **b** multi-flooded area under the study area

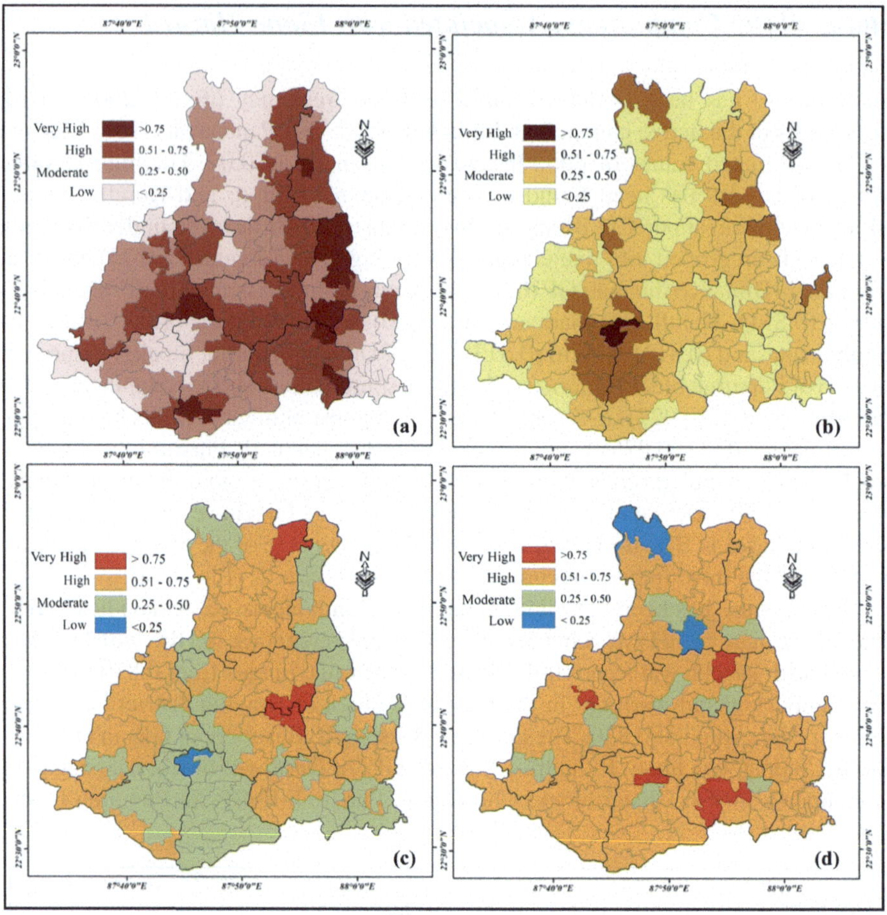

Fig. 3.7 Components for assessing the flood risk based on **a** hazard index, **b** exposure index, **c** sensitivity index and **d** adaptivity capacity index

Rampur-Dihibhusut Asanda GP in Udaynarayanpur mostly belong to the area with high exposure value ranges from 0.51 to 0.75 (Fig. 3.7b). The result has shown that the percentage of HHs with less than 1 room with kaccha walls and roof and female-headed HHs have been found more contributing components to the flood risk (Table 3.3).

The sensitivity to the flood values of GPs has been calculated based on 8 components (Table 3.3). The values range from 0.250 to 0.780. The GPs are categorized based on sensitivity to flood. The GPs namely, Malaypur-I in Arambag block, Arunda in Khanakul-I block, Ghingra in Khanakul-II block are highly sensitive to flood with a value of sensitivity of > 0.75. The high sensitivity to flood (0.51–0.75) values have been found in most of the GPS in Arambag, Khanakul-I, Khanakul-II, Ghatal, Amta-I, Udaynarayanpur blocks (Fig. 3.7c). The table has shown the very slight negative

Table 3.3 Components wise distribution of Gram Panchayats (GPs) for assessing the flood risk based on hazard, exposure index, sensitivity and adaptivity capacity

Components	Range and distribution	Very high		High		Moderate		Low	
		Act	In %	Act	In %	Act	In %	Act	In %
Hazard	Range	Above 0.60		0.41–0.60		0.20–0.40		Below 0.20	
	Distribution	40	32.00	36	28.80	28	22.40	21	16.80
Exposure	Range	Above 0.60		0.41–0.60		0.20–0.40		Below 0.20	
	Distribution	5	4.00	36	28.80	72	57.60	12	9.60
Sensitivity	Range	Above 0.60		0.46–0.60		0.30–0.45		Below 0.30	
	Distribution	32	25.60	53	42.40	34	27.20	6	4.80
Adaptive capacity	Range	Above 0.70		0.61–0.70		0.50–0.60		Below 0.50	
	Distribution	13	10.40	58	46.40	46	36.80	8	6.40

Source Computed by the authors

relation between risk and sensitivity with a value of −0.026 and the components like percentage share of working force to the total population, percentage of share of HHs casual worker, percentage of HHs owning land and percentage of share of HHs income less than RS. 5000 are the most important components that contribute to sensitivity values. It has also been revealed from the results that in the regions of very high to high risk the sensitivity values mostly belong to moderate level (0.25–0.50) although the exception is there.

The adaptive capacity to flood has been measured based on 6 components (Table 3.3). The values of adaptive capacities to flood vary from 0.157 to 0.814. In the areas under study, only a few GPs namely, Kasmoli, Jhamita in Amta-II, Tehatta-Kantaberia in Udaynarayanpur, and Tantisal in Khanakul-I have a high adaptive capacity to flood with a value of > 0.75. There are only two GPs that have a low adaptive capacity to flood with a value of < 0.25 and most of the GPS from the rest have a high adaptive capacity with values ranging from 0.51 to 0.75 (Fig. 3.7d).

Status of Risk to Flood

The risk is an amalgamation of hazard, exposure, sensitivity, and adaptive capacity as delineated in the IPCC AR5. All the GPs have been categorized into five groups based on calculated risk value. The results have shown that 4.84% area is at a very high risk of flood with a value of > 0.80. The Pursura-II GP of the block, Ghatal Municipality Rampur-Dihibhusut Asanda, Kurchi Shibpur, Harali Udaynarayanpur GPs of Udaynarayanpur CD Block, Ajabnagar-II of Ghatal block is in the very high flood risk area (Table 3.4). The result also shows that 5.65% area belongs to the high flood risk area with values ranging from 0.61 to 0.80. The Kishorepur-I, Arunda, Balipur GPs of Khanakul-I block, Rajhati-II of Khanakul-II, and Mohanpur-II of

Ghatal block are high risks prone to flood. The study also revealed that 19.35% area under the study area lies at a moderate risk to floods with a value ranging from 0.41 to 0.60 and low to very low risk to flood found over the 35.48% and 34.68% area respectively (Fig. 3.8a).

The assigned HH numbers primarily indicate the vulnerability status of Panchayats, along with the corresponding quartile distribution position (Q1, Q2, and Q3). This categorization of Panchayats proves valuable for exploratory data analysis, revealing the distribution of vulnerability indicators among homes exposed to diverse risks. Specifically, indicators strongly linked to either the family's geographic location or the inherent sensitivity of the study area is considered to exert significant influence. To analyze the risk distribution across various hazards, Violin plots were generated using SPSS 20.1 software (Mondal et al. 2022). These plots visually depict the distribution of data for the two HH groups, encompassing all fifty markers. Each indicator is associated with risk based on multiple distributions. Similar techniques were employed by Brooks et al. (2005) and Rajesh et al. (2018) in their

Table 3.4 Distribution of IPCC AR-5 based Risk of Gram Panchayats (GPs) for assessing the flood risk based on hazard, exposure index, sensitivity and adaptivity capacity

Components	Range and distribution	Very high		High		Moderate		Low		Very low	
		Act	In %	Act	In %	Act	In %	Act	In %	Act	In %
Risk	Range	Above 0.80		0.61–0.80		0.41–0.60		0.20–0.40		Below 0.20	
	Distribution	6	4.80	7	5.60	25	20.00	44	35.20	43	34.40

Source Computed by the authors

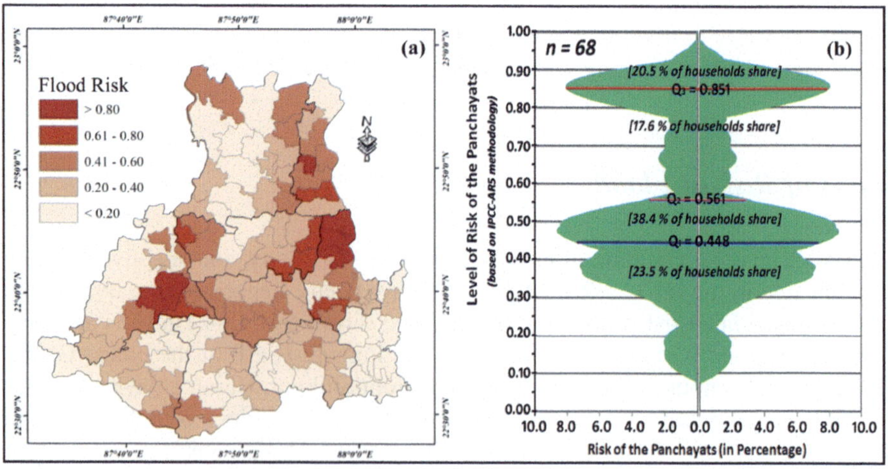

Fig. 3.8 **a** Flood hazard risk of Panchayats based IPCC-AR5 components and **b** violin plot to assess the nature of risk in different Panchayats

respective studies to identify pertinent vulnerability indicators with connections to climate-related disasters.

In numerous Panchayats, the distribution pattern clearly indicates the risk levels at the household level. Based on the IPCC-AR5 methodology, approximately 56% of households on the islands are classified as having a moderate to high risk (with Q1 = 0.448 and Q3 = 0.851). A significant portion, accounting for 20.5% of households in the Panchayats, is categorized as being at an extremely high risk, while 23.5% are considered to be at low to moderate risk. The villages on these islands have faced severe challenges due to substantial rainfall, flooding, and multiple flood events over the past fifteen years. In response to these hardships, people are compelled to transition from natural livelihoods to labour-based activities, leading to increased instances of labour migration and displacement (Fig. 3.8b).

Hotspot Analysis

A hotspot analysis was conducted to examine the spatial distribution of flood risk among all Gram Panchayats (GPs) in the lower Gangetic region of West Bengal. The results of the analysis indicate the formation of hotspots in two primary zones within the study area. The initial hotspot zone is centred around the Ghatal Municipality, while the second zone is situated on the eastern side of the region, encompassing GPs falling under the Pursurah and Udaynarayanpur blocks (Fig. 3.9). These two hotspot areas of the study region depict the sensitivity of the flood means the areas are at risk porn due to river floods. These two high-risk hotspots are formed where high flood exposures due to frequent floods are seen with located within 1 km from the active river channel and also 100% area is fully flood-affected. But here a very interesting fact is that both regions are highly flooded risk regions but the determinant factors are different likely Udaynarayanpur block is a high-risk area though having high adaptive capacity because of the high hazard exposure of the area.

Conversely, in contrast to other hotspot zones, the Ghatal municipality area shows less susceptibility to river floods. However, the significant proportion of illiterate residents in this area presents hurdles to the area's ability to adapt. Inadequate infrastructure also plagues the area, with few homes having access to electricity, poor sanitation, restricted banking options, and restricted access to safe drinking water. Despite the area being designated as having a low flood hazard, these variables lead to a reduced resilience against the negative effects of floods, meaning that the population is disproportionately vulnerable to flood hazards. On the other side, the study area has three cold spots the flood risk means fewer flood sensitive. The reason situated on is under the under one block second is the middle portion of the Daspur-I and Daspur-II and third and the last region comes under the southern portion of the Arambag CD Block. These grouping areas come under comparatively low flood hazard-exposure regions with moderate to high adaptive capacity.

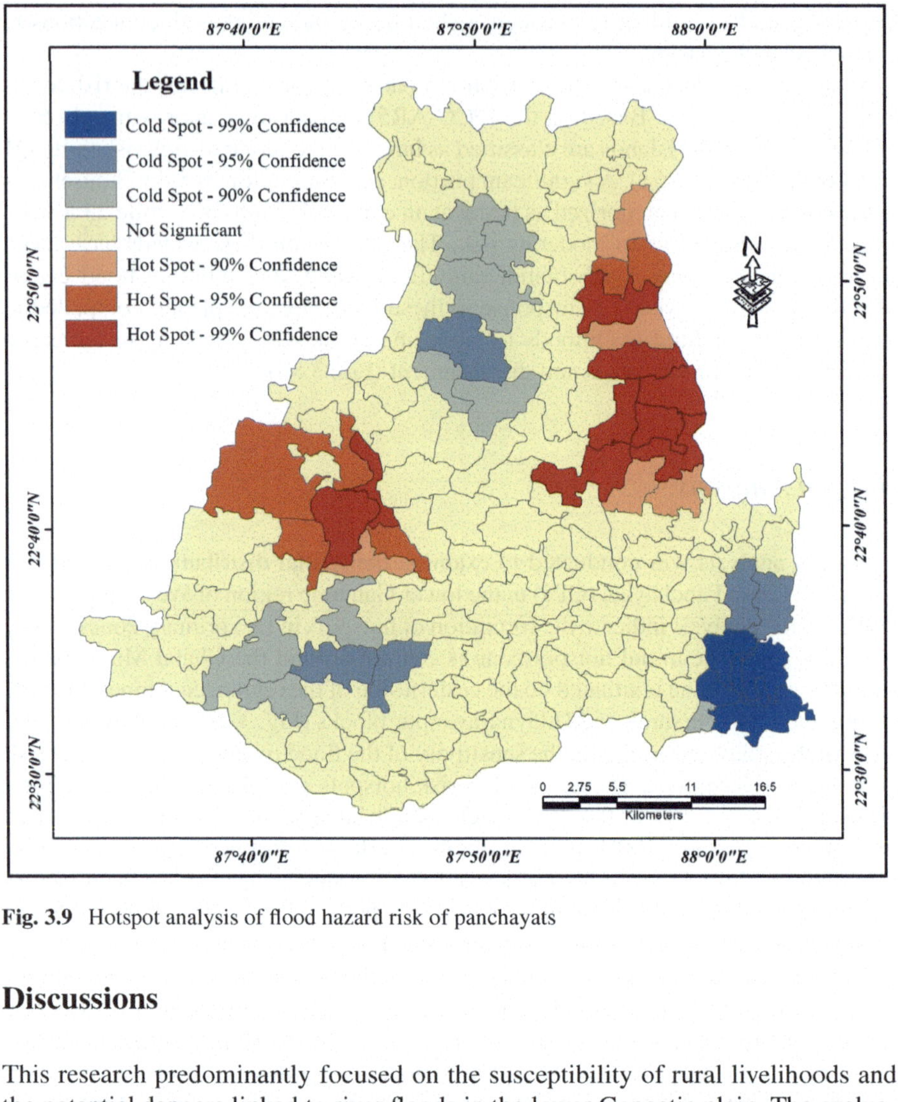

Fig. 3.9 Hotspot analysis of flood hazard risk of panchayats

Discussions

This research predominantly focused on the susceptibility of rural livelihoods and the potential dangers linked to river floods in the lower Gangetic plain. The evaluation of risk employed relevant indicators or factors associated with rural livelihoods, in accordance with the contributing factors delineated in the IPCC AR5, specifically hazards, exposure, sensitivity, and adaptive capacity. A crucial aspect lies in comprehending the impact of these components or indicators on the risk faced by the livelihoods in the study areas due to flooding. It helps in future research and to initiate coping and adaptive strategies. The study suggests that risk has a positive relationship with hazard and exposure. The hazard is the most influencing factor that enhanced the risk to the livelihoods during the food. Since the majority of the Panchayats are located near rivers, the eastern corridor and western portion are demonstrating high to very high even in terms of risk openness. The hazard risk faced by Panchayats in the study area can be illustrated through the distributional example using geomorphic units. In the southern part of the research region, low-lying seaside embankments and salt marshes at lower elevations have been identified as significantly more susceptible to erosion and storm-induced flooding. Exposure emerges as the second key factor influencing the risk to livelihoods impacted by floods. Within the study area, households led by females and those with kaccha (temporary or makeshift) rooms' exhibit the highest exposure to flood risk. Empirical observations clearly indicate that kaccha rooms are predominantly damaged during floods, leading to homelessness. The females headed HHs are exposed to flood risk. It has been well established in many types of research that during any hazard or disaster (i.e. floods, cyclones, tsunamis) women have suffered most due to gender norms that restrict women to their homesteads, their child caring role, religious bindings, their traditional dress up (like long sari) hinder fast movements, malnutrition, etc. (Ikeda 1995; Rahaman et al. 2016). In the areas of very high to high risk except the Ghatal Municipality belong to the rural area where cultivation is the predominant HH occupation. In this circumstance, floods caused extensive damage to the standing crops and lead to a high risk to livelihood. The same cases have been observed by Kamruzzaman and Shaw (2018) and Hoq et al. (2021) in the flood-prone Haor region of Bangladesh where the main source of vulnerability came due to extensive damage to standing boro rice crop by the food or flash floods. Although the result also shows the negative relationship of risk with sensitivity and adaptive capacity, it is very low. According to Smit and Wandel (2006), an effective social structure improves adaptability and reduces vulnerability and risk. Due to their proximity to rivers, Panchayats are notably susceptible to economic factors, especially in the case of impoverished households that are highly affected during floods. Livelihoods in these areas are predominantly tied to agriculture and manual labor, resulting in a lower socioeconomic resilience. The significant percentage of households earning less than Rs. 5000 and a high proportion of casual workers contribute to a heightened risk to livelihoods during floods. Casual workers often lose their jobs, and low-income households struggle to meet their basic daily needs during flood events. The research indicates that, overall, the adaptive capacity in most studied Panchayats is high and is not significantly influenced by the risk value. However, Panchayats with a high literacy rate, households with access to safe drinking water, sanitation facilities, and electricity are found to be less prone to risk. This aligns with Mansur's (2016) study, which highlighted that many urban centers in the Amazon Delta and Estuary lack essential facilities like sufficient drinking water and sanitation, exposing them to a higher risk of hazards. Another example could be cited from the Haor region of Bangladesh where, regions of high-income levels and high literacy rates were less prone to flood risk (Hoq et al. 2021).

Conclusion

This research delves into the climate-induced risks faced by Panchayats primarily involved in nature-based livelihoods like agriculture and fishery in lower Gangetic West Bengal, employing the methodology outlined in the IPCC-AR5. The study relies on secondary data and conducts a comparative analysis for 124 Panchayats. Empirical findings highlight heightened vulnerability among populations residing within a 2 km buffer from rivers in terms of risks and climatic variability. These river-proximate Panchayats exhibit a higher level of risk exposure and susceptibility compared to counterparts in other Administrative areas. The investigation extends to a detailed IPCC-AR5-based risk assessment at the household level across various geographical locations. Utilizing a violin plot, the study illustrates the quartile classification of households and the corresponding risk levels within different geographic settings. This visualization serves as an advantageous feature of the AR5 architecture. The IPCC-AR5 framework proves instrumental in pinpointing adaptive solutions within a system's current constraints and identifying climate-resilient pathways to mitigate the impacts of climate change. In an uncertain future, the IPCC-AR5 framework can be said to give a strong framework for minimizing vulnerability and risk. As a suggestion concerning with hotspot scenario, Ghatal areas have to improve their adaptive capacity likely improve infrastructure and also have to improve their educational condition which helps them to take the sudden diseases and also being ever about the flood.

Data Availability Statement (DAS) Dataset for 'Assessing the rural livelihood vulnerability and risk to river flood in lower Gangetic West Bengal applying IPCC-AR5 methodology' we accessed data Sentinal1, SAR-C for Water logging Analysis, District Disaster Management Report for flood frequency and multi flooding zones, Census of India and Socio-economic Caste Census for IPCC AR-5 Indicators.

Declaration

Author's Contribution Pintu Mandal: Methodology, Formal analysis, Writing—original draft. Jayanta Saha: Formal analysis, Software, Writing—original draft. Shyamal Santra: Writing, Formal analysis, review editing. Suman Paul: Investigation, Software Writing—review editing. Subhasis Bhattacharya: Writing—original draft and review editing. Bhaskar Samanta: Writing—review editing. Suman Paul: Methodology, Formal analysis, Writing—original draft.

Ethics Approval and Consent to Participate All the authors affirm that they have no known financial or interpersonal conflicts that would have appeared to have an impact on the research presented in this study.

Conflicts of Interests/Competing Interests Manuscripts reporting studies involving no human participants, hence no need of any Ethical Approval.

Consent for Publication Manuscripts reporting studies involving no human participants, hence no need of any Consent for publication'.

Funding This research does not promote by any funding agency.

References

Adger WN, Pulhin JM, Barnett J, Dabelko GD, Hovelsrud GK, Levy M, Vogel CH (2014) Human security. Cambridge University Press

Birkmann J (2006) Measuring vulnerability to promote disaster-resilient societies: conceptual frameworks and definitions. In: Measuring vulnerability to natural hazards: towards disaster resilient societies, vol 1, pp 9–54

Birkmann J, Cardona OD, Carreño ML, Barbat AH, Pelling M, Schneiderbauer S, Kienberger S, Keiler M, Alexander D, Zeil P, Welle T (2013) Framing vulnerability, risk and societal responses: the MOVE framework. Nat Hazards 67(2):193–211

Bohle HG (2007) Krisen, katastrophen, kollaps–Geographien von verwundbarkeit in der risikogesellschaft. na

Brooks N, Adger WN, Kelly PM (2005) The determinants of vulnerability and adaptive capacity at the national level and the implications for adaptation. Glob Environ Chang 15(2):151–163

Brouwer R, Akter S, Brander L, Haque E (2007) Socioeconomic vulnerability and adaptation to environmental risk: a case study of climate change and flooding in Bangladesh. Risk Anal Int J 27(2):313–326

Can ND, Tu VH, Hoanh CT (2013) Application of livelihood vulnerability index to assess risks from flood vulnerability and climate variability—a case study in the Mekong Delta of Vietnam. J Environ Sci Eng A 2(8A):476

Contribution of working group II to the fifth assessment report of the intergovernmental panel on climate change (IPCC). In: Climate change 2014: impacts, adaptation, and vulnerability. Part A: Global and sectoral aspects. Cambridge University Press, pp 793–832

Cutter SL (1996) Vulnerability to environmental hazards. Prog Hum Geogr 20(4):529–539

Dandapat K, Panda GK (2017) Flood hazard assessment at block level and its management strategy in Paschim Medinipur District, West Bengal, India. IOSR J Hum Soc Sci 22(6):7–14

Das S, Ghosh A, Hazra S, Ghosh T, de Campos RS, Samanta S (2020) Linking IPCC AR4 & AR5 frameworks for assessing vulnerability and risk to climate change in the Indian Bengal Delta. Progr Disaster Sci 7:100110

Das B, Bandyopadhyay A (2015) Flood risk reduction of Rupnarayana River, towards disaster management? A case study at Bandar of Ghatal Block in Gangetic Delta. J Geogr Nat Disast 5(135):2167-0587.https://doi.org/10.4172/2167-0587.1000135

Dasgupta S, Mukhopadhyay B (2014) Earthquake-landslide-flood nexus at the lower reaches of Yigong Tsangpo, Tibet: remote control for catastrophic flood in Siang, Arunachal Pradesh and upper Assam, India. J Eng Geol (A Bi-annual Journal of ISEG) 39:177–190

Dasgupta P, Morton J, Dodman D, Karapinar B, Meza F, Rivera-Ferre MG, Vincent KE (2014) Rural areas

Flood Damage Statistics (2018) (State wise and for the Country as a whole) for the period 1953 to 2016, Central Water Commission, Government of India. http://cwc.gov.in/main/downloads/statewise_flood_data_damage_statistics.pdf

Füssel HM, Klein RJ (2006) Climate change vulnerability assessments: an evolution of conceptual thinking. Clim Change 75(3):301–329

Gashaw W, Legesse D (2011) Flood hazard and risk assessment using GIS and remote sensing in Fogera Woreda, Northwest Ethiopia. Nile river basin. Springer, Dordrecht, pp 179–206

Guha-Sapir D, Hargitt D, Hoyois P (2004) Thirty years of natural disasters 1974–2003: the numbers. Presses University, de Louvain

Guha-Sapir D, Vos F, Below R, Ponserre S (2012) Annual disaster statistical review 2011: the numbers and trends

Hagen E, Lu XX (2011) Let us create flood hazard maps for developing countries. Nat Hazards 58(3):841–843. https://doi.org/10.1007/s11069-011-9750-7

Hahn MB, Riederer AM, Foster SO (2009) The livelihood vulnerability index: a pragmatic approach to assessing risks from climate variability and change—a case study in Mozambique. Glob Environ Chang 19(1):74–88

Hoq MS, Raha SK, Hossain MI (2021) Livelihood vulnerability to flood hazard: understanding from the flood-prone haor ecosystem of Bangladesh. Environ Manage 67(3):532–552

Ikeda K (1995) Gender differences in human loss and vulnerability in natural disasters: a case study from Bangladesh. Indian J Gend Stud 2(2):171–193

Imran M, Sumra K, Mahmood SA, Sajjad SF (2019) Mapping flood vulnerability from socioeconomic classes and GI data: linking socially resilient policies to geographically sustainable neighborhoods using PLS-SEM. Int J Disaster Risk Reduc 41:101288

Jamshed A, Birkmann J, Feldmeyer D, Rana IA (2020) A conceptual framework to understand the dynamics of rural-urban linkages for rural flood vulnerability. Sustainability 12(7):2894

Jha CV, Bairagya H (2013) Flood and flood plains of West Bengal, India: a comparative analysis. Revista Geoaraguaia

Jonkman SN, Kelman I (2005) An analysis of the causes and circumstances of flood disaster deaths. Disasters 29(1):75–97

Kale VS (2003a) Geomorphic effects of monsoon floods on Indian rivers. Flood problem and management in South Asia. Springer, Dordrecht, pp 65–84

Kale VS (2003b) The spatio-temporal aspects of monsoon floods in India: implications for flood hazard management. Disaster managemant. University Press, Hyderabad, pp 22–47

Kamruzzaman M, Shaw R (2018) Flood and sustainable agriculture in the Haor basin of Bangladesh: a review paper. Univ J Agric Res 6(1):40–49

Kasperson RE, Kasperson JX (2012) Climate change, vulnerability and social justice. Routledge, pp 316–336

Mansur AV, Brondízio ES, Roy S, Hetrick S, Vogt ND, Newton A (2016) An assessment of urban vulnerability in the Amazon Delta and Estuary: a multi-criterion index of flood exposure, socio-economic conditions and infrastructure. Sustain Sci 11(4):625–643

Mishra SK, Mishra N (2010) Vulnerability and adaptation analysis in flood affected areas of Orissa. Soc Change 40(2):175–193

Mollah S (2013) Types and sources of flood in Murshidabad, West Bengal. Indian J Appl Res 3(2):352–354

Mondal S (2020) Modeling the spatial pattern of household quality of living in West Bengal: an approach of hotspot and cluster analysis. Model Earth Syst Environ 6(2):833–851

Mondal M, Biswas A, Haldar S, Mandal S, Mandal P, Bhattacharya S, Paul S (2022) Rural livelihood risk to hydro-meteorological extreme events: empirical evidence from Indian Sundarban applying IPCC-AR5 and DEMATEL methodology. Int J Disaster Risk Reduction 77:103100. https://doi.org/10.1016/j.ijdrr.2022.103100

Mujiburrehman K (2015) Preparation of flood inundation map in Ganga River at Farakka Bridge, Malda, West Bengal, India. Int J Res Geogr 1(1):1–17

Mukhopadhyaya SC, Dasgupta A (2014) Bishwa o Bharater Patavumikay Pashchimbanger Nadi—Jalchitra, Ayalayed Book Agency

Nath SK, Roy D, Thingbaijam KKS (2008) Disaster mitigation and management for West Bengal, India—an appraisal. Curr Sci, 858–864

Olsson L, Opondo M, Tschakert P, Agrawal A, Eriksen S, Ma S, Perch L, Zakeldeen S (2014) Livelihoods and poverty. In: climate change 2014: impacts, adaptation, and vulnerability. Part A: global and sectoral aspects

Rahaman KM, Ahmed FRS, Islam MN (2016) Modeling on climate induced drought of north-western region, Bangladesh. Model Earth Syst Environ 2(1):45

Rajesh S, Jain S, Sharma P (2018) Inherent vulnerability assessment of rural households based on socio-economic indicators using categorical principal component analysis: a case study of Kimsar region, Uttarakhand. Ecol Ind 85:93–104

Rakib MR, Islam MN, Islam MN (2017) Flood vulnerability mapping to riverine floods: a study on the old Brahmaputra River. Curr Res Geosci 7(2):47–58

Sanyal J, Lu XX (2005) Remote sensing and GIS-based flood vulnerability assessment of human settlements: a case study of Gangetic West Bengal, India. Hydrol Processes Int J 19(18):3699–3716. https://doi.org/10.1002/hyp.5852

Sanyal J, Lu XX (2006) GIS-based flood hazard mapping at different administrative scales: a case study in Gangetic West Bengal, India. Singapore J Trop Geogr 27(2):207–220

Smit B, Wandel J (2006) Adaptation, adaptive capacity and vulnerability. Glob Environ Chang 16(3):282–292

Szabo S, Brondizio E, Renaud FG, Hetrick S, Nicholls RJ, Matthews Z, Dearing JA (2016) Population dynamics, delta vulnerability and environmental change: comparison of the Mekong, Ganges-Brahmaputra and Amazon delta regions. Sustain Sci 11(4):539–554

Tanoue M, Hirabayashi Y, Ikeuchi H (2016) Global-scale river flood vulnerability in the last 50 years. Sci Rep 6(1):1–9

Turner BL, Kasperson RE, Matson PA, McCarthy JJ, Corell RW, Christensen L, Eckley N, Kasperson JX, Luers A, Martello ML, Polsky C, Pulsipher A, Schiller A (2003) A framework for vulnerability analysis in sustainability science. Proc Natl Acad Sci 100(14):8074–8079

United Nations Development Programme. Bureau for Crisis Prevention (2004) Reducing disaster risk: a challenge for development—a global report. United Nations

Vincent K (2004) Creating an index of social vulnerability to climate change for Africa. Tyndall Center for Climate Change Research. Working Paper 56(41):1–50

Wanders N, Karssenberg D, Roo AD, De Jong SM, Bierkens MFP (2014) The suitability of remotely sensed soil moisture for improving operational flood forecasting. Hydrol Earth Syst Sci 18(6):2343–2357

Wisner B, Blaikie P, Cannon T, Davis I (2004) At risk, natural hazards. In: People's vulnerability and disasters. Routledge

WMO (2018) 2018 annual report: WMO for the twenty-first century

Yohe G, Tol RS (2002) Indicators for social and economic coping capacity—moving toward a working definition of adaptive capacity. Glob Environ Chang 12(1):25–40

Chapter 4
Impact of Rapid Urbanization on Urban Microclimate Dynamics: A Case Study on Asansol Municipal Corporation (AMC), Paschim Bardhaman, West Bengal, India

Chumki Mondal, Subham Kumar Roy, and Sanatan Ghosh

Abstract Asansol Municipal Cooperation (AMC) is one of the most important urban agglomerations in India. The fast urbanization of the AMC changes the urban environment and speeds up the high surface temperature increase over time. Using NDVI, NDWI, and NDBI to analyze the changing scenario of plant cover and built-up areas, this study aims to estimate the evolution of post-monsoonal surface temperature of AMC during the periods. This study employed LANDSAT 8 OLI/TIRS data and LANDSAT 4–5 TM data to determine the appropriate indices. The GIS and Excel programmers are used to create the maps and graphs. The amount of dense vegetation cover has reduced and the built-up area has expanded from the core to the periphery region, as can be seen clearly from evaluating all pertinent data, maps, and diagrams, which point to dynamic urbanization, expanding industrialization, and rigorous tree cutting. In 1990, Shannon's entropy reached its peak in 23 wards; nevertheless, throughout the course of 30 years, this entropy value fluctuated considerably across 50 wards of AMC. The industrial sector and the city core have a maximum surface temperature of 30 °C because they absorb heat, whereas the peripheral area has a surface temperature range of 19–24 °C because of the strong association between urbanization and Land Surface Temperature (LST). Therefore, the average temperature has risen by about 3–5 °C over the research period. The study region is divided into high-correlation and poor-correlation groups based on Pearson's correlation analysis of the measured parameters. There is a quick explanation of how LST,

C. Mondal (✉)
Department of Geography, Khandra College, Khandra (Kazi Nazrul University), Paschim Bardhaman, West Bengal 713363, India
e-mail: chumkimondal605@gmail.com

S. K. Roy
Department of Geography, Tilka Manjhi Bhagalpur University, Bhagalpur, Bihar 812007, India

S. Ghosh
Department of Geography, Krishnath College, Berhampore (University of Kalyani), Murshidabad, West Bengal 742101, India

U. Mukhopadhyay et al. (eds.), *Climate Crisis, Social Responses and Sustainability*, Climate Change Management, https://doi.org/10.1007/978-3-031-58261-5_4

NDVI, NDWI, NDBI, and correlation are calculated. Authorities can utilize the data to establish policies that reduce urban green space, highlight the most pressing issues related to land surface temperature, and reuse the data for other purposes.

Introduction

Rapidly accumulating a significant number of inhabitants in one area is the principle of underlying urbanization (Mandal 2000). Urbanization is now taking place more quickly in developing countries than in nations with advanced economies (Tah et al. 2023). The development of an urban microclimate is attributable to the extraordinary rate of urbanization and concentration of human activity. A microclimate is an isolated atmospheric zone in the heart of the city where climate is different from surroundings (Agnihotri et al. 2018). It consists of regional fluctuations in humidity, temperature, wind, and sun radiation, all of which are significantly affected by a variety of factors. Thus, the development of urban heat islands has a significant impact on the microclimate of cities (Sarkar and Kar 2023). Land use modifications resulting from unplanned urban expansion disrupted the energy balance and caused cities to be warmer than their environs. Increasing emissions of greenhouse gases into the atmosphere have been a consequence of expanding traffic and industry. These gases, in turn, absorb long-wave terrestrial radiation that is emitted, leading to elevated temperatures in urban areas. Urban heat islands occur from the city Centre becoming significantly warmer than the outer environs (Grover and Singh 2015; Tah et al. 2023; Sarkar and Kar 2023). Turner et al. (1994) assert that decreased plant cover and the expansion of developed areas have contributed to an increase in land surface temperature. This type of urban sprawl increases the surface temperature of the land because it absorbs solar radiation during the day and then reradiates it at night (Yuan and Bauer 2007). Modifications to land use and land cover (LULC) due to urbanization have impacted the energy budget and created the urban heat island (UHI) in nearly all of India's large and medium cities in recent years. As a result, gradually it raises the temperature in urban centers, creating a complex microclimate that has a significant impact on the individuals who live there (Sarkar and Kar 2023). Based on the correlations between NDVI, NDBI, and LST, numerous efforts have been carried out at the local, regional, and global levels using contemporary GIS tools. Numerous beautiful cities in India have been the subject of extensive research on LST, including Bengaluru (Govind and Ramesh 2019), Chennai (Bharath et al. 2018), Asansol–Durgapur Development Authority (Choudhury et al. 2019), Durgapur city (Tah et al. 2023), Delhi (Sharma et al. 2020), Mumbai (Taubenbock et al. 2009), Kolkata (Mandal et al. 2019) and Siliguri (Roy et al. 2022). The Asansol Municipal Corporation is a significant hub for manufacturing and mining activities in the industrial sector of West Bengal. It has experienced a rapid and dynamic urbanization process within a very short span of time. The significant transformation of previously pristine regions into urban-cultural landscapes within the city over the last few decades can also be attributed to inadequate urban expansion and economical function. Particularly concerning the establishment of heat islands in metropolitan regions, urban microclimates have produced a consequence that has a substantial effect on the living conditions and quality of life of city residents. In order to control and maintain the urban microclimate there is a need for urban vegetation cover. Therefore, in order to facilitate the expansion of the metropolitan region, it is imperative to establish a financially viable and comprehensive strategy for urban environment and land use planning that is also sustainable from a geographical and environmental standpoint. The principal aim of this research is to analyze the fluctuations in built-up area and vegetation cover within the designated study area from 1990 to 2020. Additionally, the study seeks to evaluate the Land Surface Temperature (LST) over the same time period, establish correlations among different indices, and monitor the significant urban heat island and associated microclimatic conditions in the Asansol Municipal Corporation, West Bengal. Since the primary focus of this study is urban microclimate, we take into account a roughly thirty-year time period. Long-term weather patterns have to be investigated for studying climate. Additionally, we know from the literature research that throughout this period, the percentage of built-up area has grown by around 33.87 (Maity et al. 2020; Saha and Malik 2018; Choudhury et al. 2019; Das et al. 2021).

Study Area

In terms of economic activity and urbanization, the Asansol Municipal Corporation (AMC) region is one of most important developed urban area under Asansol Durgapur Development Authority (ADDA), West Bengal. The landscape patterns have transformed during the past several decades. The urban geographer has recently turned their focus towards this field of study. AMC lies between 23°36′8″ N–23°46′45″ N and 86°49′28″ E–87°6′55″ E with a total population of 5.6 lakh, per the 2011 Census of India (Fig. 4.1). AMC is the 11th most urbanized area in India and the 42nd fastest growing city in the world, based on research conducted by the International Institute for Environment and Development in 2010 and the 2018 City Mayor's Statistics (Maity et al. 2020). Asansol Municipality was developed in the year of 1885. In the year of 1994 the city emerged as a Municipal Corporation with three municipalities (Kulti, Jamuria and Raniganj), some rural parts, collier areas and Burnpur notified area. Based on its physiographic characteristics, AMC is located on the bottom portion of the Chotanagpur Plateau, which has Precambrian-aged Meta sedimentary rocks, upper tertiary sediments, Rajmahal basalts, and Gondwana sedimentary rocks. It is 111 m above mean sea level on average. Asansol is made of conglomerate and rock fragments generated by the weathering of quartz in combination with coarse, gritty soil and some part was made by sandy soil, which is typical of granitic rocks and sandstones. The Asansol Sub-division has a large supply of mineral resources, including manganese, bauxite, iron ores, coal, silica bricks, and construction materials. The location of the Indian Iron & Steel Company (IISCO) steel mill in Burnpur was determined by the abundance of mineral resources in the area. The climate of the AMC is more tropical savanna-like. The month of April to June is the warmest, whilst December to February is the coldest. The mean yearly precipitation amounts to 1430 mm, while the average annual temperature is 26.2 °C. Given the climatic conditions, transportation and communication infrastructure, employment prospects, and overall progress in the western region of West Bengal, AMC emerges as a suitable urban locality. As a consequence, the examination of the research area's land surface temperature and pattern of development is crucial. Kolkata Urban Agglomeration is the preeminent urban entity in West Bengal, with Asansol ranking second in significance. In Asansol Municipal Corporation, the number of urban residents has grown by 18.61% over the past ten years. It is the first significant urban area in the plains of Bengal after the plateau area of Chotanagpur. Asansol is not just an urban region but also a mining and industrial hub of the western part of Bengal. As a result, throughout the past several decades, the layout of the landscape has changed drastically. LST was significantly impacted by the swift transformations in land use and land cover patterns. The population has played a crucial role in the urbanization process since the dawn of civilization (Seto et al. 2011). The fundamental necessities of human existence are clothes, food, and employment, and urban areas are the centers of those endeavors. In order to meet their most fundamental requirements, the people in the vicinity are considering moving to cities (Chatterjee and Majumdar 2022). The reasons for the urbanization process are as follows. Due to the limited quantity of land, the enormous demand for property in metropolitan areas poses a significant challenge for municipal administrators. The ecology is negatively impacted by processes of unplanned rapid urbanization (Wei 2015). It is obvious from Fig. 4.2 that Asansol Municipal Corporation (AMC) noticed steady population growth from 1901 to 2011. The total population was 14,906 in 1901 and 55,797 in 1941. The year 2001 had the greatest decadal growth rate, 81.34%, according to the Census of India (2011). Present decadal growth is low due to unhealthy physical social environment, resource destruction, losses of biodiversity, landscape change and bad impact of public health in the city core.

The territory is said to have been a part of the kingdom of Vishnupur, which was governed by the Malla dynasty for over a thousand years until the arrival of the British. As to the narrative, in the Asansol forest in 1741 AD, two men by the names of Nakdi Ray and Ramakrishna Ray, backed by troops from Lathial, repelled the Bargi assault. In 1742, King Garuda Narayan Singhadeo of Panchkot granted Nakdi Raya and Ramakrishna Raya a large tract of land in the Asansol forest area as jagir in exchange for a meagre deposit of 376 taka 4 annas 2 pies, as a token of appreciation for their valour. Following that, the two men decided to turn the Asansol forest into a place to live. Village Asansol was constructed. An upgraded route from Fort William in Calcutta to Delhi was intended to be constructed in 1831 by Governor General of India, Lord William Bentinck. It's reported to have gone farther west than Burdwan in 1836. Asansol most likely began to have a better road communication system at that time. A residential community for railway employees was constructed on Asansol Mauza in 1863, along with the extension of the East India Railway from Raniganj to Asansol. At this railway colony, Asansol started to become urbanized (Chattopadhyay 2022). One of the town's geographical benefits is

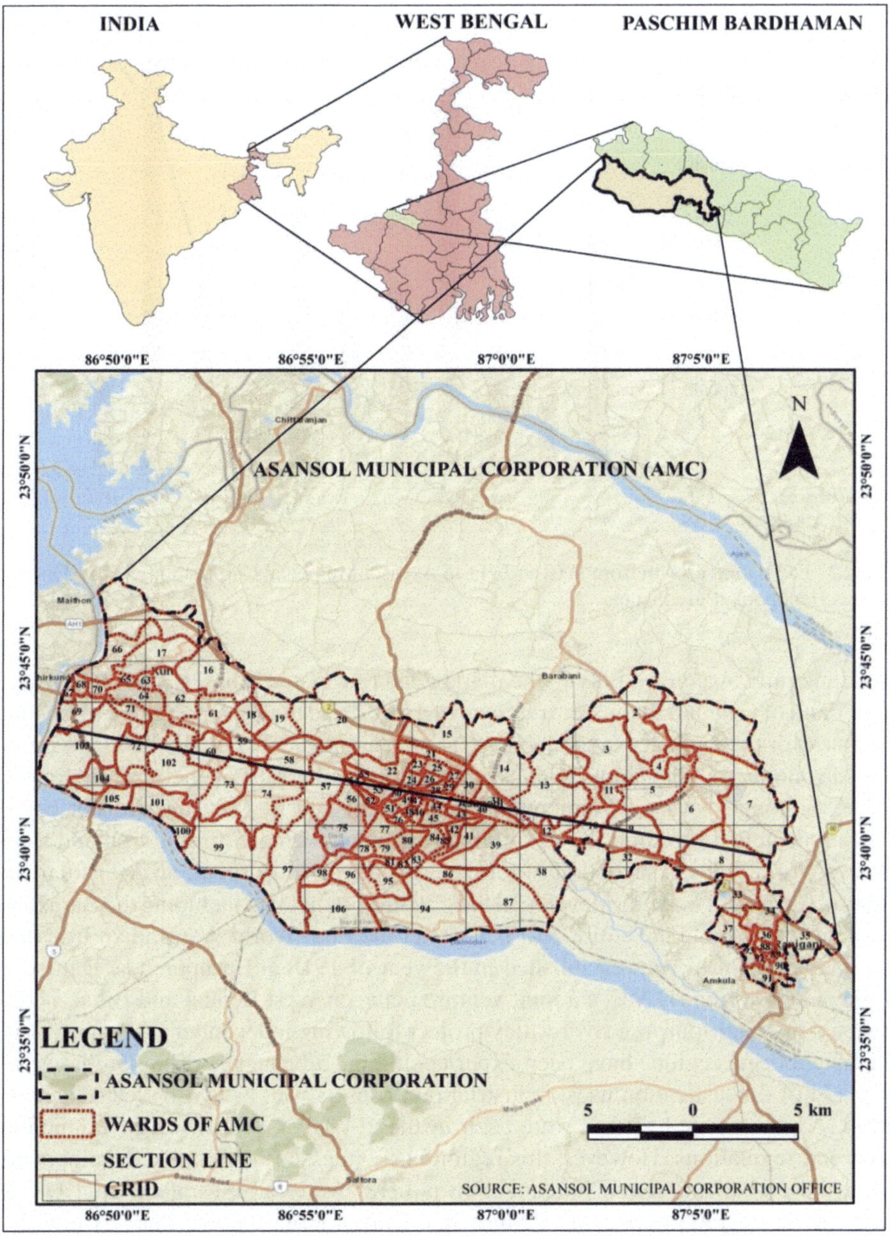

Fig. 4.1 Location map of the study area (Asansol Municipal Corporation)

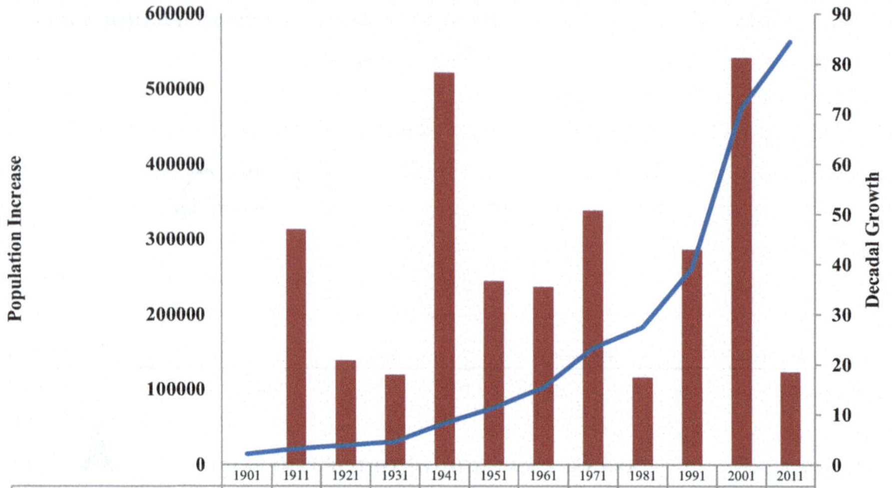

	1901	1911	1921	1931	1941	1951	1961	1971	1981	1991	2001	2011
DECADAL GROWTH	0	47.05	20.9	18.06	78.34	36.7	35.57	50.83	17.57	42.98	81.34	18.61
POPULATION INCREASE	14906	21919	26499	31286	55797	76277	103405	155968	183375	262188	475439	563917

Fig. 4.2 Population growth from 1901 to 2011 in Asansol Municipal Corporation. *Source* District Census Handbook, Barddhaman

its strong railway connectivity to the market and raw material bases, as well as to the hinterland of Calcutta. It took the removal of a forest and the construction of a housing complex for the engineers and labourers to create this plant. Tribal inhabitants from the surrounding Chotanagpur plateau made up the majority of the unskilled work force. Although Asansol Municipality was established in 1885 but did not begin operations until 1896. In order to supply the growing demand for pig iron and steel during World War I, a second blast furnace was put up in Burnpur, an Asansol town suburb. Asansol at that time developed to be an important town including the Barakar–Raniganj region and the entire coal belt. After the First World War, Indian Iron and Steel Company Ltd. was established in the year of 1918 at Burnpur. The Damodar Valley Corporation (D.V.C.), a joint venture between West Bengal and Bihar, is the nation's first multipurpose river valley project following independence. Asansol and the neighboring regions have been experiencing the advantages of this initiative in the field of domestication uses, industrial purposes as well as hydro electric generation. A few reversal effects were seen in the 1970s as a result of governmental laws and regulations. However, this region has witnessed remarkable development since the globalization and liberalization policies were put into place. In 1994, it received an upgrade to become a Municipal Corporation. Its own Police Commissionerate has existed since 2011. AMC constitutes an unique urban agglomeration that combines mining, industry, and habitation. Basically, residential, commercial, educational, medical, market, and administrative facilities are located in the core of the city. In Burnpur, Kulti, Barakar, Jamuria, Raniganj, Ikrah and Neamatpur, the majority of significant industrial parks—including those for iron, cement, soap, chemicals, silicate, endicott, and food processing—were built after the Core (Bureau of Applied Economics Statistics 2010, 2011, 2014). The open-pit coal mining fields that make up the majority of the outermost layer AMC include those at Dishergarh, Hasanpur, Kulti, Barakar, Sarakdihi, Kalla, Domohani, Kantagoria, Harabhanga, and other places.

Materials and Methods

The present study relies exclusively on secondary sources of data. In order to conduct this investigation, three Landsat photos devoid of clouds were obtained from United States Geological Survey (USGS) (Table 4.1). Additionally, collected images were prepared for radiometric and geometric modifications in order to make them useful. ArcGIS 10.5 and QGIS 2.18 software have been used for this research. For statistical analysis and illustration of the calculated data MS-Excel (2007 version) and Power Bi are used. The NDVI, NDWI, NDBI, BI and Shannon's entropy were derived using the Formulas 4.4, 4.5, 4.6, 4.7 and 4.8 to gain a more comprehensive understanding of the microclimatic effects of urban growth and changes in green space. Data from Landsat 5 (1990 and 2005) and Landsat 8 (2020) were used to determine Asansol Municipal Corporation's land surface temperature. Band 6 from Landsat 5 and Bands 10 and 11 from Landsat 8 are thermal bands, respectively (Pal and Ziaul 2017). For this reason, every process is employed to transform digital data into spectrum illumination in order to ascertain temperature using OLI TIRS and TM images, which are succinctly described below.

- Transformation of Digital value into spectrum radiance (Landsat 5 Band6)

$$L\lambda = (\text{L MAX} - \text{L MIN}) / (\text{QCAL MAX} - \text{QCAL MIN}) * (\text{QCAL} - \text{QCAL MIN}) + \text{LMIN}\lambda \quad (4.1)$$

This is where the following variables are defined: Spectral radiance (Lλ), Maximum calibrated DN (QCAL MAX), Minimum calibrated DN (QCAL MIN), Calibrated DN (QCAL), and QCALMIN, LMIN, and QCALMAX, QCALMIN, and QCAL are all scaled to QCAL.

$$\text{Radiance to BT} = T = K2/\text{In}(K1/L\lambda + 1) \quad (4.2)$$

Table 4.1 Details of used secondary data

Data type	Date	Sensor	Path row	Data source
Spatial data	05.01.1990	Landsat 5 TM	139–044	USGS earth explorer
	14.01.2005	Landsat 5 TM	139–044	
	21.01.2020	Landsat 8 OLI-TIRS	139–044	

This is where the constant values from the satellite metadata K1 and K2 are found, while T is the effective satellite temperature.

$$\text{Kelvin to Celsius} = \text{BT} - 273.15 \tag{4.3}$$

- It is quite simple to prepare Land surface temperature from Landsat 8 in QGIS. We are going to set up the RS and GIS plugins to manage and install the plugins section of QGIS in order to do this. Then we are submitting the genuine dataset as .tar data format and the actual area of interest shape file. Lastly choose LST (Celsius) from extra derived input, then click the "start processing" option and after that it takes some time and the final map is done.

- **Normalized Difference Vegetation Index (NDVI)**

Satellite images are mostly used to detect the plant cover using the NDVI. The NDVI is calculated using near-infrared (NIR) and red bands (Townshend and Justice 1986). As a result, the NDVI value is between − 1 and + 1. Values near to 0 indicate a low level of vegetation cover.

$$\text{NDVI} = (\text{NIR} - \text{RED}) / (\text{NIR} + \text{RED}) \tag{4.4}$$

- **Normalized Difference Water Index (NDWI)**

One use of the NDWI is to help locate bodies of open water in satellite photos (McFeeters 1996; Singh et al. 2014). In order to find the NDWI, the green and near-infrared (NIR) bands are utilized. The result is that the NDWI number might be anything from − 1 to + 1. If the value is close to 0, then the body of water is not present on the surface. If it is close to 1, then the surface is covered by water.

$$\text{NDWI} = (\text{GREEN} - \text{NIR}) / (\text{GREEN} + \text{NIR}) \tag{4.5}$$

- **Normalized Difference Built up Index (NDBI)**

The NDBI is an additional metric for locating man-made structures in cities using satellite imagery. The NDBI is determined by combining the results of SWIR and NIR measurements (Zha et al. 2003). The NDBI value so falls somewhere in the range of − 1 to + 1. Values between 0 and 1 denote built-up regions, while values near 1 suggest built-up areas.

$$\text{NDBI} = (\text{MIR} - \text{NIR}) / (\text{MIR} + \text{NIR}) \tag{4.6}$$

- **Built Up Index (BUI)**

BUI is another most significant measure that helps to some extent differentiate between urban areas and barren landscape (Chen et al. 2003). It is mostly used to recognize new constructions made by humans in urban landscapes. As a result, the BUI value is between − 1 and + 1. A built-up region is a greater possibility to be represented by a pixel where BUI value is high.

$$\text{BUI} = \text{NDBI} - \text{NDVI} \tag{4.7}$$

- **Shannon's Entropy**

It is common practice to use entropy as a metric for urban sprawl (Yeh and Li 2001). Trends in the accumulation or dispersion of urban areas are evaluated. This study uses Shannon's entropy to quantify patterns of urban expansion. Using entropy, we may assess the degree to which a geo-physical variable (built-up area) is concentrated or spread (homogeneous) among 'n' geographical units/wards (Jat et al. 2008).

$$\text{Shannon's Entropy } (E) = \sum_{i}^{n} Pi.\log\left(\frac{1}{Pi}\right) \tag{4.8}$$

where

Pi = Pi = $\text{Xi}/\sum_i^n Xi$ represents the quantity measured in the ith zone out of a total of n zones
n = Number of zones.

Result and Discussion

NDVI

The Normalized Difference Vegetation Index map (Fig. 4.3) explains the arrangement and distribution of plant cover from 1990 to 2020. Additionally, it shows how well a region's vegetation is doing. Our current method for NDVI computation makes use of Eq. (4.4). In 1990 (Fig. 4.3a), the highest vegetative area in AMC was basically found in eastern and north eastern part of Asansol municipality near Nigha, Damra; Girmint colliery, Jamuria, Shibpur and another patch has been found in the western portion of our area of research near Kulti, Barakar. Highest reading of NDVI varied from 0.3 to 0.68. Whereas the lowest reading of NDVI (− 0.40) was found in the IISCO steel plant which is situated in the middle part of our area of study. Negative readings of the NDVI fell between 0.02 and − 0.19 in 2005 (Fig. 4.3b), as the municipality's centre was flourishing extremely quickly. The northwestern region of the city, in particular, has the highest NDVI value (0.3–0.44) because of the abundance of flora there (Barkar, Kulti) and south eastern part of the city near Damra, Kalipahari. Major changes of

NDVI values have been seen between 1990 and 2020. Particularly detrimental to the NDVI of the AMC are the development and expansion of industrial centres, road and railway infrastructure, the coal region, and industry. This is especially true in the north-eastern, north-western, south-eastern, eastern, and southern regions of the study area, where the vegetative index value decreased from 1990 to 2020, with the exception of a few patches (Fig. 4.3c). Highest value of NDVI recorded in the year of 2020 is western part of the study area near Barakar–Kulti.

Table 4.2 summarizes the statistical data from the NDVI map, as shown. The range of NDVI varies from 0.68 to − 0.4 in the year 1990 to 0.65 to − 0.45 in the year 2020. It essentially shows that from 0.14 in 1990 to 0.1 in 2020, the proportion of vegetation in study area has decreased.

NDWI

The organization and distribution of water cover from 1990 to 2020, a period divided into 15 years, are shown in the Normalized Difference Water Index map (Fig. 4.4). Additionally, it shows how the characteristics of both natural and man-made water bodies in a certain area have been described. In order to create the NDBI Eq. (4.5) is used. In the year of 1990 (Fig. 4.4a), Maximum NDWI was seen in the western, northwestern, and south-eastern regions of the AMC, as well as in a few patches of the IISCO, Nigha, Kalapahari, Kalla. Highest value of NDWI ranged from 0.3 to 0.5. Conversely, the NDWI with the lowest value (− 0.5) was discovered in the central region of the municipality and its environs. In 2005 (Fig. 4.4b), as urban territory flourished and faced substantial changes, the core of the city was fastest growing as well as massive development in industrial sector colliery areas had negative NDWI values ranging from 0.0 to − 0.3. Major changes in NDWI values have been seen from 1990 to 2005, because of the creation of industrial setup and residential projects. However, there are certain open cast mining areas and the execution of various projects on the outskirts of the metropolitan center, that's why in 2020 water bodies had grown up. The state of balance of ground water recharge and static level in this region is greatly influenced by man-made artificial water bodies. In the year 2020 (Fig. 4.4c), The largest water value is evidently detected in the vicinity of the abandoned open cast mining regions of Eastern Coal Field and the IISCO steel factory. Due to this, the assortment of water bodies in the research region is continuously growing.

NDBI

Normalized Difference Built up Index map (Fig. 4.5) explains the arrangement and distribution of artificial structure from 1990 to 2020, with 15 years gap. It also reflects the nature and pattern of artificial structure that has been explained in a particular

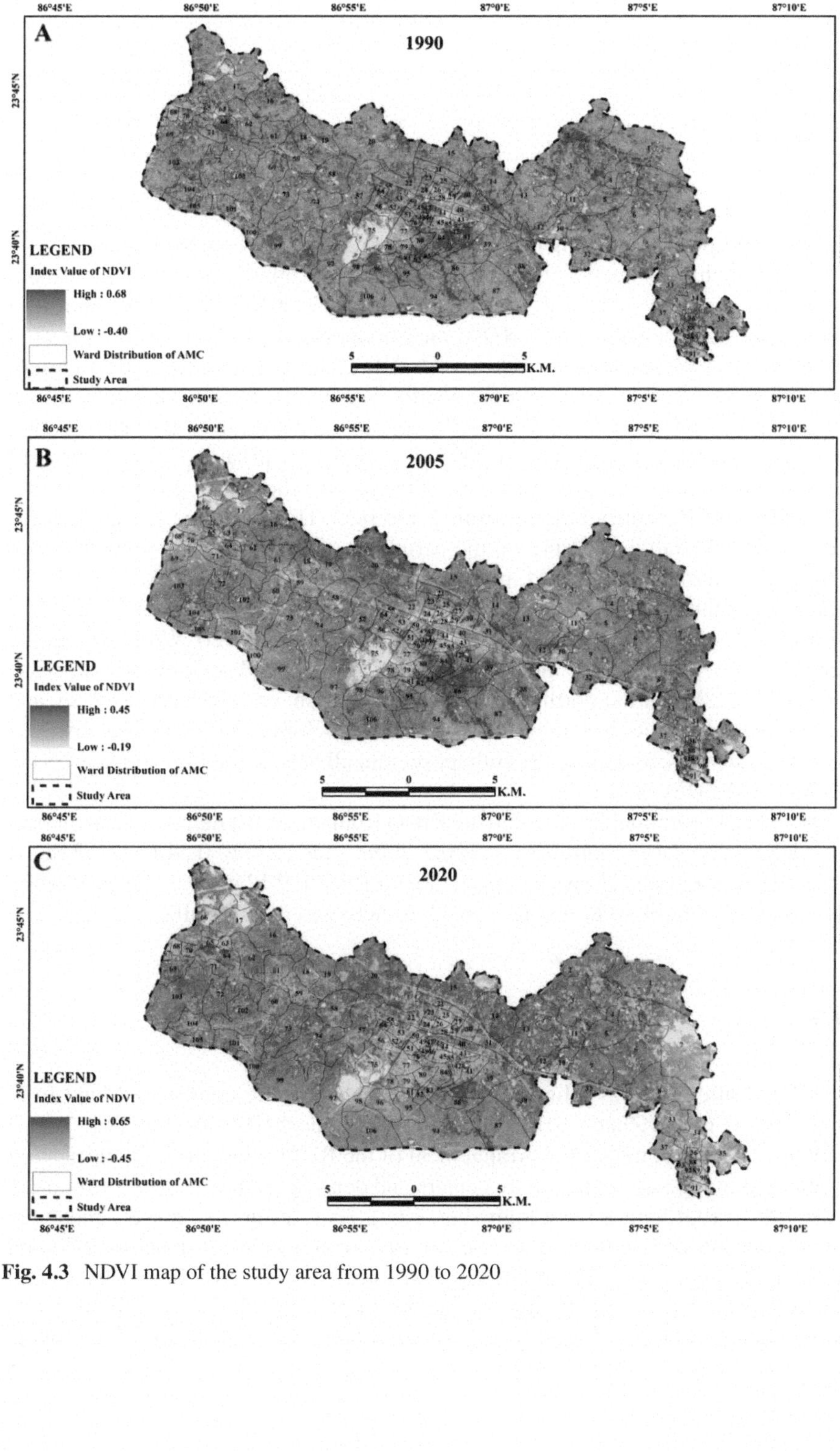

Fig. 4.3 NDVI map of the study area from 1990 to 2020

Table 4.2 Dispersion of NDVI in DMC from 1990 to 2020

Year	MAX	MIN	MEAN	SD
1990	0.68	– 0.4	0.14	0.37
2005	0.45	– 0.19	0.13	0.36
2020	0.65	– 0.45	0.1	0.31

region. For purposes of making the NDBI we have used Eq. (4.6). In the year of 1990 (Fig. 4.5a), the highest reading of NDBI in AMC was primarily observed in the main municipal towns such Asansol, Kulti, Jamuria and Raniganj and some industrial towns as Burnpur, Salanpur. The NDBI measurement that peaked was between 0.4 and 0.73. Whereas the lowest reading of NDBI (0 to -0.87) was found in eastern and north eastern part of Asansol municipality near Nigha, Damra, Girmint colliery, Shibpur and on the western portion of the research area, close to Barakar, another patch had also been discovered. In 2005 (Fig. 4.5b), as urban territory flourished and faced substantial changes, the core of the city was the fastest growing area and had positive NDBI values ranging from 0.3 to 0.55. However, due to the presence of extensive population pressure on the periphery of city, the greatest NDBI value of (0.3–0.55) has been observed, particularly in the northwestern (Asansol, Kulti, Barkar, Niyamatpur) and also in the south-eastern region of the research area, like Raniganj. NDBI has seen major changes between 1990 and 2020. With the possible exception of a few patches, the buildup index experienced a significant increase from 1990 to 2020 in the north-eastern, in the southern, eastern, northwestern, and southeasterly regions of the study area, where the development of industrial districts, roads, and railways, as well as the colliery region, all have a notable influence on the NDBI of AMC (Fig. 4.5c).

The Table 4.3 summarizes the statistical data from the NDBI map, as shown above. The NDBI value varies from 0.73 to – 0.7 in the year 1990 to 0.54 to – 0.3 in the year 2020. It essentially shows that from 0.01 in 1990 to 0.12 in 2020, the proportion that consisted of built up in research area has increased dynamically.

BUI

The BUI is often used as a line of demarcation to separate constructed land from urban areas (Liu and Zhang 2011; Ghosh and Siddique 2018; Tah et al. 2023). As a result, using Eq. 4.7, the spatial arrangement of the BUI for the area of investigation is portrayed in Fig. 4.6. In the urban centers and densely populated sections initially in the 1990s, BUI values were high (Fig. 4.6a). As a result, the densely inhabited mainly concentrated zones in the heart of city, with a built-up index of 0.70 and an extremely high range. The relatively low range of BUI (– 0.72) was frequently seen in the heavily forested regions on the outskirts of the city. Eventually in 2005, maximum range of BUI 0.5 were detected in the urban cores and industrial hubs, while

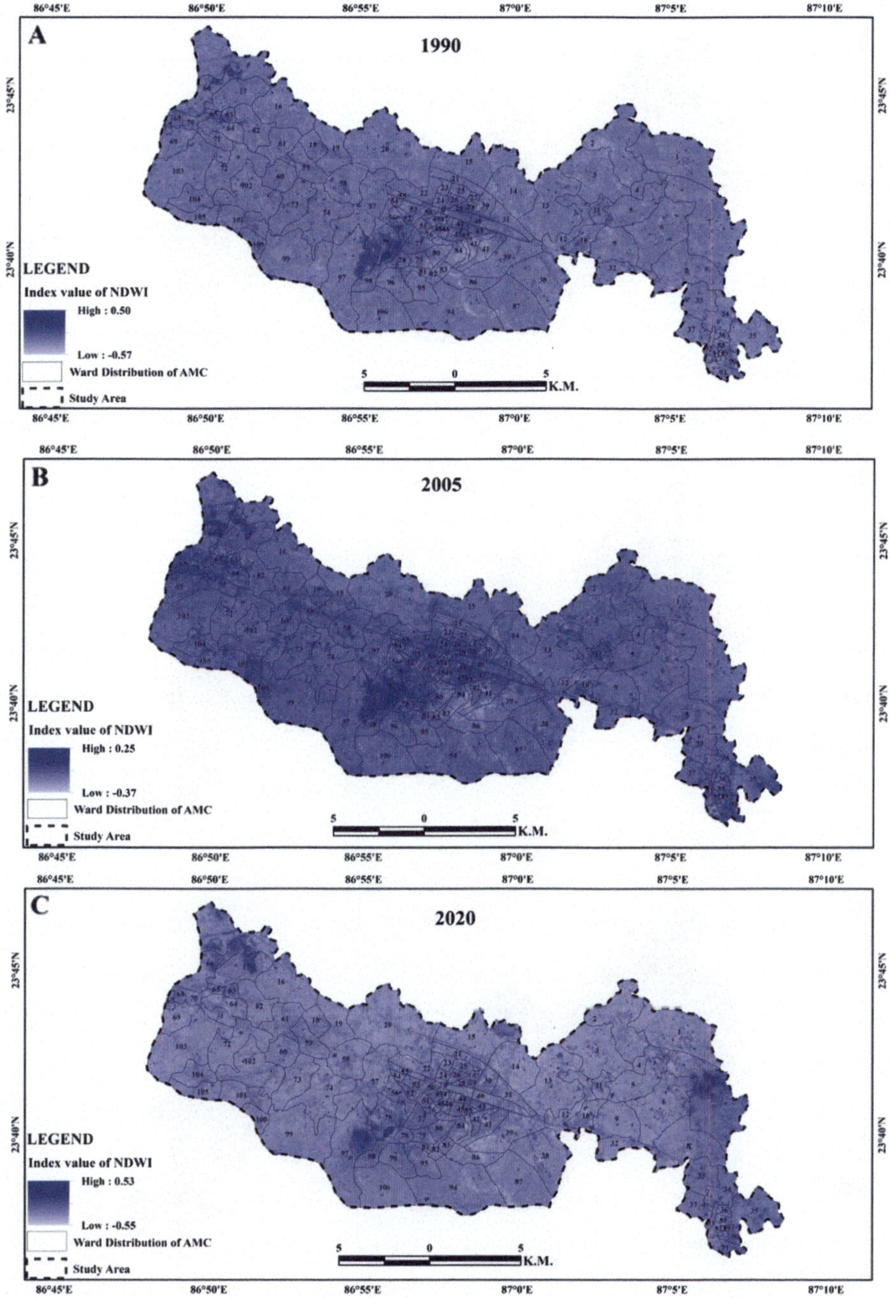

Fig. 4.4 NDWI map of the study area from 1990 to 2020

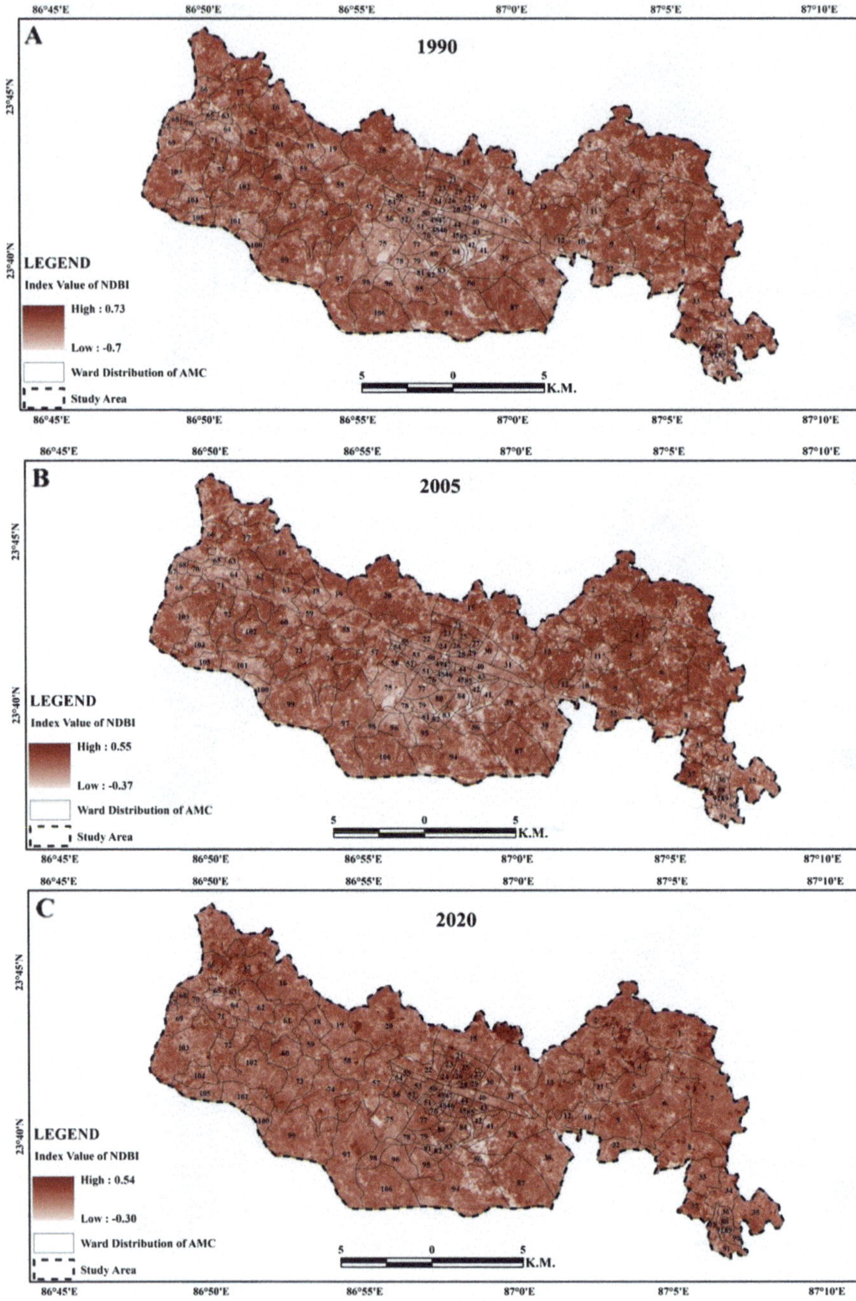

Fig. 4.5 NDBI map of the study area from 1990 to 2020

Table 4.3 Dispersion of NDBI in DMC from 1990 to 2020

Year	MAX	MIN	MEAN	SD
1990	0.73	– 0.7	0.01	0.12
2005	0.55	– 0.37	0.09	0.3
2020	0.54	– 0.3	0.12	0.355

directions to the south, south-east, south-west, and west of the investigational region showed a minimal range of BUI – 0.6 (Fig. 4.6b). The central part of the municipal area demonstrated the highest BUI values of 0.3 in 2020. The study area's highest built-up values over the thirty-year period were reported in the north, north-east, east, north-west, and south-east directions, indicating urban expansion (Fig. 4.6c). There is less urban green space in the area of study, which is indicative of the fast expansion of the urban area towards the outskirts as a result of intense population pressure and rising demand for land in the city centre, as shown by the spatial arrangement of built-up areas, which is increasingly oriented toward a specific direction.

Shannon's Entropy

The Shannon's entropy is very much useful to know how urban outgrowth has undergone change over time. As, using Eq. 4.8, Fig. 4.7 depicts the spatial layout of the Shannon's entropy for the research region. Red color is used for identification of urban outgrowth. It is generally accepted that accessibility of transportation systems and distance from the core has an impact on urban outgrowth. The distance decay parameters of urban growth were used to compute the entropy in the aforementioned scenario. With the support of the buffer in GIS, the outcomes of various geographical factors have been examined. Every row and column has been treated as an independent spatial unit for measuring Shannon's entropy. Built-up area is being applied as a variable of entropy to determine whether the arrangement of development of land in these wards is dispersed/homogenous or compact. According to the Fig. 4.7a it is clearly observed that only 23 wards of research area have the highest entropy value. It was referring to the core part of Asansol, Kulti, Jamuria, Raniganj, Burnpur, Salanpur, Ramnagar, Barakar, Nigha, etc. But within the thirty year time span urban out growth scenario is dramatically changed. The maximum value of entropy is observed in Fig. 4.7b the 50 wards of the area of research. According to the analysis, it is noticed that urban growth has concentrated in every part of research. But in Fig. 4.7c it clearly shows that old municipal towns (Asansol, Kulti, Jamuria, Raniganj) and old industrial sectors (Burnpur, Barkar) have negative change but the remaining sector emerges as a new residential, industrial sector of the study area. The alteration in entropy may be used to directly quantify the periodic shift of spatial patterns of urban development. The upward trend in the value of the entropy shows that urban growth keeps growing and development is becoming more expansive.

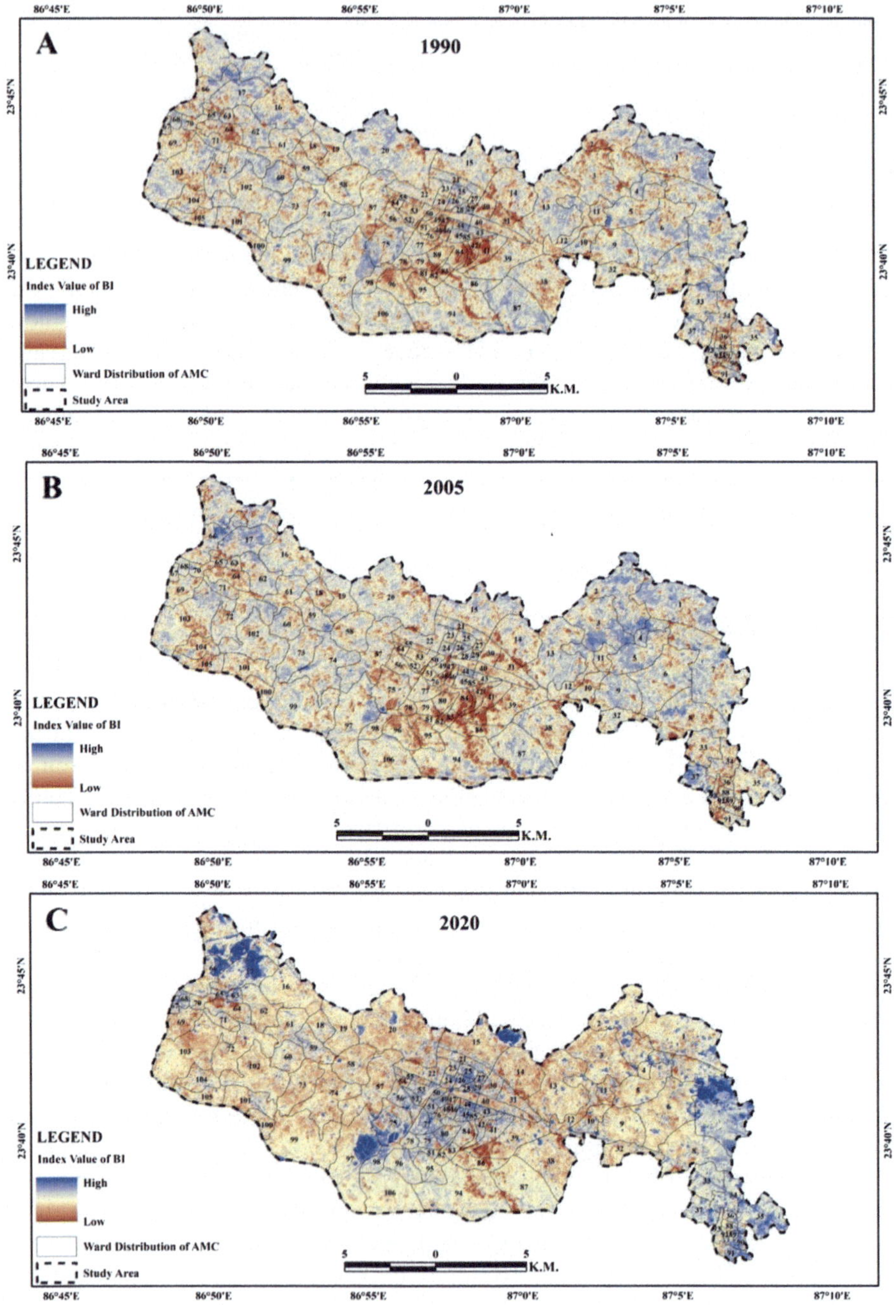

Fig. 4.6 BUI map of the study area from 1990 to 2020

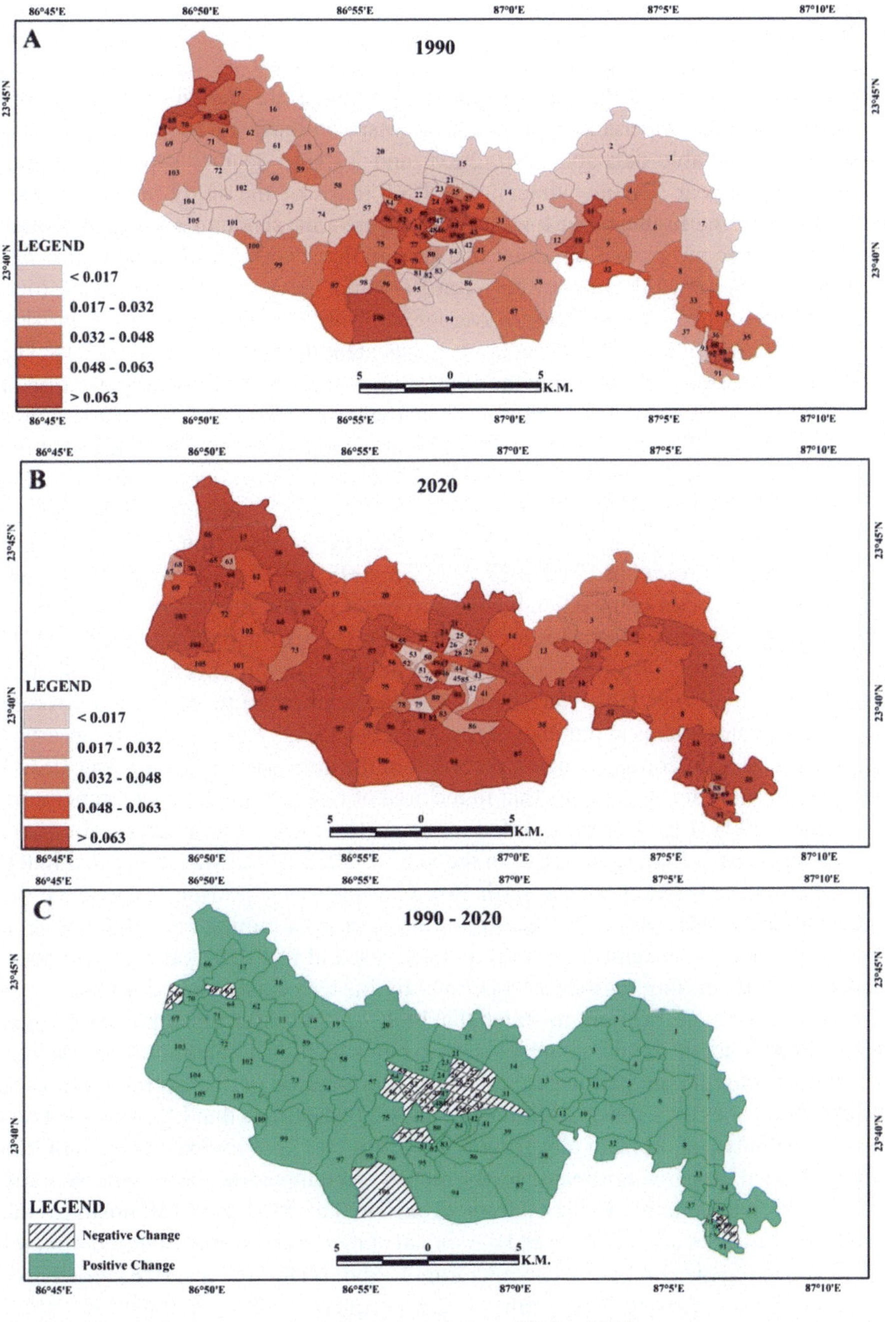

Fig. 4.7 Shannon's entropy map of the study area from 1990 to 2020

LST

Surface temperature distribution and historical change are shown on the research area's land surface temperature map. So, to show spatio-temporal oscillations in LST, we used the available data for 1990, 2005, and 2020 to acquire separate satellite photos of each year (during winter). The LST maps of the study area show that the zones with the highest and lowest temperatures are shaded dark red and blue, respectively. The program is apparent that LST in the Asansol Municipal region has been showing an upward tendency between 1990 and 2020 (Fig. 4.8) using Eqs. (4.1, 4.2, 4.3 for Landsat 5). Rising from an average of 17.66 °C in 1990 to 19.49 °C in 2005 and 20.39 °C in 2020, the LST for January has steadily increased (Table 4.4). The maximum temperature variation in the 1990 (Fig. 4.8a) is 25.8 °C, with a variation of temperatures as low as 14.7 °C. Conversely, in the year 2020 (Fig. 4.8c) temperatures range from 14.73 °C at the lowest to 29.12 °C at the highest. Thus, it is clear that the average temperature has increased by around 3 °C during the course of the 30 years.

From 1990 to 2020, the Post-Monsoonal mean surface temperature in metropolitan areas has increased from 22.07 to 25.3 °C. In the vegetative region, where temperatures rose from 15.45 °C in 1990 to 17.22 °C in 2005 and 17.53 °C in 2020, comparable trends have been seen. When it comes to water bodies, the post-monsoon mean surface temperature rose from 15.01 °C in 1990 to 16.7 °C in 2005, and then 17.1 °C in 2020, respectively. Built-up areas have demonstrated an exceptionally serious surge in post-monsoon mean surface temperature compared to vegetation-covered regions and water bodies in our area of research. From Fig. 4.8a it is clearly observed that maximum temperature is being recorded in the core part of the city and IISCO steel plant region. It also appears that in the area of research the highest temperature for the year 2005 (Fig. 4.8b) was noted in the northwestern, southeastern, southern, and western parts of the region. But in the year of 2020 which was illustrated in the Fig. 4.8c that the research area's north-east and north-west portions experience the highest temperatures, primarily because of the open coal mining zone that has been established there. near Jamuria (Ikrah) in north east and KultiBarakar region in north west, those coal mining regions are operated by Eastern Coal Limited, India.

The findings therefore demonstrated that built-up environments, highly populated places, and concrete surfaces are closely related to high temperature anomalies. Which play a significant part in guaranteeing the formation of distinct microclimates and climatic zones over time. Additionally, the study discloses that LST steadily rises during the study period from city fringes to within-city core region. As a result, the climate in a small region is affected by the increase in temperature and is dramatically different from the climate in the vicinity. From 1990 to 2020, Fig. 4.9 illustrates the correlation between urbanization and the spread of mean land surface temperature, for this we are using the comparative temperature formula (Dutta et al., 2018). Where we are collecting some sample sites from core and periphery and subtracting those values to get the value of comparative temperature. Figure 4.8 demonstrates clearly that the northwest, north, northeast, and east portions of the study area have experienced the greatest variations in surface temperature.

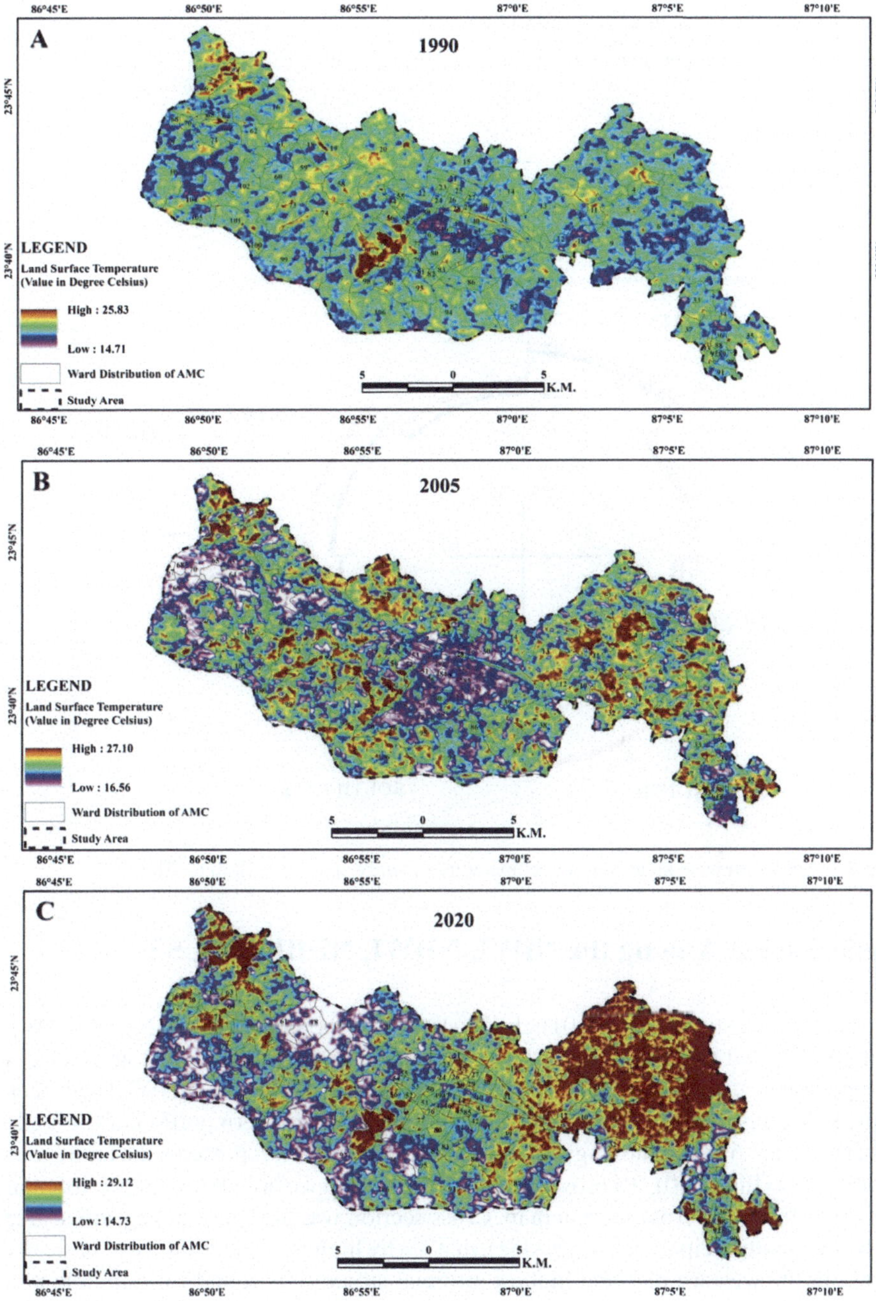

Fig. 4.8 LST map of the study area from 1990 to 2020

Table 4.4 Mean LST of study area from 1990 to 2020

Year	Vegetation area (°C)	Water body (°C)	Built up area (°C)	Mean LST (°C)
05.01.1990	15.45	15.01	22.07	17.66
14.01.2005	17.22	16.7	25.3	19.49
21.01.2020	17.53	17.1	28.41	20.39

Source The calculations were performed by the authors using Arc GIS

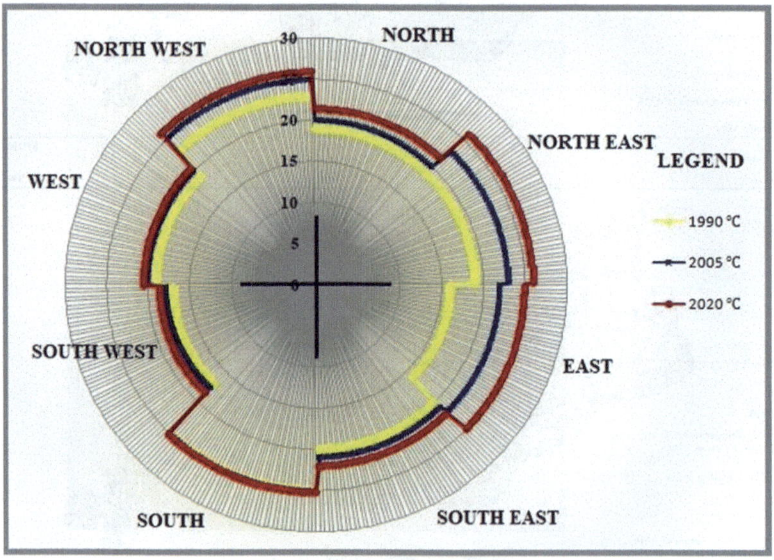

Fig. 4.9 Direction wise mean surface temperature of study area from 1990 to 2020

Relationship Among the NDVI, NDWI, NDBI and LST

Cross sectional study (Fig. 4.10) helps to show how the indices (LST, NDBI, NDVI, and NDWI) and their connection have changed over time. A–A′ for Land Surface Temperature, B–B′ for NDBI, C–C′ for NDVI, and D–D′ for NDWI have been utilized for this cross-sectional analysis. Starting in the western portion of the study area (ward no. 69) and moving eastward, this cross-section represents the entire area (ward no 7) (Fig. 4.10). To prove the interrelationship we are taking three particular positions from this cross section map. Cross section was drawn from vegetative area (left side) to municipal area (right side), that's why highest vegetation cover basically found in the western part and highest artificial structure is found from middle part of section line to right direction.

The first sample was taken in 1990. It is clearly observed that LST value is minimum when NDBI value is also minimum but NDVI value is maximum and NDWI value is moderate to low. The second sample was collected from 2005. It

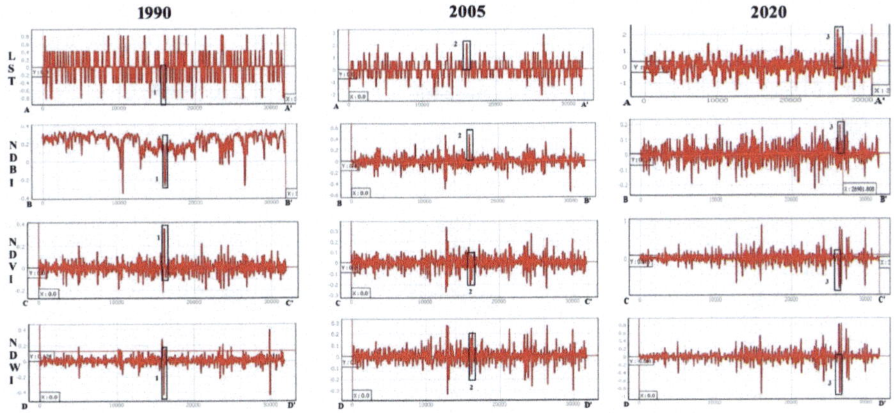

Fig. 4.10 Cross sectional studies showing the relationship among the indices

is clearly found that LST value is maximum when NDBI value is also maximum then NDVI and NDWI are minimum. Same condition was also found in the third sample, which was collected from the year 2020. Thus, the preceding diagram demonstrates that at the apex of vegetation cover, there is the least amount of built-up area and land surface temperature. However, the area of the research where plant cover is at its minimum results in the greatest documented built-up area and land surface temperature. Additionally, regression analysis and linear correlation have been utilized to investigate the relationship between LST, NDVI, NDWI, and NDBI. In order to examine the relationship between the LST, NDBI, NDWI, and NDVI in the research area, the fishnet function of the GIS programmer was used to 87 grids. Those grids were used to create a scatter plot between the LST, NDBI, NDWI and NDVI (Fig. 4.11). As a result of the study, it is established that LST and NDBI have a positive linear connection during (1990–2020), where developed areas absorb a substantial amount of sun energy, which controls LST. The correlation studies have unambiguously established a negative association between land surface temperature and plant cover (Fig. 4.11a1, a2, a3). Nevertheless, correlation investigations have demonstrated that built-up areas and land surface temperature are correlated in a substantial positive way (Fig. 4.11b1, b2, b3). It is observed that water bodies and land surface temperature exhibit a positive correlation in Fig. 4.11c1, c2, c3.

A negative correlation exists between LST and NDVI due to the fact that regions characterized by plant cover retain a lower surface temperature, thereby generating a cooling influence on the microclimate of the study area. Conversely, regions comprised of brick structures contribute to an increase in temperature (Suhali et al. 2019; Hoque and Lepcha 2019; Tah et al. 2023). Urban microclimate is greatly influenced by the distribution of vegetation, water bodies, and man-made structure in urban areas.

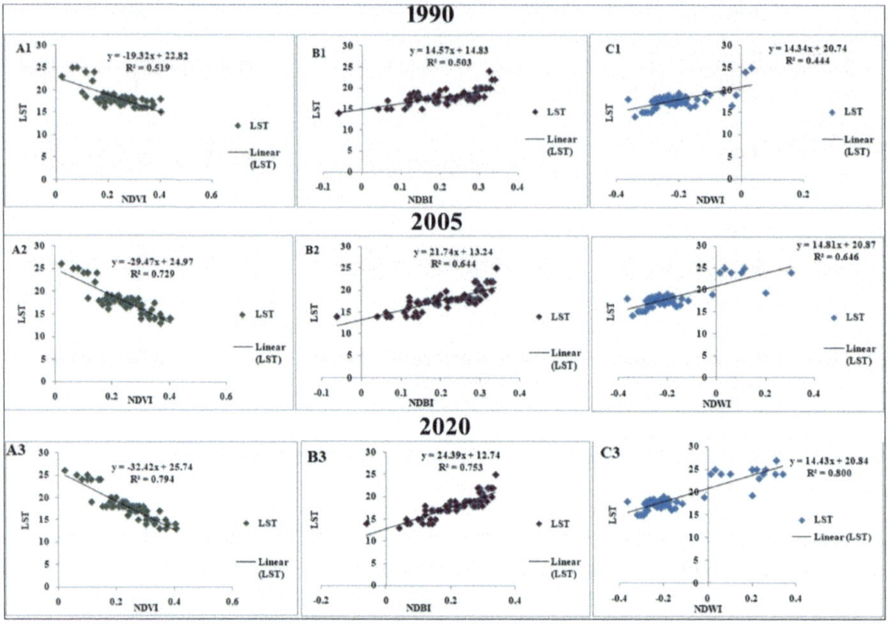

Fig. 4.11 Correlation among the indices

Conclusion

As seen by the preceding discourse, the land usc pattern of the research area, Asansol Municipal Corporation, has undergone significant transformations due to unplanned and unregulated building, in addition to rapid population expansion. A considerable extent of vegetative cover has been deforested and transformed into developed terrain, serving not only residential but also industrial and mining objectives. As a result of the formation of urban heat islands and the consequent increase in land surface temperature, the urban environment is adversely affected. The urban heat islands, which experience the highest temperatures, include the city core and industrial zones, according to the aforementioned study. Moreover, this occurrence contributes to the formation of microclimate zones throughout the research region. Rooftop gardening and the presence of urban green space exert a substantial influence on the climate of cities by mitigating the effects of urban heat islands via improved air quality, higher air humidity, and decreased land surface temperatures. As a consequence, an examination of urban microclimates is imperative for the implementation of sustainable development initiatives, mitigation of urban heat island effects, and enhancement of the living standards of urban residents. Consequently, additional newly developed towns might benefit from using the techniques and findings of this chapter. The city's planners and development authorities should remind environmental considerations above land use planning. As a result, methodical frameworks and approaches are needed to manage, plan, and monitor the newly constructed metropolitan areas as environmentally friendly as possible. A further remedial issue should be the use of high-albedo materials in building, with a specific emphasis on the regions that are absolutely necessary for the sustainable expansion of the city. Additionally, the urban green space may be essential for improving the environment of the city and making the urban microclimate healthier and more hospitable for city people.

Acknowledgements The authors express their sincere gratitude to the anonymous reviewers for their insightful feedback and recommendations.

Contribution of Authors Chumki Mondal—Investigation, Visualization and Writing Original draft.

Subham Kumar Roy—Data creation, Methodology, Data Analysis.

Dr. Sanatan Ghosh—Conceptualization, Methodology, Supervision and Visualization.

Funding No funding was received.

Conflict of Interest None of the writers have any conflicts of interest to declare.

References

Agnihotri AK, Ohri A, Mishra S (2018) Impact of green spaces on the microclimate through landsat 8 and TIRS data, in Varanasi, India. Int J Environ Sustain (IJES) 7(2):72–80

Bharath HA, Chandan MC, Vinay S, Ramachandra TV (2018) Modelling urban dynamics in rapidly urbanising Indian cities. Egypt J Remote Sens Space Sci 21(3):201–210. https://doi.org/10.1016/j.ejrs.2017.08.002

Bureau of Applied Economics & Statistics Department of Statistics & Programme Implementation West Bengal (2010, 2011, 2014) District statistical handbook, Burdwan District, West Bengal. http://wbpspm.gov.in/publications/

Census of India (2011) District census handbook, Burdwan District, West Bengal

Chatterjee U, Majumdar S (2022) Impact of land use change and rapid urbanization on urban heat island in Kolkata city: a remote sensing based perspective. Journal of Urban Management 11:59–71. https://doi.org/10.1016/j.jum.2021.09.002

Chattopadhyay D (2022) Brief history of Asansol. Anandabazar, West Bengal, Bardhaman

Chen J, Gong P, He C, Pu R, Shi P (2003) Land use/land cover change detection using improved change vector analysis. Photogramm Eng Remote Sens 69(4):369–379

Choudhury D, Das K, Das A (2019) Assessment of land use land cover changes and its impact on variations of land surface temperature in Asansol-Durgapur Development Region. Egypt J Remote Sens Space Sci 22:203–218. https://doi.org/10.1016/j.ejrs.2018.05.004

Das N, Mondal P, Sutradhar S, Ghosh R (2021) Assessment of variation of land use/land cover and its impact on land surface temperature of Asansol subdivision. Egypt J Remote Sens Space Sci 24(1):131–149. https://doi.org/10.1016/j.ejrs.2020.05.001

Dutta D, Gupta S, Kishtawal CM (2018) Linking LULC change with urban heat islands over 25 years: a case study of the urban-industrial city Durgapur, Eastern India. J Spat Sci, 501–518. https://doi.org/10.1080/14498596.2018.1537198

Ghosh S, Siddique G (2018) Change detection of built-up areas applying built-up index for Chandannagar city. Int J Technol Res Manage 5(4):1–7

Govind N, Ramesh H (2019) The impact of spatiotemporal patterns of land use land cover and land surface temperature on an urban cool island: a case study of Bengaluru. Environ Monit Assess 191. https://doi.org/10.1007/s10661-019-7440-1

Grover A, Singh RB (2015) Analysis of urban heat island (UHI) in relation to normalized difference vegetation index (NDVI): a comparative study of Delhi and Mumbai. Environments 2:125–138

Hoque I, Lepcha K (2019) A geospatial analysis of land use dynamics and its impact on land surface temperature in Siliguri-Jalpaiguri development region, West Bengal. Appl Geomatics. https://doi.org/10.1007/s12518-019-00288-1

Jat MK, Garg PK, Khare D (2008) Modeling of urban growth using spatial analysis techniques: a case study of Ajmer City (India). Int J Remote Sens 29(2):543–567

Liu L, Zhang Y (2011) Urban heat island analysis using the Landsat TM data and ASTER data: a case study in Hong Kong. Remote Sens 3:1535–1552

Maity B, Mallick SK, Rudra S (2020) Spatiotemporal dynamics of urban landscape in Asansol municipal corporation, West Bengal, India: a geospatial analysis. Geo Journal. https://doi.org/10.1007/s10708-020-10315-z.

Mandal RB (2000) Urban geography. Concept Publishing Company, New Delhi

Mandal J, Ghosh N, Mukhopadhyay A (2019) Urban growth dynamics and changing land-use land-cover of megacity Kolkata and its environs. J Indian Soc Remote Sens 47:1707–1725

McFeeters SK (1996) The use of the normalized difference water index (NDWI) in the delineation of open water features. Int J Remote Sens 17(7):1425–1432

Pal S, Ziaul S (2017) Detection of land use and land cover change and land surface temperature in English Bazar urban center. Egypt J Remote Sens Space Sci 20(1):125–145. https://doi.org/10.1016/j.ejrs.2016.11.003

Roy S, Bose A, Majumder S, Roy Chowdhury I, Abdo HZ, Almohamad H, Al Dughairi AA (2022) Evaluating urban environment quality (UEQ) for Class-I Indian city: an integrated RS-GIS based exploratory spatial analysis. Geocarto Int 38(1). https://doi.org/10.1080/10106049.2022.2153932

Saha A, Malik UM (2018) The urban social fabric of the city of Asansol, West Bengal. Indian J Spatial Sci 9(2):45–54

Sarkar S, Kar BK (2023) Assessing the impact of urban growth and green space dynamics on microclimate in urban areas: a case study of Siliguri City, West Bengal, India. In: Chatterjee U, Antipova A, Ghosh S, Majumdar S, Setiawati MD (eds) Urban environment and smart cities in Asian countries. Human dynamics in smart cities. Springer, Cham. https://doi.org/10.1007/978-3-031-25914-2_3

Seto KC, Fragkias M, Guneralp B, Reilly MK (2011) A meta-analysis of global urban land expansion. PLoSOne

Sharma S, Nahid S, Sharma M, Sannigrahi S, Anees MM, Sharma R (2020) A long-term and comprehensive assessment of urbanization-induced impacts on ecosystem services in the capital city of India. City Environ Interact 7(2):100047. https://doi.org/10.1016/j.cacint.2020.100047

Singh KV, Setia R, Sahoo S, Prasad A, Pateriya B (2014) Evaluation of NDWI and MNDWI for assessment of waterlogging by integrating digital elevation model and groundwater level. GeocartoInternational, 01–12. https://doi.org/10.1080/10106049.2014.965757

Suhail M, Khan MS, Faridi RA (2019) Assessment of urban heat islands effect and land surface temperature of Noida, India by using Landsat satellite data. MAPAN-J Metrol Soc India 34(4):431–441

Tah S, Roy SK, Mondal C (2023) Assessment of land surface temperature using Landsat images: a case study on Durgapur Municipal Corporation, West Bengal, India. In: Chatterjee U, Antipova A, Ghosh S, Majumdar S, Setiawati MD (eds) Urban environment and smart cities in Asian countries. Human dynamics in smart cities. Springer, Cham. https://doi.org/10.1007/978-3-031-25914-2_9

Taubenbock H, Wegmann M, Roth A, Mehl H, Dech S (2009) Urbanization in India—spatiotemporal analysis using remote sensing data. Comput Environ Urban Syst 33(3):179–188. https://doi.org/10.1016/j.compenvurbsys.2008.09.003

Townshend JR, Justice CO (1986) Analysis of the dynamics of African vegetation using the normalized difference vegetation index. Int J Remote Sens 7(11):1435–1445. https://doi.org/10.1080/01431168608948946

Turner B, Meyer WB, Skole DL (1994) Global land-use/land-cover change: towards an integrated study. AMBIO Stockholm 23:91–95

Wei YD (2015) Zone fever, project fever: economic transition, development policy, and urban expansion in China. Geogr Rev 105:156–177

Yeh AGO, Li X (2001) Measurement and monitoring of urban sprawl in a rapidly growing region using entropy. Photogramm Eng Remote Sens 67:83–90

Yuan F, Bauer ME (2007) Comparison of impervious surface area and normalized difference vegetation index as indicators of surface urban heat island effects in Landsat imagery. Remote Sens Environ 80(2):233–240

Zha Y, Gao J, Ni S (2003) Use of normalized difference built-up index in automatically mapping urban areas from TM imagery. Int J Remote Sens 24(3):583–594. https://doi.org/10.1080/01431160304987

Chapter 5
Impact of Urbanization on Mirco-climate and Environmental Quality in Barasat Municipality: A Geospatial Analysis

Jhoney Barui, Debabrata Chanda, Yogia Dutta, and Uttam Mukhopadhyay

Abstract Urbanization gained speed due to industrial development worldwide. The high growth of world urban population transformed the surface of the earth into concrete. The huge transformation of land helps to change the climate worldwide especially in urban spheres. The micro climatic change on urban entities has huge impact on urban environmental quality and vice versa. The research focuses on examining the negative impact of urban transformation over the past decade on the microclimate and environmental quality of Barasat Municipality. Land Surface Temperature (LST) of Barasat Municipality was quantified by using Landsat data as a tool of micro climatic change. On the other hand, different environmental parameters have been estimated to extract the changing scenario of Urban Environmental Quality. The urban environmental quality has been calculated by using multicriteria decision making like CRITIC and TOPSIS method. The evolving correlation between Urban Environmental Quality and Land Surface Temperature was assessed through the application of bivariate Moran's I and LISA (Local Indicators of Spatial Autocorrelation). The result showing the unexpected increase of temperature and the degradation of urban environmental quality and there is a strong association between them. Proper urban planning is the only way to combat this issue.

J. Barui (✉) · U. Mukhopadhyay
Department of Geography, Vidyasagar College, University of Calcutta, Kolkata 700006, India
e-mail: jonybarui9@gmail.com

D. Chanda
Department of Geography, Asutosh College, Kolkata, India

Y. Dutta
Department of Geography, Haldia Government College, East Midnapur, India

U. Mukhopadhyay et al. (eds.), *Climate Crisis, Social Responses and Sustainability*, Climate Change Management, https://doi.org/10.1007/978-3-031-58261-5_5

Introduction

Urbanization symbolizes the progress and sophistication of civilization. As human civilization is making progress, the cultural landscape of rural areas is often transformed into an urban one. Urbanization is a process by which an urban unit grows which include physical, social and cultural transformation of land surfaces (Cyril et al. 2013; Imam and Banerjee 2016; Kundu and Pandey 2020; Leeson 2018; Wenban-Smith et al. 2016). This transformation occurs when people from villages shift to urban fields. People migrated from rural to urban areas primarily because of the improved amenities available in urban settings. These include more job prospects, improved housing options, better health and educational facilities, advanced transportation and communication infrastructure, ultimately contributing to a higher quality of living. As a result, those former rural areas got dominated by secondary and tertiary activities and the population density started to increase (Nath 1986). Naturally the cities across the globe are expanding in size, whether or not there is adequate urban planning in place. Urbanization has progressed worldwide with accelerated speed after industrial revolution (Leeson 2018). World urban population was less than 3% in 1800 (Hauser and United Nations Fund for Population Activities 1982) and after 2010, urban population percentage became approximately 50 (Watson 1993). Later, it is expected to increase up to 68% by mid-twenty-first century (Kundu and Pandey 2020).

When the entire world is witnessing unprecedented urban growth, India is also not much lagging behind (Mandal 1999). With the progress of the economy, the pace of urbanization is also achieving augmented momentum (Chauhan et al. 2016). Urbanization in India experienced a significant surge after gaining independence (Nath 1986; Venkatesham 2015). Like other developing nations, urbanization in India is rapidly growing (Hasnine and Rukhsana 2020). 31.16% of Indian population resides in urban areas as per 2011 census which was 27.8% in census year 2001.

With the progress of urbanization, development and transformation occurred worldwide with the threat of environmental degradation and increased pressure on natural resources (Peng et al. 2022). Bai and Imura (2000) recognized three primary challenges confronting cities today. These are issue related to Poverty (lack of basic facilities), rapid growth-related issues (growth causes pollution and degradation of land and water) and lastly, problem caused by wealthy lifestyle (High emission of CO_2, CFC, NO_X etc.). Expansion of cities, in both the terms of population and economy are harming the natural ecosystem services as they are out of their ecological carrying capacity (Button 2002). The ongoing transformation of urban areas has led the reduction of green, open and wet spaces. Naturally, every country is now concerned about urban environment and micro-climatic change. A number of global policies promoting environmental sustainability have been implemented to uphold Urban Environmental Quality (UEQ). It is crucial to carry out a spatio-temporal assessment of UEQ to enhance comprehension and planning of urban areas.

UEQ is a multifaceted and dynamic parameter utilized to depict intercorrelation among various living and non-living factors that influence the quality of urban environment (Faisal and Shaker 2017a, b; Liang and Weng 2010; Roy et al. 2022a, b; Singh and Jain 2022). UEQ holds importance in governance; for example- land-use planning, infrastructural services, policy making etc. Collection of data of every urban environmental factor and construction of UEQ was such a hard task. In recent time, remote sensing (RS) makes UEQ modelling much easier by providing constant earth observation data, over a range of spatial, spectral and temporal resolutions (Faisal and Shaker 2017b; Liang and Weng 2010; Malah et al. 2022a). Thus, the assessment of UEQ serves not only as a valuable tool in sustainable development and planning, but also provides a deeper understanding for any urban unit. A number of works of UEQ based on RS data were recognized as instances illustrating the development and evaluation of UEQ.

The effect of urban development on micro-climate is a matter of recent concern to the researchers. LST is useful tool to measure the micro-climatic change in any region (Ali et al. 2017; Antoniadis et al. 2018; Connors et al. 2013; Shetty et al. 2022; Shimod et al. 2023; Zeren Cetin et al. 2023). The data for accruing LST is also easily available from RS and the result is accepted by scholars for understanding the changing nature of temperature of any area (Zeren Cetin et al. 2023). Surface temperature is greatly affected by land-use change (Balew and Semaw 2022). The temperature difference in inner city area and outer city promotes Urban Heat Island effect (Bhattacharjee et al. 2022). Urban heating due to higher residential area with low vegetation cover and presence of water bodies is now a common thing. The summer heating and unpleasant weather due to urban development is now in every city (Hereher et al. 2022; Khan et al. 2022; Liu et al. 2022). Urban expansion influences the environment most (Khan et al. 2022; Li et al. 2022); especially the built-up area contributes more to heating the city i.e. heat island effect.

Urban Environmental Quality Index (UEQI) serves as a potential instrument to raise public awareness about environmental condition of a specific area. The index can also assist in comprehending the threats and vulnerabilities in specific area through comparisons with others. The present work emphasizes dimension-based decomposition analysis to discern the spatial and temporal patterns of UEQ and its relationship with evolving LST. Dimensions are reflective of various themes or aspects within the environmental scenario, as evidenced by spectral indices such as NDVI, NDWI, ARVI, and NDBI, which respectively signify the quality of greenery, pollution levels, water presence and the nature of impervious surfaces. Decomposition assesses the levels of these theme-based indices and their correlation with the overall environmental condition. A class I city, Barasat Municipality was selected for the study. The city is transforming in each passing year due to the nearness of Kolkata. Thus, the study is unique in its objectives to assess, firstly the Spatio-temporal change of UEQ and LST and secondly the correlation between UEQ and LST. Finally, the spatial correlation between UEQI and LST was estimated by bivariate Moran's I, along with LISA to comprehend the spatial cluster related UEQ and LST.

Materials and Methods

Study Area

In this research, Barasat municipality has been selected as the study area. Barasat (22°40′18″–22°44′32″ N and 88°26′44″–88°31′21″ E) is situated in north-eastern direction of Kolkata city and it is the district head quarter of North 24 Parganas (Fig. 5.1). The population of this urban unit increased rapidly after partition of India due to rehabilitation of Indians living in East Pakistan which widen the town boundary over time. The municipality consists of 35 number of wards covering 34.50 km^2 area. Numerous important roads like NH 12, NH 112 (Jessore Road) and the State Highway have crossed through this municipality. The town is near to Kolkata Airport and also serves as railway junction. The fast-growing municipality has intensified the pressure of urbanization. Eventually the city has inflicted a noticeable and significant impact on its environment.

Database

The urban growth of Barasat city was examined by secondary data available in Census of India. Changing nature of UEQ and LST on spatial and temporal manner along with their correlation were analyzed through the quantification of various urban components of 2013 and 2023 by applying two Landsat-8 images (Table 5.1). Landsat-8 (OLI/TIRS) images were accrued from the United State Geological Survey (USGS). Path/Row of all the images were 138/44 with the resolution of 30 m (Visible, NIR, SWIR) and 100 m (Thermal). Images of cloud free days for the years 2013 and 2023 were used to calculate environmental indices (NDVI, ARVI, NDWI and NDBI). In case of LST also, images of both the years were used to calculate the LSTs. All the images were projected in WGS-84 datum with UTM-Zone45N. ArcGIS 10.3, Google Earth and statistical software Geoda were employed for the work.

Methodology

Urban Environmental Variables

Urban environment is interaction of human and nature. A huge number of indicators depicts the quality of urban environment. The basic environmental condition of any space can be depicted by vegetation cover, water in surface and concretization of land. Four important indices namely Normalized Difference Vegetation Index

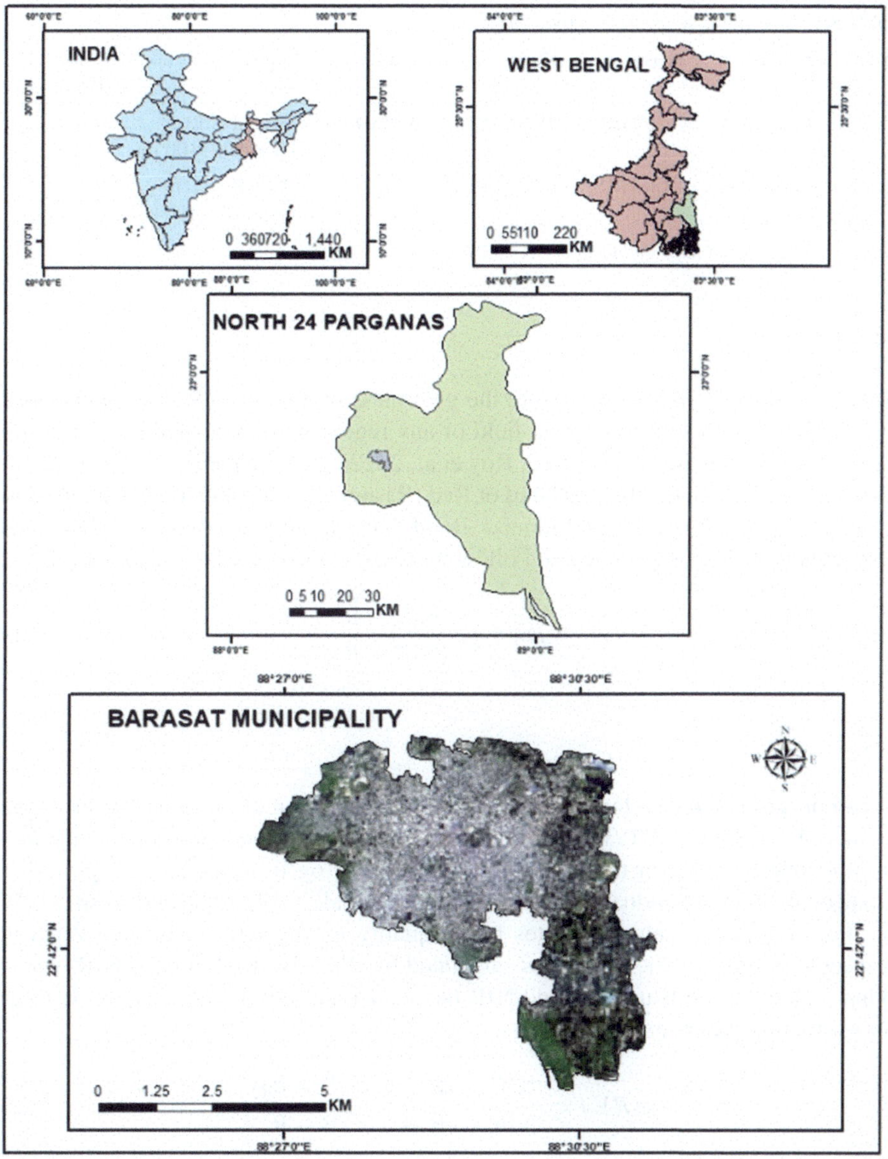

Fig. 5.1 Location of the study area

(NDVI), Atmospherically Resistant Vegetation Index (ARVI), Normalized Difference Water Index (NDWI), and Normalized Difference Built-up Index (NDBI) have been selected to assess Urban Environmental Quality Index (Malah et al. 2022b).

Table 5.1 Satellite images used in the study

Year	Satellite	Sensor	Date	Path/row	Spatial resolution	Spectral bands
2013	Landsat 8	OLI/TIRS	19.04.2013	138/044	Optical—30 m Thermal—100 m	11
2023	Landsat 8	OLI/TIRS	07.04.2023	138/044	Optical—30 m Thermal—100 m	11

Source United State Geological Survey (USGS)

NDVI

NDVI is an useful index to measure the greenness of any area (Mallick and Rudra 2021). NDVI represents the green field of any region with their quality (Guha and Govil 2021; Liang and Weng 2010; Roy et al. 2022a, b; Singh and Jain 2022; Zeren Cetin et al. 2023). The reflected band of Red (R) and Near Infrared (NIR) are used to estimate the actual vegetation biomass. Band 4 and Band 5 are Red and NIR bands respectively of Landsat 8 images. Following Eq. (5.1) was used to calculate NDVI.

$$NDVI = \frac{(\text{NIR} - \text{R})}{(\text{NIR} + \text{R})} \tag{5.1}$$

ARVI

Urban progress and development minimize the proportion of green and reduces the quality of vegetation. ARVI measures the power of vegetation to minimize the effect of atmospheric scattering due to aerosols such as rain, fog, dust, smoke or air pollution (Aziz et al. 2023; Chaudhari et al. 2023; Dhumal et al. 2017; Sinha and Basu 2023). So, the high ARVI value validates better quality of vegetation and vice versa in respect to pollution. The index was calculated by Blue (B), Red (R) and NIR bands. Band 2, 4 and 5 are Blue, Red and NIR bands of Landsat 8 images. Equation (5.2) for estimating the index is as follows.

$$ARVI = \frac{\text{NIR} - (\text{R} - \text{y} \times (\text{B} - \text{R}))}{\text{NIR} + (\text{R} - \text{y} \times (\text{B} - \text{R}))} \tag{5.2}$$

where, 'y' represents the quotient value, obtained from components of atmospheric reflectance in both blue and red channels (bands).

NDWI

Water body is an important feature of landscape as it maintains environmental stability (Javanbakht et al. 2021; Srivastava et al. 2013). Normalized Difference

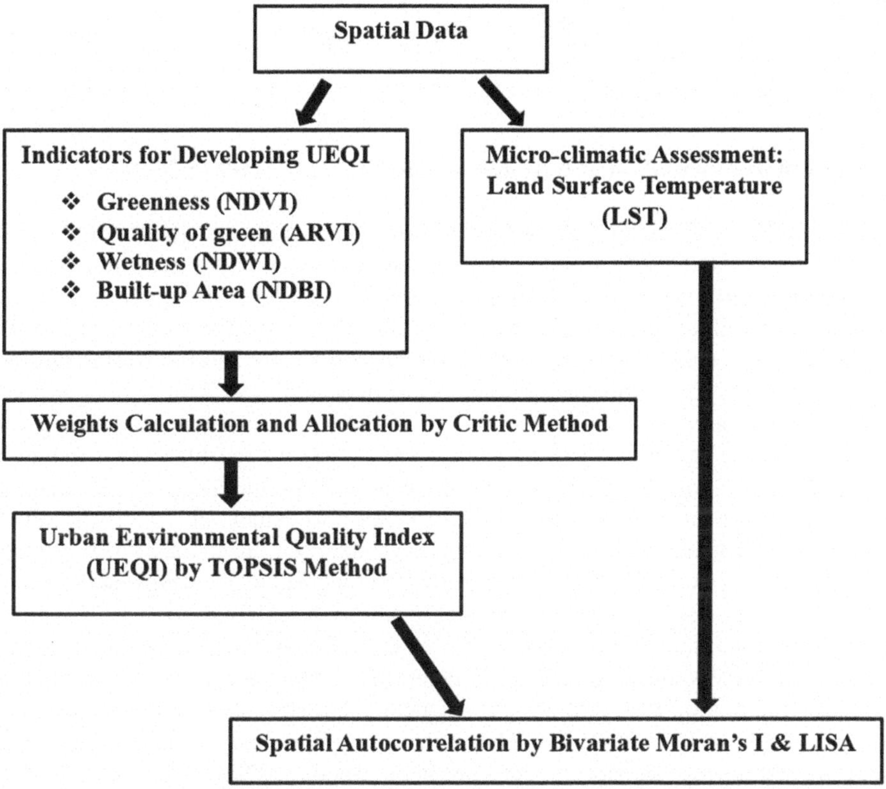

Fig. 5.2 Methodology flow chart

Water Index has been introduced to measure the moisture content over the land (Mallick and Rudra 2021; Sekertekin and Zadbagher 2021) surface using Green (G) and NIR band. Band 3 and Band 5 are Green and NIR bands respectively of Landsat 8 images. Equation (5.3) was used to calculate NDWI.

$$NDWI = \frac{(\mathrm{G} - \mathrm{NIR})}{(\mathrm{G} + \mathrm{NIR})} \tag{5.3}$$

NDBI

Proportion of built-up area of any land surface is the mark of human influence on environment (Liu et al. 2013; Xu 2010). NDBI, was employed to compute the proportion of built-up area over the land surface (Bhattacharjee and Hazra 2014). The Short-Wave Infrared (SWIR) and NIR bands were utilized for calculating NDBI. Band 6 and Band 5 are SWIR-1 and NIR bands respectively of Landsat 8 (OLI/TIRS) images. The next Eq. (5.4) was used to calculate NDBI.

$$NDBI = \frac{(\text{SWIR} - \text{NIR})}{(\text{SWIR} + \text{NIR})} \tag{5.4}$$

Urban Environmental Quality Index by CRITIC and TOPSIS Method

Urban environment constitutes an intricate structure with diversified components (Kaginalkar et al. 2021). The quality of urban environment is based on the interaction, interrelation and overlapping of all components. To evaluate the quality of urban environment with all the complexities, the multi-dimensional strategy needs to be applied (El Din et al. 2013; Fernández and Pacheco 2001; Jiří and Jan 2021; Trujillo-González et al. 2020). Multiple Criteria Decision Making (MCDM) is a branch of operational research that explicitly evaluates multiple conflicting criteria in decision making process. The urban environmental quality index was evaluated by using two methods i.e., Integrated CRICIC weight sum and TOPSIS. Initially, CRITIC method was used to determine the weights of all attributes (Adalı and Işık 2017; Alinezhad et al. 2019; Mukhametzyanov 2021; Vadivel et al. 2022; Zavadskas et al. 2006; Zhang et al. 2022). On the basis of weights determined by CRITIC, TOPSIS method was followed to estimate Urban Environmental Quality Index (UEQI).

CRiteria Importance Through Intercriteria Correlation (CRITIC) method stands out as a valuable weighting technique, determining objective weights for specific criteria. The method was introduced by Diakoulaki, Mavrotas, and Papayannakis in 1995 (Diakoulaki et al. 1995). This method amplified the contrast and conflict intensity within the decision-making problem structure (Diakoulaki et al. 1995). This method increased its significance by employing correlation analysis to identify contrasts between criteria (Yilmaz and Harmancioglu 2010). In the process of CRITIC, decision matrix has assessed and the standard deviation of normalized criterion values by columns and the correlation coefficients of all pairs of columns are used to determine the criteria contrast (Madic and Radovanovic 2015).

A number of study has been done using CRITIC method. Diakoulaki et al. (1995) adopted this method to a sample of pharmaceutical industries assessed based on three of the most notable performance indices of a firm in Greek. For developing a water resource management model for the Gediz River Basin in Turkey, Yilmaz and Harmancioglu (2010) applied this method. Financial risk assessment of large scale conglomerates was done by using this method (Kazan and Ozdemir 2014). Urban rail transit operation safety evaluation in Beijing was conducted based on improved CRITIC method and Cloud model by Wu et al. (2020). Krishnan et al. (2021) introduced by evaluating the weights of five smartphone criteria. Zafar et al. (2021) used entropy-CRITIC method to evaluate the effectiveness of 30 block chain systems and applied MCDM techniques to generate ranks. Shi et al. (2021) went with CRITIC method to evaluate a comprehensive power quality. The assessment of tide risk by integrated CRITIC weight method had been done by Chen et al. (2022). The regional distribution of power quality was examined by the technique (Zhang et al. 2022). The method was employed by Lu et al. (2022) to select a model on agricultural machinery. The method is also used to generate potential index on composite eco-tourism by Ocampo et al. (2023). Weight determined by CRITIC method for solving the problem of existing live broadcast interface of Agricultural Product (Liu et al. 2023). Vadivel et al. (2022) applied this process to find the sustainable green supplies in Indian cities. The steps for applying the method are outlined below.

The first assumption is that there is a set of m feasible alternatives A i ($i = 1, 2, 3, 4, 5 \ldots$ m) and n evaluation criteria C_j (j = 1, 2, 3, 4, 5 … n) in the problem.

At first, the decision matrix X is formed that exhibits the performance of different alternatives with respect to various criteria (Eq. 5.5).

$$X = \left[X_{ij}\right]_{m\times n} = \begin{bmatrix} x_{11} & x_{12} & x_{1n} \\ x_{21} & x_{22} & x_{2n} \\ \dot{x}_{m_1} & x_{m_2} & x_{mn} \end{bmatrix} \tag{5.5}$$

where X_{ij} stands the performance value of i th alternative on j th criterion. Here, both inter dispersion and association among variables considered in form of standard deviation (sd) and correlation. Therefore, the weightage of the j th criterion (W_j) is calculated as (Eq. 5.6):

$$W_j = \frac{C_j}{\sum_{j=1}^{n} C_j} \tag{5.6}$$

where C_j is the measure of information in j th criterion which is calculated as (Eq. 5.7):

$$C_j = \sigma_j \sum_{j'=1}^{n} (1 - r_{jj'}) \tag{5.7}$$

where σ_j is the sd of the j th criterion and $r_{jj'}$ is the coefficient of correlation between two factors.

Technique for Order of Preference by Similarity to Ideal Solution (TOPSIS) is a method of multi-criteria decision analysis, serving as a valuable tool to determine the best alternative by considering more than one criterion in the selection process (Hwang et al. 1981, 1993). Topsis calculates the optimal positive solution and identifies a performance index based on the ranking of options (Roy and Das 2018; Zavadskas et al. 2006). An ideal positive solution maximizes benefit criteria and minimizes cost criteria. On the other hand, ideal negative solution maximizes cost criteria and minimizes benefit criteria. Previous studies revealed importance of this method as performance index in various fields. For example—National Happiness Index (Niu et al. 2013), Financial Risk Assessment (Kazan and Ozdemir 2014), Road Safety Composite Index (Bao et al. 2010; Rosić et al. 2017), Financial Performance Evaluation (Roy and Das 2018) etc. The steps for calculation are as follows:

The first step is normalization of matrix which has to be avoided here as the data was already normalized as all of those were indices.

The second step is based on giving weight. In some cases, all the variables are given equal weights. The formula of weight determination is given below (Eq. 5.8).

$$W_{ij} = I_{ij} \times w_j \tag{5.8}$$

where, j indicates the weight of j th criteria. Here the step was skipped as the weight of every factor were determined earlier by CRICIC method and multiplied with index values.

The subsequent step involves calculating the ideal positive and ideal negative. The maximum values of variables (NDVI, ARVI and NDWI) are considered as ideal positive solution otherwise ideal negative solution (NDBI).

Calculation of Euclidean distance from the ideal solution (Eq. 5.9)

$$S_i^+ = \left[\sum_{j=1}^{m}\left(W_{ij} - W_j^+\right)^2\right]^{0.5} \tag{5.9}$$

Calculation of Euclidean distance from the ideal negative solution (Eq. 5.10)

$$S_i^- = \left[\sum_{j=1}^{m}\left(W_{ij} - W_j^-\right)^2\right]^{0.5} \tag{5.10}$$

Finally, Urban Environmental Quality Index (UEQI) is calculated by the following Eq. (5.11):

$$UEQI = \frac{S^-}{(S^+ + S^-)} \tag{5.11}$$

Land Surface Temperature (LST)

In the era of Globalization, every urban unit is facing the problem of micro climatic change and specially growing temperature over the years. LST characterizes the thermal behavior of land surface, which is closely associated with air temperature (Hulley et al. 2019; Li et al. 2012; Zhou et al. 2014). LST has been used here as a micro climatic tool to know about the changing climatic pattern of Barasat Municipality. The estimation of LST is a multifaceted task where LST retrieved from Thermal Infrared (TIR) data and calibrated from digital number representing multispectral image data (García 2022; Hereher et al. 2022; Kabano et al. 2021). The steps of LST calculation are synthesized below:

At first, LST was retrieved from Landsat images, using raw data (digital numbers) of the thermal bands (Band 10 in Landsat 8 OLI/TIRS) by converting them to spectral radiance using Eq. (5.12):

$$L_\lambda = M_\lambda \times Q_{cal} + A_L \tag{5.12}$$

where, L_λ is sensor spectral radiance of top of atmosphere, M_λ is radiance multiplicative scaling factor, A_L is radiance additive scaling factor and $Qcal$ is pixel values in Digital Number.

Then, brightness temperature had been calculated by following formula (5.13):

$$T_B = \frac{K2}{\ln\left(\mathrm{K1}\frac{K1}{L_\lambda} + 1\right)} \tag{5.13}$$

where, T_B is brightness temperature of sensor; K2 = Calibration constant 2, which is 1321.0789 for Landsat 8 OLI/TIRS band 10; K1 = Calibration constant 1, which is 774.8853 Landsat 8 band 10; L_λ = Spectral radiance [W/(m^2 sr μm)]; ln = Natural logarithm.

After that, Land surface emissivity had been calculated by the next formula (5.14):

$$\varepsilon = mP_V + n \tag{5.14}$$

where, $m = (\varepsilon_v - \varepsilon_s) - (1 - \varepsilon_s)F\varepsilon_v; n = \varepsilon_s + (1 - \varepsilon_s)\mathrm{F}\varepsilon_v; \varepsilon_v =$ vegetation emissivity; ε_s = soil emissivity; F = 0.55 shape parameters. Therefore, n = 0.986 and m = 0.004 (Balew and Semaw 2022).

Now, the proportion of vegetation (P_V) was necessarily computed by applying following Eq. (5.15):

$$P_V = \left(\frac{NDVI - NDVImin}{NDVImax - NDVImin}\right)^2 \tag{5.15}$$

where, the NDVImax and NDVImin are the maximum and minimum values of NDVI, calculated from raster NDVI. The calculation of NDVI has been given earlier (Eq. 5.1).

Finally, LST was retrieved from the Eq. (5.16):

$$S_T = \frac{T_B}{1 + (\lambda \times T/p)\ln \varepsilon\lambda} \tag{5.16}$$

where, λ is the emitted radiance wavelength (10.9 μm for Landsat 8 band 10), and S_T is surface temperature in Kelvin.

Finally, the last equation had applied to obtain LST values in Degree Celsius (°C) (Eq. 5.17):

$$S_T\ (°\mathrm{C}) = S_T\ (\mathrm{K}) - 273.15 \tag{5.17}$$

Spatial Autocorrelation

Bivariate Moran's I, along with LISA was applied to identify spatial correlation between LST and UEQ for both the years 2013 and 2023. Bivariate Moran's I is employed to discern the spatial clustering and correlation between two variables based on their respective spatial locations, while LISA serves as another method to characterize the nature of association between exposure and outcome variables to their location (Boots and Tiefelsdorf 2000; Griffith 2000; Hardy and Vekemans 1999; Qi and Wu 1996).

Results

Urbanization of Barasat Municipality

The urban sphere came into existence in 1869 that serves some small portion of land. Census of India depicts the urban population growth of the town with time in unplanned manner (Mukherjee et al. 2023; Mukherjee and Mondal 2022). The urban decadal growth rate between 1901 and 1911 was 5.09% which became 42.72% in 1951 and 82.7% in 1961. The population of this urban unit increased rapidly after partition of India due to shifting of people from East Pakistan to India (Mandal 1999; Mukherjee and Das 2018; Saradar 2001). There was a gradual rise and fall in urban growth till 1991. The growth rate became 115.29% in 2001 and again decreased to 20.26 between 2001 and 2011 (Fig. 5.3). The municipality has a total population of 283,443 as per Census 2011, with a density of 8216 person per sq. km.

The remarkable urban growth of the town transforming its landscape (Paul and Chatterjee 2012) where percentage of impervious surface is increasing continuously (Bhattacharjee and Hazra 2014). The growth of urban sprawl increased the LST of Barasat (Mukherjee and Das 2018). The rapid urbanization of the town reduced its vegetation cover and increased LST (Debnath et al. 2020).

Changing Urban Environmental Indices

Urban landscape is ever changing (García 2022; Li et al. 2022). Urban development in Barasat Municipality makes transformation of the landscape. Four significant indices i.e. NDVI, ARVI, NDWI, and NDBI were employed to understand the landscape transformation and to calculate Urban Environmental Quality Index for the years 2013 and 2023.

Figure 5.4 presents the spatial distribution of amount and quality of greeneries of Barasat for both the years 2013 and 2023. NDVI representing the portion of vegetation cover and ARVI showing resistance power of green spaces. The outcome

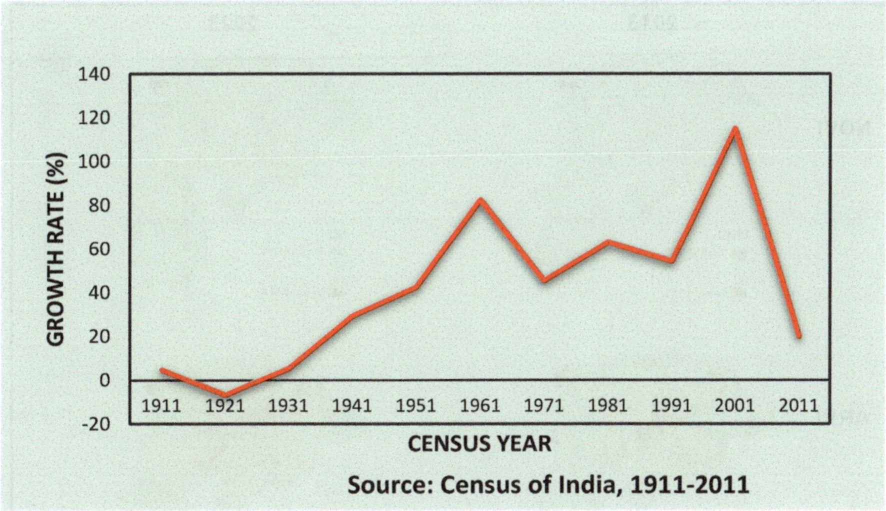

Fig. 5.3 Urban growth rate of Barasat (1911–2011)

of the calculation shows that highest values of NDVI were 0.471 and 0.416 and lowest values were − 0.023 and − 0.022 for the year 2013 and 2023 respectively. The quality of green field is also decreasing as the highest value of ARVI was 0.617 in 2013 which became 0.566 and lowest value was increased from 0.025 to 0.015 between the time frame (Table 5.2). Eastern, Western and Northern portion of the city has high amount of vegetation cover in respect to other areas (Fig. 5.4).

The proportion of wet surfaces was presented by NDWI and NDBI was calculated to understand the built-up surfaces of the city. The value of NDWI is showing the similar decreasing trend as of NDVI and ARVI. The highest values of NDWI were 0.087 and 0.082 and lowest values were − 0.418 and − 0.455 for the years 2013 and 2023 (Table 5.2). Northern and eastern corner of Barasat have higher values where the proportion of moisture contained surfaces has reduced from middle and eastern part of the city (Fig. 5.5). The result of NDBI is showing the continuous transformation of land. The highest values of NDBI were 0.233 and 0.471 for the years respectively where lowest NDBI values were − 0.323 and − 0.023 (Table 5.2). The proportion of built-up spaces continuously increasing from core to outskirt (Fig. 5.5).

Urban Environmental Quality Index

Quantification of UEQ is a hard task as urban units are full of complexities. Remote Sensing (RS) and Geographical Information System (GIS) makes the process bit easier where basic urban environment can be studied. Urban Environmental Quality Index (UEQI) of Barasat city was calculated by CRITIC and TOPSIS method with

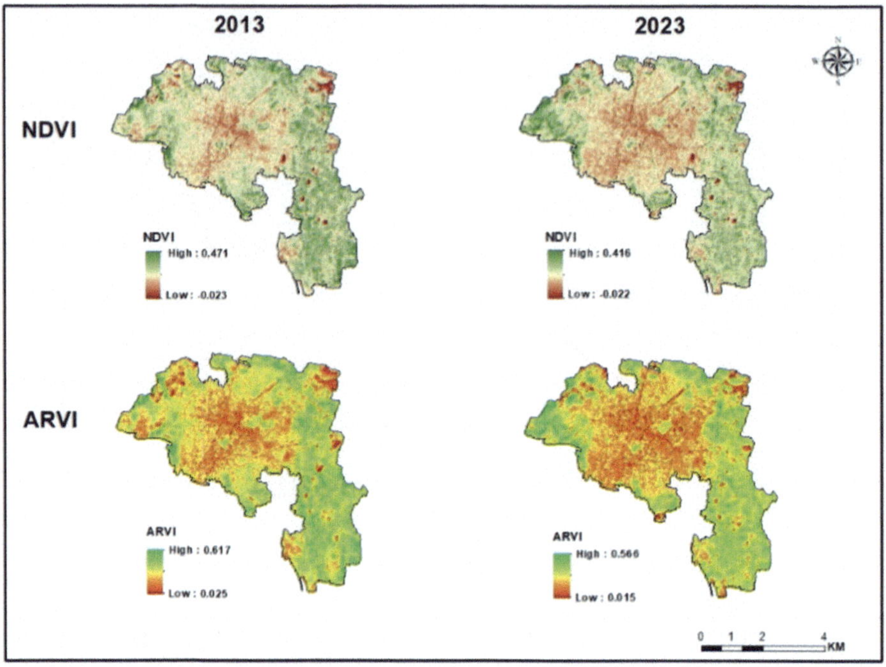

Fig. 5.4 Changing normalized difference vegetation index and atmospherically resistant vegetation index of Barasat (2013–2023)

Table 5.2 Changing urban environmental indices

Environmental Indices	2013		2023	
	Highest	Lowest	Highest	Lowest
NDVI	0.471	− 0.023	0.416	− 0.022
ARVI	0.617	0.025	0.566	0.015
NDWI	0.087	− 0.418	0.082	− 0.455
NDBI	0.233	− 0.323	0.471	− 0.023

Source Author's Calculation

the consideration of all four domains mentioned above. CRITIC method was applied earlier to the indices of NDVI, ARVI, NDWI and NDBI. In 2013, the weightage assigned from CRITIC was 0.245, 0.302, 0.234 and 0.219 for NDVI, ARVI, NDWI and NDBI respectively. The weightage became 0.247, 0.302, 0.231 and 0.220 in 2023 for those same indices. All those assigned weightage was given and TOPSIS method was applied for the better outcome. UEQI map has been generated to depict the changing quality of environment of Barasat from 2013 to 2023. According to the Fig. 5.6, Urban Environmental Quality of Barsat city is not evenly distributed and changed in last decade. In both the years, very high UEQ Index has been seen in

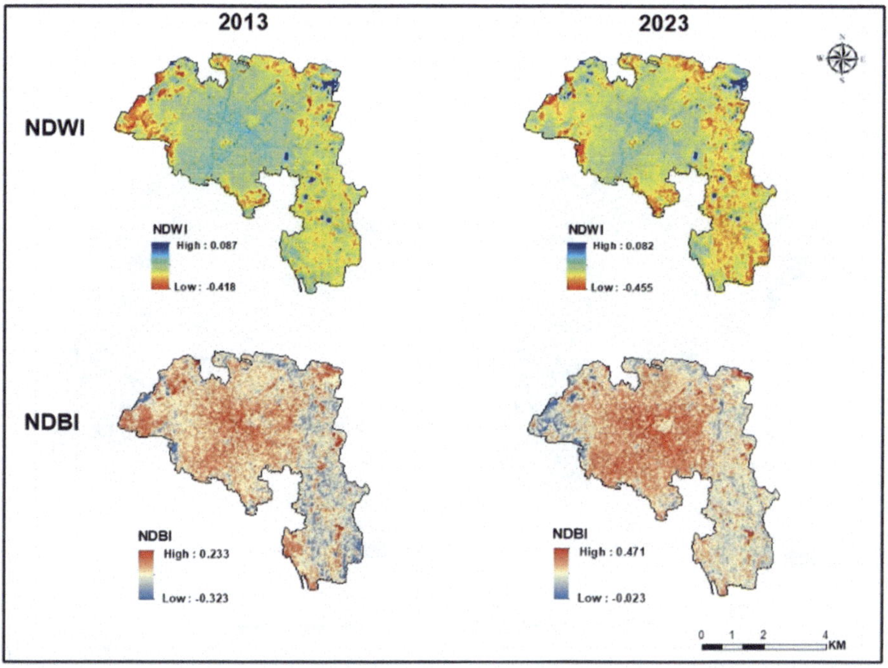

Fig. 5.5 Changing normalized difference water index and normalized difference built-up index of Barasat (2013–2023)

eastern and south-eastern parts. On the other hand, central zone of the city marked as low to very low UEQ. Moderate UEQ is in the buffer zone of middle part (Fig. 5.6). The newly formed map reveals that 14.86% of Barasat was under very low UEQI in 2013 that increased to 15.21% in 2023. Followed by 21.74% area was under low UEQI that also increased to 23.56%. Moderate UEQI zones of the city also increased in the time frame from 21.41 to 24.33%. High and very high UEQI spheres of the city show the decreasing trend between the years. The High UEQI value was 21.04% that decreased from 22.20% and Very High UEQI value decreased from 19.79 to 15.86% from 2013 to 2023 (Table 5.3).

Changing Land Surface Temperature

LST of Barasat was calculated for both the years 2013 and 2023, to know the changing scenario of surface temperature i.e., urban climate change. However, there was a remarkable change in LST from 2013 to 2023. For instance, LST were 35.46–27.35 °C in 2013. On the other hand, the LST was recorded 38.33–29.02 °C for the year 2023 (Table 5.4). So, the temperature of the city has increased over the time frame. The highest LSTs were found in central zone of the area along with some

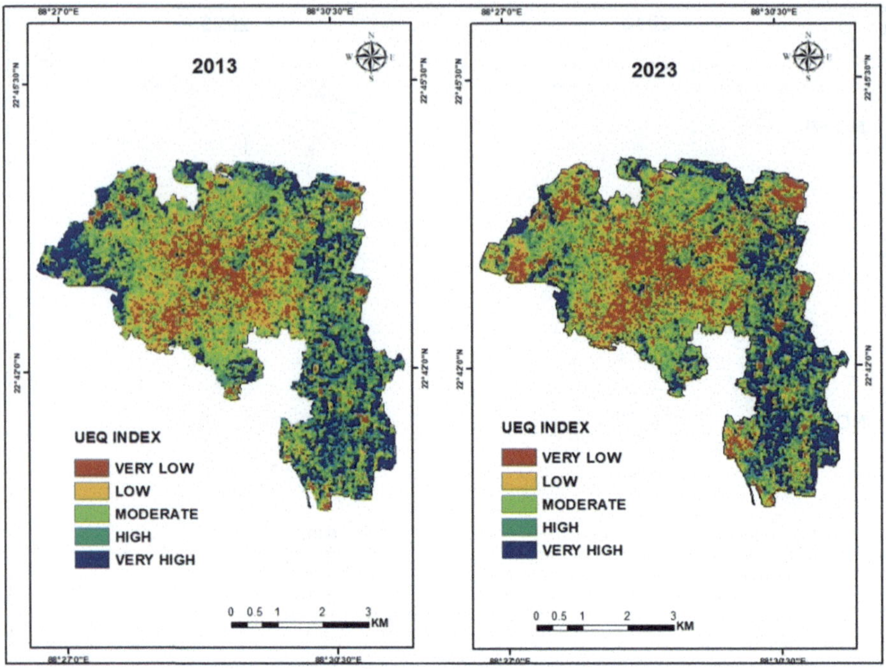

Fig. 5.6 Changing urban environmental quality index (2013–2023)

Table 5.3 Changing nature of urban environmental quality

Urban environmental quality	Percentage of area		Growth rate (in percentage)
	2013	2023	
Very low	14.86	15.21	2.36
Low	21.74	23.56	8.37
Moderate	21.41	24.33	13.64
High	22.20	21.04	– 5.23
Very high	19.79	15.86	– 19.86

Source Author's Calculation

small patches in south, North-East and Western part of the municipality in 2013 where the patches are getting larger with the time. In 2013, lowest LST were found in Eastern and Northern parts of the area but in 2023, the lowest LST parts were reduced (Fig. 5.7). It means LST on an average increased in past 10 years across the study area.

Table 5.4 Changing nature of land surface temperature

Year	Land surface temperature (°C)	
2013	Highest	35.46
	Lowest	27.35
2023	Highest	38.33
	Lowest	29.02

Source Author's Calculation

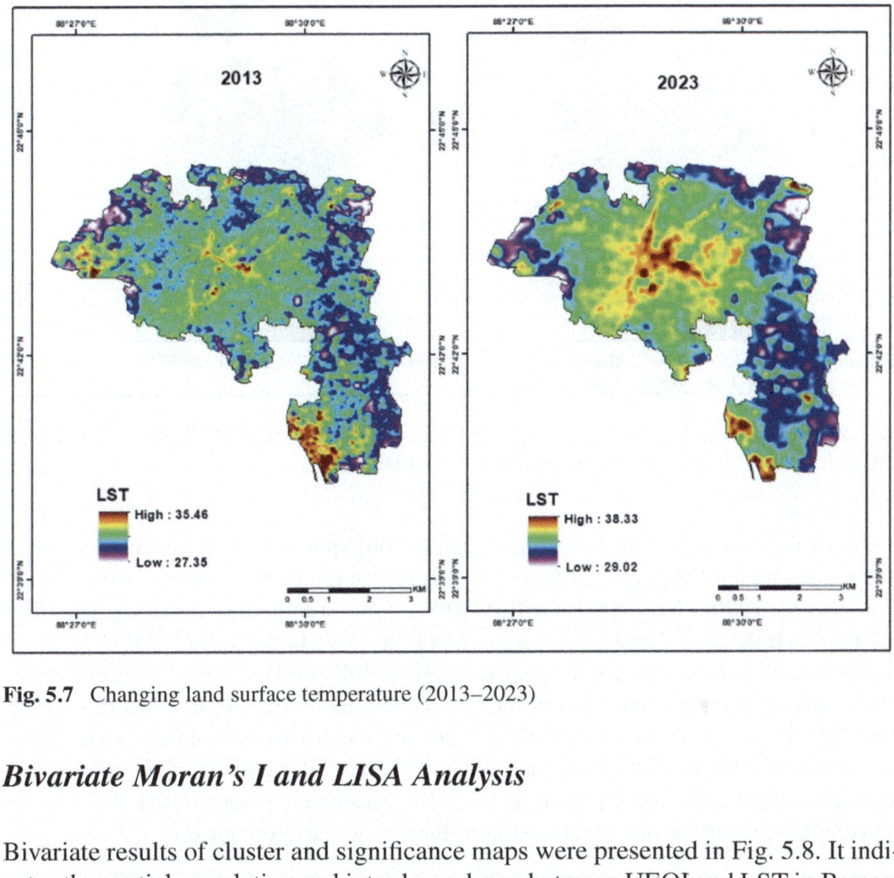

Fig. 5.7 Changing land surface temperature (2013–2023)

Bivariate Moran's I and LISA Analysis

Bivariate results of cluster and significance maps were presented in Fig. 5.8. It indicates the spatial correlation and interdependence between UEQI and LST in Barasat city for both the years 2013 and 2023. The findings present comparative features between exposure and outcome variables, offering a solid foundation to discern the spatial heterogeneity of UEQ (Gorgij et al. 2017; Hardy and Vekemans 1999; Qi and Wu 1996). Moran's I and LISA map shows negative spatial autocorrelation for both the years (Moran's I: − 0.091 and − 0.215) of LST and UEQI. The LISA map indicates hotspots in the middle portion of the city in 2013 which became larger in 2023. These hotspots are the higher LST value zones and lower UEQ zones with significance level $p < 0.001$ in both the years. The findings indicate that the core

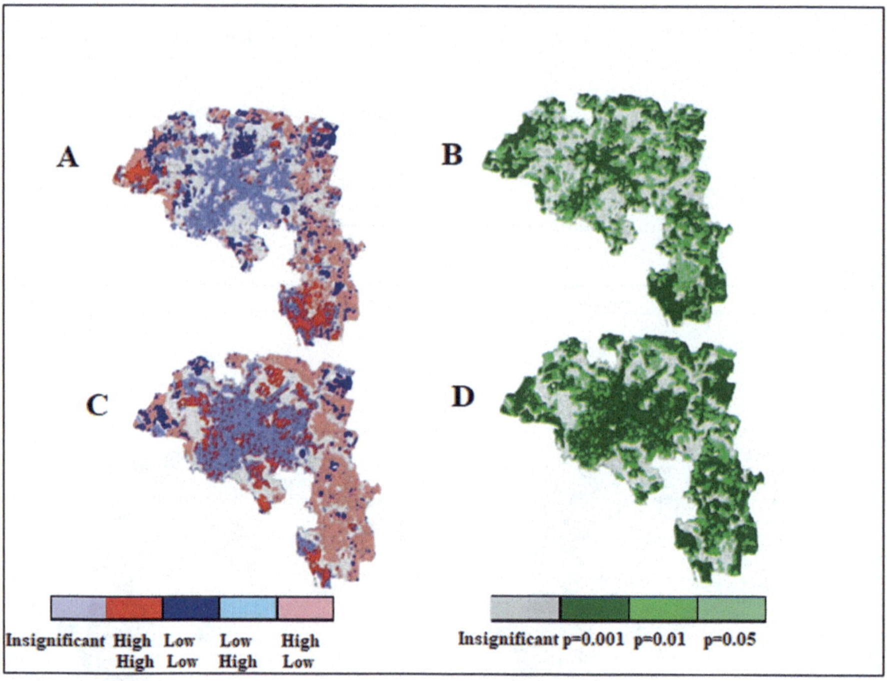

Fig. 5.8 Bivariate LISA cluster map (**A** and **C**) and significance map (**B** and **D**) of UEQ-LST domain for the years 2013 (**A** and **B**) and 2023 (**C** and **D**)

region of Barasat is hot spot and outer area is cold spot in both the years 2013 and 2023 respectively. LISA cluster map was formed on the basis of neighboring values (Fortin et al. 1990; Getis 2009; Griffith 2000). Four significant clustering are there i.e., High–High, Low–Low, Low–High and High–Low. In the year 2023, High-Low clustering had greater space with respect to others followed by Low–Low clustering. Some parts especially central zone of the city comes under High–High clustering class. The High–High clustering was in West and south-east part of the city in 2013. Low–High and High–Low clustering were accrued greater portion of the map while other clustering had smaller portion in 2013. The result suggests that LST is an important factor and it has great spatial influence on UEQ of Barasat.

Discussion

Urbanization and land transformation in Barasat is as remarkable as other Indian cities. All the indices depict the changing scenario of urban environmental factors over the time. NDVI value of Barsat resulting in lower decreasing green surfaces. The proportion of greeneries that reduced over time revealed the depletion of urban green especially in central zone of the city (Mukherjee and Mondal 2022). The outskirt and buffer areas of central part had more green spaces in both the year 2013 and 2023 because the areas still serve as agricultural field especially some parts like western and south-eastern. The value of NDVI (Fig. 5.4) in 2013 is lower in western part because the agricultural field became fallow land while crops were collected. ARVI is measuring the power of plant's resistance against pollution. The index value of ARVI (Fig. 5.4) is dependent on both the quantity and quality of green. As the central part is full of vacant greeneries, the value became lower while outskirts areas were having the better value. The overall pollution resistant power of vegetation of Barasat is not well enough which means the city needs more vegetative fields especially in the central part. Blue sphere or moisture containing lands of Barasat was represented by NDWI (Fig. 5.5) that revealed the sharp declining trend, even the proportion of water bodies has reduced in some parts like east, north-east and north-west. The unscientific and unplanned urbanization of Barasat filling the wet lands of the city makes concretization. The only increasing trend of index value is NDBI (Fig. 5.5) in Barasat city that reveals that bare surfaces are continuously increasing due to the encroachment of urban development. The built-up zone of the city is high in the middle conforming the urban concertation or location of Central Business District (CBD) of the city.

The UEQ of Barasat city has no clear pattern for both the years 2013 and 2023 but a large portion of area came under lower UEQ. The result (Fig. 5.6) indicates unscientific and unplanned constructs (narrow streets, haphazard buildings) within the city. The central part of Barasat is full of official buildings, bus terminals, railway junction and the locality in and around the zone are overcrowded and congested. Eventually, pressure in that part of Barasat negatively adversely affected the natural spaces. The outskirts of the city are full of low-rise, moderate- density housing and some portion of agricultural field, enjoying good EUQ. NDVI, ARVI and NDWI have positively influenced the UEQ since the higher value of those indices showing better UEQ. On the other hand, the NDBI is negatively influenced UEQ as the higher NDBI causes lower UEQ. The change of UEQ from 2013 to 2023 is also remarkable. The low and very low UWQ zones have increased over time from core to outer sphere of the city and the level of UEQ became lower from previous year to present in buffer zones and outskirts. The result only talks about the uninterrupted construction from core to buffer and the clearance of green and open spaces.

The selected micro-climatic tool, LST shows the increasing trend between the time frame. The impact of changing landscape pattern is reflected by LST (Dutta et al. 2019; Liu and Weng 2008; Sahana et al. 2016). Vegetation cover balances the heating effect as vegetation cover exhibited low LST (Asgarian et al. 2015; Guha and Govil 2021; Odindi et al. 2015). Vegetation cover absorbed the solar radiation before reaching the earth's surface (Abdulateef and Al-Alwan 2022; Balew and Semaw 2022; Kabano et al. 2021; Li et al. 2022; Ujang et al. 2021). So, the high vegetation cover with high canopy reduced LST in north, east and western parts of the urban fabric. Moisture contains heat for a greater time than solid particles but reduced the heat by evaporation process as heat transferred as latent heat (Srivastava et al. 2013). The proportion of moisture surfaces of Barasat Municipality reduced LST in a scattered manner mainly seen in north-east and north-west parts. Importantly the built-up area absorbs more heat and reflect less which make the environment more heated. As mentioned earlier, the LST of Barasat is showing high in the core area as bare surfaces contain more heat. Even the LST zones became greater in 2023 in many pockets of the town, especially in CBD area. All because of absence of more green and open spaces, water bodies and high congestion of bare surfaces. The outskirts have lower LST though the temperature became higher there also revealing the land transformation and degradation of greeneries over time (Fig. 5.7).

In both the years, 2013 and 2023 reveal a negative correlation of LST with UEQ showing a significant association (Fig. 5.8). Three variables of UEQ namely NDVI, ARVI and NDWI are positively associated with EQI and negatively influence LST (Ghosh et al. 2019; Guha and Govil 2021; Zeren Cetin et al. 2023). Only NDBI is positively influencing LST and is negatively associated with UEQ. The positive association within Built-up Index and LST suggests that urbanization is responsible for rise in LST. The higher LST causes heat island phenomena (Dhar et al. 2019). Higher LST and lower UEQ in central zone of the city is due to high NDBI values and the high LST zone was expanded by rapid growth of bare surfaces. The growing urban transformation activities in the study area resulted in decreasing water bodies and vegetation cover over the years. Therefore, the LST exhibited notable variations over time, correlating with various urban components i.e., urban environment.

Recommendations

Now the work provides some recommendations (Fig. 5.9) to improve and maintain the environment quality of Barasat city.

- Adopt a framework to determine urban spatial growth i.e. planned urban expansion. Continuous monitoring on urban spatial growth like land-use planning, land administration and land development should be enhanced. Some special areas like central, southern part of Barasat need more attention in this regard.
- To achieve healthier environmental setting, urban green should be increased by conserving existing vegetation and new plantation. More street plantation, gardening, creating parks etc. would be great beneficial for the environment especially the central part.
- To make the city cooler, water bodies need to be present. Existing water bodies (in the areas of northern, eastern, south-eastern and western) should be preserved. Construction of more artificial water bodies needs to be included in urban land-use planning and could be used for aquaculture.
- Preservation and promotion of urban agriculture would be a great benefit for Barasat city. Outer part of the city (especially eastern and western parts) has some agricultural field. Those lands should be protected to reduce the dependency of food on villages and also enhanced green. Roof top farming could be an effective tool to promote urban agriculture.

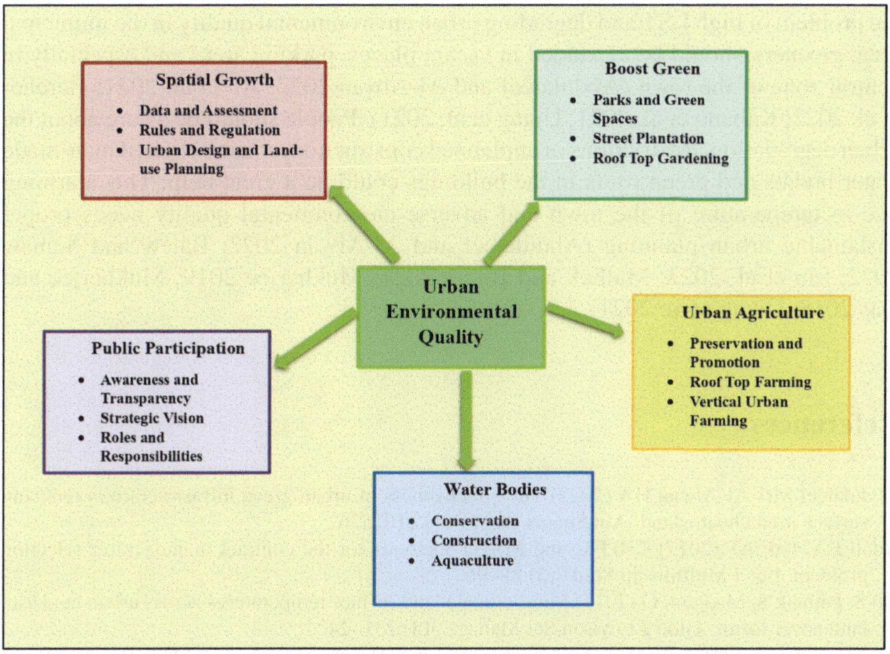

Fig. 5.9 Strategies to improve the environment quality of Barasat Municipality

- Most importantly, raising awareness among the citizen of Barasat is much needed. People should aware about the problem of environmental degradation and climate change. Urban governance should take initiation for the task and promote the participation of people in policies of environment.

Conclusion

The research has been conducted employing geospatial methodologies and statistical applications. The study unveils the connection between LST and urban environment in Barasat municipality (2013 and 2023).

The study revealed that urban environmental elements of Barasat are changing rapidly and they have a strong association with LST. Vegetation cover and Moisture content surfaces has reduced where built-up spaces have increased. On the other hand, the LST has been steadily rising each year. The higher LST could have impacted on citizens (He et al. 2021; Koç et al. 2022; Peng et al. 2022) and ecosystem services in future. The methodology introduced in the research can be applied to other cities experiencing similar conditions.

So, it can be concluded that the unplanned expansion and development of Barasat city has huge impact on its climatic condition and also the environment. To mitigate the problem of high LST and degrading urban environmental quality in the municipal area, greenery should be expanded in vacant places, parking areas and especially in central zone of the town (Abdulateef and Al-Alwan 2022; Ali et al. 2021; Hereher et al. 2022; Kabano et al. 2021; Ujang et al. 2021). People should be aware about the urban degradation. Restrictions in unplanned construction, construction of man-made water bodies and green roofs in the buildings could be a great help. This alarming rise in temperature of the town and adverse environmental quality needs proper sustainable urban planning (Abdulateef and Al-Alwan 2022; Balew and Semaw 2022; Liu et al. 2022; Mallick and Rudra 2021; Mukherjee 2019; Mukherjee and Das 2018; Ujang et al. 2021).

References

Abdulateef MF, Al-Alwan HA (2022) The effectiveness of urban green infrastructure in reducing surface urban heat island. Ain Shams Eng J 13(1):101526

Adalı EA, Işık AT (2017) CRITIC and MAUT methods for the contract manufacturer selection problem. Eur J Multidiscip Stud 2(5):88–96

Ali S, Patnaik S, Madguni O (2017) Microclimate land surface temperatures across urban land use/land cover forms. Glob J Environ Sci Manage 3(3):231–242

Ali G, Abbas S, Qamer FM, Wong MS, Rasul G, Irteza SM, Shahzad N (2021) Environmental impacts of shifts in energy, emissions, and urban heat island during the COVID-19 lockdown across Pakistan. J Clean Prod 291:125806

Alinezhad A, Khalili J, Alinezhad A, Khalili J (2019) CRITIC method. In: New methods and applications in multiple attribute decision making (MADM), pp 199–203

Antoniadis D, Katsoulas N, Kittas C (2018) Simulation of schoolyard's microclimate and human thermal comfort under Mediterranean climate conditions: effects of trees and green structures. Int J Biometeorol 62:2025–2036

Asgarian A, Amiri BJ, Sakieh Y (2015) Assessing the effect of green cover spatial patterns on urban land surface temperature using landscape metrics approach. Urban Ecosyst 18:209–222

Aziz MA, Haldar D, Danodia A, Chauhan P (2023) Use of time series Sentinel-1 and Sentinel-2 image for rice crop inventory in parts of Bangladesh. Appl Geomatics, 1–14

Bai X, Imura H (2000) A comparative study of urban environment in East Asia: stage model of urban environmental evolution. Int Rev Environ Strateg 1(1):135–158

Balew A, Semaw F (2022) Impacts of land-use and land-cover changes on surface urban heat islands in Addis Ababa city and its surrounding. Environ Dev Sustain 24(1):832–866

Bao Q, Ruan D, Shen Y, Hermans E (2010) Creating a composite road safety performance index by a hierarchical fuzzy TOPSIS approach, pp 458–463

Bhattacharjee D, Hazra S (2014) Fractional impervious surface detection of urban built-up area—a case study in Barasat Town, West Bengal. Indian Cartographer, India

Bhattacharjee A, Kamble S, Kamal N, Golhar P, Kumari V, Bhargava A (2022) Urban heat island effect: a case study of Jaipur, India. Int J Earth Sci Knowl Appl 4(1):133–139

Boots B, Tiefelsdorf M (2000) Global and local spatial autocorrelation in bounded regular tessellations. J Geogr Syst 2:319–348

Button K (2002) City management and urban environmental indicators. Ecol Econ 40(2):217–233

Chaudhari S, Sardar V, Ghosh P (2023) Drought classification and prediction with satellite image-based indices using variants of deep learning models. Int J Inf Technol, 1–10

Chauhan RK, Mohanty SK, Subramanian S, Parida JK, Padhi B (2016) Regional estimates of poverty and inequality in India, 1993–2012. Soc Indic Res 127:1249–1296

Chen Y-L, Shen S-L, Zhou A (2022) Assessment of red tide risk by integrating CRITIC weight method, TOPSIS-ASSETS method, and Monte Carlo simulation. Environ Pollut 314:120254

Connors JP, Galletti CS, Chow WT (2013) Landscape configuration and urban heat island effects: assessing the relationship between landscape characteristics and land surface temperature in Phoenix, Arizona. Landscape Ecol 28:271–283

Cyril S, Oldroyd JC, Renzaho A (2013) Urbanisation, urbanicity, and health: a systematic review of the reliability and validity of urbanicity scales. BMC Public Health 13:1–11

Debnath A, Kumar R, Singh T, Prawasi R (2020) The impact of rapid urbanization on vegetation cover and land surface temperature in Barasat Municipal Area. In: Sustainable sevelopment practices using geoinfatics, pp 1–21

Dhar RB, Chakraborty S, Chattopadhyay R, Sikdar PK (2019) Impact of land-use/land-cover change on land surface temperature using satellite data: a case study of Rajarhat Block, North 24-Parganas District, West Bengal. J Indian Soc Remote Sens 47(2):331–348

Dhumal RK, Vibhute AD, Nagne AD, Kale KV, Mehrotra SC (2017) Performance analysis of spectral features based on narrowband vegetation indices for cotton and maize crops by EO-1 Hyperion dataset, pp 581–590

Diakoulaki D, Mavrotas G, Papayannakis L (1995) Determining objective weights in multiple criteria problems: the critic method. Comput Oper Res 22(7):763–770

Dutta D, Rahman A, Paul S, Kundu A (2019) Changing pattern of urban landscape and its effect on land surface temperature in and around Delhi. Environ Monit Assess 191:1–15

El Din HS, Shalaby A, Farouh HE, Elariane SA (2013) Principles of urban quality of life for a neighborhood. HBRC Journal 9(1):86–92

Faisal K, Shaker A (2017a) An investigation of GIS overlay and PCA techniques for urban environmental quality assessment: a case study in Toronto, Ontario, Canada. Sustainability 9(3):380

Faisal K, Shaker A (2017b) Improving the accuracy of urban environmental quality assessment using geographically-weighted regression techniques. Sensors 17(3):528

Fernández I, Pacheco JM (2001) Mathematical aspects in the modeling of urban environmental quality. In: Integrative systems approaches to natural and social dynamics: systems science, pp 235–248

Fortin M-J, Drapeau P, Legendre P (1990) Spatial autocorrelation and sampling design in plant ecology. Progr Theor Vegetation Sci, 209–222

García DH (2022) Analysis of urban heat island and heat waves using Sentinel-3 images: a study of Andalusian Cities in Spain. Earth Syst Environ 6(1):199–219

Getis A (2009) Spatial autocorrelation. In: Handbook of applied spatial analysis: software tools, methods and applications. Springer, pp 255–278

Ghosh S, Chatterjee ND, Dinda S (2019) Relation between urban biophysical composition and dynamics of land surface temperature in the Kolkata metropolitan area: a GIS and statistical based analysis for sustainable planning. Model Earth Syst Environ 5:307–329

Gorgij AD, Kisi O, Moghaddam AA, Taghipour A (2017) Groundwater quality ranking for drinking purposes, using the entropy method and the spatial autocorrelation index. Environ Earth Sci 76:1–9

Griffith DA (2000) A linear regression solution to the spatial autocorrelation problem. J Geogr Syst 2:141–156

Guha S, Govil H (2021) An assessment on the relationship between land surface temperature and normalized difference vegetation index. Environ Dev Sustain 23:1944–1963

Hardy OJ, Vekemans X (1999) Isolation by distance in a continuous population: reconciliation between spatial autocorrelation analysis and population genetics models. Heredity 83(2):145–154

Hasnine M, Rukhsana (2020) An analysis of urban sprawl and prediction of future urban town in urban area of developing nation: case study in India. J Indian Soc Remote Sens 48:909–920

Hauser PM, United Nations Fund for Population Activities (1982) Population/urban future. SUNY Press

He B-J, Wang J, Liu H, Ulpiani G (2021) Localized synergies between heat waves and urban heat islands: implications on human thermal comfort and urban heat management. Environ Res 193:110584

Hereher M, Eissa R, Alqasemi A, El Kenawy AM (2022) Assessment of air pollution at Greater Cairo in relation to the spatial variability of surface urban heat island. Environ Sci Pollut Res 29(15):21412–21425

Hulley GC, Ghent D, Göttsche FM, Guillevic PC, Mildrexler DJ, Coll C (2019) Land surface temperature. In: Taking the temperature of the earth. Elsevier, pp 57–127

Hwang C-L, Lai Y-J, Liu T-Y (1993) A new approach for multiple objective decision making. Comput Oper Res 20(8):889–899

Hwang C-L, Yoon K, Hwang C-L, Yoon K (1981) Methods for multiple attribute decision making. Multiple attribute decision making: methods and applications a state-of-the-art survey, pp 58–191

Imam AU, Banerjee UK (2016) Urbanisation and greening of Indian cities: problems, practices, and policies. Ambio 45:442–457

Javanbakht M, Boloorani AD, Kiavarz M, Samany NN, Zebardast L, Zangiabadi M (2021) Spatial-temporal analysis of urban environmental quality of Tehran, Iran. Ecol Indicators 120:106901

Jiří B, Jan Š (2021) Correction to: urban environmental quality and out-migration intentions. Ann Reg Sci 67(1):243–243

Kabano P, Lindley S, Harris A (2021) Evidence of urban heat island impacts on the vegetation growing season length in a tropical city. Landsc Urban Plan 206:103989

Kaginalkar A, Ghude SD, Mohanty U, Mujumdar P, Bhakare S, Darbari H, Dwivedi AK, Gavali P, Gavhale S, Islam S (2021) Integrated urban environmental system of systems for weather ready cities in India. Bull Am Meteorol Soc

Kazan H, Ozdemir O (2014) Financial performance assessment of large scale conglomerates via TOPSIS and CRITIC methods. Int J Manage Sustain 3(4):203–224

Khan A, Carlosena L, Feng J, Khorat S, Khatun R, Doan Q-V, Santamouris M (2022) Optically modulated passive broadband daytime radiative cooling materials can cool cities in summer and heat cities in winter. Sustainability 14(3):1110

Koç A, Caf A, Koç C, Kejanli DT (2022) Examining the temporal and spatial distribution of potential urban heat island formations. Environ Sci Pollut Res 29(8):11455–11468

Krishnan AR, Kasim MM, Hamid R, Ghazali MF (2021) A modified CRITIC method to estimate the objective weights of decision criteria. Symmetry 13(6):973

Kundu D, Pandey AK (2020) World urbanisation: trends and patterns. In: Developing national urban policies. Springer, pp 13–49

Leeson GW (2018) The growth, ageing and urbanisation of our world. J Population Ageing 11:107–115

Li X, Zhou W, Ouyang Z, Xu W, Zheng H (2012) Spatial pattern of greenspace affects land surface temperature: Evidence from the heavily urbanized Beijing Metropolitan Area, China. Landscape Ecol 27:887–898

Li H, Zhou Y, Jia G, Zhao K, Dong J (2022) Quantifying the response of surface urban heat island to urbanization using the annual temperature cycle model. Geosci Front 13(1):101141

Liang B, Weng Q (2010) Assessing urban environmental quality change of Indianapolis, United States, by the remote sensing and GIS integration. IEEE J Select Top Appl Earth Observ Remote Sens 4(1):43–55

Liu H, Weng Q (2008) Seasonal variations in the relationship between landscape pattern and land surface temperature in Indianapolis, USA. Environ Monit Assess 144:199–219

Liu C, Shao Z, Chen M, Luo H (2013) MNDISI: a multi-source composition index for impervious surface area estimation at the individual city scale. Remote Sens Lett 4(8):803–812

Liu Z, Zhan W, Lai J, Bechtel B, Lee X, Hong F, Li L, Huang F, Li J (2022) Taxonomy of seasonal and diurnal clear-sky climatology of surface urban heat island dynamics across global cities. ISPRS J Photogramm Remote Sens 187:14–33

Liu D, Zhang J, He C, Wang C, Cui Y, Zhang Y (2023) Research on demand evaluation of live broadcast interface of agricultural product based on factor analysis and CRITIC method, pp 369–379

Lu H, Zhao Y, Zhou X, Wei Z (2022) Selection of agricultural machinery based on improved CRITIC-entropy weight and GRA-TOPSIS method. Processes 10(2):266

Madic M, Radovanovic M (2015) Ranking of some most commonly used nontraditional machining processes using ROV and CRITIC methods. UPB Sci Bull Ser D 77(2):193–204

Malah A, Bahi H, Radoine H, Maanan M, Mastouri H (2022a) Assessment of urban environmental quality: a case study of Casablanca, Morocco. Int Arch Photogramm Remote Sens Spat Inf Sci 46:205–210

Malah A, Bahi H, Radoine H, Maanan M, Mastouri H (2022b) Assessment of urban environmental quality: a case study of Casablanca, Morocco. Int Arch Photogramm Remote Sens Spat Inf Sci 46:205–210

Mallick SK, Rudra S (2021) Land use changes and its impact on biophysical environment: study on a river bank. Egypt J Remote Sens Space Sci 24(3):1037–1049

Mandal R (1999) Urban geography a text book New Delhi

Mukhametzyanov I (2021) Specific character of objective methods for determining weights of criteria in MCDM problems: entropy, CRITIC and SD. Decis Making Appl Manage Eng 4(2):76–105

Mukherjee K, Das P (2018) Modelling the relationship between urban growth modes and the thermal environment—a case study of the Barasat Municipality, West Bengal. J Geogr Environ Earth Sci Int 17(2):1–19

Mukherjee K, Mondal D (2022) Spatial equity in urban facility versus residents' satisfaction: challenge for the unplanned cities—a case study of Barasat City, India. GeoJournal 87(Suppl 4):931–949

Mukherjee K, Roy K, Pal B (2023) Gender parity in urbanizing city: a case study of Barasat City, India. Glob Soc Welfare, pp 1–21

Mukherjee K (2019) Delineating fringe by global diversity value: a case study on Barasat Municipality, India, pp 1–12

Nath V (1986) Urbanisation in India: review and prospects. Econ Polit Weekly, pp 339–352

Niu D, Wu Q, Fan L, Liu C (2013) The comprehensive evaluation and study based on national happiness index system, pp 204–206

Ocampo L, Aro JL, Evangelista SS, Maturan F, Casinillo L, Yamagishi K, Selerio E Jr (2023) Composite ecotourism potential index based on an integrated stochastic CRITIC-weighted sum method. Curr Issue Tour 26(15):2513–2542

Odindi J, Bangamwabo V, Mutanga O (2015) Assessing the value of urban green spaces in mitigating multi-seasonal urban heat using MODIS land surface temperature (LST) and Landsat 8 data. Int J Environ Res 9(1):9–18

Paul S, Chatterjee K (2012) Urbanisation and consistency measurement: a study on District of North 24 Parganas, West Bengal, India. Arch Appl Sci Res 4(5):2052–2067

Peng W, Wang R, Duan J, Gao W, Fan Z (2022) Surface and canopy urban heat islands: does urban morphology result in the spatiotemporal differences? Urban Climate 42:101136

Qi Y, Wu J (1996) Effects of changing spatial resolution on the results of landscape pattern analysis using spatial autocorrelation indices. Landscape Ecol 11:39–49

Rosić M, Pešić D, Kukić D, Antić B, Božović M (2017) Method for selection of optimal road safety composite index with examples from DEA and TOPSIS method. Accid Anal Prev 98:277–286

Roy S, Das A (2018) Application of TOPSIS method for financial performance evaluation: a study of selected scheduled banks in Bangladesh. J Comm Account Res 7(1):24–29

Roy AD, Gorai J, Dey R, Pal S, Mandal S (2022) Integration of field investigation and geoinformatics for urban environmental quality appraisal of Bankura Town, West Bengal, India. In: Environmental degradation in Asia: land degradation, environmental contamination, and human activities. Springer, pp 355–389

Roy S, Bose A, Majumder S, Roy Chowdhury I, Abdo HG, Almohamad H, Abdullah Al Dughairi A (2022) Evaluating urban environment quality (UEQ) for Class-I Indian city: an integrated RS-GIS based exploratory spatial analysis. Geocarto Int, 2153932

Sahana M, Ahmed R, Sajjad H (2016) Analyzing land surface temperature distribution in response to land use/land cover change using split window algorithm and spectral radiance model in Sundarban Biosphere Reserve, India. Model Earth Syst Environ 2:1–11

Saradar J (2001) Institute of Landscape Ecology and Ekistics (ILEE), ISSN 0971/4170, 2010

Sekertekin A, Zadbagher E (2021) Simulation of future land surface temperature distribution and evaluating surface urban heat island based on impervious surface area. Ecol Ind 122:107230

Shetty S, Umesh P, Shetty A (2022) Dependability of rainfall to topography and micro-climate: an observation using geographically weighted regression. Theor Appl Climatol, 1–21

Shi H, Li Y, Jiang Z, Zhang J (2021) Comprehensive power quality evaluation method of microgrid with dynamic weighting based on CRITIC. Measur Control 54(5–6):1097–1104

Shimod K, Prasad T, Vineethkumar V, Akhil R, Jayapal G (2023) Geospatial technology for analysing the dynamics in microclimate with special reference to land surface temperature of tropical cities: a case study. In: Urban commons, future smart cities and sustainability. Springer, pp 321–340

Singh S, Jain K (2022) Geospatial approach for urban environmental quality assessment. Int Arch Photogramm Remote Sens Spat Inf Sci 43:705–711

Sinha S, Basu A (2023) An assessment about the quality of life: case study of Asansol. In: Urban commons, future smart cities and sustainability. Springer, pp 427–464

Srivastava PK, Han D, Ramirez MR, Islam T (2013) Machine learning techniques for downscaling SMOS satellite soil moisture using MODIS land surface temperature for hydrological application. Water Resour Manage 27:3127–3144

Trujillo-González JM, Zapata-Muñoz YL, Torres-Mora MA, García-Navarro FJ, Jiménez-Ballesta R (2020) Assessment of urban environmental quality through the measurement of lead in bryophytes: case study in a medium-sized city. Environ Geochem Health 42:3131–3139

Ujang N, Fukuda T, Pisello AL, Vukadinovic D (2021) Resilient and responsible smart cities. Springer

Vadivel S, Shetty DS, Sequeira A, Nagaraj E, Sakthivel V (2022) A sustainable green supplier selection using CRITIC method, pp 308–315

Venkatesham V (2015) The problems and issues in urbanization in India. Indian J Res 4(8):277–278

Watson C (1993) Trends in world urbanisation. In: Proceedings of ICUP-1st international conference on urban pests, Cambridge, England

Wenban-Smith H, Faße A, Grote U (2016) Food security in Tanzania: the challenge of rapid urbanisation. Food Security 8:973–984

Wu H-W, Zhen J, Zhang J (2020) Urban rail transit operation safety evaluation based on an improved CRITIC method and cloud model. J Rail Transp Plan Manage 16:100206

Xu H (2010) Analysis of impervious surface and its impact on urban heat environment using the normalized difference impervious surface index (NDISI). Photogramm Eng Remote Sens 76(5):557–565

Yilmaz B, Harmancioglu N (2010) Multi-criteria decision making for water resource management: a case study of the Gediz River Basin, Turkey. Water SA 36(5)

Zafar S, Alamgir Z, Rehman M (2021) An effective blockchain evaluation system based on entropy-CRITIC weight method and MCDM techniques. Peer-to-Peer Netw Appl 14(5):3110–3123

Zavadskas EK, Zakarevicius A, Antucheviciene J (2006) Evaluation of ranking accuracy in multi-criteria decisions. Informatica 17(4):601–618

Zeren Cetin I, Varol T, Ozel HB (2023) A geographic information systems and remote sensing-based approach to assess urban micro-climate change and its impact on human health in Bartin, Turkey. Environ Monit Assess 195(5):540

Zhang H, Chen C, Yin X, Wang Q, Tao J (2022) Comprehensive evaluation method of power quality CRITIC-MARCOS for regional distribution network, pp 377–389

Zhou W, Qian Y, Li X, Li W, Han L (2014) Relationships between land cover and the surface urban heat island: seasonal variability and effects of spatial and thematic resolution of land cover data on predicting land surface temperatures. Landscape Ecol 29:153–167

Chapter 6
Systematic Review to Address the Effect of Urban Heat Island and Outdoor Thermal Condition on Human Comfort in Small and Medium Cities of Global South

Tirthankar Basu and Arijit Das

Abstract Urban heat island (UHI) related heat stress and its association with human comfort were mainly conducted mainly in the temperate developed countries. Lack of researches in underdeveloped and developing countries has somehow created a gap in advancing the knowledge on UHI-related issues. Further, it is well acknowledged that the issue of human comfort to the heat stress is more relevant concerning the cities of tropical climate. Therefore, the investigations on the small and medium-sized tropical cities of third world countries become essential. This chapter tried to represent a systematic investigation of the existing researches to provide guidance. This study summarized all the factors that are selected by different studies into three groups i.e. socio-economic, health, and environment for better assessment. Finally, it tried to represent the strategies that can be followed to conduct UHI-related thermal comfort assessment in the small and medium-class tropical cities of the Global South.

Introduction

In today's world, more than 50% world's population resides in urban regions and, by 2050, the proportion is likely to become around 66% (United Nations 2014). In the twenty-first century, the urbanization rate is found quite high in developing countries particularly in the Global South (Farrell 2017; Cohen 2006). Therefore, it has generated immense stress on the social, economic, and environmental aspects of the urban areas (Zhang 2016; Tacoli et al. 2015). Due to this, urbanization becomes problematic in these countries. It is observed that, due to the huge influx of rural to urban

A. Das (✉)
Department of Geography, University of Gour Banga, Malda, West Bengal 732103, India
e-mail: arijit3333@gmail.com

T. Basu
Research Scholar, Independent, Malda, West Bengal, India

U. Mukhopadhyay et al. (eds.), *Climate Crisis, Social Responses and Sustainability*, Climate Change Management, https://doi.org/10.1007/978-3-031-58261-5_6

migration and to absorb these immigrants, the cities gradually become congested. Further, this process also facilitates a high rate of loss of environmental resources (Das and Basu 2020). Overall, it culminates several adverse impacts and affects urban livability. Such adverse effects include removal of precious vegetation cover, water pollution, air pollution, biodiversity and ecosystem services loss, and urban heat island (UHI) effect (Lyu et al. 2019; Wang et al. 2020; Lee et al. 2020). Recently, the UHI effect is becoming a serious concern to policymakers as it directly affects the livability condition of the citizens (Priyankara et al. 2019). The UHI effect can be stated as the condition where the temperature of the dense urban environment remains significantly high compared to the neighbourhood rural regions (Soydan 2020). It is mainly caused because of the rapid growth in the proportion of built-up surfaces and the release of harmful gases from the industries and transport sector within the urban boundary (Ulpiani 2021; Hua et al. 2020). It alters the surface radiation, moisture, and thermal characteristics of the urban areas. Besides, the growing UHI effect in the city further leads to other urban livability issues such as air pollution and issues related to the health of citizens (Mirzaei et al. 2020). Therefore, it urges an urgent need to address a systematic approach that can help to understand and monitor the properties and effect of UHI in the cities of the developing countries of the Global South.

Now a day, the UHI effect is visible in almost all the important cities of the world. Further, the UHI effect is gradually becoming more intense with rapid urban growth. Due to the rising attention towards the UHI effect and its widespread scale, several researchers started to emphasize the different aspects of UHI such as its genesis mechanism (Athukorala and Murayama 2021), differences in spatial pattern (Lu et al. 2020), the factors that help to intensify the UHI (Dewan et al. 2021), and its mitigation strategies (Jain et al. 2020). However, the scope of the work of different researchers varies according to their disciplines. For example, an ecologist tries to observe the UHI phenomena from an ecological perspective (Westendorff 2020). An urban policymaker tries to view the situation from the perspective of urban design, building spacing, and construction materials (Kandya and Mohan 2018). A geographer observes the spatial–temporal trend of UHI and the factors associated with it (Abulibdeh 2021). In this regard, studies showed that the UHI effect is mainly visible around the city centre. It is greatly associated with congested buildings, roads, and impervious surfaces (Cao et al. 2008). However, it is also noted in the studies that some environmental elements such as green space and blue space help to reduce the UHI upshot (Aram et al. 2019).

The issue of UHI-related thermal conditions and human vulnerability is the latest and well-investigated one in urban science. However, still in the tropical underdeveloped and developing countries, the issues of the UHI effect related to urban livability and comfort have not gained significant importance. Lots of research works were available that highlighted the growing effect of urban heat islands, their spatial–temporal pattern (Kumar et al. 2021; Hassan et al. 2021). But, only a handful of research works were conducted that primarily tried to link the UHI effect and related outdoor thermal conditions with human comfort and vulnerability (Rose 2010; Dr and Kalia 2021). Further, these are mainly studied concerning the major metropolitan cities (Dewan et al. 2021; Khan 2020). However, in the twenty-first century, it is observed that the medium-class urban agglomerations are growing at a very rapid rate in the third world nations of the Global South (Tumbe 2016). These fast-growing secondary cities were considered the emerging economic hub of the countries (Roberts and Hohmann 2014). Therefore, investigations related to the sustainable development and urbanization of these cities are essential for balanced growth. In this regard, the issue of UHI-related vulnerability is a major concern to policymakers as it negatively affects the quality of life in the cities (Priyankara et al. 2019). Therefore, proper and systematic direction towards investigation in this aspect is essential for fruitful outcome.

Rationale of the Study

The investigation on the impact of UHI on urban livability concerning the small and medium-class urban agglomerations remained less priority to the researchers. The lack of a systematic direction to investigate this aspect is a major reason in this regard. Following this, this study tried to summarize the ways to link the effect of UHI-related thermal conditions with the citizens' well-being and comfort through an intense review of the existing pieces of literature.

Materials and Methodology

A systematic review is carried out by considering the existing pieces of literature related to the aspect of UHI and urban well-being. This study tried to address the following research questions: (1) What are the ways of investigation available in the existing pieces of literature to link the UHI effect and the comfort of city dwellers? (2) What are the indicators that are adopted by the researchers to explore this relation? (3) What are the possible research gaps in the available existing researches related to this aspect?. To address these research questions scientifically and comprehensively, a major focus has been given to make this study systematic, structured, comprehensive, and quantitative (Pickering and Byrne 2014). This study is systematic in the sense that all the existing researches are studied and only those are incorporated that can be applied widely. This study is structured because it tried to summarize the findings of all available researches. It is quantitative as it quantifies the existing researches to highlight the research gaps. Finally, this review is comprehensive as it tried to include available studies covering the entire globe (Hazemba and Halog 2021). This study observed that a successful review depends on the understanding of the existing methods and the approaches taken by different scholars (Iyer and Steele 2015). Following this, this study first summarized the things that needed to be addressed and examined. This is done following the three subsequent subsections.

Scope and Problem Definition

The main objective of this review is to highlight the possible ways to link the relationship between the UHI effect and the well-being of the urban resident. The priority is to select and compile the existing pieces of literature systematically to provide a basis for sustainable urban development. Therefore, this study tried to remain as unbiased as possible without giving weight to any particular method. The issue of the UHI effect is a recent and popular one in urban studies. Therefore, an initial search on UHI showed a large number of existing pieces of literature. However, to remain specific, and also for better assessment, a small number of sample articles are needed to select. This is done by the selection of search strings and screening of the existing pieces of literature (Basu and Das 2021b).

Selection of Search String

This step deals with an initial search following the objective of this review. This study deals with the UHI effect and thermal comfort matter. Therefore, keywords like heat island effect, urban, UHI, thermal comfort, and sustainable urbanization are selected. A detailed description of the database search string is shown in Table 6.1.

This study incorporated only those articles which are written in English and published between 2018 and 2021. This is to understand and highlight the most recent applied tools and techniques in this field.

Screening of Researches

An initial search using the keywords has disclosed a large number of papers (1841). These articles were sorted out into a less number of articles through title and abstract screening. In the next stage, the remaining articles were screened out by adopting a set of eligibility criteria. These criteria were as follows: (i) Studies have been carried out on field investigation; (ii) Performed the vulnerability or risk analysis related to heat; (iii) Tried to focus on all underlying factors by devising a composite index. Following the set of criteria, the total number of articles is reduced down to 60. It is further important to mention that special weight is given to the publishing outlet's quality to ensure the selection of quality research articles. Figure 6.1 represents the entire screening process of this study.

Table 6.1 Description of the research article searching procedure

Database field	String of search	Initial articles found	Date
Scopus	"Urban heat island effect + urban thermal comfort" or "heat island effect + Urban vulnerability" or "Outdoor thermal comfort + sustainable urban development" or "urban health + urban heat island effect" or "Outdoor thermal comfort + urban human vulnerability"	372	01.08.2021 to 05.08.2021
Web of science	"Urban heat island effect + urban thermal comfort" or "heat island effect + Urban vulnerability" or "Outdoor thermal comfort + sustainable urban development" or "urban health + urban heat island effect" or "Outdoor thermal comfort + urban human vulnerability"	237	06.08.2021 to 09.08.2021
Science direct	"Urban heat island effect + urban residents' vulnerability" or "heat island effect + Urban vulnerability" or "Outdoor thermal comfort + sustainable urban development" or "urban health + urban heat island effect" or "Outdoor thermal comfort + urban human vulnerability"	582	10.08.2021 to 12.08.2021
Google scholar and google search engine	"Urban heat island effect + urban thermal comfort" or "heat island effect + Urban vulnerability" or "Outdoor thermal comfort + sustainable urban development" or "urban health + urban heat island effect" or "Outdoor thermal comfort + urban human vulnerability"	650 (originally > 650)	13.08.2021 to 16.08.2021
All	Total	1841	

Results

Summary of the selected articles

Large numbers of studies are found associated with China (20.00%), followed by the United States of America (18.33%). Figure 6.2 shows the geographical distribution of the studies selected in this study. The number of articles in the developing and underdeveloped countries of the Global South is very low. Therefore, it reveals the importance of this chapter to reveal the methods, tools, and techniques in a single study, which will encourage more researches in this field in the Global South.

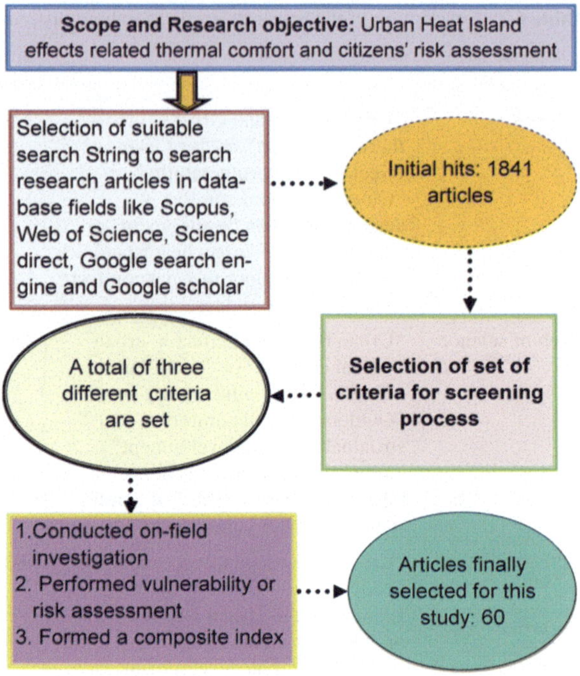

Fig. 6.1 Flow chart describing the selection process of the research articles in this study

Concerning the region, Asia possesses the highest number of articles (35.00%), followed by North America (21.67%), and Europe (13.33%).

Summary of the Research Outcomes

Formulation of Mathematical Equation Index

Details analysis and observation of the selected articles revealed that researches were mainly focused to devise different composite indexes to represent the vulnerability of humans to the UHI effect or thermal condition (Bhattacharjee et al. 2019; Cheng et al. 2020; Bassoud et al. 2021). It is observed that the models performed based on the energy balance and atmospheric data are very complex and require expert skills (Cheng and Brown 2020). Further, the availability of data that is essential to quantify the index value is also a major concern in this regard. A detailed assessment of different outdoor human thermal comfort models was presented by Coccolo et al. (2016), which can be categorized into three categories. The first group included the indexes based on energy balance parameters such as Universal Thermal Climate Index (UTCI), Comfort Formula (COMFA), Predicted Mean Vote (PMV), and Physiologically Equivalent Temperature (PET) (Manavvi and Rajasekar 2021; Ghani et al.

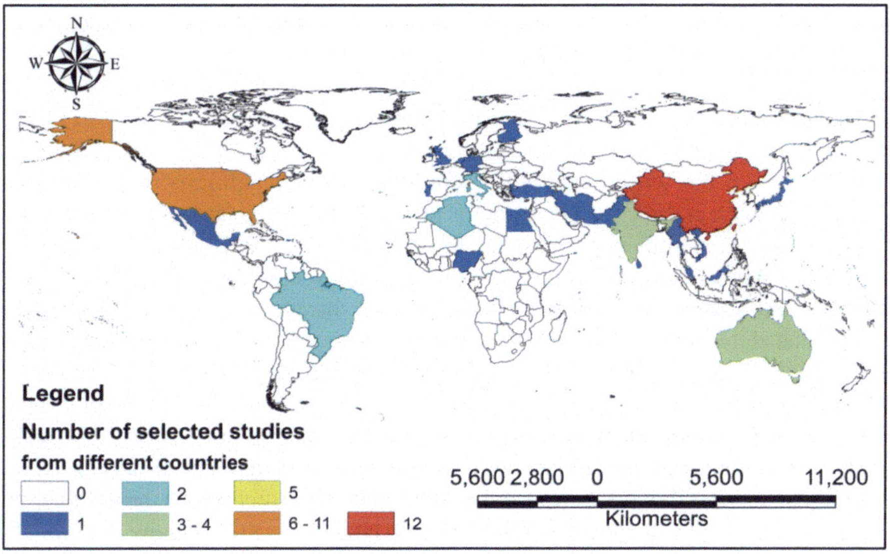

Fig. 6.2 Geographical distribution of the number of selected studies

2021; Krüger et al. 2020; Jia and Wang 2021; Lian et al. 2020). The second category of indexes belongs to the empirical group which used on-field investigations and monitoring to define human thermal comfort. In this regard, different statistical tools such as linear regressions were rigorously used to generate the outcomes (Das and Das 2020). The third group of indexes focused on atmospheric elements like air temperature, humidity, etc. It generally overlooks the parameters related to human conditions. It tried to focus on the outdoor thermal condition by representing the atmospheric situations. These indexes include operative temperature and mean radiant temperature (Tmrt), Wet Bulb Globe Temperature Index (WBGT), and Heat Index (HI) (Cheng and Brown 2020; Fang et al. 2019; Zune et al. 2020). A lot of existing researches are available that applied these indexes successfully. A general case study of these studies showed that thermal indexes following the energy balance equation were mainly useful for the analysis of the thermal sensation of humans in different climate regions (Njoku and Daramola 2019). On the other side, linear equations based on atmospheric data are mainly beneficial for meteorological predictions. It is also found useful for mapping the trend of thermal comfort over time (Antonini et al. 2020). However, this study observed that the empirical indexes are most widely used as a reliable tool to explore the perception of humans to outdoor thermal comfort (Kalisa et al. 2018; Li et al. 2021). It also enables the researchers to explore the environmental factors that affect the spatial thermal pattern.

In recent studies, the empirical models were mainly applied to identify the vulnerability of the populations to heat stress through formulating a vulnerability index model (Hammer et al. 2020; Hulley et al. 2019). A total number of 21 articles among the selected articles were found to adopt vulnerability models. In this regard, the most popular vulnerability equation to determine the vulnerability is represented as Eq. (6.1) (Zhang et al. 2018; Zuhra et al. 2019).

$$Vulnerability = Sensitivity + Exposure + Adaptive\,Capacity \tag{6.1}$$

In Eq. (6.1), the term 'sensitivity' represents those conditions or factors that increased the vulnerability of the residents to heat stress. Such factors are the proportion of the aged population, pre-existing health issues, etc. (Wilson and Chakraborty 2019). The word 'exposure' refers to those external factors that affect directly the intensity and spatial distribution of heat. Mainly the physical and environmental factors that determine the potential heat exposures are land surface temperature, green space coverage, built-up density, annual temperature growth, etc. (Zuhra et al. 2019). 'Adaptive Capacity' deals with the ability of the concerned to adapt to heat exposure. This includes different amenities or facilities such as air conditioners, technologies that can mitigate the increased heat exposure (Mallen et al. 2019). Apart from Eq. (6.1), a total of 6 articles were also observed that focused on the delineation of potential risky zones concerning heat stress (Hua et al. 2021; Huang et al. 2018; Navarro-Estupiñan et al. 2020; Zheng et al. 2020). Their applied methodology can be broadly expressed in the form of Eq. (6.2).

$$Risk = Hazard + Exposure + Vulnerability \tag{6.2}$$

In Eq. (6.2), the term 'hazard' represents extreme heat events. In this regard, the most commonly applied factor is the land surface temperature (LST) (Hua et al. 2021; Zheng et al. 2020). The definition of 'exposure' is almost the same as in Eq. (6.1). 'Vulnerability' represents the absence of effective elements to cope with the heat event. It mainly comprises the socioeconomic elements of the population (Navarro-Estupiñan et al. 2020).

Database Source and Applied Methodology

In the last four years, several studies were executed in the field of the assessment of human comfort to heat stress. It is observed that these studies included several environmental and human factors to achieve their objectives. A broad summary and comparison of these studies identified three main domains from where all the factors were primarily selected. These are socio-economic aspects, health aspects, and environmental aspects. The most trusted source to acquire information on environmental factors is satellite images. In this regard, the Landsat series satellite images are found as the most popular and widely used due to having a wide temporal span and free-of-cost availability. These set of imageries come with moderate resolution (30 m). However, it is useful enough to derive accurate land surface temperature (LST) and normalized difference vegetation index (NDVI). Some studies are also observed that have used finer remote sensing information such as LiDAR dataset or orthophotos (Mutani and Todeschi 2020; Mijani et al. 2020). However, it may incur some financial burden to the researchers. The LiDAR data is mainly used to obtain the vegetation height through the application of point clouds. Information on housing conditions and availability of different assets like an air conditioner can be availed from the Census data or the local respective offices. The majority of the studies favoured census data or official national household survey data to acquire the socio-economic data (Tran et al. 2020; Méndez-Lázaro et al. 2018). Similarly, health-related data can be obtained from the national community health survey data or national family health survey data (Zheng et al. 2020; Song et al. 2020).

Most of the studies used a wide variety of indicators to represent the vulnerability to heat stress. In this regard, researchers have provided a special emphasis on the weighting of the indicators to formulate a composite index that can capture the influence of each factor adequately. The weighting procedure of the existing studies can be broadly classified into two categories i.e. expert-based weighting (Guo et al. 2019) and weighting through the application of suitable statistical techniques (Das and Das 2020). In the case of an expert-based weighting scheme, the researchers mainly used equal-weighted (Guo et al. 2019) to all the indicators or hierarchical weighting of the indicators following the analytic hierarchy process (Tran et al. 2020). One study was also found that applied the Delphi method to obtain the expert weights (Räsänen et al. 2019). In the case of the statistical weighting procedure, most of the studies applied principal component analysis (PCA) (Song et al. 2020; Conlon et al. 2020; Nayak et al. 2018). It helps to reduce the large set of a dataset into a few representative components which adequately represent the information of the larger ones. Finally, the scores of the principal components can be summarized to obtain the composite form. After generating the composite index value, most of the studies normalized the composite score either from 0 to 1 or from 0 to 10. Then, a suitable classification scheme mainly equal interval is applied to demarcate five or seven vulnerability classes.

Discussion

This study tried to highlight the available resources that can be applied for the investigation of the effect of increased heat stress on urban communities. This research aims to deliver a state-of-the-art systematic review to focus on the current research techniques on this theme and to show the limitations and barriers that exist in the successful assessment of outdoor thermal comfort particularly in the context of urban areas. Through investigation of the existing researches revealed that the issue related to outdoor thermal condition is a complex one, wherein multiple layers along with the perception of the human played a crucial role in determining human comfort over space and time.

This literature review pointed out the application of several indexes and approaches in this field. However, it is also observed that they are mainly applied to address a particular context under some specific variables. The main limitation is observed in the consideration of the dynamic human body's adaptive response

(Li and Liu 2020). It also becomes crucial as most of the indexes emphasized the balance of energy between the human body and the environment. In recent studies, the indexes started to provide more emphasis on human behavior and perception. However, limited application and discussion for the potential outcome somehow made the studies slightly less relevant for future research (Antonini et al. 2020).

Further, the studies highlighted that the perception of the people regarding the environment varies individually from each other. Mainly, it relies on the 'pre-existing knowledge' regarding a specific situation and the psychological factors associated with the thermal perception (Nikolopoulou and Steemers 2003). Studies revealed that this mechanism is associated with four different aspects such as psychological, physical, physiological, and social (Antonini et al. 2020). This study suggests the incorporation of all these four aspects for representing the ideal framework of outdoor thermal comfort assessment.

It is observed that no significant outdoor thermal comfort model is available that included human sensitivity as an important variable with special emphasis. However, progress is also observed in this field as few studies were found that provided importance on psychological and behavioral factors and tried to explore these (Rathmann et al. 2020; Silva et al. 2020). Integration of energy balance with the human variables will surely deliver more productive models which will enhance the trustworthiness of the outcomes. Therefore, despite having several complexities in explaining this relation, studies should focus on designing an appropriate framework that can address this complex association. It will help the researchers to develop more fruitful simulation modeling tools. It will also help to attain the outdoor thermal comfort results effectively by combining the factors associated with the climatic conditions and the factors affecting human sensitivity to the environmental stimuli. Further, a deeper interdisciplinary study is essential to establish the relation between socio-economic life and thermal comfort in the outdoor space of the city.

All the existing research articles that were selected in this study examined different local aspects for inking the thermal comfort index. However, the categorization and application of different factors varied as per the objectives of the study. For example, studies were observed that used land surface temperature as a factor to measure the exposure (Zuhra et al. 2019), while the same factor is used as a measure of adaptive capacity in other studies (Zhang et al. 2018). Similarly, income and level of education are considered as two important parameters that can help to cope up with heat stress. Some studies used these two as adaptive capacity, while some used these to measure the sensitivity (Zuhra et al. 2019; Nayak et al. 2018). Building conditions such as the height of the building, asset possession (air conditioner), access to water, etc., are considered either as vulnerability or sensitivity indicators (Hulley et al. 2019; Tran et al. 2020). The bottom line is the selection of variables to assess the heat stress and thermal comfort needs to be relevant and accurate. However, their classification scheme can vary across space and time. This study also observed that further improvements can be done concerning the measurement accuracy of the climatic factors. Particularly, in the case of assessing the maximum and minimum temperature, previous studies noted the insufficiency of land surface temperature (Cheng et al. 2021). In this regard, it is suggested to use a combination of satellite data and ground station monitoring data to get the effective near-surface temperature. Further, the land surface temperature is mainly associated with terrestrial radiation. However, for the energy budget exchange model between the environment and the human, incident solar radiation is considered the most crucial factor that affects the human thermal comfort level. Therefore, it urges a more sophisticated analysis that can consider more efficient environmental elements for the energy exchange between the environment and the human body.

All the individual factors are mainly weighted following two ways for the construction of composite heat-related vulnerability or risk mapping. The first way deals with the equal or unequal weight assignment to the factors based on the judgment of the researcher/expert. However, it is important to mention that this weight assignment procedure may lead to biases due to a lack of empirical ground (Guo et al. 2019). Further, it also failed to explore the causal relationship and the role of each factor to the heat-related risks. In this regard, it is observed that statistical methods like principal component analysis were better useful to explore the causal relationship (Liu et al. 2020). PCA can deliver the variance and relative weight of each factor effectively, which helps to understand the role of each factor in the final composite index. Besides, it also reduces the probability of biases in the modeling. Some studies were also observed that tried to validate their outcome. However, no standard criteria or universal validation scheme were observed for validation. Some studies tried to show the correlation between heats related mortality rate and high heat-related risk areas. A few studies tried to show the same relation through estimation of r square value (Conlon et al. 2020). Besides, the studies were generally executed at a finer scale based on the remote sensing techniques, while the mortality-related datasets are generally associated with larger spatial scales. Further, it is very difficult to link mortality or morbidity with heat-related incidents due to the involvement of several underlying factors. Therefore, this study suggests the incorporation of a wide range of diverse datasets for validation such as hospital inpatient and outdoor patient records related to heat stress, frequency of social media posts, field surveys, and official health records.

This study highlighted some key aspects which need to be considered before taking on an effective thermal comfort assessment study. First, researchers need to understand the basic equation properly of the tools or methods for acquiring the best approach concerning the research objectives. Second, in modeling, it is observed that human sensation or their behavior plays a crucial role to determine thermal comfort. Third, a transdisciplinary approach is needed to assess the human thermal comfort and risk involving, socio-economic, behavioral, environmental, and physical elements. Fourth, understanding the outcomes and explanation to the findings including validation is very essential to get increased attention of the policymakers and urban planners.

In case of the selection of the indicators, this study recommends selection of commonly used indicators such as the proportion of vulnerable populations (young children and elderly population), socio-economic parameters like income, availability of air conditioner, environmental factors like the proportion of green space cover, and urban morphology related conditions like building height, sky view factor, etc. The main advantage of using commonly used indicators is that their causal associations with heat-related risk are already established by the existing researches. However, a deeper understanding of the energy exchanges between the environment, landscape, and humans is essential for the selection of a set of relevant local parameters. It will enable the researchers to highlight the areas where immediate spatial planning is essential and will also provide effective policy implications in this regard. In the case of assessing the role of different factors, statistical methods are better to employ. A local level statistical technique such as geographically weighted principal component analysis (GWPCA) or geographically weighted regression (GWR) is often more suitable than the global form of PCA and regression (Harris et al. 2011; Basu and Das 2021a). It helps to bring out the local spatial variation of each factor. This helps to reveal the importance of different indicators region-specific. For example, if, in any particular location of a city, environmental indicators are found more crucial in determining the outdoor thermal comfort, then environmental planning can be advised to cope with the situation in that location. Similarly, if urban morphology is found responsible, then policies should be adopted regarding urban design. This summarized direction on heat-related vulnerability and risk mapping can be adopted for future research in the small and medium-class urban agglomeration of the Global South. It may help to explore the heat related vulnerability in these regions in a simpler and lucid way.

Conclusion

All the methods for the identification of vulnerable population groups or areas to the extreme heat stress are mainly done by adopting the psychological, socioeconomic, and environmental factors. These methods were found suitable enough to take on adequate policy implementations. The main objective of these researches is to evaluate the heat-related vulnerability of the populations. For this, data were taken following theory, field observation, experiments, and survey. However, the categorization of selected indicators varied across spaces based on the research objectives and approach of the studies. Mainly two different schemes were observed to assign a weighted value to the indicators. These are expert judgment and statistical techniques. Application of statistical techniques was found more suitable due to having less probability to the biases. Among the statistical techniques, principal component analysis is found as the most commonly used. In any scientific research, validation of the outcome is considered an essential part of justifying the accuracy of results. However, only a few studies were observed that tried to validate their outcome. In this regard, the best practice followed comparison with the heat-related health issues such as mortality and morbidity. However, it possesses several limitations. This study suggested applying a diverse dataset for this purpose. Most of the researches were found beyond the tropics in both the hemisphere. Only a few studies were present that were conducted on a tropical city. Lack of studies in the hot underdeveloped and developing countries is a major concern as it can highlight new possible insights and themes in this field. This study encourages more studies in this field mainly concerning the medium class urban agglomerations as these are the emerging economical centers of these countries. Following this, the review proposed a direction for future research in this field. It will help the researchers to apply effective indexes to explore the risks related to the urban heat island effects. It in turn will assist the policymakers and urban planners to take on effective strategies to cope up with the situation.

Authors' Contribution First Author: Conceptualization, Methodology, Data Collection, Writing- Original Draft; **Second Author**: Supervision, Visualization, Validation, Writing-Formatting and Editing.

Ethical Approval and Consent to Participate Not Applicable.

Funding No funding.

Data Availability Not Applicable.

Conflict of Interest There is no potential conflict of interest.

References

Abulibdeh A (2021) Analysis of urban heat island characteristics and mitigation strategies for eight arid and semi-arid gulf region cities. Environ Earth Sci 80(7):1–26. https://doi.org/10.1007/s12665-021-09540-7

Antonini E, Vodola V, Gaspari J, De Giglio M (2020) Outdoor wellbeing and quality of life: a scientific literature review on thermal comfort. Energies 13(8):2079. https://doi.org/10.3390/en13082079

Aram F, García EH, Solgi E, Mansournia S (2019) Urban green space cooling effect in cities. Heliyon 5(4):e01339. https://doi.org/10.1016/j.heliyon.2019.e01339

Athukorala D, Murayama Y (2021) Urban heat island formation in Greater Cairo: spatio-temporal analysis of daytime and nighttime land surface temperatures along the urban-rural gradient. Remote Sens 13(7):1396. https://doi.org/10.3390/rs13071396

Bassoud A, Khelafi H, Mokhtari AM, Bada A (2021) Evaluation of summer thermal comfort in arid desert areas. Case study: old adobe building in Adrar (South of Algeria). Build Environ 205:108140. https://doi.org/10.1016/j.buildenv.2021.108140

Basu T, Das A (2021a) Formulation of deprivation index for identification of regional pattern of deprivation in rural India. Socioecon Plann Sci 74:100924. https://doi.org/10.1016/j.seps.2020.100924

Basu T, Das A (2021b) Systematic review of how eco-environmental transformation due to urbanization can be investigated in the sustainable development of Indian Cities. Environ Chall 100099. https://doi.org/10.1016/j.envc.2021.100099

Bhattacharjee S, Gerasimova E, Imbert C, Tencar J, Rotondo F (2019) Assessment of different methodologies for mapping urban heat vulnerability for Milan, Italy. IOP Conf Ser Earth Environ Sci 290(1):012162

Cao L, Li P, and Zhang L (2008) Impact of impervious surface on urban heat island in Wuhan, China. In: International conference on earth observation data processing and analysis (ICEODPA), vol 7285. International Society for Optics and Photonics, p 72855H. https://doi.org/10.1117/12.815911

Cheng W, Brown RD (2020) An energy budget model for estimating the thermal comfort of children. Int J Biometeorol 64(8):1355–1366. https://doi.org/10.1007/s00484-020-01916-x

Cheng W, Spengler JO, Brown RD (2020) A comprehensive model for estimating heat vulnerability of young athletes. Int J Environ Res Public Health 17(17):6156. https://doi.org/10.3390/ijerph17176156

Cheng W, Li D, Liu Z, and Brown RD (2021) Approaches for identifying heat-vulnerable populations and locations: a systematic review. Sci Total Environ 149417. https://doi.org/10.1016/j.scitotenv.2021.149417

Coccolo S, Kämpf J, Scartezzini JL, Pearlmutter D (2016) Outdoor human comfort and thermal stress: a comprehensive review on models and standards. Urban Clim 18:33–57. https://doi.org/10.1016/j.uclim.2016.08.004

Cohen B (2006) Urbanization in developing countries: current trends, future projections, and key challenges for sustainability. Technol Soc 28(1–2):63–80. https://doi.org/10.1016/j.techsoc.2005.10.005

Conlon KC, Mallen E, Gronlund CJ, Berrocal VJ, Larsen L, O'neill MS (2020) Mapping human vulnerability to extreme heat: a critical assessment of heat vulnerability indices created using principal components analysis. Environ Health Perspect 128(9):097001. https://doi.org/10.1289/EHP4030

da Silva I, de Almeida DS, Hashimoto EM, Martins LD (2020) Risk assessment of temperature and air pollutants on hospitalizations for mental and behavioral disorders in Curitiba, Brazil. Environ Health 19(1):1–11. https://doi.org/10.1186/s12940-020-00606-w

Das A, Basu T (2020) Assessment of peri-urban wetland ecological degradation through importance-performance analysis (IPA): a study on Chatra Wetland, India. Ecol Ind 114:106274. https://doi.org/10.1016/j.ecolind.2020.106274

Das M, Das A (2020) Exploring the pattern of outdoor thermal comfort (OTC) in a tropical planning region of Eastern India during summer. Urban Clim 34:100708. https://doi.org/10.1016/j.uclim.2020.100708

Dewan A, Kiselev G, Botje D, Mahmud GI, Bhuian MH, Hassan QK (2021) Surface urban heat island intensity in five major cities of Bangladesh: patterns, drivers and trends. Sustain Cities Soc 71:102926. https://doi.org/10.1016/j.scs.2021.102926

Dr S, Kalia A (2021) Effect of vegetation cover and shading on thermal and visual comfort in institutional campus outdoor spaces in India. Int J Recent Technol Eng 10(2):17–20. https://doi.org/10.35940/ijrte.B6078.0710221

Fang Z, Feng X, Liu J, Lin Z, Mak CM, Niu J, Tse KT, Xu X (2019) Investigation into the differences among several outdoor thermal comfort indices against field survey in subtropics. Sustain Cities Soc 44:676–690. https://doi.org/10.1016/j.scs.2018.10.022

Farrell K (2017) The rapid urban growth triad: a new conceptual framework for examining the urban transition in developing countries. Sustainability 9(8):1407. https://doi.org/10.3390/su9081407

Ghani S, Mahgoub AO, Bakochristou F, ElBialy EA (2021) Assessment of thermal comfort indices in an open air-conditioned stadium in hot and arid environment. J Build Eng 40:102378. https://doi.org/10.1016/j.jobe.2021.102378

Guo X, Huang G, Jia P, Wu J (2019) Estimating fine-scale heat vulnerability in Beijing through two approaches: spatial patterns, similarities, and divergence. Remote Sens 11(20):2358. https://doi.org/10.3390/rs11202358

Hammer J, Ruggieri DG, Thomas C, Caum J (2020) Local extreme heat planning: an interactive tool to examine a heat vulnerability index for Philadelphia, Pennsylvania. J Urban Health 97:519–528. https://doi.org/10.1007/s11524-020-00443-9

Harris P, Brunsdon C, Fotheringham AS (2011) Links, comparisons and extensions of the geographically weighted regression model when used as a spatial predictor. Stoch Env Res Risk Assess 25(2):123–138. https://doi.org/10.1007/s00477-010-0444-6

Hassan T, Zhang J, Prodhan FA, Pangali Sharma TP, Bashir B (2021) Surface Urban heat islands dynamics in response to LULC and vegetation across South Asia (2000–2019). Remote Sens 13(16):3177. https://doi.org/10.3390/rs13163177

Hazemba M, Halog A (2021) Systematic review of how environmental management policies are incorporated into national development plans in order to achieve sustainable development. Environ Chall 100041

Hua L, Zhang X, Nie Q, Sun F, Tang L (2020) The impacts of the expansion of urban impervious surfaces on urban heat islands in a coastal city in China. Sustainability 12(2):475. https://doi.org/10.3390/su12020475

Hua J, Zhang X, Ren C, Shi Y, Lee TC (2021) Spatiotemporal assessment of extreme heat risk for high-density cities: a case study of Hong Kong from 2006 to 2016. Sustain Cities Soc 64:102507. https://doi.org/10.1016/j.scs.2020.102507

Huang KT, Yang SR, Matzarakis A, Lin TP (2018) Identifying outdoor thermal risk areas and evaluation of future thermal comfort concerning shading orientation in a traditional settlement. Sci Total Environ 626:567–580. https://doi.org/10.1016/j.scitotenv.2018.01.031

Hulley G, Shivers S, Wetherley E, Cudd R (2019) New ECOSTRESS and MODIS land surface temperature data reveal fine-scale heat vulnerability in cities: a case study for Los Angeles County, California. Remote Sens 11(18):2136. https://doi.org/10.3390/rs11182136

Iyer D, Steele A (2015) UNDP-UNEP poverty-environment initiative—2015—stg-wedocs.unep.org

Jain S, Sannigrahi S, Sen S, Bhatt S, Chakraborti S, Rahmat S (2020) Urban heat island intensity and its mitigation strategies in the fast-growing urban area. J Urban Manag 9(1):54–66. https://doi.org/10.1016/j.jum.2019.09.004

Jia S, Wang Y (2021) Effect of heat mitigation strategies on thermal environment, thermal comfort, and walkability: a case study in Hong Kong. Build Environ 107988. https://doi.org/10.1016/j.buildenv.2021.107988

Kalisa E, Fadlallah S, Amani M, Nahayo L, Habiyaremye G (2018) Temperature and air pollution relationship during heatwaves in Birmingham, UK. Sustain Cities Soc 43:111–120. https://doi.org/10.1016/j.scs.2018.08.033

Kandya A, Mohan M (2018) Mitigating the Urban Heat Island effect through building envelope modifications. Energy Build 164:266–277. https://doi.org/10.1016/j.enbuild.2018.01.014

Khan M (2020) Assessment of intra-city urban heat island effect in relation to vulnerable stakeholders in Karachi, Pakistan via LCZ classification, land surface temperature analysis and traverse surveys. http://urn.fi/URN:NBN:fi:amk-2020102221540

Krüger EL, Silva TJV, da Silveira Hirashima SQ, da Cunha EG, Rosa LA (2020) Calibrating UTCI'S comfort assessment scale for three Brazilian cities with different climatic conditions. Int J Biometeorol 1–10. https://doi.org/10.1007/s00484-020-01897-x

Kumar A, Agarwal V, Pal L, Chandniha SK, Mishra V (2021) Effect of land surface temperature on urban heat island in Varanasi City. India J 4(3):420–429. https://doi.org/10.3390/j4030032

Lee K, Kim Y, Sung HC, Ryu J, Jeon SW (2020) Trend analysis of urban heat island intensity according to urban area change in Asian mega cities. Sustainability 12(1):112. https://doi.org/10.3390/su12010112

Li J, Liu N (2020) The perception, optimization strategies and prospects of outdoor thermal comfort in China: a review. Build Environ 170:106614. https://doi.org/10.1016/j.buildenv.2019.106614

Li K, Zhang H, Zhao L (2021) Thermal comfort interventions of landscape elements in a humid and subtropical residential area in China. J Asian Archit Build Eng 1–18. https://doi.org/10.1080/13467581.2021.1900857

Lian Z, Liu B, Brown RD (2020) Exploring the suitable assessment method and best performance of human energy budget models for outdoor thermal comfort in hot and humid climate area. Sustain Cities Soc 63:102423. https://doi.org/10.1016/j.scs.2020.102423

Liu X, Yue W, Yang X, Hu K, Zhang W, and Huang M (2020) Mapping urban heat vulnerability of extreme heat in Hangzhou via comparing two approaches. Complexity. https://doi.org/10.1155/2020/9717658

Lu L, Weng Q, Xiao D, Guo H, Li Q, Hui W (2020) Spatiotemporal variation of surface urban heat islands in relation to land cover composition and configuration: a multi-scale case study of Xi'an, China. Remote Sens 12(17):2713. https://doi.org/10.3390/rs12172713

Lyu R, Clarke KC, Zhang J, Jia X, Feng J, Li J (2019) The impact of urbanization and climate change on ecosystem services: a case study of the city belt along the Yellow River in Ningxia, China. Comput Environ Urban Syst 77:101351. https://doi.org/10.1016/j.compenvurbsys.2019.101351

Mallen E, Stone B, Lanza K (2019) A methodological assessment of extreme heat mortality modeling and heat vulnerability mapping in Dallas, Texas. Urban Clim 30:100528. https://doi.org/10.1016/j.uclim.2019.100528

Manavvi S, Rajasekar E (2021) Evaluating outdoor thermal comfort in "Haats"—the open air markets in a humid subtropical region. Build Environ 190:107527. https://doi.org/10.1016/j.buildenv.2020.107527

Méndez-Lázaro P, Muller-Karger FE, Otis D, McCarthy MJ, Rodríguez E (2018) A heat vulnerability index to improve urban public health management in San Juan, Purerto Rico. Int J Biometeorol 62(5):709–722. https://doi.org/10.1007/s00484-017-1319-z

Mijani N, Alavipanah SK, Firozjaei MK, Arsanjani JJ, Hamzeh S, Weng Q (2020) Modeling outdoor thermal comfort using satellite imagery: a principle component analysis-based approach. Ecol Ind 117:106555. https://doi.org/10.1016/j.ecolind.2020.106555

Mirzaei M, Verrelst J, Arbabi M, Shaklabadi Z, Lotfizadeh M (2020) Urban heat island monitoring and impacts on citizen's general health status in Isfahan metropolis: a remote sensing and field survey approach. Remote Sens 12(8):1350. https://doi.org/10.3390/rs12081350

Mutani G, Todeschi V (2020) The effects of green roofs on outdoor thermal comfort, urban heat island mitigation and energy savings. Atmosphere 11(2):123. https://doi.org/10.3390/atmos11020123

Navarro-Estupiñan J, Robles-Morua A, Díaz-Caravantes R, Vivoni ER (2020) Heat risk mapping through spatial analysis of remotely-sensed data and socioeconomic vulnerability in Hermosillo, México. Urban Clim 31:100576. https://doi.org/10.1016/j.uclim.2019.100576

Nayak SG, Shrestha S, Kinney PL, Ross Z, Sheridan SC, Pantea CI, Hsu WH, Muscatiello N, Hwang SA (2018) Development of a heat vulnerability index for New York State. Public Health 161:127–137. https://doi.org/10.1016/j.puhe.2017.09.006

Nikolopoulou M, Steemers K (2003) Thermal comfort and psychological adaptation as a guide for designing urban spaces. Energy Build 35(1):95–101. https://doi.org/10.1016/S0378-7788(02)00084-1

Njoku CA, Daramola MT (2019) Human outdoor thermal comfort assessment in a tropical region: a case study. Earth Syst Environ 3(1):29–42. https://doi.org/10.1007/s41748-019-00090-4

Pickering C, Byrne J (2014) The benefits of publishing systematic quantitative literature reviews for Ph.D. candidates and other early-career researchers. High Educ Res Dev 33(3):534–548. https://doi.org/10.1080/07294360.2013.841651

Priyankara P, Ranagalage M, Dissanayake D, Morimoto T, Murayama Y (2019) Spatial process of surface urban heat island in rapidly growing Seoul metropolitan area for sustainable urban planning using Landsat data (1996–2017). Climate 7:110. https://doi.org/10.3390/cli7090110

Räsänen A, Heikkinen K, Piila N, Juhola S (2019) Zoning and weighting in urban heat island vulnerability and risk mapping in Helsinki, Finland. Reg Environ Change 19(5):1481–1493. https://doi.org/10.1007/s10113-019-01491-x

Rathmann J, Beck C, Flutura S, Seiderer A, Aslan I, André E (2020) Towards quantifying forest recreation: exploring outdoor thermal physiology and human well-being along exemplary pathways in a central European urban forest (Augsburg, SE-Germany). Urban for Urban Green 49:126622. https://doi.org/10.1016/j.ufug.2020.126622

Roberts B, Hohmann R (2014) Secondary Cities: managing urban land governance systems. Paper presented at the world bank and land poverty 2014: uses of spatial data in urban management

Rose AL (2010) Impact of urbanization on the thermal comfort conditions in the hot humid city of Chennai, India. In: Recent advances in space technology services and climate change 2010 (RSTS & CC-2010). IEEE, pp 262–267. https://doi.org/10.1109/RSTSCC.2010.5712856

Song J, Huang B, Kim JS, Wen J, Li R (2020) Fine-scale mapping of an evidence-based heat health risk index for high-density cities: Hong Kong as a case study. Sci Total Environ 718:137226. https://doi.org/10.1016/j.scitotenv.2020.137226

Soydan O (2020) Effects of landscape composition and patterns on land surface temperature: urban heat island case study for Nigde, Turkey. Urban Clim 34. Article 100688. https://doi.org/10.1016/j.uclim.2020.100688

Tacoli C, McGranahan G, Satterthwaite D (2015) Urbanisation, rural-urban migration and urban poverty. Human Settlements Group, International Institute for Environment and Development. London. http://pubs.iied.org/16574IIED

Tran DN, Van Quang D, Nguyen VT, Khan A, Thai PK, Huang C, Cordia C, Schak E, Dung P (2020) Spatial patterns of health vulnerability to heatwaves in Vietnam. Int J Biometeorol 64(5):863–872. https://doi.org/10.1007/s00484-020-01876-2

Tumbe C (2016) Urbanization, demographic transition and the growth of cities in India. International Growth Centre. Reference number: C-35205-INC-1

Ulpiani G (2021) On the linkage between urban heat island and urban pollution island: three-decade literature review towards a conceptual framework. Sci Total Environ 751:141727. https://doi.org/10.1016/j.scitotenv.2020.141727

United Nations (2014) World urbanization prospects: the 2014 revision. Highlights, New York

Wang W, Wu T, Li Y, Xie S, Han B, Zheng H, Ouyang Z (2020) Urbanization impacts on natural habitat and ecosystem services in the Guangdong-Hong Kong-Macao "megacity". Sustainability 12(16):6675. https://doi.org/10.3390/su12166675

Westendorff VE (2020) Role of trees in mitigating urban heat island in Charlotte, North Carolina, USA. In: WIT transactions on ecology and the environment, vol 245, pp 73–83. https://doi.org/10.2495/EID200081

Wilson B, Chakraborty A (2019) Mapping vulnerability to extreme heat events: lessons from metropolitan Chicago. J Environ Planning Manage 62(6):1065–1088. https://doi.org/10.1080/09640568.2018.1462475

Zhang XQ (2016) The trends, promises and challenges of urbanisation in the world. Habitat Int 54:241–252. https://doi.org/10.1016/j.habitatint.2015.11.018

Zhang W, McManus P, Duncan E (2018) A raster-based subdividing indicator to map urban heat vulnerability: a case study in Sydney, Australia. Int J Environ Res Public Health 15(11):2516. https://doi.org/10.3390/ijerph15112516

Zheng M, Zhang J, Shi L, Zhang D, Pangali Sharma TP, Prodhan FA (2020) Mapping heat-related risks in Northern Jiangxi Province of China based on two spatial assessment frameworks approaches. Int J Environ Res Public Health 17(18):6584. https://doi.org/10.3390/ijerph17186584

Zuhra SS, Tabinda AB, Yasar A (2019) Appraisal of the heat vulnerability index in Punjab: a case study of spatial pattern for exposure, sensitivity, and adaptive capacity in megacity Lahore, Pakistan. Int J Biometeorol 63(12):1669–1682. https://doi.org/10.1007/s00484-019-01784-0

Zune M, Rodrigues L, Gillott M (2020) The vulnerability of homes to overheating in Myanmar today and in the future: a heat index analysis of measured and simulated data. Energy Build 223:110201. https://doi.org/10.1016/j.enbuild.2020.110201

Chapter 7
Evaluating the Sensitivity of Saffron Yield to Climate Change in Western Himalaya, India. A Study from Kashmir Valley

Iqra Binti Ayoub, Shoukat Ara, and Suhail A. Lone

Abstract Agriculture stands as a basis of economy in some major regions of the Himalaya, contributing substantially to the livelihoods of millions. However, the persistent alterations in meteorological variables, attributed to climate warming, have profoundly impacted agricultural productivity, disrupting the economic stability of the region. Kashmir Himalayas emerge prominently as a key global producer of saffron (*Crocus sativus* L.) cultivated within a specific and delicate micro-climate, is found vulnerable to the ramifications of evolving climatic conditions. In this study, we employed meteorological data (temperature and precipitation), coupled with saffron production statistics, to unravel the intricate relationship between changing climate patterns and saffron cultivation. The findings unveiled a noteworthy escalation in temperature, progressing from an increment of 0.1 °C per decade during the period spanning 1901 to 2014, to a more pronounced rise of about 0.3 °C per decade from 1951 to 2021. This temporal shift in climatic variables stands as a compelling factor exerting a profound influence on saffron production. Further analysis of the results underscored a discernible decline in saffron production over the past two decades. Specifically, the saffron yield dwindled from 15.95 metric tons in 1997 to 9.6 metric tons in 2015. The reduction was mirrored in the cultivated area, which contracted from 5707 ha in 1997 to 3674 ha in 2015. These findings illuminate the role of changing climate in the substantial decrease witnessed in saffron production. The study also endeavours to dissect the underlying issues impeding crop cultivation and consequently, the economic vitality of the region.

I. B. Ayoub (✉) · S. Ara
Division of Environmental Sciences, SKUAST-K, Shalimar, Srinagar 190025, India
e-mail: iqrabintiayoub@gmail.com

S. A. Lone
Department of Earth Sciences, University of Kashmir, Srinagar 190006, India

U. Mukhopadhyay et al. (eds.), *Climate Crisis, Social Responses and Sustainability*, Climate Change Management, https://doi.org/10.1007/978-3-031-58261-5_7

Introduction

Agriculture is one of the major sources of economy and plays an important role in socio-economic development of the region. Agricultural production has observed a laudable growth after 1948 (Bosworth and Collins 2008; Makwana 2013; Gulati et al. 2017; Pathak et al. 2022; Rao et al. 2022). India is one of the leading producers of agricultural crops in the world (Suresh and Mathur 2016; Narayan and Bhattacharya 2019; Mariappan and Zhou 2019; Dhulipala and Patil 2020). However, the agricultural production and productivity is critically affected by climatic variability. Due to recent climate warming, changes in meteorological variables including rainfall, ambient temperature, relative humidity, wind speed and wind direction have serious impact on agricultural production across globe (Huo et al. 2013; Zhang et al. 2017; Vicente-Serrano et al. 2020; Jain et al. 2022). Climate change has profound effect on various processes including: transpiration, evaporation and photosynthesis thereby affecting the most crop yields. Jammu and Kashmir a part of western Himalayas is a region enriched with various agricultural products, which is also the predominant sector of its economy supporting nearly 80% of the population (Dixit et al. 2014; Malik et al. 2018; Peerzada et al. 2021; Kumar et al. 2021). The Kashmir valley in India is renowned for its high quality saffron. Saffron, originated from the word "Zafaran," means yellow, is a charismatic spice with a rich historical background (Kafi et al. 2018; Pandey et al. 2020; Pandita 2021). The spice is grown across Europe and Asia. It serves dual purposes as both a flavour enhancer and a natural coloring agent in culinary applications, while also playing a pivotal role in industries related to dyeing, perfumery, and flavoring. Furthermore, saffron boasts numerous biological properties, including its potential as an anticancer, antimutagenic, and antioxidant agent. In the Kashmir region, saffron cultivation takes place in elevated areas locally called "wudur," characterized by glacio-fluvio-lacustrine deposits situated at altitudes ranging from 1585 to 2300 m above sea level, within temperate climatic conditions (Kanth et al. 2008). However, this region is not resistant to the impact of climate change, which present substantial challenges to the cultivation of saffron.

Climate change is altering the environmental conditions necessary for successful production affecting the yield, quality and overall sustainability in Kashmir Valley. Himalayan environment is very fragile to changing climatic conditions and from last few decades Himalaya has experienced significant effect on weather patterns, including acute temperature variability, changes in precipitation form and rise in extreme weather events (Lone et al. 2022a). Based on time series temperature and precipitation data, the studies have suggested that temperature has increased and precipitation has decreased during last 100 years (Lone et al. 2019, 2022b). The socio-economic development which is highly dependent upon the agriculture activities are dominantly controlled by water resources generated from snow and glacier cover. The studies have postulated that ~ 70% annual stream water discharge and groundwater recharge is sustained by water derived from the winter accumulated snow across western Himalayas (Lone et al. 2021, 2022a, b; Lone and Jeelani

2023). Thus, decrease in precipitation particularly during the winter season can have profound effect on the regional hydrology of the basin (Lone et al. 2021). These hydrological and hydrometeorological changes will ultimately impact the agriculture in general and saffron production in particular. Being, a source of economy to major section of people living in Kashmir Himalayas and keeping in view the importance of Saffron in medical and other uses the dynamics and response of saffron to changing climatic variables is least understood and remained a huge research gap. It becomes prerequisite that the regular and detailed monitoring of saffron production is critically important for sustaining the economic stability of the region. In order to fill this gap, the main aim of the present study is to evaluated the current declining rate of saffron cultivation in Kashmir valley using decadal saffron production data and meteorological parameters. The record of monitoring the saffron production can help us to better understand the response of saffron to changing climate on a long-term basis. The study can be useful to construct past climatic conditions and will add more insights to agriculture management and ecosystem restoration policies across the region.

Study Area

The Kashmir Valley, situated in the western Himalayas and forming a significant part of the Union Territory of Jammu and Kashmir, lies within the geographical coordinates ranging from 33° 20′ N to 34° 35′ N Latitude and 74° 10′ N to 75° 50′ N Longitude, covering an area of approximately 15,520 km^2 (Fig. 7.1).

This region experiences a temperate climate characterized into four seasons including; spring, summer, autumn, and winter. The climatic conditions in the Kashmir Valley is dominantly controlled by Western disturbances and Indian summer monsoons. receiving an average precipitation of around 686 mm (Lone et al. 2022a, b). The highest temperature and precipitation levels are typically recorded in July, with temperatures reaching about 24 °C and precipitation averaging 165 mm. Conversely, the lowest temperature and precipitation occur in January, with temperatures dropping to approximately – 4 °C and precipitation dwindling to around 12 mm in November. The Kashmir Valley is blessed with verdant forests at higher altitudes and terrestrial vegetation in the plains. The local hydrology plays a critical role in sustaining agricultural and horticultural activities, which, in turn, drive the region's economy. This hydrology heavily relies on the accumulation of winter snowfall, ensuring a continuous flow of streams and groundwater recharge throughout the year. Any alteration in the form or quantity of precipitation can have detrimental effects on the hydrological system, ultimately impacting the socio-economic development of the valley.

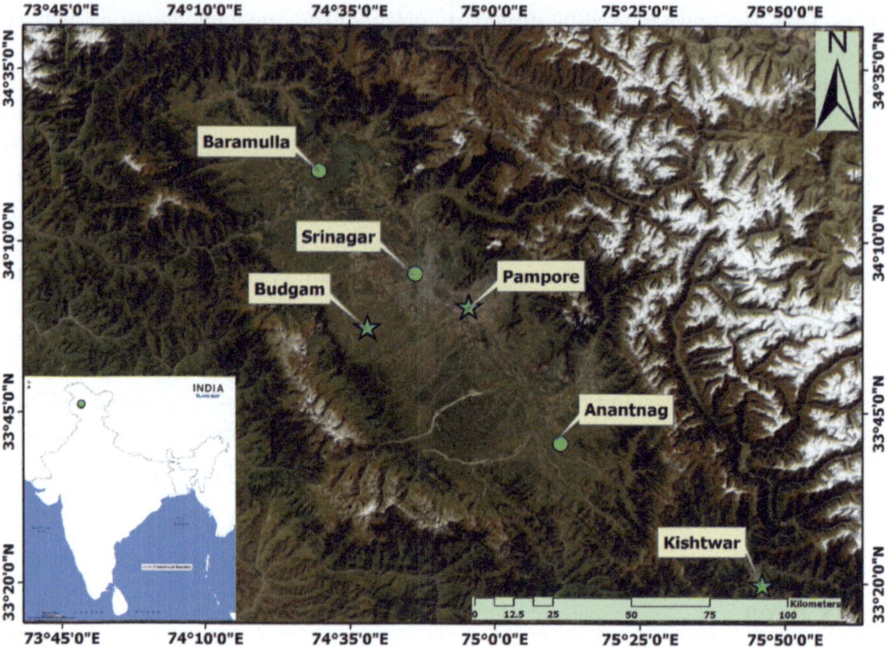

Fig. 7.1 Study area map. The stars are depicting the location of saffron producing areas in the region. The yellow circles are depicting the location of major cities in Kashmir Valley

Materials and Methods

The examination of meteorological variables, specifically long-term precipitation and temperature data spanning from 1901 to 2022, aimed to discern the evolving climatic patterns prevalent in the western Himalayas (Fig. 7.2).

The data utilized for this investigation was sourced from the Indian Meteorological Department (IMD) in Pune and the National Ocean and Atmospheric Administration (NOAA). Simultaneously, comprehensive data on saffron yield and cultivation practices over an extended period were procured from the Department of Agriculture in the Union Territory of Jammu & Kashmir. The analytical process involved conducting statistical assessments, data visualization through plotting, and accuracy evaluations of both meteorological and saffron cultivation and yield data. These tasks were accomplished using Microsoft Excel, version 2021, and Statistics 2014. The integration of statistical analyses and graphical representation facilitated a nuanced understanding of the long-term trends in precipitation and temperature, allowing for a more in-depth exploration of the relationship between climatic variables and saffron cultivation dynamics. This comprehensive approach not only shed light on the changing climatic patterns in the western Himalayas but also provided insights into their potential impacts on saffron cultivation in the region.

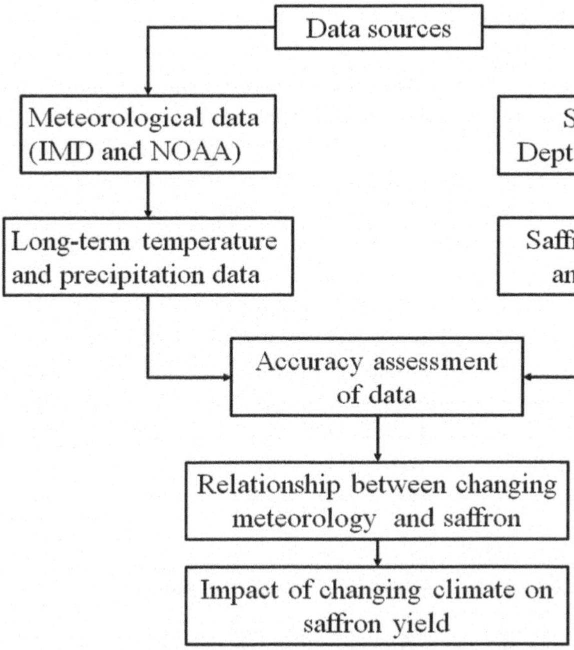

Fig. 7.2 Flowchart depicting the methodology adopted to understand the relationship between meteorological variable with saffron

Results and Discussion

Over an extended temporal span, an in-depth examination of atmospheric variables, specifically temperature and precipitation, uncovered notable fluctuations in the western Himalayas. Our investigation relied on climate data derived from two key sources: the Indian Meteorological Department (IMD) and the National Oceanic and Atmospheric Administration (NOAA). This comprehensive dataset served as the foundation for our analysis of various meteorological parameters in the region.

The temperature trends spanning from 1901 to 2022 unveiled intriguing patterns. While there was an overall average temperature increase of approximately 0.166 °C, it's crucial to recognize the non-linear nature of this progression. Initially, there were two distinct periods, from 1901 to 1930 and from 1974 to 1990, during which temperatures exhibited an upward trend at a rate of 0.02 °C per year. Intriguingly, there was a temporary deviation with a decline of 0.0002 °C per year from 1931 to 1973, followed by a significant resurgence at a rate of 0.10 °C per year from 1991 to 2022 (Fig. 7.3). Similarly, precipitation levels displayed a variable pattern over the years, with an average decrease of about 0.032 mm per year. However, a noteworthy exception occurred between 1910 and 1930, marked by an increase in annual precipitation by 0.8 mm per year (Fig. 7.3). Beyond these overarching trends, our analysis delved into distinct seasonal variations within the western Himalayas for both temperature and precipitation. In terms of temperature, the time series revealed marginal increases in summer temperatures by 0.006 °C, while winter temperatures exhibited a more substantial uptick of 0.08 °C. Conversely, precipitation patterns showed a nuanced picture across seasons. Winter witnessed a decrease of 0.18 mm per year, while other seasons, including spring (0.21 mm per year), summer (0.09 mm per year), and autumn (0.18 mm per year), experienced increases in precipitation. This detailed exploration not only sheds light on the long-term climatic dynamics but also highlights the intricate seasonal variations that contribute to the overall understanding of the meteorological landscape in the western Himalayas. It's worth noting that these findings align with conclusions drawn from other studies conducted in various parts of the Himalayas and in regions beyond, all of which emphasize the dynamic nature of meteorological parameters, specifically temperature and precipitation, in this ecologically sensitive area (Auer and Böhm 1994; Diaz and Bradley 1997; Sharma et al. 2000; Jamelli et al. 2004; Bhutiyani et al. 2007; Shekhar et al. 2010; Kohler et al. 2014; Rao et al. 2016; Bari et al. 2016; Ferreira et al. 2018; Shrestha et al. 2019; Lone et al. 2019, 2022a, b).

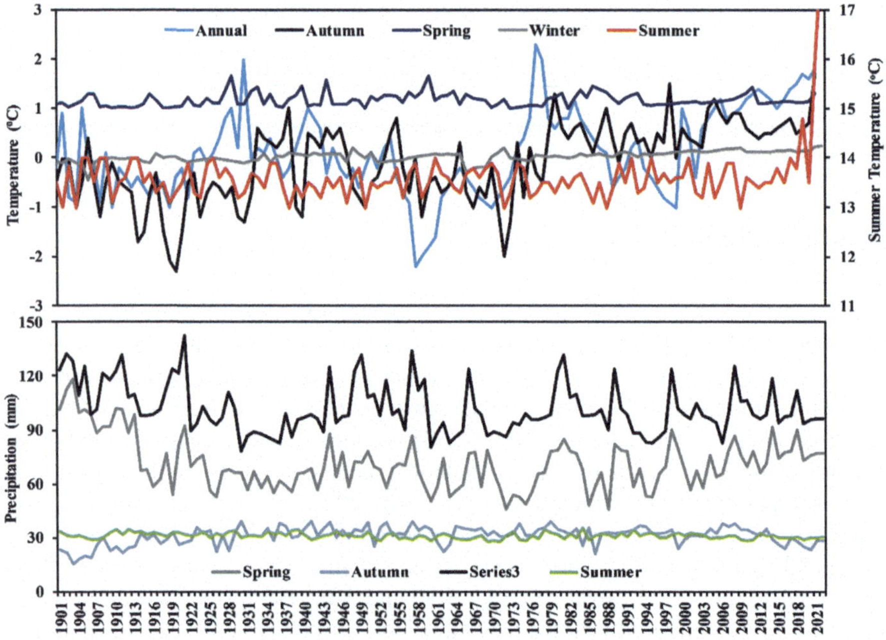

Fig. 7.3 Time-series plot showing the variability of mean annual temperature and precipitation from 1901 to 2022. The plots depicted that the mean annual temperature has increased by 1.5 °C decade^{-1} and mean annual precipitation has decreased by 0.032 mm year^{-1}

Saffron Cultivation and Yield

In our quest to comprehend the temporal dynamics of Saffron cultivation and yield data in response to changing climatic conditions, we meticulously gathered information from the Directorate of Agriculture, Jammu & Kashmir, spanning the last three decades. The statistical analysis, focusing on the years 1996 to 2019, revealed notable shifts in saffron cultivation trends. During this period, the saffron cultivation area witnessed a significant decline, plummeting from 5707 to 1323 ha, marking a reduction of over 33% in cultivation (Table 7.1; Fig. 7.4). Concurrently, the saffron production also experienced a decline, diminishing from 15.95 tonnes in 1996 to 9.46 tonnes in 2010. This dual decline in both cultivation area and production underscores the vulnerability of saffron cultivation to the changing climate conditions. In 2012, the saffron production in the Kashmir Valley covered a total area of 2469 ha, yielding an annual output of 10 metric tons. In stark contrast, approximately fifteen years earlier, in 1997, the saffron cultivation area was substantially larger at 5707 ha, resulting in an annual production of 15.95 metric tons. The year 2003–04 marked a period of exceptionally low productivity, with only 1.57 kg per ha, attributed to a severe drought experienced between 1999 and 2003 (Alam 2007). Examining the earlier years from 1990 to 1996, the data unveiled a relatively stable saffron cultivation area in Kashmir, ranging from 4036 to 4496 ha. During this period, annual production remained consistent, fluctuating between 13.0 and 14.1 metric tons, with productivity ranging from 2.90 to 3.21 kg per ha. This historical perspective emphasizes the resilience of saffron cultivation during a more stable climatic period and underscores the potential impacts of climate change on the cultivation dynamics observed in recent decades.

Numerous factors have contributed to the decline of the saffron industry in Kashmir. These include the scarcity of high-quality corms for use as seed material, suboptimal soil fertility, the absence of reliable irrigation systems, issues with rodent and disease infestations, inadequate postharvest management practices, insufficient marketing infrastructure, growing urbanization encroaching on saffron lands, widespread adulteration concerns, and the illicit trafficking of lower-grade saffron. However, the most pressing and formidable challenge threatening the survival of the saffron industry is the adverse impact of climate change. The changing climatic conditions in the region, as evidenced by fluctuations in temperature and precipitation patterns, directly affect saffron cultivation. Variability in weather conditions, including temperature extremes and altered precipitation regimes, disrupts the delicate balance required for successful saffron growth. These climatic shifts can impact the flowering and yield cycles, leading to reduced productivity and increased vulnerability to pests and diseases.

The quantity of rainfall plays a pivotal role in shaping saffron production, with specific climatic conditions being particularly conducive to optimal cultivation. According to reports, the most favourable locations for saffron cultivation are those with well-distributed rainfall ranging from 100 to 150 cm, coupled with snowfall during the winter months. This climatic combination is essential for meeting the

Table 7.1 Statistical summary of Saffron yield and production in Kashmir Valley from 1996 to 2019

Year	Area (ha)	Production (MT)	Yield (kg/ha)
1996–97	5707	15.9	2.80
1997–98	4618	13.3	2.90
1998–99	4116	12.8	3.13
1999–2000	3997	7.65	1.89
2000–01	2831	3.59	1.27
2001–02	2713	0.30	0.095
2002–03	2825	6.50	2.28
2003–04	2742	5.15	1.88
2004–05	3143	6.86	2.23
2005–06	3010	6.5	2.15
2006–07	3280	8.2	2.50
2007–08	3280	7.7	2.34
2008–09	3715	9.4	2.50
2009–10	3715	10.4	2.75
2010–11	2479	9.5	2.55
2011–12	2469	10.0	2.64
2012–13	2441	12.4	2.75
2013–14	1913	14.0	2.8
2014–15	1756	5.5	1.5 (floods)
2015–16	1580	16.1	2.8
2016–17	1497	16.4	2.85
2017–18	1323	5.2	1.4
2018–19	1323	5.2	1.4
2019–20	1323	5	0.9

water requirements of saffron corms, crucial for achieving optimum flower yields. The significance of rainfall in September is highlighted, as it ensures the corms receive the necessary moisture for healthy growth and development. In the hilly regions of the valley, there has been a noticeable decline in precipitation, with a substantial decrease of 10.3 mm/year. In contrast, flood plains have experienced a relatively less dramatic decline of 3.6 mm/year. The foothills and Karewa, situated between the hilly regions and the plains, have observed a more moderate decline, with decreases of 6.3 mm/year and 5.8 mm/year, respectively. Winter plays a vital role in saffron cultivation, as snow covering the fields provides continuous moisture to the corms throughout the dormant season, extending up to the following spring. This ensures the corms remain healthy and well-moisturized, critical for preventing infections and facilitating proper outgrowth. The maintenance of ideal moisture levels during the winter season is imperative for the successful cultivation of saffron. As spring approaches, the initial hoeing of the fields becomes crucial for soil aeration.

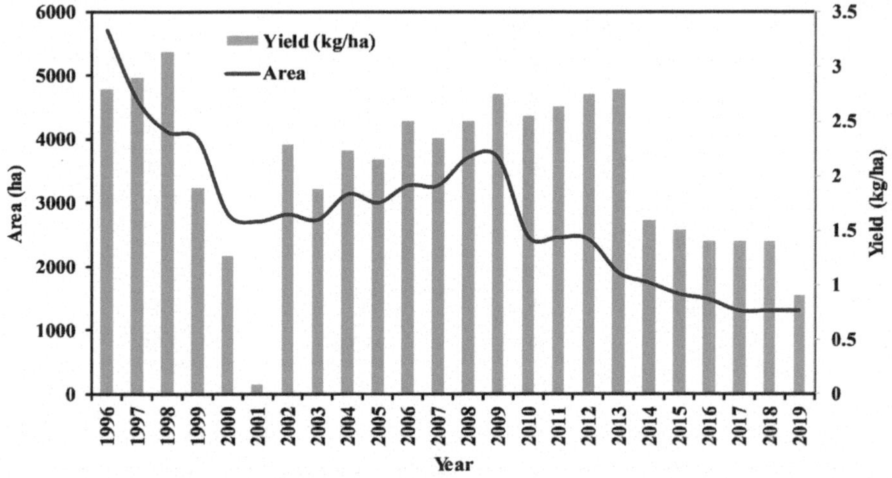

Fig. 7.4 The bivariate plot depicting the change in Area (ha) and Saffron yield from 1996 to 2019. The plot depicts that Saffron cultivation and net yield has drastically decreases from last few decades

Subsequent to this activity, intermittent rain showers are necessary until the blossoming period. The decline in precipitation, particularly in steep areas of the valley, poses a challenge to saffron cultivation, with a notable decrease of 10.3 mm/year. Flood plains, experiencing a somewhat slower rate of decline at 3.6 mm/year, and foothills and the Karewa Mountains, undergoing a more moderate decline at 6.3 mm/year and 5.8 mm/year, respectively, underscore the nuanced impact of changing precipitation patterns on saffron cultivation in different topographical regions of the valley. Addressing these variations becomes crucial for devising adaptive strategies to sustain and enhance saffron production in the face of evolving climatic conditions.

The Union Territory has seen its lowest productivity in the previous two to three decades due to a lack of rainfall. It was observed in the UT of Jammu and Kashmir, where productivity fell from 3.12 to 1.57 kg/ha between 1999 and 2003 due to an extreme drought (Alam 2007). However, favourable rainfall in 2004–05 increased output to 2.96 kg/ha. It implies that without rain, saffron production suffers a severe decline. The farmers' observations that if there is less snowfall in the winter season, it immediately affects production and yield in the following year have been linked to differences in the timing of precipitation in the valley and the decreasing tendency of saffron. As in the winter, snow covers the saffron fields and continuously supplies moisture to the corms up to the following spring season, ensuring that the corms do not suffer and continue to be healthy. Corms must have the ideal moisture content during the winter (dormant) season to avoid infection and failure to develop the outgrowth. When the first hoeing is done in the spring, the soil is aerated. After the first hoeing, the fields require intermittent rain showers till the time of blossoming. In terms of precipitation, steep areas of the valley have experienced a sharp decline (10.3 mm/year), whereas flood plains have experienced a somewhat slower rate of decline (3.6 mm/year), and foothills and the Karewa Mountains have experienced a

Table 7.2 Major Saffron producing regions of the world

S. No.	Country	Area (ha)	Production (MT)	Yield (kg/ha)
1	Iran	43,408	174	4
2	India	4265	7.50	2.29
3	Greece	1000	4.30	4.30
4	Azerbaijan	675	3.70	5.48
5	Spain	600	5	2
6	Morocco	500	1	2
7	Italy	29.4	0.24	8.16
Total		49,477.4	195.74	3.96

more moderate decline (6.3 and 6.6 mm/year, respectively). It is important to mention here that Pulwama district of Kashmir Valley occupies major Saffron cultivation area (3200 ha) followed by Budgam (300 ha), Srinagar (165 ha) and Kishtwar (120) districts. Apart, from Kashmir Valley the other major Saffron producing areas of the world are given in Table 7.2. Similar conclusions highlighting the impact of climate changing on reducing saffron production is also put forward by many studies from different parts of the world (Kumar et al. 2008; Colla and Rouphael 2009; Husaini 2014; Khanali et al. 2017; Mukhter et al. 2020).

Future of Agricultural Production Under Changing Climate

The vitality of numerous mountainous basins across the globe, particularly in Southeast Asia, hinges critically upon the precious meltwaters that emanate from the Himalayan cryosphere. This cryospheric reservoir, comprising snow and glaciers, constitutes a pivotal component of the Himalayan hydrological cycle (Gruber et al. 2017; Lone et al. 2022a, b, 2023). These invaluable water resources play a monumental role in sustaining livelihoods and agro-economics within the major Himalayan basins. However, the adverse effects of a warming climate have profoundly disrupted the customary ebb and flow of cryospheric advance and retreat. This, in turn, has disturbed both aquatic and terrestrial ecosystems (Lone et al. 2021), and concurrently, has adversely impacted agricultural productivity. Consequently, this upheaval reverberates through the agro-economy of the region.

The variability observed in long-term meteorological parameters strongly indicates that the discernible alterations in the hydrological patterns and agricultural production of Himalayas can be directly attributed to historical and current climatic trends. It is anticipated that these patterns may undergo further changes due to the ongoing rise in winter temperatures and the decrease in precipitation, particularly in the western Himalayas. This is a region where local and regional domestic, agricultural, and hydropower activities are predominantly reliant on meltwaters (Miller et al. 2012; Nepal et al. 2021; Shah and Lone 2019; Lone et al. 2021, 2022a, b; Lone and Jeelani 2022). Both observed and projected variations in meteorological parameters underscore the vulnerability of the Himalayan cryosphere to the adverse impacts of climate change. Climate change exerts significant and far-reaching effects on agricultural production worldwide. These impacts are contingent on factors such as geographical region, specific crops, and livestock, as well as the severity of climate change.

Escalating temperatures can influence the duration of growing seasons and disrupt temperature-sensitive crop development. While some crops may thrive in extended growing seasons, others, such as wheat, maize, and saffron, may suffer under extreme heat. In regions like the Kashmir Valley, climate change can intensify water scarcity, posing challenges for crop irrigation and livestock hydration. This, in turn, leads to a reduction in agricultural output and heightens competition for limited water resources. Furthermore, the mounting frequency and intensity of extreme weather events, such as hurricanes, droughts, and heatwaves, exact a toll on crops and infrastructure. This results in crop losses and disruptions in the food supply chain. Shifts in temperature and precipitation can disturb ecosystems, resulting in biodiversity loss. This, in turn, indirectly impacts agriculture by compromising crucial services like pollination, natural pest control, and soil health. Disruptions in agricultural production due to climate change can contribute to food insecurity and market price volatility. Crop failures or reduced yields can lead to an increase in food prices, affecting both producers and consumers. The Himalayan mountainous catchments, home to the world's largest irrigated agriculture system, relies heavily on cryospheric waters. Consequently, these exceptional externalities render the public water supply and agro-economy of this region highly vulnerable, potentially leading to economic and political instability, particularly in mountainous communities.

Recommendations

Introduction and promotion of climate-resilient farming practices that can adapt to changing temperature and precipitation patterns. Investment in research and development to create saffron varieties that are more resilient to climate variations. This could involve breeding programs focused on developing saffron plants that thrive under changing conditions. Implementation of efficient water management strategies to address variations in precipitation. This may involve the development of reliable irrigation systems, rainwater harvesting, and efficient water-use practices to ensure optimal soil moisture for saffron cultivation. Conducting awareness programs and educational initiatives for saffron farmers about climate change impacts and adaptation strategies. Also, Conduct research on the impact of changing climate on saffron pests and diseases. Develop effective strategies for pest and disease management to mitigate potential threats to saffron production. Implementing a combination of these recommendations requires concerted efforts from farmers, government agencies, research institutions, and the community. By adopting a holistic and collaborative approach, the resilience of saffron cultivation in the Kashmir Valley can be enhanced, contributing to the economic well-being of the region.

Conclusion

In this study, a time series meteorological data and Saffron yield/production data were used to assess and understand the response of Saffron production and yield to changing climatic scenarios. The time series temperature and precipitation data of the Himalayas indicated that an increase of temperate by 0.2 °C decade^{-1} and a decrease in precipitation by 0.032 mm year^{-1} was observed from 1901 to 2022. The Saffron production and yield data also showed that areal extent of Saffron production has declined significantly during the observation period (1996–2019). The study further suggested total Saffron production has declined from 15.96 MT in 1996 to 5.2 MT in 2019 with an aerial decline of 5707 ha to 1323 ha from 1996 to 2019. We concluded that change in meteorological variables due changing climate is one of the major factors that declined the Saffron yield and production across Kashmir Valley. The study also suggested that the continuous increase in temperature and decrease in precipitation will negatively affect the agriculture/horticulture in general and Saffron production and yield in particular which would affect the socio-economic development of the Kashmir Valley. The result of this study would be highly substantial for agricultural policy development and will serve as primary information to different organizations to sustainably manage the agricultural production in the country.

Acknowledgements The authors are highly thankful to Indian Meteorological Department (IMD Pune) and National Oceanic and Atmospheric Administration (NOAA) for open access to Satellite and meteorological datasets.

References

Alam A (2007) Status and prospects of mechanisation in saffron cultivation in Kashmir. Acta Hortic 739:383–388

Auer I, Böhm R (1994) Combined temperature-precipitation variations in Austria during the instrumental period. Theoret Appl Climatol 49(3):161–174

Bari SH, Rahman MTU, Hoque MA, Hussain MM (2016) Analysis of seasonal and annual rainfall trends in the northern region of Bangladesh. Atmos Res 176:148–158

Bhutiyani MR, Kale VS, Pawar NJ (2007) Long-term trends in maximum, minimum and mean annual air temperatures across the Northwestern Himalaya during the twentieth century. Clim Change 85(1–2):159–177

Bosworth B, Collins SM (2008) Accounting for growth: comparing China and India. J Econ Perspect 22(1):45–66

Colla G, Rouphael Y (2009) Evaluation of saffron (*Crocus sativus* L.) production in Italy: effects of the age of saffron fields and plant density. J Food Agric Environ 7(1):19–23

Dhulipala S, Patil GR (2020) Freight production of agricultural commodities in India using multiple linear regression and generalized additive modelling. Transp Policy 97:245–258

Diaz HF, Bradley RS (1997) Temperature variations during the last century at high elevation sites. Clim Change 36:253–279

Dixit J, Sharma S, Ali M (2014) Present status, potential and future needs for mechanization of agricultural operations in Jammu and Kashmir state of India. Agric Eng Int CIGR J 16(3):87–96

Ferreira RN, Nissenbaum MR, Rickenbach TM (2018) Climate change effects on summertime precipitation organization in the Southeast United States. Atmos Res 214:348–363

Gruber S, Fleiner R, Guegan E, Panday P, Schmid MO, Stumm D, Wester P, Zhang Y, Zhao L (2017) Inferring permafrost and permafrost thaw in the mountains of the Hindu Kush Himalaya region. Cryosphere 11(1):81–99

Gulati A, Roy R, Hussain S (2017) Getting Punjab agriculture back on high growth path: sources, drivers and policy lessons

Huo Z, Dai X, Feng S, Kang S, Huang G (2013) Effect of climate change on reference evapotranspiration and aridity index in arid region of China. J Hydrol 492(7):24–34

Husaini AM (2014) Challenges of climate change: omics-based biology of saffron plants and organic agricultural biotechnology for sustainable saffron production. GM Crops Food 5(2):97–105

Jain P, Castellanos-Acuna D, Coogan SC, Abatzoglou JT, Flannigan MD (2022) Observed increases in extreme fire weather driven by atmospheric humidity and temperature. Nat Clim Change 12(1):63–70

Jamelli V, Pech VP, Chochillan C, Brunstein D (2004) Geomorphic variability of debris flows and recent climate changes in the French Alps. Clim Change 64:77–102

Kafi M, Kamili AN, Husaini AM, Ozturk M, Altay V (2018) An expensive spice saffron (*Crocus sativus* L.): a case study from Kashmir, Iran, and Turkey. In: Global perspectives on underutilized crops, pp 109–149

Kanth RH, Khanday BA, Tabassum S (2008) Crop weather relationship for saffron production. In: Nehvi FA, Wani SA (eds) Saffron production in Jammu and Kashmir, Directorate of Extension Education. SKUAST-K, India, pp 170–188

Khanali M, Shahvarooghi Farahani S, Shojaei H, Elhami B (2017) Life cycle environmental impacts of saffron production in Iran. Environ Sci Pollut Res 24:4812–4821

Kohler T, Wehrli A, Jurek M (2014) Mountains and climate change: a global concern. Sustainable Mountain Development Series Centre for Development and Environment (CDE), Swiss Agency for Development and Cooperation (SDC) and Geographica Bernensi, Bern, Switzerland, p 136

Kumar R, Singh V, Devi K, Sharma M, Singh MK, Ahuja PS (2008) State of art of saffron (*Crocus sativus* L.) agronomy: a comprehensive review. Food Rev Int 25(1):44–85

Kumar P, Sharma PK, Kumar P, Sharma M, Butail NP (2021) Agricultural sustainability in Indian Himalayan region: constraints and potentials. Indian J Ecol 48(3):649–661

Lone SA, Jeelani G (2022) Appraising the groundwater potential of Liddar Sub-Basin (Western Himalayas) using geospatial techniques. In: Climate change impact on groundwater resources. Springer, Cham, pp 329–344

Lone SA, Jeelani G (2023) Spatio-temporal dynamics of Groundwater recharge in Dras sub-basin of upper Indus River Basin, Western Himalayas. Impacts of urbanization on hydrological systems in India. Springer International Publishing, Cham, pp 313–322

Lone SA, Jeelani G, Deshpande RD, Mukherjee A (2019) Stable isotope (δ^{18}O and δD) dynamics of precipitation in a high-altitude Himalayan cold desert and its surroundings in Indus river basin, Ladakh. Atmosp Res 221:46–57

Lone SA, Jeelani G, Mukherjee A, Coomar P (2021) Arsenic fate in upper Indus river basin (UIRB) aquifers: controls of hydrochemical processes, provenances, and water-aquifer matrix interaction. Sci Total Environ 795:148734

Lone SA, Jeelani G, Alam A, Bhat MS, Farooq H (2022a) Effects of changing climate on the water resources of upper Jhelum basin (UJB), India. In: Mukherjee A (ed) Riverine systems: understanding the hydrological, hydrosocial and hydroheritage connections. Springer Capital Publishing, pp 139–154

Lone SA, Jeelani G, Padhya V, Deshpande R (2022b) Identifying and estimating the sources of river flow in the cold arid desert environment of Upper Indus River Basin (UIRB), western Himalayas. Sci Total Environ 795:154964

Lone SA, Jeelani G, Deshpande RD, Bhat MS, Padhya V (2023) Assessing the hydrological controls on spatio-temporal patterns of streamwater in glacierized mountainous Upper Indus River Basin (UIRB), western Himalayas. J Hydrol 619:129310

Makwana AK (2013) Agricultural education in India: challenges and prospects. Voice of research working papers, 2013-12-25

Malik SH, Kumar K, Kumar A (2018) Agricultural Sustainability in North-Western Himalayan Region: issues and challenges. Univers Rev 7(11):65–70

Mariappan K, Zhou D (2019) A threat of farmers' suicide and the opportunity in organic farming for sustainable agricultural development in India. Sustainability 11(8):2400

Miller JD, Immerzeel WW, Rees G (2012) Climate change impacts on glacier hydrology and river discharge in the Hindu Kush-Himalayas. Mt Res Dev 32(4):461–467

Mukhter I, Ganie AU, Ahad B, Ahmad S, Chowdhary R, Seerat S (2020) Qualitative analysis of factors influencing the decline of Saffron (*Crocus Sativus*) cultivation in the Valley of Kashmir, India: need of regaining its legacy. J Soc Polit Sci 3(3)

Narayan S, Bhattacharya P (2019) Relative export competitiveness of agricultural commodities and its determinants: some evidence from India. World Dev 117:29–47

Nepal S, Khatiwada KR, Pradhananga S, Kralisch S, Samyn D, Bromand MT, Jamal N, Dildar M, Durrani F, Rassouly F, Azizi F (2021) Future snow projections in a small basin of the Western Himalaya. Sci Total Environ 795:148587

Pandey DK, Nandy S, Mukherjee A, Dey A (2020) Advances in bioactive compounds from *Crocus sativus* (saffron): structure, bioactivity and biotechnology. Stud Nat Prod Chem 66:273–304

Pandita, D (2021) Saffron (*Crocus sativus* L.): phytochemistry, therapeutic significance and omics-based biology. In: Medicinal and aromatic plants. Academic Press, pp 325–396

Pathak H, Mishra JP, Mohapatra T (2022) Indian agriculture after independence, vol 110(001). Indian Council of Agricultural Research, New Delhi, p 426

Peerzada IA, Chamberlain J, Reddy M, Dhyani S, Saha S (2021) Policy and governance implications for transition to NTFP-based bioeconomy in Kashmir Himalayas. Sustainability 13(21):11811

Rao VB, Franchito SH, Gerólamo ROP, Giarolla E, Ramakrishna SSVS, Rao BRS, Naidu CV (2016) Himalayan warming and climate change in India. Am J Clim Change 5(04):558

Rao CR, Raju BMK, Josily, S, Rao, AVMS, Kumar, RN, Rao, MS, Swapna, N, Siva, GS, Meghana, YL, Prabhakar, Singh, VK (2022) Impact of climate change on productivity of food crops: a sub-national level assessment for India. Environ Res Commun 4(9):095001

Shah RA, Lone SA (2019) Hydrogeomorphological mapping using geospatial techniques for assessing the groundwater potential of Rambiara river basin, western Himalayas. Appl Water Sci 9:64

Sharma KP, Moore B III, Vorosmarty CJ (2000) Anthropogenic, climatic and hydrologic trends in the Kosi Basin. Himalaya Clim Change 47:141–165

Shekhar MS, Chand H, Kumar S, Srinivasan K, Ganju A (2010) Climate-change studies in the western Himalaya. Ann Glaciol 51(54):105–112

Shrestha S, Yao T, Adhikari TR (2019) Analysis of rainfall trends of two complex mountain river basins on the southern slopes of the Central Himalayas. Atmos Res 215:99–115

Suresh A, Mathur VC (2016) Export of agricultural commodities from India: performance and prospects. Indian J Agric Sci 86(7):876–883

Vicente-Serrano SM, McVicar TR, Miralles DG, Yang Y, Tomas-Burguera M (2020) Unraveling the influence of atmospheric evaporative demand on drought and its response to climate change. Wiley Interdiscip Rev Clim Change 11(2):e632

Zhang P, Zhang J, Chen M (2017) Economic impacts of climate change on agriculture: the importance of additional climatic variables other than temperature and precipitation. J Environ Econ Manag 83:8–31

Chapter 8
Global Climate Change Crisis: Lessons Learned from COVID-19 in the Context of Solid Waste Management and Allied Sectors in India

Tinku Casper D'Silva, Sunitha Vijayan, Saptashish Deb, and Ram Chandra

Abstract The human existence with the recent spread of COVID-19 disease infected by the coronavirus novel variant SARS-CoV-2 was uncertain. The nations together abided to overcome the challenges posed by the pandemic. However, during those testing times, the major activities were disrupted abruptly; one among them was solid waste management. This has put up a question mark on the safe and hygienic surroundings, an essential social responsibility towards a sustainable, hygienic, and safe environment. The socio-economic scenario during the pandemic wave has triggered concern for proper waste management, which has varied from nation to nation and regionally. A convenience online survey conducted in India during the pandemic suggested that about 72 billion face masks were used, further contributing to solid waste. Further, the possible solutions to achieve a sustainable solid waste management framework are discussed specifically for rural and urban setups in India. These solutions seem to align with the United Nations 17 sustainable development goals, depicting that they can be continued to combat climate change suitable for developing countries. Finally, the importance of circular economy and social inclusiveness with responsibilities are highlighted to promote and achieve sustainability within the solid waste management sector.

T. C. D'Silva (✉) · S. Deb · R. Chandra
Centre for Rural Development and Technology, Indian Institute of Technology Delhi, Hauz Khas, New Delhi 110016, India
e-mail: casperdsilva23@gmail.com

S. Deb
e-mail: sapta121@gmail.com

T. C. D'Silva · S. Vijayan
Kerala State Council for Science, Technology and Environment—National Transportation Planning and Research Centre (KSCSTE—NATPAC), Government of Kerala, Thiruvananthapuram, Kerala 695011, India
e-mail: sunitha.natpac@gmail.com

U. Mukhopadhyay et al. (eds.), *Climate Crisis, Social Responses and Sustainability*, Climate Change Management, https://doi.org/10.1007/978-3-031-58261-5_8

Introduction

COVID-19, caused by SARS-CoV-19, has caused tremendous losses in regard to life, economy, and resources in the global context and has drastically changed the lifestyle and individual behaviour of humans (Botzen et al. 2021). It has partitioned the recent times into pre-COVID and post-COVID periods. The pre-COVID period implies a normal life before the COVID-19 pandemic hit across the globe. At the same time, after COVID hit globally, it needed a broader sense of understanding of disruptive human lifestyles, social responses, and sustainability.

In those testing times, there was a vast shift in the composition of solid waste in both urban and rural regions. A considerable difference has been observed during the pre-COVID and COVID periods, owing to food insecurity and a disrupted supply chain. The imposition of the lockdown resulted in the closure of public places, including shops, restaurants, canteens, cafeterias, etc. This led to an increase in additional supply chain requirements to transport food items, triggering an exponential growth in plastic waste generation (Tenenbaum 2020). The intensified panic buying of packed food items during the pandemic owing to the concern of food security has subsequently influenced the waste characterization in all regions, be it urban or rural. This resulted in the littering of expired/unused food into the waste (Layton 2020; Neel 2020). Similarly, a possible surge in the percentage of plastic waste in urban areas was reported due to a rise in medical items and food packaging waste during the pandemic (Tripathi et al. 2020; Geyer et al. 2017). Thus, the overall scenario can be summed up as a combination of the possible inflation of non-biodegradable waste and deflation of biodegradable waste in urban regions, and inflation of both biodegradable and non-biodegradable wastes in rural areas during the COVID and post-COVID period.

Subsequently, during COVID-19, proper collection, disposal, and recycling of these wastes have become an added burden for the waste management sector across the nations (Klemeš et al. 2020; Kaufman and Chasan 2020; Ferronato and Torretta 2019). Thus, there is a need to assess the status of different regions at a regional or national level in investigating the effect of the pandemic on waste management and allied sectors to list the possible challenges and solutions.

Further elaborating, when the world came almost to a standstill during COVID-19, nature recuperated slowly with reports of its revival with clean water resources, better air quality, improved soil fertility, and resurgence of lost biodiversity from different parts of the world (Gardiner 2020). In different to that, the situation was entirely opposite in the case of solid waste management. The pandemic has changed waste generation and its composition. A rapid rise in biomedical wastes and other personal protection equipment (PPE) contaminated the regular municipal solid waste (MSW), increasing the risk of virus transmission (UNEP 2020). It further disrupted the standard operating procedures involved in solid waste management, thereby posing a serious risk to the people involved at the ground level (Mallapur 2020), which ultimately threatens the socio-ecological conditions. Thus, induced capacity building of the solid waste management infrastructure has become essential for dispensing proper training and knowledge to effectively combat COVID-19 (AcrPlus 2020) and its further consequences in terms of the climate crisis and socio-ecological sustainability.

Thus, the present study addresses the issues faced during COVID and post-COVID times in India, one of the most populous countries in the world. The lessons learned during the COVID-19 pandemic, the contribution of PPE to solid waste generation, possible strategies for proper solid waste management in urban and rural areas, its challenges and the importance of social inclusiveness are highlighted, and how it can be further continued to achieve the United Nations sustainable development goals and mitigate the climate change crisis are discussed.

Global Climate Change and Future Pandemics: Importance of Learning from COVID-19 Pandemic

As explained, COVID-19 has been recognized as one of the biggest health threats that affected the world very badly. The chances of a COVID-19 pandemic outbreak were predicted much earlier from its first report from China (Fan et al. 2019). However, the lack of preparedness made its impact very detrimental for all the nations. There are many lessons that the COVID-19 crisis can provide to humanity, be it how vulnerable a region is with regard to social adhesiveness, inequalities, poor infrastructure for health and sanitation, and many more. High population density and population growth, water shortage, and inadequate liquid/solid waste treatment and disposal facilities are some of the critical factors that will definitely promote the spread of infectious diseases rather than controlling them. All these will contribute to an impact at the ground level and should be addressed at the regional level.

Even though the impact is not as quick as that of COVID-19, climate change has also been enlisted in the list due to its catastrophic characteristics (Salas 2020), with CO_2 rising at an alarming rate in the atmospheric environment (Subbarao et al. 2023). In the case of COVID-19, the world has been slowly recuperating from thereon, while the risk is aggravating from the latter. Climate change is one of the primary global health threats to humanity, particularly in South Asian and African countries (Cramer et al. 2018; Watts et al. 2019). The atmospheric temperature is on the rise, directly influencing the heat waves and intensity, and a reduction in precipitation and unpredictable changes in seasons are expected as a result of the climate change (Hochman et al. 2019). The factors that limit the adaptation to mitigate climate change at a regional level are similar to those observed during COVID-19. These can be enlisted as population density, water scarcity, economic and social inequalities, poor health infrastructure, limited data/resource availability and sharing, and restricted trade-offs between the regions, locally or internationally. All these factors ultimately depend on the political cohesion or conflicts. Combining the uncontrolled urbanization rate with these effects calls for adaptive measures to combat the climate change impacts (Frumkin and Haines 2019; Negev et al. 2021).

A holistic approach to innovation, research, and development is important to tackling climate change. Preventive measures to control the excess greenhouse gases going into the atmosphere are considered the most common strategy being adopted across different nations. Recent research suggests that the levels of greenhouse gases in the atmosphere can be controlled and even reduced by adopting scalable and economically viable technologies, especially in the energy and food sector (Drawdown 2020). For example, shifting to renewable energy sources such as bioenergy, lowering food wastage, optimizing the food supply chain and waste management sector with a local circular economy approach, improving and supporting carbon sinks with more natural and organic agricultural practices, and restoring local natural ecosystems are few to list out (Drawdown 2020). Thus, adaptation in a timely manner: intersectoral, regional, national, and global tie-ups are crucial for controlling the short and long-term impacts of climate change. In addition, being vigilant for future pandemic outbreaks and climate change shares seems to have similar features and can act as a motivational force for reforming intersectoral activities (Negev et al. 2021).

From a circular economy point of view, solid waste management should focus on reducing the usage of natural resources and sustainable reuse of waste materials as secondary raw materials. This symbiotic approach is expected to be a part of renewable energy production, enhance products' life without hampering the ecology energy and add value to the products (Negrete-Cardoso et al. 2022; Islam et al. 2021; Velenturf et al. 2019; Parajuly et al. 2020). However, the circular economy requires a robust approach through effective policies and good governance to shift the economy from a linear flow to a circular flow model. 10 R is the most prominent ideology proposed for the circular economy transition, which consists of recycling, recovering, refurbishing, repurposing, repairing, remanufacturing, rethinking, reducing, refusing and re-using (Morseletto 2020).

Contribution of Personal Protection Equipment to Solid Waste Generation During COVID-19: A Case Evidence from India

Survey

A convenient survey (n = 456) was conducted to understand the ground situation during the lockdown in June 2020 belonging to different parts of India (46% urban and the rest from rural regions). The participants in this survey were from various sectors of society, including students, faculties, government/private employees, and common people. The survey was preferred online due to the COVID-related movement/lockdown restrictions. The marginal equal proportion of the sampling population from urban and rural contexts eliminated any biases towards any specific community, which is essential to understanding the ground-level situation in both circumstances. A questionnaire consisting of 10 questions was used for carrying out this survey, as given in Table 8.1.

Results and Discussion

Population density is one of the major parameters that influence the spreading of airborne and waterborne pathogens because people stay together in a very close space and have larger contact groups and poor hygiene (Saadat et al. 2020). In this study, about 68% of the sampling population was from moderately populated areas (500–800 persons per km^2). However, 20% of the population belonged to densely populated areas (> 800 persons per km^2), which raises concern.

Solid wastes can be a major source for spreading the pandemic if proper disposal is not done. Generally, landfilling, incineration, composting, and anaerobic digestion are the major solid waste management strategies adopted in developing countries (Miezah et al. 2015; Srivastava et al. 2015). However, eliminating viruses and bacteria requires a minimum temperature of 70 °C, which cannot be accomplished using these adopted technologies. Even though biochemical waste reduction methods such as composting and anaerobic digestion reduce the pathogen concentration during treatment (Zhao and Liu 2019; McCarthy et al. 2011), these methods cannot achieve complete pathogenic resilience. Nonetheless, integrating thermal pretreatment methods (> 100 °C) with existing anaerobic digestion systems (Subbarao et al. 2023; Ahmed et al. 2020) could improve the pathogenic stability of the anaerobic digestion process in the future. In rural areas, burying, burning, and littering of solid wastes are the commonly adopted solid waste management methods, irrespective of the nature of solid wastes (D'Silva et al. 2022). In urban areas, landfilling (21.30%) was the preferred standard disposal method, followed by incineration (8.50%). However, during the COVID-19 pandemic, no changes in the waste management strategies (81.60%) were experienced by the sampling population, while 18.40% felt amendments in terms of sanitation and hygiene.

The survey also assessed the use of PPEs among the sampling population. While 96.10% of the sampling population used face masks of different materials, including N95, cloth-based, or surgical, cloth-based masks were the most commonly used (about 76.50%). Other kinds of face masks were majorly from urban areas. Other PPEs, such as gloves and face shields, were also used by some participants (< 4%). Interestingly, the use of face masks was limited to less than 6 h by most participants (75.70%), mainly in the working environment or outside the home. Another exciting fact observed was that 88.90% of the participants were using reusable masks for up to a month, about 32.10% only up to a week and 28.80% of them for up to two weeks too. Using these results, quantification of PPEs waste generation for a month was calculated for the sampling population, and then magnification of these results to the total urban and rural population for the country was done with an 80% acceptance rate for the rural population and 90% for urban population (Nzediegwu and Chang 2020). Table 8.2 shows the estimated total face mask generation that could contribute

Table 8.1 Questionnaire used for carrying out the survey

S. No.	Question asked	Options given for answering
1.	Which type of area you are residing in?	(a) Urban (b) Rural
2.	In which category does your area come in terms of population density?	(a) Very dense (> 800 people per km^2) (b) Moderate (in between 500 and 800 people per km^2) (c) Less dense (< 500 people per km^2)
3.	What waste management practices were being practiced in your area during the pre-COVID-19 period?	(a) Landfilling (b) Incineration (c) Composting (d) Anaerobic digestion (e) No proper waste management strategy was applied
4.	Is there any change in these waste management practices during the lock-down or during the CoVID-19 period?	(a) Yes (b) No (c) Maybe
5.	What kind of precautions have you taken daily to tackle COVID-19?	(a) Disinfecting your home frequently (b) Use of hand sanitizers or handwash at regular intervals (c) Use of face masks (d) Use of gloves
6.	What type of personal protection equipment (PPE) are you using?	(a) N-95 mask (b) Surgical mask (c) Cloth mask (d) Surgical gloves (e) Cloth gloves (f) Face shield
7.	For how long do you use the PPEs in one run?	(a) Less than 6 h (b) 6–12 h (c) More than 12 h
8.	Do you reuse the PPEs?	(a) Yes (b) No
9.	Do you wash it or sanitize it before reusing it?	(a) Yes (b) No
10.	If yes, for how long do you use the same masks or gloves again and again?	(a) Less than a week (b) 1–2 weeks (c) 2–3 weeks (d) Up to a month

Table 8.2 Estimated monthly face mask waste generated that could be disposed off from households in India (lockdown period: March–June 2020)

	Approximate monthly face masks generation (in numbers)
Rural India	23,188,734,720
Urban India	48,464,407,566
Total	71,653,142,286

as a surplus to the usual waste generated in the country. The population of India (both rural and urban) was taken from https://www.worldometers.info. Considering that a face mask weighs an average of 3 g (Cornelio et al. 2022), the quantification of this surplus waste generated would be about a total of 0.22 megatonnes across India. The next section gives an overview of how urban and rural setups in developing countries can adapt to the "new normal" conditions without disrupting the sustainability of solid waste management technologies.

Possible Solid Waste Management Strategies During Pandemics in Urban and Rural Regions

In New Delhi (an example of an urban region), the existing waste-to-energy incineration plants utilize nearly 50% of municipal solid wastes (MSW) for energy recovery. Only 13% of generated MSW possesses high calorific value, according to the Centre for Science and Environment (CSE 2019). A minimum calorific value (CV) of 0.07 kW/kg is essential for satisfactory performance of incineration (Korai et al. 2016). However, the higher moisture content in MSW reduces the plant efficiency, and the chances of toxic gas emission are high. The presence of more non-biodegradable wastes with higher calorific value (National Green Tribunal 2017) and dry nature improves the gross calorific value of generated MSW, improving the incineration efficiency than before.

Solid waste handling and disposal need significant attention, with varied lifespans of the SARS-CoV-2 virus on different materials (24 h for cardboard, 3 d for plastics, 4 h—3 d for metals) (Munshi et al. 2022; Chin et al. 2020). Nghiem et al. (2020) have proposed a strategy for non-biodegradable wastes in urban regions having home quarantine systems as follows: (a) sorting and 3 d storage before collection, (b) increased collection and transportation frequency, and (c) incineration for the urban home quarantine facilities. Similar strategies are also discussed by Kulkarni and Anantharama (2020). This strategy could be replicated with a slight modification in urban regions of developing countries with the integration of proper waste handling, as Di Maria et al. (2020) reported. The waste collection to treatment chain must implement the process of waste collection (i.e., sealed double-layered polythene bags by trained waste workers with proper PPEs) through sanitation transport vehicles to

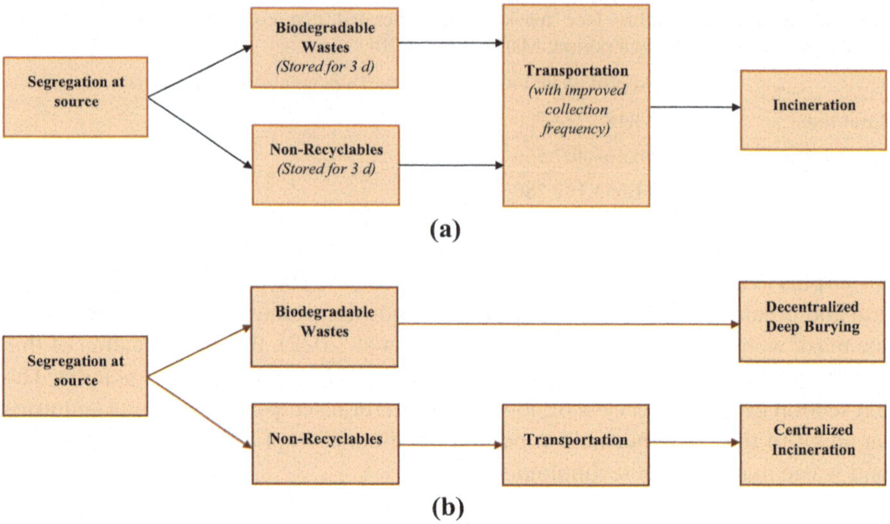

Fig. 8.1 **a** Proposed solid waste management strategy in urban regions of developing countries during COVID and post-COVID period. **b** Proposed solid waste management strategy in rural regions of developing countries during COVID and post-COVID period

incineration (Di Maria et al. 2020). Figure 8.1a shows the suggestions for possible solid waste management during a pandemic.

However, the rural region needs to be strategized differently (Fig. 8.1b). Though the agricultural waste could be composted or anaerobically digested for resource recovery; however, the organic waste from domestic households (especially quarantined homes) must have a suitable waste management strategy since the chances of viral contamination are high. Deep burial is one of the scientific strategies for the disposal, be it dead bodies of pathogen-infected patients or biomedical waste disposal from the healthcare system in rural areas, where suitable site selection and land availability are not a concern (WHO 2014; BMWM Guidelines 2016). Deep burial can be considered an extended version of pit composting, in which the organic wastes are buried in soil pits of 2 ft × 2 ft × 2 ft (D'Silva et al. 2022). A similar strategy could also be applied to the disposal of organic waste from domestic households in rural areas in a decentralized manner. The organic waste could be disposed of in a pit or trench (1.5–3.0 m deep). The pit area must be far from any water bodies with a minimum of 200–350 m distance (WHO/WEDC Technical Report 2013). According to the report, an aqueous calcium hydroxide ($Ca(OH)_2$) solution must be sprinkled on the pit, and a layer of soil (0.5 m thick) must be covered. The SARS-CoV-2 viruses are resistant to a pH range of 3–10 (Tripathi et al. 2020). Hence, the application of $Ca(OH)_2$ increases the pH (> 11) inside the pit and eliminates harmful bacteria, viruses and coliforms (Grabow et al. 1978). In addition, the area might be covered using galvanized iron or wire mesh until the pit is filled to avoid human or animal activity.

Though the "deep burial" of organic wastes might reduce the burden on overall solid waste management, incineration is found to be the most suitable approach for managing non-biodegradable wastes. A low-cost incinerator made of bricks and locally available raw materials could be a solution. De Montfort University has proposed a low-cost incineration system recommended by WHO (Batterman 2004), which is affordable for the rural community. A centralized approach for the incineration of biomedical wastes has been successfully reported in Wuhan, China, during the peak of the COVID cases (Singh et al. 2020). Hence, localized incineration using the low-cost system would be the best solution for managing non-biodegradable and biomedical wastes.

Challenges in the Solid Waste Management Sector During the Post-COVID-19 Pandemic

According to Sharma and Jain (2019), 82% of solid waste is collected in India, and only 28% of the waste undergoes any scientific treatment, while the remaining 72% is disposed of in landfills. Waste segregation at the source, door-to-door collection, reuse and recycling, adoption of scientific treatment technologies, availability of land, and proper disposal are some of the prime challenges that require attention. Most urban local bodies cannot manage such a large amount of solid waste due to financial debilities and inadequate infrastructure. Several nations reported improper functioning and lack of returns from recycling sectors during the lockdown period, creating a severe financial and economic crisis (Tripathi et al. 2020). Thus, there is a need further to strengthen the infrastructure for proper solid waste management to maintain safety and hygiene at a community level in developing countries. However, this requires local-level planning and designing.

During the COVID-19 lockdown, about 93% of the global population lived in countries that have implemented movement restrictions in place (Connor 2020). Amidst the lockdown, the availability of essential services such as food items, medicine, sanitation, and hygiene resulted in panic buying of these items worldwide. In a study by Di Renzo et al. (2020), increased panic buying and food consumption during the lockdown period was reported in an Italian community, which ultimately increased solid waste generation, especially non-biodegradable wastes. Such a lifestyle could negatively impact developing countries like India. Mmereki et al. (2016) pointed out some major challenges in developing countries: lack of necessary plans and precautions for collecting and disposing of this solid waste, inadequate institutional capacities, lack of collection facilities etc. Since the traces of COVID-19 cases are reported to prevail until 2025 (Scudellari 2020), a long-term solid waste management approach is essential (Tripathi et al. 2020), which will sustain the waste management sector during a future crisis. Thus, the engineering, scientific and organizational challenges must be addressed sustainably and adequately for the waste management sector.

Moreover, in the post-COVID period, shifting from the traditional linear economic model to a sustainable circular model in solid waste management is crucial for transforming the economy into a green economy and achieving sustainable development goals (Negrete-Cardoso et al. 2022; Ibn-Mohammed et al. 2021; Sharma et al. 2021). Hence, the time calls for restructuring the SWM practices to make the system more effective and resilient to adversities. This pinpoints the development of an economy based on zero-waste and zero-carbon targeting, lowering the waste management expenditure. Adopting appropriate methods for sorting and disposing of MSW should be the main focus of solid waste management (Islam et al. 2021). Otherwise, the world will be heading to a catastrophe of 'waste disaster' (Kothari et al. 2021).

Social Inclusiveness and Responsibilities for Sustainable Solid Waste Management

Social inclusiveness is a factor that determines the harmony in which individuals and groups take part in society, improving opportunities, dignity, and mutual respect based on their identity. In solid waste management, ragpickers are essential in completing at least partial recycling and resource recovery at the source itself or from landfills in developing countries. E.g., ragpickers recycle almost 20% of waste generated in India. About 4 million ragpickers are estimated to work in India (Dandapani 2017). Generally, this trade has no employer-employee hierarchy, even though contractors may organize some waste-picking activity. Most ragpickers are usually self-employed with no formal linkages with the local authorities or waste management industries. Moreover, ragpickers commonly suffer from musculoskeletal, respiratory and gastrointestinal ailments. Also, these workers are not given any benefits of social security.

Especially during the COVID period and while the vaccination programs were underway, biomedical waste generation escalated exponentially, and the need for proper management was at a peak. Several vaccines have been tested and made available for the public to prevent and mitigate COVID-19. Under the Massive Immunisation Programme, India vaccinated more than 250 million people (500 million vaccine doses) by mid-2021 (Biswas 2020). In the biggest vaccination drive against corona, frontline workers such as health workers were prioritized. However, there is a need to include sanitation workers and others involved in similar work in the category equivalent to health workers, as their working environment is conducive to viral transmissions.

From the waste management point of view and future possible unprecedented events, managing additionally generated gloves, sanitizer bottles, syringes, vials, and cotton swabs are crucial and are expected during the vaccination programs. As per CPCB data, around 33,000 tonnes of COVID-19 biomedical waste were generated (CPCB 2020). For about 30 crore frontline workers, India procured about 60 crore syringes for the first phase of the vaccination programs (Kadidal 2020), which is humungous.

Thus, with the changing pandemic scenario, the inclusiveness of ragpickers and other sanitation workers from their safety point of view might be highly preferred for a hygienic and safe environment. This is essential considering these workers' ground-level participation and contribution to society. Incidents have also been reported where proper personal masks and gloves were not distributed to these workers, or there were shortages of gloves and masks in Delhi and Kerala, India (Dasgupta 2020; Karunakaran 2020). The social inclusion of sanitation workers and the informal waste sector in a solid waste management policy could go a long way for developing countries (Nzeadibe and Ejike-Alieji 2020). In India, guidelines for the safe disposal of biomedical waste from hospitals and COVID-19 care centers were issued by the national authority (CPCB 2020). However, these do not cover the handling of home-quarantined generated waste. Hence, it can be concluded that education and awareness are important factors regarding the social acceptance of these workers. The governing bodies must consider this a challenge and an opportunity to make aware and educate the people.

Interlinkages Between COVID-19 Management, Achieving Sustainable Development Goals and Climate Change Mitigation

The roadmap to achieve the United Nations' sustainable development goals (SDGs), initiated to curb the global climate change crisis, is very much linked with other regional/global societal goals, including health and sanitation. COVID-19 was a reminder that the nations must focus on identifying the regional challenges, taking advantage of the sustainable solutions, and the possibilities of trade-offs. Uncertainties during COVID-19 were evident that the measures to control the spread of COVID-19 were very slow, with clarity on proper management unclear for the authorities. However, in the case of climate change mitigation, SDGs can play a pivotal role in finding appropriate solutions to protect the global population from unforeseen future crises and subsequent resurgence.

Progressing toward achieving the enlisted SDGs, such as climate action, sustainable cities and communities, affordable and clean energy, life on land and life underwater, etc., will be crucial for the fight against climate change. Also, SDG goals such as good health and well-being, clean water, and sanitation can minimize the unexpected outbreaks of future pandemics and present a better management option at the authority level. The Paris Agreement 2016 reestablished the goal of controlling the global temperature increase below 2 °C above pre-industrial levels, substantially reducing risks from climate change. However, there is a need to pick up the pace to achieve the SDGs as quickly as possible. Developed countries must aid other developing and underdeveloped countries to achieve their goals within the targeted timeframe through financial and technical support.

Conclusion

The COVID-19 pandemic has hit the world very severely. The solid waste management sector has been considered one of the most vulnerable sectors in context with the possibility of transmission of viral infections. This article elucidated that the impact of the pandemic put a huge burden on solid waste composition, emphasizing both urban and rural regions of India. Population density is one of the most critical factors that is decisive in the waste generation potential and also contributes to the spreading of the pandemic. In India, approximately 72 billion face masks were generated from preventing COVID-19 during the pandemic wave. Existing solid waste management technologies are vulnerable to pathogenic infections. However, several strategies could be applied during a pandemic or emergency for effective waste management, specifically to urban and rural setups. It was found that these strategies strategized for the pandemic crisis could also be adopted to achieve the United Nations sustainability goals owing to mitigating climate change. Establishing a circular economy, social awareness, responsibilities, and inclusiveness in the context of humanity, and a proper scientific solid waste management technique is essential to pave a new way to deal with the climate change and pandemic crises in urban and rural regions.

Conflict of Interest The authors declare that there is no conflict of interest.

Informed Consent Informed consent was obtained from all individual participants involved in the convenient survey.

References

AcrPlus (2020) Association of cities and regions for sustainable resource management, municipal waste management and COVID-19. https://www.acrplus.org/en/municipalwaste-management-covid-19

Ahmed B, Tiwari SB, Naseem A, Kazmi AA, Khursheed A, Tyagi VK (2020) Thermal pre/treatment of organic fraction of municipal solid waste. In: Biorefineries: a step towards renewable and clean energy. Springer, Singapore, pp 105–120. https://doi.org/10.1007/978-981-15-9593-6_5

Batterman S (2004) Assessment of small-scale incinerators for health care waste. Environmental Health Sciences, University of Michigan. Accessed from: https://www.who.int/immunization_safety/publications/waste_management/en/assessment_SSIs.pdf

Biswas S (2020) India coronavirus: how do you vaccinate a billion people? BBC News Article. Available from: https://www.bbc.com/news/world-asia-india-55048925

BMWM Guidelines (2016) Retrieved from: https://dhr.gov.in/sites/default/files/Bio-medical_Waste_Management_Rules_2016.pdf

Botzen W, Duijndam S, van Beukering P (2021) Lessons for climate policy from behavioral biases towards COVID-19 and climate change risks. World Dev 137:105214

Chin A, Chu J, Perera M, Hui K, Yen HL, Chan M, Peiris M, Poon, L (2020) Stability of SARS-CoV-2 in different environmental conditions. medRxiv. https://doi.org/10.1101/2020.03.15.20036673

Connor P (2020) More than nine-in-ten people worldwide live in countries with travel restrictions amid COVID-19. Available from: https://www.pewresearch.org/

Cornelio A, Zanoletti A, Federici S, Ciacci L, Depero LE, Bontempi E (2022) Environmental impact of surgical masks consumption in Italy due to COVID-19 pandemic. Materials 15(6):2046. https://doi.org/10.3390/ma15062046

CPCB (2020) Guidelines for handling, treatment and disposal of waste generated during treatment/diagnosis/quarantine of COVID-19 patients. Available from: https://ncdc.gov.in/WriteReadData/l892s/63948609501585568987.pdf

Cramer W, Guiot J, Fader M, Garrabou J, Gattuso JP, Iglesias A, ... and Xoplaki E (2018) Climate change and interconnected risks to sustainable development in the Mediterranean. Nat Clim Change 8(11):972–980

CSE (2019) Most of Delhi's garbage unfit for waste-to-energy plants: CSE report. Hindustan Times News Article. Available from: https://www.hindustantimes.com/delhi-news/most-of-delhi-s-garbage-unfit-for-waste-to-energy-plants-cse-report/story-4BAf2tyrX2iIcbF0M7mn3K.html

Dandapani S (2017) Unpaid and undervalued, how India's waste pickers fight apathy to keep our cities clean. The News Minute News Article. Available from: https://www.thenewsminute.com/article/oppressed-and-unrecognised-life-waste-pickers-crucial-india-s-sanitation-72426#:~:text=available%20to%20them

Dasgupta S (2020) Not enough masks & gloves, sanitation workers fear infection at Delhi's Covid hospitals. The Print News Article. Available from: https://theprint.in/india/not-just-doctors-sanitation-workers-also-run-risk-of-infection-without-protective-gear/406818/

Di Maria F, Beccaloni E, Bonadonna L, Cini C, Confalonieri E, La Rosa G, Milana MR, Testai E, Scaini F (2020) Minimization of spreading of SARS-CoV-2 via household waste produced by subjects affected by COVID-19 or in quarantine. Sci Total Environ 140803. https://doi.org/10.1016/j.scitotenv.2020.140803

Di Renzo L, Gualtieri P, Pivari F, Soldati L, Attinà A, Cinelli G, Leggeri C, Caparello G, Barrea L, Scerbo F, Esposito E, De Lorenzo A (2020) Eating habits and lifestyle changes during COVID-19 lockdown: an Italian survey. J Transl Med 18(1):1–15. https://doi.org/10.1186/s12967-020-02399-5

Drawdown (2020) The drawdown review. Available from: https://drawdown.org/publications/the-drawdown-review?q=drawdown-review

D'Silva TC, Verma S, Magdaline RM, Chandra R, Khan AA (2022) Environmental resilience and sustainability through green technologies: a case evidence from rural coastal India. Environ Eng Res 27(5):210262. https://doi.org/10.4491/eer.2021.262

Fan Yi, Zhao K, Shi Z-L, Zhou P (2019) Bat coronaviruses in China. Viruses 11(3):210

Ferronato N, Torretta V (2019) Waste mismanagement in developing countries: a review of global issues. Int J Environ Res Public Health 16(6):1060

Frumkin H and Haines A (2019) Global environmental change and noncommunicable disease risks. Ann Rev Public Health 40:261–282

Gardiner B (2020) Pollution made COVID-19 worse. Now, lockdowns are clearing the air. Natl Geogr. https://www.nationalgeographic.com/science/2020/04/pollution-made-the-pandemic-worse-but-lockdowns-clean-the-sky/

Geyer R, Jambeck JR, Law KL (2017) Production, use, and fate of all plastics ever made. Sci Adv 3(7):e1700782. https://doi.org/10.1126/sciadv.1700782

Grabow WO, Middendorff IG, Basson NC (1978) Role of lime treatment in the removal of bacteria, enteric viruses, and coliphages in a wastewater reclamation plant. Appl Environ Microbiol 35(4):663–669

Hochman A, Alpert P, Harpaz T, Saaroni H, & Messori G (2019) A new dynamical systems perspective on atmospheric predictability: eastern Mediterranean weather regimes as a case study. Sci Adv 5(6):eaau0936

Ibn-Mohammed T, Mustapha KB, Godsell J, Adamu Z, Babatunde KA, Akintade DD, Koh SCL et al (2021) A critical analysis of the impacts of COVID-19 on the global economy and ecosystems and opportunities for circular economy strategies. Resour Conserv Recycl 164:105169

Islam A, Kalam MA, Sayeed MA, Shano S, Rahman MK, Islam S, Hassan MM et al (2021) Escalating SARS-CoV-2 circulation in environment and tracking waste management in South Asia. Environ Sci Pollut Res 1–18

Kadidal A (2020) Post-vaccination, tackling mounds of biomedical waste an uphill task for Karnataka. Deccan Herald News Article. Available from: https://www.deccanherald.com/state/post-vaccination-tackling-mounds-of-biomedical-waste-an-uphill-task-for-karnataka-928292.html

Karunakaran B (2020) Coronavirus in Kerala: no masks, only cotton towels for health workers. Times of India News Article. Available from: https://timesofindia.indiatimes.com/city/kochi/no-masks-only-cotton-towels-for-health-workers/articleshow/74782749.cms

Kaufman L, Chasan E (2020) Cities wonder whether recycling counts as essential during the virus. Bloomberg Green. https://www.bloomberg.com/news/articles/2020-03-27/cities-wonder-whether-recycling-counts-as-essential-during-the-virus

Klemes JJ, Van Fan Y, Tan RR, Jiang P (2020) Minimising the present and future plastic waste, energy and environmental footprints related to COVID-19. Renew Sustain Energy Rev 127(2020):109883

Korai MS, Mahar RB, Uqaili MA (2016) Optimization of waste to energy routes through biochemical and thermochemical treatment options of municipal solid waste in Hyderabad, Pakistan. Energy Conv Manag 124:333–343. https://doi.org/10.1016/j.enconman.2016.07.032

Kothari R, Sahab S, Singh HM, Singh RP, Singh B, Pathania D, Tyagi VV et al (2021) COVID-19 and waste management in Indian scenario: challenges and possible solutions. Environ Sci Pollut Res 28(38):52702–52723

Kulkarni BN, Anantharama V (2020) Repercussions of COVID-19 pandemic on municipal solid waste management: challenges and opportunities. Sci Total Environ 140693. https://doi.org/10.1016/j.scitotenv.2020.140693

Layton J (2020) Out-of-date food 'being dumped by panic-buyers'. Coventry Telegraph News Article. Available from: https://www.coventrytelegraph.net/news/coventry-news/out-date-food-being-dumped-18009419

Mallapur C (2020) Sanitation workers at risk from discarded medical waste related to COVID-19, IndiaSpend. https://www.indiaspend.com/sanitation-workersat-risk-from-discarded-medical-waste-related-tocovid-19/

Mc Carthy G, Lawlor PG, Coffey L, Nolan T, Gutierrez M, Gardiner GE (2011) An assessment of pathogen removal during composting of the separated solid fraction of pig manure. Biores Technol 102(19):9059–9067

Miezah K, Obiri-Danso K, Kádár Z, Fei-Baffoe B, Mensah MY (2015) Municipal solid waste characterization and quantification as a measure towards effective waste management in Ghana. Waste Manage 46:15–27. https://doi.org/10.1016/j.wasman.2015.09.009

Mmereki D, Baldwin A, Li B (2016) A comparative analysis of solid waste management in developed, developing and lesser developed countries. Environ Technol Rev 5(1):120–141

Morseletto P (2020) Restorative and regenerative: Exploring the concepts in the circular economy. J Ind Ecol 24(4):763–773

Munshi M, Deb S, Malakar S, Pou KJ, Chakraborty S (2022) Novel coronavirus disease (COVID-19): emergence, early infection rate, and deployment strategies for preventive solutions. In: Lessons from COVID-19. Academic Press, pp 213–240

National Green Tribunal (2017) Judgement order. Retrieved from: http://www.indiaenvironmentportal.org.in/files/Almitra%20Patel%20Corrected%20Order%202%20Jan%202017%20NGT.pdf

Neel (2020) Punjab farmers start dumping vegetables due to curfew, Chandigarh News (Times of India). https://timesofindia.indiatimes.com/city/chandigarh/punjab-farmers-start-dumping-vegetables-due-tocurfew/articleshow/748 01554.cms

Negev M, Dahdal Y, Khreis H, Hochman A, Shaheen M, Jaghbir MT, ... and Davidovitch N (2021) Regional lessons from the COVID-19 outbreak in the middle east: from infectious diseases to climate change adaptation. Sci Total Environ 768:144434

Negrete-Cardoso M, Rosano-Ortega G, Álvarez-Aros EL, Tavera-Cortés ME, Vega-Lebrún CA, Sánchez-Ruíz FJ (2022) Circular economy strategy and waste management: a bibliometric analysis in its contribution to sustainable development, toward a post-COVID-19 era. Environ Sci Pollut Res 29(41):61729–61746

Nghiem LD, Morgan B, Donner E, Short MD (2020) The COVID-19 pandemic: considerations for the waste and wastewater services sector. Case Stud Chem Environ Eng 100006. https://doi.org/10.1016/j.cscee.2020.100006

Nzeadibe TC, Ejike-Alieji AU (2020) Solid waste management during COVID-19 pandemic: policy gaps and prospects for inclusive waste governance in Nigeria. Local Environ 1–9. https://doi.org/10.1080/13549839.2020.1782357

Nzediegwu C, Chang SX (2020) Improper solid waste management increases potential for COVID-19 spread in developing countries. Resour Conserv Recycl 161:104947

Parajuly K, Fitzpatrick C, Muldoon O, Kuehr R (2020) Behavioral change for the circular economy: a review with focus on electronic waste management in the EU. Resour Conserv Recycl X 6:100035

Saadat S, Rawtani D, Hussain CM (2020) Environmental perspective of COVID-19. Sci Total Environ 138870

Salas RN (2020) Lessons from the COVID-19 pandemic provide a blueprint for the climate emergency. BMJ 370

Scudellari M (2020) How the pandemic might play out in 2021 and beyond. Nature 22–25

Sharma KD, Jain S (2019) Overview of municipal solid waste generation, composition, and management in India. J Environ Eng 145(3):04018143

Sharma HB, Vanapalli KR, Samal B, Cheela VS, Dubey BK, Bhattacharya J (2021) Circular economy approach in solid waste management system to achieve UN-SDGs: solutions for post-COVID recovery. Sci Total Environ 800:149605

Singh N, Tang Y, Zhang Z, Zheng C (2020) COVID-19 waste management: effective and successful measures in Wuhan, China. Resour Conserv Recycl 105071

Srivastava V, Ismail SA, Singh P, Singh RP (2015) Urban solid waste management in the developing world with emphasis on India: challenges and opportunities. Rev Environ Sci Biotechnol 14(2):317–337

Subbarao PMV, D'Silva TC, Adlak K, Kumar S, Chandra R, & Vijay VK (2023) Anaerobic digestion as a sustainable technology for efficiently utilizing biomass in the context of carbon neutrality and circular economy. Environ Res 116286

Tenenbaum (2020) The amount of plastic waste is surging because of the coronavirus pandemic. Forbes. https://www.forbes.com/sites/lauratenenbaum/2020/04/25/plastic-waste-during-the-time-of-covid-19/#7c4e661f7c48

Tripathi A, Tyagi VK, Vivekanand V, Bose P, Suthar S (2020) Challenges, opportunities and progress in solid waste management during COVID-19 pandemic. Case Stud Chem Environ Eng 2:100060

UNEP (2020) BASEL: waste management an essential public service in the fight to beat COVID-19. The United Nations Environment Programme (UNEP) and the Basel Convention. http://www.basel.int/Implementation/PublicAwareness/PressReleases/WastemanagementandCOVID19/tabid/8376/Default.aspx

Velenturf APM, Purnell P, Macaskie LE, Mayes WM, Sapsford DJ (2019) A new perspective on a global circular economy. In: Resource recovery from wastes: towards a circular economy, pp 1–22

WHO/WEDC (2013) Technical notes on drinking water, sanitation and hygiene in emergencies. Retrieved from: https://www.who.int/water_sanitation_health/publications/2011/WHO_TN_08_Disposal_of_dead_bodies.pdf?ua=1

Watts N, Amann M, Arnell N, Ayeb-Karlsson S, Belesova K, Boykoff M, ... and Montgomery H (2019) The 2019 report of The Lancet Countdown on health and climate change: ensuring that the health of a child born today is not defined by a changing climate. The Lancet 394(10211):1836–1878

World Health Organization (2014) Field situation: how to conduct safe and dignified burial of a patient who has died from suspected or confirmed Ebola virus disease (No. WHO/EVD/Guidance/Burials/14.2). World Health Organization

Zhao Q, Liu Y (2019) Is anaerobic digestion a reliable barrier for deactivation of pathogens in biosludge? Sci Total Environ 668:893–902

Part III
Climate Crisis, Social Responses and Sustainability on Sociological Perspectives

Chapter 9
Harnessing the Power of Climate Activism: Insights from Psychological Perspectives on Climate Change Engagement—A Systematic Review

Dipanjan Bagchi, Akancha Srivastava, and Bhawna Tushir

Abstract Scientific evidence has validated the inevitability of global warming and its effect in the form of climate change. There has been an increase in climate strikes and other forms of climate activism in recent years. It is important to understand the research landscape in psychological literature with regards to climate change and climate activism, to help guide future researchers. The databases of PubMed (Keywords: "climate activism", "climate change", "psychology", n = 1), Google Scholar (Keywords = "climate activism", "climate change", "psychology", n = 200) and Scopus database (Keywords: "climate activism" AND "climate change" AND "psychology", n = 160) were searched to create the pool of research documents. This was further filtered according to the inclusion and exclusion criteria. In the first section of this article, we have tried to explore the temporal and geographic growth trends of climate change research and collaborations using R (Bibliometric package). In the second section, we have used a text-mining approach to identify the research topics being explored in the climate change literature. R package tm along with associated packages were used to do the processing and subsequent grouping of the themes. In order to refine the classification the identified groupings were supervised by the authors. The final documents have been scoured to extract an overall understanding of the existing concepts explored so far and gauge their impact in the realm of climate change research. This systematic study casts light on the psychological views on climate activism and offers insightful information about the underlying causes that affect people's involvement in the fight and struggle against climate change. The creation of more effective techniques for encouraging climate activism and utilizing

D. Bagchi · B. Tushir
Department of Psychology, CHRIST (Deemed to be University) Delhi NCR, Ghaziabad, India
e-mail: dipanjan.bagchi@res.christuniversity.in

B. Tushir
e-mail: bhawna@christuniversity.in

A. Srivastava (✉)
Jindal School of Psychology and Counseling, O.P. Jindal Global University, Sonipat, Haryana, India
e-mail: akancha.srivastava@jgu.edu.in

U. Mukhopadhyay et al. (eds.), *Climate Crisis, Social Responses and Sustainability*,
Climate Change Management, https://doi.org/10.1007/978-3-031-58261-5_9

its capacity to inspire significant action to address climate change can be influenced by an understanding of these elements. In order to address the complex issues of climate change, this chapter emphasizes the value of multidisciplinary collaboration amongst psychologists, policymakers, educators, and activists.

Introduction

Disruption within Earth's ecological landscape is an enduring occurrence that has played a significant role in shaping our planet over centuries. We commonly refer to this persistent disruption as 'environmental disruption.' It continuously challenges the delicate equilibrium of our world, manifesting as ongoing environmental degradation. This degradation encompasses habitat destruction, wildlife extinction, and widespread pollution. Such indicators are telltale signs of environmental disruption, a phenomenon intricately linked to shifts in weather patterns and global warming. The consequences of environmental disruption are far-reaching, affecting both human and non-human life. It leads to the loss of biodiversity, disrupts ecosystems, and jeopardizes essential resources such as clean water and food. Moreover, environmental disruption can exacerbate social and economic inequalities, disproportionately impacting marginalized communities. Woven into the fabric of environmental disruption is climate change, a transformation in the Earth's climate system characterized by sustained alterations in weather patterns and temperatures. Although historically there have been natural factors that have played a role in influencing climate, we largely attribute the current rate of warming, as shown by rising global temperatures and polar ice melt, to human activities (NASA: Climate Change Evidence 2023).

A vast body of data indisputably points to these human activities as key contributors to changes in our climate, highlighting the enormous influence of environmental disruption on the Earth's climate system (Steffen et al. 2015; Woodward 2019). This scientific understanding highlights the importance of taking swift action to address environmental disruption and mitigate its widespread effects on global climate patterns (Steffen et al. 2015; Datta and Krishnamoorti 2022).

The implications of this global climate change are manifold and extend into the economic, health, social, and political spheres. These consequences encompass shifts in agricultural productivity, a heightened frequency and intensity of natural disasters, population displacement, and geopolitical tensions over resources (IPCC 2014; Nelson et al. 2007). Additionally, climate change has contributed to the emergence of new health risks and the spread of infectious diseases (Haines et al. 2006; Watts et al. 2018).

Climate activism is a complex phenomenon that has been the subject of scholarly investigation across multiple disciplines. While environmental sciences, political science, economics, and sociology have shed light on various aspects of climate change and activism through their respective lenses—political action and societal structures—a deeper understanding necessitates incorporating psychological perspectives. Political science has focused on increasing participation and awareness, emphasizing long-term benefits (Sherman et al. 2021). Sociology, on the other hand, has analyzed social movements and their advocacy for policy changes (Weckroth and Ala-Mantila 2022). These fields have extensively explored socioeconomic determinants, geographic variations, and attitudes towards climate change (Dietz et al. 2020; Weckroth and Ala-Mantila 2022). However, compared to these established fields, the field of psychology has relatively fewer and scattered studies, often focusing solely on individual or organizational perspectives. Thus, the current chapter examines the research landscape of climate change and climate activism from a psychological perspective by using systematic review along with bibliometric analysis and text mining. This chapter combines the traditional systematic review methodology with the quantitative and qualitative insights from text mining and bibliometric analysis, respectively, to provide a more comprehensive understanding of the research landscape.

Navigating Climate Change: Challenges and Opportunities

Environmental disruptions resulting in climate change exemplify the intricate interplay between natural processes, human activities, and the Earth's climate system. These disruptions stem from human activities, manifesting as temperature increases and unpredictable weather patterns. Present climate change poses substantial threats to both the planet and human civilization, but it also presents opportunities for positive transformation. Consequently, we must reevaluate our current operations, processes, and policies, viewing them as a call to embrace sustainable, long-term solutions that mitigate and enhance climate change responses.

Woodward (2019) emphasizes that environmental disruptions serve as catalysts for change, motivating individuals and authorities to adapt and pursue a more sustainable, resilient future. Recognizing the urgency of addressing climate change and implementing sustainable practices not only mitigates its impacts but also fosters innovation in renewable energy, conservation, and green technologies. This, in turn, paves the way for future generations to enjoy a sustainable future.

The momentum of climate activism mirrors the imperative outlined above to address climate change. Climate change crises, including those caused by environmental disruptions, drive a collective commitment to sustainable solutions. Consequently, climate activists tirelessly advocate for sustainable practices, renewable energy advancements, conservation initiatives, and green technology development. Climate activism emerges from an acknowledgement of the climate crisis and the shared aspiration to transition toward a sustainable and resilient future for generations to come.

On the flip side, an opposing perspective emerges. While there is a growing demand for sustainable solutions, the current climatic situation also presents opportunities for positive transformation. However, the unpredictable forces of nature and their complex interaction with climate systems and human activities create formidable challenges when addressing environmental disruptions, as the consequences can be irreversible. Moreover, despite the urgency surrounding climate change, mitigation and adaptation efforts often fall short of their intended impact. Inequities in resource distribution, access to technology, and policy implementation create substantial barriers to achieving a sustainable and resilient future. The existing body of literature highlights a multitude of inconsistencies concerning mitigation and adaptation challenges and leads to an important question on how to develop a comprehensive understanding of the complexities related to the development of effective solutions for a sustainable future and their implementation at the individual and social level to achieve the desired result. Moreover, in order to comprehend and reconcile the gap that exists between intention and action with regard to fostering a resilient and sustainable ecosystem, it is crucial to explore the underlying factors that hinder the successful implementation of mitigation and adaptation strategies.

From Climate Change Process to Climate Activism Outcomes: An Evolution

Climate change has precipitated the emergence of numerous climate movements, which embody loosely organised groups of individuals united by a common goal: advocating for government policy changes that promote environmental sustainability. The tangible outcome of these collective efforts is known as climate activism. Climate activism encompasses a spectrum of deliberate actions and activities aimed at effecting change in the realms of politics, society, and the environment. These actions primarily serve to raise awareness and mobilize individuals or groups to support an issue or challenge an existing norm in pursuit of the desired change.

Climate activism is a sequential process that begins with the identification of a challenge or issue. The process of identifying a challenge begins with the identification of a hazard that represents a substantial loss for humanity. Change requires a collaborative endeavor from a large number of individuals, resulting in community building. Community building occurs when a large cross-section of society comprehends the gravity of the obstacles and dangers involved in the process. After the threat has been successfully identified, people spring into action, and behavior interventions are devised using a serious and committed Action Plan. This results in a measurable improvement. This process culminates in a collective movement in the form of activism, which ultimately results in the creation of permanent changes to the environment (Fig. 9.1).

Fisher and Nasrin (2020) classify climate activism into two primary categories: direct and indirect. Direct climate activism is a proactive effort to achieve immediate environmental change. It goes beyond mere encouragement, imploring group members to embrace sustainable lifestyle choices, modify consumer behaviours, and champion pro-environmental practices that prioritize coexistence with nature, leaving no detrimental footprint on our ecosystems (Haenfler et al. 2012; Olzak and

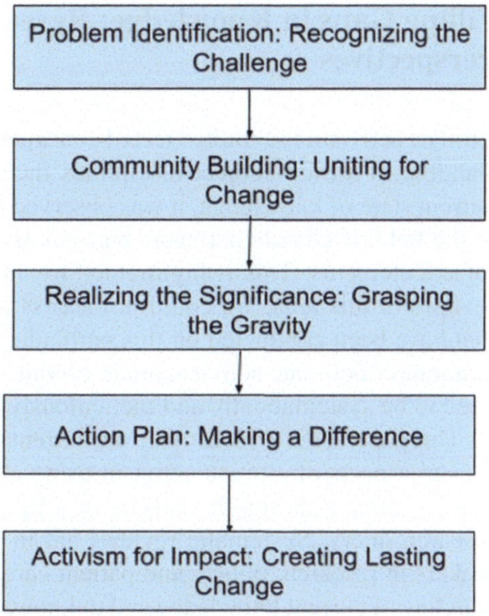

Fig. 9.1 Figure showing the sequential process of climate activism

Soule 2009; Steinhardt and Wu 2016; Fisher and Nasrin 2020). In this context, direct activism, as a facet of civic engagement, is dedicated to motivating individuals within society to make minor and major lifestyle choices and participate in identity-based movements that promote sustainable living, such as the adoption of eco-friendly habits and dietary changes like veganism (Woodward 2019; Büchs et al. 2015).

Conversely, indirect climate activism represents a form of civic engagement not directly focused on immediate environmental changes; rather, it targets governments and policymakers to instigate political and economic transformations (Woodward 2019). These civic engagement initiatives are most prominent in democratic countries where individuals have the right to voice their opinions (Fisher and Nasrin 2020). This type of activism often involves activities such as lobbying, advocating for policy changes, and participating in peaceful protests. These efforts aim to influence decision-makers and create systemic change that can have long-term effects on environmental sustainability and social justice (Haenfler et al. 2012; Saunders et al. 2013). Additionally, indirect climate activism recognizes the interconnectedness of environmental issues with other social and economic challenges, such as poverty and inequality, highlighting the need for comprehensive solutions that address multiple dimensions of sustainability (Carleton and Hsiang 2016).

Filling Gaps in Knowledge: Revealing Innovative Perspectives

Climate activism is a multi-faceted concept which could be deciphered by the understanding of varied subject disciplines incorporating a few perspectives. With the current state of knowledge, it was observed that there is a substantial knowledge gap in the field of climate activism, particularly about the understanding of its psychological elements. This is highlighted by the fact that due to the limited number of reviews in this area, we could not access any other systematic review evaluations that have been conducted on this particular subject matter of climate activism. The dynamics of climate activism are intertwined with psychological intricacies that still need to be systematically and meticulously investigated.

This paper aims to investigate the current state of knowledge regarding the psychological aspects of climate activism using a systematic review. A systematic review aims to offer a thorough analysis of all pertinent papers about a specific topic or research query. Systematic reviews aid in evaluating large data sets for decision-makers in research, policy, and patient care, offering a more comprehensive understanding of current knowledge and unknowable aspects (Owens 2021.) Conducting a systematic literature review is a necessary measure in this regard as it offers a comprehensive examination of the current body of literature through a meticulous process of selection, analysis, and synthesis of relevant available literature. By employing a systematic review of the literature, we aim to present an accurate representation of the current understanding of the issue under investigation.

Psychological research is essential in the realm of climate activism as it provides insights into motives, behavioral modification, coping strategies, social interaction, community resilience and development. It promotes the development and understanding of robust social networks, adaptive coping mechanisms, and sustainable behavioral approaches. The psychological viewpoint identifies the characteristics and intricacies of climate activism which requires an examination of individual involvement, including psychological processes such as cognition, emotion, and behavior. It is imperative for psychologists to surpass expectations and demonstrate conclusively that the incorporation of psychological concepts into viable interventions can substantially diminish greenhouse gas emissions compared to the absence of such integration (Nielsen et al. 2021). This will facilitate the advancement of practical and environmentally sound strategies to alleviate the difficulties presented by the climate crisis. Furthermore, a psychological standpoint delves into the determinants that impact the endorsement of environmentally mindful behavior, acting as catalyst between conscious intent and tangible results. Therefore, it is necessary to conduct research on the psychological underpinnings of climate activism to identify its unresolved human factors.

In an attempt to fill the existing gap, the current paper has utilized three different methodological approaches which are systematic review, bibliometric analysis, and text mining. The mix of these approaches enhances our understanding by presenting wide-ranging analysis and grasp of the psychological aspects inherent in climate activism. Through the utilization of a systematic review, the study carefully combines and assesses previous research, providing a detailed summary of the pertinent psychological investigations and understandings in the field of climate activism. Similarly, the bibliometric analysis enables a quantitative evaluation, highlighting the academic canvas, identifying significant publications, and highlighting notable authors, journals, and theme patterns relevant to the psychological aspects of climate action. In addition to these methodologies, the text mining approach utilizes computational techniques to extract, analyze, and interpret textual data from various sources. Together, these three review approaches combine to provide a detailed and comprehensive viewpoint, and psychological aspect of activism. The purpose of the current paper is to examine the research landscape of climate activism from a psychological perspective.

Objective

- To understand the publication of documents by year, affiliations, country, document type, and journals.
- To understand the research done and emerging topics associated with climate change and activism.
- To classify the retrieved documents in terms of common themes.

Method

Context

To explore the research landscape in the psychological literature on climate change and activism.

Applications

It is important to understand the research done in this domain and identify emerging research and associated topics for future researchers.

Inclusion and Exclusion Criteria

The selection of studies followed a two-stage process, commencing with an initial screening guided by the subsequent criteria:

- Studies concentrating on the theme of climate change and climate activism in the psychology discipline.
- Studies disseminated in peer-reviewed and indexed academic journals.
- Empirical investigations employing quantitative and mixed research methodologies.
- Studies published in the English language.
- Inclusion of documents available in their entirety to facilitate comprehensive comprehension.
- The choice to restrict inclusion to English-language studies was motivated by the aim of achieving a global perspective on this subject matter. Additionally, it served to overcome potential linguistic limitations encountered by researchers, given the predominance of English-language publications in this domain.

  Exclusion of studies will be warranted in cases where:

- The information provided within the study is deemed insufficient.
- Studies are published in non-peer-reviewed journals.
- The research constitutes commentaries, editorials, or review articles.

Planning the Review

This review is organized into 4 phases. These phases are organized to enhance the effectiveness and comprehensiveness of the review (Fig. 9.2).

Phase 1

A two-stage process was systematically employed for the organization and screening of the gathered data. The initial stage involved data extraction and compilation. The data for this review was retrieved between June 25 and June 26. Three sources were searched to gather data on topics relevant to this chapter, namely, Scopus, Google Scholar and PubMed. Among these three sources, a subscription is needed for Scopus and its access is provided by Elsevier and the rest are open to public use. In this present chapter the researchers connected through Google Meet to discuss, clear documents and share details of the study at regular intervals.

The documents related to climate change and climate activism were scoured through three databases. The search strings used in the Scopus database were: "climate activism" AND "climate change" AND "psychology" (n = 160). The search

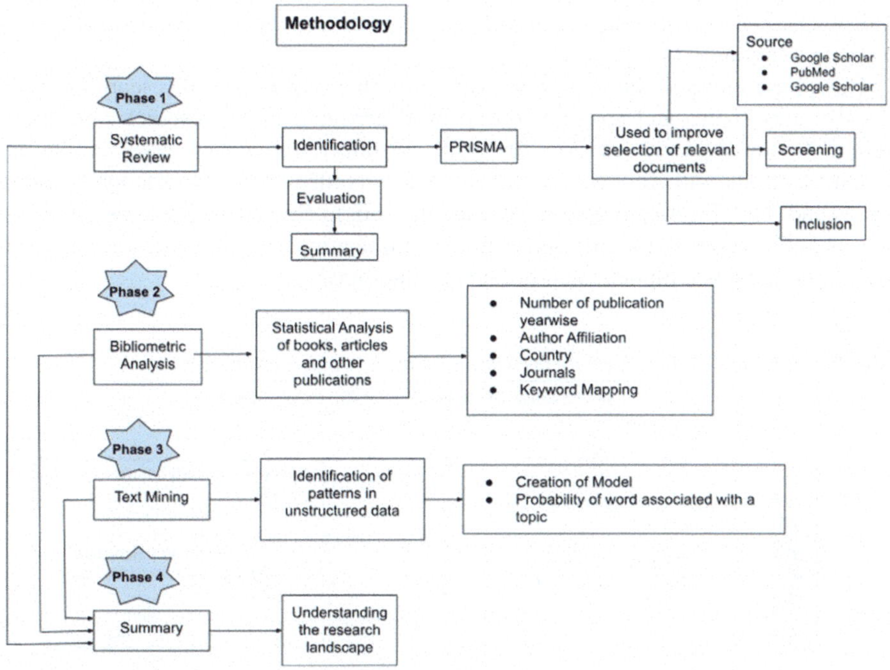

Fig. 9.2 The four phases of the review

string used in Google Scholar was "climate activism", "climate change", "psychology" (n = 71) and the same for PubMed was "climate activism", "climate change", "psychology" (n = 1). A total of 232 documents were retrieved from these 3 sources (Table 9.1).

Subsequently, the second stage encompassed the screening process, which was undertaken through collaborative consultation with fellow authors via Google Meet. In this phase the 'Preferred Reporting Items for Systematic Reviews and Meta-Analysis' (PRISMA) guidelines (Sarkis Onofre et al. 2021; Page et al. 2022; Sohrabi

Table 9.1 Details of the database used in the present study

Indicators	Description
Database source	Scopus, Google Scholar, PubMed
Publication type	Research papers, book chapters
Record volume	232
Record content	Journal, publication type, title, abstract, keywords, authorship, institution, country/region, and citation frequency
Retrieval time	June 25 and 26 June 2023

et al. 2021). The primary purpose of using this guideline is to plan, organize and report the review literature (Table 9.2).

This phase entailed the comprehensive assessment of both full research papers and abstracts. Titles and abstracts that proved pertinent to the research objectives were meticulously reviewed. Any disparities or uncertainties were resolved during virtual meetings with fellow researchers. The recorded data was efficiently catalogued and shared among the research team via Google. Spreadsheet for ease of reference and utilization. To identify and eliminate duplications, a systematic arrangement of authors' names in alphabetical order was employed.

Table 9.2 Prisma 2020 flow diagram presenting the process of systematic review

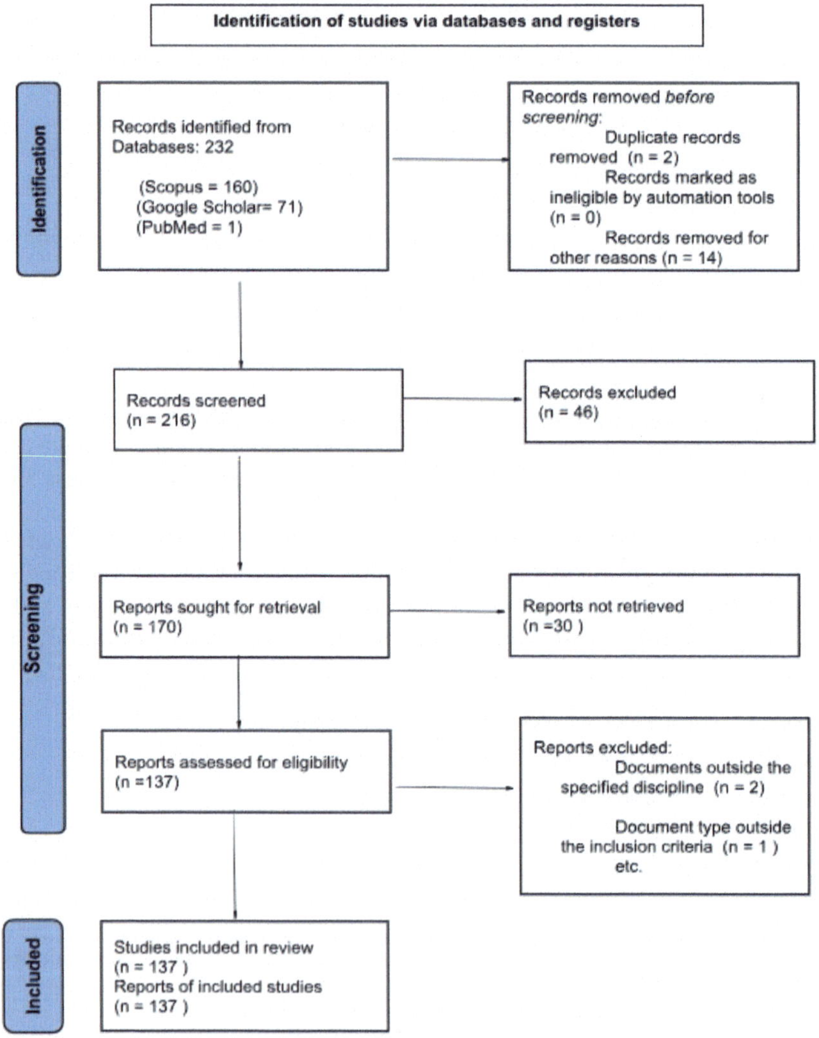

After completion of the screening process, the data was systematically reviewed and arranged according to the common themes identified by the researchers.

Phase 2

In the second phase, the Bibliometric analysis was used to understand the preferred journals that are publishing literature on climate change. To conduct bibliometric analysis a minimum of 50–100 documents are required (Seglen 1994). The present chapter meets this requirement and was used to assess several documents produced over the years.

Performance analysis involves describing the research contributors' input (Cobo et al. 2011; Ramos-Rodrígue and Ruíz-Navarro 2004). We employed Biblioshiny, a tool within the R Bibliometrics package, to gain insights into these contributors in the context of our research theme, which is Moral Intelligence. The search results were scrutinized based on the number of documents by year, affiliations, country, document type, subject area, and journals. Furthermore, we examined authors' productivity, their impact, and identified the most influential authors. We discussed data on publications from single and multiple countries, providing insights into both local and international collaborations. This data also facilitates our comprehension of the most frequently cited documents, both at a local and global scale.

Science mapping explores the connection among research components (Donthu et al. 2021). To conduct this analysis, we exported citation, bibliographical, and author keyword data to VOSviewer (version 1.6.18, Center for Science and Technology Studies, Leiden University, The Netherlands), a software tool used for visualizing network maps. The analysis included co-authorship, and we examined it in terms of authors, organizations, and countries. Link strength in this analysis reflects the frequency of collaborations among authors, organizations, and countries. Additionally, we employed co-occurrence analysis of author keywords, where the link strength between keywords indicates how often they appear together in various publications.

Phase 3

The grouping of the documents was done by utilizing the text-mining techniques. Text mining was done by using the R package 'tm' to process the corpus (137 documents). The processing includes stripping white space, converting the corpus to lowercase, and removing punctuations, stopwords and numbers. Furthermore, lemmatization was done on the corpus. Lemmatization plays a pivotal role in enhancing language modelling techniques, ultimately leading to superior document retrieval performance, where the implementation of lemmatization yields the most favourable results (Balakrishnan and Lloyd-Yemoh 2014). The R package UDPipe was used for lemmatization purposes in the present study. After lemmatization of the corpus topic modelling was done. Topic modelling is an unsupervised machine learning process where groups of similar words are present. It is also straightforward to employ the model for predicting the topic to which a new document pertains. Topic models' package was used to serve this purpose.

Phase 4

The findings from the three phases were summarized to get an understanding of the research landscape on climate activism.

Results and Discussion

Publication Output Through Time

The documents retrieved after screening span a noteworthy 11 years, during which time the volume of publications has increased, reaching a total of 51 by the year 2022. In 2023 the number of publications has reached 22 and is expected to increase as the data mining has been conducted in June 2023. It is interesting to note that Greta Thunberg founded the movement known as Fridays for Future in 2018. The impetus received from these movements also reflects on the publications in the field of psychology as will be discussed in the later sections (Fig. 9.3).

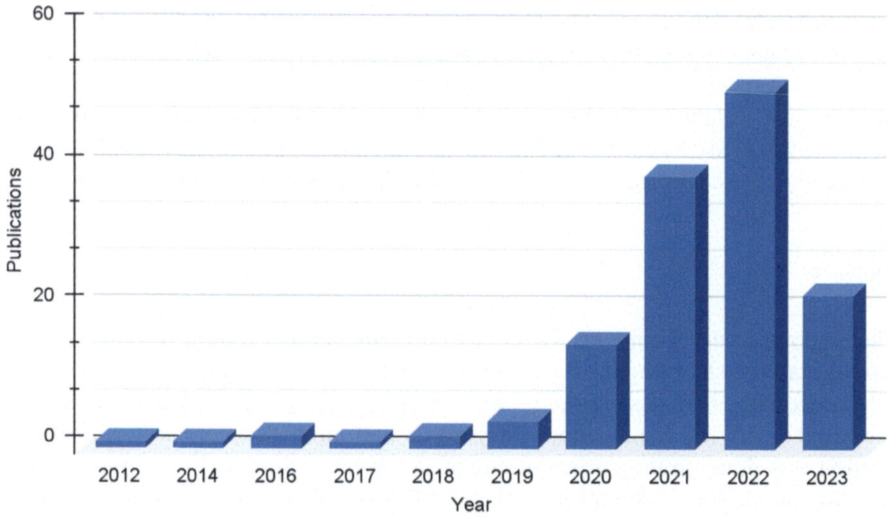

Fig. 9.3 Year-wise publication of bibliometric papers. *Note(s)* This figure illustrates the publication trend of bibliometric papers from 2012 to 2023

Preferred Journals

The analysis of most relevant sources related to climate change and climate actions were analyzed on R (bibliometric package). The list includes 8 journals with a minimum of 3 documents on the selected topic. The journal Sustainability (Switzerland) published by MDPI tops the list with 14 documents. This journal's coverage has been discontinued in Scopus since 2020. The journal Frontiers in Psychology which is indexed in Scopus comes next with 9 publications. The Journal of Environmental Psychology which is indexed in Scopus with a cite score of 10.7, has 6 publications. The next journal is Frontiers in Political Science, also indexed in Scopus with a cite score of 1.6 and has 5 publications. The Journal of Environmental Education Research also indexed in Scopus with a cite score of 7.1 and has a total of 4 publications on climate change and climate activism. Lastly, two journals have a minimum of 3 publications each, namely Children's Geographies and Frontiers in Communication; both indexed in Scopus having a cite score of 4.4 and 2.7 respectively. It should be noted that the journal's most cited publications have started in 2018. This coincides with the timeline of the establishment of Fridays for Future and the increase in research publications (Table 9.3).

It should be noted that the cite score mentioned in the above table is an alternative to the Impact factor. Although cite score is exclusively used for journals indexed in Scopus it can be all the journals except one that is not indexed in Table 9.3. Before publishing an article, selecting a journal is important. The selection process is influenced by different factors such as cite score, language of the journal and even reaching the target audience. In the above table, it can be seen that the article with the highest citation has been published in a non-indexed journal. Hence, fellow researchers should choose the journal carefully keeping in mind the primary purpose of the article.

Author Keyword

- **Terminology and Concept**. The total number of keywords in the 137 documents was 378. These keywords were further filtered for cogeneric words and the minimum number of occurrences of a keyword was set to 2. This reduced the number of author keywords to 62. The reason for setting the occurrence of authors' keywords to 2 was to identify the emerging topics related to climate change and climate activism. The keywords with a minimum of 2 occurrences were 33 (53%), 12 keywords (social identity, ecological grief, climate change anxiety, risk perception, education for sustainable development, psychology, eco-anxiety, anxiety disorders, empowerment, resilience, climate crisis, behaviour change) occurred thrice (19%), 8 keywords (social movements, agency, anxiety, coping, sustainability, young people, climate change education, social media), occurred 4 times (13%), 3 keywords (emotion, transformative learning, pro-environmental

Table 9.3 The top 7 most productive journals on climate change and climate activism research with their most cited article

	Journal	PY start	TC	Cite score	The most cited article (reference)	Times cited	Publisher
1	Sustainability (Switzerland)	2020	245	NA	Anxiety and the ecological crisis: an analysis of eco-anxiety and climate anxiety	141	MDPI
2	Frontiers in Psychology	2021	12	4.5	Climate change and emotions: analysis of people's emotional states in Southern Ecuador	5	Frontiers
3	Journal of Environmental Psychology	2021	16	10.7	Climate anxiety, wellbeing and pro-environmental action: correlates of negative emotional responses to climate change in 32 countries	10	Science Direct
4	Environmental Education Research	2022	14	7.1	Educating for hope and action competence: a study of secondary school students and teachers in England	8	Taylor and Francis
6	Children's Geographies	2021	12	1.2	To trust or not to trust? Young people's trust in climate change science and implications for climate change engagement	16	Taylor and Francis
7	Frontiers in Communication	2018	23	2.7	Climate change marches as motivators for bystander collective action	21	Frontiers

Note PY start: Publication year start: Total publications; *TC* total citations

behaviour) have occurred 5 times (5%), 1 keyword (hope) has occurred only 1 time (2%), 4 keywords (Friday's for future, climate activism, climate anxiety and emotions) occurred 7 times(6%), 1 keyword (collective action) have occurred 8 times (2%) and 1 keyword (youth) occurred 14 times (2%).

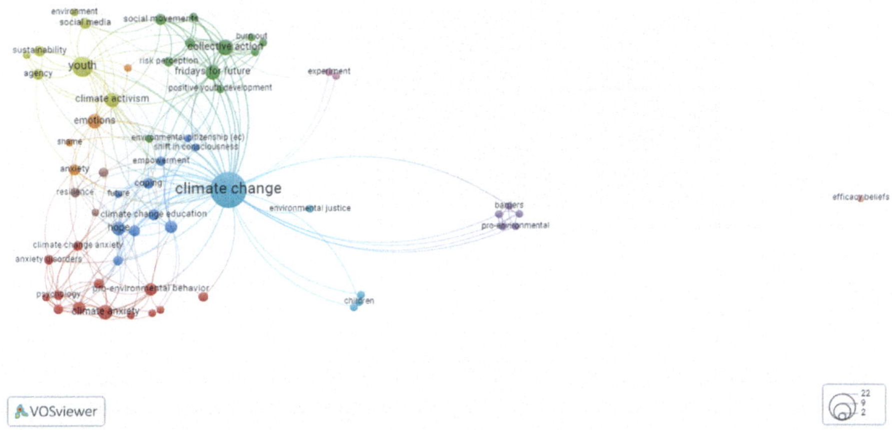

Fig. 9.4 A screenshot of the bibliometric map created based on author keywords co-occurrence with the overlay visualization model. *Note* Minimum occurrences of a keyword are set to two. The following URL can be used to open Fig. 9.5 in VOSviewer: https://tinyurl.com/yqaks9dh

- ***Emerging Topics***. The emerging topics are the ones with a minimum occurrence of 2. These include barriers, facilitators, older persons, pro-environmental, responsibility, youth climate strikes, affect, environmental citizenship, shift in consciousness, burnout, collective activism, extended parallel processing model, positive youth development, future, pro-environmental behaviour, climate emotions, protest, shame, sustainability science education, children, community engagement, experiment, environment, environmental justice, fight shame, Greta Thunberg, political participation, efficacy beliefs and environmental communication (Fig. 9.4).

Text Mining

- ***Topic Modelling***. "Topic modelling is part of a class of text analysis methods that analyze "bags" or groups of words together—instead of counting them individually–in order to capture how the meaning of words is dependent upon the broader context in which they are used in natural language" (Bail n.d.). Although there are other methods of doing this same analysis like k-means clustering, topic modelling has certain advantages. First, it does not require binary classification of documents and texts. It assigns probability values to each document, hence with a large data set the accuracy increases. Secondly, this method 'employs more sophisticated iterative Bayesian techniques'.

  In, this present chapter the Latent Dirichlet Allocation (LDA) method was used. The R package named 'topicmodels' was used to do the analysis. The selection of an optimal number of topics depends on a number of factors for example the

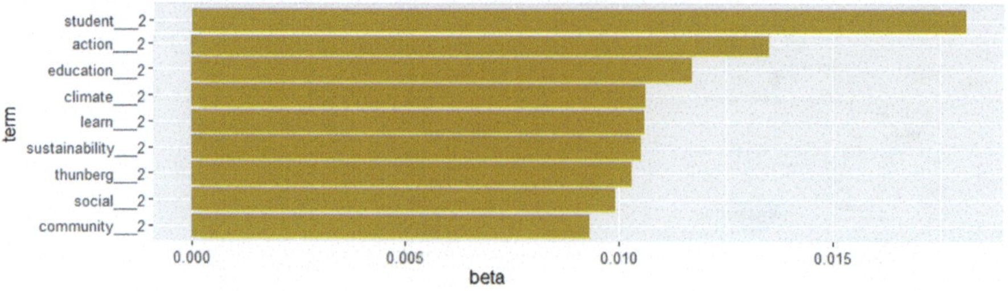

Fig. 9.5 Topic 1: Greta Thunberg

purpose of the study, the interpretability of the result and one can take the help of perplexity values (Gan and Qi 2021). In this chapter, the authors discussed and explored the results of various numbers of topics. It was decided to model the data on 6 topics for the purpose of interpretability in spite of a high perplexity value of 328.9576.

- ***Identified Topics***. The primary aim of this section is to give an overview of the documents present in the chapter according to the topics identified by the LDA method. The overview is related to

Topic 1. The first topic consists of the word among others Thunberg. Greta Thunberg is a well-known climate activist. She has been a tremendous influence on how climate change is being perceived in today's world. Other words associated with this topic are student, action, education, climate, learn, sustainability, social and community. In this present study, we have used the most basic approach to text analysis: the unigram approach. For further understanding of this theme, correlation was conducted to understand the context of the keywords identified in topic 1 (Fig. 9.5; Table 9.4).

It is quite clear from the above table that this theme is related to student activism initiated by Greta Thunberg and the call for collective action. It is also noteworthy that the word social is related to media. This group of documents discusses about the power of social media to build community resilience. Researcher Susan Clayton, professor of psychology at The College of Wooster, is also associated with this topic as her research work encompasses conservation psychology, climate anxiety, psychology of justice and identity.

Topic 2. The words associated with the second topic include people, young, child, future, youth, climate, hope, think and strike. For further understanding, the words associated with these keywords are explored. The associate words with a minimum of 0.30 correlation coefficient are included in the list (Fig. 9.6; Table 9.5).

The above table demonstrates that the associated words are linked to young children's participation in strikes. Fridays for Future is organized throughout the world by school children which was started in 2018 by Greta Thunberg (BBC 2023). Mention of Australia is also there in the table. Further analysis could shed light on

Table 9.4 Words with a minimum of 0.30 correlation with each of the keywords in Topic 1

Keywords	Associated words							
Student	School (0.48)	Teacher (0.46)	Course (0.34)	Classroom (0.31)				
Action	Collective (0.57)	Climate (0.31)						
Education	Transformative (0.32)	Learn (0.33)						
Clayton	Change (0.78)	Anxiety (0.42)	Clayton (0.35)	Action (0.31)				
Thunberg	Collective (0.63)	Mediation (0.47)	Ideology (0.47)	Stronger (0.46)	Familiar (0.45)	Intention (0.35)	Liberal (0.34)	Action (0.31)
Social	Media (0.34)							
Community	Project (0.36)	Resilience (0 36)	Local (0.32)					

Note No words are associated with 'sustainability', 'learn' is already associated with education. The correlation coefficient values are mentioned in brackets for each associated word

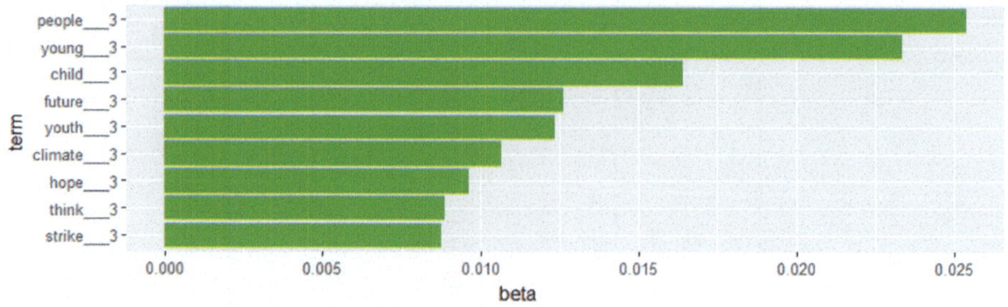

Fig. 9.6 Topic 2: climate anxiety and school strikes

Table 9.5 Words with a minimum of 0.30 correlation with each of the keywords in Topic 2

Keywords	Associated words				
People	Young (0.46)				
Young	Youth (0.46)	Adult (0.44)			
Child	Food (0.52)	Animal (0.49)	Parent (0.41)	Family (0.35)	
Future	Hope (0.33)				
Climate	Anxiety (0.63)	Clayton (0.47)	Action (0.47)		
Think	Complexity (0.55)	Competence (0.46)			
Strike	School (0.36)	Motivator (0.36)	Participation (0.32)	September (0.32)	Australian (0.31)

Note The correlation coefficient values are mentioned in brackets for each associated word

its correlation with strikes. On running further analysis on the words Australian it was found that the words motivator, school and strike were associated with it. This points out the student activism for climate change. September was significantly associated with the word strike pointing towards the September strike in Australian schools. So documents included in these topics talk about climate anxiety, hope and school strikes in Australia.

Third Topic. The third topic has words like emotion, climate, eco-anxiety, emotional, hope, change, people, ecological, and feel. Further analysis of correlated words will shed light on their distinctiveness from other topics (Fig. 9.7; Table 9.6).

This analysis provides insight into how this group is different from the other two mentioned above. This group of documents deals with the relationship between

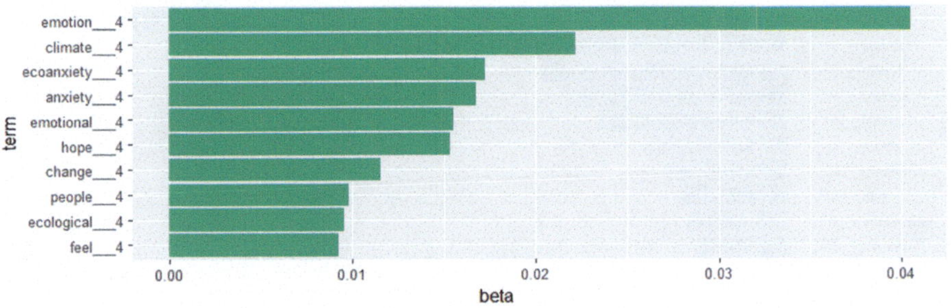

Fig. 9.7 Topic 3: eco-anxiety and hope

Table 9.6 Words with a minimum of 0.30 correlation with each of the keywords in Topic 3

Keywords	Associated words		
Emotion	Sadness (0.32)	Ecological (0.31)	
Anxiety	Adaptive (0.35)	China (0.31)	Impairment (0.30)
Exo-anxiety	Ecological (0.47)	Phenomena (0.35)	Interdisciplinary (0.34)
Feel	Guilt (0.32)		

Note The correlation coefficient values are mentioned in brackets for each associated word

humans and between humans and their physical world (ecological). Also, the eco-anxiety and sadness associated with it. The analysis also shows guilt is associated with the word feel. Interdisciplinary is associated with eco-anxiety as an interdisciplinary approach is required to tackle the current climate crisis. This topic is also related to documents associated with China. Further analyses were run to find words associated with China. The associated words are impairment (0.53), functional (0.51), harm (0.40), observe (0.38) and anxiety (0.31).

Fourth Topic. Words like climate and change are common as the search keywords include these terms. The words that make this topic unique are health, mental, country, risk, experience and impact (Fig. 9.8; Table 9.7).

These words associated with health are mental, physical, wellbeing, determinant and depression indicating these documents deal with determinants of mental and physical wellbeing. Also impact of mental distress on health. Interestingly the word risk is associated with disaster, reduction, perception and flood. Overall it can be said that documents associated with these topics deal specifically with the risk of disaster and flood and the impact it has on mental and physical well-being. It should be noted this analysis gives an estimate of the topics that are being explored by researchers.

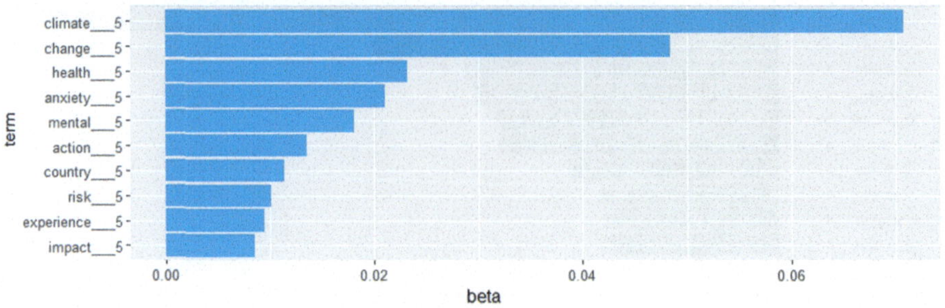

Fig. 9.8 Topic 4: risk of disaster and mental health

Table 9.7 Words with a minimum of 0.30 correlation with each of the keywords in Topic 4

Keywords	Associated words				
Health	Mental (0.84)	Physical (0.42)	Wellbeing (0.42)	Determinant (0.41)	Depression (0.30)
Mental	Distress (0.35)	Impact (0.35)			
Country	Harm (0.33)	Observe (0.31)	Anxiety (0.31)	Impairment (0.30)	
Risk	Disaster (0.66)	Reduction (0.42)	Perception (0.40)	Flood (0.30)	

Note The correlation coefficient values are mentioned in brackets for each associated word. No words are associated with 'experience' and 'impact'

Fifth Topic. The keywords that define this topic are effect, model, efficacy, collective and belief. The prominence of these words will be further explained by the words associated with them (Fig. 9.9).

The first term 'model' is associated with the fit, variance, mediation, test, variables and regression indicating the statistical analysis associated with this group of studies. The word efficacy is associated with self-efficacy. This can be explained by

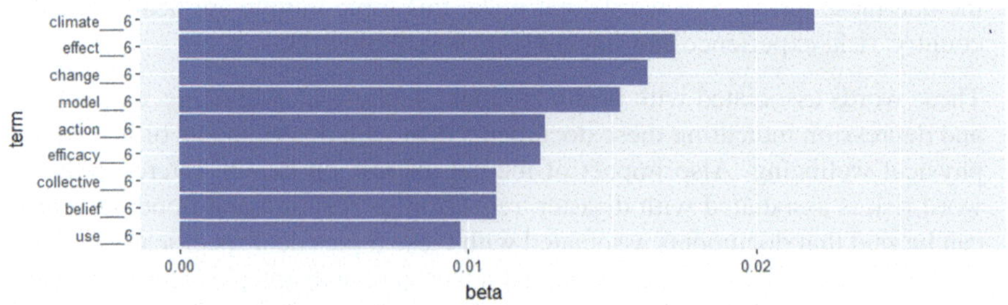

Fig. 9.9 Topic 5 self-efficacy and beliefs

Table 9.8 Words with a minimum of 0.30 correlation with each of the keywords in Topic 5

Keywords	Associated words				
Model	Fit (0.49)	Variance (0.44)	Mediation (0.41)	Test (0.39)	Variables (0.38)	Regression (0.31)
Efficacy	Belief (0.55)	Collective (0.54)	Predictor (0.51)	Self-efficacy (0.40)	Collaboration (0.33)	
Collective	Greta (0.63)	Action (0.57)	Efficacy (0.54)	Ideology (0.46)	Support (0.33)	
Belief	Agreement (0.47)	Linden (0.45)	Consensus (0.40)	Perceive (0.39)	Leiserowitz (0.33)	

Note The correlation coefficient values are mentioned in brackets for each associated word. No words are associated with 'experience' and 'impact'

the unigram approach. Since this approach takes into account individual words the word self-efficacy has been separated into two words. The word efficacy is also associated with Greta, belief, collective, predictor and collaboration. It is interesting to note that the words Linden and Leiserowitz. Sander van der Linden is a professor of psychology in the Department of Psychology, at the University of Cambridge (Google Scholar n.d). His area of research encompasses social psychology, social influence, persuasion, misinformation Judgement and decision-making. On further investigation, it was found words like consensus, agreement, and message were associated with 'linden' indicating his line of research in climate change. Another researcher identified through word association was Anthony Leiserowitz, Senior Research Scientist, Lecturer, and Director of the Yale Program on Climate Change Communication (YPCCC). It should be noted that these two researchers were identified for their notable contribution to this field of research through word association. Although on further investigation other researchers could be identified, but that is outside the scope of this study (Table 9.8).

Topic 6. The most notable keywords unique to this group are political and issue. The other keywords are already discussed above. Hence these two words will be focused upon (Fig. 9.10; Table 9.9).

These groups of articles focus on political ideology and activism. The other word 'issue' revolves around public, membership, concrete, spark and distance. Due to the lack of association, it can be concluded that these documents are linked to political action and social media.

Summary of the Result

This paper presents a comprehensive picture of the research landscape and highlights the various research gaps that exist in climate change research. The current

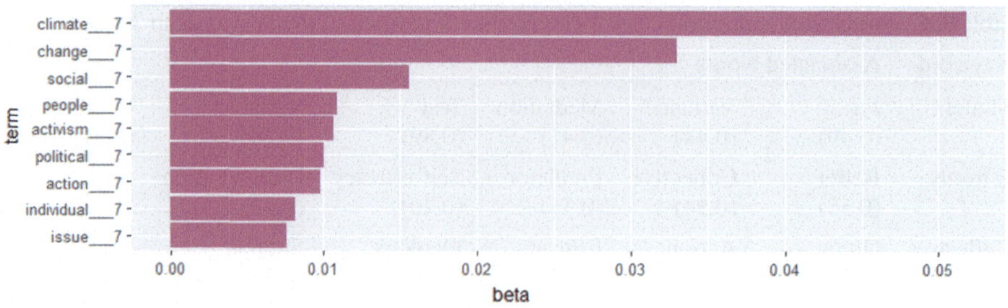

Fig. 9.10 Topic 6 political action and social media

Table 9.9 Words with a minimum of 0.30 correlation with each of the keywords in Topic 6

Keywords	Associated words				
Political	Politics (0.42)	Ideology (0.33)			
Issue	Public (0.55)	Membership (0.50)	Concrete (0.41)	Spark (0.37)	Activism (0.33)	Distance (0.31)

paper has utilized a systematic review as a methodological approach (Synder 2019). Any researcher willing to investigate this domain will get a comprehensive understanding of the emerging topics, and relevant journals and connect the dots through future research. The literature was investigated in three phases. In the first phase, bibliometric analysis helped in identifying Sustainability, Frontiers in Psychology and Journal of Environmental Psychology as the top 3 journals with relevant articles on climate activism. Emerging topics in climate activism research include collective activism, shame, and environmental justice efficacy beliefs among others. The systematic review demonstrates the necessity of investigating and comprehending individual attitudes, emotions, and behaviours regarding climate change (Kleres and Wettergren 2017; Steg 2023). These studies establish emotions, such as hope and anxiety, as powerful indicators of climate action and involvement in climate activism (Nabi et al. 2018; Brosch 2021). Moreover, engagement in such activities may be associated with the development of one's identity (Martiskainen et al. 2020).

The literature on psychological aspects influencing climate activism highlights important factors such as collective action triggers, affective involvement, self-efficacy, and collective efficacy (Castiglione 2020, 2022). They place further focus on the role of methods employed by citizen activists in influencing decision-makers, such as affirmation and cost–benefit framing. These studies highlight the necessity of thoroughly investigating and comprehending the psychological factors associated with climate action to better incorporate psychological findings.

In the third phase through text mining, additional information about climate activism was procured. The first topic that was identified was Greta Thunberg and her role in initiating climate activism cannot be undermined. Researchers have tried to understand this phenomenon by understanding her role in climate activism as is evident from the associated words like collective action, intention, media and resilience. The role of social media in political action came up as a new topic that needs further investigation. The role of social influence, persuasion, and misinformation in climate activism and its effect were identified as the major areas that warrant further research. Hence, using text mining helped to understand this topic in more detail in unique ways that would not have been possible otherwise.

Furthermore, research acknowledges the necessity of a thorough and unified approach to climate change interventions and policies that integrate psychological considerations. Furthermore, numerous studies have highlighted the significance of different psychological factors in shaping successful climate change interventions and policies. These studies primarily focus on comprehending human perceptions and behaviours, and incorporating psychological concepts into practical interventions (Clayton et al. 2015). The research findings collectively demonstrate the potential for psychological views to improve the efficacy of climate change interventions and policies. Furthermore, there is an urgent need for more comprehensive and cohesive research that takes into account both human and organizational viewpoints to enhance the effectiveness of climate change treatments and policies.

Limitations

Apart from the databases of Scopus, Google Scholar and PubMed no other databases were searched. Although Scopus is a credible database with over 69 million records and a total of 23,000 indexed journals and it is bigger than the Web of Science (Falagas et al. 2008), a database like the Web of Science could be searched for a comprehensive review. This limitation increases the probability of excluding other prospective studies.

Another limitation of this study is the use unigram approach. This means every conjoined word is separated. In this process, a single word loses the power of explaining the correlated keywords.

It should be noted that due to the use of different databases identification of authors with significant contributions towards this topic was not possible through the biblometrix package (namely Biblioshiney). Also, identification of Multi-Country Publication and Single Country Publication was not possible as due to the use of different databases.

Lastly, in the topic modelling section, the number of topics is usually selected on the basis of certain guidelines. Perplexity value is one of them. The value is said to be acceptable if it is less than 100. But in this study, it was more than 100 (Gan and Qi 2021) but still accepted to ensure the interpretability of the topics. For a more accurate review, full-text screening is required.

Scope and Future Directions

This chapter offers a comprehensive overview of the research landscape of climate change and climate activism research. It can act as a quick reference for researchers new to this area of research. Researchers interested in working on this topic can refer to emerging topics to find an area that is still evolving but has the potential to provide more insights into climate change and activism.

The importance of bibliometric analysis and text mining approaches to understand a corpus of data is also highlighted through this chapter. Text mining techniques using supervised and unsupervised machine learning techniques have the potential to give a quick overview of large datasets. But for a more detailed analysis of full-text screening and classification is recommended. Future researchers are encouraged to conduct a systematic review using the same keywords used in this study.

The aim of this chapter is to stimulate discussion of climate change and climate activism. Understanding the determinants of activism, and its significance in slowing climate change is necessary this is one of the most pressing topics of the twenty-first century.

Conclusion

Based on 137 documents, this chapter offers an overview of research trends in climate change and climate activism research in the psychology discipline. The number of publications has been consistently increasing since 2012, and this upward trend is anticipated to persist. We identified journals having a significant volume of publications related to this subject. This can guide researchers in the selection of journals based on the aim of their study. The existing research work was divided into 6 major topics: Greta Thunberg, Climate Anxiety and School Strikes, Eco-Anxiety and Hope, Risk of Disaster and Mental Health, Self-Efficacy and Beliefs and Political Action and Social Media. Emerging topics were also highlighted to guide researchers in navigating the research landscape. Lastly, this article ended with future directions for researchers interested in this domain of climate change and climate activism in the domain of psychology.

References

Bail C (n.d.) Topic modeling. Github. Retrieved 17 Oct 2023, from https://cbail.github.io/SICSS_Topic_Modeling.html

Balakrishnan V, Lloyd-Yemoh E (2014) Stemming and lemmatization: a comparison of retrieval performances

BBC (2023) Greta Thunberg: who is the climate activist and what has she achieved?. BBC News. https://www.bbc.com/news/world-europe-49918719

Brosch T (2021) Affect and emotions as drivers of climate change perception and action: a review. Curr Opin Behav Sci 42:15–21. https://doi.org/10.1016/j.cobeha.2021.02.001

Büchs M, Saunders C, Wallbridge R, Smith G, Bardsley N (2015) Identifying and explaining framing strategies of low carbon lifestyle movement organisations. Glob Environ Chang 35:307–315. https://doi.org/10.1016/j.gloenvcha.2015.09.009

Carleton TA, Hsiang SM (2016) Social and economic impacts of climate. Science 353(6304). https://doi.org/10.1126/science.aad9837

Castiglione A (2020) Climate activism: what we know and what more we need to learn. A psychological and sociological review. https://doi.org/10.31234/osf.io/7brz3

Castiglione A, Brick C, Holden S, Miles-Urdan E, Aron AR (2022) Discovering the psychological building blocks underlying climate action—a longitudinal study of real-world activism. R Soc Open Sci 9(6). https://doi.org/10.1098/rsos.210006

Clayton S, Devine-Wright P, Stern PC, Whitmarsh L, Carrico A, Steg L, Swim J, Bonnes M (2015) Psychological research and global climate change. Nat Clim Change 5(7):640–646. https://doi.org/10.1038/nclimate2622

Cobo MJ, López-Herrera AG, Herrera-Viedma E, Herrera F (2011) Science mapping software tools: review, analysis, and cooperative study among tools. J Am Soc Inform Sci Technol 62(7):1382–1402

Datta A, Krishnamoorti R (2022) Understanding the greenhouse gas impact of deforestation fires in Indonesia and Brazil in 2019 and 2020 frontiers in climate. Front Media. https://doi.org/10.3389/fclim.2022.799632

Dietz T, Shwom RL, Whitley CT (2020) Climate change and society. Ann Rev Sociol 46(1):135–158. https://doi.org/10.1146/annurev-soc-121919-054614

Donthu N, Kumar S, Mukherjee D, Pandey N, Lim WM (2021) How to conduct a bibliometric analysis: an overview and guidelines. J Business Res 133:285–296. https://doi.org/10.1016/j.jbusres.2021.04.070

Falagas ME, Pitsouni EI, Malietzis GA, Pappas G (2008) Comparison of PubMed, scopus, web of science, and google scholar: strengths and weaknesses. FASEB J 22(2):338–342. https://doi.org/10.1096/fj.07-9492lsf

Fisher DR, Nasrin S (2020) Climate activism and its effects. WIREs Clim Change 12(1). https://doi.org/10.1002/wcc.683

Gan J, Qi Y (2021) Selection of the optimal number of topics for LDA topic model—taking patent policy analysis as an example. Entropy 23(10):1301

Google (n.d.) Google Scholar. Google scholar profiles. https://www.google.com/intl/en/scholar/citations.html

Haenfler R, Johnson B, Jones E (2012) Lifestyle movements: exploring the intersection of lifestyle and social movements. Soc Mov Stud 11(1):1–20. https://doi.org/10.1080/14742837.2012.640535

Haines A, Kovats R, Campbell-Lendrum D, Corvalan C (2006) Climate change and human health: impacts, vulnerability and public health. Public Health 120(7):585–596. https://doi.org/10.1016/j.puhe.2006.01.002

IPCC (n.d.) AR5 synthesis report: climate change 2014. https://www.ipcc.ch/report/ar5/syr/

Kleres J, Wettergren S (2017) Fear, hope, anger, and guilt in climate activism. Soc Mov Stud 16(5):507–519. https://doi.org/10.1080/14742837.2017.1344546

Leiserowitz A (n.d.) Yale center for business and the environment. https://cbey.yale.edu/our-community/anthony-leiserowitz#:~:text=Anthony%20Leiserowitz%2C%20Ph.,climate%20change%20and%20the%20environment

Martiskainen M, Axon S, Sovacool BK, Sareen S, Furszyfer Del Rio D, Axon K (2020) Contextualizing climate justice activism: knowledge, emotions, motivations, and actions among climate strikers in six cities. Glob Environ Change 65:102180. https://doi.org/10.1016/j.gloenvcha.2020.102180

Nabi RL, Gustafson A, Jensen R (2018) Framing climate change: exploring the role of emotion in generating advocacy behavior. Sci Commun 40(4), 442–468. https://doi.org/10.1177/1075547018776019

NASA (2023) Climate change evidence: how do we know? (n.d.) Climate change: vital signs of the planet. NASA. https://climate.nasa.gov/evidence/#otp_history

Nelson DR, Adger WN, Brown K (2007) Adaptation to environmental change: contributions of a resilience framework. Annu Rev Environ Resour 32(1):395–419. https://doi.org/10.1146/annurev.energy.32.051807.090348

Nielsen KS, Clayton S, Stern PC, Dietz T, Capstick S, Whitmarsh L (2021b) How psychology can help limit climate change. Am Psychol 76(1):130–144. https://doi.org/10.1037/amp0000624

Olzak S, Soule SA (2009) Cross-cutting influences of environmental protest and legislation. Soc For 88(1):201–225. https://doi.org/10.1353/sof.0.0236

Owens JK (2021) Systematic reviews: brief overview of methods, limitations, and resources. Nursing Open. Advance Online Publication. https://doi.org/10.1111/nae2.28

Page MJ, Moher D, McKenzie JE (2022) Introduction to PRISMA 2020 and implications for research synthesis methodologists. Res Synth Methods 13(2):156–163

Ramos-Rodrígue AR, Ruíz-Navarro J (2004) Changes in the intellectual structure of strategic management research: a bibliometric study of the Strategic Management Journal, 1980–2000. Strateg Manag J 25(10):981–1004

Sarkis-Onofre R, Catalá-López F, Aromataris E, Lockwood C (2021) How to properly use the PRISMA Statement. Syst Rev 10(1):117. https://doi.org/10.1186/s13643-021-01671-z. PMID: 33875004; PMCID: PMC8056687

Saunders C, Büchs M, Papafragkou A, Wallbridge R, Smith G (2013) Beyond the activist Ghetto: a deductive blockmodelling approach to understanding the relationship between contact with environmental organisations and public attitudes and behaviour. Soc Mov Stud 13(1):158–177. https://doi.org/10.1080/14742837.2013.832623

Seglen PO (1994) Causal relationship between article citedness and journal impact. J Am Soc Inf Sci 45(1):1–11. https://doi.org/10.1002/(sici)1097-4571(199401)45:1%3c1:aid-asi1%3e3.0.co;2-y

Sherman DK, Shteyn MF, Han H, Van Boven L (2021) The exchange between citizens and elected officials: a social psychological framework for citizen climate activists. Behav Publ Policy 5(4):576–605. https://doi.org/10.1017/bpp.2020.41

Snyder H (2019) Literature review as a research methodology: an overview and guidelines. J Bus Res 104:333–339

Sohrabi C, Franchi T, Mathew G, Kerwan A, Nicola M, Griffin M, Agha M, Agha R (2021) PRISMA 2020 statement: what's new and the importance of reporting guidelines. Int J Surg 88. Article 105918. https://doi.org/10.1016/j.ijsu.2021.105918

Steffen W, Broadgate W, Deutsch L, Gaffney O, Ludwig C (2015) The trajectory of the anthropocene: the great acceleration. Anthr Rev 2(1):81–98. https://doi.org/10.1177/2053019614564785

Steg L (2023) Psychology of climate change. Annu Rev Psychol 74:391–421. https://doi.org/10.1146/annurev-psych-032720-042905

Steinhardt HC, Wu F (2016) In the name of the public: environmental protest and the changing landscape of popular contention in China. China J 75:61–82. https://doi.org/10.1086/684010

Watts N, Amann M, Arnell N, Ayeb-Karlsson S, Belesova K, Berry H, Bouley T, Boykoff M, Byass P, Cai W, Campbell-Lendrum D, Chambers J, Daly M, Dasandi N, Davies M, Depoux A, Dominguez-Salas P, Drummond P, Ebi KL, Costello A et al (2018) The 2018 report of the Lancet Countdown on health and climate change: shaping the health of nations for centuries to come. Lancet 392(10163):2479–2514. https://doi.org/10.1016/s0140-6736(18)32594-7

Weckroth M, Ala-Mantila S (2022) Socioeconomic geography of climate change views in Europe. Glob Environ Change 72:102453.https://doi.org/10.1016/j.gloenvcha.2021.102453

Woodward A (2019) Climate change: Disruption, risk and opportunity. Glob Trans 1:44–49. https://doi.org/10.1016/j.glt.2019.02.001

Chapter 10
A Review on Traditional Knowledge: A Sustainable Solution for the Climate Crisis

Sharaniya Vijitharan

Abstract Nature-based solutions are crucial to climate change mitigation and adaptation. Traditional knowledge is a credible resource based on nature-based solutions through understanding and predicting environmental and climatic changes by direct observation, practices, and information passed over generations verbally. The local and indigenous people's lifestyle is to survive with nature without disturbing the balance of ecosystem services. This chapter aimed to explore the role of traditional knowledge as a sustainable solution for predicting, mitigating, and adapting to climate change, as it was recently recognized and started to be incorporated into international climate agreements. This chapter reviewed 63 research papers collected from Web of Science, Scopus, and Google Scholar databases for examining traditional knowledge across different cultures and regions. Indigenous and local communities use various indicators to forecast the weather. These traditional indicators denoted the indication of weather events to develop their adaptive responses against changes and variations in climate. Types of local indicators were related to vegetation, animals, insects, birds, astronomy, and atmosphere. Conventional climate projection and adaptive measures play a crucial role in developing nations with limitations in modern technology. Documentation and records of such prediction methods and adaptation mechanisms are still lacking globally. A lack of documentation records, modernization, and conflict with Western science hinders traditional knowledge. Integrating this old-aged knowledge into modern practices can help develop climate-smart agriculture and sustainable natural resource management, and biodiversity conservation to overcome impacts and design appropriate adaptation strategies by scaling up from the local level. Therefore, this chapter provides in-depth insights into how traditional knowledge can tackle climate change.

S. Vijitharan (✉)
Department of Bio-science, Faculty of Applied Science, University of Vavuniya, Vavuniya, Sri Lanka
e-mail: sharaniya@vau.ac.lk

U. Mukhopadhyay et al. (eds.), *Climate Crisis, Social Responses and Sustainability*, Climate Change Management, https://doi.org/10.1007/978-3-031-58261-5_10

Introduction

Climate change has emerged as a worldwide challenge, affecting people from all corners of life, regardless of their socio-economic status. It is a key driver for many natural calamities. However, its effects can vary depending on geographical location, frequency, and intensity of climate-related hazards. Integrating sustainable strategies for mitigating and adapting to climate change is imperative to address climate change and adhere to the Paris Climate Agreement's aim of limiting the earth's surface temperature to 1.5 °C (United Nations 2013). Therefore, we are facing an escalating climate crisis globally by joining hands with dedicated and responsible efforts. However, the integration of traditional knowledge in international and national-level assessments has been poorly recognized and exercised in practice, including Millenium Ecosystem Assessment, the Intergovernmental Panel on Climate Change (IPCC), and the Economics of Ecosystems and Biodiversity (Turnhout et al. 2012). These global initiatives are focused on the sustainable use and management of ecosystems and various ecosystem services, and biodiversity. As a result, renewed attention has been gained in the traditional knowledge for its potential to extend sustainable climate solutions as the indigenous people and their traditional systems were recognized by the Paris Agreement (Hosen et al. 2020). Mainly in developing countries, traditional knowledge advances excellent help in decision-making on agricultural and natural resources governance, sustainable use, and conservation of ecosystems and biodiversity in addition to adapting to climate change, where immature modern technologies still exist (Niang et al. 2014; Tengo et al. 2014; IPBES 2018; Mekonnen et al. 2021; IPCC 2021, 2022).

Currently, the acknowledgement of traditional knowledge can be seen in many inetenational platforms. Notably, there is no universally agreed single definition available for traditional knowledge. It indicates that traditional knowledge can be defined differently since it is specific to the community and geographic locations associated with the environment. However, the World Intellectual Property Organization (WIPO) defines traditional knowledge is "knowledge encompassing the skills, know-how, and practices that are developed, sustained and passed from generation to generation within a local communities or indigenous peoples, often forming part of its cultural or spiritual identity (WIPO 2020). According to the WIPO, the terms traditional, indigenous, and local knowledge can be interchangeable (WIPO 2012) and have overlapping definitions. The traditional wisdom was not documented in a penned tradition because it has been passed in diverse forms, including songs, folklore, narratives, cultural values, beliefs, proverbs, the local language, customs, and agricultural practices (Ohajunwa and Miji 2018; Brondizio et al. 2021). This knowledge system is in the stage of abandoned usage and lack of generational transmission by indigenous and local communities due to decreaded reliability in weather forcasting as a result of amplified variability in weather (Kalanda-Joshua et al. 2011; Roncoli et al. 2002). Also, the poor utilization and accessibility of modern forcasting systems and skills related to access, lack of language competencies, interpretation capacities with scientific data can lead to the loss of traditional knowledge (Shoko and Shoko 2013) and inconsistencies in weather calculation. Therefore, it is crucial in promoting sustainable actions by strengthening climate mitigation and adaptation efforts through integration of sciences (Makondo and Thomas 2018) and aimed at achieving the Sustainable Development Goals (SDGs) outlined within the framework under the 2030 Agenda.

Filho et al. (2022) mentioned that traditional knowledge is vital in identifying and forecasting climate risks, predicting weather, and providing viable adaptation options to enhance responses to climate change. Various indigenous and local communities use different indicators for forecasting weather to mitigate and adapt to the impacts of climate change. The repeated observations from the environment can extend the understanding of spatiotemporal dynamics of each indicator (Gagnon and Berteaux 2009). These indicators form the environment have been used to observe the changes, which primarily affect the alternatives of livelihood (Nyong et al. 2007). Indicators are based on plant phenology, animals, birds, insects, meteorological, and astronomical factors (Zuma-Netshiukhwi et al. 2013; Jiri et al. 2015; Soropa et al. 2015; Phu and De 2016; Iticha and Husen 2018; Nyakaisiki et al. 2019; Radney et al. 2019; Son et al. 2019; Chuong et al. 2020; Filho et al. 2022). These indicators are closely rooted in the surrounding environment and can be sustainable solutions regardless of the rapidly changing weather patterns due to increased anthropogenic causes.

Recently, worldwide organizations, scientists, researchers, naturalists, environmentalists, and policy-makers have widely recognized traditional knowledge of climate change and variability. Many previous researches were conducted in the African continent compared to other continents. However, there are gaps in collecting and documenting the application of traditional knowledge to abate the climate crisis, which is still lacking and challenging to integrate with Western science. Therefore, this review chapter aimed to identify indicators for forecasting climate across different countries and regions, how indigenous and local communities used traditional indicators to interpret weather events, and how climate change is mitigated and adapted, by reviewing the literature systematically. This systematic review chapter serves as evidence highlighting the usage of traditional knowledge by various countries. It showcases reliance on environmental observation, where the absence of technological advancements, aims to preventatively mitigate climate-related adversities and promote sustainable, environmentally friendly solutions.

Material and Methods

This study searched and derived the academic articles from Web of Science (WoS) and Scopus databases, and other sources (google searches, reference lists, policy documents, reports, and grey literature), to perform the systematic literature review. Scopus and WoS were selected due to their widely recognized standards, as most of the official institutions and researchers trusted and valued the accuracy and reproducibility of these databases. From the journal articles, we considered only the peer-reviewed articles, which were in the English language, with the availability of full text for the period between 2012 and 2022. The keywords used in driving the articles were "traditional knowledge, indigenous knowledge, local knowledge, indigenous ecological knowledge, traditional ecological knowledge, climate change, climate change impacts, adaptation, and mitigation." Totally 618 full-text articles were obtained through the databases of WoS, Scopus, and Google Scholar. We applied the Preferred Reporting Items for Systematic Reviews and Meta-Analyses (PRISMA) as a manual to conduct this review (Moher et al. 2009). Figure 10.1 shows the detailed process of identification through databases, screening to select the most appropriate papers based on the study objectives, and inclusion of chosen papers for systematic review using the PRISMA flow diagram. Firstly, the duplicate articles were removed from the collected articles. Secondly, we excluded the articles, which were not retrievable and relevant to the objectives. Finally, this chapter considered only 63 published articles and reports for the above-stated study period for this review. The scope of this study covered only the full-text articles, and also the review papers, not the abstracts. Since original research articles were scarce in some countries. The limitation of this review was the articles and reports before 2012 were not considered. This chapter explored the trend of traditional knowledge utilization in climate predictions and adaptations based on countries, including the key terms used to interpret the traditional knowledge in literature, weather forecast indicators, adaptation measures, and the application of traditional insight to the breadth of the sustainability paradigm.

Results and Discussion

Overview of the Reviewed Articles

A detailed description of this systematic is represented in Figs. 10.2, 10.3 and 10.4. Of 63 articles, more than half were from the African continent, 17 were from Asia, and four were from the Oceania and American continents, as shown in Fig. 10.2. The selected articles were narrowly downed based on traditional knowledge and climate change adaptation rather than a broader scope. Figure 10.3 displays the counties and the number of studies. More studies on traditional knowledge of climate change were recorded in Africa, followed by the Asian region.

Furthermore, among the reviewed articles, the most commonly used term was indigenous knowledge (IK), employed by 45 of the 63 articles. Eight articles used traditional knowledge (TK) and local knowledge (LK), and one used traditional ecological knowledge (TEK) and folk knowledge (FK) (Fig. 10.4).

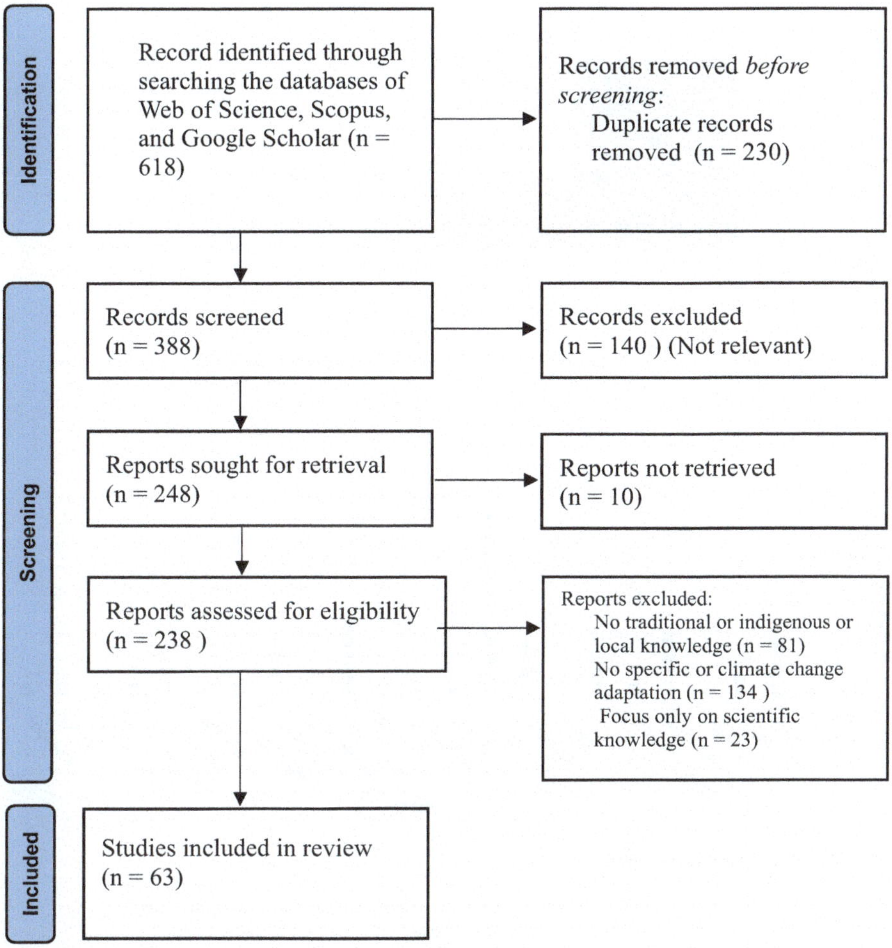

Fig. 10.1 Systematic review process diagram

Climate Change and Traditional Knowledge

The United Nations Framework Convention on Climate Change defines climate change as " alteration in the composition of the global atmosphere is attributed directly or indirectly by anthropogenic activities, which is in addition to natural climate variability observed over a comparable period" (IPCC 2014, 2018). The impacts of climate change are inevitable and unprecedented challenges in various sectors due to intensified temperature rise and decreased rainfall patterns. Climate change threatens ecosystem stability, biodiversity, water resources, agriculture, the livelihood of marginalized communities, the economy, human settlements, and health due to increased atmospheric greenhouse gases (Imoro et al. 2021). Traditional

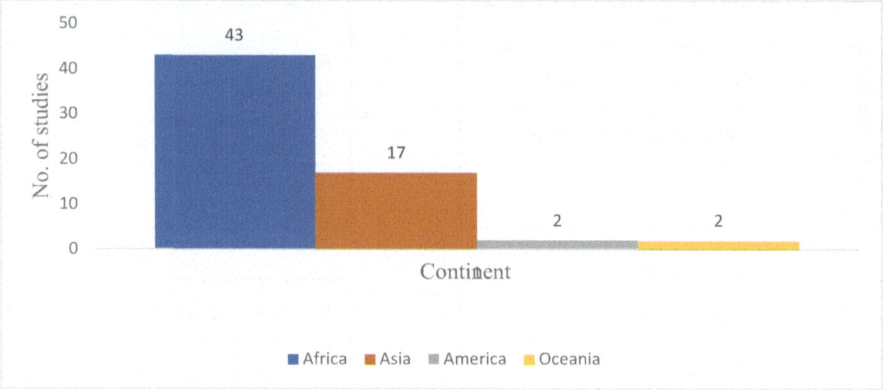

Fig. 10.2 Distribution by continent

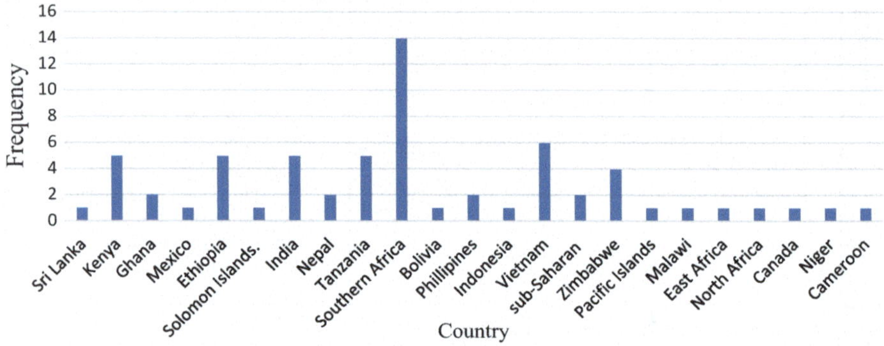

Fig. 10.3 Distribution by country

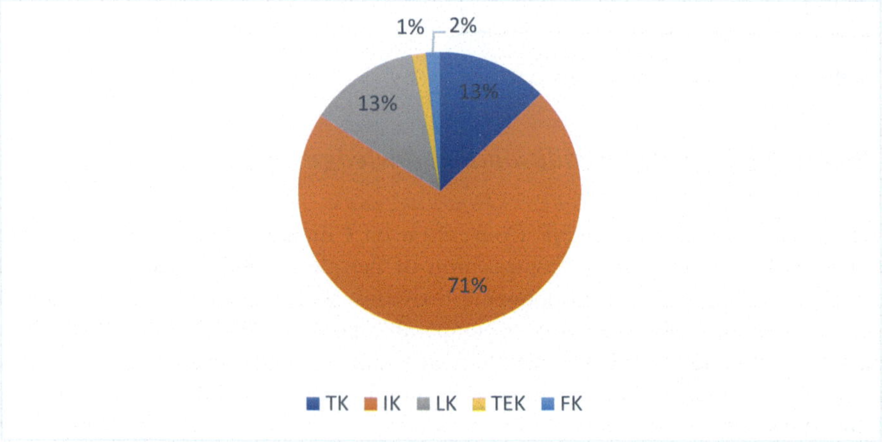

Fig. 10.4 Term used to indicate traditional knowledge

knowledge has the power to determine the future of the earth. The Indigenous and local communities in remote areas used the location and culture-specific traditional indicators to predict climate extremes in advance, reduce the effects, and increase the adaptation mechanisms.

Traditional knowledge is uneven across the diverse indigenous and local communities globally. Also, it built a strong and thick bond between the people and the living environment. Over many decades, it was passed within the communities, and people adapted to survive with nature by observing the changes using the indicators that were present locally. Traditional knowledge and practices in different communities can vary based on the experiences of day-to-day practices. Figure 10.5 describes the features of traditional knowledge. The main features of this knowledge: no systematic documentation, generated and passed with communities, location-specific, culture-specific, oral tradition, oral transmission, decision-making, experimentation, adaptation (Sharma et al. 2020), no individual ownership, nature-based solution, dynamic, and closed surveillance with nature. The role of traditional knowledge is vital for countries to attain sustainable development. This knowledge system uses observation, practices, and skills to predict weather situations (Jiri et al. 2015). Previous studies found that plant phenology, animal, bird, and insect behaviors, and atmospheric conditions were used as indicators to forecast weather and seasonal quality. The description of climate effects is based on these indicators, such as observing the environment (trees, moon, stars, and clouds), the reaction of animals and birds (movement and sounds), and signals from atmospheric conditions.

Traditional Indicators for Weather Prediction

Table 10.1 describes the traditional plant phenological indicators used by different countries. Morphological characteristics of plants projected prediction of the rainy season, dry season, and windstorms (Balehegn et al. 2019; Chuong et al. 2020; Iticha and Husen 2018; Jiri et al. 2015; Kijazi et al. 2013; Mugi-Ngenga et al. 2021; Mahoo et al. 2015; Matsa and Mukoni 2013; Risiro et al. 2012; Shoko 2012; Soropa et al. 2015; Zuma-Netshiukhwi et al. 2013). Fruiting, flowering, and flower budding of specific plant species, the growth of bamboo shoots, and the occurrence of mosses and mushrooms were sourced for rainfall projection. On the other hand, due to lack of rainfall, the dry season was forecasted by heavy flowering and continuous flowering of mango trees, Colophospermum mopane, and *Sclerocarya birrea.* The findings of this chapter reveal that usage of traditional and local plant indicators was highly recorded in the African continent. The trees located near the farming field and house are used to predict rainfall based on long-term observation, and the phenology of trees signifies good rain or drought (Mafongoya and Ajayi 2017). It sheds light on how indigenous and local communities survive with nature by observing its changes with their long-term experience. However, these traditional predictions differed from scientific forecasting.

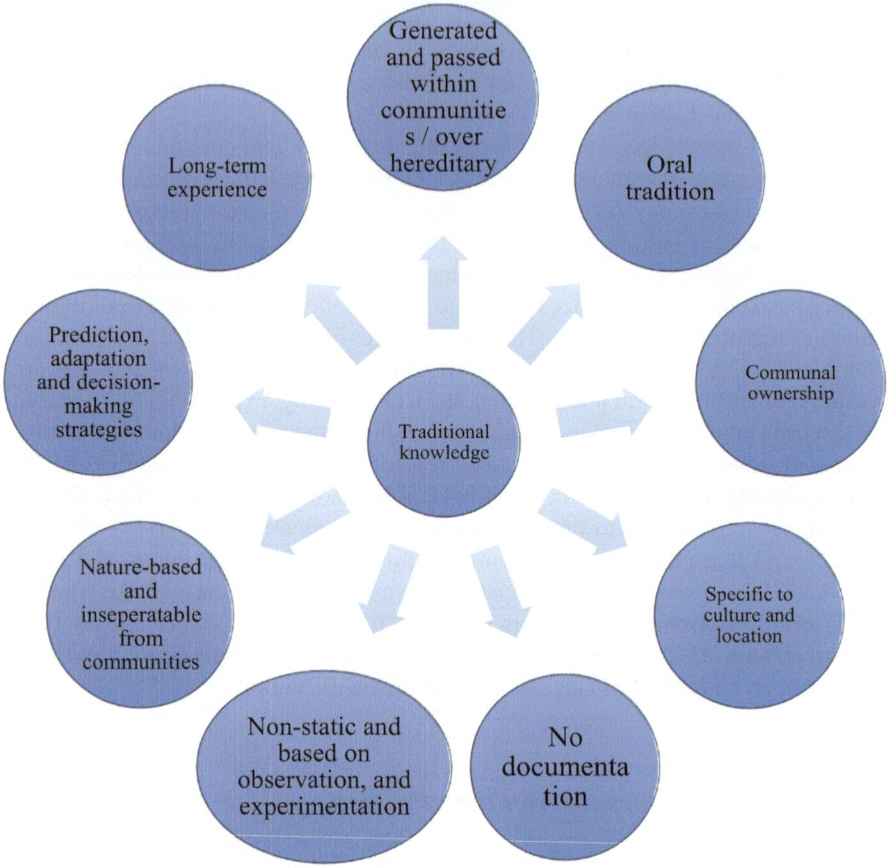

Fig. 10.5 Features of traditional knowledge in climate change

Table 10.2 illustrates how the occurrence and behavior of various birds are used as indicators for weather prediction (Jiri et al. 2015; Soropa et al. 2015; Chuong et al. 2020; Nyadzi et al. 2021; Phu and De 2016; Radeny et al. 2019). The crying and singing sounds of certain birds and the appearance of sparrows and cicadas were the traditional indicators based on bird behavior used in Zimbabwe, Zambia, and Kenya. The singing behavior of the birds illustrates that birds were happy and welcoming the rainy season. The flight pattern of birds, specifically when birds moved from the ocean to high-altitude areas, was indicated by windstorms. Noticeably, Vietnam, Zambia, and Kenya shared common indicators of heading of swallow and storks for flood and onset of rain

The relationships between the behavior of certain insects and weather forecasts are listed in Table 10.3. Ants, bees, termites, butterflies, grasshoppers, scorpions, worms, and dragonflies were used as indicators (Kijazi et al. 2013; Elia et al. 2014;

Table 10.1 Traditional plant phenological indicators used for various countries

Indicator	Indication	Location	Source
Shedding off of leaves, dropping fruits before maturing, and delays in fruiting (*Erythrina abyssinica*, *Ficus ovata*, *Ficus natalensis*, and *Ficus sycomorus*)	Drought/dry season	Zimbabwe, Zambia Southern Ethiopia Kenya	Jiri et al. (2015), Zuma-Netshiukhwi et al. (2013), Iticha and Husen (2018) and Mugi-Ngenga et al. (2021)
Fruiting of wild trees or heavy fruiting (Ebony trees, *Periary capensis*, *Vapaca kikiana*, *Facourtia indica*, *Vangueria infausta*, *Ximenia arnericana*, *Securinga virosa*, *Peppae capensis*, *Arecaceae* spp., *Uapaca kirkiana*, *Solanumin cunum*, *Acacis tortillis*, *Adansonia digitata*, and *Commiphora erythraea*)	Rainy season	Zimbabwe Vietnam Zambia Tanzania Ethiopia	Shoko (2012) Soropa et al. (2015) Chuong et al. (2020), Risiro et al. (2012), Matsa and Mukoni (2013), Zuma-Netshiukhwi et al. (2013), Mahoo et al. (2015) and Balehegn et al. (2019)
Budding and (early) flowering of *Acacia* spp., Peach tree (*Prunus persica*), Apricot (*Prunus armeniaca*) *Sclerocarya cattra*, *Brachystegia spiciformis*, and *Julbenardia globiflora*	Start of the Rainy season	Southern Ethiopia; Zimbabwe	Iticha and Husen (2018), Soropa et al. (2015) and Mafongoya and Ajayi (2017)
Heavy flowering of mango tree (*Mangifera indica*)	Potential drought season	Zambia Tanzania	Risiro et al. (2012) and Kijazi et al. (2013)
Initial heavy-flowing in *Colophospermum mopane*	Good rains	Zimbabwe	Jiri et al. (2015)
Subsequent heavy flowering in *Colophospermum mopane* and *Sclerocarya birrea*	Drought	Zimbabwe	Jiri et al. (2015)
Appearance of mushrooms	Rainy season	Vietnam	Son et al. (2019)
Alternating size of banana buds	Windstorm	Vietnam	Chuong et al. (2020)
Floating moss in waterbodies	Rainy season	Vietnam	Son et al. (2019)
Bamboo shoots grow higher than older bamboo shoots	Flood	Vietnam	Phu and De (2016)

Mahoo et al. 2015; Phu and De 2016; Balehegn et al. 2019; Son et al. 2019; Mugi-Ngenga et al. 2021). The movement pattern of ants was correlated with drought and rainy seasons, such as if ants move in a straight line, representing good rain, while ants move in a disturbed line with food, depicting the onset of drought (Mafongoya and Ajayi 2017). Similarly, the dry season was indicated by moving bee colonies from the original place to another place. The low-flying pattern of dragonflies, the visibility of earthworms near the soil surface, and termites flying into the house interpreted the rainy season's occurrence and the season's quality (Phu and De 2016; Chuong et al. 2020; Mugi-Ngenga et al. 2021).

Table 10.2 Traditional Indicators based on birds

Indicator	Indication	Location	Source
Crying sound of birds (Southern ground hornbill (*Bucorvus leadbeateri*), Vigors sunbird, and Rain cuckoo)	Rainy season	Zimbabwe	Shoko (2012) Jiri et al. (2015)
Singing and movement of birds	Rainy season	Zimbabwe	Soropa et al. (2015)
The appearance of birds (Chochomela and Kwarakwara)	Rainy season	Zimbabwe	Jiri et al. (2015)
The flight pattern of birds [moving from sea to mountainous areas; Example: White-rumped Shama (*Copsychus malabaricus*)]	Windstorms	Vietnam	Chuong et al. (2020)
Presence of many butterflies (Amblyscirtes)	The rain starts soon	Zambia Ghana	Zuma-Netshiukhwi et al. (2013), Mahoo et al. (2015) and Nyadzi et al. (2021)
Hearding of swallow and storks	Flood and rain starts soon	Vietnam Zambia Kenya Ethiopia Tanzania	Phu and De (2016), Zuma-Netshiukhwi et al. (2013), Mahoo et al. (2015), Mugi-Ngenga et al. (2021) and Radeny et al. (2019)
Movement of cicadas	The rain starts soon	Zambia	Zuma-Netshiukhwi et al. (2013)
Chicken moving with outstretched wings	The rain starts soon	Zambia	Zuma-Netshiukhwi et al. (2013) and Mahoo et al. (2015)
First appearance of sparrows	Good rain	Zambia	Zuma-Netshiukhwi et al. (2013)

On the other hand, no earthworms near the ground surface indicated drought season since earthworms survive in moist and swampy areas and respire through the skin. If there are no humid conditions, earthworms may die out. Indicators of bees, ants, termites, and dragonflies were common indicators for African countries and Vietnam. Primarily, all indicators represent and predict the timing of prospective weather. The listed descriptions of indicators were locally relevant based on the available resources and biodiversity surrounding the communities.

Table 10.4 displays the traditional indicators based on certain animal behavior and movements. It is based on the crying sound of frogs, vocalizing squirrels, crabs moving outside water streams, the appearance of pangolins, rats digging burrows in

Table 10.3 Indicators based on insects

Indicator	Description	Indication	Location	Source
Ants	Appearance with increased size Moves in a straight line Appearance of red ants, increased anthills Crawling in the pillars of the house, or crossing roads, or moving to a higher place Appearance of winged ants Avoid crossing-through ditches	Good rain	Southern Ethiopia Zimbabwe Zambia (South Africa) Vietnam Malawi Ethiopia	Iticha and Husen (2018) and Mafongoya and Ajayi (2017) Zuma-Netshiukhwi et al. (2013) Son et al. (2019) and Phu and De (2016) Phu and De (2016), Mahoo et al. (2015), Mugi-Ngenga et al. (2021) and Balehegn et al. (2019)
The appearance of termites (*Ancistrotermes* spp.)	It flies into the house Occurrence in maize field	Onset rainfall will start Dry spell	Vietnam Zimbabwe Tanzania Malawi	Son et al. (2019), Phu and De (2016), Shoko (2012), Kijazi et al. (2013) and Mugi-Ngenga et al. (2021)
The presence of many butterflies	The rain starts soon	The rain starts soon	Zambia	Zuma-Netshiukhwi et al. (2013) and Mahoo et al. (2015)
Flies and gadflies	Attach paddy field		Vietnam	Phu and De (2016)
Bees	Hives located in a high position Hives located in a lower position Moving colonies	Drought or hot weather Cold temperature or flooding Rain deficient	Vietnam Vietnam Kenya	Son et al. (2019) Chuong et al. (2020) Mugi-Ngenga et al. (2021)
Grasshoppers (*Hesperotettix speciosus*)	Abundance and hatching of new grasshoppers Flying within the vicinity of households	Good rain Drought	Zimbabwe Malawi	Jiri et al. (2015), Mahoo et al. (2015) Nkomwa et al. (2014)
Appearance of black scorpion (*Arachnida* spp.)		Good rain	Tanzania Burkina Faso	Kijazi et al. (2013) and Elia et al. (2014)
Millipedes	Big millipedes making sounds in the soil	Onset of rains	Tanzania Zimbabwe Burkina Faso	Jiri et al. (2015) and Elia et al. (2014)

(continued)

Table 10.3 (continued)

Indicator	Description	Indication	Location	Source
Appearance of Morpani worms (*Imbrassia belina*)		Rainy season	Zimbabwe Tanzania	Soropa et al. (2015) and Kijazi et al. (2013)
Earthworms (*Lumbricina* spp.)	There is no appearance near the ground surface Occurrence of more earthworms	Drought Onset of rain	Vietnam Ghana	Chuong et al. (2020) Nyadzi et al. (2021)
The flight pattern of Dragonfly	Flying low Flying middle Flying high	Rainy season Shady Sunny	Vietnam, Kenya	Phu and De (2016) and Mugi-Ngenga et al. (2021)

elevated areas, wandering tortoises, and increased courtship behavior of domestic livestock denoted by the rainy season (Shoko 2012; Kijazi et al. 2013; Okonya and Kroschel 2013; Zuma-Netshiukhwi et al. 2013; Elia et al. 2014; Jiri et al. 2015; Mahoo et al. 2015; Phu and De 2016; Iticha and Husen 2018; Nyakaisiki et al. 2019; Son et al. 2019; Mugi-Ngenga et al. 2021). The drought was indicated by invading snakes to living areas due to increased environmental temperature and cattle not feeding their calves

Table 10.5 explains traditional astronomic and atmospheric indicators and their indicators, including sky, moon, stars, and wind (Shoko and Shoko 2013; Elia et al. 2014; Jiri et al. 2016; Balehegn et al. 2019; Mugi-Ngenga et al. 2021). The results showed that African countries applied almost all atmospheric and astronomic indicators. It included the direction of the wind, the color of the star and the moon, the appearance of the morning stars and fog, the rainbow, the wind blowing from west to east, the southern side, and frequent wind swirls signifying the good rainy season ahead. Also, indigenous communities recognized the dark cloud, the appearance of a long rainbow, a dark and cloudy sky, the sky without stars, moving clouds from the north to the south side, and the moon's crescent facing outward for forecasting rain and the onset of rainfall. Based on previous studies, Table 10.5 confirmed that most South African countries have relied on astrological and atmospherical indicators

Climate vulnerabilities and related phenomena were forecasted in the above-listed countries using traditional knowledge through interpreting the conditions and behavior of plants, animals, birds, and insects, and meteorological and astronomical conditions using wind, moon, and stars. However, this information was not appropriately documented (Sharma et al. 2020) and disappeared due to gaps in intergeneration (Fernández-Llamazares et al. 2015; Rankoana 2022) and a lack of awareness about using traditional knowledge towards self-sufficiency economy and sustainable development. Chisadza et al. (2015) highlighted that projections using traditional indicators have been proven to correlate with scientific findings by supporting the results of Zimbabwean catchment in the Limpopo basin. It shows a great signal in

Table 10.4 Indicators based on animals

Indicator	Indication	Location	Source
Restless animals	Rainy season	Uganda	Nyakaisiki et al. (2019) and Okonya and Kroschel (2013)
Hearding and croaking of frogs	Rainy season	Zimbabwe Vietnam	Shoko (2012), Son et al. (2019), Jiri et al. (2015) and Mahoo et al. (2015)
Vocalizing squirrel	Rainy season	Vietnam	Son et al. (2019)
Snakes invading the living areas	Drought	Kenya	Mugi-Ngenga et al. (2021)
Crabs move outside from the streams	Flood	Vietnam	Son et al. (2019)
Appearance of pangolin (scaly anteaters)	Rainy season	Tanzania Burkina Faso Zimbabwe	Elia et al. (2014) and Kijazi et al. (2013)
Rats burrow in elevated locations	Flood	Vietnam	Phu and De (2016)
Intestines of animals			
• Clear fluid • Turbid fluid	Good rain Flood	Southern Ethiopia	Iticha and Husen (2018)
Behavior of livestock			
• Frequent courtship • Lack of interest in entering the barn after grazing • Cattle stand for a long time after having water • Not milking their calves	Rainy season Rainy season Drought Drought	Southern Ethiopia	Iticha and Husen (2018)
Increased libido and frequent courtship of goats and sheep	Good rainy season	Zambia	Zuma-Netshiukhwi et al. (2013)
Frequent appearance of wandering tortoise	Good rainy season	Zambia	Zuma-Netshiukhwi et al. (2013)

including the importance of traditional knowledge for predicting and developing adaptation strategies at the local level to scale up.

The study conducted by Soropa et al. (2015) discovered that distinct weather-predicting indicators rooted in indigenous knowledge were employed at the district level in Zimbabwe. Those indicators included trees, wild fruits, fauna, birds, and dew, which were not commonly observed in the selected districts (Chiredzi, Murehwa, and Tsholotsho). The timing of delays and early flowering indicated forthcoming dry and wet seasons and correlated with more and lower fruit production levels, respectively. It signifies that the choice of indicators by local communities is influenced by their sociocultural background and the availability of natural resources in

Table 10.5 Indicators based on astrology and atmospheric conditions

Indicator	Indication	Location	Source
Wind direction			
• West to east • Blowing from the Southeastern side • Heavy wind at the end of October • High frequency of wind swirls	Good rainy season Rainy season Poor rainy season Good rain	Tanzania Ethiopia Zimbabwe Kenya Zambia Uganda	Nyakaisiki et al. (2019) Iticha and Husen (2018) and Zuma-Netshiukhwi et al. (2013) Jiri et al. (2015) Mugi-Ngenga et al. (2021), Balehegn et al. (2019) and Radeny et al. (2019)
The moon and star appear in red color	Rain expected in 1–3 days	Tanzania	Nyakaisiki et al. (2019) and Mafongoya and Ajayi (2017)
Stars			
• Morning star appears for 240 days with no eclipse • Morning star does not appear for 60 days, and 7 stars with one morning star • Invisible start or fewer star • Appearance of 3–5 starts	Lack of rainfall Rainy season Rainy season Rainy season	Southern Ethiopia Vietnam Ethiopia Kenya	Iticha and Husen (2018) and Balehegn et al. (2019) Chuong et al. (2020), Phu and De (2016) and Mugi-Ngenga et al. (2021)
Sun			
• Halo around the sun • Red sunset	Strom, erratic rain, and cyclones Humidity	India Bolivia	Salim and Monolisha (2019) Riva et al. (2013)
Moon			
• Dark night after full moon eclipse • Dense halo • Moon crescent facing outwards • Yello moon • Oval shape with a tiny black spot • Positioned sideways • Full moon without tilt	Good rainy season Drought	Southern Ethiopia Zimbabwe; Zambia Kenya Bolivia Tanzania Philippines South Africa	Iticha and Husen (2018) and Kijazi et al. (2013) Jiri et al. (2015), Zuma-Netshiukhwi et al. (2013), Mugi-Ngenga et al. (2021), Riva et al. (2013) and Elia et al. (2014) Nelson et al. (2019) and Ubisi et al. (2019)
Cloud			

(continued)

Table 10.5 (continued)

Indicator	Indication	Location	Source
• Dark clouds • Dark clouds with strong winds • Moving from north to south direction • Fish scales like clouds and move from south to north • The appearance of large pale blue clouds in the daytime	Heavy rainfall Rainfall with thunderstorms Heavy rain Sunny	Southern Ethiopia, Ghana, Zambia; Cameroon, Indonesia, Philippines, Timore-Leste Zimbabwe; Zambia Vietnam	Iticha and Husen (2018) and Zuma-Netshiukhwi et al. (2013) Jiri et al. (2015) Son et al. 2019 Chuong et al. (2020) and Phu and De (2016) Phu and De (2016), Radeny et al. (2019), Nyadzi et al. (2021), Hiwasaki et al. (2014) and Tume et al. (2019)
Rain			
• Low rain in summer • High rain in summer	Warm winter Cold winter	South Africa	Ubisi et al. (2019)
Water surface			
• Warm surface water • Appearance of bubbles	Abundance of fish Strom	India	Salim and Monolisha (2019)
Rainbow			
• Long rainbow • Short rainbow • Red color dominates • Blue color dominates	Extended rainy season Lack of rainfall Indication of rain Rain has passed	Southern Ethiopia Zambia	Iticha and Husen (2018) Zuma-Netshiukhwi et al. (2013)
Temperature			
• Clod temperature • High temperature • Extended cold winter • Very high temperatures in mid of November	Low rainfall Good rain Onset of drought Drought	South Africa South Africa Southern Ethiopia Malawi	Ubisi et al. (2019) Iticha and Husen (2018) Nkomwa et al. (2014)
Appearance of fog in the morning	Indication of rain	Zimbabwe	Mafongoya and Ajayi (2017)

their specific locality. Traditional weather projection indicators, such as birds, trees, clouds, and frogs, were shown to have a high-reliability level compared to science-based indicators among the smallholder farmers in Ghana. The reliability index score for traditional predictors (0.72) and scientific predictions (0.88) did not differ much. Therefore, traditional knowledge in predicting climate vulnerability is imperative, and equal attention and recognition need to be given to developing sustainable locally specific solutions (Ankrah et al. 2022).

This chapter expounds on local and indigenous communities that applied almost all indicators. Remarkably, these indicators played a substantial role in how these traditions indicators are considered for foreseeing climate-related risks, planning agricultural activities, managing crops, soil, livestock, water, forest, and wildlife, conserving biodiversity and medicinal plants, and creating sustainable lifestyles (Mekonnen et al. 2021). Research by Mahoo et al. (2015) found that 56% of farmers stated that traditional forecasts were reliable, while 22% mentioned that only scientific climate forecasts were reliable in Tanzania. Young people (below 40 years old) showed less interest and belief in scientific prediction. In contrast, people above 40 years old have strong opinions and still follow nature for anticipating weather and planning farming activities based on existing traditional knowledge. The elderly farmers have more conventional forecasting knowledge of weather, including upcoming seasons, than young farmers, which was supported by the findings of Elia et al. (2014) and Mafongoya and Ajayi (2017). This gap interprets that traditional knowledge in climate can disappear if this is not detailed and incorporated into international and national policies to safeguard the invaluable knowledge system that was followed over generations.

A study by Shali (2016) indicated that fishing communities in Tanzania also prioritized meteorological conditions such as the appearance of clouds, specifically dark clouds, clouds with thunder and strong wind, reddish clouds at dusk, and the sky filled with cloud cover reported to predict heavy onset rainfall in a next few hours or days. Traditional indicators are nature-based solutions to reduce the impacts of climate change that led to agricultural and crop production loss, food insecurity, lowering the severity of hydrological disasters and biological disasters, biodiversity loss, and ecosystem imbalance. These findings confirm that local communities and indigenous people still relied on their way of observation based on day-to-day life routine rather than depending on scientific-based projections and early warning.

Traditional Knowledge and Climate Change Adaptation

Traditional knowledge plays an incredible role in adapting to climate change by adjusting to the adverse effects of climate change (IPCC 2018). The adaptation measures reduce vulnerability, make changes in response to existent and anticipated climate changes, and enhance resilience. Adaptation measures can include natural resource management, forest management, land management, livestock management, agricultural and crop management practices (Mekonnen et al. 2021), and water management to cope with shifting weather patterns. These adaptatives' strategies aim to minimize the adverse impact of human and natural systems.

Indigenous and local communities follow Various traditional adaptation strategies, as discussed in the following sections. The farming practices, water conservation and irrigation practices, native crop and tolerant crop and animal varieties, soil improvement practices, food storage, pest and disease control practices, resilient infrastructures, and livelihood and economic diversification are identified in this review chapter.

Crop and Livestock Farming Practices

Farmers, pastoralists, and local communities were applied the traditional to adapt to the vulnerability of weather by adjusting crop palnting calenders (Abeywardana et al. 2019; Huynh et al. 2020a, b): early planting and harvesting, delayled palnting in the rainy season (Schilling et al. 2012), shortening growing season of crops (Phu and De 2016; Chuong et al. 2020; Mugi-Ngenga et al. 2021), intercropping during the drout by intercropiing dry rice with *Acacia*, planting banana around the paddy field to maintain moisture (Son et al. 2019; Chuong et al. 2020; Schilling et al. 2012), and growing indigenous animal breeds and crops, such as dry rice, soynean, banana, cassava, maize, cinnamon, and areca, can increase the level of tolerance to drought and cold (Nkomwa et al. 2014; Chaudhary et al. 2021), reduce illness of livestocks (Ayal et al. 2015) and crops, and cultivation cinnamon and areca to reduce soil erosion (Schilling et al. 2012; Van Dien and Van 2014; Nalau et al. 2018; Iticha and Husen 2018; Chuong et al. 2020; Huynh et al. 2020a, b). Native plant species and livestock breeds resist drought, require less water during drought season, and are resilient to pests and diseases. Also, crop spacing (Mugi-Ngenga et al. 2021), crop diversification (Son et al. 2021; Huynh et al. 2020a, b), avoiding planting in valleys during heavy rainfall and flooding (Iticha and Husen 2018), planting commercial crops (Hiwasaki et al. 2014), dryland agriculture (Padal et al. 2013) mixed cropping, increase plants densities when growing.

Water Management and Irrigation Practices

Rainwater, flood water, and roof water harvest are widespread practices by storing water in dug ponds and boreholes, storage tanks, and deeper wells to conserve water for the upcoming drought season (Ofoegbu et al. 2016; Iticha and Husen 2018; Mugi-Ngenga et al. 2021). The water harvested using different water harvesting methods can be used for drinking, plant watering, and other purposes. Traditional water dams were built for storing water and fish generation (Kihila 2018). In addition, plant native and drought-resistant crops to reduce water consumption (Nkomwa et al. 2014; Chaudhary et al. 2021), cultivating crops near water wells (Snorek et al. 2014) to reduce the water evaporation and absorption by surfaces when passing to long distances, changing irrigation practices, namely bamboo drip irrigation (Das et al. 2012; Chuong et al. 2020) and planting cover crops (Ncube and Lagardien 2015) are adopted to fight against climate change.

Soil Fertility and Improvement Practices

The following practices were collected by reviewing previous studies. These are mulching (Rawat and Kharwal 2016)—improving the soil by adding a layer of litter or crop residues, crop rotation (Ncube and Lagardien 2015), crop diversification (Huynh et al. 2020a, b; Son et al. 2021) to expand the quality of soil, conservation of agrobiodiversity, and enhance nutritional uptake from soil. Growing multipurpose trees—*Moringa stenopetalla* (Mugi-Ngenga et al. 2021), agroforestry and terraces, fodder banks reduce land degradation (Ncube and Lagardien 2015) and storing fodder for drought season (Iticha and Husen 2018), planting cover crops to control soil moisture, zero tillage, and surface seeding (Smith et al. 2013), application of organic fertilizers and manures—cow dung and compost (Phu and De 2016) are adopted by indigenous peoples to climate change by improving soil management practices (Amwata 2013), crop productivity, crop resilience (Taremwa et al. 2016), and food insecurity.

Climate Resilient Structures

Indigenous people built stone revetment, walls using coconut leaf and woods, seawalls using stones, concrete, and sandbags, tree planting in the coastal areas, and coastal forests to reduce soil and coastal erosion and flooding triggered by climate change (Arkhurst et al. 2022). Besides, the houses built using bamboo and stone acted as natural air conditioners to survive during high temperatures because of increased heat-trapping greenhouse gases in the atmosphere (Mugi-Ngenga et al. 2021).

A SWOT Analysis of Traditional Knowledge

Recently, a growing interest in traditional weather and climate forecasting, locally specific and acceptable traditional adaptation strategies, reliability, integration with contemporary science, and policies denoted the increased attention due to its practical reality and intrinsic heritage values (Filho et al. 2021; Mugambiwa 2018; Audefroy and Cabrera Sánchez 2017). A SWOT of traditional knowledge was tabulated in Table 10.6 to showcase its strengths, weaknesses, opportunities, and threats. The numerous strong points of traditional knowledge, such as a strong relationship with nature and cultural beliefs, contributing to resources management, climate-resilient local adaptation, enhancing sustainable livelihood, and field-tested adaptation strategies to reduce climate change vulnerabilities (Sharma et al. 2020; Mekonnen et al. 2021). The weaknesses are a lack of documentation specific to geographic areas (Mafongoya and Ajayi 2017; Sharma et al. 2020), no availability in a common language, and limited scientific validation. Threats are lack of intergeneration transmission (Fernández-Llamazares et al. 2015; Rankoana 2022), cultural erosion, and language loss.

The Role of Traditional Knowledge—A Sustainable Solution for the Climate Crisis

Traditional knowledge system has a unique link between human survival and their nearby natural resources. As a result, it provides information and insights that can be applied to local climate change adaptation (Chisadza et al. 2015). This can lay a foundation for technological developments and contributions to modern science. Previous studies have unearthed a positive correlation between the traditional indicators used

Table 10.6 A SWOT analysis of traditional knowledge in climate change

Strengths (S)	Weakness (W)
• Strong relationship with environment, culture, and spiritual beliefs • Deep connections with Mother Nature (environment, culture, and spiritual beliefs) • Give nature-based sustainable solutions through a different lens • Provide climate-resilient mitigation and adaptation measures • Contribute to biodiversity protection • Promote sustainable land management • Promotes resources management • Location-specific knowledge • Offer sustainable livelihood practices with food security • Offer community-based and solutions conservation-oriented solutions • Provide field-tested or practically oriented adaptation measures • Local relevance or specific to geographic regions • Developed based on day-to-day observation of climate conditions	• Lack of documentation • Language barriers • Limited to geographic areas • Some indicators were no longer • Vulnerable to climate change • Possible uncertainties due to rapid shifts in climate change • Limited scientific validation
Opportunities (O)	**Threats (T)**
• The paring of traditional knowledge and modern science through a complementary approach to developing sustainable solutions • Integration of traditional language in national and international policies • Knowledge transmission through studies • Preservation of traditional knowledge through nature-based and climate-concerned institutions or governmental and non-governmental organizations	• Possibility of lost or diluted due to lack of systematic documentation and climate-induced displacement • Cultural erosion as a result of a lack of interest and awareness among the young generation in their cultural beliefs, practices, and environmental knowledge • Possibilities of neglegation • Issues in intellectual property • Inadequate legal protection • Language loss • Lack of intergenerational transmission

by indigenous communities and Western science (Chisadza et al. 2015). Further, the positive relationship shows the benefits of integrating traditional knowledge with modern science and its tremendous potential to reduce the risks.

Table 10.7 illustrates the benefits of traditional adaptation strategies in climate change and their dimension toward sustainable development and being a sustainable solution. Natural and cultural resources management, crop and livestock management, sustainable land management, forest management, water conservation, ecological pests and disease management, and livelihood diversification have the potential to address the three pillars of sustainable development.

Recognizing the value of traditional wisdom for incorporating with contemporary science both knowledge systems (Zuma-Netshiukhwi et al. 2013; Jiri et al. 2016;

Table 10.7 Benefits of traditional adaptation and sustainability dimension

Benefits of traditional adaptation	Sustainability dimension
Natural and cultural resource management	Social, economic, and environmental
Crop and livestock management	Social, economic, and environmental
Sustainable land management	Social, economic, and environmental
Water conservation	Social, economic, and environmental
Forest management	Social, economic, and environmental
Ecological pest and disease management	Social, economic, and environmental
Livelihood diversification and adaptation	Social, economic, and environmental

Source (Author)

Armatas et al. 2016) is essential to synergize the local relevance and accuracy of conventional and contemporary approaches, respectively, for scaling up (Table 10.7). The co-production of traditional knowledge and modern science can be used to predict the quantity and timing of rainfall together. This consensus forecasting system was successfully applied in Tanzanian and Uganada (Plotz et al. 2017). It could benefit the local community by enhancing the usefulness of combining the traditional indicators, interpretation of indicators, and the meteorological station records as a value addition to finding sustainable solutions for climate vulnerabilities.

Another striking piece of evidence from the Climate Change 2023: Synthesis report released by the Intergovernmental Panel on Climate Change (IPCC), leveraging a variety of knowledge and cultural perspectives and inclusive engagement, including indigenous and local knowledge, enhancing capacity and promoting climate-resilient development to enable socially appropriate and locally relevant solutions (high confidence). Also, these knowledge systems' inclusive processes can lead to long-term sustainable growth by addressing adaptation gaps and eluding maladaptation practices (high confidence). The information received through community approaches (local and indigenous) can hasten planning and behavioral changes toward combating climate change chance risks (high confidence) (IPCC 2023). Therefore, documenting and understanding indicators using indigenous traditional knowledge is imperative to validating and developing integrated traditional and scientific adaptations to tackle climate catastrophes.

Community-based adaptation plans were also crucial in resource management and disaster risk reduction (Basel et al. 2020). Based on the findings of this chapter, the precious traditional knowledge of local communities and indigenous people gained from observations and passed through generations is an incredible weapon to fight against climate change through sustainable usage of natural resources and ecosystem management. Around 122 million people in Africa and South Asia could fall into intense poverty and hunger due to climate change by 2030 (FAO 2016). Scientific knowledge alone is insufficient for solving the climate crisis, while traditional knowledge can provide appropriate and location-specific adaptation strategies with room to develop new approaches and offer solutions for food insecurity and poverty (Hussein and Suttie 2016). Its recognition and growing awareness influenced that considering scientific knowledge alone is inadequate for resolving the climate crisis and strengthening adaptation measures and strategies (Mafongoya and Ajayi 2017; Petzold et al. 2020; Huynh et al. 2020a, b; Mugambiwa and Makhubele 2021; Nyadzi et al. 2021). Combining scientific and traditional knowledge provides the best solutions and is especially important in agriculture, food security, and water management. Importantly, locally adapted tactics can be appropriate and efficient for making innovative, applicable strategies. Importantly, this knowledge system needs to be amalgamated with national institutional and legal frameworks to reduce gaps (Filho et al. 2022) between the government and local and indigenous communities. Traditional knowledge is essential in developing sustainable development (Magni 2017) to reduce the impacts of climate change.

Based on the overall study results, this chapter underscored how traditional knowledge plays a role in addressing the contemporary climate crisis by analyzing the existing prediction mechanisms and adaptation measures in various countries worldwide. The collection and documentation of traditional knowledge on climate change showed which indicators are used to predict the weather, and how adaptation measures are applied to lessen the effects of climate change. The current practices represented that still developing countries relied on this precious knowledge. However, many traditional communities are victims of climate calamities due to a lack of scientific interventions and integration. Crucially, in-depth studies on the relationship between plant indicators and animal behavior with weather parameters can provide a strong foundation for effective projections at local levels. These future directions can be a scientific path to integrate traditional knowledge with modern sciences.

Conclusions

This review study investigated the traditional indicators for projecting weather and climate with its timing and quality, adaptation measures, and how traditional knowledge can contribute to proposing sustainable solutions. In this chapter, we identified indicators, and signs of these indicators were recorded from their field observations and information passed over generations. The indicators used for weather forecasting were flora, fauna, astronomical, and atmospheric conditions. These indicators are specific to geographic regions. The adaptation measures based on traditional knowledge have a strong capacity for reducing impacts and enhancing resilience to climate-related disasters. Undoubtedly, these measures followed by indigenous and local communities can be a local, sustainable solution by addressing the social, economic, and environmental dimensions of sustainable development. Therefore, this chapter provides in-depth insights to policymakers, decision-makers, the scientific community, and the general public, including youth, about traditional knowledge since it is a tremendous resource for informing sustainable climate solutions to ensure the well-being of ecosystems and human communities.

Further, this chapter recommends that proper attention and interaction mechanisms be considered at the subnational and national level policies and regulations and combined with scientific knowledge to scale up by performing validation studies and further climate-smart innovations. Additionally, increasing awareness among younger generations and incorporating traditional knowledge into the education system can be more beneficial. The co-production of integrated knowledge systems can offer a new partnership path and empower innovative thinking among organizations, governments, experts, and scientists for national-level projects and rural development.

References

Abeywardana N, Schütt B, Wagalawatta T, Bebermeier W (2019) Indigenous agricultural systems in the dry zone of Sri Lanka: management transformation assessment and sustainability. Sustainability 11(3):910. https://doi.org/10.3390/su11030910

Amwata DA (2013) The influence of climate variability and change on land-use and livelihoods in Kenya's southern rangelands. Doctoral dissertation. Available online: repository.mut.ac.ke. Accessed on 12 Apr 2021

Ankrah DA, Kwapong NA, Boateng SD (2022) Indigenous knowledge and science-based predictors reliability and its implication for climate adaptation in Ghana. Afr J Sci Technol Innov Dev 14(4):1007–1019. https://doi.org/10.1080/20421338.2021.1923394

Arkhurst BM, Poku-Boansi M, Adarkwa KK (2022) Perception on coastal erosion: an assessment of how national level coastal resilience strategies promote indigenous knowledge and affect local level adaptation in Ghanaian communities. Environ Sci Policy 137:290–300. https://doi.org/10.1016/j.envsci.2022.08.019

Armatas CA, Venn TJ, McBride BB, Watson AE, Carver SJ (2016) Opportunities to utilize traditional phenological knowledge to support adaptive management of social-ecological systems vulnerable to changes in climate and fire regimes. Ecol Soc 21(1). https://www.jstor.org/stable/26270350

Audefroy JF, Cabrera Sánchez BN (2017) Integrating local knowledge for climate change adaptation in Yucatán, Mexico. Int J Sustain Built Environ 6(1):228–237. https://doi.org/10.1016/j.ijsbe.2017.03.007

Ayal DY, Desta S, Gebru G, Kinyangi J, Recha J, Radeny M (2015) Opportunities and challenges of indigenous biotic weather forecasting among the Borena herders of southern Ethiopia. Springer Plus 4:617

Balehegn M, Balehey, S, Fu C, Wu L (2019) Indigenous weather and climate forecasting knowledge among Afar pastoralists of north eastern Ethiopia: role in adaptation to weather and climate variability. Pastoralism 9(1). https://doi.org/10.1186/s13570-019-0143-y

Basel B, Goby G, Johnson J (2020) Community-based adaptation to climate change in villages of Western Province, Solomon Islands. Mar Pollut Bull 156:111266

Brondízio ES, Aumeeruddy-Thomas Y, Bates P, Carino J, Fernández-Llamazares Á, Ferrari MF, Galvin K, ReyesGarcía V, McElwee P, Molnár Z, Samakov A, Shrestha UB (2021) Locally based, regionally manifested, and globally relevant: indigenous and local knowledge, values, and practices for nature. Annu Rev Environ Resour 46(1):481–509. https://doi.org/10.1146/annurev-environ-012220-012127

Chaudhary BR, Acciaioli G, Erskine W, Chaudhary P (2021) Responses of the Tharu to climate change-related hazards in the water sector: indigenous perceptions, vulnerability and adaptations in the western Tarai of Nepal. Clim Dev 13(9):816–829. https://doi.org/10.1080/17565529.2021.1889947

Chisadza B, Tumbare MJ, Nyabeze WR, Nhapi I (2015) Linkages between local knowledge drought forecasting indicators and scientific drought forecasting parameters in the Limpopo River Basin in Southern Africa. Int J Dis Risk Reduct 12:226–233

Chuong HV, Le QNP, Nguyen MTH, Tran PT, Tân NQ, Pham TG, Nguyen L, Nguyen LTD, Trinh HN (2020) Indigenous knowledge in relation to climate change: adaptation practices used by the Xo Dang people of central Vietnam. Heliyon 6(12):e05656. https://doi.org/10.1016/j.heliyon.2020.e05656

Das A, Ramkrushna GI, Choudhury BU, Patel DP, Ngachan SV, Ghosh PV, Tripathi AK, Das S, Kumar M (2012) Natural resource conservation through indigenous farming systems: wisdomalive in North East India. Indian J Tradit Knowl 11(3):505–513

de la Riva MV, Lindner A, Pretzsch J (2013) Assessing adaptation–climate change and indigenous livelihood in the Andes of Bolivia. J Agric Rural Dev Trop Subtrop JARTS 114:109–122

Elia EF, Mutula S, Stilwell C (2014) Indigenous knowledge use in seasonal weather forecasting in Tanzania: the case of semi-arid central Tanzania. S Afr J Libr Inf Sci 80:18–27

Fernández-Llamazares A, Díaz-Reviriego I, Luz AC, Cabeza M, Pyhälä A, Reyes-García V (2015) Rapid ecosystem change challenges the adaptive capacity of local environmental knowledge. Glob Environ Chang 31:272–284. https://doi.org/10.1016/j.gloenvcha.2015.02.001

Filho WL, Matandirotya NR, Lütz JM, Alemu EA, Brearley FQ, Baidoo AA, Kateka A, Ogendi GM, Adane GB, Emiru N, Mbih RA (2021) Impacts of climate change to African indigenous communities and examples of adaptation responses. Nat Commun 12(1):1–4. https://doi.org/10.1038/s41467-021-26540-0

Filho WL, Barbir J, Gwenzi J, Ayal DY, Simpson NP, Adeleke L, Tilahun B, Chirisa I, Gbedemah SF, Nzengya DM, Sharifi A, Theodory TF, Yaffa S (2022) The role of indigenous knowledge in climate change adaptation in Africa. Environ Sci Policy 136:250–260. https://doi.org/10.1016/j.envsci.2022.06.004

Food and Agriculture Organization (FAO) (2016) The state of food and agriculture: climate change, agriculture and food security. Food and Agriculture Organization of the United Nations (FAO), Rome, Italy

Gagnon CA, Berteaux D (2009) Integrating traditional ecological knowledge and ecological science: a question of scale. Ecol Soc 14(2). https://www.jstor.org/stable/26268305

Hiwasaki L, Luna EM, Syamsidik S, Shaw R (2014) Process for integrating local and indigenous knowledge with science for hydro-meteorological disaster risk reduction and climate change adaptation in coastal and small island communities. Int J Dis Risk Reduct 10:15–27. https://doi.org/10.1016/j.ijdrr.2014.07.007

Hosen N, Nakamura H, Hamzah A (2020) Adaptation to climate change: does traditional ecological knowledge hold the key? Sustainability 12(676). https://doi.org/10.3390/su12020676

Hussein K, Suttie D (2016) Rural-urban linkages and food systems in sub-Saharan Africa: the rural dimension, vol 5. IFAD, Rome

Huynh CV, Le QNP, Nguyen MTH, Tran PT, Nguyen TQ, Pham TG, Nguyen LHK, Nguyen LTD, Trinh HN (2020) Indigenous knowledge in relation to climate change: adaptation practices used by the Xo Dang people of central Vietnam. Heliyon 6. https://doi.org/10.1016/j.heliyon.2020.e05656

Imoro ZA, Imoro AZ, Duwiejuah AB, Abukari A (2021) Harnessing indigenous technologies for sustainable management of land, water, and food resources amidst climate change. Front Sustain Food Syst 5. https://doi.org/10.3389/fsufs.2021.691603

Intergovernmental Panel on Climate Change (IPCC) (2014) Summary for policy makers. climate change 2014: impacts, adaptation, and vulnerability. Part A: global and sectoral aspects. Contribution of working group II to the fifth assessment report of the IPCC. Cambridge University Press, Cambridge, pp 1–32

IPBES (2018) In: Archer E, Dziba, LE, Mulongoy, KJ, Maoela, MA, Walters, M, Biggs, R, Cormier-Salem, M-C, DeClerck, F, Diaw, MC, Dunham, AE, Failler, P, Gordon, C, Harhash, KA, Kasisi, R, Kizito, F, Nyingi, WD, Oguge, N, OsmanElasha, B, Stringer, LC, Tito de Morais, L, Assogbadjo, A, Egoh, BN, Halmy, M W, Heubach, K, Mensah, A, Pereira, L, Sitas, N (eds) Summary for policy-makers of the regional assessment report on biodiversity and ecosystem services for Africa of the intergovernmental science-policy platform on biodiversity and ecosystem services. IPBES Secretariat, Bonn, p 49

IPCC (2018) Global warming of 1.5 °C: an IPCC special report on impacts of global warming of 1.5 °C above pre-industrial levels and related global greenhouse gas emission pathways, in the contex of strengthening the global response to the thereat of blimate change, sustainable development, and efforts to eradicate poverty

IPCC (2021) In: Masson-Delmotte V, Zhai P, Pirani A, Connors SL, Péan C, Berger S, Caud N, Chen Y, Goldfarb L, Gomis MI, Huang M, Leitzell K, Lonnoy E, Matthews JBR, Maycock TK, Waterfield T, Yelekçi O, Yu R, Zhou B (eds) Climate change 2021: the physical science basis. Contribution of working group I to the sixth assessment report of the intergovernmental panel on climate change. Cambridge University Press

IPCC (2022) In: Pörtner H-O, Roberts DC, Tignor M, Poloczanska ES, Mintenbeck K, Alegría A, Craig M, Langsdorf S, Löschke S, Möller V, Okem A, Rama B (eds) Climate change 2022: impacts, adaptation, and vulnerability. contribution of working group II to the sixth assessment report of the intergovernmental panel on climate change. Cambridge University Press

IPCC (2023) Climate change 2023: synthesis report. In: Core Writing Team, Lee H, Romero J (eds) Contribution of working groups I, II and III to the sixth assessment report of the intergovernmental panel on climate change. IPCC, Geneva, p 184. https://doi.org/10.59327/IPCC/AR6-9789291691647

Iticha B, Husen A (2018) Adaptation to climate change using indigenous weather forecasting systems in Borana pastoralists of southern Ethiopia. Clim Dev 11(7):564–573. https://doi.org/10.1080/17565529.2018.1507896

Jiri O, Mafongoya PL, Chivenge P (2015) Indigenous knowledge systems, seasonal "quality" and climate change adaptation in Zimbabwe. Climate Res 66:103–111. https://doi.org/10.3354/cr01334

Jiri O, Mafongoya PL, Mubaya C, Mafongoya O (2016) Seasonal climate prediction and adaptation using indigenous knowledge systems in agriculture systems in Southern Africa: a review. J Agric Sci 8:156

Kalanda-Joshua M, Ngongondo C, Chipeta L, Mpembeka F (2011) Integrating indigenous knowledge with conventional science: enhancing localised climate and weather forecasts in Nessa, Mulanje, Malawi. Phys Chem Earth, Parts A/B/C 36(14–15):996–1003. https://doi.org/10.1016/j.pce.2011.08.001

Kihila JM (2018) Indigenous coping and adaptation strategies to climate change of local communities in Tanzania: a review. Clim Dev 10(5):406–416

Kijazi AL, Chang'a LB, Liwenga ET, Kanemba A, Nindi SJ (2013) The use of indigenous knowledge in weather and climate prediction in Mahenge and Ismani Wards, Tanzania. J Geogr Reg Plann 6(7):274–280

Mafongoya PL, Ajayi OC (eds) (2017) Indigenous knowledge systems and climate change management in Africa. CTA, Wageningen, 316pp. https://cgspace.cgiar.org/handle/10568/91189

Magni G (2017) Indigenous knowledge and implications for the sustainable development agenda. Eur J Educ 52(4):437–447. https://doi.org/10.1111/ejed.12238

Mahoo H, Mbungu W, Yonah I, Recha J, Radeny M, Kimeli P, Kinyangi J (2015) Integrating indigenous knowledge with scientific seasonal forecasts for climate risk management in Lushoto District in Tanzania. CCAFS working paper no. 103. CGIAR Research Program on Climate Change, Agriculture and Food Security (CCAFS). Copenhagen, Denmark. Available online at: www.ccafs.cgiar.org

Makondo CC, Thomas DSG (2018) Climate change adaptation: Linking indigenous knowledge with western science for effective adaptation. Environ Sci Policy 88:83–91. https://doi.org/10.1016/j.envsci.2018.06.014

Matsa W, Mukoni M (2013) Traditional science of seed and crop yield preservation: exploring the contributions of women to indigenous knowledge systems in Zimbabwe. Int J Humanit Soc Sci 3(4):234–245

Mekonnen Z, Kidemu M, Abebe H, Semere M, Gebreyesus M, Worku A, Tesfaye M, Chernet A (2021) Traditional knowledge and institutions for sustainable climate change adaptation in Ethiopia. Curr Res Environ Sustain 3:100080. https://doi.org/10.1016/j.crsust.2021.100080

Moher D, Liberati A, Tetzlaff J, Altman DG, PRISMA Group (2009) Preferred reporting items for systematic reviews and meta-analyses: the PRISMA statement. Ann Intern Med 151(4):264–269

Mugambiwa SS (2018) Adaptation measures to sustain indigenous practices and the use of indigenous knowledge systems to adapt to climate change in Mutoko rural district of Zimbabwe. Jàmbá 10(1):388. https://doi.org/10.4102/jamba.v10i1.388

Mugambiwa SS, Makhubele JC (2021) Indigenous knowledge systems based climate governance in water and land resource management in rural Zimbabwe. J Water Clim Change. https://doi.org/10.2166/wcc.2021.183

Mugi-Ngenga E, Kiboi MN, Mucheru-Muna M, Mugwe JN, Mairura F, Mugendi DN, Ngetich FK (2021) Indigenous and conventional climate-knowledge for enhanced farmers' adaptation to climate variability in the semi-arid agro-ecologies of Kenya. Environ Chall 5:100355. https://doi.org/10.1016/j.envc.2021.100355

Nalau J, Becken S, Schliephack J, Parsons M, Brown C, Mackey B (2018) The role of indigenous and traditional knowledge in ecosystem-based adaptation: a review of the literature and case studies from the Pacific Islands. Weather Clim Soc 10(4):851–865. https://doi.org/10.1175/wcas-d-18-0032.1

Ncube B, Lagardien A (2015) Insights into indigenous coping strategies to drought for adaptation in agriculture: a Karoo scenario. WRC report no. 2084/1/15. Water Research Commission, Pretoria

Nelson GL, Zamora O, de Guzman LE, Tatlonghari R, Espaldon MV, Brillon J (2019) The indigenous practices and climate change responses of ati and suludnon farmers in Iloilo, Philippines. J Environ Sci Manag 22:87–98

Niang I, Ruppel OC, Abdrabo MA, Essel A, Lennard C, Padgham J, Urquhart P (2014) Africa. In: Barros VR, Field CB, Dokken DJ, Mastrandrea MD, Mach KJ, Bilir TE, Chatterjee M, Ebi KL, Estrada YO, Genova RC, Girma B, Kissel ES, Levy AN, MacCracken S, Mastrandrea PR, White LL (eds) Climate change 2014: impacts, adaptation, and vulnerability. Part B: regional aspects. Contribution of working group II to the fifth assessment report of the intergovernmental panel on climate change. Cambridge University Press, Cambridge, pp 1199–1265

Nkomwa EC, Joshua M, Ngongondo C, Monjerezi M, Chipungu F (2014) Assessing indigenous knowledge systems and climate change adaptation strategies in agriculture: a case study of Chagaka Village, Chikhwawa, Southern Malawi. Phys Chem Earth, Parts A/B/C 67–69:164–172. https://doi.org/10.1016/j.pce.2013.10.002

Nyadzi E, Werners SE, Biesbroek R, Ludwig F (2021) Techniques and skills of indigenous weather and seasonal climate forecast in northern Ghana. Clim Dev 13(6):551–562

Nyakaisiki K, Mugume I, Ngailo T, Nakabugo R (2019) The use of indigenous knowledge in predicting changes in seasonal rainfall by smallholder farmers of Ruteete Subcounty, Kabarole district. J Geosci Environ Prot 7:13–22. https://doi.org/10.4236/gep.2019.71002

Nyong A, Adesina FA, Elasha BO (2007) The value of indigenous knowledge in climate change mitigation and adaptation strategies in the African Sahel. Mitig Adapt Strat Glob Change 12(5):787–797. https://doi.org/10.1007/s11027-007-9099-0

Ofoegbu C, Chirwa PW, Francis J, Babalola FD (2016) Assessing forest-based rural communities' adaptive capacity and coping strategies for climate variability and change: the case of Vhembe district in South Africa. Environ Dev. https://doi.org/10.1016/j.envdev.2016.03.001

Ohajunwa C, Mji G (2018) The African indigenous lens of understanding spirituality: reflection on key emerging concepts from a reviewed literature. J Relig Health 57(6):2523–2537. https://doi.org/10.1007/s10943-018-0652-9

Okonya JS, Kroschel J (2013) Indigenous knowledge of seasonal weather forecasting: a case study in six regions of Uganda. Agric Sci 4:641–648

Padal SB, Raju JB, Chandrasekhar P (2013) Traditional knowledge of Konda Dora Tribes, Visakhapatnam District, Andhra Pradesh, India. IOSR J Pharm 3(4):22–28
Petzold J, Andrews N, Ford JD, Hedemann C, Postigo JC (2020) Indigenous knowledge on climate change adaptation: a global evidence map of academic literature. Environ Res Lett 15(11):113007. https://doi.org/10.1088/1748-9326/abb330
Phu PX, De NN (2016) The situation and solutions for using indigenous knowledge of local people in adaptation to floods in an Giang Province, Vietnam. Asia Pac J Rural Dev. https://doi.org/10.1177/1018529120160204
Plotz R, Chambers LE, Finn CK (2017) The best of both worlds: a decision-making framework for combining traditional and contemporary forecast systems. J Appl Meteorol Climatol 56(8):2377–2392. https://doi.org/10.1175/jamc-d-17-0012.1
Radeny M, Desalegn A, Mubiru D, Kyazze F, Mahoo H, Recha J et al (2019) Indigenous knowledge for seasonal weather and climate forecasting across East Africa. Clim Change 156:509–526
Rankoana SA (2022) Indigenous knowledge and innovative practices to cope with impacts of climate change on small-scale farming in Limpopo Province, South Africa. Int J Clim Change Strateg Manag 14(2):180–190. https://doi.org/10.1108/IJCCSM-04-2021-0040
Rawat DS, Kharwal AD (2016) Folk knowledge on indigenous practices in North-West Himalaya with special reference to Himachal Pradesh (H.P.), India. Int J Adv Sci Res 1(8):6–12
Risiro J, Mashoko D, Tshuma D, Rurinda E (2012) Weather forecasting an indigenous knowledge systems in Chimanimani District of Manicaland, Zimbabwe. J Emerg Trends Educ Res Policy Stud 3(4):561–566
Roncoli C, Ingram KT, Kirshen P (2002) Reading the rains: local knowledge and rainfall forecasting in Burkina Faso. Soc Nat Resour 15(5):409–427. https://doi.org/10.1080/08941920252866774
Salim SS, Monolisha S (2019) Indigenous traditional ecological knowledge of Tamil Nadu Fisher folks: to combat the impact of climate and weather variability. Indian J Tradit Knowl 18:781–792
Schilling J, Freier KP, Hertig E, Scheffran J (2012) Climate change, vulnerability and adaptation in north Africa with focus on Morocco. Agr Ecosyst Environ 156:12–26. https://doi.org/10.1016/j.agee.2012.04.021
Sharma S, Hamal K, Khadka N, Joshi BB (2020) Dominant pattern of year-to-year variability of summer precipitation in Nepal during 1987–2015. Theord Appl Climatol 142(3):1071–1084. https://doi.org/10.1007/s00704-020-03359-1
Shoko K (2012) Indigenous weather forecasting systems: a case study of the biotic weather forecasting indicators for wards 12 and 13 in Mberengwa district Zimbabwe. J Sustain Dev Africa 14:92–114
Shoko K, Shoko N (2013) Indigenous weather forecasting systems: a case study of the abiotic weather forecasting indicators for wards 12 and 13 in Mberengwa District Zimbabwe. Asian Soc Sci 9:285
Smith WN, Grant BB, Desjardins RL, Kroebel R, Li C, Qian B, Drury CF et al (2013) Assessing the effects of climate change on crop production and GHG emissions in Canada. Agric Ecosyst Environ 179(0):139–150. https://doi.org/10.1016/j.agee.2013.08.015
Snorek J, Renaud FG, Kloos J (2014) Divergent adaptation to climate variability: a case study of pastoral and agricultural societies in Niger. Glob Environ Chang 29:371–386. https://doi.org/10.1016/j.gloenvcha.2014.06.014
Son HN, Linh DT, Kingsbury A (2019) Indigenous knowledge and climate change adaptation of ethnic minorities in the mountainous regions of Vietnam: a case study of the Yao people in Bac Kan Province. Agric Syst 176:102683. https://doi.org/10.1016/j.agsy.2019.102683
Son H, Kingsbury A, Hoa H (2021) Indigenous knowledge and the enhancement of community resilience to climate change in the Northern Mountainous Region of Vietnam. Sustain Food Syst 45:499–522
Soropa G, Gwatibaya S, Musiyiwa K, Rusere F, Mavima G, Kasasa P (2015) Indigenous knowledge system weather forecasts as a climate change adaptation strategy in smallholder farming systems of Zimbabwe: case study of Murehwa, Tsholotsho and Chiredzi districts. Afr J Agricul Res 10(10):1067–1075. https://doi.org/10.5897/ajar2013.7205

Taremwa NK, Gashumba D, Butera A, Ranganathan T (2016) Climate change adaptation in rwanda through indigenous knowledge practice. J Soc Sci 46:165–175

Tengo M, Brondizio ES, Elmqvist T, Malmer P, Spierenburg M (2014) Connecting diverse knowledge systems for enhanced ecosystem governance: the multiple evidence base approach. Ambio 43:579–591

Tume SJP, Kimengsi JN, Fogwe ZN (2019) Indigenous knowledge and farmer perceptions of climate and ecological changes in the Bamenda Highlands of Cameroon: insights from the Bui Plateau. Climate 7(12):138. https://doi.org/10.3390/cli7120138

Turnhout E, Bloomfield B, Hulme M, Vogel J, Wynne B (2012) Listen to the voices of experience. Nature 488(7412):454–455

Ubisi NR, Kolanisi U, Jiri O (2019) The role of indigenous knowledge systems in rural smallholder farmers' response to climate change: case study of Nkomazi local municipality, Mpumalanga, South Africa. J Asian Afr Stud 55:273–284

United Nations Framework Convention on Climate Change (UNFCCC) (2013) Report of the subsidiary body for scientific and technological advice on its thirty-eighth session. Germany: Bonn. held in Bonn from 3 to 14 June 2013

Van Dien T, Van DX (2014) Agricultural production model adapt to climate change based on indigenous knowledge of ethnic minorities in Bac Kan Province, Vietnam. Earth Sci 3(1):1–8

Van Huynh C, Phuong L, Hong Nguyen M, Tran P, Nguyen T, Pham T et al (2020b) Indigenous knowledge in relation to climate change: adaptation practices used by the Xo Dang people of central Vietnam. Heliyon 6:e05656

WIPO (World Intellectual Property Organization) (2012) Traditional knowledge. https://www.wipo.int/tk/en/tk/. Accessed 07 Oct 2023

WIPO (2020) Climate action and sustainability: indigenous peoples are part of the solution (nd). https://www.wipo.int/wipo_magazine/en/2020/01/article_0007.html

Zuma-Netshiukhwi G, Stigter K, Walker S (2013) Use of traditional weather/climate knowledge by farmers in the south-western free state of South Africa: agrometeorological learning by scientists. Atmosphere 4:383–410. https://doi.org/10.3390/atmos4040383

Chapter 11
The Intersection of Climate Crisis and Disease Outbreaks: Cataclysmic Consequences

Debangshu Banerjee, Susanta Nath, and Biplob Kumar Modak

Abstract A few of the negative effects of climate change that are threatening ecosystems include rising temperatures, extreme weather and sea level rise. The conditions under which vectors carrying the disease are present, such as ticks, mosquitoes and fleas, result from rising temperatures and precipitation patterns. The wide range of *Aedes* mosquitoes aid in the spread of the Zika virus and dengue fever. Lyme disease has expanded northward in Quebec due to climate change, which was previously considered a low-risk zone for this disease. It is important to note that the spread of malaria to previously low-risk areas, in particular the highlands and the hills of northern India, has been facilitated by rising temperatures and changing precipitation. The ability of cholera-causing organisms to multiply in vulnerable areas is further aided by changes in rising temperatures. The introduction of zoonotic diseases like Ebola, SARS, and COVID-19 is facilitated by ecological changes brought on not only by climate change but by loss of biodiversity also. Population displacement caused by climate-related disasters can also escalate the risk of disease transmission. In most cases, victims of these outbreaks of diseases are the poor people of developing countries. To minimise or stop the spread of disease, effective public health policies and combating climate change are indispensable. Interdisciplinary collaboration, policy interventions, and public awareness are necessary to combat all of these. This communication aims to discuss the impact of climate change on epidemics and to stress the urgent need for comprehensive and coordinated measures to address the problem.

D. Banerjee
Department of Zoology, Tarakeswar Degree College, Hooghly, WB 712410, India

S. Nath
Department of Zoology, Government General Degree College Singur, Hooghly, WB 712409, India

B. K. Modak (✉)
Department of Zoology, Sidho-Kanho-Birsha University, Purulia, WB 723104, India
e-mail: bkmodak09@gmail.com

U. Mukhopadhyay et al. (eds.), *Climate Crisis, Social Responses and Sustainability*, Climate Change Management, https://doi.org/10.1007/978-3-031-58261-5_11

Introduction

Climate change is a process in which, primarily due to human activities that produce greenhouse gases on an ongoing basis, Earth's climate has been gradually changing. Climate change is a complex issue caused by human activities such as burning fossil fuels, deforestation, and certain farming practices. It affects different aspects of the planet, including the air, oceans, land, and living organisms. This, in turn, has significant impacts on humans, such as rising sea levels, extreme weather patterns, and changes in the availability of food and water (Dhiman et al. 2010).

As we confront the growing danger of climate change and recurring outbreaks of diseases, we are starting to see how these global challenges converge to bring about catastrophic consequences that impact ecosystems, societies, and economies. The intricate interplay between the destabilizing effects of climate change and the dynamics of infectious diseases has brought about an unprecedented level of complexity, where the health of our planet is closely tied to the well-being of humans (McMichel and Woodruff 2008). This synthesis of ecological and epidemiological realms has resulted in a complex web of interconnected effects that range from changes in disease vectors to strained healthcare infrastructures and from vulnerable communities to the intricate web of international relationships undergoing profound shifts (Smith et al. 2005). This communication explores the intersection of the climate crisis and disease outbreaks, highlighting the urgent need for comprehensive solutions.

Earth's climate is rapidly changing due to increasing temperatures, shifting precipitation patterns, and rising sea levels (Mimura 2013). These changes have profound impacts on the spread of diseases that have plagued humanity for centuries (Shivanna 2022). Climate change can have a significant impact on disease-carrying insects, also known as vectors, in areas that are already hot and humid (Ma et al. 2022). This is because the hot and humid conditions make it easier for these insects to breed and survive (Annel et al. 2010). As climate change progresses, the ranges of these vectors are expected to expand, which means they'll have the potential to infect and spread diseases to more people (Dhara et al. 2012). With the rise in global temperatures, the ranges of disease-carrying vectors such as mosquitoes and ticks are expanding, exposing new populations to infections like malaria, dengue fever, Lyme disease, and more (Mojahed et al. 2022). Altered rainfall patterns lead to waterborne illness outbreaks like cholera and leptospirosis (Levy et al. 2018). It is concerning to observe that climate change is leading to a loss of biodiversity (Zhao et al. 2023). As habitats are destroyed and ecosystems are disrupted, many animal species are at risk of extinction (Jung et al. 2023). However, some species may have the chance to thrive in new environments (Chu and Karr 2017). The result is the emergence of zoonotic diseases, which are transmitted from animals to humans, as a result of this disturbance in the natural balance of things (Ellwanger and Chies 2021; Barbier 2021). The increasing frequency of zoonotic disease outbreaks highlights the consequences of environmental degradation and encroachment on wild habitats (Bavinger et al. 2020).

Tropical and subtropical climates provide favourable conditions for parasites and their vectors due to warm temperatures and high humidity (Almberg et al. 2012). This supports their growth, reproduction, and complex life cycles, which often require specific environmental parameters for survival (Bronson 2009). Climate-driven disasters, such as droughts, floods, etc. can trigger mass displacements of populations due to cascading effects (Ebi et al. 2021). Displacements of people from their homes often lead to overcrowded living conditions, inadequate sanitation, and limited access to healthcare. Such conditions promote the expansion of diseases (Macrogliese 2008). The Pandemic COVID-19 has demonstrated how easily healthcare systems can become overwhelmed by rapidly spreading infectious diseases (Filip et al. 2022). It can be imagined the impact of climate disasters and disease outbreaks, where resources are stretched to their limits, exacerbating suffering (Djalante et al. 2020). This chapter aims to investigate the complex relationship between the increasing climate crisis and the heightened risk of disease outbreaks (Seventers and Hochberg 2017). It recognizes the potentially catastrophic consequences of these phenomena on global health and societal structures (Lacetera 2018). We are trying to understand how climate change affects infectious diseases by analyzing changes in the behavior of disease carriers, the way pathogens behave, and how susceptible hosts are to them (Eldred and Reilly 2014). The primary objective is to explain the direct and indirect links between climate change and the emergence, spread, and intensity of infectious diseases (Confalonieri et al. 2015). Additionally, the research aims to assess the vulnerability of communities and healthcare systems to these threats, highlighting the disparities in impact across socio-economic strata (Niekerk 2020). By analyzing historical precedents and current trends, the study aims to develop a comprehensive understanding of the dynamics driving the convergence of climate crisis and disease outbreaks. This chapter aims to examine the cascading effects of health crises on social, economic, and political systems, analyzing the pathways through which environmental degradation and disease can amplify existing vulnerabilities, and contribute to societal collapse. Through this investigation, the study seeks to provide valuable insights to inform adaptive strategies, policy interventions, and global cooperation necessary to mitigate the catastrophic consequences of this intersection. The research will also look into future scenarios and how they could affect disease patterns and society as a whole. The goal is to provide insights that can guide policymakers and international collaborations in addressing the challenges brought about by climate change and disease outbreaks (Paul and Mity 2013). By combining epidemiology, ecology, and social sciences, the study hopes to provide a comprehensive understanding of the threats at the intersection of climate change and infectious diseases.

The present chapter aims to uncover the mechanisms that contribute to the amplification of diseases under the influence of climate change and its impact on society.

Methodology

Literature Review and Search Strategy

We conducted a thorough search on reputable databases, such as PubMed and Google Scholar, to identify relevant publications exploring the link between climate change and disease outbreaks. The search queries were strategically designed to cover literature containing the terms "climate crisis" or "climate variability" and "disease outbreaks," "epidemics," or "pandemics." We gave special attention to studies that focused on the interdependence of climate change and the occurrence, spread, or aggravation of diseases. To obtain a detailed understanding of the subject, the search was narrowed down to studies that specifically focused on the effects of climate change and disease outbreaks in different regions across the globe. The emphasis was placed on gaining a global perspective, with particular attention to areas that are vulnerable to both climate-related problems and significant disease outbreaks.

Criteria for Inclusion

The papers found during the literature search were thoroughly screened in a three-step process that involved the evaluation of their titles, abstracts, and full texts. The inclusion criteria were defined to ensure that the papers were relevant to the intersection of the climate crisis and disease outbreaks. Only articles that presented substantial empirical evidence, novel insights, or theoretical contributions to understanding the catastrophic consequences of the interplay between climate change and disease dynamics were considered (Fig. 11.1).

Framework

The study utilized a conceptual framework to guide the analysis and focused on answering central questions regarding the relationship between climate change and disease outbreaks. Some of the questions included: How does climate change affect the frequency and severity of disease outbreaks? What are the mechanisms that explain the interaction between climate-related factors and disease dynamics? Finally, how can societies effectively respond to the health challenges arising from the intersection of climate crisis and disease outbreaks?

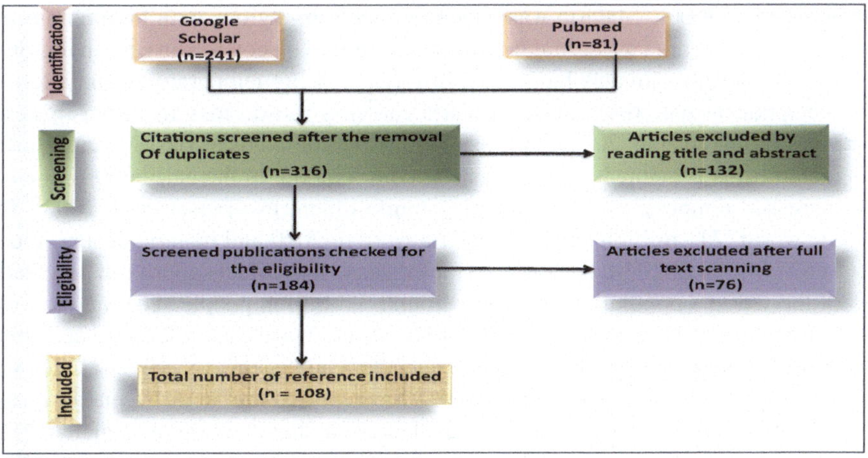

Fig. 11.1 Flow diagram summarizing the approach used in the selection of literature

Data Analysis

The study utilized a conceptual framework to guide its analysis, intending to answer key questions such as how climate change affects disease outbreaks, how climate-related factors interact with disease dynamics, and how societies can respond effectively to the health challenges arising from both climate crisis and disease outbreaks. A systematic process was used to extract information, including targeted queries to explore key aspects such as the influence of climate variables on disease vectors, the role of environmental changes in disease emergence, and the societal vulnerabilities that contribute to the amplification of health risks. The analysis aimed to uncover patterns, correlations, and potential causal relationships between climate crisis and disease outbreaks. Overall, this approach rigorously investigates the intersection of climate crisis and disease outbreaks, aiming to uncover insights that contribute to our understanding of their consequences.

Instances of Disease Spread with Climatic Fluctuations

Climate variations have a significant impact on the transmission of diseases, with changes in weather patterns often leading to outbreaks. Warmer temperatures and increased humidity facilitate the growth of mosquitoes (JanosiIm et al. 2023). Climate changes also increase the range of diseases, exposing new populations to them (Eldred and Reilly 2014; Tzanakakis et al. 2020). Extreme weather events, which are exacerbated by climate change, compromise sanitation and healthcare, amplifying the transmission of waterborne diseases during floods or droughts (Tabari 2020).

(a) Before the 2000s, Malaria was endemic in certain regions with high humidity and rainfall in the 1990s. Malaria was not a significant concern in certain regions with temperate climates but after a few years, rising temperatures and altered precipitation patterns allowed malaria-carrying mosquitoes to thrive in new areas (Tonnevi et al. 2014). For instance, in the highlands of East Africa, where malaria was previously rare, cases started to emerge in the early 2000s due to increased temperatures and changing rainfall patterns (Tanser et al. 2003). In the world, 241 million cases of malaria were in 2020 and malaria deaths stood at 627,000 in 2020.Malaria is a serious disease that affects approximately half of the world's population, which is around 3.3 billion people. The majority of malaria cases are found in Sub-Saharan Africa, but it also affects Asia, Latin America, and some parts of Europe and the Middle East (Xiang et al. 2018). The World Health Organization (WHO) has reported that in 2010, malaria was present in 106 countries and territories. That year, there were approximately 216 million cases of malaria, resulting in nearly 0.7 million deaths, with most of the fatalities occurring among children living in Africa (https://www.unicef.org). In India, new areas in states like Himachal Pradesh and Jammu and Kashmir, which were previously malaria-free, have reported cases due to changing climate conditions (Patz et al. 2005).

(b) Prior to 1990s, Dengue fever was primarily confined to tropical and subtropical regions. But as temperatures rose, the range of the *Aedes aegypti* mosquito, which carries the dengue virus, expanded. New outbreaks occurred in areas previously unaffected, like several areas of the United States South as well as southern Europe (Hales et al. 2002). Dengue fever was limited to certain parts of India with tropical and subtropical climates in the 1990s. But rising temperatures and increased rainfall created favourable conditions for the breeding of *Aedes* mosquitoes, leading to a significant rise in dengue cases across various states in India from 2000 to 2017 (Fig. 11.2). Cities like Delhi, Mumbai, and Kolkata have witnessed a surge in dengue fever cases due to changing climate patterns (Dhiman et al. 2010). Due to climate change, the potential for the spread of the disease is expected to increase in the southern, central, and eastern regions of India, leading to potential outbreaks, unless appropriate measures are taken to control the disease. Due to high temperatures, the western regions of India may not be suitable for transmitting the dengue virus (Hui 2006). In light of various climate change scenarios, a Climate Driven Dengue Model has been developed to anticipate vulnerable areas in India where dengue occurs (Kakarla et al. 2020). In the case of public health preparedness and vector control operations in endemic areas, this model approach provides essential information (Oladipo et al. 2022).

(c) Prior to 2000s, Schistosomiasis was predominantly found in tropical and subtropical regions with suitable snail hosts (Gupta et al. 2010). However, the habitats of schistosome snails that are hosts to this parasite have been extended as a result of changes in temperature and precipitation patterns post-2000s (Brooks and Hobeng 2007). Cases of schistosomiasis have therefore started to be reported in new areas.

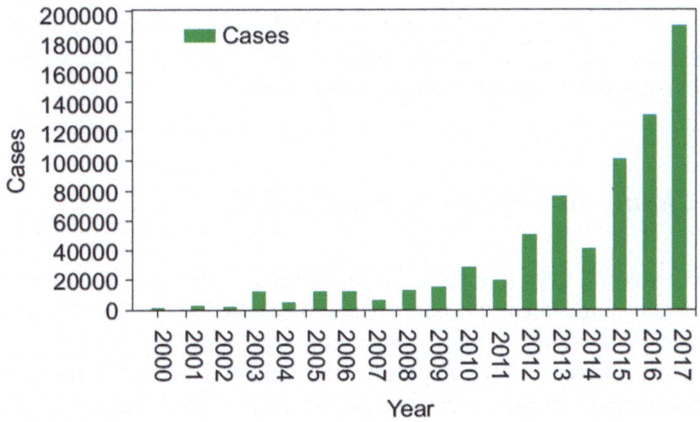

Fig. 11.2 Dengue cases by year in India from 2000 to 2017 (Modified from Kakarla et al. 2020)

(d) Lyme disease was primarily confined to some parts of Europe, North America, and Asia in the 1970s. But later Warmer temperatures allowed ticks, which carry the bacteria causing Lyme disease, to thrive in new regions (Cabral 2010; Stiasny et al. 2021). The disease began to emerge in areas where it was previously rare, such as parts of Canada and northern Europe (Monaghan et al. 2015).

(e) Zika virus was limited to equatorial regions in Africa and Asia in the 1990s. But later, *Aedes* mosquitoes, which transmit the Zika virus, expanded their habitat due to rising temperatures (Mittal et al. 2017). Zika outbreaks occurred in new areas, including parts of South America and the southern United States (Bogoch et al. 2016).

(f) *Vibrio* bacteria, including *Vibrio cholerae*, were mainly a concern in warm coastal waters in the early 2000s (de Magny et al. 2012). Rising sea surface temperatures created favourable conditions for *Vibrio* bacteria in new regions (Kumar et al. 2005). Consequently, areas previously unaffected, including parts of northern Europe and North America, started reporting cases of Vibrio infections (Patz et al. 2008).

Climate Change and Disease Outbreak

Climate change significantly amplifies the risk of disease outbreaks worldwide. Mostly the diseases are vector-borne but they can be water-borne also (Cam et al. 2013). As temperatures rise, it creates a more favourable environment for disease-carrying mosquitoes, which, in turn, increases the prevalence of diseases like malaria, dengue fever, and Zika virus. Climate change also affects the pattern of water resources, leading to the spread of waterborne diseases such as cholera (Walton et al. 2023). Extreme weather events and altered ecosystems facilitate the emergence of new infectious diseases, posing challenges for public health systems. Climate change-induced migration further accelerates disease transmission as displaced populations often live in overcrowded and unsanitary conditions.

Climate Change and Vector-Borne Diseases

According to studies, ticks that spread Lyme disease across North America are now found in a wider range due to climate change (Ostfeld 2008). Originally, the Northeastern states of the United States were home to many cases of Lyme disease. However, due to increasing temperatures, the ticks that transmit the disease have expanded their range further north into states such as Vermont, New Hampshire, and Maine (Bouchard et al. 2019). Due to the rising temperatures caused by climate change, the tick population is expanding and moving into new areas, thriving in warmer regions, and as a result, increasing the risk of exposure and bites (JanosiIm et al. 2023). As the climate warms, the tick season is getting longer, providing ticks with more time to feed and transmit Lyme disease to humans. The location and kind of flora are also changing due to climate change, which may have an impact on tick and other wildlife populations' access to food. They may then relocate as a result, raising the possibility of disease transmission (Gray and Ogden 2021).

Our world is experiencing a rise in global temperatures due to climate change. This warming trend is causing changes in temperature patterns, particularly in regions that were once cooler, such as highland areas. Unfortunately, this has led to an increase in malaria transmission in regions where the disease was previously uncommon. Malaria cases have been reported in highland regions that were previously too cold for transmission. In the East African highlands, such as parts of Kenya, Uganda, Rwanda, and Ethiopia, the malaria pathogen and mosquito vectors are benefiting from higher temperatures, which is increasing the rate of transmission. The disease is also spreading to higher altitudes and new areas previously thought to be free of it. An increase in the number of breeding sites for Anopheles mosquitoes can result from frequent and heavy rainfalls, increasing the risk that malaria will be transmitted (Mares et al. 2017). In addition, the spread of malaria to new areas can be made possible by migration and urbanization. This is particularly true when people from regions with high malaria prevalence move to highland areas and introduce the disease to populations that were previously unaffected. Moreover, rising temperatures have caused *Aedes* mosquitoes, which are responsible for spreading Zika virus and Dengue fever, to expand their geographical range (Ryan et al. 2019). The habitat of *Aedes* mosquitoes in the United States has expanded to include more northern areas, such as Florida, Texas, and parts of the Gulf Coast. This expansion has increased the likelihood of Zika virus and Dengue fever transmission in areas that were previously less exposed to these diseases (Rezza 2014). Due to the presence of *Aedes* mosquitoes, local outbreaks of dengue fever have been reported in southern European cities such as Spain, Italy, and Greece. The occurrence of these outbreaks has been linked to favourable climatic conditions in these regions. In Asia and South America, where

Zika virus and Dengue fever are already prevalent, there is grave concern about their worsening due to climate change altering rainfall patterns, which create mosquito breeding grounds and increase transmission rates. *Aedes* mosquitoes are the carriers of chikungunya. Climate change can expand *Aedes* mosquitoes' geographic range, allowing them to survive and reproduce in new areas (Ryan et al. 2019).

The prevailing climate conditions have a significant impact on the spread of dengue. The biology of *Aedes* mosquitoes is greatly influenced by factors such as temperature, precipitation, and relative humidity (Curriero et al. 2001). Dengue virus optimally requires 18 °C or more to stay alive within an *Aedes mosquito* (Alto and Bettinardi 2013). Immature stages of the *Aedes* mosquito are unable to survive in temperatures above 42 °C. Dengue is currently prevalent in southern parts of India, and with climate change, it is predicted to spread even further. A study has shown that there may be a 1-week increase in transmission in New Delhi when the temperature rises to 2 °C. However, transmission can be reduced to only 34 weeks with a temperature increase of 4 °C. There has been a 44-week period of transmission of dengue in Kolkata. Overall, the increase in temperature by 4 °C could lead to an additional 2–5 times higher incidence of dengue disease with new transmission areas across India's northern sub-Himalayas and southern regions (Ye et al. 2023). Based on global climate models, with a 2 °C or 4 °C increase in temperature, transmission could continue for up to 53 weeks. In addition, the potential for an epidemic of Dengue has been assessed.

The West Nile virus is a globally present pathogen that mosquitoes transmit. It belongs to the encephalitic Flavivirus spp. family, which is the most widely distributed virus of this family. The *Culex* genus, which belongs to the Culicidae family, serves as the primary amplification and bridge vector for this virus. The distribution and abundance of these mosquito species can be affected by climate change, which may lead to changes in the spread of the West Nile virus. A more favourable breeding environment for mosquitoes, which may lead to an expansion of their geographic range, can be created by warmer temperatures and changing weather patterns. This increases the risk that the West Nile virus will spread to new areas (Ahmed et al. 2019). The West Nile virus has a cycle that involves the transmission of the virus from adult mosquitoes to birds through their blood meals, which results in the multiplication of the virus. Some birds can carry the virus without showing any symptoms. These birds act as reservoir hosts. However, climate change can affect bird migration patterns, which can disrupt the movement and distribution of bird populations that serve as virus reservoirs (Duan et al. 2016). These changes can impact the virus's spread to different areas. Leishmaniasis is a disease caused by the leishmania parasite which infects females with phlebotomine spp (Horn and Luong 2018). According to research, *Phlebotomous papatasi*, a type of sand fly that can lead to leishmaniasis, thrives in temperatures ranging from 32 to 36 °C. Sand flies tend to be less active in temperatures ranging from 11–20 °C to 37–40 °C. The study has also discovered that sand flies are most active during the wettest months of the dry season, which are May and November, despite being active throughout the dry season (May to November). An estimated 50,000 to 90,000 new cases of Visceral Leishmaniasis occur worldwide annually.

Effect of Climatic Factors on Vector-Borne Diseases

Climatic factors, in terms of the prevalence and distribution of vector-borne diseases, are playing an important role (Fig. 11.3). Life cycles and behaviour of the vectors that carry diseases, e.g. mosquitos, ticks, or flies, have a direct influence on temperature, humidity conditions, and precipitation patterns. Climate change may speed up the development of pathogens in these vectors, which can lead to increased transmission of diseases such as malaria and dengue fever. Changes in precipitation patterns can create breeding grounds for mosquitoes in stagnant water, fostering the spread of diseases. Additionally, alterations in climate can expand the geographical range of vectors, exposing new populations to diseases previously uncommon in those areas. To predict and control outbreaks of vector-borne diseases, it is important to understand and monitor these climate influences, which illustrate the crucial link between weather conditions and public health.

Effect of Temperature

The impact of temperature fluctuations on vectors and pathogens, as well as the change in their life cycle or geographical distribution is considerable. The growth of pathogens in vectors could be accelerated by warmer temperatures, thereby increasing transmission rates for infectious diseases such as Zika and Lyme disease.

(a) On Vector

i. Activity and Feeding Behaviour: Generally, higher temperature increases the metabolic rate of vectors leading to their being more active and prone to seek out hosts for blood meal. As a result of higher temperatures, feeding frequency and duration may be increased which can facilitate the spread of pathogens (D'Amato et al. 2016).

ii. Flight and Dispersal: The temperature may affect vectors' flight capabilities and movement patterns. Vectors, like mosquitoes, flies, and biting insects, are likely to be more active when the temperature rises (Bellone and Failloux 2020). With the increased activity, they may move further and spread to wider areas which could raise their risk of transmitting diseases.

iii. Reproduction and Development: The temperature plays an important role in the reproduction and development of vectors. They can produce more eggs in hotter weather and increase their population by speeding up the reproductive cycle (Agyekum et al. 2021). In the development period of young adults, such as larvae and pupae, temperature also plays a key role. Growth of young adults like larvae and pupae, can accelerate with warmer temperatures, leading to shorter generation times and the potential for a higher population of vectors.

iv. Vector Distribution and Disease Spread: Temperature can influence the distribution of vectors as well as their diseases. The evolution of vectors' habitats into areas where they have not yet existed can be caused by changes in climate and temperature patterns (Rocklov and Dubrow 2020). This could lead to new

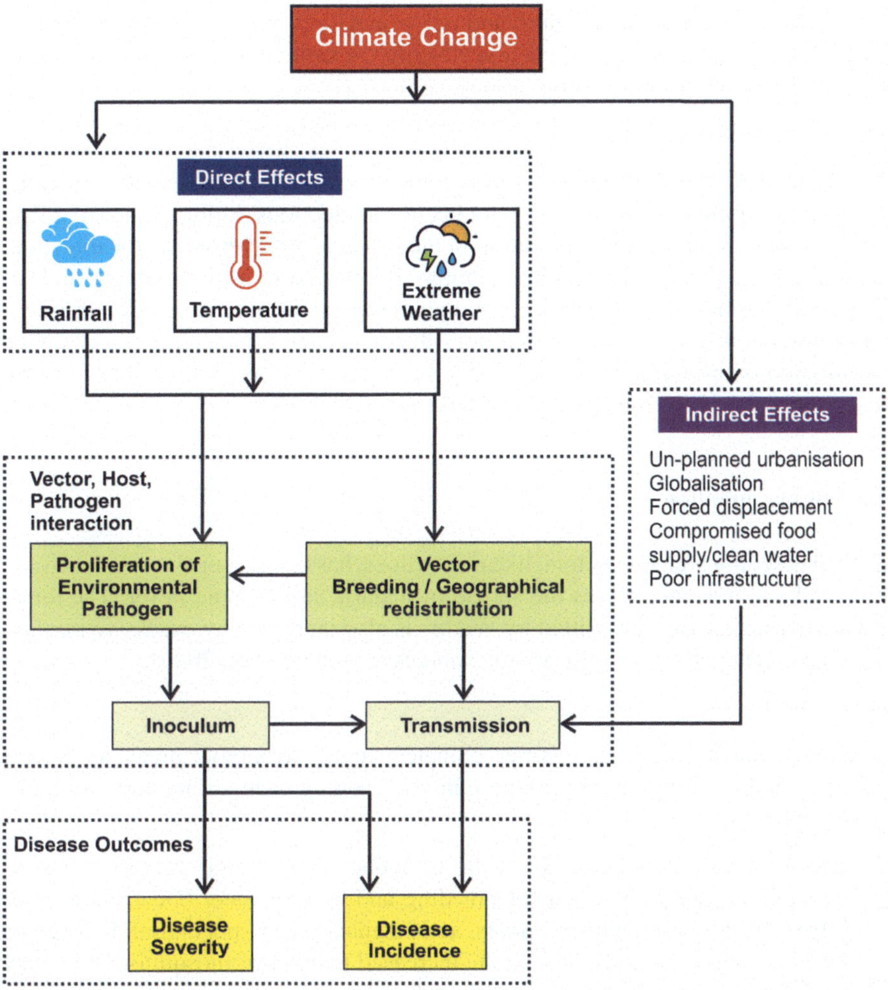

Fig. 11.3 Chart showing the severity and incidence of infectious diseases that could be directly and indirectly affected by climate change (Modified from Dennis and Fisher 2018)

diseases, possibly leading to an outbreak or epidemic in the population that was not previously exposed.

Insect vectors can become infected by taking pathogens from an infected host's blood. However, vectors need time to become infectious. Until the Vector has become infectious, pathogens will multiply and spread throughout the insect's body (Semenza et al. 2022). The duration of this multiplication has been described as an intrinsic incubation period, which is greatly affected by temperature in the case of arbovirus transmission (Christofferson et al. 2016). As a result, an increase in the spread of Arboviral Diseases is associated with climate change. These infections occur most often during summer months when the mosquito and tick populations are in full bloom (Rosenthal 2009). It results in a lengthened vector life cycle at cold temperatures, raising mortality and decreasing fertility.

(b) On the pathogen

After ingestion of the pathogen through the blood meal of the infected host, the pathogen transmitted by the insect vector becomes infectious (Smith et al. 2019). The pathogens are amplified and circulating in the insect's body before vector infection. The duration of this amplification is commonly referred to as the intrinsic incubation period and has a significant influence on temperature in case of viral transmission. Consequently, the climatic conditions fluctuate having a head-on impact on the transmission of arboviral diseases. Mostly these infections are noticed during warm weather months, when mosquitoes and ticks are active.

Effect of Precipitation

An ideal breeding area for vectors, like mosquitoes, has been set up due to an increase in rainfall and this encourages the spread of malaria and Dengue Fever. The threat of waterborne disease transmitted by vectors is also increased by excessive rainfall, which can flush pathogens into aquatic sources (Lymbery et al. 2020).

(a) On Vector

As eggs hatch earlier in warmer climates, more favourable areas for vector breeding can be created by increasing rainfall, flooding or humidity and will make it easier to breed.

i. Breeding and Population Growth: In moist areas, some vectors, such as mosquitoes, are well-suited for breeding and require water (Geatachew et al. 2015). By filling containers, pools, and stagnant water spots, rainfall creates a lot of breeding grounds. Due to the increased number of mosquitoes, this may lead to an increase in the transmission of vector-borne diseases such as malaria, dengue fever, and Zika virus.
ii. Distribution and Movement: Precipitation can affect vectors, affecting their distribution and movement. In the event of severe rainfall and flooding, vectors may be dislocating from their original habitats which leads them to look for new sites where they can survive (Vorobevskii et al. 2020). This migration may bring the vectors and diseases they carry to new locations, increasing their range and leading to outbreaks in places that were not previously affected (Caminade et al. 2019).
iii. Habitat Modification: Since precipitation creates a favourable environment for their survival, the action of vectors can be influenced (Socha et al. 2022). For instance, floods may result in the creation of transitional or permanent water bodies such as swamps and marshes which are a good breeding ground for specific vectors. Furthermore, more vegetation growth may result from

increased precipitation which is sufficient to provide vectors with a further resting place and resources.

(b) On the pathogen

i. Survival and Reproduction: The ability of pathogens to adapt and develop can be influenced by the amount of rainfall. During heavy rains, the number of pathogens present in waters may be washed off or diluted and their population reduced. Some pathogens, on the other hand, that can develop protective structures such as pollens or cysts if they are capable of doing so, may persist and live in humid environments or water according to their characteristics (Swick et al. 2016).

ii. Vector Population dynamics: Rain affects the vector population's behaviour, as well as influences its transfer of pathogens (Mackey et al. 2014). If more rain comes, it may lead to perfect breeding conditions for the vectors leading to an increase in their populations. It is possible to increase the probability of interactions between vectors and pathogens that could lead to the spread of disease.

iii. Host availability and behaviour: The change in behaviour and availability of host populations can influence the transfer of pathogens due to rainfall or flooding (Levy et al. 2016). A large amount of rain may affect the habitat of the animals or reservoirs that host them, leading to reduced chances of contact between the vectors and their hosts. Host behaviour may also change when they are looking for refuge and changing their movements during a thunderstorm. These changes may affect the ability to feed vectors and transmit pathogens.

iv. Disease Outbreaks and Epidemics: It is important to point out how rainfall falls may have a bearing on the likelihood of disease outbreaks and epidemics. A favourable environment for breeding vectors and the transfer of pathogens can be created due to prolonged rains or excess rainfall which may lead to increased disease transmission, as well as a larger outbreak (Campbell-Lendrum et al. 2015). Moreover, temporary water bodies can form during dry periods, followed by heavy rains, which may lead to an increase in vector populations and disease (Raimondi et al. 2023).

Effects of Extreme Weather Events on Vectors

Extreme weather events can have a significant impact on vector species, e.g. mosquitoes and ticks, by modifying their habitats and breeding grounds. This change may lead to an increase in the incidence of vector-borne diseases, which pose a higher risk to human and animal populations.

(a) Hurricanes, Heavy rainfall, and Displacement

Vectors can be forced to leave their natural habitats while a hurricane is in force, which may result in disturbance of the breeding habitat and reduced populations at ground level. In contrast, the aftermath of hurricanes usually produces an increase in rain and standing water that could result in new breeding grounds and possibly a larger number of vector populations. Humidity and precipitation can also have a significant role in vector activity. Humidity and vector activity have a rectilinear relation. The rise in relative humidity can enhance the vector activity.

(b) Water Accumulation

Flooding can result from hurricanes or other severe storms that are accompanied by a lot of rain, which leaves stagnant water in the pools. As a result, there are more places for mosquitoes and other water-dependent vectors to reproduce, which ultimately increases their populations. As a result, the danger of vector-borne infections is increased (Nosrat et al. 2021).

(c) Floods

Floods may lead to the displacement of vectors and their host, including animals or humans, leading to an increased risk of contact with vulnerable individuals (Okaka and Odhiambo 2018). If they are forced to move, it may result in the migration of the vectors into new regions and possible transmission of diseases from uninhabited areas.

(d) Heatwaves

i. Accelerated Development: During the heat wave, certain insects known as vectors may grow and breed more rapidly as temperatures rise. For example, mosquitoes can grow more quickly and have less breeding season in warmer climates, which means their population is growing at a quicker rate (Ahmed et al. 2019).
ii. Extended Activity Period: The duration of vector activity may be prolonged due to warmer temperatures to increase the potential for transmitting disease caused by vectors.
iii. Altered Distribution: Extreme temperatures may lead the vectors to expand their range, which could result in new populations of vector-borne diseases.

Dynamics of Vector-Borne Diseases

Climate change is playing a key role in increasing the risk of vector-borne diseases to people. Climate change has both direct and indirect consequences, affecting physical conditions like temperature and rainfall, as well as the biological conditions of plants and animals. The distribution of ticks and mosquito vectors may be affected by these changes, leading to a higher risk of vector-borne diseases. Most of these diseases are transmitted by mosquitoes and ticks to humans through their bites (Socha et al. 2022). Climate change influences the dynamics of these diseases in several key ways:

The Expansion of Suitable Habitats for Vectors

A rise in world temperature has had an important role to play in the geographical distribution of vectors. Climate warming may encourage the survival and reproduction of diseases carried by these organisms in previously unsuitable areas. In warmer and humid environments, for instance, mosquitoes and ticks which have been identified as vectors of a variety of diseases are thriving (Campbell-Lendrum et al. 2015). This rate will be significantly increased by a rise of 2 °C in the lower end of the temperature range, which enables malaria parasites to spread (Mordae et al. 2013). These vectors may find new habitats in previously cooler areas due to climate change, posing a greater risk for the spread of diseases.

Impact of Precipitation

Changes in rainfall patterns are also a cause of the spread of vector-borne diseases. The formation of stagnant water bodies on which mosquitoes breed may be caused by an increase in rainfall. Paradoxically, pools of water from previously flowing sources may be created during periods of drought, which may in turn encourage vector breeding (Norat et al. 2021). Public health efforts to control and eradicate communicable diseases resulting from vectors are being severely hampered by these changes in vector animal distribution and breeding behaviour (Wilson et al. 2020).

Prolonged Disease Transmission Seasons

Climate change is extending the active transmission period of vector-borne diseases (Rocklow and Dubrow 2020). The warmer climate creates favourable conditions for both the vectors and pathogens carrying them, which is why they are more infectious during a longer period. This implies that individuals in the affected regions are at greater risk of exposure to these diseases during the course of the year, not on a seasonal basis (Grassly and Fraser 2006). The healthcare system could be strained and the burden of diseases increased as a result of this extended disease transmission season.

Behavioural Changes in Vectors

Climate change temperature variations can have a significant impact on the behaviour of disease vectors. For instance, mosquitoes can change the way they bite as their temperature rises (Caminadae et al. 2019). If people are less likely to be protected with bed nets or insect repellents, they might become more active during the day and night. This behavioural change may lead to reduced effectiveness of prevention measures and an increase in the risk of disease transmission. Climate change is also changing the dynamics of vector borne diseases, not only as a result of environmental changes (Campbell-Lendrum et al. 2015). The increase in the threat posed by these diseases is due to their expansion into suitable habitats, extended disease transmission periods, and altered vector behaviour (Lindahl and Grace 2015).

Climate Change and Water-Borne Diseases

Water quality and the spread of waterborne diseases, such as cholera, may be affected by changes in temperature and precipitation patterns (Colwell et al. 2008), Amoebiasis, and Giardiasis. The contamination of water sources with pathogens may be caused by heavy rainfall and floods (Jung et al. 2023). In warm, salt water, the cholera bacterium *Vibrio* which causes cholera is thriving. As temperatures rise, their growth rate in aquatic environments increases, leading to a potential rise in their overall population. Climate change can further exacerbate the situation by altering precipitation patterns, resulting in more rainfall and more frequent and severe storms. Heavy rainfall can cause flooding, which can contaminate water sources with cholera bacteria from human waste, thereby posing a serious health risk to humans. Coastal freshwater sources can also be contaminated with salt water, resulting in unsafe drinking water and the risk of cholera outbreaks due to increased sea levels caused by climate change.

Climate Change and Lifestyle Diseases

The World Health Organization's (WHO) 2002 report estimated that chronic diseases, particularly cardiovascular diseases, were responsible for 60% of all reported deaths and 46% of the global disease burden (Boutayeb 2010). Heat-related illnesses are becoming more common as temperatures rise, even though they are not infectious. Climate change is causing a rise in global temperatures, which is leading to an increase in extreme heat events that can cause heat-related illnesses like heat exhaustion and heatstroke. These illnesses are often classified as lifestyle diseases because they are usually caused by prolonged exposure to hot temperatures, especially in urban areas with limited green spaces (Pathak and Mckinney 2021). Increasing temperatures due to climate change can increase the concentration of ground-level ozone, a major component of smog, which worsens air pollution (Ebi and McGorgre 2008). Respiratory disorders such as asthma and COPD may be caused by prolonged exposure to air pollution. Climate change may hurt food production and availability, which may lead to a variety of health problems, including obesity, diabetes, and cardiovascular disease.

Effects of Climate Change on Psychological Disorders

Climate change can be the reason of mental health problems due to the stressful situations it creates. While not traditionally considered lifestyle diseases, anxiety, depression, and other mental health issues can develop due to severe weather events, job loss, and displacement caused by climate-related factors. The effects of climate change are becoming increasingly apparent, affecting our physical and biological systems which play an essential role in the health of humans. These impacts include changes in aeroallergens, increased levels of ground-level ozone pollution, and declining air quality. People who suffer from allergies and asthma are particularly vulnerable to the worsening of their condition and a decline in their overall quality of life due to these changes in the environment (D'Amato et al. 2016). Climate change's influence on environmental exposures may increase the occurrence of allergic diseases and asthma. The extent of the disease burden is directly proportional to the magnitude of climate change. To reduce the burden of disease, physicians should use a combination of disease adaptation and mitigation factors that exacerbate climate change. Some strategies include optimizing medical management, providing guidance on interpreting air quality indices and pollen counts, and educating patients on minimizing exposure and symptoms, which can enhance the quality of life for at-risk individuals. Proper public health tracking and collaboration with planners and policymakers are crucial in implementing appropriate responses and health-protective measures.

Socioeconomic Factors and Vulnerability to Infectious Diseases

Social factors play a key role in shaping the vulnerability thread about climate crises and outbreaks of disease. Poverty increases the effect and prevents communities from investing in resilient infrastructure or adequately funded healthcare (Douthit and Alemu 2016). Due to lack of funds, preparedness and response efforts in the event of climate disasters are hampered resulting in a disproportionate impact on poor regions. Overcrowded living conditions in economically deprived areas become breeding grounds for diseases, spreading swiftly and making containment challenging. Misinformation is promoted through limited access to education and prevention measures are hindered by lack of health services, thereby exacerbating the incidence of epidemics.

Elderly people and persons who earn less often have limited means of access to essential health care, thus putting them at increased risk for both climate-related disasters and disease outbreaks. To address these socioeconomic disparities, an integrated policy that builds on climate resilience while also ensuring equal opportunities for access to education, health care, and welfare systems is needed (Van 2020). Society must achieve effective solutions to both climate crisis and disease outbreaks through these integrated approaches, to secure the greatest vulnerability among us.

Outbreak of Diseases and Societal Collapse

Climate change worsens parasitic diseases, particularly in developing countries, by providing perfect conditions for parasites to thrive (Patz et al. 2000). This is mainly due to the temperature rise that expands the habitats of parasite vectors like mosquitoes and ticks, leading to the increased spread of diseases. These conditions disproportionately affect vulnerable populations that lack access to healthcare. Therefore, addressing the relationship between climate change and parasitic diseases is critical in mitigating their impact on global public health, particularly in areas where these diseases are prevalent.

Parasitic diseases are often spread through contaminated food and water or the bites of vectors. However, the common people of this region lack access to safe food and water, proper waste disposal, and adequate health care infrastructure. Inadequate sewage systems and piles of garbage create a favourable environment for the growth of vectors (Olds et al. 2018). Though the causes and cures of all parasitic diseases are well known, parasitic diseases with malnutrition are the primary cause of death in most tropical countries. Seemingly harmless parasitic infections like *Pneumocystis, Toxoplasma,* and *Cryptosporidium* can take a severe form in immune-compromised patients. Since 1980, with the emergence of HIV, the parasitic disease burden has increased severalfold, without restricting its geographical boundaries. Several parasitic diseases are now also common in developed countries, particularly in AIDS and immune-compromised patients (Larry et al. 2009). In addition, tropical countries have incurred significant financial losses due to several parasitological diseases like fascioliasis, trypanosomiasis, and coccidiosis, among veterinary important animals (Narladkar 2018). In 1796, Edward Jenner developed the first successful vaccine against smallpox. However, despite our advancements in medicine, we have yet to develop any successful and safe vaccine against parasitic diseases. The issue is that many pharmaceutical companies and developed nations are hesitant to invest in parasitology research as most of the victims of these diseases are from impoverished tropical countries and cannot afford expensive treatments. In 2003, \$543 and \$32 were spent per person on research for cancer and cardiovascular diseases, respectively. In comparison, the World Health Organization only spent \$0.004 per case on research for schistosomiasis in 2002 (US NIH 2006). Most parasitic diseases, excluding malaria, are now considered "tropical neglected diseases" (Table 11.1) (Mackey et al. 2014). Visceral leishmaniasis (VL) is reported to occur each year in the range of 50,000 to 90,000, while cutaneous leishmaniasis (CL) affects 600,000 to 1 million people each year (Tidman et al. 2021). Filaria had 51 million infections as of 2018. Chagas disease, caused by *T. cruzi*, affects 6–7 million people worldwide. Schistosomiasis affects nearly 240 million individuals and has varying death estimates ranging from 24,072 to 200,000 annually. Hookworm continues to burden around 500 million people globally, with 5.1 billion at risk. Dracunculus, also known as dracunculiasis, had over 3.5 million cases in the mid-1980s, but the number has significantly declined, with only 27 cases reported in 2020.

Table 11.1 Table showing the distribution, prevalence and deaths due to different infectious diseases, especially in developing countries (Adapted from cdc.gov, who.int, datacatalog.worldbank.org and Georgiadou et. al. 2015)

Name of the diseases	Most impacted regions	Global Prevalence (year-over-year)	Population at risk	Death (As on year)
Chagas disease	Latin America	6–7 million	28 million	10,000
River blindness	Sub-Saharan Africa	14.6 million (2017)	220 million (2017)	
Malaria	The WHO African Region carries a disproportionately high share of the global malaria burden	241 million (2020)	3.3 billion people	627,000 (2020)
Visceral Leishmaniasis	Brazil, East Africa and India	50,000 to 90,000 new cases of VL occur worldwide annually	350 Million	20,000–30,000
Cutaneous leishmaniasis	Afghanistan, Algeria, Brazil, Colombia, Iraq, Libya, Pakistan, Peru, the Syrian Arab Republic and Tunisia	600,000 to 1 million new cases occur worldwide annually		
Amebiasis (due to *Entamoeba histolytica)*	Worldwide	500 million		40,000–110,000
Schistosomiasis (caused by Blood fluke)	Africa, the Middle East, the Caribbean, Brazil, Venezuela and Suriname, China, Indonesia, the Philippines, Africa, the Middle East, Corsica (France)	200 million	236.6 million (2019)	200,000 (2000)
African sleeping sickness	Angola, Central African Republic, Chad, Congo, Gabon, Guinea, Malawi and South Sudan, Cameroon, Côte d'Ivoire, Equatorial Guinea, Uganda, United Republic of Tanzania, Zambia and Zimbabwe	663 New Cases in 2020	55 million people for the period 2016–2020; with only 3 million people at moderate or higher risk	

(continued)

Table 11.1 (continued)

Name of the diseases	Most impacted regions	Global Prevalence (year-over-year)	Population at risk	Death (As on year)
Lymphatic filariasis	Throughout the tropics and sub-tropics of Asia, Africa, the Western Pacific, and parts of the Caribbean and South America	51 million (2020)	863 million	
Hookworm infection	Tropics and Subtropics	740 million		65 thousand
Ascariasis (due to *Ascaris lumbricoides)*	Tropical and subtropical regions of the world	1221 million		60 thousand

The impact of climate change on the spread of parasitic diseases has profound effects on both human and animal populations. A comprehensive approach is necessary to mitigate impacts, considering the complex interplay of climate patterns, parasites, hosts, and vectors (Short et al. 2017). Changes in temperature, precipitation, and ecological conditions have a significant impact on the geographical range and behaviour of disease vectors like mosquitoes, ticks, and mollusks (Caminadae et al. 2019). As a result, the areas where parasitic diseases pose a risk can expand. For example, climate change can extend the breeding period of disease vectors, introducing new populations into regions that were previously unaffected by diseases such as malaria or dengue fever. Climate change may alter host habitats, increasing cross-species parasite transmission risk between animals and humans (Byers 2020). Changes in temperature and humidity can affect the behaviour and migration patterns of wild animals. This can increase the risk of zoonotic disease transmission to domestic animals and humans. A well-known example is schistosomiasis, which is a waterborne parasite infection. The larvae of the parasite develop in snails, and insulated water temperature facilitates their growth. This increases the risk of human infection upon contact with contaminated waters (Short et al. 2017). This underscores how climate affects disease transmission.

Conclusion

The research highlights the urgent need for a comprehensive and integrated approach involving experts from various fields such as climatology, epidemiology, and healthcare. It advocates for strategies to strengthen the resilience of communities at risk for parasitic diseases under changing climatic conditions, including monitoring and controlling disease vectors. The study also explores the intersection of the climate crisis and disease outbreaks, highlighting the catastrophic consequences, especially for marginalized populations in developing countries. It identifies rising temperatures, extreme weather events, and loss of biodiversity as factors that facilitate the spread of disease vectors, leading to outbreaks of diseases like Zika virus, dengue fever, Lyme disease, malaria, cholera, and various zoonotic diseases (Ellwanger and Chies 2021). Effective public health policies, concerted efforts to combat climate change, interdisciplinary collaboration, policy interventions, and public awareness are proposed as indispensable measures to mitigate the alarming spread of diseases. The research underscores the urgency for coordinated global actions to address the intricate interplay between climate change and epidemics. Adaptation strategies include early warning systems, improved health infrastructure, vector control programs, and vaccination campaigns. By considering future scenarios, the study provides actionable insights for adaptive strategies, policy formulation, and international cooperation. Its interdisciplinary approach and focus on societal implications make it a pioneering effort to unravel the complexities at the intersection of the climate crisis and disease outbreaks. One area for future research is predictive modelling to anticipate how parasitic diseases might spread in different parts of the world and at different times, given various climate change scenarios. Another area is investigating how well different adaptation strategies, such as ecosystem-based approaches or community-driven interventions, might work to reduce the impact of climate change on disease transmission. It is important to understand how socioeconomic factors affect vulnerability and resilience to climate-induced disease outbreaks, especially in marginalized communities. In addition, exploring innovative technologies such as remote sensing and genomic surveillance for early detection and monitoring of disease emergence could improve readiness and response efforts.

References

Agyekum TP, Botwe PK, Arko-Mensah J, Issah I, Acquah AA, Hogarh JN, Dwomoh D, Robins TG, Fobil JN (2021) A systematic review of the effects of temperature on *Anopheles* mosquito development and survival: implications for malaria control in a future warmer climate. Int J Environ Res Public Health 18(14):7255. https://doi.org/10.3390/ijerph18147255

Ahmed T, Hyder MZ, Liaqat I, Scholz M (2019) Climatic conditions; Conventional and Nano technology-based methods for the control of mosquito vectors causing human health issues. Int J Environ Res 16(17):3165. https://doi.org/10.3390/ijerph16173165

Almberg FS, Cross PC, Dobson AP, Smith DN, Hudson PJ (2012) Parasite invasion following host reintroduction: a case study of Yellowstone's wolves. Philos Trans R Soc Lond 367(1604):2840–2851

Alto BW, Bettinardi D (2013) Temperature and dengue virus infection in mosquitoes: independent effects on the immature and adult stages. Am J Tropical Med Hygiene 88(3):497–505

Annel NW, Love JA, Challinor AJ, Osbron TJ (2010) Global and regional impacts of climate change at different & levels of temperature increase. Clim Change 155:377–301

Barbier EB (2021) Habitat loss and the risk of disease outbreak. J Environ Econ Manag 108:102451. https://doi.org/10.1016/j.jeem.2021.102451

Bavinger JC, Shantha JG, Yeh S (2020) Ebola, Covid 19, and emerging infectious lessons: learned and future preparedness. Curr Opin Ophthalmol 31(5):416–422

Bellone R, Failloux AB (2020) The role of temperature in shaping mosquito-borne viruses transmission. Front Microbiol 11:584846

Bogoch II, Brady OJ, Kraemer MUG, German M, Creatore MI, Kulkarni MA, Brownstein JS, Mekaru SR, Hay SI, Groot E, Watts A, Khan K (2016) Anticipating the international spread of Zika virus from Brazil. Lancet 387(10016):335–336. https://doi.org/10.1016/S0140-6736(16)00080-5

Bouchard C, Dibernardo A, Koffi J, Wood H, Leighton PA, Lindsay LR (2019) N Increased risk of tick-borne diseases with climate and environmental changes. Can Commun Dis Rep 45(4):83–89

Boutayeb A (2010) The burden of communicable and non-communicable diseases in developing countries. Handb Dis Burdens Qual Life Measures 2010:531–546. https://doi.org/10.1007/978-0-387-78665-0_32

Bronson FH (2009) Climate change and seasonal reproduction in mammals. Philos Trans R Soc Lond 364(153):3331–3340

Brooks DR, Hobeng EP (2007) How will global climate change affect parasite host assemblages? Trends Parasitol 23(12):571–574. https://doi.org/10.1016/j.pt.2007.08.016

Byers JE (2000) Effects of climate change on parasites and disease in estuarine and nearshore environments. PLOS Biol 18 (11). https://doi.org/10.1371/journal.pbio.3000743

Cabral JPS (2010) Water microbiology. Bacterial pathogens and water. Int J Environ Res Public Health 7(10):3657–3703

Cam KF, Thomas DR, Salmon RL, Wyn-Jones AP, Kay D (2013) Extreme water-related weather events and water borne disease. Epidemiol Infection 191(4):671–686

Caminadae C, Melntyre KM, Jones AF (2019) Impact of recent and future climate change on vector-borne diseases. Anal N Y Acad Sci 11236(1):157–173

Caminade C, Mcln-tyre KM, Jones AF (2019) Impact of recent and climate change on vector-borne diseases. Ann N Y Acad Sci 1436(1):157–173

Campbell-Lendrum D, Manga L, Bagayoko M, Sommerfeld J (2015) Climate change and vector-borne diseases: what are the implications for public health research and policy? Philos Trans R Soc B 370(1665)

Christofferson RC, Mones CN (2016) Potential for entrinsic incubation temperature to alter interplay between transmission potential and mortality of dengue infected *Aedesaegypti*. Environ Health Insights 10:119–123

Chu EW, Karr JR (2017) Environmental impact: concept, consequences, measurement reference module in life sciences. https://doi.org/10.1016/B978-0-12-809633-8.02380-3

Colwell RK, Brehm G, Cardelús CL, Gilman AC, Longino JT (2008) Global warming, elevational range shifts, and lowland biotic attrition in the wet tropics. Science 322(5899):258–261

Confalonieri UE, Menezes JA, Margonari de Souza C (2015) Climate change and adaptation of the health sector: the case of infectious diseases. Virulence 6(6):554–557

Curriero FC, Patz JA, Rose JD, Lele S (2001) The association between extreme precipitation and waterborne disease outbreaks in the United States 1948–1984. Am Public Health Assoc 91(8):1194

D'Amato G, Pawankar R, Vitale C, Lanza M, Molino A, Stanziola A, Sanduzzi A, Vatrella A, D'Amato M (2016) Climate change and air pollution: effects on respiratory allergy. Allergy Asthma Immunol Res 8(5):391–395. https://doi.org/10.4168/aair.2016.8.5.391

de Magny GC, Thiaw W, Kumar V, Manga NM, Diop BM, Gueye L, Kamara M, Roche B, Murtugudde R, Colwell RR (2012) Cholera outbreak in Senegal in 2005: was climate a factor? PLoS One 7(8). https://doi.org/10.1371/journal.pone.0044577

Dennis S, Fisher D (2018) Climate change and infectious diseases. Ann Acad Med Singapore 47(10):401–404

Dhara VR, Schramm PJ, Luber G (2012) Climate change and infectious diseases in India: implication for health Care providers. Indian J Med Res 138(6):847–852

Dhiman RC, Pahwa S, Dhillon GPS, Dash AP (2010) Climate change and threat of vector-borne diseases in India: are we prepared? Parasitol Res 106(4):763–773

Djalante R, Nurhidayah L, Van Minh H, Mehindradhata NTN, Trias A, Lassa J, Miller MA (2020) COVID-19 and ASEAN responses: comparative policy analysis. Progr Disaster Sci 8:100129. https://doi.org/10.1016/j.pdisas.2020.100129

Douthit NT, Alemu HK (2016) Social determinants of health: poverty, national infrastructure and investment. Br Med J Case Rep. https://doi.org/10.1136/bcr-2016-215670

Duan RY, Kong XQ, Huang MY, Varela S, Ji X (2016) The potential effects of climate change on amphibian distribution, range fragmentation and turnover in China. PeerJ 4

Ebi KL, Vanos J, Baldwin JW, Bell JF, Hondala DM, Emmett NA, Hayes K, Reid CF, Saha S, Spector J, Beray P (2021) Extreme weather and climate change: population health and health system implications. Ann Rev Public Health 42:1293–1315

Ebi KL, McGregor G (2008) Climate change, tropospheric ozone and particulate matter, and health impacts. Environ Health Perspect 116(11):1449–1455

Ebi Kl, Vanos J, Baldwin JW, Bell JE, Hondula DM, Errett NA, Hayes K, Reid CE, Saha S, Spector J, Berry P (2021) Extreme weather and climate change: population health and health system implication. Ann Rev Public Health 42:293–315

Eldred BD, Reilly JR (2014) Warmer temperatures increase disease transmission and out break intensity in a host-pathogen system. J Animal Ecol 83(4):838–840

Ellwanger JH, Chies JAB (2021) Zoonotic Spillover: understanding basic aspects for better prevention. Genetics Mol Biol 44(1). https://doi.org/10.1590/1678-4685-GMB-2020-0355

Filip R, Puscaselu RG, AnchidinNorocel L, Dimian M, Savage WK (2022) Global challenges to public health care systems during the CoviD-19 pandemic: a review of pandemic measures and problems. J Personalized Med 12(8):1295. https://doi.org/10.3390/jpm12081295

Georgiadou SP, Makaritsis KP, Dalekos GN (2015) Leishmaniasis revisited: current aspects on epidemiology, diagnosis and treatment. J Transl Med 3(2):43–50

Getachew D, Takie H, Gebre Michael T, Balkow M, Mesfin A (2015) Breeding sites of aedesaegypti: potential Dengu vectors in Dine Dawa East Ethiopia. Interdisc Perspect Infectious Dis 70:6276

Grassly NC, Fraser C (2006) Seasonal infectious disease epidemiology. Proc Biol Sci 34(1):2541–2550

Gray JS, Ogden NH (2021) Ticks, human babesiosis and climate change. Pathogens 10(11):1430

Gupta P, Vishnudas PK, Ramakrishnan U, Robin VV, Dhammarajan G (2010) Geographical and host species barriers differentially affect generalist and specialist parasite community structure in a tropical sky-island archipelago. Proc Biol Sci 286(100):20100439

Hales S, de Wet N, Maindonald J, Woodward A (2002) Potential effect of population and climate changes on the global distribution of dengue fever: an empirical model. Lancet 20360(9336):830–834

Horn CJ, Luong LT (2018) Proximity to parasites reduces host fitness independent of infection in *Drosophila marches* system. Parasitology 15:1564–1560

https://datacatalog.worldbank.org/search/dataset/0039831

https://www.unicef.org/press-releases/ten-things-you-didnt-know-about-malaria

https://www.cdc.gov/globalhealth/topics/chagas/index.html

https://www.who.int/health-topics/chagas-disease

https://www.who.int/news-room/fact-sheets/detail/onchocerciasis

https://www.who.int/news-room/fact-sheets/detail/malaria

https://www.who.int/news-room/fact-sheets/detail/schistosomiasis

https://www.who.int/news-room/fact-sheets/detail/trypanosomiasis-human-african-(sleeping-sickness)

https://www.who.int/news-room/fact-sheets/detail/lymphatic-filariasis

Hui EK (2006) Reasons for the increase in emerging and reemerging viral infectious diseases. Microbes Infection 8(3):905–916

JanosiIm, Biro T, Lakatos Bo, Gallas JAC, Szollosi Nagy A (2023) Changing waterunder a warming climate: tendencies in the Carpathian basin. Climate 11(6):118

Jung Y-J, Kant NA, Kim H, Namkoong S (2023) Impact of climate change. On water borne diseases: directions towards sustainability. Water 15(7):1298

Kakarla SG, Bhimala KR, Kadiri MR, Kumaraswamy S, Muthenemi SR (2020) Dengue. Dengue situation in India: suitability and transmission potential model for present and projected climate change scenarios. Sci Total Environ 739

Kumar K, Gupta SC, Baidoo SK, Chander Y (2005) Antibiotic uptake by plants from soil fertilized with animal manure. J Environ Qual 34:2082–2085. https://doi.org/10.2134/jeq2005.0026

Lacetera N (2018) Impact of climate change on animal health and welfare. Anim Front 9(1):26–31

Larry SR, John Janovy Jr, Gerald D, Schmidt, Larry S (2009) Roberts' Foundations of parasitology, 8th edn. McGraw-Hill, 1221 Avenue of the Americas, New York, NY 10020, pp 1–701

Levy K, Smith SM, Carlton EJ (2018) Climate change impacts on waterborne diseases: moving toward designing interventions. Curr Environ Health Rep 5(2):272–282

Lindahl JE, Grace D (2015) The consequences of human actions on risks for infectious diseases: a review. Infection Ecol Epidemiol 5:30048

Lymbery AJ, Lymbery SJ, Beatty SJ (2020) Fish out of water. Aquatic parasites in a drying world. Philos Trans R Soc Lond 12:300–307

Ma J, Guo Y, Gao J, Liu Q, Xu L (2022) Climate change drives the transmission and spread of vector-borne diseases: an perspective. Biology 11(11):1628

Mackey TK, Liang BA, Cuomo R, Hafen R, Brouwer Key Lee DE (2014) Emerging and reemerging neglected tropical diseases: a review of key characteristics, risk factors, and the policy and innovation environment. Clin Microbiol Rev 27(4):949–979

Macrogliese DJ (2008) The impact of climate change on the parasites and infectious diseases of gouatic animals. Revue Scientifique Technique 2:667–684

Mares R, Deutralant C, Paquet M, Spottiswoode CN, Cevas R (2017) Breeding decisions and output are Compelated with both temperature and rain fall in an arid-region passerine, the sociable weavers. R Soc Open Sci 4(9):170835

McMichael AJ, Woodruff RE (2008) Climate change and infectious diseases. Soc Ecol Infectious Dis 378–407

Mimura N (2013) Sea-level rise caused climate change and its implications Society. Proc Jpn Acad Ser B Phys Biol Sci 89(7):281–301

Mittal R, Nyuyen D, Debs LH, Patel AP, Liu G, Thavenivm S, Kay S, Mittal J, Bandstra E, Younis RT, Chapagain P, Jayaweera DT, Liu XZ (2017) Zika virus: an emerging global health threat. Front Cellular Infection Microbiol 7:486

Mojahed N, Mohammadkhani MA, MohamadKhani A (2022) Climate crises and developing vector—borne diseases. Iran J Public Health 51(12):2664–2673

Monaghan AJ, Moore SM, Sampson KM, Beard CB, Eisen RJ (2015) Climate change influences on the annual onset of Lyme disease in the United States. Ticks Tick-Borne Dis 6(5):615–622

Mordecai EA, Paaijmans KP, Johnson LR, Balzer C, Ben Horin T, de Moor E, McNally A, Pawar S, Ryan SJ, Smith TC, Lafferty KD (2013) Optimal temperature for malaria transmission is dramatically lower than previously predicted. Ecol Lett 16(6):29–30

Narladkar BW (2018) Projected economic losses due to vector and vector-borne parasitic diseases in livestock of India and its significance in implementing the concept of integrated practices for vector management. Vet World 11(2):151–160

Niekerk AJV (2020) Inclusive economic sustainability: SDGS and global inequality. Sustainability 12(13):5427. https://doi.org/10.3390/su12135427

Norart C, Altraminano J, Anymba A, Caldwell JM, Domoah R, Mutuku F, Ndenga B, Le Beaud AD (2021). Impact of recent climate extremes on mosquito borne disease transmission in Kenya. PLoS Neglected Tropical Dis 15(3):2009182. https://doi.org/10.1371/journal.pntd.0009182

Okaka FO, Odhiambo BDO (2018) Relationship between flooding and out break of infectious disease in kenya: a review of the literature. J Environ Public Health. https://doi.org/10.1155/2018/5452938

Oladipo HJ, Tajudden YA, Oladujoye LO, Yusuff SI, Yusuf RO, Oluwaseyi EM, Abdul Basit MO, Adebisi YA, El-Sherbini MS (2022) Increasing challenges of Malanic Control in Sub-Saharan Africa: priorities for public health research and policymakers. Ann Med Surg 81:104–366

Olds HT, Corsi SR, Dila DK, Halmo KM, Bootsma MJ, McLellan SL (2018) High levels of sewage contamination released from urban areas after storm events: a quantitative survey with sewage specific bacterial indicators. J Public Lib Sci Med 15(7). https://doi.org/10.1371/journal.pmed.1002614

Ostfeld RS (2008) Parasite as weapons of mouse destruction. J Anim Ecol 77(2):201–204

Pandey PK, Kass PH, Soupir ML, Biswas S, Singh VP (2014) Contamination of water resources by pathogenic bacteria. AMB Express 4:51

Pathak N, Mckinney A (2021) Planetary health, climate change and lifestyle medicine: threats and opportunities. Am J Lifestyle Med 15(5):541–552

Patz JA, Graczy K, Geller N, Vittor AY (2000) Effects of environmental change on emerging parasitic diseases. Int J Parasitol 30(12–13):1395–405

Patz J, Campbell-Lendrum D, Holloway T (2005) Impact of regional climate change on human health. Nature 438:310–317

Patz JA, Olson SH, Uejio CK, Gibbs HK (2008) Disease emergence from global climate and land use change. Med Clin North Am 92(6):1473–1491

Paul AR, Mity R (2013) Future projection, of climate extremes across contiguous northeast India and Bangladesh. Sci Rep 13(1):15616

Raimondi A, Quinn R, Abhijith GR, Becciu G, Ostfeld A (2023) Rainwater harvesting and treatment: state of the art and perspectives. Water 15(8):1518. https://doi.org/10.3390/w15081518

Rezza G (2014) Dengue and chikungunya: long-distance spread and outbreaks in naïve areas. Pathogen Global Health 108(8):349–355

Rocklov J, Dubrow R (2020) Climate change: an enduring challenge for vector-borne disease prevention and control. Nat Immunol 21(5):479–483

Rosenthal J (2009) Climate change and the geographic distribution of infectious diseases. EcoHealth 6(4):489–495

Ryan SJ, Carlson J, Mordecai FA, John Son LR (2019) Global expansion and redistribution of Aedes-borone virus transmission risk with Climate Change. PLOS Neglected Tropical Dis 13(3). https://doi.org/10.1371/journal.pntd.0007213

Scheraga JD, Ebi KL, Furlow J, Moreno AR (2003) Climate change and human health: risks and responses

Semenza JC, Rocklöv J, Ebi KL (2022) Climate change and cascading risks from infectious disease. Infectious Dis Therapy 11(4):1371–1390

Shivanna KR (2022) Climate change and its impact on biodiversity and human welfare. Proc Indian Natl Sci Acad 88(2):160–171

Short EE, Caminade C, Thomas BN (2017) Climate change contribution to the emergence or re-emergence of parasitic diseases. Res Treatment 10:117. https://doi.org/10.1177/1178633617732296

Smith KF, Dobson AP, Mckenzie FE, Real LA, Smith DL, Wilson ML (2005) Ecological theory to enhance infectious disease control and public health policy. Front Ecol Environ 3(1):29–37

Smith KM, Machalaba CC, Seifman R, Feferholtz Y, Karesh WB (2019) Infectious disease and economics: the case for considering multi-sectoral impacts. One Health 7:100080. https://doi.org/10.1016/j.onehlt.2018.100080

Socha N, Kwasnik M, Larska M, Rola J, Rozek W (2022) Vector-bonne viral diseases as a current threat for human and animal health—one health perspective. J Clin Med 11(11):3026

Stiasny K, Santonja I, Holzomann H, Essi A, Stanek G, Kundi M, Hema FL (2021) The regional decline and rise of tick-bone encephalitis incidence do not correlate with lyme borreliosis, Austria 2005 to 2018. Eurosurveillance 26(35):2002108. https://doi.org/10.2807/1560-7917.ES.2021.26.35.2002108

Swick MC, Koehler TM, Driks A (2016) Surviving between hosts: sporulation and transmission. Microbiol Spectrum 4(4):10

Tabari H (2020) Climate change impact on flood and extreme precipitation increases with water availability. Sci Rep 10(1):13768. https://doi.org/10.1038/s41598-020-70816-2

Tanser FC, Sharp B, le Sueur D (2003) Potential effect of climate change on malaria transmission in Africa. Lancet 362(9398):1792–1798

Tidman R, Abela-Ridder B, Castaneda RRD (2021) The impact of Climate change on neglected tropical diseases: a systematic review. Trans R Soc Tropical Med Hygine 115(2):147–168

Tonnevi A, Bergstedt O, Fonsberg B (2014) Precipitation effects on microbial pollution in a river: lag structures and seasonal effect modification. Public Lib Sci 9(5). https://doi.org/10.1371/journal.pone.0098546

Tzanakakis VA, Pananychiamaks NV, Angelakis AN (2020) Water supply and water scarcity. Water 12(9):2347. https://doi.org/10.3390/w12092347

U.S. National Institutes of Health (2006) Estimates of funding for various diseases, conditions, research areas. National Institutes of Health, Bethesda, 10 pp. http://www.nih.gov/news/fundingresearchareas.htm

Van Seventers JM, Hochberg NS (2017) Principles of infectious diseases: transmission, diagnosis; prevention and control. Int Enclopedia Public Health 22–39. https://doi.org/10.1016/B978-0-12-803678-5.00516-6

Vorobevskii I, Janabi FA, Schneebeck F, Bellera J, Krebs P (2020) Urban floods: linking the overloading of a strom water sewer system to precipitation parameters. Hydrology 7(2):35. https://doi.org/10.3390/hydrology7020035

Walton MG, Cubillejo I, Nag D, Withey JH (2023) Advances in cholera research: from molecular biology to public health initiatives. Front Microbiol 14

Wilson AL, Courtenay O, Kelly Hope LA, Scott TW, Takken W, Torn SJ, Lindsay SW (2020) The importance of vector control for the control and elimination of vector-borne diseases. PLoS Neglected Tropical Dis 14(1). https://doi.org/10.1371/journal.pntd.0007831

Xiang J, Hansen A, Liv Q, Tong MX, Liu X, Sun Y, Cameron S, Hanson-Easey S, Han GS, Williams C, Weinstein P, Bi P (2018) Association between malaria incidence and meteorological factors: a multi-location Study in China, 2005–2012. Epidemiol Infection 146(1):89–99

Ye G, Xu Z, Yang M, Wang J, Liang J, Yin J, Yang Y, Xia H, Liu Y (2023) Clinical features and transmission risk analysis of dengue virus infections in Shenzhen, during 2014–2019. Comput Struct Biotechnol J 21. https://doi.org/10.1016/j.csbj.2023.07.001

Zhao M, Duan Q, Shen X, Zhang S (2023) Climate change influences the population density and suitable area of *Hippotisous dorsalis* (Hemiptera: Pentatomidae) in China. Insects 14(2):135. https://doi.org/10.3390/insects14020135

Chapter 12
Socio-economic Challenges of Environmental Refugees: An Overview of Sagar Island of Sundarban, India

Semanti Das and Chandan Surabhi Das

Abstract This article examines the traits and strategies of environmental refugees who moved to the Sagar Block in the Sundarbans from coastal villages in Ghoramara, Lohachara, and Khasimara because their livelihoods were becoming more and more vulnerable to climate change. Three sample villages in the Sagar Block, where the refugees were gathered, underwent a household survey. Based on financial capital, a Sustainable Livelihood Security Index (SLSI) was developed, and a Livelihood Asset Status Tracking (LAST) matrix was created using assets, nutritional status, education, community engagement, water, sanitation, primary health, and reproductive health. Based on a combined evaluation of the SLSI and LAST Index, the sample villages were deemed to be at the vulnerable end of sustainable livelihood security. There is a weak positive correlation ($R = 0.450$) between their ability to adapt and their actual adaptation. Their lives can be made better by using resilient farming techniques, adaption strategies based on rural communities, and diversification of livelihood practices.

Introduction

According to Myers (2005), the phrase "environmental refugee" describes a relatively recent phenomena in the world. Lester Brown of the World Watch Institute first used the term "environmental refugee" in the 1970s. Despite the fact that such changes frequently occur without proof or empirical evidence, environmental change is the main cause of environmental refugees (Castles 2002; Findlay and Geddes 2011; Kibreab 1997; McGregor 1993). Following the publication of El-UNEP Hinnawi's

S. Das (✉)
Department of Geography, Chandrakona Vidyasagar Mahavidyalaya, Chandrakona Town, Paschim Medinipur, West Bengal, India
e-mail: semantidas11@gmail.com

C. S. Das
Department of Geography, Barasat Government College, North 24 Parganas, Barasat, West Bengal, India

U. Mukhopadhyay et al. (eds.), *Climate Crisis, Social Responses and Sustainability*, Climate Change Management, https://doi.org/10.1007/978-3-031-58261-5_12

report 'Environmental Refugees' in 1985, environmental problems gained increased attention in research work related to migration (El-Hinnawi 1985; Westing 1992; Lonergan 1998). "Persons or groups of people who, for compelling reasons of sudden or progressive change in the environment that adversely affects their lives or living conditions, are forced to leave their habitual homes, or choose to do so, either temporarily or permanently, and who move either within their country or abroad," is what the International Organization for Migration (IOM) defined as environmental migrants in 2007.... IOM (2009). Complex systems known as deltas serve as the home for populations that depend on natural resources and are prone to changes in the social, political, economic, and environmental spheres. Numerous individuals are undertaking migration in reaction to the aforementioned variables, which are consequences of climate change (Seto 2011). According to research carried out in the Sundarban region over the last three decades, an estimated 7000 people have been forcibly removed from their original residences due to coastal erosion, cyclones, sea level rise, and embankment breaches (Hazra et al. 2010). According to Sward and Codjoe (2012) and Gemenne (2011), climate change will most likely cause temporary, localized human migration, particularly for socioeconomically disadvantaged populations. Hajra et al. (2017) claim that migration raises the risk of injury and losses in the event of a natural disaster. Conversely, Black et al. (2013) and McLeman and Smit (2006) assessed migration as a coping strategy for sustained vulnerability. Environmental migration can be a "transformational" reaction to such changes, helping to build long-term resilience, so it's not simply a small problem (Foresight 2011). Occasionally, it is possible to see migration or displacement as the result of people's inability to adapt (IPCC 2007). "Displacement" describes circumstances in which individuals are evicted from their houses as a direct result of events that may occur slowly or quickly. 2017; ODI and UNDP). Documenting the susceptibility and sustainability of livelihoods can help understand the socioeconomic issues faced by environmental migrants in deltas. Because livelihood approaches to migration can just describe the true causes of migration, they have been more and more popular since the 1990s (Moser 1998). "A livelihood involves the capabilities, assets (including both material and social resources), and activities required for a means of life," Carney (1998) stated. A means of subsistence is deemed sustainable if it can sustain itself through shocks and pressures, recover from them, and retain or increase its resources going forward without jeopardizing the foundation of natural resources. People can build "layers of resilience" to withstand "waves of adversity" and turn many adversities into opportunities when they have a robust and sustainable livelihood system (Glavovic et al. 2003). 'Adaptation' in socio-ecological systems, according to Smit and Pilifosova (2001), is a moderate-to long-term adjustment. Conversely, "coping strategies" describe temporary actions taken by households to mitigate the negative effects of climate variability on their means of subsistence (Engle 2011). One of the most crucial methods for assessing the degree to which climate hazards affect ecological and human systems and for recording victims' reactions to threats is vulnerability assessment (Adger et al. 2007).

Two helpful methods for evaluating vulnerability are the Sustainable Livelihood Security Index (SLSI) and the LAST matrix. Boateng (2013) developed the LAST technique to assess how a development initiatives affects agricultural productivity and poverty reduction in Ghana. It is based on the sustainable lifestyles analytical framework. Through a formative process evaluation methodology, the LAST technique has been shown by Bond and Mukherjee (2002) to be a helpful method for rapidly assessing program outcomes related to livelihood development. Social vulnerability, or the stress that groups or people experience as a result of unforeseen social and environmental change, is inescapably related to household level vulnerability (Neil Adger 1999). The sustainable livelihoods approach, as per Chambers and Conway (1992), places emphasis on five categories of household assets, namely natural, social, financial, physical, and human capital. This method aids in the creation of community-based development programs (United Nations General Assembly 1997). With the aid of the aforementioned five capitals, the Department for International Development (DFID) evaluated the Sustainable Livelihoods (SL) framework for property accumulation of the poor in rural areas (DFID 2000). "The ability of systems, institutions, humans, and other organisms to adjust to potential damage, to take advantage of opportunities, or to respond to consequences" is the definition of adaptive capacity given by the IPCC (2014). The SLSI approaches were developed in 1993 by the MS Swaminathan Research Foundation (MSSRF). CARE's program analysis in Kenya, India, and Sri Lanka is now based on the Household Livelihood Security (HHLS) model (Lindenberg 2002). Rain-fed agriculture in India is significantly impacted by climate change. Raymond (2007) used a 40-year panel data set with more than 200 districts to evaluate the impact of yearly weather fluctuations on agricultural output in India. The catastrophic implications of climate change on livelihoods depending on agriculture, especially in developing countries like Bangladesh, were illustrated by Hoque et al. (2019).

Climate change is putting strain on communities through forced migration, relocation, and displacement, but it can be challenging to understand the scope of the issue (Doevenspeck 2011; Black et al. 2011). Coastal hazards now threaten the Ghoramara, Lohachara, and Khasimara Islands in the Sundarbans due to climate change, sea level rise, and regular storms. According to Ghosh et al. (2014), people were compelled to relocate to Sagar Island, which is nearby, as a result of the villages of Khasimara, Khasimara Char, Lakshmi Narayanpur, Bagpara, and Baishnabpara on Ghoramara Island submerging. Low-lying tidal sections of Sagar Island were being recovered by local government officials in the 1970s as a means of resettling islanders because of their close proximity to Lohachara. The locations with the highest people concentrations are Bankimnagar, Gangasagar, and Jibantala-Kamalpur colony areas (Harms 2015; Chakma 2014).

Numerous research on vulnerability and adaptability overlook real adaptation behaviors in favor of focusing just on adaptive capacity. The populations that have recovered in the delta have not received much attention (Samling et al. 2015). The goal of this research is to highlight the socioeconomic vulnerability of the Islands' coastal community to climate change utilizing the LAST matrix and SLSI tool, as well as the relationship between adaptability and adaptive capacity.

Materials and Methods

Study Area

The study region on Sagar Island consists of three villages: Gangasagar (Gangasagar GP), Kamalpur (Rudranagar GP), and Bankimnagar (DhasparaSumatinagar II GP) (Fig. 12.1). Sagar Island lies in the Sundarban Delta at the mouth of the Hugli estuary. The Hooghly River borders the island on the north and west, the Muriganga River borders it on the east, and the Bay of Bengal borders it on the south. For the selection of villages of the Sagar Block, at the first stage Villages were ordered in an ascending manner in respect to the distance from the sub district Headquarter Rudranagar. At the second stage, villages were divided into some strata based on the distance from the sub district Headquarter Rudranagar. Finally, from these strata villages along the coastline were selected where the refugees were gathered. Three villages, Bankimnagar, Kamalpur, and Gangasagar, with respective areas of 3.08, 5.72, and 12.26 km^2 and populations of 3885, 6602, and 10,340 (Census 2011).

Methodology

Primary data was collected through semi-structured interviews with randomly selected families (n = 352) as part of a door-to-door household survey. The quality of life varied very little, hence this sample size was considered enough. Based on a formula with a 50% response distribution and a 5% margin of error (95% confidence level), the sample size for the household survey was determined. To find the houses of respondents, a multi-stage sampling strategy was employed. To locate rehabilitated individuals, clusters were created in the initial phase. Households chosen by stratified random selection were sampled in the second phase. Both left- and right-hand rules were used to choose households in the third stage. The average age of the participants was 52 years old, and the oldest respondent was 85 years old. 50% of Jibantala-Kamalpur's population, 10% of Gangasagar Colony's population, and 0% of Bankimnagar Colony's population were female responders (Fig. 12.2).

An Analytical Framework for the Livelihood Asset Status Tracking (LAST) Matrix

Qualitative evaluations of capital assets can be converted into quantitative ratings using the LAST matrix, which can then be used for various temporal and spatial analysis. The primary instrument for collecting and rearranging data in this case is word images, or quality of life indexes. Respondents assist in illustrating the "best case" (100% response score) and "worse case" (0% response score) scenarios of their

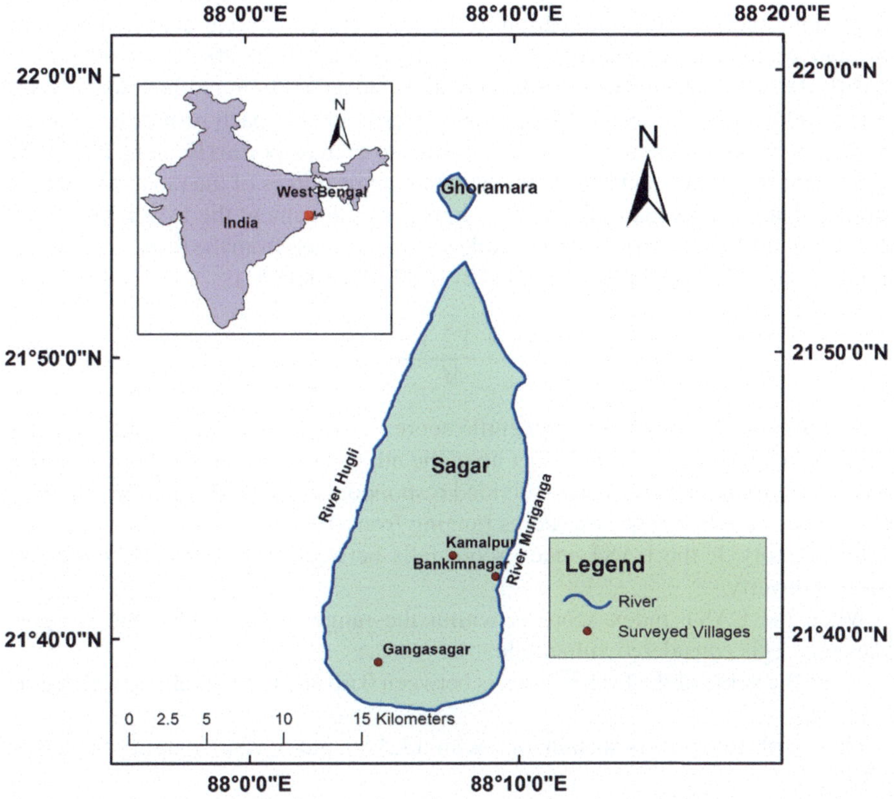

Fig. 12.1 Study area showing surveyed villages of Sagar block of Sundarban, India

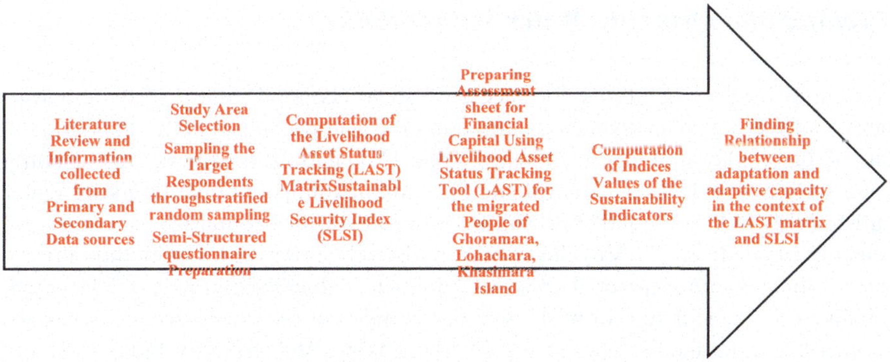

Fig. 12.2 Schematic diagram demonstrating the steps followed this Methodological Flowchart

living situations, or any combination of the two. The point total deducted for both affirmative and negative answers. The score intervals of 0–20, 20–40, 40–60, 60–80, and 80–100 were selected in accordance with Boateng's (2013) methodology. Based on the midpoint of the interval score that corresponded to each response, a precise absolute score and score interval were awarded to each response (Boateng 2013). An assessment sheet was made based on the financial resources of the randomly chosen families that participated in the survey. In order to document the LAST Matrix for each household, the cumulative percentile scores obtained from the responses of each family in the LAST sheet were divided by 1.00. (Constant M).

$$\frac{\sum(Y1 + Y2 + \cdots Yn)}{M}$$

In this case, *Y* denotes the percentile score of answers for every indicator that makes up a household's capital asset up to the nth response score for that asset. *M* is an acronym for constant. Boateng divided respondent households into the following four groups based on response scores ranging from 0 to 100.

In Category 1, the LAST index score falls between 0.01 and 0.30, indicating extreme poverty.

When the LAST index score is within the range of 0.31–0.59, the category (Category 2) is considered vulnerable.

When the value of the LAST index is between 0.60 and 0.79, it falls into Category 3: Viable.

The fourth category is sustainable, with a LAST index value ranging from 0.80 to 1.00.

Dwelling and Drinking Water Vulnerability

Household level vulnerability for accessing socio-economic status of the environmental refugees were analyzed by computing the average values of the standardized data of the condition of Census houses (Table 12.1) and the location of the drinking water sources (Table 12.2). To evaluate the combined impact of the Drinking Water Vulnerability Index (VI) and Dwelling at Village Level, the geometric mean of several vulnerability (V1, V2, …, Vn) classes with rank has been derived. The methodology's idea originated with Hajra et al. 2021's Composite Vulnerability Index (CVI), which is built using the square root of the geometric mean of the prioritized variables. To evaluate the combined impact of the Drinking Water Vulnerability Index (VI) and Dwelling at Village Level, the geometric mean of several vulnerability (V1, V2, … Vn) classes with rank has been derived. The methodology's idea originated with Hajra et al. 2021's Composite Vulnerability Index (CVI), which is built using the square root of the geometric mean of the prioritized variables.

$$\mathrm{VI} = 3\sqrt{(\mathrm{V}_1 * \mathrm{V}_2 * \mathrm{V}_3)}$$

Sustainable Livelihood Security Index (SLSI)

The development of the SLSI (Socioeconomic Livelihood Security Index) aims to identify rural poor households that possess the capacity to effectively manage and adapt to natural hazards. Aggregated indices that employ normalized and combined measures of socio-ecological risk are useful tools for identifying places where multiple vulnerabilities exist (Abson et al. 2012). To compute the SLSI, we utilized the formula proposed by Singh and Hiremath (2009).

$$SLSI_{ij} = \frac{X_{ij-\min_j X_{ij}}}{\max_j X_{ij-} \min_j X_{ij}} \quad i = 1, 2 \ldots, \mathrm{I}, \quad j = 1, 2 \ldots, \mathrm{J}$$

where, *Xij* is the value of the Sustainable Livelihood Security Index related to the *j*-th entity of the Bankimnagar, Kamalpur and Gangasagar village. *Xij* is the value of the variable representing the *i*-th component of SLSI related to the *j*-th entity.

Standardization of the indicators should adhere to the methodology employed by the United Nations Development Programme (UNDP) in its computation of the Human Development Index (HDI), given that each indicator possesses a unique unit. In the same way that Hahn et al. (2009) established this index, indicators were identified and assigned equivalent significance. The SLSI comprised eight dependent variables or components of livelihood security: income and assets, nutrition and food security, education, community involvement, access to potable water, sanitation and hygiene, primary and reproductive health. As areas of livelihood security, these factors were subdivided into the following five subcomponents: health security, economic security, nutritional status assessment (based on the availability of nutritional centers), and educational security evaluation. In order to evaluate the subsistence security index of villages, CARE employed an ordinal scale ranging from one to five (Lindenberg 2002). Scores were assigned to each ingredient according to its availability. The Village Index, which was subsequently devised using a scale of one to five, was the result. In consideration of SLSI and the LAST Matrix, the relationships between Adaptive Capacity and Adaptation were computed utilizing SPSS (PASW Statistics 18.0.0).

Results and Discussion

Implementation of LAST Tool and SLSI

Researchers use five capital models—natural, financial, physical, human, and social—to implement the LAST tool for sustainable livelihood frameworks. However, the current study's implementation of LAST relied solely on financial resources. The LAST indices are 0.045, 0.045, and 0.058 in the villages of Bankimnagar, Kamalpur, and Gangasagar, respectively (Table A.1) Boateng (2013) states

Table 12.1 Dwelling Vulnerability based on condition of Census House

Village	Condition of census house						Dwelling Vulnerability (DV)	Rank DV (based on geometric mean of different DV)
	Good (%)	Standardized Value	Livable (%)	Standardized Value	Dilapidated (%)	Standardized Value		
Bankimnagar	36.3	0.912	51.9	0	11.4	0.673	0.850	2
Kamalpur	37.2	1.000	57.4	1.000	4.0	0	1.000	1
Gangasagar	27.1	0	55.2	0.600	15.0	1.000	0.843	3

Source Office of the Registrar General & Census Commissioner, India, Census of India 2011

Table 12.2 Drinking Water vulnerability based on Location of Drinking Water Source

	Location of Drinking water sources							
Village	Within* Premises (%)	Standardized Value	Near **Premises (%)	Standardized Value	Away*** (%)	Standardized Value	Drinking Water Vulnerability (DWV)	Rank DWV (based on geometric mean of different DWV)
Bankimnagar	0.2	0	29	0.069	70.8	1.000	0.410	2
Kamalpur	1.2	0	55.9	1.000	42.9	0.762	0.913	1
Gangasagar	2.4	0	58.6	1.000	39.0	0.065	0.402	3

Source Office of the Registrar General & Census Commissioner, India, Census of India 2011

that the respondent families are classified as belonging to category 1, which indicates extremely poor living conditions. Based on the state of Census houses (Table 12.1) and the locations of drinking water sources (Table 12.2), the villages of Gangasagar (DV 0.843, DWV 0.402) and Bankimnagar (DV 0.850, DWV 0.410) and Kamalpur (DV 1.000, DWV 0.913) are the most vulnerable in terms of dwelling (DV) and drinking water vulnerability (DWV). The educational security index obtained the maximum score (1.000) subsequent to the economic security index (0.125), participation and empowerment index (0.250), health security index (0.344), and nutritional potential index (0.375), subsequent to the standardization of sustainability measures. Bankimnagar village achieved the highest ranking with a standardized SLSI value of 0.375. Kamalpur village ranked third with an SLSI value of 0.281, and Gangasagar village ranked second with an SLSI value of 0.369 (Table A.2). The idea of a continuum from relief to rehabilitation to development was embraced by CARE. The total score of the amenities offered was used to compute the village indices of Bankimnagar, Kamalpur, and Gangasagar, which were 2.25, 1.875, and 2.875, respectively. The village scores, then, fall between well-protected livelihood security (village index = 5) and substantial threats to livelihood security (village index = 1). Villages are therefore in the rehabilitation phase.

Relationship Between Adaptation and Adaptive Capacity in the Context of the LAST Matrix and SLSI

Political and societal interference can occasionally undermine adaptation, impacting people's experiences, attitudes, and behaviors. As a result, adaptation may grow unpredictable and complex (Adger et al. 2013; Dilling et al. 2019). If adaptive capacity has the ability to explain adaptive behaviors, then there should be a positive correlation between adaptation and adaptive capacity (Mortreux et al. 2020). At the village level, there was a weakly positive connection (R = 0.450) found between the SLSI and LAST Index (Table 12.3). Capacity for adaptation accounts for just 20% of the diversity in actual adaption. Almost 27 percent of adaptation is guaranteed if adaptive capacity stays constant. The sample has positive autocorrelation, as indicated by the Durbin Watson value of 1.006. Consequently, the adaptive capability indicated that the security of the villages' livelihoods was in jeopardy. Households lived in appalling conditions due to their true adaptive behavior.

Table 12.3 Relationship between adaptive capacity and adaptation at village level through SLSI and LAST index

Village	SLSI	LAST	R	R^2	Sig	Unstandardized coefficient	Eigen value	Durbin-Watson
Bankimnagar	0.375	0.045	0.450	0.202	0.703	0.272	1.992	1.006
Kamalpur	0.281	0.045						
Gangasagar	0.369	0.058						

Source Computed by author

Social Security Status of Sagar Block

National Social Assistance Programme (NSAP)

Active performances of NSAP, a Centrally Sponsored Scheme of the Government of India, help to portray how the elderly, widows, and people with disabilities get financial assistance in the form of social pensions. The total beneficiaries in the Sagar Block were 4418 in 2016–2017, and 3807 in 2017–2018. However, the number of bank beneficiaries increased from 2616 in 2016–2017 to 3566 in 2017–2018. The number of postal beneficiaries decreased from 1802 in 2016–2017 to 241 in 2017–2018.

State Disaster Response Fund (SDRF)

In terms of the year, the number of SDRF beneficiaries in Sagar block increased from 2289 to 37 261 between 2014–2015 and 2016–2017. During the Boro (winter cropping season) season, the amount of compensation was Rs. 4,854,258 (US $739.11) in 2014–2015. At the time of the Kharif (monsoon cropping season) season, the number of beneficiaries increased from 31,294 in 2015–2016 to 37,261 in 2016–2017, and the amount of compensation increased from Rs. 102,277,072 (the US $1562,233.32) in 2015–2016 to Rs. 307,261,000 (the US $4,581,062.14) in 2016–2017. This compensation hike assisted victims in combating natural disasters.

Financial Benefit Scheme

Special Component Plan (SCP)

The SCP initiative was introduced by the West Bengal Scheduled Castes and the Scheduled Tribe Development and Finance Corporation to improve the lives of scheduled caste members who are below the poverty line. SCP lends money to the following industries: (i) small-scale irrigation and agriculture; (ii) animal husbandry;

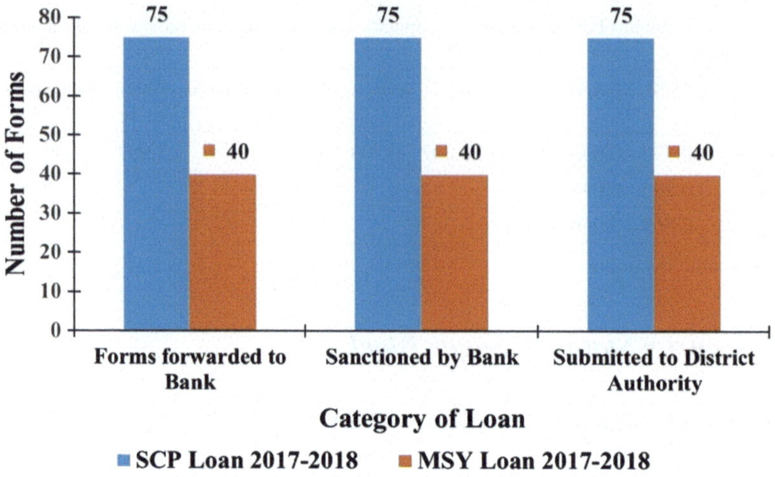

Fig. 12.3 Financial benefit scheme

(iii) fisheries; (iv) village industries; (v) businesses; and (vi) small-scale transportation. The maximum amount of the subsidy is Rs. 10,000 (US $131.90), which is half of the project's total cost. According to the Employment Bank, Government of West Bengal Initiative (2012), the margin money loan is offered up to the project cost of Rs. 12,000 (US $158.27) at a rate of 20% of the project cost or Rs. 2000 (US $26.38), whichever is less.

Mahila Samridhi Yojana (MSY)

State channelizing agency (SCAs), regional rural banks (RRBs), and nationalised banks provide loans to target communities (National Safai Karmacharis Finance and Development Corporation 2017). The Financial Benefit Scheme in 2017–18 is summarised in (Fig. 12.3).

Discussion

Livelihood Status of Environmental Refugees After Migration

The project to resettle the Ghoramara, Khasimara, and Lohachara Islands started in Bankimnagar in 1972. It moved to the Gangasagar colony in 1981, Jibantala-Kamalpur colony in 1983, and finally, at Bankimnagar in 1972 (Chakma 2014). Through the local administrative authority, the West Bengal State Government initiated the project (Harms 2013, 2015; Chakma 2014; Mukherjee 2014). There is a range of 4000–7000 displaced people overall. (Mukherjee 2014; Ghosh et al. 2014; CSE 2007). The SagarPanchayat report states that the government gave individual environmental refugees patta, or grants, totaling 0.20067 ha. Many Jibantala-Kamalpur colony residents, however, did not acquire title or patta after rehabilitation. At first, there were enough rehabilitation packages available, but as the number of resettlers increased, fewer and fewer packages were provided (Samling et al. 2015). Patta was first made available by the current administration as a part of a special package called "NijoBhumiNijoGriha" (own land, own house). When displaced people with legal title (patta) were surveyed by Harms (2015), she found that they are still worried about losing their land.

Daily Hardship of the Displaced People

Respondents from the Khasimara and Lohachara Islands migrated to the Bankimnagar colony and engaged in fishing activities in the Battala River, a tributary of the Chemaguri River that ultimately empties into the Muriganga River. Daily profits varied from Rs. 100 to Rs. 150 (about $1.32 to $1.98 USD). 45.45% of respondents were BPL (Below Poverty Line) cardholders. PMAY (Pradhan MantriAbasYojana) did not provide housing for all of them. However, the economic status of this colony is significantly superior to the other two.

The infrastructure in Jibantala colonies 1 and 2 is deplorable. Insufficient quantities of secure potable water wells are available. Their residences were around 1.5 km away from tube wells. The vast majority of individuals lack access to basic housing. Though the village Kamalpur (Jibantala colony is a part of village Kamalpur) is less vulnerable in terms of Census housing condition and location of drinking water supply (Tables 12.1 and 12.2). The presence of unpaved roads and inadequate sanitation posed significant challenges to sustaining a satisfactory standard of living. The majority of responders from Adibashi Mahalla engage in the collection of prawn spawn from the Chemaguri River, yielding a daily income of Rs. 50–100 (US $0.66–US $1.32).

The poor, young, old, sick, and other marginalized populations are disproportionately affected by climate change (Kasperson et al. 2001). As a result, they frequently require the assistance of benefactors for quick help. Only non-governmental organizations, such as the Tagore Society for Rural Development and SagarMangal, were carrying out charitable activities in Jibantala-Kamalpur, such as programs for maternal and child development, sanitation, testing facilities for drinking water, vocational training, and the establishment of self-help groups. However, in the Gangasagar colony, just 16.13% of homes were BPL, indicating a relatively acceptable performance in SLSI (0.369).

It is debatable if adaptive ability can be used to infer adaptation (Engle 2011; Juhola and Kruse 2015; Malone and Engle 2011; Siders 2018). If persons with poor adaptive capacities can adjust while people with great adaptive capacities cannot. Similarly, it appears that adaptive capacity cannot fully explain adaptation (Mortreux et al. 2020). In this case, adaptive ability can explain 20% ($R^2 = 0.0202$) of the variation in adaptation, while the remaining 80% is unexplained (Table 12.3).

Migrants' Struggle for Existence: An Overview

The two practical strategies for mitigating the effects of climate change are migration and displacement. The optimum site of destination for migrants in response to both abrupt and slow-onset environmental shocks is ignored in the majority of research on the environmental dimensions of migration, which focuses on location-specific mobility drivers (Findlay 2011). The paradox is that Sagar Island takes in and partially rehabilitates displaced persons even though it is vulnerable to coastal erosion (Bandhopadhyay 1997, 2000; Kumar et al. 2001; Hazra et al. 2002; Gopinath and Seralathan 2005; Purkait 2008). Migrant discontent stems from issues with land allocation, unemployment, health care, sanitation, education, transportation, electrification, and drinking water due to the shortcomings in the rehabilitation program (Samling et al. 2015).

The marginal people of Sagar Island are particularly vulnerable due to their poor economic security, nutritional status, community engagement, educational security, and health security (particularly primary and reproductive health). A scarcity of resources causes clashes between early settlers and newcomers (Ghosh et al. 2014; Mukherjee 2014). Government administrations in India are always apathetic when it comes to environmental and migratory issues. No proper records are kept to manage migration or to develop programmes for their rehabilitation (Hazra and Bakshi 2003).

Management Strategy

Local governments have already developed some management strategies to ensure the Sagar Block's long-term survival. According to the 2011 Master Plan, brick block pitching was underway, with the goal of saving the island for the next thirty years. According to the 2011 Master Plan, brick block pitching was underway, with the goal of saving the island for the next thirty years. According to the World Bank (2014), the "embankment system, which was originally built to safeguard the residents, has become a liability: it creates a false sense of security while being increasingly prone to erosion and failure." Furthermore, resettlement was only successful when the resettlers' livelihoods were rebuilt (Voutira and Barbara 2000) through the establishment of community networks, without ignoring the economic and social costs involved (Tabucanon 2012).

Displaced individuals were deprived of access to tangible assets. Therefore, it is crucial to consistently advocate for the promotion of resilient agricultural methods. It is advisable to promote mechanized agricultural practices, including the cultivation of chile known as "Sagar-beauty" and watermelon known as "Sagar-sweet," in this context. Microfinance should be accessible for activities such as fishing, pond and canal excavations for irrigation, poultry farming, animal husbandry, small-scale cottage industries, and individuals engaged in non-formal business sectors. The effect of diversity differs according on geographical location, assets, family income, economic possibilities, and social networks (Davies 1996; Ellis 1998; Barrett et al. 2001). The Sagar Block demonstrates effective coping mechanisms. Despite this, the displaced persons have not yet reached the Rehabilitation Stage. There may be several underlying causes that explain the disparities between adaptive capacity calculations and real adaption scenarios.

Limitations of the Study

Proper identification of environmental refugees who moved to the Sagar Block in the Sundarbans from coastal villages in Ghoramara, Lohachara, and Khasimara are quiet difficult due to unavailability of the record. Proportional distribution of schemes related to social security status at village level should be considered for getting adaptation status of the rural communities. Rigorous research work will open up new vistas in future.

Conclusion

The investigation reveals two contrasting scenarios. First, the local administration has placed environmental refugees in the villages of Kamalpur, Gangasagar, and Bankimnagar, where their livelihoods are in terrible condition. Second, the overall coping mechanisms of the Sagar Block are quite good. The administration's social security efforts include the active implementation of NSAP, SDRF, and Financial Benefit Schemes such as SCP for scheduled caste people and MSY Loan, all of which contribute to the development of coping and adapting mechanisms in the face of climate-related disasters. These strategies, however, are not without flaws. Migrants face challenges in obtaining adequate accommodation, sanitation, safe drinking water, education, and employment possibilities. If a livelihood system's assets are depleted and institutions fail to cope, the effect is diminished resilience and greater vulnerability to disasters (Glavovic et al. 2003). To assist settlers in becoming economically self-sufficient and overcoming climate change-related livelihood difficulties, decentralized planning and community-based adaption methods are urgently needed.

Acknowledgements The authors express their gratitude to all the villages for their valuable aid and backing throughout the field study. We express our gratitude to the authors for the numerous official documents that were generously utilized in the development of this paper.

Funding The research did not receive any financing from public, commercial, or not-for-profit sources.

Declaration of Interests The authors affirm that they have no known conflicting financial interests or personal affiliations that could potentially influence the findings presented in this study.

References

Abson DJ, Dougill AJ, Stringer LC (2012) Using principal component analysis for information-rich socio-ecological vulnerability mapping in Southern Africa. Appl Geogr 35(1–2):515–524 (online). https://doi.org/10.1016/j.apgeog.2012.08.004

Adger WN, Agrawala S, Mirza MMQ, Conde C, O'Brien K, Pulhin J, Pulwarty R, Smit B, Takahashi K (2007) Assessment of adaptation practices, options, constraints and capacity. Climate change: impacts, adaptation and vulnerability, contribution of working group II to the fourth assessment report of the intergovernmental panel on climate change. Cambridge University Press, Cambridge, pp 717–743

Adger WN, Barnett J, Brown K, Marshall N, O'Brien K (2013) Cultural dimensions of climate change impacts and adaptation. Nat Clim Change 3(2):112–117. https://doi.org/10.1038/NCLIMATE1666

Bandyopadhyay S (1997) Coastal erosion and its management in Sagar Island, South 24-Parganas, West Bengal. Indian J Earth Sci 24(3–4):51–69

Bandyopadhyay S (2000) Coastal changes in the perspective of long term evolution of an estuary: Hugli, West Bengal, India. Quat Sea Level Var Shoreline Displacement Coast Environ 103–115

Barrett CB, Reardon T, Webb P (2001) Nonfarm income diversification and household livelihood strategies in rural Africa: concepts, dynamics, and policy implications. Food Policy 26:315–331

Black R, Adger WN, Arnell NW, Dercon S, Geddes A, Thomas D (2011) The effect of environmental change on human migration. Glob Environ Change 21:S3–S11

Black R, Arnell NW, Adger WN, Thomas D, Geddes A (2013) Migration, immobility and displacement outcomes following extreme events. Environ Sci Policy 27(1):32–43. https://doi.org/10.1016/j.envsci.2012.09.001

Boateng PA (2013) A livelihood asset status tracking method for the assessment of the effects of a development programme on agricultural productivity and poverty reduction: evidence from the Ejisu-Juaben District, Ghana. J Agric Environ Int Dev JAEID 107(2):267–296. https://doi.org/10.12895/jaeid.20132.181

Bond R, Mukherjee N (2002) Livelihood asset status tracking: an impact monitoring tool? J Int Dev 14:805–815. https://doi.org/10.1002/jid.926

Carney D (ed) (1998) Sustainable rural livelihoods: what contribution can we make? Papers Presented at the DFID Natural Resources Advisers' Conference, July 1998. DFID, London, p 4

Castles S (2002) Environmental change and forced migration: making sense of the debate. New issues in refugee research, Working Paper No. 70, UNHCR Evaluation and Policy Analysis Unit Geneva

Census (2011) Complete village directory. https://data.gov.in/Catalog/Complete-Villages-directory-indiastatedistrictsub-district-level-Census2011/. Accessed 17 Dec 2018

Census (2011) Office of the Registrar General & Census Commissioner, India. https://censusindia.gov.in/tables_published/H-Series/houselist_main.html. Accessed 17 Dec 2020

Chakma N (2014) Inter Island relocation due to natural hazards and socio-economic challenges: a case study on Hugli Estuary Islands. Golden Res Thoughts 4(5):1–7

Chambers R, Conway G (1992) Sustainable rural livelihoods: practical concepts for the 21st century. Institute of Development Studies, UK

CSE (2007, Jan 15) Bad tidings: land erosion, sea level rise threaten Sundarbans. Down to Earth, pp 25–34

Davies S (1996) Adaptable livelihoods. Coping with food insecurity in the Malian Sahel. St. Martin's Press, New York, NY

DFID (2000) Sustainable livelihoods guidance sheets. https://www.ennonline.net/attachments/871/dfid-sustainable-livelihoods-guidance-sheet-section1.pdf. Accessed 12 May 2020

Dilling L, Prakash A, Zommers Z, Ahmad F, Singh N, de Wit S, Nalau J, Daly M, Bowman K (2019) Is adaptation success a flawed concept? Nat Climate Change 9(8):572–574. https://doi.org/10.1038/s41558-019-0539-0

Doevenspeck M (2011) The thin line between choice and flight: environment and migration in rural Benin. Int Migr 49:e50–e68

El-Hinnawi E (1985) Environmentalrefugees. UNEP. http://www.deccma.com. Accessed 18 Feb 2021

Ellis F (1998) Household strategies and rural livelihood diversification. J Dev Stud 35(1):1–38

Employment Bank, Government of West Bengal Initiative (2012) Special Component Plan (SCP). http://employmentbankwb.gov.in/tsp.php/. Accessed 20 Mar 2020

Engle N (2011) Adaptive capacity and its assessment. Glob Environ Change 21(2):647–656. https://doi.org/10.1016/j.gloenvcha.2011.01.019

Engle NL (2011b) Adaptive capacity and its assessment. Glob Environ Chang 21(2):647–656. https://doi.org/10.1016/j.gloenvcha.2011.01.019

Findlay AM (2011) Migrant destinations in an era of environmental change. Glob Environ Chang 21(Supplement 1):S50–S58

Findlay A, Geddes A (2011) Critical views on the relationship between climate change and migration: some insights from the experience of Bangladesh. Migration and climate change Cambridge. UNESCO Publishing, Cambridge University Press, Cambridge, pp 138–159

Foresight (2011) Migration and global environmental change. Final Project Report. The Government Office for Science, London

Gemenne F (2011) Why the numbers don't add up: a review of estimates and predictions of people displaced by environmental changes. Global. https://doi.org/10.1016/j.gloenvcha.2011.09.005

Ghosh T, Hajra R, Mukhopadhyay A (2014) Island erosion and afflicted population: crisis and policies to handle climate change. In: International perspectives on climate change, climate change management, pp 217–225. https://doi.org/10.1007/978-3-319-04489-7_15

Glavovic BC, Scheyvens R, Overton J (2003) Waves of adversity, layers of resilience: exploring the sustainable livelihoods approach. Contesting development: pathways to better practice. In: Proceedings of the third biennial conference of the Aotearoa New Zealand International Development Studies Network (Dev Net), Massey University, 5–7 Dec 2002, Institute of Development Studies, pp 289–293

Gopinath G, Seralathan P (2005) Rapid erosion of the coast of Sagar Island, West Bengal, India. Environ Geol 48(8):1058–1067. https://doi.org/10.1007/s00254-005-0044-9

Hahn MB, Riederer AM, Foster SO (2009) The livelihood vulnerability index: a pragmatic approach to assessing risks from climate variability and change, a case study in Mozambique. Center for Sustainability and the Global Environment, University of Wisconsin-Madison, pp 74–88. https://doi.org/10.1016/j.gloenvcha.2008.11.002

Hajra R, Szabo S, Tessler Z, Ghosh T, Matthews Z, Foufoula-Georgiou E (2017) Unravelling the association between the impact of natural hazards and household poverty: evidence from the Indian Sundarban delta. Sustain Science, Springer Japan. https://doi.org/10.1007/s11625-016-0420-2

Hajra R, Ghosh A, Ghosh T (2021) Micro level vulnerability assessment of Estuarine Islands: a case study from Indian Sundarban. Springer Nature Switzerland AG. https://doi.org/10.1007/978-3-319-54612-4_11

Harms A (2013) Dwelling in loss: environment, displacement and memory in the Indian Ganges Delta. PhD thesis, University of Berlin

Harms A (2015) Leaving Lohachara: on circuits of emplacement and displacement in the Ganges Delta. Glob Environ 8:62–85. https://doi.org/10.3197/ge.2015.080104
Hazra S, Bakshi A (2003) Environmental refugees from Vanishing Islands. In: Environment and human security. Lancer Books, pp 219–227
Hazra S, Ghosh T, DasGupta R, Sen G (2002) Sea level and associated changes in the Sundarbans. Sci Culture 68(9–12):309–321
Hazra S, Samanta K, Mukhopadhyay A, Akhand A (2010) Temporal change detection (2001–2008) Study of Sundarban (Final Report). School of Oceanographic Studies, Jadavpur University
Hoque ZM, Cui S, Xu L, Islam I, Tang J, Ding S (2019) Assessing agricultural livelihood vulnerability to climate change in coastal Bangladesh. Int J Environ Res Public Health 16:4552. https://doi.org/10.3390/ijerph16224552
IOM (2009) Migration, environment and climate change: assessing the evidence, Geneva. International Organization for Migration. In collaboration with United Nations University UNU-EH S Institute for Environment and Human Security, p 19
IPCC (2007) Climate Change. Impacts, adaptation and vulnerability. Cambridge University Press, Cambridge
IPCC (2014) Annex II: Glossary. In: Barros VR, Field CB, Dokken DJ, Mastrandrea MD, Mach KJ, Bilir TE, Chatterjee M, Ebi KL, Estrada YO, Genova RC, Girma B, Kissel ES, Levy AN, MacCracken S, Mastrandrea PR, White LL (eds) Climate Change 2014: impacts, adaptation, and vulnerability. Part B: regional aspects. Contribution of Working Group II to the Fifth Assessment Report of the Intergovernmental Panel on Climate Change. Cambridge University Press, Cambridge, pp 1757–1776
Juhola S, Kruse S (2015) A framework for analysing regional adaptive capacity assessments: challenges for methodology and policy making. Mitig Adapt Strat Glob Change 20(1):99–120. https://doi.org/10.1007/s11027-013-9481-z
Kasperson RE, Kasperson JX (2001) Climate change, vulnerability and social justice. Risk and Vulnerability Programme, Stockholm Environment Institute, Stockholm
Kibreab G (1997) Environmental causes and impact of refugee movements: a critique of the current debate. Disasters 21(1):20–38. https://doi.org/10.1111/1467-7717.00042
Kumar P, Meenakumari B, Bandyopadhyay JK (2001) Geomorphological changes in Sagar Island—a remote sensing study. In: Proceedings of the national seminar on riverine and reservoir fisheries—challenges and strategies, 23–24 May 2001, Cochin, India
Lindenberg M (2002) Measuring household livelihood security at the family and community level in the developing world. World Dev 30(2):301–318
Lonergan S (1998) The role of environmental degradation in population displacement. Environmental change and security project report, p 4
Malone EL, Engle NL (2011) Evaluating regional vulnerability to climate change: purposes and methods. Wiley Interdisc Rev Clim Change 2(3):462–474. https://doi.org/10.1002/wcc.116
McGregor J (1993) Refugees and the environment. Geography and refugees: patterns and processes of change. Belhaven, London, pp 157–170
McLeman R, Smit B (2006) Migration as an adaptation to climate change. Clim Change 76:31–53. https://doi.org/10.1007/s10584-005-9000-7
Mortreux C, O'Neill S, Barnett J (2020) Between adaptive capacity and action: new insights into climate change adaptation at the household scale. Environ Res Lett 15(7). https://doi.org/10.1088/1748-9326/ab7834
Moser C (1998) The asset vulnerability framework: reassessing urban poverty reduction strategies. World Dev 26(1):1–19. https://doi.org/10.1016/S0305-750X(97)10015-8
Mukherjee B (2014) Coastal erosion at Ghoramara and its rehabilitation programme in Jibantala. Indian J Landsc Syst Ecol Stud 37(1):159–172
Myers N (2005) Environmental refugees: an emergent security issue. Session III—Environment and Migration, 13th Economic Forum, Prague, 23–27 May 2005, EF.NGO/4/05

National SafaiKarmacharis Finance and Development Corporation (2017) Mahila SamridhiYojna. http://nskfdc.nic.in/en/content/schemes-programmes/mahila-samridhi-yojna-msy/. Accessed 20 Mar 2020

Neil Adger W (1999) Social vulnerability to climate change and extremes in coastal Vietnam. World Dev 27(2):249–269. https://doi.org/10.1016/s0305-750x(98)00136-3

ODI, UNDP (2017) Climate change, migration and displacement: the need for a risk-informed and coherent approach. https://www.odi.org/publications/10977-climate-change-migration-and-displacementneed-risk-informed-and-coherent-approach. Accessed 18 Feb 2021

Purkait B (2008) Coastal erosion in response to wave dynamics operative in Sagar Island, Sundarban delta, India. Front Earth Sci China 3(1):21–33. https://doi.org/10.1007/s11707-009-0001-0

Raymond G (2007) The impact of climate change on Indian agriculture. Job market paper draft, pp 1–31

Samling CL, Ghosh AK, Hazra S (2015) Resettlement and rehabilitation: Indian scenario. DECCMA Working Paper, Deltas, Vulnerability and Climate Change: Migration and Adaptation, IDRC Project Number 107642

Seto KC (2011) Exploring the dynamics of migration to mega-delta cities in Asia and Africa: contemporary drivers and future scenarios. Glob Environ Change 21(Supplement 1):S94–S107

Siders AR (2018) Adaptive capacity to climate change: a synthesis of concepts, methods, and findings in a fragmented field. Wires Clim Chang 10(3):e573. https://doi.org/10.1002/wcc.573

Singh PK, Hiremath BN (2009) Sustainable livelihood security index in a developing country: a tool for development planning. Ecol Ind 10:442–451. https://doi.org/10.1016/j.ecolind.2009.07.015

Smit B, Pilifosova O (2001) Adaptation to climate change in the context of sustainable development and equity. Climate change 2001: impacts, adaptation and vulnerability. Cambridge University Press, Cambridge, pp 877–912

Sward J, Codjoe S (2012) Human mobility and climate change adaptation policy: a review of migration. National Adaptation Programmes of Action (NAPAs), Migrating out of Poverty RPC Working Paper 6. Migrating out of Poverty Consortium, University of Sussex, Brighton, UK, p 44

Tabucanon GM (2012) The Banaban resettlement: implications for Pacific environmental migration. PacificStudies 35:343–370

United Nations General Assembly (1997) S/19-2. Programme for the further implementation of agenda 21. 19th Special Session, 23–27 June, New York

Voutira E, Barbara H-B (2000) Successful refugee settlement: are past experiences relevant? In: Michael C, McDowell C (eds) Risks and reconstruction: experiences of resettles and refugees. World Bank, Washington, pp 56–76

West Bengal forest Department (2019) Sabujsree Scheme. https://westbengalforest.gov.in/sabujsri.php/. Accessed 20 Mar 2020

Westing AH (1992) Environmental refugees: a growing category of displaced persons. Environ Conserv 19:201–207

World Bank (2014) Building resilience for sustainable development of the Sundarban. Strategy report. Report No 88061, South Asia Region, Sustainable Development Department, Environment and Water Resources Management Unit

Chapter 13
Socio-economic Dimensions of Climate Change in Urban Bangladesh: A Focus on the Initiatives of Local Governing Agencies

Imran Hossain, A. K. M. Mahmudul Haque, and S. M. Akram Ullah

Abstract This chapter examines the socio-economic impacts of climate change in non-coastal cities in Bangladesh. It aims to identify the severity of climate change on various socio-economic dimensions, assess the challenges faced by governing agencies, and provide recommendations for effective adaptation and mitigation measures. The study employs a combination of quantitative and qualitative methods to assess the specific impact of climate change on each dimension. A structured survey method is employed, incorporating a five-point Likert scale, to quantify the severity of climate change impacts. Additionally, qualitative methods such as Key Informant Interviews (KII), Focus Group Discussions (FGD), and observation techniques enrich the data collection process, providing a holistic understanding of the challenges faced by governing agencies. The findings reveal that climate change significantly affects the socio-economic well-being of communities in non-coastal cities. Professional security, health security, income security, economic security, and social security are all negatively impacted by climate change, with varying degrees of severity. Furthermore, the research identifies major challenges faced by governing agencies, including unawareness, corruption, poverty, political factors, religious factors, the apathy of the authority, a lack of international aid, and population growth. These challenges hinder effective climate change adaptation and mitigation efforts. Based on the research findings, recommendations are provided to address these challenges and enhance the capacity of governing agencies. By implementing the recommendations, governing agencies can enhance their ability to address climate change challenges and promote resilience in these cities.

I. Hossain (✉)
Department of Political Science, Varendra University, Rajshahi 6204, Bangladesh
e-mail: imranbd.ru@gmail.com; imran.pol@vu.edu.bd

I. Hossain · A. K. M. Mahmudul Haque · S. M. Akram Ullah
Department of Political Science, University of Rajshahi, Rajshahi 6205, Bangladesh
e-mail: akmmahmudul@ru.ac.bd

S. M. Akram Ullah
e-mail: akram@ru.ac.bd

U. Mukhopadhyay et al. (eds.), *Climate Crisis, Social Responses and Sustainability*,
Climate Change Management, https://doi.org/10.1007/978-3-031-58261-5_13

Introduction

Bangladesh, a country with a population of more than 150 million, faces a number of difficulties, such as environmental dangers, socio-political unrest, developmental crises, and the effects of climate change (Spillan and Rahman 2023). Climate change and its associated hazards, such as temperature and rainfall fluctuations, intensified floods, droughts, cyclones, storm rises, and salinity intrusion, are already affecting communities, ecosystems, and national infrastructure (Sesana et al. 2021). This impact is particularly profound due to Bangladesh's geophysical location, hydrological influences of monsoons, regional water flow patterns, the population's limited resilience, inadequate policies, and institutional responses (Parven et al. 2022). These factors combine to create a perfect storm of vulnerability for Bangladesh. The country's low-lying deltaic geography makes it highly susceptible to sea-level rise, resulting in increased coastal erosion and the displacement of coastal communities (Mojid 2020). Additionally, the monsoon season brings heavy rainfall, which can lead to devastating floods that destroy homes, crops, and infrastructure (Kabir et al. 2023). The population's limited resilience and lack of access to resources exacerbate the consequences of climate change, leaving communities struggling to adapt and recover. Without adequate policies and institutional responses, Bangladesh will continue to face significant challenges in mitigating and adapting to the effects of climate change (Isfat and Raihan 2022). The monsoon-induced floods and dry-season rain shortages trigger flooding in Bangladesh. This situation is exacerbated by the warming climate, which causes frequent and intensified floods and severe droughts. Moreover, alarming projections anticipate sea level rises of 30 cm and 50 cm by 2030 and 2050, respectively, which could lead to severe socio-economic and environmental crises (Hossen and Sultana 2023). The Sundarbans, a vital ecosystem, faces 40% loss with a mere 25 cm sea-level rise and complete submersion with over 60 cm rise (Waheduzzaman and Mizanuzzaman 2021). A meter's rise could submerge significant portions of productive land.

However, the term "climate change" refers to evolving climate attributes over time (Celik 2020). Researchers across disciplines study its implications, driven by heightened concerns (Agrawala and van Aalst 2008). Notably, the Intergovernmental Panel on Climate Change (IPCC) and the United Nations Framework Convention on Climate Change (UNFCCC) present different definitions. IPCC denotes any climatic shift over time, natural or human-induced (IPCC 2007), while UNFCCC identifies climate change as human-induced alterations in the global atmosphere's composition beyond natural variability (IPCC 2007). The IPCC, established by the World Meteorological Organization and the United Nations Environment Programme, evaluates climate change's scientific, technical, and socioeconomic dimensions. The differing definitions and focus of the IPCC and UNFCCC highlight the complexities and debates surrounding climate change. A more thorough knowledge of the causes of climate change, including both natural and human-caused ones, is made possible by the IPCC's expanded definition. On the other hand, the UNFCCC's focus on changes brought about by humans emphasizes the need for collective action to lessen the effects of climate change and prepare for them. These differing perspectives impact policy-making and international negotiations as countries grapple with the urgency and responsibility to address climate change.

In this aspect, Bangladesh experiences inevitable climate change consequences, marked by unprecedented human tragedies (Glennon 2018). Despite being among the least energy-intensive nations, Bangladesh ranks among the ten most climate-affected countries, posing a significant contemporary challenge (Mahmood 2012). According to the most current study, Bangladesh ranked seventh among the countries most negatively affected by climate change as of 2021 (The Climate Risk Index 2021). A range of natural and quasi-natural hazards brought on by climate change highlight the dynamics of these sensitivities. Bangladesh's varied geography is reflected in the various threats that different sections of the country face. The north faces significant challenges from cyclones and tidal rises; the center faces river erosion and flooding; and the north faces droughts and floods (Das 2010). A symptom of the persistent advance of climate change is the intensification and frequency of hydro-meteorological risks. The lives and livelihoods of millions are severely impacted by the cumulative effects of excessive temperatures, floods, droughts, tropical cyclones, lightning, tidal surges, and saltwater incursions (Ministry of Disaster Management and Relief 2020). The devastating rate of river erosion is a key indicator of the effects of climate change. According to Rahman and Gain (2019), climate change has a significant impact on the acceleration of riverbank erosion, which in turn causes massive land loss, widespread homelessness, and forced migration. These harsh facts highlight how important it is to comprehend the particular circumstances and events that pushed Bangladesh into its difficult position as one of the most climate-affected countries. In this regard, governments and non-governmental organizations prioritize climate issues, despite varying successes and setbacks. Particularly concerning are predictions that climate change could exacerbate existing poverty, pushing vulnerable countries like Bangladesh into greater distress. The IPCC forecasts a temperature increase of 2.5–10 °F over the next century due to human-induced greenhouse gases, as corroborated by over 1300 scientists (Kumari et al. 2020). The consequences of inaction are already becoming apparent, with extreme weather events becoming more frequent and severe. The urgency to address climate change is not only based on scientific evidence but also on the ethical responsibility to protect future generations and ensure a resilient future for all. In this aspect, human activities exacerbate climate change's impacts, especially in economically vulnerable countries like Bangladesh. Coastal regions, though prioritized, are not exclusive victims; non-coastal regions, including the north and east, face detrimental effects. The northern area already grapples with drought and dropping surface water levels, leading to an increase in climate refugees in cities like Dhaka, triggering significant socio-economic disruption (Salehin 2017).

Various studies analyze Bangladesh's climate trends. For instance, Karmakar and Nessa (1997) and Karmakar (2003) study the impact of climate change on natural disasters, while Warrick et al. (1994), Karmakar and Shrestha (2000), and

Debsarma (2003) evaluate rainfall and temperature fluctuations. Projected temperature increases by Haque et al. (1992) underscore concerns, especially winter temperature rises due to monsoon-induced precipitation, potentially causing severe floods. Additionally, warming trends affect Bangladesh's lowest minimum temperatures. Warrick et al. (1994) investigated temperature and rainfall variations, indicating increasing temperatures. Since the twentieth century's turn, the average global temperature has risen by 0.5 °C, paralleling observed global warming. Karmaker and Nessa (1997) found rising decadal mean annual temperatures post-1961–1970 over Bangladesh. Bangladesh's Climate Change Strategy Action Plan 2009 acknowledges climate change's hindrance to achieving middle-income status, emphasizing disaster risk reduction, adaptation, low-carbon development, mitigation, and international funding provision. It is clear that Bangladesh recognizes the urgency and importance of addressing climate change. The country's efforts to reduce natural catastrophe risk and adapt to changing conditions are crucial to protecting its citizens and ensuring sustainable development. The inclusion of low-carbon development and international funding provision demonstrates Bangladesh's commitment to global efforts in mitigating climate change and creating a more resilient future. The Bangladesh Climate Change Strategy Action Plan (BCCSAP 2008), updated in 2009, outlines goals and governmental involvement.

The UNFCCC and SDG[1] processes share climate change mitigation and adaptation goals. A multitude of studies address Bangladesh's climate vulnerability. Approaches include production-function and Ricardian methods for assessing impacts on agriculture and economics (Hossain et al. 2020). Coastal areas, annually battered by cyclones and storms due to climate change (Mojid 2020), receive significant attention, but non-coastal regions also grapple with issues like land degradation, salinity, and agricultural loss (Roy et al. 2022; Sikder 2010). Climate-induced land and soil loss jeopardizes food security (Bhandari 2021). Additionally, rising sea levels and increased precipitation lead to flooding, further exacerbating the challenges faced by Bangladesh. The country's vulnerability to climate change is also heightened by its high population density and limited resources for adaptation.

A comprehensive review of existing literature, as presented in the earlier sections of this manuscript, underscores the disproportionately limited attention given to the socio-economic dimensions of climate change in non-coastal regions (Hossen and Sultana 2023; Salehin 2017; Mojid 2020; Kabir et al. 2023; Isfat and Raihan 2022; Glennon 2018; Karmakar and Nessa 1997). Despite the wealth of existing literature, a comprehensive examination reveals a shortage of studies focusing on the complex socio-economic consequences of climate change in regions away from the shoreline. This research endeavors to address this critical gap by delving into the specific dynamics of non-coastal areas, shedding light on their unique vulnerabilities and resilience mechanisms, with a particular emphasis on the pivotal role played

[1] The term "Sustainable Development Goals" (SDG) is used. These 17 global objectives were created by the UN in 2015 as a component of the 2030 Agenda for Sustainable Development. The objectives are to address a variety of global social, economic, and environmental issues in order to build a more just and sustainable future for all.

by local governing agencies. By conducting an in-depth analysis of the existing literature, this study aims to not only identify the research gap but also contribute significantly to the understanding of climate change impacts in non-coastal areas. The core objective is to understand climate change's socio-economic impact on non-coastal areas and to investigate the role of local agencies in this aspect. The study also sought to investigate the major challenges faced by the governing agencies regarding climate issues. Through illustrating impacts, identifying measures, and evaluating priorities, this study aims to contribute to effective climate change management.

Profile of Study Area

Rajshahi City Corporation (RCC) and Sylhet City Corporation (SCC) represent two distinct urban areas in Bangladesh, each with its unique geographical, demographic, and socio-economic-political characteristics. Located in the northwest region, Rajshahi is positioned at approximately 24.3636° N latitude and 88.6241° E longitude (Clemett et al. 2006). Sylhet, on the other hand, is situated in the north-eastern part of the country, marked by coordinates approximately 24.8949° N latitude and 91.8687° E longitude. Rajshahi, often referred to as the Silk City, stands as a major urban center with a population mix reflective of its historical and cultural significance (Clemett et al. 2006). The city is positioned along the northern banks of the Padma River, contributing to its socio-economic vibrancy through trade and commerce (Clemett et al. 2006). In contrast, Sylhet, known for its picturesque landscapes, lies nestled amidst hills and tea gardens, offering a unique demographic composition shaped by its distinct geographical features (Siddique 2021). Demographically, Rajshahi and Sylhet exhibit variations in population density and composition. Rajshahi, as a populous urban center, boasts a diverse demographic makeup with a blend of urban and rural migrants. Its socio-economic landscape is marked by a mix of traditional and modern economic activities. Sylhet, renowned for its tea industry, is characterized by a demographic fabric influenced by its hilly terrain, with a substantial population engaged in tea cultivation (Siddique 2021).

The socio-economic-political situation of these areas plays a crucial role in understanding their vulnerability to climate change. Ranked highly vulnerable, experts identify Rajshahi and neighboring districts within the Barind region as particularly impacted (Rahman et al. 2022). The Barind region's water supply challenges stem from its height above sea level, leading to deepening underground water reservoirs and socio-economic implications (Jamat 2010). These challenges have prompted the government and various organizations to focus on implementing measures to address the water scarcity issue. Furthermore, rapid urbanization, infrastructural development, and economic activities in both RCC and SCC contribute to their exposure to climate-related challenges. Additionally, the political landscape, governance structures, and the effectiveness of local governing bodies are vital elements influencing the regions' adaptive capacity. Understanding these contextual factors is paramount for comprehending the nuanced vulnerabilities and crafting effective

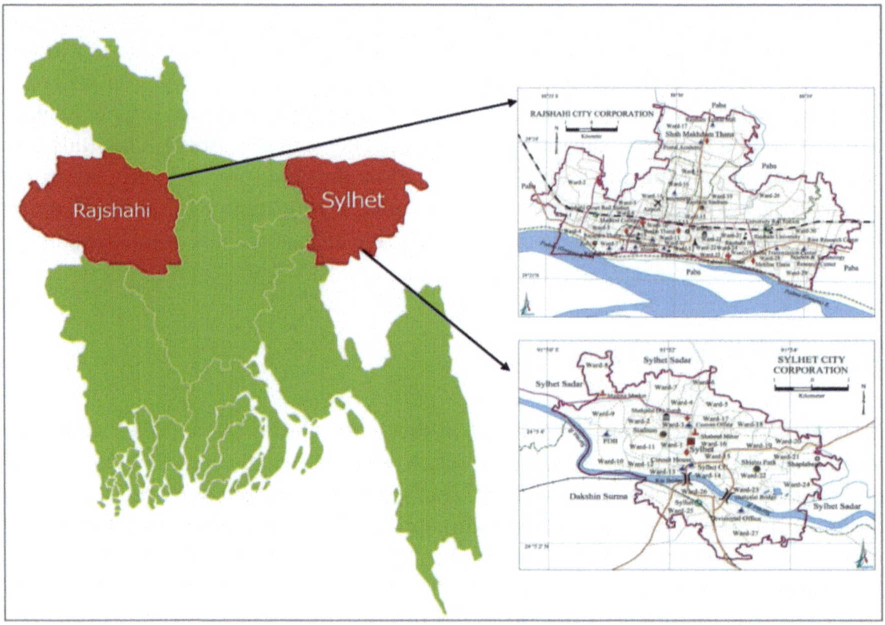

Fig. 13.1 Map of the study area

climate change adaptation and mitigation strategies tailored to the specific challenges faced by Rajshahi and Sylhet.

Our study seeks to improve the relationship between the discussed aspects and the primary goals by weaving these contextual threads. The foundation for comprehending the complex vulnerabilities of Rajshahi and Sylhet to climate change is laid by their distinct geographical, demographic, and socio-economic-political features. By exploring these variations, we may better understand the difficulties that these areas experience, which is a critical first step in developing policies for mitigation and adaptation to climate change (Fig. 13.1).

Rationale of the Study

The rationale for conducting research on the socio-economic dimensions of climate change in urban Bangladesh, with a focus on the initiatives of local governing agencies, is multifaceted and grounded in several key considerations. Bangladesh stands as one of the most climate-vulnerable nations globally, frequently encountering devastating natural disasters like cyclones, floods, and droughts that disproportionately affect its urban communities (Seddiky et al. 2023). The swift pace of urbanization in the country has resulted in escalated population density, urban expansion, and heightened strain on already limited resources, thereby exacerbating its susceptibility to the adverse impacts of climate change. The significance of local governing agencies in addressing these challenges is paramount. Situated at the closest administrative level to communities, these agencies shoulder the responsibility of ensuring fundamental service provision, responsible natural resource management, and the promotion of sustainable development (Meyer and Auriacombe 2019). Their critical role in climate change adaptation and mitigation in urban areas is evident. Hence, the driving rationale of this study is to delve into the proactive measures adopted by local governing agencies to confront the socio-economic ramifications of climate change in urban Bangladesh. Through a meticulous examination of these agency actions, the research aims to offer valuable insights into both the obstacles and opportunities associated with creating resilient and sustainable urban centers in the face of climate change. Additionally, the findings of this study carry the potential to inform policy formulation, strategic planning, and effective implementation at both local and national levels. By doing so, the research seeks to foster the adoption of robust climate change adaptation and mitigation strategies across urban Bangladesh. The nation grapples with considerable challenges in adapting to climate change impacts, particularly given its unique topographical and geographical features that render it susceptible to severe climatic events.

However, while these non-coastal areas grapple with heightened vulnerability, they tend to receive comparatively less climate financing. This disparity in resource allocation underscores the urgent need to focus on these regions, given their significant vulnerability and the severe socio-economic dimensions of climate change they face. The lack of awareness and preparedness among local authorities in these non-coastal regions to address climate change impacts is concerning. Despite the evident socio-economic impacts, some governing agencies fail to acknowledge climate change's presence beyond coastal cities. Nonetheless, the engagement of local authorities is deemed pivotal in effective climate change mitigation, and this research underscores the necessity to investigate the socio-economic impacts of climate change comprehensively in non-coastal Bangladesh. It also stresses evaluating the adequacy of initiatives taken by governing agencies in light of the intensity of these socio-economic impacts, highlighting a critical gap in current practices. In sum, this research is propelled by the urgent need to address the socio-economic dimensions of climate change in urban Bangladesh's non-coastal regions and to examine the role of local governing agencies in this crucial endeavor.

Materials and Methods

The research methodology employed in this study adopts an empirical approach with a blend of both qualitative and quantitative techniques, ensuring data validity and reliability. The data is sourced from primary and secondary sources. Primary sources encompass field studies, responses from stakeholders and city governance bodies, and the opinions of individuals knowledgeable about climate issues or directly affected by climate change. Secondary data includes research reports, articles, official statistics, newspapers, and relevant literature.

A wide range of variables affecting job security and stability are included in professional security. Based on field data, we identified certain sub-parameters, such as changes in employment patterns, job losses related to climate-induced events, and variations in the demand for various professions. The justification for choosing these sub-parameters is that they are closely related to climate change, wherein extreme occurrences and environmental changes can have a substantial impact on livelihoods and career paths. Furthermore, our sub-parameters also take into account the significant effects of climate change on public health, such as the frequency of diseases linked to climate change, the accessibility of medical services during extreme events, and the effects of stresses linked to climate change on mental health. Based on empirical observations of health vulnerabilities in the research areas, which emphasize the complex relationship between climate change and health outcomes, these sub-parameters were selected. To further account for the financial effects of climate change, sub-parameters related to income security, such as shifts in sources of income and economic pursuits, were chosen. These decisions are based on field-level analyses of changes in local economies, especially in industries that are vulnerable to changes in the climate. In the context of economic security, the focus is on things like how flexible economic structures are and how vulnerable local industries are to climate-related hazards. This is because real-world evidence shows that economies are weak when climate-related problems happen. The vulnerability of marginalized groups, social support networks, and community integration were the final sub-parameters of social security that we selected based on our observations of social dynamics in the research locations.

The study's population comprises mayors, councilors, government officials, policymakers, civil society members, and city residents across two city corporations in Bangladesh. The unit of analysis comprises those well-versed in climate issues or those directly or indirectly impacted by climate change. The research focuses on two non-coastal city corporations chosen through purposive sampling, with a total of 424 respondents. Sampling procedures encompass a mix of simple random, purposive, and key informant interviews (KII), while data collection techniques include questionnaire surveys, KII, focused group discussions (FGD), observation, and document study. The survey includes four sets of questionnaires adapted to different respondent categories, combining open-ended structured and closed-ended unstructured questions. The questionnaire covers socio-economic data, climate change mitigation measures, and challenges faced by city corporations. Responses are coded for computer tabulation, and a 5-point Likert scale is used to measure variables, with identical response categories for uniformity (Table 13.1).

Key informant interviews involve government officials responsible for climate policy at different levels. FGDs are conducted in both city corporations with local elites and public representatives as participants, aimed at cross-checking data and exploring real-life climate governance scenarios. The study also employs observation to empirically assess measures taken by institutions to mitigate climate change impacts. Furthermore, document studies supplement primary data with secondary

Table 13.1 Sampling and data collection techniques

Respondents	Sampling procedure	Data collection techniques	Sample size	
Government officials	Purposive	KII	5X2 CC	10
Residents	Simple random	QS	192X2 CC	394
	Purposive	IDI	10X2 CC	
Local elites and public representatives	Purposive	FGD	5X2 FGDs	10
Total respondents				424

[*Note* In the table, CC means City Corporations, KII means Key Informant Interview, QS means Questionnaire Survey, FGD means Focus Group Discussion and IDI means In-depth Interview]

sources like publications, journals, research reports, and newspapers, enhancing the reliability and validity of the research. The collected quantitative data undergoes inferential analysis using techniques such as frequency distribution, statistical regression, and chi-square test through tools like SPSS V.22. Qualitative data is analyzed descriptively with logical reasoning. The research findings are presented through various visual aids, including graphs, tables, and narrative explanations.

Results and Discussion

The study sought to investigate the socioeconomic effects of climate change in Bangladesh's non-coastal cities. The study evaluated the effects of climate change on various socioeconomic aspects and looked at the initiatives taken by local governing bodies. The increasing frequency and intensity of natural disasters have resulted in loss of lives, damage to infrastructure, and disruption of economic activities (Chaudhary and Piracha 2021; Rahman and Gain 2019; Karmakar 2003).

The study reveals that there is a significant impact of climate change on various socio-economic dimensions in both Rajshahi and Sylhet City Corporations in Bangladesh. In terms of professional security, a small percentage of respondents in the RCC (2.6%) stated being affected by climate change, indicating that there are certain vulnerabilities and risks to professional livelihoods (Table 13.2). During FGD, respondents says,

> Our everyday activities are disturbed by the intense heat waves. It is difficult for people in many occupations, particularly those in construction and agriculture, to work outside due to the extreme heat. It threatens our well-being and professional security (FGD, RCC, Bangladesh).

However, in SCC, no respondents reported being affected, which may be attributed to the different geographical and climatic conditions. The study suggests that climate change has a significant impact on various socio-economic dimensions in both the RCC and SCC in Bangladesh. While professional security is a concern for some

Table 13.2 Socio-economic impact of climate change

Variables		RCC		SCC	
		f	%	f	%
Professional security	Yes	5	2.6%	0	0.0%
	No	187	97.4%	192	100.0%
Health security	Yes	37	19.3%	1	0.5%
	No	155	80.7%	191	99.5%
Income security	Yes	19	9.9%	0	0.0%
	No	173	90.1%	192	100.0%
Economic security	Yes	18	9.4%	0	0.0%
	No	174	90.6%	192	100.0%
Social security	Yes	15	7.8%	0	0.0%
	No	177	92.2%	192	100.0%

[*Source* Field Survey, 2022]

respondents in RCC, SCC seems to be less affected due to different geographical and climatic conditions.

In terms of health security, more RCC respondents (19.3%) than SCC respondents (0.5%) reported being impacted by climate change. This suggests that climate change impacts in non-coastal cities have significant health implications, such as increased vulnerability to diseases and health risks. In terms of income security, a greater proportion of respondents in RCC (9.9%) than in SCC (0.0%) reported being impacted by climate change (Table 13.2). This indicates that climate change can have adverse effects on income generation and economic stability in non-coastal areas, potentially leading to livelihood challenges and financial insecurity (Kabir et al. 2023; Ministry of Disaster Management and Relief 2020). So, this study demonstrates that climate change has an adverse effect on income security. The changing climatic patterns, including erratic rainfall and prolonged droughts, have significantly impacted our agricultural activities, which are the primary source of income for many people in this region (Kabir et al. 2023). The unpredictability of the weather has resulted in crop failures, reduced yields, and increased vulnerability to pests and diseases (Bajaj and Kumria 2023; Isfat and Raihan 2022). People heavily rely on agriculture for their livelihood and have been hit hard by these challenges. The decline in agricultural productivity has not only affected the income of the residents but has also caused a ripple effect on the entire local economy, leading to job losses and reduced employment opportunities (Aryal et al. 2020). Similarly, in the context of economic security, a higher percentage of respondents in the RCC (9.4%) reported being affected by climate change compared to the SCC (0.0%). This implies that non-coastal cities may experience economic disruptions and vulnerabilities due to climate change impacts, affecting sectors such as agriculture, infrastructure, and businesses. In comparison to SCC (0.0%), a higher percentage of respondents in RCC (7.8%) reported being impacted by social security (Table 13.2). This suggests that climate change can pose social challenges and affect community well-being in non-coastal areas, including issues related to displacement, social cohesion, and access to basic services (Roy et al. 2022; Sikder 2010).

Climate Change Severity on Socio-economic Dimensions

The severity scale of climate change is derived from a combination of vulnerability assessments and the acknowledged seriousness of threats affecting different security aspects. This strategy is consistent with the more general discussion of methods for assessing the impact of climate change (IPCC 2014; Cutter et al. 2003).

The study reveals that climate change has significant implications for professional security in non-coastal cities in Bangladesh. In terms of the RCC, 25.0% of respondents reported feeling severely unsafe, while 27.1% considered their professional security to be unsafe. Similarly, in the SCC, 5.2% felt severely unsafe, and 28.6% perceived their professional security as unsafe (Table 13.3). This discrepancy can be attributed to the specific vulnerabilities of professions in RCC, such as agriculture, which is more susceptible to climate-induced disruptions like floods and erratic weather patterns (Das 2010). The intensity and frequency of hydro-meteorological hazards in the north, where RCC is situated, contribute to the exacerbated professional insecurities (Ministry of Disaster Management and Relief 2020). These findings suggest that climate change events, such as extreme weather conditions and natural disasters, pose risks to employment opportunities, livelihoods, and occupational safety (Rahman and Gain 2019). Furthermore, it is important to note that these risks are not limited to just the two aforementioned cities. Climate change affects communities across the globe, and its impact on employment and occupational safety is a growing concern (Leisner 2020). As temperatures continue to rise and weather patterns become more unpredictable, industries such as agriculture, construction, and transportation are particularly vulnerable (Roy et al. 2022; Sikder 2010; Karmakar 2003). Additionally, the displacement of communities due to climate change-related events can lead to job loss and economic instability (Leisner 2020).

The study also demonstrates that climate change has implications for health security in non-coastal cities. In RCC, 35.4% of respondents reported feeling severely unsafe, and 39.6% perceived their health security as unsafe. In SCC, 17.7% felt severely unsafe, and 70.8% considered their health security to be unsafe (Table 13.3). The disparities in health security vulnerabilities can be linked to the unique climate-related health challenges faced by each city. For instance, RCC contends with an increased prevalence of climate-related diseases due to the frequency of floods, affecting water and sanitation conditions (Rahman and Gain 2019). These findings indicate that climate change-related factors, such as heatwaves, floods, and changing disease patterns, pose risks to public health, including increased morbidity and mortality rates. So, climate change has far-reaching impacts on public health, including the spread of waterborne illnesses and malnutrition due to changes in food production and distribution (Salvador et al. 2023). The increasing frequency and

Table 13.3 Climate change severity on socio-economic dimensions

		Professional security					Chi-square statistics (N, df) Significant value $p < 0.05$
		Severely unsafe	Unsafe	Neither safe nor unsafe	Safe	Completely safe	
RCC	n	48	52	43	41	8	34.354[a] (384, 4) 0.000
	%	25.0%	27.1%	22.4%	21.4%	4.2%	
SCC	n	10	55	76	41	10	
	%	5.2%	28.6%	39.6%	21.4%	5.2%	
	Health Security						45.845[a] (384, 4) 0.000
RCC	n	68	76	26	17	5	
	%	35.4%	39.6%	13.5%	8.9%	2.6%	
SCC	n	34	136	6	16	0	
	%	17.7%	70.8%	3.1%	8.3%	0.0%	
	Income Security						110.909[a] (384, 4) 0.000
RCC	n	73	45	46	24	4	
	%	38.0%	23.4%	24.0%	12.5%	2.1%	
SCC	n	0	29	89	64	10	
	%	0.0%	15.1%	46.4%	33.3%	5.2%	
	Economic Security						141.398[a] (384, 4) 0.000
RCC	n	81	56	31	21	3	
	%	42.2%	29.2%	16.1%	10.9%	1.6%	
SCC	n	1	119	7	65	0	
	%	0.5%	62.0%	3.6%	33.9%	0.0%	
	Social Security						53.920[a] (384, 4) 0.000
RCC	n	38	74	40	31	9	
	%	19.8%	38.5%	20.8%	16.1%	4.7%	
SCC	n	0	58	65	50	19	
	%	0.0%	30.2%	33.9%	26.0%	9.9%	

[*Source* Field Survey, 2022]

intensity of extreme weather events have taken a toll on the physical and mental well-being of the residents (Dhimal et al. 2021). Health concerns are made worse by inadequate infrastructure and access to lowering facilities, especially for vulnerable populations like the elderly, children, and those with pre-existing medical issues. In addition, irregular rainfall patterns and protracted droughts have led to a shortage of water and contaminated water sources (Dixit et al. 2022). During FGD, a school teacher states,

> Droughts disrupt the normal functioning of schools. With water scarcity, maintaining proper hygiene becomes a challenge. This impacts the health and attendance of students, creating obstacles in our efforts to provide quality education (FGD, RCC, Bangladesh).

This has contributed to the spread of waterborne diseases and compromised overall hygiene and sanitation standards. Access to clean drinking water and proper sanitation facilities is essential for maintaining good health, but it remains a significant challenge for many residents in the selected cities. However, Income security is also negatively impacted by climate change in Bangladesh's non-coastal cities. In RCC, 23.4% and 38.0% of respondents thought that their income stability was somewhat unsafe. 15.1% of SCC residents think their financial stability is insecure (Table 13.3). The higher susceptibility in RCC can be elucidated by the reliance on agriculture, an occupation profoundly affected by climate-induced events like floods and droughts (Ministry of Environment and Forests 2015). These findings suggest that the financial stability of people and households have been adversely impacted by climate change-related events such as crop failures, business interruptions, and income variations. Floods and droughts are examples of extreme weather conditions that can destroy agricultural production and cause food scarcity (Das 2010). Rising sea levels can also result in evictions and property losses, which can have long-term economic repercussions (Hossen and Sultana 2023). For vulnerable populations, such as low-income households, the effects of climate change on income security are particularly severe.

Climate change has a noticeable impact on economic security in non-coastal cities in Bangladesh. The analysis revealed that in RCC, 42.2% of respondents considered their economic security severely unsafe, and 29.2% felt it was unsafe. Similarly, in SCC, 62.0% felt their economic security was unsafe (Table 13.3). The impacts of climate change on vulnerable industries, such as agriculture and small-scale enterprises, contribute significantly to economic insecurity in RCC (Start Network Bangladesh 2019). These findings highlight the vulnerability of local economies to climate change-induced disruptions, including damage to infrastructure, agricultural productivity, and business operations. Furthermore, the economic impact of climate change is not limited to local communities. Global markets are also affected by extreme weather events and natural disasters, leading to supply chain disruptions and price volatility (Davis et al. 2021).

Climate change affects social security in non-coastal cities. The study indicates that in RCC, 19.8% of respondents reported feeling severely unsafe, and 38.5% perceived their social security as unsafe. In SCC, 30.2% felt unsafe regarding social security (Table 13.3). These differences can be traced back to the socio-demographic composition of the cities, where RCC experiences higher vulnerability due to a concentration of marginalized communities (Rahman and Gain 2019). These results suggest that climate change events can disrupt community cohesion, access to basic services, and social well-being, leading to increased vulnerabilities and social inequalities (Sesana et al. 2021).

Initiatives of Local Governing Agencies

In order to assess the governing process of Bangladesh considering climate issues, it should be measured whether the existing governance of Bangladesh is environmental or not. On the other hand, the sustainability and inclusiveness of government schemes and aid programs should be justified as well.

Table 13.4 shows how residents of two selected city corporations have reacted to natural disasters in terms of how many have been impacted by them and how many have gotten aid after or during the disaster. A cross-tabulation between the two sets of variables has been performed. Following the investigation, it appears that 58.3% of those affected by temperature need some kind of help. During KIIs, a senior officer from the Ministry of Disaster Management and Relief (MoDMR) district office states,

> We typically don't provide direct assistance to anyone having issues with high temperatures. You understand that this is an international issue. While the heat is also getting worse, the global temperature is rising quickly. Undoubtedly, those who cannot afford electricity or air conditioning will suffer. Very few homes lack power, with the exception of those who reside in severe slums. The government is working hard to guarantee an equitable supply of electricity. Additionally, during the hotter months, we want to pay special attention to patients in hospitals who have diarrhea (KII, SCC, Bangladesh).

Despite the fact that the cold wave is not a significant issue in the chosen city corporations, many of the respondents received help, primarily in the form of winter

Table 13.4 Inclusiveness in Government programs addressing the climate induced phenomenon

			Received Help after Calamities						
			Temp	Cold W	Strom	Flood	Thunder	Heavy R	Draught
Affected by calamities	Temperature	*f*	7						
		%	58.3%						
	Cold wave	*f*		0					
		%		0.0%					
	Strom	*f*			2				
		%			28.6%				
	Flood	*f*				2			
		%				33.0%			
	Thunder	*f*					–		
		%					–		
	Heavy rainfall	*f*						1	
		%						2.4%	
	Draught	*f*							–
		%							–

[*Source* Field Survey, 2022]

clothing. During a storm, 29% of people received assistance from the authorities. In the event of a flood, 33% of respondents provided unfavorable feedback, and 3% did so in the event of heavy rain, with the majority of their grievances centered on water maroon (Table 13.4). The city corporations of Bangladesh did not perceive storms and draughts as threats.

Similar calculations were made considering the livelihood aspects of the respondents and whether or not the authorities had addressed them. Only 50% of individuals who believe their food security is at risk due to climate change have received assistance, according to the tabulation. The outcome for health security is good; 100.0% of respondents have received assistance from the authorities. However, the situation is different when it comes to economic security. Only 30% of respondents who reported being affected by climate change have received any kind of support from the authorities (Table 13.5). This is concerning because many people have lost their jobs as a result of natural calamities brought on by climate change, like floods and droughts. In the long run, a lack of aid in this area can result in increased hunger and poverty. To ensure sustainable development and a better future for everybody, it is essential for authorities to address these challenges and offer sufficient support to individuals impacted by climate change. As a result, it is critical for the government to intervene and support those impacted by disasters brought on by climate change. Inaction could have negative effects on sustainable development, such as increasing food insecurity and poverty. Individuals must put the welfare of those impacted by climate change as a top priority and seek to improve everyone's future.

The RCC has implemented measures to address professional, economic, social, health, and income security in response to climate change. However, there is a significant proportion of respondents who still perceive their security in these dimensions as unsafe, indicating the need for further improvements and interventions. The SCC has also taken measures to address security concerns, but there is room for enhancing their efforts to ensure the safety and well-being of the population.

Challenges in Tackling and Mitigating Climate Issues

The research aimed to identify the major challenges faced by governing agencies in non-coastal cities in Bangladesh in tackling and mitigating climate issues. The government's limited financial resources often hinder their ability to implement effective climate change policies and programs. Additionally, many citizens were unaware of the severity of climate change and its potential impact on their daily lives. This lack of awareness made it difficult for governing agencies to mobilize public support for their initiatives.

The study demonstrates that 54.7% of respondents in the RCC reported a lack of awareness regarding climate change, while 45.3% claimed to be aware. In SCC, 48.4% of respondents identified unawareness as a challenge, whereas 51.6% reported being aware (Table 13.6). Both RCC and SCC face the challenge of unawareness, although the prevalence is slightly higher in RCC. This indicates the need for

Table 13.5 Role of authority regarding climate change

			Addressed by the authority								
			Food	Health	Income	Econo	Disaster	Health	Comm	Infras	W. & C
Affected livelihood sectors	Food security	*f*	1								
		%	50.0%								
	Health security	*f*		4							
		%		100.0%							
	Income security	*f*			1						
		%			100.0%						
	Economic security	*f*				0					
		%				0.0%					
	Disaster management	*f*					0				
		%					0.0%				
	Health sector	*f*						4			
		%						80.0%			
	Communication system	*f*							0		
		%							0.0%		
	Social Security	*f*								0	
		%								0.0%	
	Women and child	*f*									2
		%									100.0%

[*Source* Field Survey, 2022]

increased awareness campaigns and educational programs to enhance knowledge and understanding of climate change among the population in non-coastal cities. Furthermore, it is important to note that the consequences of climate change are not limited to coastal areas. In fact, inland cities are also vulnerable to the impacts of extreme weather events such as droughts and floods. Therefore, it is crucial for all cities, regardless of their proximity to the coast, to take action to mitigate and adapt to climate change.

The study emphasizes that while addressing climate change challenges, corruption appears as a major barrier for governmental authorities in both the RCC and the SCC. Only a small number of respondents in both regions did not identify any corruption, while the majority categorically classified corruption as a significant concern. It is helpful to consult the most recent corruption perceptions indicator, such as the World Corruption indicator, to support this statement. Bangladesh's Corruption Perceptions Index in 2022 has a score of 25, which is representative of the country's current level of corruption. According to data from the World Corruption Index, its score increased from 13 in 2003 to 25 in 2022, expanding at an average yearly rate of 3.76% (Transparency International 2019).

The high rates of corruption in both the RCC and SCC highlight how pervasive corruption is and how urgently robust anti-corruption measures must be put in place. The available research indicates that combating corruption is essential to promoting

Table 13.6 Major challenges faced by governing agencies in non-coastal cities

Variables		RCC		SCC	
		f	%	f	%
Unawareness	Yes	105	54.7%	93	48.4%
	No	87	45.3%	99	51.6%
Corruption	Yes	153	79.7%	165	85.9%
	No	39	20.3%	27	14.1%
Poverty	Yes	74	38.5%	58	30.2%
	No	118	61.5%	134	69.8%
Political Issue	Yes	129	67.2%	82	42.7%
	No	63	32.8%	110	57.3%
Religious	Yes	10	5.2%	1	0.5%
	No	182	94.8%	191	99.5%
Apathy of the authority	Yes	99	51.6%	103	53.6%
	No	93	48.4%	89	46.4%
Lack of international aid	Yes	38	19.8%	80	41.7%
	No	154	80.2%	112	58.3%
Large Population	Yes	54	28.1%	56	29.2%
	No	138	71.9%	136	70.8%

[*Source* Field Survey, 2022]

good governance and guaranteeing the effective execution of adaptation and mitigation strategies for climate change. The widespread influence of corruption presents a significant barrier to advancement in both areas. Proactive action is desperately needed to address corruption, as indicated by the high proportion of respondents who cited it as a major challenge. During KIIs, a government official says,

> When it comes to climate change policies and initiatives, corruption poses serious problems for us. Our efforts to address the effects of climate change and create a resilient and sustainable city are significantly hampered by corruption. One of the primary concerns is the misallocation and embezzlement of funds meant for climate change adaptation and mitigation projects. Corruption diverts resources away from essential initiatives, resulting in inadequate funding and implementation. Another form of corruption that affects climate change policies is bribery and nepotism in decision-making processes. Unethical practices and favoritism can lead to the selection of projects and policies that benefit certain individuals or groups rather than addressing the pressing climate change challenges (KII, RCC, Bangladesh).

The study also revealed that a significant portion of respondents in the RCC and SCC regarded poverty as a significant obstacle in light of climate change. Thirty-two percent of respondents in SCC and 38.5% of respondents in RCC agreed that poverty was an urgent problem. Interestingly, though, most respondents in both regions—61.5% in RCC and 69.8% in SCC—did not think poverty posed a serious threat to climate change. It is relevant to consult the most recent poverty index, such as the World Multi-dimensional Poverty Index, in order to have a deeper understanding of the state of poverty. The most recent survey data available for Bangladesh's MPI estimation in 2019 indicates that 24.6% of the population is multidimensional poor, or 41,730 thousand people in 2021, and an additional 18.2% is vulnerable to multidimensional poverty, or 30,834 thousand people in 2021 (Multidimensional Poverty Index 2023). While poverty is recognized as a significant issue in both the RCC and SCC, the new data highlights the complex aspects of poverty in relation to climate change. For a more thorough knowledge of the socioeconomic difficulties entwined with climatic issues in the locations under study, it is imperative to take these numbers into account.

According to this study, in RCC, 24.8% of respondents reported inadequate access to healthcare as a challenge, while 20.3% identified inadequate access to education. In SCC, 19.5% of respondents reported inadequate access to healthcare as a challenge, while 16.9% identified inadequate access to education (Table 13.6). Additionally, both RCC and SCC face challenges related to environmental degradation and natural disasters.

RCC cities are more affected by political difficulties than SCC cities are. For resource allocation, decision-making, and the successful implementation of climate change policies and programs, these political obstacles must be removed. The study emphasizes how political obstacles have a big impact on climate change initiatives and policies. One interesting conclusion is that over two-thirds of respondents identified political obstacles as more common in RCC. This unexpected discovery suggests that political concerns significantly obstruct efficient decision-making and resource distribution in RCC cities. It is relevant to consult the most recent World Political

Stability Index in order to offer a modern viewpoint on political stability. Bangladesh is currently ranked 172nd out of 171 countries, with an index of −1.15 (Rahman and Rashid 2018). The difficulties caused by Bangladesh's political unrest are highlighted by this indicator. In contrast, the study finds that SCC appears to be less impacted by political impediments, with just over 40% of respondents mentioning such difficulties. The incorporation of the most recent World Political Stability Index enhances our comprehension of the current political terrain, illuminating the distinct obstacles presented by political instability in Bangladesh and its possible consequences for the governance of climate change.

However, it is important to note that even in SCC, nearly half of the respondents reported no political challenges. This suggests that there is still room for improvement in terms of creating a conducive environment for climate change policies and initiatives. During KIIs, a government official argues,

> We are facing significant political challenges when it comes to climate change policies and initiatives. The lack of political will and commitment to address climate change is hindering our progress in mitigating its effects and building a sustainable future. One of the key challenges we face is the limited prioritization of climate change issues by our governing agencies. Climate change is a global crisis that requires urgent attention, yet political interests and other pressing issues often take precedence over environmental concerns. This lack of prioritization leads to inadequate funding, resources, and policy measures to effectively tackle climate change at the local level (KII, SCC, Bangladesh).

The study also suggests that religious challenges regarding climate change are not a significant concern for the majority of respondents in both the RCC and SCC. Religious challenges do not appear to be significant concerns in either the RCC or SCC, indicating that religious factors are less likely to hinder climate change initiatives. However, cultural and religious sensitivities should still be considered while formulating and implementing strategies to ensure inclusivity and community engagement. It is encouraging to see that religious challenges are not a major obstacle in the fight against climate change in both the RCC and SCC.

The study also demonstrates that a significant portion of respondents in both the RCC and SCC identified the apathy of the authorities as a challenge for climate change. In RCC, 51.6% of respondents acknowledged this challenge, while in SCC, 53.6% of respondents recognized the apathy of the authority. On the other hand, a minority of respondents in both regions reported no such challenge, with 48.4% in RCC and 46.4% in SCC stating that the apathy of the authority is not a challenge for climate change (Table 13.6). Overall, it can be inferred that a significant number of respondents in both regions perceive the apathy of the authorities as a hindrance to addressing climate change. The presence of apathy among governing authorities is a shared challenge in both the RCC and SCC. This highlights the importance of fostering a sense of responsibility and urgency among governing agencies to prioritize climate change issues and take proactive measures. Furthermore, engaging local stakeholders and communities is crucial to spreading awareness of the effects of climate change and the need for action. Nonetheless, a significant difference was noted in SCC, as a greater proportion of participants recognized the absence of foreign assistance as a more significant obstacle in contrast to RCC. This suggests that the SCC will have greater difficulty obtaining foreign assistance to combat climate change. To the extent that governing agencies in non-coastal cities confront obstacles, it is imperative that they enhance their international collaboration and obtain sufficient support. Based on the most recent World Bank data, Bangladesh was provided with significant foreign aid totaling 5.04 billion US dollars for development assistance as of 2021 (World Bank 2021). The study's conclusions, however, emphasize that despite this foreign assistance, a real problem still exists, especially in SCC, underscoring the necessity for sophisticated approaches to guarantee that international aid is used effectively in addressing the climate change issues that governing bodies in non-coastal cities must deal with.

The comparative analysis highlights several major challenges faced by governing agencies in non-coastal cities of Bangladesh, including unawareness, corruption, poverty, political barriers, apathy of the authority, a lack of international aid, and population-related issues. These findings underscore the need for targeted interventions, policy reforms, capacity building, and increased collaboration among various stakeholders to overcome these challenges and ensure effective climate change governance in non-coastal areas. Due to Bangladesh's high population density, subpar infrastructure, and limited capacity for adaptation, its urban population is especially vulnerable to climate change.

However, Bangladesh was identified as being particularly vulnerable to the effects of climate change, the People's Republic of Bangladesh's government began a number of plans and projects to ensure sustainable and inclusive measures to prevent and mitigate the effects of climate change, as well as building capacity to adapt to the changing climate, taking into consideration all the issues. Programs like Low Carbon Development (LCD), Bangladesh Climate Change Strategy Action Plan (BCCSAP), and National Sustainable Development Strategy (NSDS) have been modified as a consequence of combining the efforts of all stakeholders. The climate change crisis is given high attention in this program, which also contains a forceful call to action.

These initiatives have so far made good progress. The Prime Minister herself established an environmental council, which now has members at the divisional, district, and Upazilla levels. The laws and regulations governing the environment have been updated and modified. The capabilities of the Department of Environment (DoE) have increased. The government has primarily implemented poverty-sensitive programs to shield the most vulnerable populations from the worst types of natural disasters. The government is also working to guarantee that the victims can take advantage of all international rights. A campaign against less persistent pollutants is currently underway. 17 clean development mechanisms have already received national government approval for the entire country. Overall, the government is taking steps to address environmental concerns and protect vulnerable populations from natural disasters. In addition to implementing poverty-sensitive programs, they are also working to ensure that victims have access to international rights. The campaign against less persistent pollutants and the approval of Clean Development Mechanisms demonstrate a commitment to sustainable development.

The discussion above sheds light on a few issues, including the fact that Bangladesh's urban areas are feeling the effects of climate change less quickly than its coastal areas. The government is not giving citizens' concerns about climate change and its effects the proper attention. The governing bodies do not, however, feel compelled to alter the way they are currently handling environmental governance. The current policies implemented by the governing bodies are, above all, not at all long-lasting, inclusive, environmentally friendly, or SDG-focused. During KIIs, the deputy director of the Sylhet Environmental Office has the exact opposite perspective. He argued that when it comes to the environment, we don't always think realistically or practically. People are all too familiar with how fetal a brick kiln can be, for instance. But what if we needed to convince a business owner who had made a sizable financial investment to remove or destroy his infrastructure? It is quite harsh to think that a country that is already experiencing many problems could suddenly have its entire system upended by the UN's novel suggestion. By the way, we are working extremely hard. We no longer dig for stones, though it was unthinkable a few years ago. Since it will take time, I believe things will eventually improve.

Even though it has been acknowledged as one of the greatest threats facing all of humanity, certain parts of Bangladesh do not treat it seriously. Moreover, the people of Bangladesh should be seriously conscious of this climate issue. The traditional model of government has barely been modified. Despite the introduction of digitalization in governance, environmental concerns still overlap. People do not receive adequate education or guidance from the authorities on how to lead innovative and energy-efficient lives on a daily basis. However, there aren't many government initiatives to produce energy or water supplies that could mitigate the effects of climate change. The sustainability of government facilities is still in jeopardy, which is the most important point. This lack of action from the government has led to a rise in public awareness and activism towards environmental issues. People are taking matters into their own hands by implementing sustainable practices in their daily lives, such as reducing waste and using renewable energy sources. However, more needs to be done on a larger scale to address the root causes of climate change. Governments must prioritize the development of renewable energy sources and invest in infrastructure that supports sustainable living. Education on environmental issues should also be made a priority, so that people can make informed decisions about their impact on the planet. Ultimately, it is up to individuals, businesses, and governments to work together towards a more sustainable future for our planet.

Conclusion

This chapter has shed light on the socio-economic impacts of climate change in non-coastal cities in Bangladesh. The findings and analysis have provided valuable insights into the severity of climate change on various dimensions, including professional security, health security, income security, economic security, and social security. The chapter has also highlighted the severity of different natural calamities experienced in these cities. Furthermore, the study has identified key challenges faced by governing agencies in addressing climate change issues. These challenges include unawareness, corruption, poverty, political factors, religious factors, apathy of the authority, a lack of international aid, and population growth. These challenges pose significant barriers to effective climate change adaptation and mitigation efforts. To address these challenges, the study recommends a multi-faceted approach. There are several critical needs that must be addressed to tackle the challenges faced by cities. First and foremost, anti-corruption measures and initiatives must be implemented to alleviate poverty. These measures should be tailored to the specific socio-economic context of each city. Secondly, political challenges can be addressed by promoting transparency, accountability, and inclusive decision-making processes. Strategies that encourage these values should be put in place. Thirdly, it is important to increase awareness and understanding of climate change among the general population through targeted awareness campaigns, educational programs, and information dissemination. This can help build a knowledgeable and engaged community that actively participates in climate change adaptation and mitigation efforts. In addition to this, local communities must be empowered by involving them in decision-making processes, project planning, and implementation. Community-based initiatives, knowledge-sharing platforms, and capacity-building programs must be encouraged to enhance local resilience and adaptive capacities.

In light of the Sustainable Development Goals (SDGs), these recommendations align with Goal 16 (Peace, Justice, and Strong Institutions) by advocating for corruption-free governance, and Goal 1 (No Poverty) by addressing poverty-related challenges. Furthermore, Goal 13 (Climate Action) is central to the study's focus, emphasizing the urgency of climate change adaptation and mitigation measures. In essence, this research contributes to the ongoing discourse on climate change resilience in non-coastal cities, providing insights that can inform policy decisions and interventions. The localized nature of the study allows for a targeted and context-specific approach, essential for crafting effective strategies to enhance the socio-economic well-being of communities in the face of climate change challenges.

References

Agrawala S, Van Aalst M (2008) Adapting development cooperation to adapt to climate change. Climate Policy 8:183–193

Aryal JP, Sapkota TB, Khurana R, Khatri-Chhetri A, Rahut DB, Jat ML (2020) Climate change and agriculture in South Asia: adaptation options in smallholder production systems. Environ Dev Sustain 22(6):5045–5075

Bajaj S, Kumria R (2023) Crop improvement under climate change. In: Global climate change and plant stress management, pp 9–12

Bhandari MP (2021) Climate change impacts on agriculture, a case study of Bangladesh, India, Nepal, and Pakistan

Celik S (2020) The effects of climate change on human behaviors. In: Environment, climate, plant and vegetation growth, pp 577–589

Chaudhary MT, Piracha A (2021) Natural disasters—origins, impacts, management. Encyclopedia 1(4):1101–1131

Clemett A, Amin MM, Ara S, Akan MMR (2006) Background information for Rajshahi City, Bangladesh. WASPA Asia Project Rep 2:1–29

Climate Change Risk Index (2021) Global climate risk index 2021. https://www.germanwatch.org/en/19777

Das PK (2010) Climate change and education, Bangladesh. UKaid. https://www.unisdr.org/fles/16355_climatechangeedbangladesh.pdf

Davis KF, Downs S, Gephart JA (2021) Towards food supply chain resilience to environmental shocks. Nat Food 2(1):54–65

Debsarma SK (2003) Intra-annual and inter-annual variation of rainfall over different regions of Bangladesh. In: Proceedings of SAARC seminar on climate variability in the South Asian Region and its impacts, SMRC, 20–24 Dec 2002, Dhaka, Bangladesh. SAARC Meteorological Research Centre

Dhimal M, Bhandari D, Dhimal ML, Kafle N, Pyakurel P, Mahotra N, Müller R et al (2021) Impact of climate change on health and well-being of people in Hindu Kush Himalayan region: a narrative review. Front Physiol 12:651189

Dixit A, Madhav S, Mishra R, Srivastav AL, Garg P (2022) Impact of climate change on water resources, challenges and mitigation strategies to achieve sustainable development goals. Arab J Geosci 15(14):1296

Glennon R (2018) The unfolding tragedy of climate change in Bangladesh. The New Nation. https://thedailynewnation.com/news/181928/The-unfolding-tragedy-of-climate-change-in-Bangladesh

Hossain MS, Arshad M, Qian L, Kächele H, Khan I, Islam MDI, Mahboob MG (2020) Climate change impacts on farmland value in Bangladesh. Ecol Ind 112:106181

Hossen MF, Sultana N (2023) Shoreline change detection using DSAS technique: case of Saint Martin Island, Bangladesh. Remote Sens Appl Soc Environ 30:100943

IPCC (2007) Fourth assessment report. November. New York, United Nations Environment Program (UNEP), 368, 397

Isfat M, Raihan A (2022) Current practices, challenges, and future directions of climate change adaptation in Bangladesh. Journal homepage: www.ijrpr.com ISSN, 2582, 7421

Jamat MK (2010) Convener of Rajshahi protection committee, Article on the Daily Prothom Alo, Jolobayu Poribortoner Shikar Brihottoro Rajshahi Onchol (the great Rajshahi Region is on the grasp of Climate Change, published date: 14 Jan 2010

Kabir MH, Nawaz S, Hossen MN, Hossain ML, Habiba SU (2023) Impacts of floods on infrastructures in developing countries. Handbook of flood risk management in developing countries, p 2

Karmakar S, Shrestha ML (2000) Recent climate change in Bangladesh. SMRC Series, no 4. Dhaka, SAARC Meteorological Research Centre

Karmakar S (2003) Trends in the annual frequency of cyclonic disturbances and storms in the Bay of Bengal. In: Proceedings of SAARC seminar on climate variability in the South Asian Region and its impacts, SMRC, 20–24 Dec 2002, Dhaka, Bangladesh. SAARC Meteorological Research Centre

Karmakar S, Nessa J (1997) Climate change and its impacts on natural disasters and south-west monsoon in Bangladesh and the Bay of Bengal. J Bangladesh Acad Sci 21(2):127–136

Kumari B, Solanki H, Kumar A (2020) Climate change: a burning issue for the world. Medicine 35:501–507

Leisner CP (2020) Climate change impacts on food security-focus on perennial cropping systems and nutritional value. Plant Sci 293:110412

Mahmood SAI (2012) Impact of climate change in Bangladesh: the role of public administration and government's integrity. J Ecol Nat Environ 4(8):223–240

Meyer N, Auriacombe C (2019) Good urban governance and city resilience: an afrocentric approach to sustainable development. Sustainability 11(19):5514

Ministry of Disaster Management and Relief (2020) National Plan for Disaster Management (2021–2025). https://modmr.portal.gov.bd/sites/default/fles/fles/

Mojid MA (2020) Climate change-induced challenges to sustainable development in Bangladesh. In: IOP conference series: earth and environmental science, vol 423, no 1. IOP Publishing, p 012001

Multidimensional Poverty Index (2023) United Nations Development Programme. https://hdr.undp.org/sites/default/files/Country-Profiles/MPI/BGD.pdf

Parven A, Pal I, Witayangkurn A, Pramanik M, Nagai M, Miyazaki H, Wuthisakkaroon C (2022) Impacts of disaster and land-use change on food security and adaptation: evidence from the delta community in Bangladesh. Int J Disaster Risk Reduction 78:103119

Rahman MS, Gain A (2020) Adaptation to river bank erosion induced displacement in Koyra Upazila of Bangladesh. Progr Disaster Sci 5:100055

Rahman R, Rashid MM (2018) Political instability and economic growth in Bangladesh. In: Peer-reviewed academic journal innovative issues and approaches in social sciences

Rahman SM, Mori A, Rahman SM (2022) How does climate adaptation co-benefits help scale-up solar-powered irrigation? A case of the Barind Tract, Bangladesh. Renew Energy 182:1039–1048

Roy S, Tandukar S, Bhattarai U (2022) Gender, climate change adaptation, and cultural sustainability: insights from Bangladesh. Front Climate 4:841488

Salehin A, Haque M (2019) Climate change in Bangladesh: effect versus awareness of the local agencies of Rajshahi City. Rajshahi Univ Soc Sci J 22:119–140

Salvador C, Nieto R, Vicente-Serrano SM, García-Herrera R, Gimeno L, Vicedo-Cabrera AM (2023) Public health implications of drought in a climate change context: a critical review. Annu Rev Public Health 44:213–232

Seddiky DMA, Ara DE, Karim MA (2023) Climate change-induced hazard risks and migration in Bangladesh: a case study. Available at SSRN 4423261. https://doi.org/10.2139/ssrn.4423261

Sesana E, Gagnon AS, Ciantelli C, Cassar J, Hughes JJ (2021) Climate change impacts on cultural heritage: a literature review. Wiley Interdisc Rev Climate Change 12(4):e710

Siddique A (2021) Forest governance transitions in Bangladesh. Forest, 06

Sikder MT (2010) The impacts of climate change on the coastal belt of Bangladesh: an investigation of risks & adaptations on agricultural sector. In: Proceedings of international conference on environmental aspects of Bangladesh, Japan, September 2010. Hokkaido University, Sapporo, pp 26–28

Spillan JE, Rahman M (2023) The political climate in Bangladesh. Doing business in Bangladesh and Sri Lanka: challenges and opportunities. Springer International Publishing, Cham, pp 109–130

Transparency International (2019) Bangladesh. Transparency. https://www.transparency.org/en/countries/bangladesh

Waheduzzaman M, Mizanuzzaman M (2021) Impact of salinity on the livelihoods of the coastal people in Bangladesh: an assessment on post Amphan situation. J Glob Ecol Environ 11(4):6–24

Warrick RA, Bhuiya AH, Mirza MQ (1994) The greenhouse effect and climate change: briefing document No. 1. Dhaka, Bangladesh Unnayan Parishod (BUP), pp 17–20

World Bank (2021) World Bank Open Data. https://data.worldbank.org/indicator/DT.ODA.ODAT.CD?end=2021&locations=BD&start=1971&view=chart

Chapter 14
Examining Mediating Effect of Climate Change Worry Between Climate Change Anxiety and Divergent Migration: A Scale to Measure Climate-Induced Post-migration Anxiety and Stress

Sarthak Dash, Sugyanta Priyadarshini, and Sukanta Chandra Swain

Abstract The agricultural sector is predominantly a climate-sensitive sector. Sadly, the extreme climatic variation in the recent years has forced farmers to migrate resulting in non-marginal cognitive adjustments due to economic losses, displacement and lack of adaptability to climate shocks in post-migration location. The current study aims at analysing the mediating role of climate change worry (CCW) between climate change anxiety (CCA) and divergent migration (DM) among farmers of cyclone-prone Ganjam district, Odisha. Further, this chapter intends to develop a novel single scale in order to map the extent of anxiety and stress among the migrated farmers in the post-migration phase. Based on the prior literature, this study has performed a path analysis keeping CCW as a mediator between CCA and DM. The hypothesis is tested using AMOS 23 on a data set of 552 farmers in pre-migration stage. Further, in developing 28-item Climate-Induced Post-Migration Anxiety-Stress (CIPMAS) scale, factor analysis including exploratory and confirmatory factor analysis has been used to investigate the factorial structure, construct reliability, and model adequacy for a sample of additional 510 migrated farmers. The results indicate that CCW has a significant influence on CCA anxiety and DM. This study can provide micro-evidence of the contribution of CCA and CCW to farmer's migration, which will benefit analyses of climate-induced migration in other developing nations with greater agricultural dependence. In addition, valuable interpretations derived from CIPMAS scale can help in policy implication for mapping farmer's cognitive vulnerability to condense the extent of anxiety and worry in post-migration period.

S. Dash (✉) · S. Priyadarshini · S. C. Swain
School of Humanities, KIIT Deemed to be University, Bhubaneshwar, India
e-mail: 2181018@kiit.ac.in

S. Priyadarshini
e-mail: sugyanta.priyadarshini@kiit.ac.in

S. C. Swain
e-mail: sukanta.swainfhu@kiit.ac.in

U. Mukhopadhyay et al. (eds.), *Climate Crisis, Social Responses and Sustainability*, Climate Change Management, https://doi.org/10.1007/978-3-031-58261-5_14

Introduction

Climate change is one of the greatest furies for the agricultural sector and a major setback for its development. The farming sector is vulnerable to climatic changes as it is heavily dependent on climate patterns for its growth and thus, the impact of climatic changes has a direct impression on agricultural yield (Ochieng et al. 2016). However, it is evident that the climatic distortions mostly remain unmitigated by the farmers and cultivators resulting in change in occupation, increased poverty level, development of suicidal ideation, and migration (Nguyen and Sean 2021). Thus, migration in case of farmers is considered to be a functional alternative to natural distortions (environmental degradation, natural calamities, and diseases), socio-cultural changes (change in population, communication, knowledge) and economic and technical factors (loss of livelihood, change in technique of production) (Raitzer et al. 2015). Farmers often migrate due to the adverse impact of climatic variations on agricultural productivity leading to lack of adequate crop production and loss of livelihood (Falco et al. 2019). Further, due to lack of adequate knowledge over mitigation or adaptive measures, the adverse impact of climatic distortions ruins the dependency of rural farmers on agricultural sector (Alam et al. 2017) pushing them to migrate in order to seek alternative livelihood as a way of adaptation (McLeman 2013a).

It has been recognized that variations in climate have induced migration among rural farmers as migration has turned out to be an adaption strategy or consequence of failure to adapt (Tacoli 2009). However, the entire phenomenon of migration due to climatic transformation and pressure creates climatic worry and anxiety among rural farmers for resettlement. The expected future extreme climatic distortions influence the adaptive and risk perception behaviour (worry and anxiety) of rural farmers driving them to migrate. Recent empirical investigations into the relationship between climate change worry (Searle and Gow 2010; Sundblad et al. 2007), climate change anxiety (Pihkala 2020; Doherty and Clayton 2011) and divergent migration (Garip 2008) have dragged attention towards the a clearly delineated and inevitable link between them.

In order to avert the catastrophic effects of climate change, adaptation is considered to be a significant policy option, and migration is one of those possible adaptation strategies (Nordås and Gleditsch 2007; Tacoli 2009; Scheffran et al. 2012). The agricultural sector is majorly affected by climate change, and as a result, it is more likely to experience the highest risk and uncertainty resulting in loss of livelihood. In this context, Stojanov et al. (2016a, b) discovered that farmers initially discover changes in climatic conditions and then decide to relocate as a means of coping with these changes. In the agricultural sector, farmers are the main performers, and their perception regarding the change in climatic conditions which hampers their livelihood can act as the major driving force behind migrating to other places. In response to the climate-related repercussions, farmers often choose to relocate in order to find alternate means of subsistence as they are constantly under danger of crop failure or low crop yield.

Taking into consideration this theoretical foundation, it is necessary to investigate the relationship between migration and climate change in a manner that is more context-specific by incorporating several psychological factors. Furthermore, there is a pressing need to uncover the ways in which psychological factors such as worry and anxiety about climate change have impacted migration decisions in order to establish proper regional policies to cope with problems associated with migration. In addition, this research aims to establish a novel scale that may be used to quantify the post-migration stress experienced by farmers who are willing to migrate.

Significantly, these analyses have highlighted the consideration of migration as an array of potential responses to environmental change, and as being inextricably linked to other psychological dynamics. While these studies have provided considerable insight into the relationship between psychological impact of migration due to climate change, they have tended to rely on global sectors, while perceptions from marginalised sector such as "agricultural sector" have remained untouched. In an attempt to address this shortcoming, the current study has made an attempt to understand the impact of climate change distortions on psychological behaviour of migrating farmers of cyclone-prone Ganjam district, Odisha by

(i) analysing the mediating role of climate change worry (CCW) between climate change anxiety (CCA) and divergent migration (DM)
(ii) developing a novel single scale in order to map the extent of anxiety and stress among the migrated farmers in the post-migration phase

Conceptual Framework

Relationship Between Climate Change Worry and Divergent Migration

The concept of 'climate change worry' has emerged from the massive research on psychological impact of climate change (Searle and Gow 2010; Sundblad et al. 2007), thereby revolving around the theme of chronic fear of environmental doom (Gregersen et al. 2020). Climate change worry involves trepidation about unprecedented climatic negative event associated with tension, nervousness, irritability and stress (Stewart 2021; Sundblad et al. 2007). From few of these investigations, the degree of concern of people about climate change was reported (Fetzek and Mazo 2014) which evidentially demonstrated that climate change worry is an additional driver of divergent migration, both in terms of interior and cross-borders. Divergent migration is the functional alternative to social change ranging from socio-cultural, economic and environmental factors (Garip 2008). Unfortunately, it is dejecting to note that the sector that employs majority of workers in India is the most vulnerable segment to climate change i.e., agricultural sector. In context of climatic devastation comprising of climate uncertainties affecting quality of soil, quality and quantity of water, and yield of crop (Nguyen and Sean 2021), typical seasonal changes, and natural climatic events create CCW in farmers pushing them for rural–urban migration (McLeman 2013a). When livelihood of farmers is compromised and their survival is at stake, they prefer migrating in search of better opportunities. The anticipation of negative impact due to climatic changes results in CCW, thereby making migration a necessity, not a choice. The potential positive association between climate change worry and divergent migration has become noticeable and with supportive arguments it can be hypothesised that

H1: Climate change worry is directly influencing divergent migration

Relationship Between Climate Change Worry and Climate Change Anxiety

In general, worry is considered as a key feature of anxiety issues (Weems 2008) tending to manifest as future-focused thought (McEvoy et al. 2013). The concept of 'climate change anxiety' has been obtained from the investigation on eco-anxiety and the psychological impact of climate change (Pihkala 2020; Doherty and Clayton 2011). Eco-anxiety is conceptualized as a long-term terror of ecological doom. Researchers have defined climate change anxiety as a type of anxiety disorder which is strongly associated to anthropogenic climate change (Pihkala 2020) or a type of anxiety that revolves around the perception derived from the concern among the people, even among those who are not directly impacted from the climate change (Clayton 2020). Climate change worry resulting from adverse climatic distortions creates climate change anxiety among the people who are specifically dependent on agriculture for their livelihood (Daghagh Yazd et al. 2019). Additionally, few scholars have already investigated that the observed pattern of climate change distortions have already worsened the worries of the farmers contributing to the chronic anxiety disorder regarding the upcoming unfavourable distortions to be caused from climatic changes (Ellis and Albrecht 2017; Berry et al. 2018). The worry of crisis generates anxiousness among farmers regarding financial catastrophe due to probable crop failure and reduced market prices for untimely delivery originating from disruptive climatic regulations (Rudolphi et al. 2020). The current analysis views climate change worry as a significant influencer of climate change anxiety and with supportive arguments of assessing the association between Climate change worry and climate change anxiety it can be hypothesised that

H2: Climate change worry is directly influencing climate change anxiety

Relationship Between Climate Change Anxiety and Divergent Migration

In the context of climate change, divergent migration is universally observed as an adaptation failure (Avis 2017). Prior research made it evidential that divergent migration is an adaptive step undertaken by people who have developed anxiety due to adversely impacting climatic changes (de Sherbinin et al. 2022). Agricultural sector is mostly vulnerable to climate change and most of the farmers are likely unable to mitigate the climate change anxiety induced by climate change and hence migrate due to the risk of livelihood loss (Nguyen and Sean 2021). The anxiety developed in farmers due to climate change trigger them to migrate that changes the rural landscape due to rural–urban migration. However, the CCA not only affects the behaviour of the farmers but also their migration patterns induced from climatic distortions such as droughts, floods, sea level rise, heatwaves, other natural disasters and food insecurity (McLeman 2013b). Threatening their household livelihood loss by adversely impacting agricultural yield, farmers are vulnerable to climate change anxiety, thereby forcing them to migrate for seeking an alternate livelihood as no other option of adaptation (Falco et al. 2019; Alam et al. 2017). Hence, with these supportive arguments it can be hypothesised that

H3: Climate change anxiety is directly influencing divergent migration

Relationship Between Climate Change Anxiety, Climate Change Worry and Divergent Migration

Climate change worry is fundamentally a concern about change in climate resulting in climate change anxiety, thereby manifesting intrusive thoughts and feelings of anxiousness about future time disasters, including one's own descendants (Innocenti et al. 2023). Worrying about change in climate is not a matter of concern as it can bring favourable changes by growing a sense of concern among humans about the environment, but it can only be problematic when the worry related to climate change creates anxiety among people as their livelihood and life gets adversely affected resulting in overwhelming and debilitating thoughts about future time (Goodwin et al. 2017). Changing climatic patterns disrupts the lives of the farmers who are predominantly dependent on agriculture for their livelihood, as they are not acquainted to practice agriculture in intensive rainfall shower during the rainy season and frequent droughts throughout the dry time of the year resulting in significant crop yield reduction and hence, are subjected to CCA (Nguyen and Sean 2021).The supportive arguments believe that the farmer's experience of climatic disruptions results in climate change worry resulting in climate change anxiety, thereby ultimately triggering their migration motivation. Hence, it can be hypothesised that

H4: Climate change anxiety acts as a mediator between climate change worry and divergent migration

Rationale of the Study

The study attempts to establish the nexus between climate change and its impact on mental health of farmers leading to migration. From the existing literature, it is found that climate change worry and climate change anxiety are the significant drivers of divergent migration, both in terms of interior and cross-borders (Fetzek and Mazo 2014). In order to analyze the relationship, this study considered Ganjam as the research area as it is the most cyclonic-prone area and is evident for the highest out-migration (Rajakumari et al. 2022; Mishra and Sundaray 2021). Ganjam is a

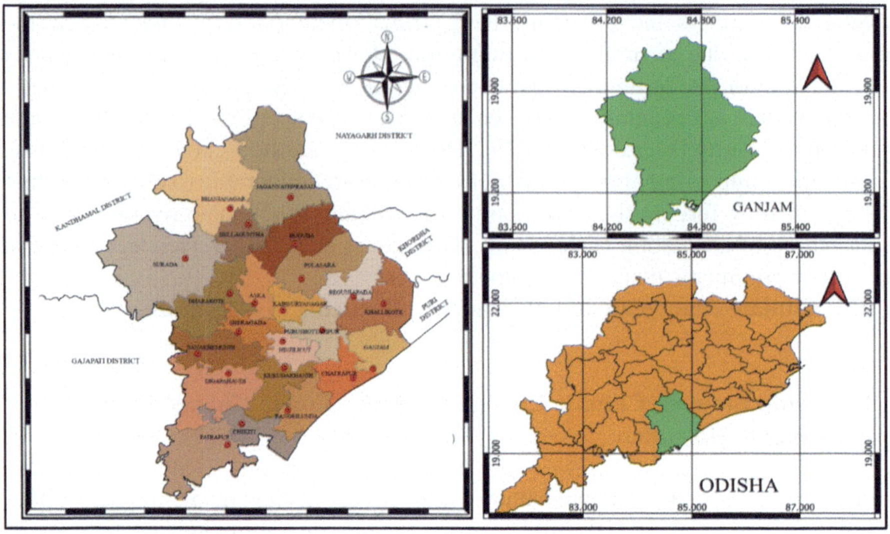

Fig. 14.1 Study area map. *Source* Computed by Authors

coastal district of Odisha and in the last 10 years, more than 10 cyclones have hit this coastal district and have tragically claimed a large number of human lives and their livelihood. Due to this, many farmers are migrating to other places in search of better resettlement. Hence, this calls for an immediate attention to determine how climate change impacts the mental health of the farmers and is ultimately leading to out-migration or the development of migration intention among them. However, for the current study the four blocks of Ganjam district nearer to the coastal line named Ganjam, Chhatrapur, Chikiti, and Rangeilunda are taken into consideration (Fig. 14.1).

Materials and Methods

Data Collection

The current study aims to analyze the impact of CCW on the relationship between CCA and DM. Further, the study aims to formulate a scale that can measure migration-generated stress and anxiety. So, the data is collected from two targeted population spectrums of the Ganjam district. Ganjam district is taken into consideration through purposive sampling as it is witnessed as one of the highest natural disaster-prone districts and the highest out-migration zone of Odisha (Rajakumari et al. 2022; Mishra and Sundaray 2021). Further, four blocks named Ganjam (n = 254), Chhatrapur (n = 262), Chikiti (n = 275), and Rangeilunda (n = 271) are

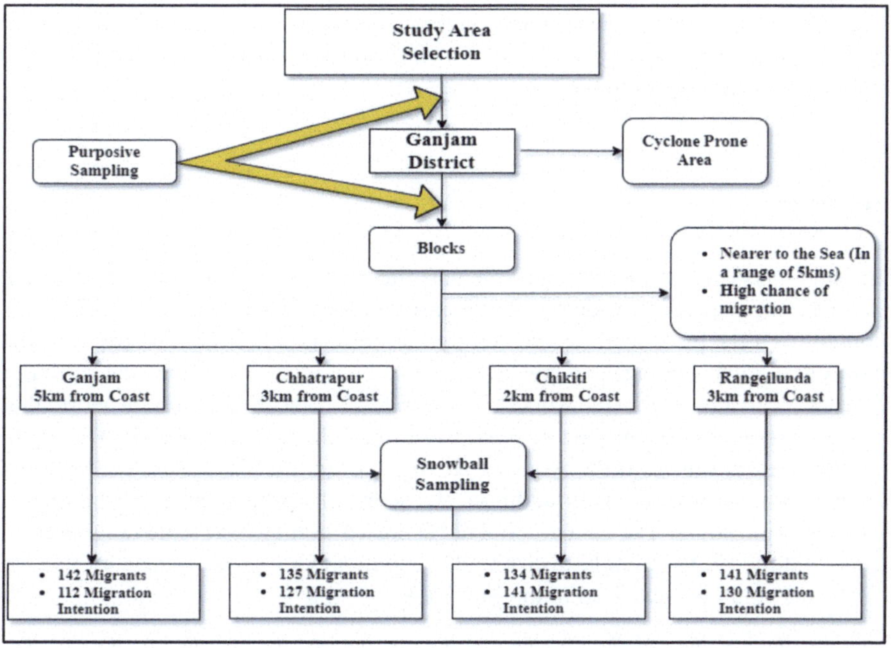

Fig. 14.2 Framework for study area selection. *Source* Computed by Authors

considered (Fig. 14.2) based on their distance from the coastline. These four blocks are under 5 km from the coastal line hence they experienced most of the cyclonic impact in the past.

Further, in the second phase, the data is collected via face-to-face interviews and telephonic interviews, targeting migrants or families of migrants through exponential non-discriminative snowball sampling where each referral has guided the investigator with many more potential research samples (Makwana et al. 2023). For the sample selection technique, the exponential non-discriminative snowball sampling method was used as the members of the research were difficult to locate and there is no trace of secondary source of record regarding the migrated farmers and farmers intending to migrate in the given study area. Consequently, the research investigators could reach 510 migrants and 552 farmers intending to migrate in the near future from the aforementioned four blocks. The demographic details from the collected sample are tabulated in Table 14.4.

Before the data collection, a structured close-ended questionnaire is designed and submitted to practitioners and subject experts. Adjustments to the structure and content of the questionnaire were made accordingly after a thorough evaluation of the comprehensiveness, clarity, and reliability of the questionnaire. Further, a two-phase pilot survey is conducted comprising of 57 and 63 samples respectively, and hence it is observed that there is no ambiguity. The pilot survey provided a clear picture regarding the feasibility of the study protocol and the expected time frame for data collection which was considered satisfactory. Based on the successful pilot study, the final survey is conducted and a total of 552 and 510 data is collected from two different targeted populations.

Measures

This study has extensively reviewed the past literature in order to conceptualize the theoretical constructs considered in the current study. The attributes of CCA are adopted from the prior work conducted by Larionow et al. (2022) that explains the anxiety that occurs among the population due to climate change. Further, the attributes of the CCW are taken from the works of Stewart (2021) and it measures the worry that develops among the population affected by climate change. The items of Divergent Migration are taken from the work of Hoong and Soon (2012) which talks about the intention for migration. In addition, the work of Malm et al. (2020) explains post-migration stress. The data is collected using a 5-point Likert scale ranging from strongly agree (5) to strongly disagree (1). The methodological framework used in the current study is picturized in the Fig. 14.3.

Data Analysis

The present investigation makes use of AMOS.23 for hypothesis and model fitness testing. Before diving into the data given in below Table 14.1, SPSS 26 is used to check for normality through skewness and kurtosis. The result indicates that the data is normal since both the skewness and kurtosis values are less than the threshold of $\pm$ 2 (George and Mallery 2010).

Non-response Bias and Common Method Variance

The use of common method variance (CMV) is employed to assess the impact of measurement error by illustrating the distorted association across measurement constructs (Podsakoff et al. 2003). Therefore, Harman's single-factor tests were undertaken in order to assess the presence of common method variance (CMV). The findings indicate that there were five factors with eigenvalues over 1.0, collectively accounting for 76.84 of the overall variances. The component with the highest eigenvalue accounted for 29.345 of the total variation. Therefore, it can be inferred that the presence of common technique bias does not pose a significant risk for the present investigation.

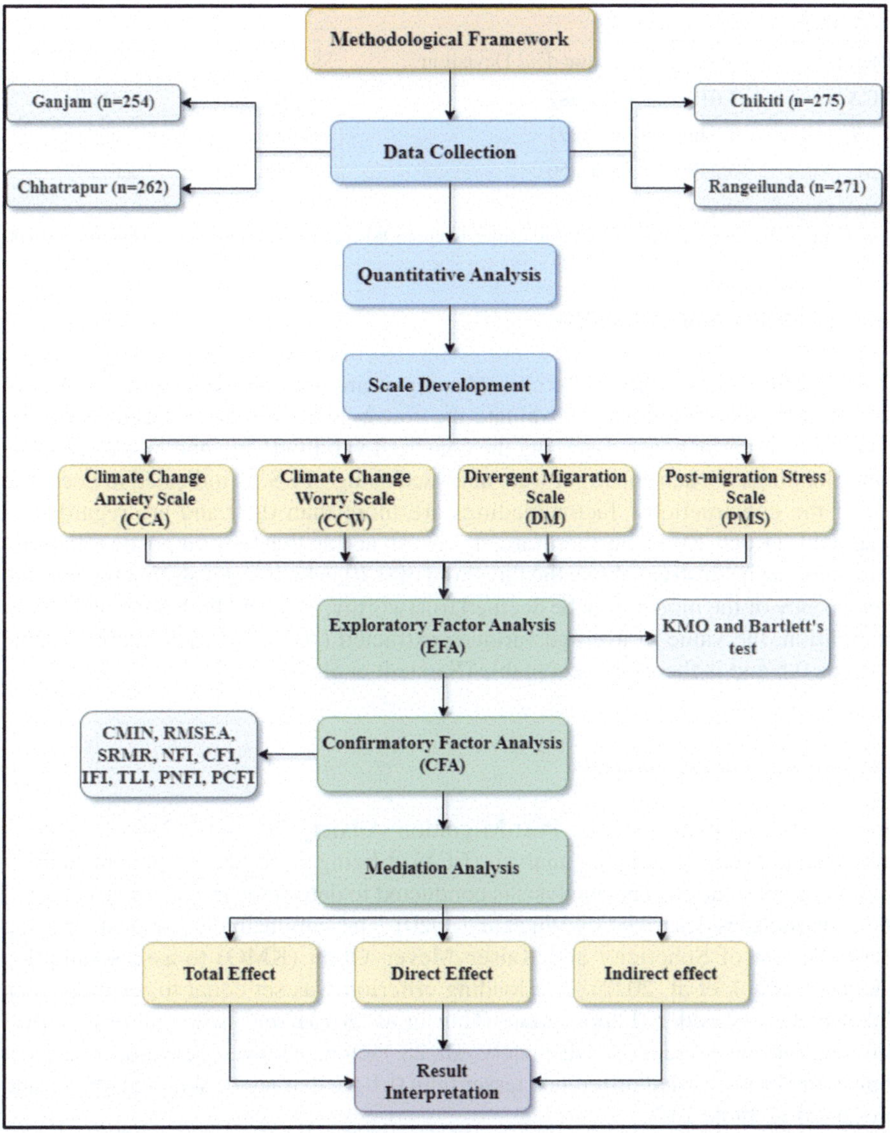

Fig. 14.3 Phases of research methodology. *Source* Computed by Authors. *Note* Comparative Fit Index (CFI), Normed Fit Index (NFI), Incremental Fit Index (IFI), Tucker-Lewis Coefficient (TLI), Parsimony Normed Fit Index (PNFI), and Parsimony Comparative Fit Index (PCFI), Root Mean Square Error of Approximation (RMSEA), and Standardized Root Mean Square Residual (SRMR)

Table 14.1 Descriptive statistics

Factors	Mean	Standard Deviation	Skewness	Kurtosis
CCA	3.019	1.082	−0.263	−1.466
CCW	3.345	1.293	−0.379	−1.529
DM	3.247	1.356	−0.380	−1.671

Source Computed by Authors
Note Climate Change Anxiety (CCA), Climate Change Worry (CCW), Divergent Migration (DM)

Model Validity and Reliability

The reliability and validity of the current analysis are shown in Tables 14.2 and 14.3. The present research aimed to evaluate the convergent validity of the measures by using three criteria, hence assessing the construct validity (Chin and Yao 2014). The factor loadings of each construct were analysed using SPSS, as indicated in Sect. 3.2. All of the constructions' factor loadings are more than 0.60 and are regarded as acceptable (Kock 2015). Furthermore, it is worth noting that the composite reliability values for each construct above the threshold of 0.70, which suggests that the internal consistency of the model may be deemed trustworthy, as stated by Hair et al. (2019). In addition, the value of average variance extracted (AVE) for each latent variable exceeds 0.5 and is therefore acceptable (Sarstedt et al. 2014).

Exploratory Factor Analysis

Initially, the Climate-Induced Post-Migration Anxiety-Stress (CIPMAS) scale is subjected to exploratory factor analysis (EFA) utilizing principal component analysis with varimax rotation. This analysis is conducted to determine the factor structure of the constructs, as described by Wu et al. (2020). The sample undergoes testing using Bartlett's test of Sphericity and Kaiser–Meyer–Olkin (KMO) to assess sampling adequacy (Seok et al. 2022). The loading criterion was set equal to, or more than 0.6, which is considered appropriate (Hair et al. 2013), and hence items less than this loading were reduced. Ultimately, all 28 factors whose eigenvalue is greater than 1 and cross-loaded difference lesser than 0.1 are extracted (Wu et al. 2020) and presented in Table 14.3.

Confirmatory Factor Analyses

In the study, confirmatory factor analysis is performed in order to ensure the model's fitness, validity, and reliability of the two proposed models which is represented in Table 14.6. The CFA result shows that the model has an acceptable fit in terms of $\chi 2/df$ as the values are 2.167 and 2.759 (Hair et al. 2019). In addition, the values of the Comparative Fit Index (CFI), Normed Fit Index (NFI), Incremental Fit Index

Table 14.2 Factor loadings and reliability of CCA, CCW, DM

Construct	Items	Factor loadings	CR	AVE	Cronbach's Alpha
Climate Change Anxiety	It is hard for me to pay attention when I am thinking about climate change. (CCA 1)	0.809	0.975	0.695	0.986
	I have trouble sleeping when I think much about climate change. (CCA 2)	0.863			
	Climate change gives me bad dreams. (CCA 3)	0.828			
	Stress about climate change makes it challenging for me to enjoy time with family and friends. (CCA 9)	0.888			
	It's hard for me to worry about sustainability while also taking care of my family's needs. (CCA 10)	0.889			
	It's hard for me to get work or school done because I'm worried about climate change. (CCA 11)	0.883			
	My worries about climate change limit my capacity to perform to my full potential. (CCA 12)	0.809			
Divergent Migration	How often do you think about searching for better job prospects outside this state? (DM 1)	0.856	0.984	0.751	0.992
	How often do you think about setting up a business in another state or place? (DM 2)	0.889			

(continued)

Table 14.2 (continued)

Construct	Items	Factor loadings	CR	AVE	Cronbach's Alpha
	How often do you think about working and living in another state or place for an extended period of time? (DM 3)	0.855			
	How often do you think about immigrating to another state or place to live there permanently? (DM 4)	0.868			
	I will actively examine the possibility of emigrating to another state or place within the next 5 years. (DM 5)	0.841			
	An increase in migration is inevitable as Ganjam gets more stressful and competitive. (DM 13)	0.859			
	Ganjam's future has become too unpredictable and migration is the best option. (DM 14)	0.858			
	Compared to Ganjam, many other states or places can offer a better environment to raise a family. (DM 15)	0.832			
Climate Change Worry	I have a higher level of concern over climate change compared to other people. (CCW 1)	0.900	0.977	0.669	0.964

(continued)

Table 14.2 (continued)

Construct	Items	Factor loadings	CR	AVE	Cronbach's Alpha
	Contemplations over climate change evoke concerns about the potential outcomes of the future. (CCW 2)	0.760			
	I actively pursue knowledge on climate change through various media platforms such as television, newspapers, and the internet. (CCW 3)	0.787			
	I experience a tendency to feel anxious when I come across information on climate change, even in cases when the consequences of climate change may not be imminent. (CCW 4)	0.838			
	I am concerned that occurrences of extreme weather events may be attributed to climate change (CCW 5)	0.848			
	I am deeply concerned about cllmate change to the extent that I feel immobilised in my ability to take any action to address it (CCW 6)	0.860			

(continued)

Table 14.2 (continued)

Construct	Items	Factor loadings	CR	AVE	Cronbach's Alpha
	I am concerned about my ability to effectively manage the impacts of climate change. (CCW 7)	0.797			
	It is difficult for me to cease worrying about climate change once I have begun to do so (CCW 9)	0.766			

Source Computed by Authors
Note Composite Reliability (CR), Average Variance Extracted (AVE)

(IFI), Tucker-Lewis Coefficient (TLI), Parsimony Normed Fit Index (PNFI), and Parsimony Comparative Fit Index (PCFI), Root Mean Square Error of Approximation (RMSEA), and Standardized Root Mean Square Residual (SRMR) fall under the acceptable range as presented in Table 14.6.

Mediation Analysis

The bootstrapping technique is used to conduct a mediation study in order to investigate the function of CCW as the mediator between CCA and DM, as seen in Fig. 14.5. The research used mediation analysis to get insight into the statistical significance of both the direct and indirect associations among the variables under investigation (Zhao et al. 2011). The present investigation used bootstrapping in multiple parallel meditation studies to mitigate the non-normality concerns associated with the sample distribution and to evaluate potential indirect effects among the variables (Zhao et al. 2010). The present study used a bootstrapping technique with 2000 resamples and replacements (Omar et al. 2021) to examine the presence of a statistically significant mediation pathway. This was achieved by calculating the indirect effect on the samples and constructing a 95% confidence interval.

Table 14.3 Factor loadings and reliability of CIPMAS scale

Construct	Items	Factor loadings	CR	AVE	Cronbach's Alpha
Climate Change Anxiety	It is hard for me to pay attention when I am thinking about climate change. (CCA 1)	0.794	0.989	0.677	0.986
	I have trouble sleeping when I think much about climate change. (CCA 2)	0.850			
	Climate change gives me bad dreams. (CCA 3)	0.824			
	I ponder, "Why am I unable to effectively manage the challenges posed by climate change?" (CCA 5)	0.824			
	I take some time to reflect on why I have these feelings regarding climate change once I am alone. (CCA 6)	0.836			
	I ask myself, "Why do I feel this manner regarding climate change?" (CCA 8)	0.900			
	Stress about climate change makes it challenging for me to enjoy time with family and friends. (CCA 9)	0.898			
	It's hard for me to worry about sustainability while also taking care of my family's needs. (CCA 10)	0.888			
	It's hard for me to get work or school done because I'm worried about climate change. (CCA 11)	0.885			

(continued)

Table 14.3 (continued)

Construct	Items	Factor loadings	CR	AVE	Cronbach's Alpha
	My worries about climate change limit my capacity to perform to my full potential. (CCA 12)	0.827			
	Some of my friends think I spend too much time dwelling on climate change. (CCA 13)	0.840			
Post Migration Stress	I experience discrimination in school or at work. (PMS 2)	0.896	0.997	0.772	0.991
	People around me making racist remarks towards me. (PMS 4)	0.876			
	Difficulties in understanding how ordinary life activities work here (shopping, buying tickets, traveling, etc.) (PMS 6)	0.863			
	I am worried about an unstable financial situation. (PMS 8)	0.890			
	I feel frustrated for not being able to support myself financially. (PMS 9)	0.878			
	I am worried about my all debts. (PMS 10)	0.870			
	I am missing my social life in my home town. (PMS 11)	0.838			
	Longing for my hometown. (PMS 12)	0.903			

(continued)

Table 14.3 (continued)

Construct	Items	Factor loadings	CR	AVE	Cronbach's Alpha
	I am missing all the activities that I used to do earlier in my hometown. (PMS 13)	0.869			
	I am worried about my family members that I am separated from them. (PMS 14)	0.880			
	I am sad because I have not been reunited with my family. (PMS 15)	0.847			
	For some reason, I feel like I don't belong here. (PMS 16)	0.872			
	I feel frustrated due to the loss of status in this new society. (PMS 17)	0.841			
	I feel frustrated because I am not able to make use of my competencies here. (PMS 18)	0.856			
	In my family, there are some really upsetting disagreements. (PMS 19)	0.830			
	I am experiencing a lack of respect within my family. (PMS 20)	0.801			
	I am experiencing a sense of insignificance within my family. (PMS 21)	0.840			

Source Computed by Authors
Note Composite Reliability (CR), Average Variance Extracted (AVE)

Table 14.4 Demographic details of migrants

	Group A		Group B	
Characteristics	Respondents (N = 552)	Observed %	Respondents (N = 510)	Observed %
Gender				
Male	480	87	472	93
Female	72	13	38	7
Age Group				
18–29	336	61	246	48
30–39	127	23	134	26
40–49	62	11	84	16
50–64	27	5	46	9
Marital Status				
Unmarried	353	64	267	52
Married	254	46	243	48
Level of Education				
Primary	99	18	64	12
Secondary	77	14	101	20
Intermediate	127	23	124	23
Graduation	144	26	123	24
Masters and Higher	105	19	98	19
Years of migration for Group B				
0–5 Years			211	41
6–10 Years			243	48
More than 10 years			56	11

Note Group A: Farmers Intending to Migrate; Group B: Migrated Farmers
Source Computed by Authors

Results and Discussion

Descriptive Statistics

The demographic details of the 552 farmers intending to migrate and 510 migrated farmers from the four blocks of Ganjam districts are described in Table 14.4. It is observed that out of the total respondents in Group A (farmers intending to migrate), 87% are male and 13% are female. Further, around 61% of respondents belong to the age group 18–29, and around 64% are unmarried. Out of these 552 respondents, the majority are having intermediate (23%) and graduation (26%) levels of education. Group B shows the demographic features of 510 migrants from the four blocks of Ganjam district. From the analysis, it is found that out of these 510 migrants, 93% are male and the majority are unmarried (52%) and belong to the 18–29 age group (48%). Further, around 20%, 23%, and 24% of respondents have intermediate, graduation, and master's and higher levels of education. Out of these 510 migrants, the majority of respondents have left their homeland for 6–10 years. It is observed that for both the groups, the higher the level of education greater the intention and probability to migrate as they have comparatively more options to shift their source of livelihood in case of any inevitable scarcity. Additionally, it is found that unmarried farmers have migrated or intend to migrate in a greater proportion in comparison to the married respondents as resettlement with the family in the new location could be a critical task, which may not be suitable for the family members.

Factor Analysis Using Exploratory and Confirmatory Factor Analysis

The rotated components matrix for both the scales are presented in Table 14.5, which indicates that the majority of the indicators possess factor loadings exceeding 0.5. This indicates that all items meet the criteria for further hypothesis testing techniques (Gour et al. 2017), as all factors are within a valid range and at an appropriate threshold level. Consequently, these data may be utilised in subsequent testing methodologies. The absence of cross-loading in the rotated components matrix establishes the data's dependability.

Primarily, KMO and Bartlett's sphericity test methods are used to evaluate the adequacy and suitability of the sample for EFA. The value of KMO varies between 0 and 1, and the ideal value must be higher than 0.50 for a better adequacy level of the sample (Bulent Gurbuz and Ozkan 2021). Additionally, if the p-value is < 0.05, the data set is suitable for factor analysis (Bulent Gurbuz and Ozkan, 2021). The KMO value for the CIPMAS scale is 0.981 and Bartlett's test is significant at the $p = 0.000$ level. Thus, the obtained data is suitable for the EFA. Further, the principal component method with varimax rotation is used to depict the constructs for EFA analysis.

Further, Structural validity is examined by both EFA and CFA. After item reduction in EFA, CFA is performed on the proposed CIPMAS scale to examine the construct more closely and to establish construct validity for each of the most preferred items determining post-migration stress among the migrants (Fig. 14.4). The model fitness assessed various fit indices including relative Chi-square ($\chi 2/\text{df}$), Comparative Fit Index (CFI), Normed Fit Index (NFI), Incremental Fit Index (IFI), Tucker-Lewis Coefficient (TLI), Parsimony Normed Fit Index (PNFI), and Parsimony Comparative Fit Index (PCFI), Root Mean Square Error of Approximation (RMSEA), and Standardized Root Mean Square Residual (SRMR). The model fit indices showed that the values under CIPMAS are closer to 1, representing a better fit for the proposed model (Table 14.6).

Table 14.5 Rotated component matrix

Rotated component matrix						
Items	Component 1	Component 2	Items	Component 1	Component 2	Component 3
CCA1	0.663		CCW1	0.863		
CCA2	0.786		CCW2	0.776		
CCA3	0.708		CCW3	0.636		
CCA5	0.659		CCW4	0.792		
CCA6	0.685		CCW5	0.803		
CCA8	0.712		CCW6	0.701		
CCA9	0.748		CCW7	0.686		
CCA10	0.704		CCW8	0.701		
CCA11	0.815		CCW9	0.501		
CCA12	0.751		CCW10	0.718		
CCA13	0.761		CCA1		0.591	
PMS2		0.738	CCA2		0.729	
PMS4		0.735	CCA3		0.586	
PMS6		0.692	CCA4		0.680	
PMS8		0.679	CCA5		0.658	
PMS9		0.728	CCA6		0.668	
PMS10		0.672	CCA7		0.631	
PMS11		0.660	CCA8		0.648	
PMS12		0.734	CCA9		0.654	
PMS13		0.752	CCA10		0.658	
PMS14		0.803	CCA11		0.680	

(continued)

Table 14.5 (continued)

Rotated component matrix						
Items	Component 1	Component 2	Items	Component 1	Component 2	Component 3
PMS15		0.785	CCA12		0.589	
PMS16		0.786	CCA13		0.618	
PMS17		0.754	MI1			0.617
PMS18		0.777	MI2			0.668
PMS19		0.702	MI3			0.623
PMS20		0.658	MI4			0.668
PMS21		0.702	MI5			0.658
Extraction Method: Principal Component Analysis			MI13			0.713
Rotation Method: Varimax with Kaiser Normalization			MI14			0.749
			MI15			0.755
			MI16			0.800
			MI17			0.761
			MI18			0.764
			MI19			0.717
			MI20			0.649
			MI21			0.701
			MI22			0.747
			MI23			0.710
			MI24			0.704

Source Computed by Authors

Table 14.6 Model fit indices of proposed models

Global adjustment measures	Fit indices	Acceptable range	Reference	Result CIPMAS	Result CCA, CCW, DM
Absolute Fit Indices	CMIN	≤ 3	Daniel-González et al. (2020)	2.167	2.759
	RMSEA	≤ 0.08		0.073	0.074
	SRMR	≤ 0.05		0.044	0.041
Relative Fit Indices	NFI	≥ 0.90	Daniel-González et al. (2020)	0.95	0.92
	CFI	≥ 0.90		0.97	0.95
	IFI	≥ 0.95	Brosseau-Liard and Savalei (2014)	0.98	0.95
	TLI	≥ 0.95	Sharma et al. (2005)	0.96	0.95
Parsimonious Fit Indices	PNFI	> 0.80	Daniel-González et al. (2020)	0.86	0.84
	PCFI	≥ 0.80		0.84	0.86

Source Computed by Authors
Note Comparative Fit Index (CFI), Normed Fit Index (NFI), Incremental Fit Index (IFI), Tucker-Lewis Coefficient (TLI), Parsimony Normed Fit Index (PNFI), and Parsimony Comparative Fit Index (PCFI), Root Mean Square Error of Approximation (RMSEA), and Standardized Root Mean Square Residual (SRMR)

Relation of Climate Change Anxiety with Divergent Migration and Climate Change Worry

In line with the proposed hypothesis (H1), the current study validates the direct relationship between CCA and DM as significant based on the analysis (Table 14.7). The possible reason for the significant direct relationship is that divergent migration is an adaptive step undertaken by people who have developed anxiety due to adversely impacting climatic changes (de Sherbinin et al. 2022). The agricultural sector is mostly vulnerable to climate change and most of the farmers are likely unable to mitigate the CCA induced by climate change and hence migrate due to the risk of livelihood loss (Nguyen and Sean 2021). In this similar context, Taylor (2020) argued that due to CCA, there is forced migration and people migrate to different areas in search of food, and shelter. Further, Watts et al. (2019) argued that due to climatic change physical and mental health of the human being is impacted, hence people prefer to migrate to other places.

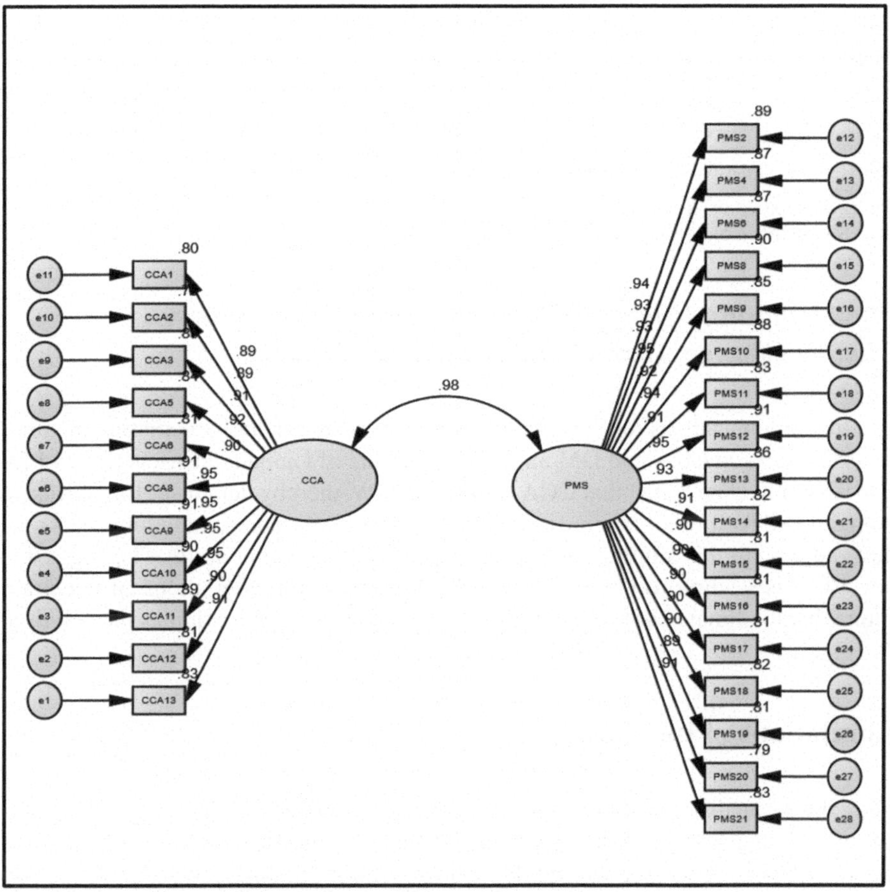

Fig. 14.4 CFA of CIPMAS scale. *Source* Computed by Authors

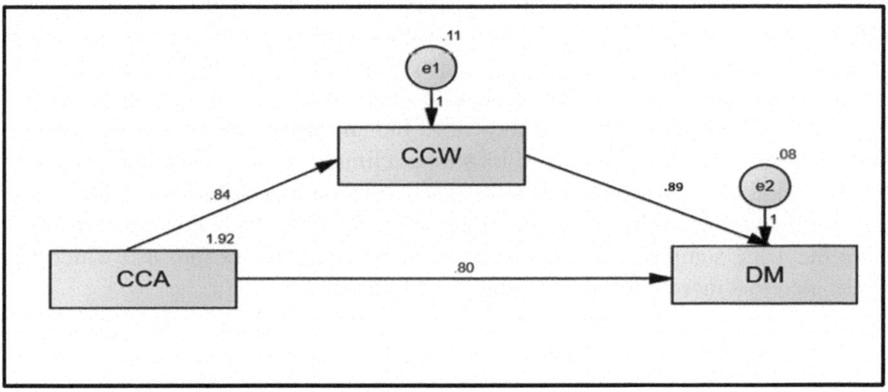

Fig. 14.5 Mediation analysis between the constructs. *Source* Computed by Authors

Table 14.7 Hypothesis results for direct relationships between constructs

Hypothesis	Constructs	Estimates (β)	t-value	Decision
H1	CCA → DM	0.424^{***}	11.645	Supported
H2	CCA → CCW	0.675^{***}	12.824	Supported
H3	CCW → DM	0.429^{***}	09.792	Supported

Source Computed by Authors
Note Significance of correlation: $^{***}p \leq 0.001$

The Mediating Effect of Climate Change Worry Between Climate Change Anxiety and Divergent Migration

The mediation analysis results explain that CCW is significantly mediating the relationship between CCA and DM among the farmers of Ganjam Districts (Fig. 14.5). The reason may be stated that CCA results in CCW, thereby demonstrating intrusive thoughts and feelings of anxiousness about future time disasters, including one's own descendants. In this similar context, the work of Innocenti et al. (2023), suggested that CCA is leading to CCW and thereby leading to thinking about an uncertain future due to climate change and thus leading to choosing migration as the solution.

Limitations of the Study

The current analysis is subjected to be considered under the light of certain limitations. First, while performing the factor analysis, the Pearson matrix is applied but it is important to note that while dealing with original data, the use of Pearson correlation is not recommended (Velicer et al. 2000). To rectify this apprehension, a secondary analysis is run over the data collected by using polychoric correlations instead of Pearson correlations. However, upon re-evaluation, it is evident that the results obtained from utilizing Pearson correlation retains while using polychoric correlations with the same number of factors as used previously while running Pearson correlation. Second, it is crucial to mention that the sample collected for the study may not fully represent the entire Indian farmers of diverse sociodemographic levels and are subjected to locational climatic issues concerned. Third, in the regression model use of a large sample size is recommended with a response rate displayed asymmetrically in the distribution chart. To address this concern while using the 1062 sample, a quantile regression approach is put into use which has contributed to a more interpretable analysis of the results.

Recommendations

The current study has several policy recommendations to be followed up. First, as it is observed that the migration intention of farmers strongly increases under the scenario of climate change worry and anxiety (McLeman 2013b; Duijndam et al. 2023), policymakers should come up with how to effectively address mental health risk factors keeping in mind how future adverse impacts from climate change can be addressed. This widespread area of research can further be built upon the existing findings where the mediation effect of climate change anxiety between climate change worry and farmer's divergent migration is revealed. The current research will assist the scholars and other researchers to analyze the prevalence of any other mental issues among the farmers due to climate change forcing them for out migration on the basis a larger and more representative sample and its correlations with any psychological behaviors. However, it is understood while data collection that better data is required for rigorous research on internal migrations and displacement in connection with psychological behaviour. If adequate local based data can be provided regarding the same then it will be easy for the policy makers to come forward with a complete assessment of local resilience and adaptation programs across diverse climatic distorted zones (Abdi et al. 2023). The current study recommends for an agenda to be set forth a practical engagement of both policy makers and academic researchers in identifying the risk factors affecting mental issues of farmers and outlining possible suggestions for preventing migration by farmers due to climatic distortions.

Conclusions

This study has offered a pertinent insight into the impact of CCW and CCA on farmers' DM. The result of the study revealed that climatic variation creating CCA has led to forced migration among the farmers of the Ganjam district. Additionally, it has also been deduced that CCW is successfully mediating the relationship between CCA and DM. The use of the measures scaling CCA, CCW, and DM is crucial in the assessment and quantification of the impact of the climate crisis on farmers' mental well-being. Additionally, they provide the examination of the interconnections between different emotions associated with climate change, as shown in our research. This study has developed a single scale which has the potential to not only measure climate-induced post-migration stress but also can provide valuable insights for developing therapies that can effectively address the mental health consequences of farmers by future researchers and healthcare professionals.

Based on the results, it is crucial to note that climate change adaptation strategies are very essential to overcome their level of worry and anxiety to reduce their rate of migration. For this change, agro-knowledge, tertiary agricultural education, and technical investment (Ma et al. 2023) in agricultural sector should be enhanced by increasing agricultural productivity, branching out farming methods, coping with climatic distortions, accessing rural and urban vegetable markets, introducing farm diversification and agricultural entrepreneurship, and refining eco-sustainability which can help them to adapt with the climatic changes instead of relocating out of worry and anxiety.

Acknowledgements The authors are deeply indebted to the Kalinga Institute of Industrial Technology, Deemed to be University for the tremendous support in terms of accessing information and data collection. The authors are also grateful to the farmers and their families for their voluntary participation in the data collection process.

References

Abdi AH, Mohamed AA, Sugow MO (2023) Exploring the effects of climate change and government stability on internal conflicts: evidence from selected sub-Saharan African countries. Environ Sci Pollut Res 30(56):118468–118482. https://doi.org/10.1007/s11356-023-30574-w

Alam GMM, Alam K, Mushtaq S (2017) Climate change perceptions and local adaptation strategies of hazard-prone rural households in Bangladesh. Clim Risk Manag 17:52–63. https://doi.org/10.1016/j.crm.2017.06.006

Avis WR (2017) Scoping study on defining and measuring distress migration. FAO Governance, Social Development, Humanitarian and Conflict (GSDRC) Applied Knowledge Services. Food and Agriculture Organization of the UN, Rome

Berry HL, Waite TD, Dear KBG, Capon AG, Murray V (2018) The case for systems thinking about climate change and mental health. Nat Clim Change 8(4):282–290. https://doi.org/10.1038/s41558-018-0102-4

Brosseau-Liard PE, Savalei V (2014) Adjusting incremental fit indices for nonnormality. Multivar Behav Res 49(5):460–470. https://doi.org/10.1080/00273171.2014.933697

Bulent Gurbuz I, Ozkan G (2021) A holistic approach in explaining farmers' intentional behaviour on manure waste utilization. New Medit 20(4). https://doi.org/10.30682/nm2104g

Chin C-L, Yao G (2014) Convergent validity. In: Michalos AC (ed) Encyclopedia of quality of life and well-being research. Springer Netherlands, pp 1275–1276. https://doi.org/10.1007/978-94-007-0753-5_573

Clayton S (2020) Climate anxiety: psychological responses to climate change. J Anxiety Disord 74:102263. https://doi.org/10.1016/j.janxdis.2020.102263

Daghagh Yazd S, Wheeler SA, Zuo A (2019) Key risk factors affecting farmers' mental health: a systematic review. Int J Environ Res Public Health 16(23):4849. https://doi.org/10.3390/ijerph16234849

Daniel-González L, Moral De La Rubia J, Valle De La OA, García-Cadena CH (2020) Structure analysis of subjective well-being. Salud Mental 43(3):119–127. https://doi.org/10.17711/SM.0185-3325.2020.017

de Sherbinin A, Grace K, McDermid S, van der Geest K, Puma MJ, Bell A (2022) Migration theory in climate mobility research. Front Climate 4. https://doi.org/10.3389/fclim.2022.882343

Duijndam SJ, Botzen WJW, Hagedoorn LC, Bubeck P, Haer T, Pham M, Aerts JCJH (2023) Drivers of migration intentions in coastal Vietnam under increased flood risk from sea level rise. Clim Change 176(2):12. https://doi.org/10.1007/s10584-022-03479-9

Falco C, Galeotti M, Olper A (2019) Climate change and migration: is agriculture the main channel? Glob Environ Chang 59:101995. https://doi.org/10.1016/j.gloenvcha.2019.101995

Fetzek S, Mazo J (2014) Climate, scarcity and conflict. Survival 56(5):143–170. https://doi.org/10.1080/00396338.2014.962803

Garip F (2008) Social capital and migration: how do similar resources lead to divergent outcomes? Demography 45(3):591–617. https://doi.org/10.1353/dem.0.0016

George D, Mallery M (2010) SPSS for windows step by step: a simple guide and reference, 17.0 Update, 10th edn. Pearson, Boston

Goodwin H, Yiend J, Hirsch CR (2017) Generalized anxiety disorder, worry and attention to threat: a systematic review. Clin Psychol Rev 54:107–122. https://doi.org/10.1016/j.cpr.2017.03.006

Gour L, Maurya, SB, Koutu, GK, Singh SK, Shukla SS, Mishra DK (2017) Characterization of rice (*Oryza sativa* L.) genotypes using principal component analysis including scree plot & rotated component matrix. Int J Chem Stud 975:975–983

Gregersen T, Doran R, Böhm G, Tvinnereim E, Poortinga W (2020) Political orientation moderates the relationship between climate change beliefs and worry about climate change. Front Psychol 11:1–12. https://doi.org/10.3389/fpsyg.2020.01573

Hair JF, Black WC, Babin BJ, Anderson RE (2013) Multivariate data analysis. Pearson Education Limited. https://www.drnishikantjha.com/papersCollection/Multivariate%20Data%20Analysis.pdf

Hair JF, Risher JJ, Sarstedt M, Ringle CM (2019) When to use and how to report the results of PLS-SEM. Eur Bus Rev 31(1):2–24. https://doi.org/10.1108/EBR-11-2018-0203

Hoong LC, Soon D (2012) A study on emigration attitudes of young Singaporeans (2010). Working Papers, Article id: 4742. https://ideas.repec.org//p/ess/wpaper/id4742.html

Innocenti M, Perilli A, Santarelli G, Carluccio N, Zjalic D, Acquadro Maran D, Ciabini L, Cadeddu C (2023) How does climate change worry influence the relationship between climate change anxiety and eco-paralysis? A moderation study. Climate 11(9), Article 9. https://doi.org/10.3390/cli11090190

Kock N (2015) Common method bias in PLS-SEM: a full collinearity assessment approach. Int J e-Collaboration (IJeC) 11(4):1–10. https://doi.org/10.4018/ijec.2015100101

Larionow P, Sołtys M, Izdebski P, Mudło-Głagolska K, Golonka J, Demski M, Rosińska M (2022) Climate change anxiety assessment: the psychometric properties of the Polish version of the climate anxiety scale. Front Psychol 13:870392. https://doi.org/10.3389/fpsyg.2022.870392

Ma R, Abid N, Yang S, Ahmad F (2023) From crisis to resilience: strengthening climate action in OECD countries through environmental policy and energy transition. Environ Sci Pollut Res 30(54):115480–115495. https://doi.org/10.1007/s11356-023-29970-z

Makwana D, Engineer P, Dabhi A, Hardik C (2023) Sampling methods in research: a review. Int J Trend Sci Res Dev (IJTSRD) 7:762–768

Malm A, Tinghög P, Narusyte J, Saboonchi F (2020) The refugee post-migration stress scale (RPMS)—development and validation among refugees from Syria recently resettled in Sweden. Confl Heal 14(1):2. https://doi.org/10.1186/s13031-019-0246-5

McEvoy PM, Watson H, Watkins ER, Nathan P (2013) The relationship between worry, rumination, and comorbidity: evidence for repetitive negative thinking as a transdiagnostic construct. J Affect Disord 151(1):313–320. https://doi.org/10.1016/j.jad.2013.06.014

McLeman RA (2013a) Developments in modelling of climate change-related migration. Clim Change 117(3):599–611. https://doi.org/10.1007/s10584-012-0578-2

McLeman RA (2013b) Climate and human migration: past experiences, future challenges. Cambridge University Press, Cambridge. https://doi.org/10.1017/CBO9781139136938

Mishra A, Sundaray BK (2021) A study on aftermath of COVID-19 on migrant workers in Odisha with special reference to Ganjam District. Train Dev J 12(1):1–8. https://doi.org/10.5958/2231-069X.2021.00001.9

Nguyen TPL, Sean C (2021) Do climate uncertainties trigger farmers' out-migration in the Lower Mekong Region? Curr Res Environ Sustain 3:100087. https://doi.org/10.1016/j.crsust.2021.100087

Nordås R, Gleditsch NP (2007) Climate change and conflict. Polit Geogr 26(6):627–638. https://doi.org/10.1016/j.polgeo.2007.06.003

Ochieng J, Kirimi L, Mathenge M (2016) Effects of climate variability and change on agricultural production: the case of small scale farmers in Kenya. NJAS Wageningen J Life Sci 77:71–78. https://doi.org/10.1016/j.njas.2016.03.005

Omar S, Mohsen K, Tsimonis G, Oozeerally A, Hsu J-H (2021) M-commerce: the nexus between mobile shopping service quality and loyalty. J Retail Consum Serv 60:102468. https://doi.org/10.1016/j.jretconser.2021.102468

Pihkala P (2020) Anxiety and the ecological crisis: an analysis of eco-anxiety and climate anxiety. Sustainability 12(19), Article 19. https://doi.org/10.3390/su12197836

Podsakoff PM, MacKenzie SB, Lee J-Y, Podsakoff NP (2003) Common method biases in behavioral research: a critical review of the literature and recommended remedies. J Appl Psychol 88(5):879–903. https://doi.org/10.1037/0021-9010.88.5.879

Raitzer DA, Bosello F, Tavoni M, Orecchia C, Marangoni G, Samson JNG (2015) Southeast Asia and the economics of global climate stabilization. Asian Development Bank

Rajakumari S, Minnu A, Sarunjith KJ (2022) Determination of vulnerable zones along Brahmapur coast, Odisha using AHP and GIS with validation against multiple cyclones. Environ Monit Assess 194(4):278. https://doi.org/10.1007/s10661-022-09886-w

Rudolphi JM, Berg RL, Parsaik A (2020) Depression, anxiety and stress among young farmers and ranchers: a pilot study. Community Ment Health J 56(1):126–134. https://doi.org/10.1007/s10597-019-00480-y

Sarstedt M, Ringle CM, Smith D, Reams R, Hair JF (2014) Partial least squares structural equation modeling (PLS-SEM): a useful tool for family business researchers. J Fam Bus Strat 5(1):105–115. https://doi.org/10.1016/j.jfbs.2014.01.002

Scheffran J, Marmer E, Sow P (2012) Migration as a contribution to resilience and innovation in climate adaptation: social networks and co-development in Northwest Africa. Appl Geogr 33:119–127. https://doi.org/10.1016/j.apgeog.2011.10.002

Searle K, Gow K (2010) Do concerns about climate change lead to distress? Int J Climate Change Strat Manage 2(4):362–379. https://doi.org/10.1108/17568691011089891

Seok CB, Cosmas G, Hashmi SI, Ading C (2022) Psychometric and gender invariance analysis of the flourishing scale in the malaysian context. SAGE Open 12(2):215824402210964. https://doi.org/10.1177/21582440221096447

Sharma S, Mukherjee S, Kumar A, Dillon WR (2005) A simulation study to investigate the use of cutoff values for assessing model fit in covariance structure models. J Bus Res 58(7):935–943. https://doi.org/10.1016/j.jbusres.2003.10.007

Stewart AE (2021) Psychometric properties of the climate change worry scale. Int J Environ Res Public Health 18(2):494. https://doi.org/10.3390/ijerph18020494

Stojanov R, Duží B, Kelman I, Němec D, Procházka D (2016a) Local perceptions of climate change impacts and migration patterns in Malé, Maldives. Geogr J 183(4):370–385. https://doi.org/10.1111/geoj.12177

Stojanov R, Kelman I, Ullah AA, Duží B, Procházka D, Blahůtová KK (2016b) Local expert perceptions of migration as a climate change adaptation in Bangladesh. Sustainability 8(12), Article 12. https://doi.org/10.3390/su8121223

Sundblad E-L, Biel A, Gärling T (2007) Cognitive and affective risk judgments related to climate change. J Environ Psychol 27:97–106. https://doi.org/10.1016/j.jenvp.2007.01.003

Tacoli C (2009) Crisis or adaptation? Migration and climate change in a context of high mobility. Environ Urban 21(2):513–525. https://doi.org/10.1177/0956247809342182

Taylor S (2020) Anxiety disorders, climate change, and the challenges ahead: introduction to the special issue. J Anxiety Disord 76:102313. https://doi.org/10.1016/j.janxdis.2020.102313

Velicer WF, Eaton CA, Fava JL (2000) Construct explication through factor or component analysis: a review and evaluation of alternative procedures for determining the number of factors or components. In: Goffin RD, Helmes E (eds) Problems and solutions in human assessment. Springer US, pp 41–71. https://doi.org/10.1007/978-1-4615-4397-8_3

Watts N, Amann M, Arnell N, Ayeb-Karlsson S, Belesova K, Boykoff M, Byass P, Cai W, Campbell-Lendrum D, Capstick S, Chambers J, Dalin C, Daly M, Dasandi N, Davies M, Drummond P,

Dubrow R, Ebi KL, Eckelman M, Montgomery H et al (2019) The 2019 report of The Lancet countdown on health and climate change: ensuring that the health of a child born today is not defined by a changing climate. The Lancet 394(10211):1836–1878. https://doi.org/10.1016/S0140-6736(19)32596-6
Weems C (2008) Developmental trajectories of childhood anxiety: identifying continuity and change in anxious emotion☆. Dev Rev 28(4):488
Wu S, Feng X, Sun X (2020) Development and evaluation of the health belief model scale for exercise. Int J Nursing Sci 7:S23–S30. https://doi.org/10.1016/j.ijnss.2020.07.006
Zhao X, Lynch JG Jr, Chen Q (2010) Reconsidering Baron and Kenny: myths and truths about mediation analysis. J Consum Res 37(2):197–206. https://doi.org/10.1086/651257
Zhao X, Huo B, Selen W, Yeung JHY (2011) The impact of internal integration and relationship commitment on external integration. J Oper Manag 29(1):17–32. https://doi.org/10.1016/j.jom.2010.04.004

Chapter 15
Eco-anxiety in Adolescents: A Meta-Analysis of the Impact of Climate Change on Mental Health

Subhajit Mondal, Jhikmik Kar, Emon Bagchi, and Uttam Mukhopadhyay

Abstract The constant perceptual dread of environmental disasters and catastrophes caused by allegedly irreversible impacts of climate change (climate change) with corresponding fear for future generations is called eco-anxiety. Yet it is not considered a psychological disease, but a frequent increase in eco-anxiety can result into mental illness. It ranges from mild stress to clinical disease viz; anxiety, depression, self-harm tendencies along with maladaptive coping practices i.e. substance abuse. Adolescents develop their personality very rapidly with higher magnitude of substantial development and changes. Presently, they are far more conscious about the adverse impact of climate change on environment and human health. The present chapter aims to identify the existing literature, primarily focused on identification of the impact of climate change on adolescents' mental health and pointing out various knowledge gaps in the existing literature and future research considerations. The review has synthesized and evaluated the existing scholarly literatures regarding eco-anxiety while using Preferred Reporting Items for Systematic Reviews and Meta-Analyses (PRISMA) for reporting. It has been evident from existing literature that, adolescents are facing eco-anxiety along with depression, extreme emotion, sadness, fear, etc. due to emotional attachment with the environment and awareness about climate change which also leads to the use of maladaptive substances to cope with eco-anxiety. Most of the available researches are conducted in American and European countries. Children, adolescents and indigenous inhabitants are identified as the most vulnerable group. Methodological framework to analyze grassroot experiences of eco-anxiety is still not satisfactory.

S. Mondal (✉) · J. Kar · U. Mukhopadhyay
Department of Geography, University of Calcutta, Kolkata, West Bengal, India
e-mail: subhajitmondal.vb@gmail.com

E. Bagchi
Department of Geography, Heramba Chandra College, Kolkata, West Bengal, India

U. Mukhopadhyay et al. (eds.), *Climate Crisis, Social Responses and Sustainability*, Climate Change Management, https://doi.org/10.1007/978-3-031-58261-5_15

Introduction

Today, the phenomena of Climate change has emerged as one of the urgent global concern. Consequences of increasing temperatures, extreme weather events, habitat destruction, and other environmental shifts have garnered considerable focus. Scientific consensus unequivocally supports the notion that these transformations stem from human activities, particularly the emission of greenhouse gases (IPCC 2022a). The implications of climate change extend beyond its physical effects; it also represents a growing menace to mental health, notably among adolescents. The Intergovernmental Panel on Climate Change (IPCC), recently released a comprehensive report elucidating numerous potential future scenarios and the inevitable consequences humanity is poised to face in the forthcoming years (Climate Change 2021 2021; Costello et al. 2009). They forecasted that climate change will increasingly present significant threats for natural ecosystems and global human health within 2040, contingent upon the extent of global warming along with a projected temperature increases of 1.5 degrees Celsius by 2040, using 1850–1900 as the reference period (IPCC 2022a). Consequently, a consensus within the scientific community affirms that climate change is primarily anthropogenic in nature as supported by (Climate Change 2021 2021; Costello et al. 2009; IPCC 2022b; Orru et al. 2017). Media and Climate Change Observatory (MeCCO) heightened attention directed towards interdependency of climate change on cognitive health of the young people (Katzung et al. 2021). This report underscored the media's profound focus on issues like escalating sea levels, the proliferation of wildfires globally, and the ongoing climate deliberations by the United Nations. Simultaneously, an increasing awareness was observed regarding the mounting mental health implications of climate change with prevailing consensus regarding undeniable physical effects on humanity, biodiversity, and the planet (Ballew et al. 2019). Moreover, the world currently grapples with not one but two major crises, both of which significantly impact mental health. Discussions surrounding anxiety in the context of ecological concerns have predominantly gained momentum since 2007. Cossman offers a valuable retrospective of the early dialogues on eco-anxiety spanning from 2007 to 2013 (Cossman 2013), while Maiteny delves into the broader topic (Maiteny 2018). In the 2010s, attention shifted toward exploring the psychological repercussions of ecological crises, with climate change taking center stage in discussions and research, generating substantial interest. Among the evolving domain of climate psychology, various facets of anxiety have been extensively deliberated upon in numerous sub-disciplines (Clayton 2020a; Doherty 2011a). Notably, in 2017, the American Psychological Association (APA) and Eco America introduced a working definition of "eco-anxiety" in their publication, "Mental Health and Our Changing Climate" (Clayton 2018a) sparking considerable discourse on the subject. Numerous scholars, including Australian environmental philosopher Glenn Albrecht (Albrecht 2011, 2019a), have made significant contributions to the growing awareness of eco-anxiety. In addition, publications providing self-help strategies and recommendations for collective action to mitigate eco-anxiety, particularly climate anxiety, began to emerge in 2020 (Ray 2020b),

However, there are numerous approaches to discussing anxiety in a broad sense, and this diversity extends to the discourse surrounding eco-anxiety. The term seems to primarily encompass anxiety, despite the phenomenon involving a spectrum of emotions, several scholars have questioned whether "eco-anxiety" is the most fitting expression for this (Ojala 2012b; Pihkala 2020b; Searle 2010).

Adolescence is a critical developmental phase marked by rapid transformations in one's physical, cognitive, and emotional aspects. During this transitional period, individuals transition from childhood to adulthood, heightening their awareness of their environment and the global issues that impact their lives. Adolescents, owing to the ongoing maturation of their cognitive and emotional processes, are notably receptive to pressures and traumatic experiences (Doherty 2011a). When discussing climate change and its influence on mental health, it is imperative to consider adolescents as a significant demographic factor due to their heightened sensitivity to environmental challenges and the psychological repercussions these challenges entail. Adolescence is distinguished by profound transformations in brain structure and functions. This is the period when the prefrontal cortex, responsible for emotional regulation and decision-making, undergoes significant development (Blakemore 2008). These changes in the brain have a notable impact on an adolescent's ability to process and respond to environmental stimuli. Adolescents are simultaneously navigating the complex process of identity formation. Their self-concept, values, and beliefs are in a state of evolution. Their comprehension of environmental issues, such as climate change, becomes intertwined with their emerging identities, rendering them more susceptible to eco-anxiety. Adolescents assign significant value to peer relationships and are significantly shaped by social norms and group dynamics. Their susceptibility to climate change concerns is amplified due to the prevailing social environment, which can either exacerbate or alleviate their apprehensions about it (Rutland et al. 2010). Furthermore, research has unveiled the rising awareness of climate change and its potential consequences among adolescents (Ojala 2012b). Their heightened sensitivity stems from their growing comprehension, which in turn amplifies their concerns about the future. The distinct vulnerability of adolescents to environmental pressures and the potential psychological ramifications underscores the necessity to scrutinize how climate change impacts their mental well-being. Climate change exerts three types of impacts on mental health i.e. direct impact, indirect impact, and vicarious impact (Berry et al. 2010). Research predominantly focuses on the direct consequences, triggered by extreme weather events, leading to posttraumatic stress disorder, anxiety, depression, substance use disorders, and suicidal ideation (Berry et al. 2010; Cianconi 2020; Hayes 2018). Awareness of the global environmental crisis induces distress irrespective of direct exposure (Clayton 2018a; Pihkala 2018). While heightened emotional responses may catalyse positive action and sustainable habits, eco-anxiety can also immobilize individuals in the face of environmental challenges (Albrecht 2019b; Pihkala 2020b; Wolf 2011; Wolfe 2018). Climate change awareness can promote pro-environmental behaviours but also elicit paralyzing anxiety and denial (Albrecht 2019b).

A survey conducted among 600 individual aged 10–14 in Australia revealed that "one quarter of children worry that the world will end before they grow older," and

"44% of they are concerned about the future impact of climate change" (Tucci et al. 2007). The absence of coping mechanisms to deal with the emotional repercussions of learning about climate change can lead to feelings of hopelessness and denial (Ojala 2012a). Nonetheless, despite their pronounced levels of worry and apprehension, young individuals seem to possess limited general knowledge about climate change (Erkal, n.d.) and (Corner 2015). Eco-anxiety could potentially evolve into a future public health concern, highlighting the importance of ongoing vigilance in staying current with research outcomes. In recent years, there has been a notable proliferation of research on teenage eco-anxiety in response to the climate crisis, marking the emergence of this burgeoning field. However, the current state of research remains fragmented and diverse, with numerous studies delving into different dimensions of this intricate issue. This fragmented nature of the research makes it challenging to attain a comprehensive understanding of the prevalence and repercussions of eco-anxiety among young individuals. Consequently, a meta-analysis is imperative to consolidate the findings from existing studies, offering a more comprehensive insight into this phenomenon and providing coherence to the current body of research.

A notable research gap is apparent in the current literature concerning the comprehensive synthesis of data on eco-anxiety among adolescents. This inadequacy extends to the lack of a systematic and quantitative methodology for accurately assessing impact sizes and prevalence rates of eco-anxiety among adolescents. Along with the identification of the overall prevalence, contributory factors, and potential consequences of eco-anxiety on the adolescents. Addressing these gaps necessitates the amalgamation of data from diverse studies, employing a comprehensive approach that involves identifying patterns and trends across various demographic groups, cultural contexts, and spatial diversification. The use of meta-analysis in this context impedes the development of a more comprehensive understanding with enhanced generalizability.

This chapter aimed to accomplish the following objectives: (1) scrutinize the concept of eco-anxiety in the adolescents, (2) assess the existing evidence of eco-anxiety among them (3) identify the trend and contribution in the field of eco-anxiety (4) identify knowledge gaps in "eco-anxiety" research with a comprehensive understanding of the present scenario and offer potential avenues for future research. Eco-anxiety exhibits an intrinsic connection to the overarching theme of "Climate Crisis, Social Responses, and Sustainability: Socio-Ecological Study." This interdisciplinary exploration delves into the intricate web of relationships between environmental concerns, societal responses, and sustainable practices, embodying a comprehensive socio-ecological perspective. Within this framework, eco-anxiety emerges as a significant psychological dimension, reflecting the emotional and mental impact of the ongoing climate crisis on individuals, particularly in the context of adolescents. By interweaving the psychological facets of eco-anxiety with broader socio-ecological dynamics, this chapter intended to provide insights about the scientific contribution to the intricate interplay among human emotions, societal actions, and sustainable practices amidst contemporary climate crises.

Rationale of the Study

The terms "climate anxiety" and "eco-anxiety" are increasingly prevalent in both traditional and digital media discussions with a simultaneous growing research interest (Cunsolo 2018). However, a significant knowledge gap persists regarding the diverse expressions of these anxieties and their psychological interplay in ecological crises. While a substantial portion of the population grapples with eco-anxiety, specifics about the concerns it entails and appropriate societal and medical responses remain underexplored. Urgently needed is further investigation into the mental health implications of ecological emergencies, particularly climate change, given their anticipated significance (Berry et al. 2010; Bourque 2014). Comparative to the burgeoning literature on eco-anxiety in adults, very limited has been learned about the emotional experiences of adolescents regarding climate change awareness. Growing up amidst media narratives emphasizing climate change's negative aspects (Tucci et al. 2007), children face uncertainty. Numerous studies explore teenage eco-anxiety, covering emotional repercussions (Munn et al. 2018), the effectiveness of environmental education, and media's impact on adolescent emotional responses (Milfont et al. 2012). Therefore, diverse eco-anxiety literature required for a systematic synthesis of findings.

Materials and Methods

This chapter follows a systematic scoping review approach for meta-analysis which systematically compiles and synthesizes extant literature concerning eco-anxiety in adolescents in Web of Science. The methodological framework employed in this inquiry comprises three distinct stages which are succinctly illustrated in Fig. 15.1.

Stage-1

Data Source, Search Strategy and Data Extraction

From the earliest available date up to the end of August 2023, a comprehensive search for scientific articles was conducted across the Web of Science with following markers and extracted from the same.

"Eco-anxiety" [Title/Abstract], Alternative Title/Abstract: "Ecoanxiety", Alternatively, "Climate Change Anxiety" [Title/Abstract], Alternative Title/Abstract: "Climate Anxiety", OR "Climate Change Worry" [Title/Abstract], Alternative Title/Abstract: "Environmental Worry" [Title/Abstract], Alternatively, "Environmental Distress" [Title/Abstract], Alternative Title/Abstract: "Ecological Grief", OR "Ecological Stress" [Title/Abstract].

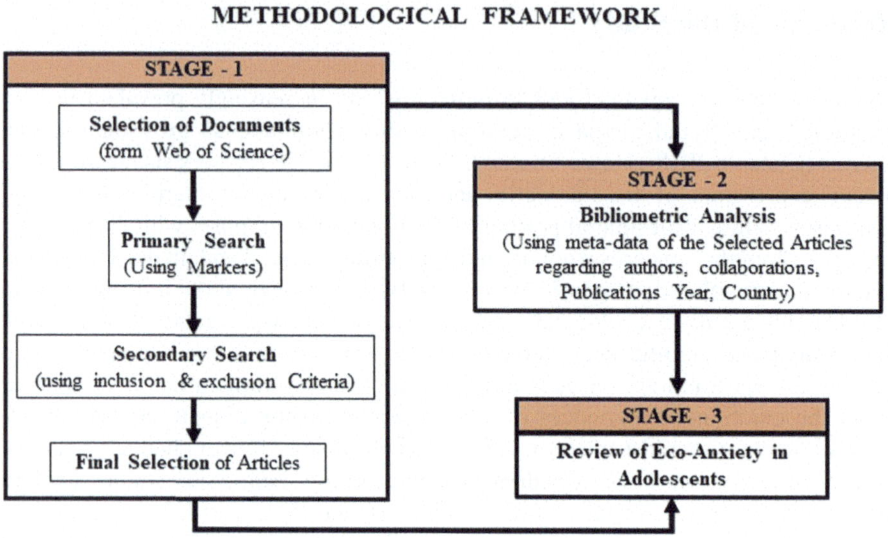

Fig. 15.1 Methodological framework

Protocol for Inclusion and Exclusion Criteria

The developed methodology by Arksey (2005) and enhanced by Levac (2010) for Preferred Reporting Items for Systematic Reviews and Meta-Analyses for Scoping Reviews (PRISMA-ScR) (Fig. 15.1a) introduced by (Léger-Goodes et al. 2022) served as the foundational basis of this review protocol.

This review encompasses the studies conducted across the globe and published till 31st August, 2023 with availability in English only. This time frame was selected due to the prominence of eco-anxiety and the gravity of the climatic issues since the onset of the 21st century (Pihkala 2020a). The focus of this scoping study was on adolescents as the target population defined as "individuals between 10 to 19 years (*Adolescent Health : National Health Mission*, n.d.). The review methodology drew inspiration from the work of Pihkala, who conducted comprehensive investigation for elucidating the nature and occurrence of the phenomenon. Pihkala's research encompassed articles on eco-anxiety from the domains of social sciences and political sciences, psychodynamic perspectives, existential psychology, and various anxiety-related theories (Pihkala 2020a). In accordance with the established inclusion criteria, we also applied exclusion criteria. Eco-anxiety is portrayed as an overwhelming emotional response, particularly among the selected group of young individuals in relation to climate change. Hence, the articles included in this review must satisfy three key conditions: (1) they must pertain to adolescents or young, (2) they should investigate the impact of climate change on mental health, and (3) these consequences must be elicited by climate change awareness. Consequently, articles that exclusively focused on adults aged 19 and above, articles discussing the physical, economic, or social consequences without addressing mental health, or articles only providing brief mentions of the impacts of climate-related natural disasters such as tornadoes or tsunamis were excluded as they did not align with the specified criteria.

This review encompasses the studies conducted across the globe and published till 31st August, 2023 with availability in English only. This time frame was selected due to the prominence of eco-anxiety and the gravity of the climatic issues since the onset of the 21st century (Pihkala 2020a). The focus of this scoping study was on adolescents as the target population defined as "individuals between 10 to 19 years (*Adolescent Health : National Health Mission*, n.d.). The review methodology drew inspiration from the work of Pihkala, who conducted comprehensive investigation for elucidating the nature and occurrence of the phenomenon. Pihkala's research encompassed articles on eco-anxiety from the domains of social sciences and political sciences, psychodynamic perspectives, existential psychology, and various anxiety-related theories (Pihkala 2020a). In accordance with the established inclusion criteria, we also applied exclusion criteria. Eco-anxiety is portrayed as an overwhelming emotional response, particularly among the selected group of young individuals in relation to climate change. Hence, the articles included in this review must satisfy three key conditions: (1) they must pertain to adolescents or young, (2) they should investigate the impact of climate change on mental health, and (3) these consequences must be elicited by climate change awareness. Consequently, articles that exclusively focused on adults aged 19 and above, articles discussing the physical, economic, or social consequences without addressing mental health, or articles only providing brief mentions of the impacts of climate-related natural disasters such as tornadoes or tsunamis were excluded as they did not align with the specified criteria.

Study Selection

Following the removal of duplicates, 83 sources remained from the initial 133 sources identified in the entire search process. These identified articles were imported into Covidence (version v2532 1ea49715) and stored in Zotero (version 6.0.23) to facilitate separate assessments by two reviewers. The software included predefined inclusion and exclusion criteria applied during both level 1 and level 2 screenings, with no conflicts emerging. The flowchart, in adherence to PRISMA guidelines (Tricco 2018), is depicted in Fig. 15.2, and all 83 articles were consistently marked as "N."

Stage-2

The ultimate set of relevant documents selected for this study was exported from the Web of Science (WoS) database in the form of Research Information System (RIS) and Comma Separated Values (CSV) files. The Biblioshiny package in R was subsequently employed for executing a comprehensive bibliometric analysis, delving into the most impactful publications, distinguished authors, prominent countries of origin, and discernible trends over the years within the specific subject.

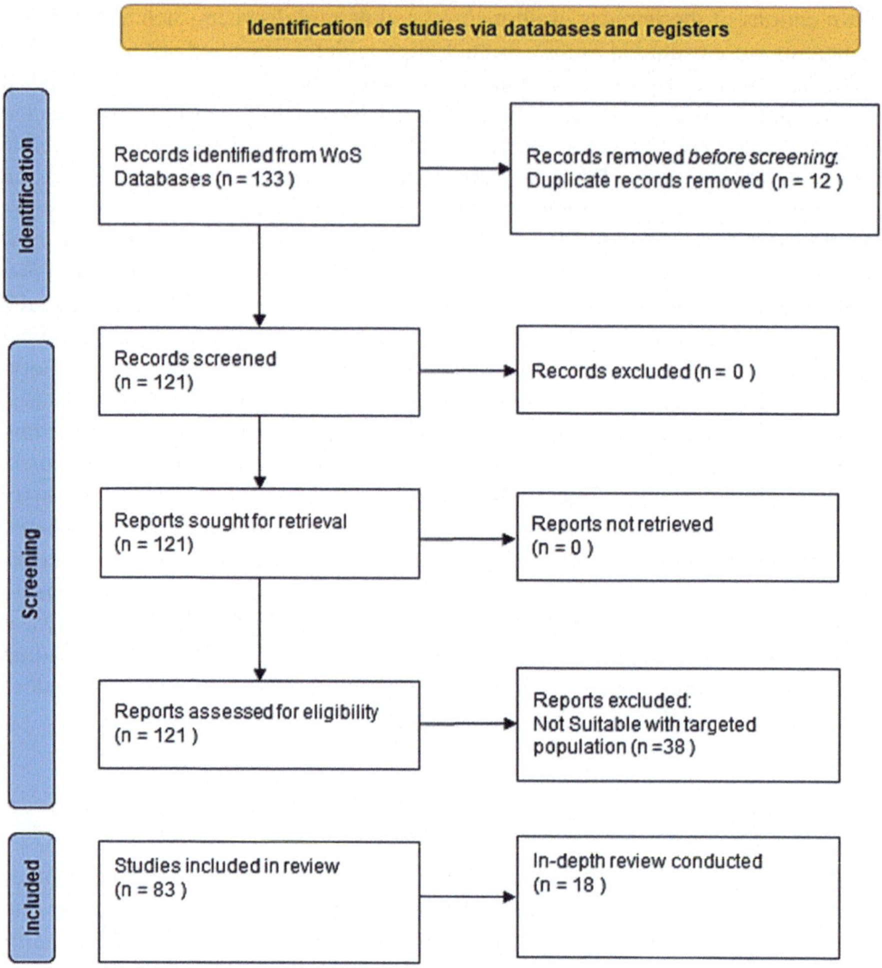

Fig. 15.2 A PRISMA flow chart

Stage-3

Moreover, this chapter identified and analyzed the most pertinent themes encompassing concepts, types, scholarly contributions, trends, prevalence and perspectives of eco-anxiety in adolescents along with vulnerable groups and demographic characteristics to provide a comprehensive view regarding the present scenario and future research prospects in particular in this stage.

Result and Discussion

This chapter focused on the meta-analysis of the available scholarly literature on the interplay of negative emotions among the adolescents regarding the consequences of the climate change which is termed as "eco-anxiety" and the outcome of the chapter are as explained followingly.

Concept of Eco-anxiety

Environmental awareness and climate change evoke distressing emotions, leading to anxiety, particularly among those not directly impacted (Helm 2018a). This anxiety extends to broader environmental concerns (Doherty 2011b) and is characterized as a powerful response to a dynamic natural environment (Helm 2018b). (Clayton 2020b) and (Pihkala 2020a) highlight a spectrum of challenging emotions and mental states arising from environmental factors. Eco-anxiety, often an indirect consequence of environmental issues, prompts individuals to employ defense mechanisms and coping strategies. Described as a widespread reaction to the planet's ecological state, eco-anxiety is synonymous with climate anxiety, denoting unease about the environmental situation and the need for structural changes (Clayton 2020b). Albrecht (2019b) address eco-anxiety as a general apprehension about our interactions with environmental support systems. Searle (2010) and Gow (2010) term it environmental distress—an excessive concern and worry about climate change, including reactions to climate-related issues. It is the experience of anxiety in response to the ecological catastrophe (Stanley 2021).

These various definitions collectively reflect the complex nature of eco-anxiety as a multifaceted emotional response to environmental challenges and climate change. However, there are ambiguities in the usage of the term "eco-anxiety." For instance, when discussing topics related to eco-anxiety, phrases such as "climate change anxiety" (Clayton 2020b) and "ecological stress" (Helm 2018a) have been employed. Any anxiety related to environmental issues is encompassed within the term "eco-anxiety," with "climate anxiety" representing a closely related but distinct form. This diversity is reflected in the wide array of terminology and phrases in current literature that illustrate the various concepts related to eco-anxiety. "Climate change caused by humans" (Clayton 2018b) is another term used interchangeably with climate anxiety and is also used in discussions of solastalgia.

Scientific Contribution in Eco-anxiety

This chapter identified significant disparity (Figs. 15.3 and 15.4) in the scientific contributions to the knowledge domain of eco-anxiety. The prominence of research undertaken mostly in developed nations, such as the USA and Australia, compared to their rarity in less developed nations. It depicts the inequal distribution of research and possibility of an imbalance in the direction of scientific inquiry. Developed nations have taken the lead in exploring the complex components of eco-anxiety, including the psychological, social, and environmental aspects. They are supported by more solid research infrastructure and plentiful financing mechanisms. Their significant contributions have significantly enriched the burgeoning body of knowledge on this subject. Conversely, developing countries may contend with resource limitations, a lack of prioritization, or divergent research agendas, which pose hindrances to their active involvement in scientific endeavours related to the phenomenon of eco-anxiety. Recognising the significance of fostering research capacities and supporting multidisciplinary studies in developing nations is essential to redressing this imbalance. By doing this, it is possible to get a larger and more representative view on eco-anxiety, taking into consideration the particular difficulties and vulnerabilities that communities in these areas confront. Such an inclusive strategy will not only promote a more equitable global response to the environmental and mental health issues brought on by climate change, but it will also improve the scientific knowledge of eco-anxiety. It is an appeal to close the gap and guarantee that voices from wealthy and poor countries are equally heard and respected in the conversation on eco-anxiety. In order to provide a more inclusive and thorough understanding of eco-anxiety that considers the varied experiences and contexts across the globe, there is a need for international collaboration, knowledge exchange, and assistance. This asymmetry in scientific contributions is a good example of this need.

The "Timeline in Scientific Contribution in Eco-Anxiety" in (Fig. 15.5) provides an insightful look at the chronological development of research in this important area. An interesting finding from the data is that there has been a substantial concentration of research in recent years, with 2022 and 2023 showing as the most productive years.

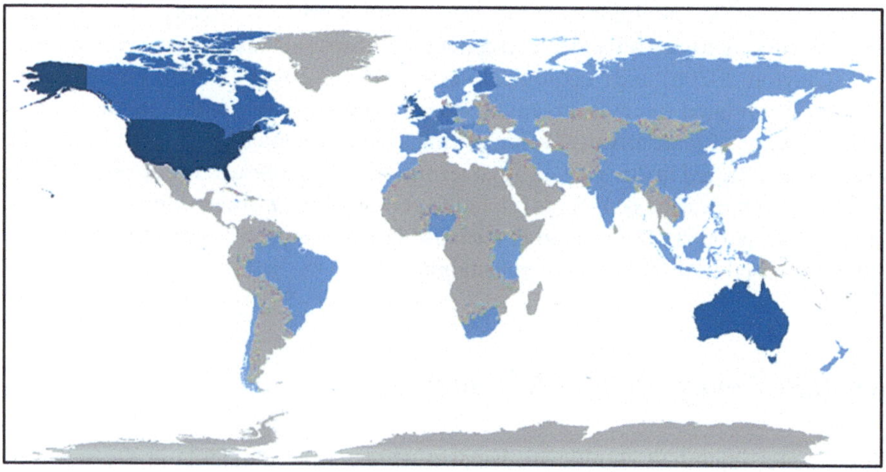

Fig. 15.3 Major contributing countries

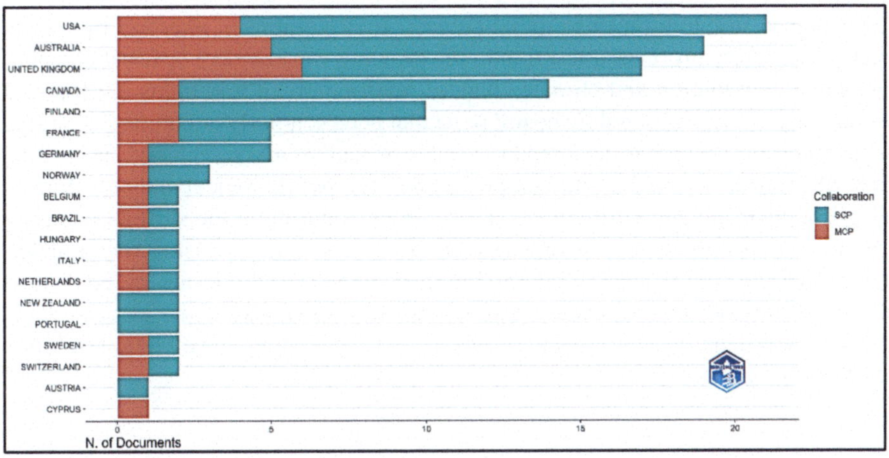

Fig. 15.4 Corresponding authors and collaborations

61 records, or a significant 45.865% of the total, were devoted to the study of eco-anxiety in 2022. This shows that as communities deal with escalating concerns about climate change and its effects, there is an increasing need to address the psychological, social, and environmental elements of eco-anxiety. The following year, 2023, maintains this growing trend with 33 records, making up 24.812% of the whole, demonstrating ongoing scholarly interest in the subject.

On the other hand, older years, such 2013 and 2018, make less significant contributions, making up just 0.752% and 1.504% of all articles, respectively. These differences in temporal distribution demonstrate the dynamic character of scientific engagement with eco-anxiety, which has seen a recent increase in attention and

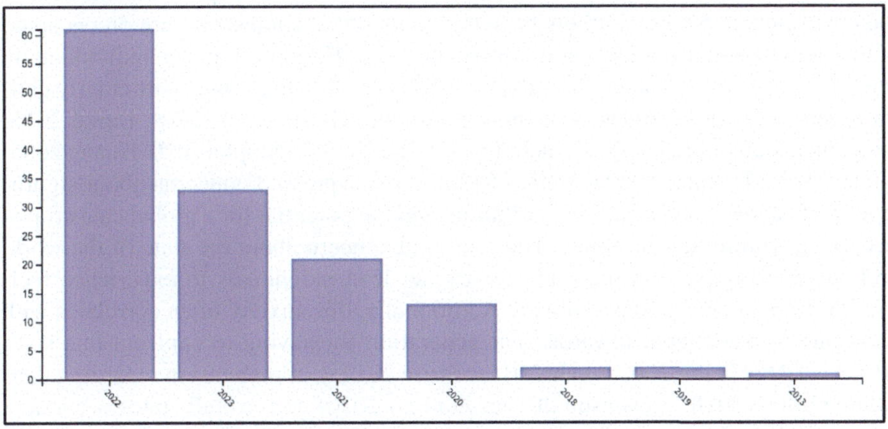

Fig. 15.5 Trend in eco-anxiety literature

dedication to research on this crucial subject. It is possible to interpret the recent focus of research efforts as a reaction to the growing worldwide worries about environmental degradation and climate change. Additionally, it highlights the fact that eco-anxiety is a growing public health issue that need scholarly attention. According to the timeframe, which is depicted in the table, scientists and researchers are putting more effort into examining and comprehending the psychological and sociological effects of environmental difficulties. These findings highlight the shifting nature of scientific contributions to eco-anxiety and could be a sign of a rising social understanding of the significance of addressing these mental health issues in the context of a dynamic and unpredictable environment. It is evidence of the proactive role science plays in addressing current problems and establishing the direction of future eco-anxiety research.

Prevalence of Eco-anxiety

The present chapter also recognised that, before "eco-anxiety" widely adopted as a specific term, numerous papers explored the emotions expressed by adolescents and teenagers regarding environmental concerns. Which denoted the prevalence of the eco-anxiety or the anxiety developed due to the catastrophes caused by the climate change is prevailed in the adolescents back in the distant past. To identify these emotions, a content analysis was conducted, examining the reported psychological states in the literature (Erlingsson 2017). The authors of the featured papers acknowledged that young individuals often grapple with various emotions as they learn about climate change, including anger, hopelessness, and sadness, while also experiencing optimism and anxiety. Among these emotions, concern (26.9%) and hope (20.8%) were the two most frequently reported in the selected articles. Numerous indications suggest that young people, especially adolescents, are deeply concerned about the environment. Several authors have highlighted that adolescents are preoccupied with environmental matters on a consistent basis. Furthermore, for individuals in regions already experiencing the adverse effects of climate change, particularly children, this concern becomes even more prominent (Berry et al. 2018; Burke 2018; Chalupka 2020). Other authors, including (Burke 2018; Chalupka 2020; Huang 2005; Ojala 2012a; Ratinen 2020a; Strife 2012a), have expressed concerns about significant biodiversity loss, escalating pollution, and the potential for a global catastrophe occurring within their lifetimes. The profound concern about the state of the world and its future impact on their adult lives can lead individuals to experience feelings of despondency and pessimism. Additionally, this anxiety often correlates with emotions of resentment towards their generation for having to confront this issue (Boggs 2016; Huang 2005; Ojala 2012a; Strife 2012a), apprehension regarding their future (Burke 2018; Stevenson 2015a; Zummo 2020) and overall states of anxiety (Ojala 2012c). Contrastingly, evidences with significant indication on young people and children harbour their hope regarding climate change. Notably, two separate studies (Li 2017; Stevenson 2015a) have highlighted the coexistence of concern and hope. These studies found that hope was intricately linked to awareness (Li 2017) and that a positive perspective was often associated with taking action (Burke 2018; Ojala 2012c; Stevenson 2015a). The discussion within these articles regarding children's responses to climate change represents a noteworthy outcome. Notably, (Ojala 2012c) explores three distinct forms of coping mechanisms can be employed, including: (1) Problem-Focused Coping, which entails actively addressing climate change through tangible actions; (2) Emotion-Focused Coping, that formulates measures to control the negative thoughts and emotions caused by the climate change; and (3) Meaning-Focused Coping, where positive emotions are elicited with the help of beliefs and values. According to (Ojala 2012c) adolescents' most commonly used coping mechanisms include:

- Emotional distancing was a coping strategy adopted by young adults and teenagers to emotionally deal with the effects of climate change.
- However, the most effective coping method was meaning-focused coping, as it has the highest potential for fostering optimism. It was the most successful coping technique because it engages individuals in a positive emotional context while being mindful of the negative aspects. This approach was associated with an overall sense of well-being and a commitment to environmental advocacy (Ojala 2012c)

This chapter also highlighted various publications with the evidence of potential factors that could instill hope in adolescents. The authors have provided their insights into the strengths and vulnerability factors of the prevalence of eco-anxiety are as follows:

- According to Hickman, significant portion of adolescents experienced profound psychological responses which includes feelings of concern, fear, sadness, anxiety, dread, powerlessness, helplessness, and grief due to degradation of climate which impact on their daily lives, and those who believed that government responses were inadequate reported higher distress levels (Hickman 2021a).
- In contrast, Chalupka explored that, how young individuals in high-income developed nations grappled with climate change awareness, which often resulted in substantial anxiety and internal conflicts, particularly among those who perceived themselves as privileged. Interestingly, children from indigenous or subsistence communities or those deeply connected to the land displayed higher levels of concern. However, this study did not identify specific protective factors in this context (Chalupka 2020).
- Ratinen explored the significance of imparting age-appropriate information about climate change to adolescents. This research reveals that providing incomplete or inappropriate information can trigger anxiety and distress among youngsters. The study underscores the importance of allowing children to openly express these emotions and take proactive steps. He also highlighted that, girls tended to feel a greater sense of responsibility and had lower confidence in the efficacy of scientific interventions, rendering them more susceptible to worry and pessimism compared to the boys. Additionally, he underlined that, promoting climate action

in primary schools can be used as a protective action for the future. (Ratinen 2020b).

- On the other hand, it has been investigated that the amplified stress levels, experienced by adolescents due to climate change, can result in intense emotional distress and feelings of helplessness. These emotions encompass anxiety, grief, apprehension for the future, and a sense of injustice. Notably, this chapter reveals that the emotional sensitivity of the adolescents is positively correlated with a greater comprehension of the significant obstacles and repercussions related to climate change.
- In 2020, Zummo conducted a study to examine the influence of media discourse on adolescents' emotional responses to climate change. The findings suggested that doomsday narratives tend to evoke feelings of fear and apprehension, ultimately leading to inaction, a loss of hope, and a shift in focus towards negative aspects. Regrettably, this study did not identify specific protective or vulnerable factors (Berry 2015; Kurth, 2018; Searle 2010; Verplanken 2020).
- Meanwhile, Li delved into the relationship between students' hope levels and their concerns regarding climate change. They uncovered a positive correlation, indicating that heightened awareness and concern can lead to the development of coping strategies and enhanced effectiveness (Li 2017).
- Children often express concerns about their personal environmental impact and the extent to which they can contribute to combating climate change. While this study did not identify specific vulnerability or protective factors, it implies that children may experience worry, fear, and anger (Boggs 2016).
- Stevenson emphasized that concern about climate change can be a catalyst for positive action, particularly among young individuals who will bear the long-term consequences. However, their research stressed that this process is effective only when feelings of worry and anxiety are coupled with hope. Notably, it was observed that girls tend to exhibit higher levels of both concern and hope (Stevenson 2015b).
- Strife observed that adolescents often exhibit emotions such as despair, anxiety, anger, pessimism, and a sense of being overwhelmed when confronted with the realities of climate change. These emotional responses include crying, nightmares, profound despair, and even thoughts about the potential end of the world. Notably, the study found that higher levels of hope were positively associated with trust in technology. Nevertheless, this research did not identify specific vulnerability or protective factors (Huang 2005; Strife 2012).

These studies collectively illustrate that adolescents' awareness of climate change can evoke a multitude of emotional expressions encompassing feelings of concern, anxiety, fear, sorrow, and, in certain instances, hopelessness. Age, gender, and coping strategies emerge as pivotal vulnerable factors significantly influencing individuals' reactions. Nonetheless, protective factors such as encouragement, empowerment, and support have the potential to mitigate the adverse impacts of climate change awareness on the psychological well-being of the adolescents. A comprehensive understanding of these intricate dynamics is crucial to addressing the mental health of the younger generation amid the global challenges posed by climate change.

Demographic Characteristics and Vulnerable Group

Younger individuals consistently identified with higher magnitude of eco-anxiety, aligning with prior researches (Berry et al. 2018b; Searle 2010). Adolescents 'exhibited similar, or occasionally even higher, magnitude of concerns in environmental issues compared to the adults in accordance to various surveys (Diamond et al. 2014). Conversely, older age groups appeared to express lesser concern regarding climate change. However, the fact remains that there are and will continue to be children growing up in a society where they will inevitably confront climate change issues. Unless we can significantly mitigate its impact, which, regrettably, seems more likely than not, younger individuals will bear a more substantial burden than their older counterparts. As eloquently expressed by a 16-year-old in the study by (Hickman 2021b), "I think it's different for young people. For us the destruction of the planet is personal". Moreover, when compared to adults, they demonstrate a stronger propensity to place trust in the scientific accord about human-induced causes of climate change. Furthermore, female respondents consistently reported elevated levels of eco-anxiety, a pattern that resonates with previous findings suggesting that females are usually express a higher magnitude of concerns about climate change than their male counterparts (Berry et al. 2015; Kurth 2018; Searle 2010; Verplanken 2020). Their analysis of affective and cognitive risk evaluations revealed that both men and women predicted similar likelihoods of severe negative consequences arising from climate change though women expressed greater apprehension regarding these potential repercussions. Regional disparities in the prevalence of eco-anxiety were also observed, with wealthier nations indicating less apprehension about climate change and its influence on everyday life in contrast to less affluent countries in the Global South (Hickman 2021b). This variance may stem from differences in the population's susceptibility to climate change and their adaptability, as highlighted by the Intergovernmental Panel on Climate Change in 2022. Therefore, there is an imperative to provide assistance to populations that face higher susceptibility to climate change, particularly individuals with pre-existing medical issues and economically weak, as their lives, well-being, and opportunities for advancement are in jeopardy. A national Australian study pinpointed the populations most vulnerable to the mental health ramifications associated with climate change by assessing their direct exposure to the events associated with change in climatic phenomena. A significant portion of the respondents reported firsthand encounters with climate change-related incidents. The study revealed that women exhibited a higher likelihood of experiencing both pre- and post-traumatic stress disorders compared to men, and individuals from economically disadvantaged regions were more prone to eco-anxiety (Doherty 2011b; Patrick 2022).

Notably, none of the studies we reviewed directly eco-anxiety from an Indigenous perspective. In two of the studies, the ethnic backgrounds of the participants were documented: 83% identified as white, 11 as American Indians, 18 as Asian Americans, 38 as Black or African Americans, and 5 as individuals of other racial backgrounds. Moreover, less than 10% of the participants in the studies identified as African, Asian Latinx, or from other racial backgrounds. While some general insights into the cultural connections of Indigenous peoples to the land were provided in four of the reviewed articles (Barlow 1988). Indigenous communities are particularly susceptible to heightened levels of climate anxiety due to their geographic vulnerability, profound interdependence on the environment, and cultural practices, as well as the irreplaceable loss of their spiritual and cultural heritage intertwined with the natural world (Ray 2020a). The probability of experiencing eco-anxiety is greater among individuals who hold a deep concern for environmental matters (Ray 2020a), feel a strong connection to nature, whether for cultural or personal reasons (Grose 2020), or describe having a close and personal relationship with the natural world (Grose 2020). Individuals who have undergone physical alterations in their surroundings due to climate change, including damages or distress caused by extreme weather events (Barlow 1988), climate change-induced displacement or homelessness (Salamon and Gage 2023), or have been impacted by droughts, rising sea levels, or erratic weather patterns (Ray 2020a), are at a higher risk of experiencing elevated stress levels and increased susceptibility to climate change-related anxiety. Naturalists and climate scientists, with their profound familiarity with and emotional attachment to the natural world, also encounter eco-anxiety (Clayton 2020c).

Knowledge Gap

For future research, a comprehensive understanding of the social context, age-specific factors, and the influence of environmental concerns on immediate and long-term behaviours in adolescents should be explored, in line with recommendations from (Ojala 2012c) and (Strife 2012). In 2018 Burke emphasized on the development of sustainability in adolescents to prevent the impact of climate change in near future and also stressed on expanding the concerned research to underdeveloped, developing and non-Western countries (Burke 2018). This chapter reveals a notable gap in the availability of empirically grounded data with rigorous methodological designs. This gap is evident, given that our review exclusively identified studies employing descriptive research designs. Furthermore, there exist significant knowledge gaps concerning the direct impact of climate change awareness on adolescents who are already grappling with mental health challenges. As Hickmn assert, a holistic comprehension of climate anxiety among this demographic necessitates consideration of a wide array of factors, including aspects of inter-personal relationships, psychology of individual, cultural practices, ethics, law, and politics (Hickman 2021a).

Recommendations

An initial review and meta-analysis of the literature indicating that evidence related to eco-anxiety predominantly revolves around emerging concepts, providing a definitive guidance at this juncture may be premature. Instead, it is imperative to emphasize the provision of mental health and behavioural support by focusing on following recommendations by various experts. Parents play a pivotal role in providing a secure environment for children to express climate concerns and maintaining a balance between positive and negative information (Berry et al. 2018). Additionally, parents and teachers should establish an open channel of conversation for children who frequently watch television, allowing them to freely communicate their perspectives, apprehensions and affective responses (Strife 2012; Zummo 2020). Student leadership in climate action with a realistic yet optimistic approach to climate education is crucial (Pinto and Grove-White 2020). They stressed on incorporating students' backgrounds and socio-political contexts into discussions and adopting a holistic Learn-Think-Act framework in the school curriculum. Using literature in climate education and incorporation of students' perspectives with dialogue is important to mitigate eco-anxiety in near future (Boggs 2016; Ojala 2012c). On the other hand, Sobel recommended changes in the school curriculum and introduction of climate education with nature-focused programs (Sobel 2007). Zummo emphasised on creation of open spaces for students to discuss climate-related thoughts and concerns to encourage more open discussions and engagement regarding climate change and eco-anxiety (Zummo 2020). Addressing climate anxieties requires collective adult responsibility (Hickman 2021b), displaying personal commitment to environmental care (Ojala 2012c), and adopting a systems-based approach that integrates mental health services and community design (Chalupka 2020). Vulnerable communities need resources, support, and skills for sustainable service provision, including psychological first aid training (Burke 2018). Nevertheless, it is crucial to acknowledge that we are currently in a stage where recommendations stem from incomplete knowledge, necessitating a cautious approach in their assessment. As a result, we propose that more transdisciplinary research should be conducted on eco-anxiety to comprehend this phenomenon from a multidimensional perspective. Unlikely the present scenario, research concentration should be proportionate to the representative population while both developing and underdeveloped nations should be considered.

Limitations of the Study

Eco-anxiety research, especially constituted with children and adolescents appears to be progressing at a notably slower pace than the parallel public discourse on this topic. The quality and quantity of existing studies significantly influenced our meta-analysis, specifically the inherency and insufficiency of the available data. Also, limited studies are focused on eco-anxiety in adolescents compared to eco-anxiety in adults. Heterogeneity across selected studies in methodologies, sample populations, and measurement tools poses a challenge to generalizability and comparability. Demonstrating correlations and establishing causal relationships between climate change and eco-anxiety was intricate for us due to the observational differences among the available scholarly literature. Temporal factors, cross-cultural variations, age and developmental stage differences, and the lack of longitudinal studies further constrain the analysis. The definition and measurement challenges associated with eco-anxiety contribute to the complexity and emphasized on the cautious interpretation.

Conclusion

This chapter unveiled the presence of eco-anxiety in adolescents. Eco-anxiety is a term used to describe the relationship between anxiety related to beliefs of the negative effects of climate change and anxiety related to climate change. It is evident that this group experiences a spectrum of emotions associated with eco-anxiety, encompassing feelings such as anger, grief, remorse, and hopelessness. Nevertheless, none of the examined articles offered adolescent specific metrics on this subject, underscoring the need for further research focused on adolescence experiences. Most of the available researches are conducted in American and European countries till now which defiantly create a knowledge divide. Children, adolescents and indigenous inhabitants are identified as the most vulnerable group. Furthermore, to mitigate the adverse impacts of climate change on the mental health of the young individuals, this assessment underscores the pivotal roles played by parents, educators, mental health professionals, school systems, adults, and other influential figures. While the existing research forms a foundational basis for studying children's eco-anxiety, it also highlights numerous research gaps that merit exploration in future studies.

References

Adolescent Health: National Health Mission (n.d.) Retrieved October 14, 2023, from https://nhm.gov.in/index1.php?lang=1&level=2&sublinkid=818&lid=221

Albrecht G (2011) Chronic environmental change: emerging 'psychoterratic' syndromes. In: International and cultural psychology, pp 43–56. https://doi.org/10.1007/978-1-4419-9742-5_3

Albrecht G (2019a) Earth emotions: new words for a new world

Albrecht G (2019b) Earth emotions: new words for a new world

Arksey HO, Lisa (2005) Scoping studies: towards a methodological framework. Int J Soc Res Methodol 8(1):19–32. https://doi.org/10.1080/1364557032000119616

Ballew MT, Leiserowitz A, Roser-Renouf C, Rosenthal SA, Kotcher JE, Marlon JR, Lyon E, Goldberg MH, Maibach EW (2019) Climate change in the American mind: data, tools, and trends. Environ Sci Policy Sustain Dev 61(3), Article 3. https://doi.org/10.1080/00139157.2019.1589300

Barlow DH (1988) Anxiety and its disorders: the nature and treatment of anxiety and panic

Berry HL, Bowen K, Kjellstrom T (2010) Climate change and mental health: a causal pathways framework. Int J Public Health 55(2), Article 2. https://doi.org/10.1007/s00038-009-0112-0

Berry HL, Dominic P (2015) Worrying about climate change: is it responsible to promote public debate? BJPsych Int 12(2):31–32. https://doi.org/10.1192/s2056474000000234

Berry HL, Waite TD, Dear KBG, Capon AG, Murray V (2018a) The case for systems thinking about climate change and mental health. Nat Climate Change 8(4):282–290. https://doi.org/10.1038/s41558-018-0102-4

Blakemore S-J (2008) The social brain in adolescence. Nat Rev Neurosci 9(4):267–277. https://doi.org/10.1038/nrn2353

Boggs GL, Nance S, Ackland RT, Danna S, Grant KB (2016) Beyond the lorax. Reading Teacher 69(6):665–675. https://doi.org/10.1002/trtr.1462

Bourque FÇOW, Cunsolo A (2014) Climate change: the next challenge for public mental health? Int Rev Psychiatry (Abingdon, England) 26(4):415–422. https://doi.org/10.3109/09540261.2014.925851

Burke SEL, Sanson AV, Van Hoorn J (2018) The psychological effects of climate change on children. Curr Psychiatry Rep 20(5):35–35. https://doi.org/10.1007/s11920-018-0896-9

Chalupka S, Anderko L, Pennea E (2020) Climate change, climate justice, and children's mental health: a generation at risk? Environ Justice 13(1):10–14. https://doi.org/10.1089/env.2019.0034

Cianconi P, Betrò S, Janiri L (2020) The impact of climate change on mental health: a systematic descriptive review. Front Psychiatry 11:74–74. https://doi.org/10.3389/fpsyt.2020.00074

Clayton S (2018a) Mental health risk and resilience among climate scientists. Nat Climate Change 8(4), Article 4. https://doi.org/10.1038/s41558-018-0123-z

Clayton S (2018b) Mental health risk and resilience among climate scientists. Nat Climate Change 8(4), Article 4. https://doi.org/10.1038/s41558-018-0123-z

Clayton S (2020a) Climate anxiety: psychological responses to climate change. J Anxiety Disorders 74, Article NA. https://doi.org/10.1016/j.janxdis.2020.102263

Clayton S (2020b) Climate anxiety: psychological responses to climate change. J Anxiety Disorders 74, Article NA. https://doi.org/10.1016/j.janxdis.2020.102263

Clayton S (2020c) Climate anxiety: psychological responses to climate change. J Anxiety Disorders 74:102263. https://doi.org/10.1016/j.janxdis.2020.102263

Climate change 2021: The physical science basis : summary for policymakers : working group I contribution to the sixth Assessment report of the Intergovernmental Panel on Climate Change (2021) IPCC

Corner A, Roberts O, Chiari S, Völler S, Mayrhuber ES, Mandl S, Monson K (2015) How do young people engage with climate change? The role of knowledge, values, message framing, and trusted communicators. WIREs Climate Change 6(5), Article 5. https://doi.org/10.1002/wcc.353

Cossman B (2013) Anxiety governance. Law Soc Inquiry 38(4), Article 4. https://doi.org/10.1111/lsi.12027

Costello A, Abbas M, Allen A, Ball S, Bell S, Bellamy S, Friel S, Groce N, Johnson A, Kett M, Lee M, Maria; Levy MAC, McCoy PH, McGuire B, Montgomery MD, Napier D, Pagel C, Patel J, de Oliveira JAP, Redclift N, Rees H, Rogger D, Scott J, Stephenson J, Twigg J, Wolff J, Patterson C (2009 Managing the health effects of climate change. Lancet (London, England) 373(9676):1693–1733. https://doi.org/10.1016/s0140-6736(09)60935-1

Cunsolo AE, Neville (2018) Ecological grief as a mental health response to climate change-related loss. Nat Climate Change 8(4):275–281. https://doi.org/10.1038/s41558-018-0092-2

Diamond IR, Grant RC, Feldman BM, Pencharz PB, Ling SC, Moore AM, Wales PW (2014) Defining consensus: a systematic review recommends methodologic criteria for reporting of Delphi studies. J Clin Epidemiol 67(4):401–409. https://doi.org/10.1016/j.jclinepi.2013.12.002

Doherty TJC, Susan (2011a) The psychological impacts of global climate change. Am Psychol 66(4), Article 4. https://doi.org/10.1037/a0023141

Doherty TJ, Clayton S (2011b) The psychological impacts of global climate change. Am Psychol 66(4):265–276. https://doi.org/10.1037/a0023141

Erkal S (n.d.) Comparison of environmental attitudes of University Students determined via the new environmental paradigm scale according to the students' personal characteristics
Erlingsson C, Brysiewicz P (2017) A hands-on guide to doing content analysis. Afr J Emergency Med Revue Africaine de La Medecine d'urgence 7(3):93–99. https://doi.org/10.1016/j.afjem.2017.08.001
Grose A (2020) A guide to eco-anxiety: how to protect the planet and your mental health. Watkins
Hayes K, Blashki G, Wiseman J, Burke S, Reifels L (2018) Climate change and mental health: risks, impacts and priority actions. Int J Mental Health Syst 12(1):28–28. https://doi.org/10.1186/s13033-018-0210-6
Helm SV, Pollitt A, Barnett MA, Curran MA, Craig ZR (2018a) Differentiating environmental concern in the context of psychological adaption to climate change. Glob Environ Change 48, Article NA. https://doi.org/10.1016/j.gloenvcha.2017.11.012
Helm, SV, Pollitt A, Barnett MA, Curran MA, Craig ZR (2018b) Differentiating environmental concern in the context of psychological adaption to climate change. Glob Environ Change 48:158–167. https://doi.org/10.1016/j.gloenvcha.2017.11.012
Hickman C, Marks E, Pihkala P, Clayton S, Lewandowski RE, Mayall EE, Wray B, Mellor C, van Susteren L (2021a) Climate anxiety in children and young people and their beliefs about government responses to climate change: a global survey. Lancet Planetary Health 5(12), Article 12. https://doi.org/10.1016/s2542-5196(21)00278-3
Hickman C, Marks E, Pihkala P, Clayton S, Lewandowski RE. Mayall EE, Wray B, Mellor C, van Susteren L (2021b) Climate anxiety in children and young people and their beliefs about government responses to climate change: a global survey. Lancet Planetary Health 5(12):e863–e873. https://doi.org/10.1016/s2542-5196(21)00278-3
Huang H-PY, Larry D (2005) A comparative study of Canadian and Taiwanese grade 5 children's environmental behaviors, attitudes, concerns, emotional dispositions, and knowledge. Int J Sci Math Educ 1(4):419–448. https://doi.org/10.1007/s10763-005-1098-6
IPCC (2022a) Global warming of 1.5°C: IPCC special report on impacts of global warming of 1.5°C above pre-industrial levels in context of strengthening response to climate change, sustainable development, and efforts to eradicate poverty, 1st edn. Cambridge University Press. https://doi.org/10.1017/9781009157940
IPCC (2022b) Global warming of 1.5°C: IPCC special report on impacts of global warming of 1.5°C above pre-industrial levels in context of strengthening response to climate change, sustainable development, and efforts to eradicate poverty, 1st edn. Cambridge University Press. https://doi.org/10.1017/9781009157940
Katzung J, Boykoff M, Nacu-Schmidt A (2021) Media and climate change observatory special issue 2020: a review of media coverage of climate change and global warming in 2020 [Xlsx]. University of Colorado Boulder. https://doi.org/10.25810/2NC9-BJ43
Kurth C (2018) The anxious mind. https://doi.org/10.7551/mitpress/11168.001.0001
Léger-Goodes T, Malboeuf-Hurtubise C, Mastine T, Généreux M, Paradis P-O, Camden C (2022) Eco-anxiety in children: a scoping review of the mental health impacts of the awareness of climate change. Front Psychol 13:872544. https://doi.org/10.3389/fpsyg.2022.872544
Levac D, Colquhoun H, O'Brien KK (2010) Scoping studies: advancing the methodology. Implementation Sci 5(1):69–69. https://doi.org/10.1186/1748-5908-5-69
Li CJM, Martha C (2017) Exploring the essential psychological factors in fostering hope concerning climate change. Environ Educ Res 25(6):936–954. https://doi.org/10.1080/13504622.2017.1367916
Maiteny P (2018) Longing to be human: evolving ourselves in healing the earth. In: Vital signs, pp 69–82. https://doi.org/10.4324/9780429484728-11
Milfont TL, Wilson J, Diniz P (2012) Time perspective and environmental engagement: a meta-analysis. Int J Psychol 47(5):325–334. https://doi.org/10.1080/00207594.2011.647029
Munn Z, Peters MDJ, Stern C, Tufanaru C, McArthur A, Aromataris E (2018) Systematic review or scoping review? Guidance for authors when choosing between a systematic or scoping review approach. BMC Med Res Methodol 18(1):143. https://doi.org/10.1186/s12874-018-0611-x

Ojala M (2012a) Hope and climate change: the importance of hope for environmental engagement among young people. Environ Educ Res 18(5), Article 5. https://doi.org/10.1080/13504622.2011.637157

Ojala M (2012b) Regulating worry, promoting hope: how do children, adolescents, and young adults cope with climate change? Int J Environ Sci Educ 7(4), Article 4

Ojala M (2012c) Regulating worry, promoting hope: how do children, adolescents, and young adults cope with climate change? Int J Environ Sci Educ 7(4), Article 4

Orru H, Ebi KL, Forsberg B (2017) The interplay of climate change and air pollution on health. Curr Environ Health Rep 4(4), Article 4. https://doi.org/10.1007/s40572-017-0168-6

Patrick R, Snell T, Gunasiri H, Garad R, Meadows G, Enticott J (2022) Prevalence and determinants of mental health related to climate change in Australia. Austr N Z J Psychiatry 57(5):710–724. https://doi.org/10.1177/00048674221107872

Pihkala P (2020a) Anxiety and the ecological crisis: an analysis of eco-anxiety and climate anxiety. Sustainability 12(19):7836. https://doi.org/10.3390/su12197836

Pihkala P (2020b) Eco-anxiety and environmental education. Sustainability 12(23), Article 23. https://doi.org/10.3390/su122310149

Pihkala PP (2018) Eco-anxiety, tragedy, and hope: psychological and spiritual dimensions of climate change. Zygon® 53(2):545–569. https://doi.org/10.1111/zygo.12407

Ratinen I, Uusiautti S (2020a) Finnish students' knowledge of climate change mitigation and its connection to hope. Sustainability 12(6):2181. https://doi.org/10.3390/su12062181

Ratinen I, Uusiautti S (2020b) Finnish students' knowledge of climate change mitigation and its connection to hope. Sustainability 12(6):2181. https://doi.org/10.3390/su12062181

Ray SJ (2020a) A field guide to climate anxiety. https://doi.org/10.1525/9780520974722

Ray SJ (2020b) A field guide to climate anxiety: how to keep your cool on a warming planet

Rutland A, Killen M, Abrams D (2010) A new social-cognitive developmental perspective on prejudice: the interplay between morality and group identity. Perspect Psychol Sci 5(3):279–291. https://doi.org/10.1177/1745691610369468

Salamon MK, Gage M (2023) Facing the climate emergency, 2nd edn. New Society Publishers

Searle K, Gow K (2010) Do concerns about climate change lead to distress. Int J Climate Change Strategies Manage 2(4), Article 4. https://doi.org/10.1108/17568691011089891

Stanley SK, Hogg TL, Leviston Z, Walker I (2021) From anger to action: differential impacts of eco-anxiety, eco-depression, and eco-anger on climate action and wellbeing. J Climate Change Health 1. https://doi.org/10.1016/j.joclim.2021.100003

Stevenson K, Peterson N (2015a) Motivating action through fostering climate change hope and concern and avoiding despair among adolescents. Sustainability 8(1):6. https://doi.org/10.3390/su8010006

Stevenson K, Peterson N (2015b) Motivating action through fostering climate change hope and concern and avoiding despair among adolescents. Sustainability 8(1):6. https://doi.org/10.3390/su8010006

Strife S (2012) Children's environmental concerns: expressing ecophobia. J Environ Educ 43(1):37–54. https://doi.org/10.1080/00958964.2011.602131

Tricco AC, Lillie E, Zarin W, O'Brien KK, Colquhoun H, Levac D, Moher D, Peters MDJ, Horsley T, Weeks L, Hempel S, Akl EA, Chang C, McGowan J, Stewart LA, Hartling L, Aldcroft A, Wilson MG, Garritty C, Lewin S, Godfrey C, Macdonald M, Langlois EV, Soares-Weiser K, Moriarty J, Clifford T, Tunçalp Ö, Straus SE (2018) PRISMA extension for scoping reviews (PRISMA-ScR): checklist and Explanation. Ann Internal Med 169(7):467–473. https://doi.org/10.7326/m18-0850

Tucci J, Mitchell J, Goddard C (2007) Children's fears, hopes and heroes: modern childhood in Australia (Australia) [Report]. Australian Childhood Foundation. https://apo.org.au/node/8255

Verplanken BM, Elizabeth D, Alexandru I (2020) On the nature of eco-anxiety: how constructive or unconstructive is habitual worry about global warming? J Environ Psychol 72:101528. https://doi.org/10.1016/j.jenvp.2020.101528

Wolf JM, Susanne C (2011) Individual understandings, perceptions, and engagement with climate change: insights from in-depth studies across the world. Wires Clim Change 2(4):547–569. https://doi.org/10.1002/wcc.120
Wolfe SE, Tubi A (2018) Terror management theory and mortality awareness: a missing link in climate response studies? WIREs Climate Change 10(2). https://doi.org/10.1002/wcc.566
Zummo L, Gargroetzi E, Garcia A (2020) Youth voice on climate change: using factor analysis to understand the intersection of science, politics, and emotion. Environ Educ Res 26(8):1207–1226. https://doi.org/10.1080/13504622.2020.1771288

Chapter 16
Navigating Academic Challenges and Psychological Well-Being: A Study Among Students of HEIs Amidst COVID-19

Nanigopal Kapasia, Margubur Rahaman, Avijit Roy, and Pradip Chouhan

Abstract COVID-19 pandemic has devastated the entire education system of the globe. Academic institutions have faced an immense challenge about teaching and learning process in lockdown phases. Students often face a multitude of psychological stresses in the form of sleeplessness, heightened anxiety, feelings of frustration, and struggles with maintaining concentration on their academic pursuits. Therefore, the present chapter aimed to investigate and navigate the educational challenges as well as psychological well-being of students in higher education. A total of 570 students from higher education institutions adopted the snowball sampling method. The models of Multivariate Ordered Logistic Regression (MOLR) were run for envisaging predictors of challenges and psychological well-being among students of higher education. Fourth of ten participants (40%) were academically moderate satisfied amidst COVID-19. Literally, 41.4% respondents perceived a higher psychological stress in their academic careers. The chapter highlights a pronounced adverse effect of COVID-19 on students' academic satisfaction as well as psychological well-being. The chapter suggests that it is necessary to improve the technology infrastructure and develop a new strategic model effectively to navigate such unexpected situations. The chapter provides valuable insights, suggesting the need for more in-depth research on this critical issue.

N. Kapasia (✉) · A. Roy
Department of Geography, Malda College, Malda, West Bengal 732101, India
e-mail: nanigopal.kapasia@gmail.com

M. Rahaman
Department of Migration and Urban Studies, International Institute for Population Sciences (IIPS), Mumbai, India

P. Chouhan
Department of Geography, University of Gour Banga, Malda, West Bengal 732103, India

U. Mukhopadhyay et al. (eds.), *Climate Crisis, Social Responses and Sustainability*, Climate Change Management, https://doi.org/10.1007/978-3-031-58261-5_16

Introduction

COVID-19 pandemic exploded during the climate change crisis, forcing individuals to adapt abruptly to a new environment and also triggered to be changes in everyone's routine life and works as a sustainable way (Ruiu et al. 2022). Apparently there are no linkages and association between climate change and COVID 19 but both involves in interconnected factors and impacts especially on health, air quality, disease transmission, global response and collaboration to address its complexities and interconnected challenges. COVID-19 affects every aspect of the society including economy, social, education, industry, health infrastructures, and technological advancements. Among them, the far reaching as well as extensive impact mostly evidenced on global education system irrespective of developed and developing nations and global heath (Servant-Miklos, 2022; Rashid and Yadav 2020; Dhawan 2020). Abrupt closure of educational institutions, mandated by social distancing measures to curb infection spread, posed a significant challenge to the teaching as well as learning process (Sahoo et al. 2021). This led to movement of traditional classes to online learning platforms (Rashid and Yadav 2020). Even in lockdown, students perceived stresses of encountered heightened psychological stress, manifesting as sleeplessness, headaches, anxiety, frustration, and irritation that may led a difficulties for studies (Sreeram & Mundada, 2021; Kumalasari & Akmal, 2021). The educational institutions were compelled to adapt swiftly, addressing concerns related to the quality of education, delivery modes, accessibility, and the imperative of lifelong learning facilitated by ICT tools among others (Kapasia et al., 2020; Mondal et al. 2021; Saha et al., 2021).

Students engaged inpractical-oriented courses express concernsabout missing hands-on learning opportunities (Muthuprasad et al. 2021). The previous studies show that fear of infection, uncertainty about isolation (Sharma et al. 2021) and disruption of daily routine (Drane 2020), anxious with self-health, unavailability of proper study spaces (Wathelet et al. 2020) causes of negativities on online learning. During online classes, learners are disappointed with the problems of internet connectivity, absence of gadgets for study (Drane 2020; Mondal et al. 2021), lack of proper guidance (Surahman and Sulthoni 2020). Some studies reveal learners are more spontaneous about studying in physical classes than in virtual classes because of more interaction (Chakraborty et al. 2021), study gestures, and joyful learning (Dhawan, 2020). Indeed, many studies in different countries (Hussein et al. 2020; Hashemi et al. 2021; She et al. 2021; Hassan et al. 2021) found a positive linkage between learner's satisfaction and e-learning amidst covid-19.

Pandemic of COVID-19 unprecedentedly disrupted to entire education system, necessitating a movement of conventional classrooms to virtual learning. Existing research has provided substantial insights into comprehending the nexus between the COVID-19 crisis and students' mental well-being on a broader scale. For instance, numerous studies underscore the challenges posed to learners, impacting both academic performance and mental health (Dhawan, 2020; Razami and Ibrahim 2021). However, it is acknowledged that the association between the COVID-19 crisis and mental health well-being exhibits variations based on spatial, socio-demographic, and temporal backgrounds (Sreeram and Mundada 2021). Considering the spatial aspect, there is a limited body of literature that specifically delves into the well-being of students in West Bengal during the pandemic. Therefore, this chapter aims to address this gap by exploring academic satisfaction and psychological stress among students in West Bengal. The contribution of this chapter lies in two key aspects. Firstly, it presents the context of perceived academic satisfaction and psychological stress among students, considering their background characteristics. Secondly, it provides an in-depth understanding of these perceptions in West Bengal by drawing parallels with similar studies conducted in different locations. The findings of this chapter are intended to be valuable for researchers and policymakers, offering insights into site and situation-specific changes in perceived academic satisfaction and psychological stress during the pandemic. This, in turn, can inform the formulation of policies that incorporate socio-ecological approaches, tailored to the unique site and situation dimensions.

Materials and Methods

The online survey (through social media i.e. WhatsApp, email, and Face book) was conducted in July–August, 2021 to assess the academic satisfaction and psychological stress among students in higher education institutions. We received 570 responses from students in different higher education levels including undergraduates, postgraduates, and research scholars. In a structured online questionnaire, students were provided both close-ended and open ended questions to elucidate their existing situation. We selected the snowball sampling method for collecting data. The details description of the survey procedure and flow of research is presented in Fig. 16.1. We provided consent to assure their confidentiality.

Two outcomes e.g. (a) Respondent's perceived academic satisfaction and (b) Psychological stress had been analysed in this chapter. The perceived academic satisfaction level (Hassan et al. 2021; Subashini et al. 2022) was assessed by using thirteen questions related to online learning. The responses were recorded on 4 points of Likert scale: 1 = not happy at all, 2 = happy to some extent, 3 = moderately happy and 4 = extremely happy. The psychological stress (Rabaan et al. 2021; Sundarasen et al. 2020) was measured from ten questions related to learning stress and stress related to mental health. The responses were recorded on 4 4-point Likert scale: 1 = not at all stressed; 2 = feeling stressed to some extent; 3 = moderately feeling stressed; 4 = extremely feeling stressed. For framing the structures of questionnaire, previous relevant studies were also followed (AlOmari et al. 2020; Sundarasen et al. 2020). The social, demographic and economic factors as independent variables were taken into consideration. These include respondent's age, sex (male and female), educational level (undergraduate, postgraduate and researchers), residential area (rural and urban) family income (Rs.), and family structure (nuclear, joint and broken).

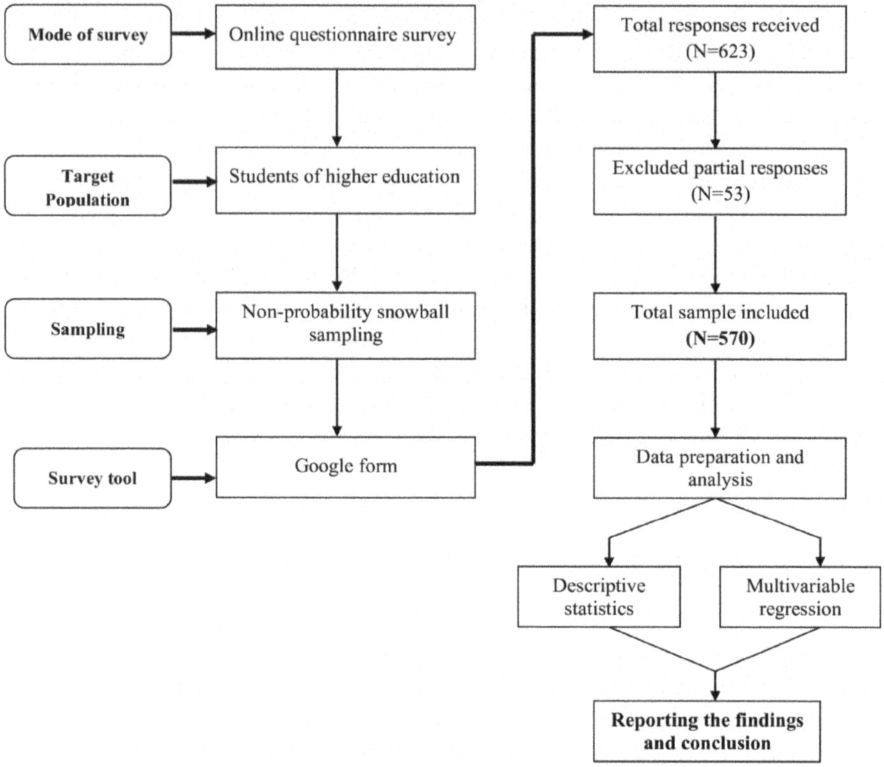

Fig. 16.1 Survey procedure and flow of research

After creating a spreadsheet, the responses were exported into Stata 12.1 (Stata Corp., USA) for the statistical analyses. In summary statistics, frequencies and percentages were used for categorical variables while mean was used for continuous variables. For the assessment of predictors regarding respondent's academic satisfaction as well as psychological stress, model of multiple-ordered logistic regression was performed using STATA (Version: 12.1).

Results and Discussion

Learning Status of the Students During Covid-19

During the lockdown period, along with disruption of traditional learning, students faced challenges and dilemmas in remote learning (Bodrud-Doza et al. 2020). Many educational institutions have tried quickly to adapt virtual and online learning comfortably for the stakeholders (Insigne et al. 2021). Since academic calendar

Table 16.1 Learning progress amidst Covid-19

Learning modalities	Frequency	Percentage
Through textbook and online classes	345	60.53
Only through online classes	129	22.63
Reading textbooks with own effort	96	16.84
Completion of study Syllabus		
Below 50%	188	32.98
Above 50%	327	57.37
Just exam completed	13	2.28
Not yet completed exam	42	7.37
Visit e-pathshala for learning materials		
Yes	302	52.98
No	114	20.00
Don't know	154	27.02
Time spent on the study		
Below of routine time	318	55.79
Above of routine time	145	25.44
Same like routine time	107	18.77

Source Online questionnaire survey, 2021

has been disrupted due to the closures of institutions for lockdown. The learners learned online to cope with this situation. In this situation, proportionately most of the learners (60.53%) were learning through text book and online classes and only 16.84% of learners were learning only through reading of their textbooks. There exists a slight variation in the duration of academic sessions for participants' enrolment in various universities. Majority of participants (57.37%) reported more than half of their concerned syllabus has been covered. Along with quick adaptation to e-learning, learners also followed e-pathsala for the purpose of study-materials where more than half (52.98%) reported they had visited the e-pathsala website for learning materials along with online classes. During the pandemic, most of the learners spent below of their routine time for study purposes because of the monotonicity in online classes, problems related to internet connectivity, not having their own devices for e-learning, and the absence of a separate and proper study room (Table 16.1).

Academic Challenges During Covid-19

At the time of COVID 19, all the countries are faced adverse effect of COVID 19 with enduring climate change crisis (Liwanag and Lee 2021). Since, COVID-19 pandemic

Table 16.2 Incidence of contentment with academics among students

Satisfaction level	Frequency	Percentage
Low	199	34.91
Moderate	228	40.00
High	124	21.75
Very high	19	3.34

Source Online questionnaire survey, 2021

spurred the advancement of technology on educational field to address difficult situation. Digital solutions like virtual meeting for communication, online learning platforms have become essential tools for navigating the challenges faced by the learners amidst pandemic (Kaukko et al. 2021). Table 16.2 displays the level of academic satisfaction, perceived by higher education students during Covid-19 pandemic. A total of 570 students, more than one-third of learners (34.91%) had low levels of academic satisfaction and faced academically strong challenges whereas proportionately 40% of students had moderate levels of academic satisfaction. The possible reasons behind academically challenges of students lie on educational inequality due to properly not access of technology, low economic status and pre-existing health problem as a cause of climate change. Approximately 21.75% of students reported a high level of satisfaction with their academics during COVID-19.

The lockdown phases due to COVID-19 have changed the landscape of higher education (Singha et al. 2020). Where the conventional teaching–learning process has shifted into mode of online, certain evaluations and examinations are also conducted through online platforms. So the conventional academic institutions pose a significance challenge for both teachers and learners (Mondal et al. 2021). Previous research revealed that students are faced with challenges for physically practical subjects, anxious about the completion of the syllabus, and experiencing eye strain after prolonged periods of e-learning (Sasmal and Roy 2021). Contrarily, greatest advantages of online learning lies it time and cost-effectiveness (Rahiem et al. 2021) along with more flexible and wider access (Rashid and Yadav 2020).

Based on ordered logistic regression models, Table 16.3 shows the likelihood of satisfaction regarding academic performance among participants during COVID-19.The findings indicate that, in this pandemic scenario, students older than 25 years were less happy with their academic performance (OR: 0.66; CI: 0.17, 3.06).Interestingly, in online mode of learning, Female students showed a higher likelihood of satisfaction (OR: 1.24; CI: 0.81, 1.90) compared to their male counterparts. Participants who are pursuing research tended to have a poor likelihood of academic satisfaction (OR: 0.38; CI: 0.21, 0.54) as compared to other students. Regarding location of residence, learners of urban areas were less happy than those participants who are residing in rural settings (OR: 0.87; CI: 0.56, 1.42).

Table 16.3 Likelihood of academic satisfaction among participants by OLR models

Variables	Academic satisfaction index	
	OR	95% CI
Respondent's age (years)		
< 20®		
21–25	0.89*	0.56, 2.14
> 25	0.66**	0.17, 3.06
Sex of respondents		
Male®		
Female	1.24*	0.81, 1.90
Level of education		
UG level®		
PG level	0.84	0.47, 1.33
Researchers	0.38***	0.21, 0.74
Family structures		
Nuclear®		
Joint and extended	0.74	0.47, 1.11
Broken	0.23***	0.11, 1.79
Monthly family income (Rs.)		
Below 20,000®		
20,000–50,000	1.00	0.58, 1.69
Above 50,000	1.22*	0.59, 2.51
Residential area		
Rural®		
Urban	0.87*	0.56, 1.42

Note *** $p < 0.01$, ** $p < 0.05$, * $p < 0.1$; *OR* odds ratio; *CI* confidence interval; ® reference category
Source Online questionnaire survey, 2021

Challenges Associated with Online Learning Amidst COVID-19

Previous research studies reported that the climate changes and COVID pandemic negatively influenced on mental health especially among elderly, children, women, people with pre-existing health problems and low economic status (Marazziti et al. 2021; Perkins et al. 2021). Poor economic conditions the family for low-income and family members infected by COVID-19 were also significant contributors to the psychological stress of learners was perceived by 18.48% and 5.28% pupils respectively (Table 16.4). It is considered that pupil belongs to poor income family because of climate change (rising temperature and global warming), loss of job in COVID phase are felt stressful at the time of learning, besides chronically ill person at home,

Table 16.4 Problems related to learning and others during Covid-19

Various problems	Frequency(n)	Percentage (%)
Related to learning		
The feeling of stress, depression, and anxiety due to isolation	124	20.46
Don't have your gadgets for online learning	98	16.17
Absence of a favourable learning environment at home	73	12.05
Limited access to the internet	87	14.36
Not interested in teaching	51	8.41
Related to other than learning		
Stress due to financial crisis at home	112	18.48
Stress due to chronically ill person at home	29	4.79
Stress because of a family member infected by COVID-19	32	5.28

Source Online questionnaire survey, 2021

family member infected by COVID all articulate to learners into a academically challenges environment at the time of online classes. It must be established that pupil's adaptation in online learning is not a technical issue rather a pedagogical and instructional challenges (Ali and Kour 2020; Insigne et al. 2021). Along with these, learners felt a lot of psychological stress with studies and families. The chapter revealed that proportionately most of the learners (20.46%) suffered feelings of stress, depression, and anxiety due to remaining isolated. The lack of interaction within class room as well as technological problems during online study has provoked more stress for learners (Joshi 2021; Mondal et al. 2021). The pupils facing challenges for the lack of their gadgets for online study (16.17%) and limited access to the internet (14.36%), insufficient interaction with their teachers (8.41%). Learners faced lots of academic challenges to learn in virtual environment, to mediating these challenges, teachers and facilitators needs to have basic ICT tools effectively, access to its application and learning platforms along with acknowledgement of learners readiness.

•

Psychological Well-Being of Students During Covid-19

The COVID pandemic and climate change negatively affect the well being of the pupils. In COVID pandemic and climate crisis, learners are associated with greater anxiety, frustration, distress, depression that hamper significantly their psychological well being (Lawrance et al. 2022). Like climate change, COVID 19 issues presenting significant challenges to global health. As a greatest public health crisis, COVID pandemic mostly affects the learner's mental health due to its uncertainty, learner's loneliness, social isolation and financial crisis of the family (Van der Ven and Sun 2021; Manzanedo and Manning 2020). Table 16.5 reveals nearly one-fourth

Table 16.5 Prevalence of psychological stress among students

Psychological stress levels	Psychological stress	
	n	%
Low	14	2.46
Moderate	166	29.12
High	154	27.02
Very high	236	41.40

Source Online questionnaire survey, 2021

of students (41.40%) had a high level of stress. Around 27.02% of learners faced high levels of psychological stress in their daily lives during the COVID-19 phase.

Previous research works in different settings found that psychological well being of higher-educated learners are agitated by feeling of loneliness, hopelessness (Rahiem et al. 2021), social anxiety (Stockinger et al. 2021), depression, and unhealthy behaviours (Eden et al. 2020). Literally, COVID 19 pandemic led psychological distress and poor mental health (Joshi 2021; Stockinger et al. 2021). During the lockdown, isolated environment, unable to access physical interaction with peers and relatives influence their emotional well-being (Drane et al. 2021). To preserve the psychological well being and a good mental health of learners, there needs to watch entertaining media content, conducting active virtual group discussion session, to arrange mental health counselling programme to enhance psychological well being (Singh et al. 2021; Ebrahim et al. 2022).

The results (Table 16.6) show students' psychological stress significantly associated with age of the participants. Similar line of result found in previous relevant studies of different countries (Lawrance et al. 2022; Ebrahim et al. 2022; Azizet al. 2023) amidst COVID 19. Students involved in research work tended to have a greater probabilities of psychological stress (OR: 3.58; CI: 2.36, 4.14) than who were enrolled in UG and PG. The probable reason of greater distress among high educated learners are not to getting job in uncertain pandemic environment, increasing age having enough academic qualification and strongly isolated within concrete wall at urban areas. Mental stress was more prevalent among students belonging to broken family backgrounds. Similarly, a higher probability of mental stress was significantly observed among students lived in urban counterparts (OR: 1.25; CI: 0.76, 1.81) than rural.

Limitations of the Study

We adopted the method of snowball sampling to collect a desirable sample size on a circumstance. Since our chapter did not represent the entire population. So, we cannot make statistical inferences about the entire population and are not able to suggest

Table 16.6 Probability of psychological stress among students amidst COVID-19

Variables	Psychological stress index	
	OR	95% CI
Respondent's age (years)		
< 20®		
21–25	1.83	1.11, 2.27
> 25	4.63***	2.89, 5.36
Sex of respondents		
Male®		
Female	0.84	0.21, 1.42
Level of education		
UG level®		
PG level	1.35**	0.95, 1.83
Researchers	3.58*	2.36, 4.14
Family structures		
Nuclear®		
Joint and extended	0.67	0.33, 1.06
Broken	5.71***	3.38, 8.09
Monthly family income (Rs.)		
Below 20,000®		
20,000–50,000	0.87	0.51, 1.47
Above 50,000	0.37***	0.31, 1.27
Residential area		
Rural®		
Urban	1.25**	0.76, 1.81

Note *** $p < 0.01$, ** $p < 0.05$, * $p < 0.1$; *OR* odds ratio; *CI* confidence interval; *Ref*. reference category
Source Online questionnaire survey, 2021

any policy perspective orientation. But we can generalize the situation which may be helpful for better understanding and further in-depth research about this situation.

Conclusion

COVID pandemic has come in light at the period of climate change crisis. Although both are distinct phenomena, yet they shared commonalities in terms of their impacts on human health and environmental sustainability. To address these interconnected challenges requires holistic approaches to build more resilient and sustainable societies in the face of emerging threats and uncertainties. In the context of COVID pandemic, the chapter explored more or less all the learners are affected by Covid-19 pandemic. They felt worries in their studies, physical health, and safety remaining isolation in their home. It is true that online learning enables students to explore various opportunities to enhance skill and knowledge that ensured to navigate the academic challenges and psychological well being of the pupils. To navigate the psychological well being of learners, teachers also need to be more interactive and encourageous to online teaching along with upgrade technological infrastructures using ICT tools. In addition, parent and guardians need to take proactive interest in their children's education and fulfil their study required needs. Since, the chapter comprises a situational study, so future researchers are suggested to validate the findings related to academic challenges and psychological well being of the higher educated pupils using more balance sample from different types of higher educational institutions along with longitudinal studies.

Acknowledgements The authors would like to thank each and every participant for contributing their valuable time, opinion, and thinking for their perceived academic challenges and psychological stress during COVID-19.

References

AlOmari O, Al Sabei S, Al Rawajfah O, Abu Sharour L, Aljohani K, Alomari K, Alhalaiqa F (2020) Prevalence and predictors of depression, anxiety, and stress among youth at the time of COVID-19: an online cross-sectional multicountry study. In: Depression research and treatment

Aziz ARA, Ashaari RNSR, Ali MZM (2023) The impact of stress to student's psychological well-being amidst Covid-19 pandemics: Impak Tekanan terhadap Kesejahteraan Psikologi Pelajar semasa Pandemik Covid-19. Sains Insani 8(2):179–189

Bodrud-Doza M, Shammi M, Bahlman L, Islam ARMT, Rahman MM (2020) Psychosocial and socio-economic crisis in Bangladesh due to COVID-19 pandemic: a perception-based assessment. Front Public Health 341

Chakraborty P, Mittal P, Gupta MS, Yadav S, Arora A (2021) Opinion of students on online education during the COVID-19 pandemic. Human Behav Emerg Technol 3(3):357–365

Dhawan S (2020) Online learning: a panacea in the time of COVID-19 crisis. J Educ Technol Syst 49(1):5–22

Drane CF, Vernon L, O'Shea S (2021) Vulnerable learners in the age of COVID-19: a scoping review. Austr Educ Research 48(4):585–604

Ebrahim AH, Dhahi A, Husain MA, Jahrami H (2022) The psychological well-being of university students amidst COVID-19 pandemic: scoping review, systematic review and meta-analysis. Sultan Qaboos Univ Med J 22(2):179

Hassan SUN, Algahtani FD, Zrieq R, Aldhmadi BK, Atta A, Obeidat RM, Kadri A (2021) Academic self-perception and course satisfaction among university students taking virtual classes during the COVID-19 pandemic in the Kingdom of Saudi-Arabia (KSA). Educ Sci 11(3):134

Hashemi A (2021) Effects of COVID-19 on the academic performance of Afghan students' and their level of satisfaction with online teaching. Cogent Arts Human 8(1):1933684

Hussein E, Daoud S, Alrabaiah H, Badawi R (2020) Exploring undergraduate students' attitudes towards emergency online learning during COVID-19: a case from the UAE. Child Youth Serv Rev 119:105699

Joshi S (2021) Rising importance of remote learning in India in the wake of COVID-19: issues, challenges and way forward. World J Sci Technol Sustain Develop 18(1):44–63

Kapasia N, Paul P, Roy A, Saha J, Zaveri A, Mallick R et al (2020) Impact of lockdown on learning status of undergraduate and postgraduate students during COVID-19 pandemic in West Bengal, India. Children Youth Serv Rev 116:105194

Kaukko M, Kemmis S, Heikkinen HL, Kiilakoski T, Haswell N (2021) Learning to survive amidst nested crises: can the coronavirus pandemic help us change educational practices to prepare for the impending eco-crisis? Environ Educ Res 27(11):1559–1573

Kumalasari D, Akmal SZ (2021) Less stress, more satisfied in online learning during COVID-19 pandemic: the moderation role of academic resilience. Psychol Res Urban Soc 4(1):36–44

Lawrance EL, Jennings N, Kioupi V, Thompson R, Diffey J, Vercammen A (2022) Psychological responses, mental health, and sense of agency for the dual challenges of climate change and the COVID-19 pandemic in young people in the UK: an online survey study. Lancet Planetary Health 6(9):e726–e738

Manzanedo RD, Manning P (2020) COVID-19: Lessons for the climate change emergency. Sci Total Environ 742:140563

Marazziti D, Cianconi P, Mucci F, Foresi L, Chiarantini I, Della Vecchia A (2021) Climate change, environment pollution, COVID-19 pandemic and mental health. Sci Total Environ 773:145182

Mondal BK, Sahoo S, Paria P, Chakraborty S, Alamri AM (2021) Multi-sectoral impact assessment during the 1st wave of COVID-19 pandemic in West Bengal (India) for sustainable planning and management. Arab J Geosci 14:1–26

Muthuprasad T, Aiswarya S, Aditya KS, Jha GK (2021) Students' perception and preference for online education in India during COVID-19 pandemic. Social Sci Human Open 3(1):100101

Perkins KM, Munguia N, Ellenbecker M, Moure-Eraso R, Velazquez L (2021) COVID-19 pandemic lessons to facilitate future engagement in the global climate crisis. J Clean Prod 290:125178

Rabaan AA, Al-Ahmed SH, Muhammad J, Khan A, Sule AA, Tirupathi R et al (2021) Role of inflammatory cytokines in COVID-19 patients: a review on molecular mechanisms, immune functions, immunopathology and immunomodulatory drugs to counter cytokine storm. Vaccines 9(5):436

Rahiem MD, Krauss SE, Ersing R (2021) Perceived consequences of extended social isolation on mental well-being: narratives from Indonesian university students during the COVID-19 pandemic. Int J Environ Res Public Health 18(19):10489

Rashid S, Yadav SS (2020) Impact of Covid-19 pandemic on higher education and research. Indian J Human Develop 14(2):340–343

Razami HH, Ibrahim R (2021) Distance education during COVID-19 Pandemic: the perceptions and preference of university students in Malaysia towards online learning. Int J Adv Comput Sci Appl 118–126

Ruiu G, Ruiu ML, Ragnedda M (2022) How the COVID-19 pandemic impacted the perception of climate change in the UK. Am Behav Sci

Saha T, Das PP, Singh R (2021) Challenges in higher education during and after COVID-19 pandemic in India. J Phys Conf Ser 1797(1):012065

Sahoo KK, Muduli KK, Luhach AK, Poonia RC (2021) Pandemic COVID-19: an empirical analysis of impact on Indian higher education system. J Stat Manag Syst 24(2):341–355

Sasmal S, Roy M (2021) Perception of undergraduate nursing students regarding e-learning during COVID-19 pandemic in West Bengal. Int J Commun Med Public Health 8(4):1892–1898

Servant-Miklos V (2022) Environmental education and socio-ecological resilience in the COVID-19 pandemic: lessons from educational action research. Environ Educ Res 28(1):18–39

She L, Ma L, Jan A, Sharif Nia H, Rahmatpour P (2021) Online learning satisfaction during COVID-19 pandemic among Chinese university students: the serial mediation model. Front Psychol 12:743936

Singh J, Steele K, Singh L (2021) Combining the best of online and face-to-face learning: hybrid and blended learning approach for COVID-19, post vaccine, & post-pandemic world. J Educ Technol Syst 50(2):140–171

Sreeram M, Mundada MA (2021) Survey to assess perceived causes of stress, its manifestations and coping strategies among dental college students in Maharashtra during COVID-19 lockdown. Biomed Pharmacol J 14(1):53–60

Stockinger K, Rinas R, Daumiller M (2021) Student adaptability, emotions, and achievement: navigating new academic terrains in a global crisis. Learn Individ Differ 90:102046

Subashini N, Udayanga L, De Silva LHN, Edirisinghe JC, Nafla MN (2022) Undergraduate perceptions on transitioning into E-learning for continuation of higher education during the COVID pandemic in a developing country: a cross-sectional study from Sri Lanka. BMC Med Educ 22(1):1–12

Sundarasen S, Chinna K, Kamaludin K, Nurunnabi M, Baloch GM, Khoshaim HB, Sukayt A (2020) Psychological impact of COVID-19 and lockdown among university students in Malaysia: implications and policy recommendations. Int J Environ Res Public Health 17(17):6206

Surahman E, Sulthoni S (2020) Student satisfaction toward quality of online learning in Indonesian higher education during the Covid-19 pandemic. In: 2020 6th International conference on education and technology (ICET). IEEE, pp 120–125

Van der Ven H, Sun Y (2021) Varieties of crises: comparing the politics of COVID-19 and climate change. Global Environ Politics 21(1):13–22

Wathelet M, Duhem S, Vaiva G, Baubet T, Habran E, Veerapa E, D'Hondt F (2020) Factors associated with mental health disorders among university students in France confined during the COVID-19 pandemic. JAMA Netw Open 3(10):e2025591–e2025591

Chapter 17
Socio-economic, Health, and Psychological Impacts of Covid-19 on Human Well-Being in East Africa

Kelvin Ngongolo and Doreen Doreen Mrimi

Abstract The Covid-19 pandemic has significantly disrupted society around the world and had an impact on many facets of human welfare. This chapter offers a thorough analysis of how Covid-19 has affected people's quality of life in East Africa. The study highlights the socio-economic, health, and psychological effects of the pandemic in the area, such as elevated mortality rates, difficulties with mental health, a downturn in the economy, and disruptions in social interactions by looking at a wide range of scholarly articles, reports, and data sources. It also looks at regional responses, difficulties encountered, and viable solutions to lessen the consequences and improve people's well-being. It also considers prospective benefits including enhanced cleanliness standards, technology breakthroughs, and societal resiliency. The chapter's findings advance knowledge of the multiple effects of COVID-19 on human wellbeing and offer policymakers, researchers, and practitioners' useful information for formulating plans to lessen the adverse effects and advance the general welfare of people and communities in the post-pandemic era.

Introduction

The outbreak of the novel coronavirus disease (COVID-19) in late 2019 marked the beginning of an unprecedented global health crisis. What began as a localized health concern swiftly evolved into a global crisis, transcending borders, and profoundly impacting societies worldwide. For instance, by the end of July 2023, the total case of COVID-19 was about 692,169,025 worldwide (World Health Organization 2023; Worldometer 2023), The virus, caused by the severe acute respiratory syndrome coronavirus 2 (SARS-CoV-2), rapidly spread across continents, challenging healthcare systems, economies, and societies at large (Chaturvedi 2020; Nuwematsiko et al.

K. Ngongolo (✉) · D. D. Mrimi
Department of Biology, College of Natural and Mathematical Sciences, University of Dodoma, Dodoma, Tanzania
e-mail: kelvin.ngongolo@udom.ac.tz

U. Mukhopadhyay et al. (eds.), *Climate Crisis, Social Responses and Sustainability*, Climate Change Management, https://doi.org/10.1007/978-3-031-58261-5_17

2022). In East Africa, a region characterized by its diverse cultures, vibrant communities, and unique challenges, the COVID-19 pandemic brought forth a myriad of consequences that profoundly affected human well-being (Atzrodt et al. 2020; Chaturvedi 2020; Nuwematsiko et al. 2022; Solymári et al. 2022; Tarimo et al. 2022).

Among the affected economic activites is Agriculuter sector. COVID-19 disrupted global agriculture, causing labor shortages, supply chain disruptions, and fluctuating demand. Market closures, export challenges, and financial strain on farmers ensued (Barman et al. 2021a, 2021b). Livestock and perishable goods suffered, while technology adoption accelerated. Government support became vital. Agriculture faced unprecedented challenges, emphasizing the sector's need for resilience and innovation. The Agriculture sector in East Africa is the socio-economic practised by major and was affected by COVID-19. In addition to agriculture and supply chains, COVID-19 has been noted to impact the global tourism sector, infrastructure, electricity consumption, industrial and small business, and overall GDP growth. It has significantly affected markets and the marketing of various products worldwide (Barman et al. 2021a; Ngongolo 2023).

To address the multifaceted impacts of COVID-19 in Africa, several strategies have been recommended (Petersen et al. 2020). Strengthening public health systems is imperative, involving enhanced healthcare infrastructure, testing capabilities, and medical supplies. Public health awareness campaigns, especially regarding preventive measures, should be prioritized through various channels. Education adaptation, vaccination campaigns, and socio-economic support, including social protection programs and food security initiatives, are crucial (Jahn et al. 2021). Strengthening digital innovation, community engagement, and international collaboration are also emphasized (El Otmani Dehbi et al. 2021; Stewart et al. 2022). This comprehensive review provides valuable insights for policymakers and stakeholders, offering a basis for targeted interventions and future research in the African context (Ngongolo 2023).

This chapter delves into the multifaceted impacts of the COVID-19 pandemic on human well-being in East Africa (Louis et al. 2022). The east Africa coutries in this chapter are represented by four countries whicu are Tanzania, Kenya, Uganda and Rwanda. Well-being encompasses not only physical health but also social, economic, and psychological dimensions. As the virus took its toll on the region, it exposed vulnerabilities and inequalities while revealing the resilience and resourcefulness of East African communities in the face of adversity.

The chapter begins by providing a brief overview of the early stages of the COVID-19 pandemic in East Africa. It will cover the reported cases from 2020 to 2023 for three quarters (Q1,Q2,and Q3) in each year whereby Q1 = January–April, Q2 = May–August Q3 = September–December. Also, the chapter exploired the effects in terms of Health Impact on Human Well-being that includes the strain on healthcare systems, access to medical services, the burden on healthcare workers, and the impacts on vulnerable populations, Socio-Economic Disruption such as job losses, food insecurity, disruptions in supply chains, and challenges faced by the informal sector. Also, the chapter highlighted the Psychological and Mental Health Effects, Gender and Social Impact.

Rationale of the Study

The COVID-19 pandemic has significantly impacted global health, caused socio-economic disruptions and affected vulnerable regions like East Africa. This book chapter investigates the impact on human well-being in East Africa, focusing on the unique challenges faced by the population due to diverse cultures, economies, and poverty. The pandemic's effects are multifaceted, impacting public health, social fabric, economy, and governance structures. This study aims to address the knowledge gap on COVID-19's impact on human well-being in East Africa. Despite extensive research, there is a lack of comprehensive studies on this region. By analyzing aspects of human well-being, such as physical, mental, social, economic, and healthcare, the study aims to identify areas for improvement and develop strategies to strengthen community resilience for future challenges. This study contributes to academic and policy discussions on pandemics and their socio-economic impacts by providing evidence-based insights from the East African context. It helps policymakers understand challenges and develop context-specific responses. The findings inform decision-making and help design targeted interventions to mitigate the pandemic's adverse effects on human well-being.

Objective of the Book Chapter

The aim of this book chapter is to perform a thorough analysis of the various ways in which the COVID-19 epidemic has affected human welfare in East Africa. It aims to provide insights into the resilience of communities by placing the socioeconomic, health, and psychological effects within the particular regional context. Through comparing experiences across East African countries, evaluating long-term effects, and offering policy recommendations, the chapter seeks to advance academic understanding. It hopes to raise public awareness of the interrelated difficulties people have been facing since the pandemic and promote a comprehensive approach to well-being in the area through this investigation.

Materials and Methods

The book chapter involved conducting a comprehensive literature search using relevant keywords and databases. The study focused on the following question; What are the socio-economic, health, and psychological impacts of the COVID-19 pandemic on human well-being in East Africa? Specifically, how do these impacts manifest across diverse populations, geographic regions, and socio-economic strata in the East African context? Additionally, what interventions or coping strategies have

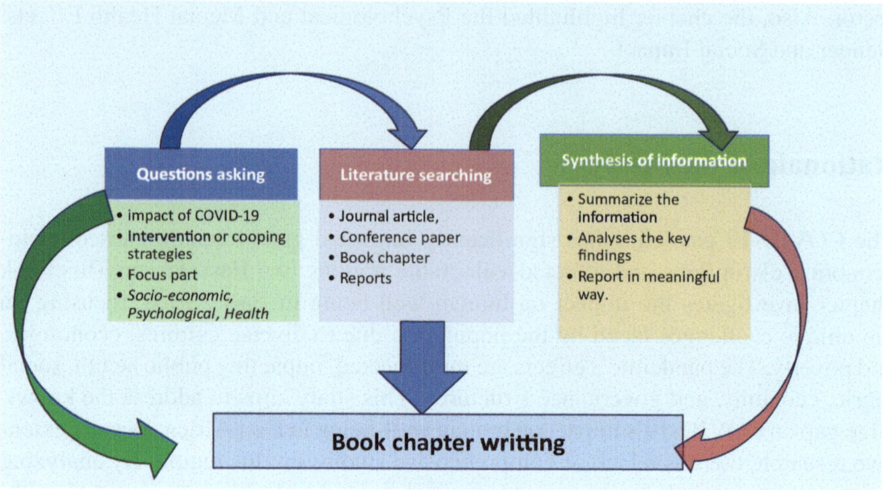

Fig. 17.1 Methodological flowchart for this book chapter

been employed, and what are the long-term consequences of these impacts on the overall well-being of individuals in East Africa? Inclusion Criteria: Empirical studies from January 2020 to present, conducted in East African countries, exploring socio-economic, health, and psychological impacts of COVID-19 on individuals of all demographics. Publication types include peer-reviewed articles, conference papers, reports, and grey literature available in English or with accessible English translations. Exclusion Criteria: Studies outside East Africa, non-original research, non-human studies, non-English studies without translation, irrelevant outcomes, pre-COVID-19 studies, and studies lacking essential data (Fig. 17.1).

A search strategy was performed in databases like PubMed and Google Scholar where key words like COVID-19, well-being, socio-economic, health, psychology, and East Africa. to enhance search in the database, the key words were combined using Boolean operators (AND, OR) to narrow or broaden the search. This make the following searchable strings; PubMed Search String:n ("COVID-19" OR "coronavirus") AND ("socio-economic" OR "health" OR "psychological impact") AND ("human well-being") AND ("East Africa"), Google Scholar Search String: "COVID-19" OR "coronavirus" "socio-economic" OR "health" OR "psychological impact" "human well-being" "East Africa" To comprehend the trajectory, diverse data on COVID-19 cases were gathered from authoritative health sources, including the World Health Organization (WHO). This facilitated a comprehensive analysis of the pandemic's progression and impact.

Trend of COVID -19 in East Africa

The trend of covid varied across the three countries in in Eastern Africa. By July 2023 the highest xases and death due to COVID-19 was reported in Kenya followed by Uganda, Rwanda and Tanzania Table 17.1). The trend of cases from 2020 to 2023 was increasing in all the countries (Fig. 17.2). The cases can increase due to various factors. Some of the common causes of COVID-19 cases increasing include (Kimani et al. 2020). Delta Variant or other variants of concern where by New and more transmissible variants of the virus can lead to a rapid increase in cases as they spread more easily among the population. Another factor may include reduced immunity. Over time, immunity acquired from previous infections or vaccinations may wane, making individuals more susceptible to infection (Weitz et al. 2020). Relaxation of preventive measures such as mask-wearing, social distancing, and hand hygiene can facilitate viral transmission. Also, Large gatherings and events Mass gatherings or events without proper safety protocols can lead to super-spreader situations. Travel and mobility whereby the Increased travel, especially from areas with higher transmission rates, can introduce the virus to new locations and contribute to its spread.

Other factors which can promote the increase of cases for COVID-19 includes; Vaccine hesitancy of which when a significant portion of the population is hesitant or refuses to get vaccinated, it can impede achieving herd immunity and allow the virus to circulate more freely. Seasonal factors whereby Some respiratory viruses, including coronaviruses, exhibit seasonal patterns of transmission, which might lead to increases during certain times of the year (Aabed and Lashin 2021). Asymptomatic and presymptomatic spread. People who are infected but have no symptoms (asymptomatic) or not yet showing symptoms (presymptomatic) can unknowingly spread the virus to others. Public health capacity and response. Inadequate testing, contact tracing, and isolation measures can lead to uncontrolled outbreaks. The last possible factors Mutations and viral evolution. The virus can continue to mutate and

Table 17.1 The overall cases of COVID-19 for three countries of East Africa (Tanzania, Kenya, Uganda and Rwanda) by 27th July 2023 (*Source* World Health Organization (2023), Worldometer (2023))

	Tanzania	Kenya	Uganda	Rwanda
CCT	43,078	343,915	171,729	133,194
DCT	187	5689	3632	1468
VTDA	65.95	44.17	57.72	210.94
VPVCP	53.85	20.63	28.55	80.29
VPVB	0	3.72	1.55	32.62

Note CCT Cases cummurative total, *DCT* Death cummulative total, *VTDA* vaccines—total doses administered per 100 population, *VPVCP* Vaccines—persons vaccinated with a complete primary series per 100 population, *VPVB* Vaccines—persons vaccinated with at least one booster or additional dose 100 population

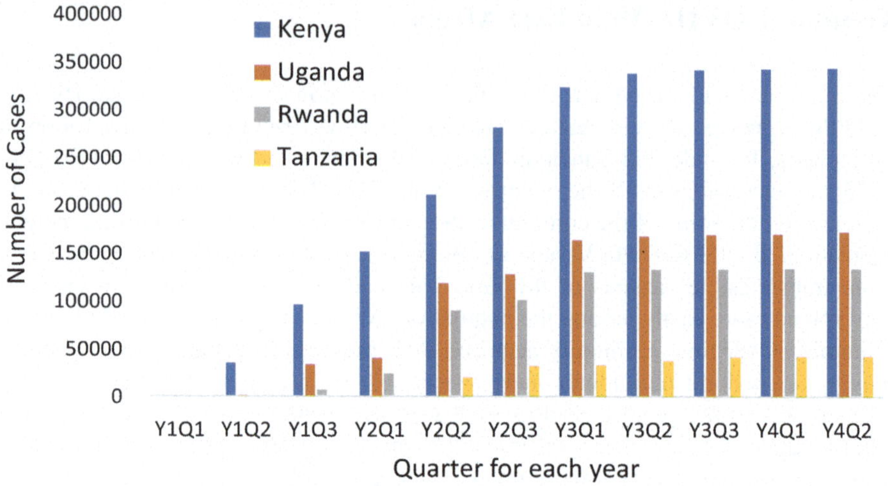

Fig. 17.2 Trend of COVID-19 in three African countries from 2020 to 2023. Note. Q1-January-April (2020), Q2-May–August, Q3-September–December, Y1-2020, Y2-021, Y3-2022, Y4-2023

adapt, potentially affecting its transmissibility or virulence (Jindal et al. 2021). It's important to address these factors through a combination of measures, including vaccination campaigns, public health education, adherence to preventive guidelines, and timely responses to localized outbreaks. Keep in mind that different regions and countries may have unique circumstances contributing to COVID-19 case increases. Always consult reputable sources for the latest information and guidelines in your specific area.

Results and Discussion

Socio-economic on COVID-19 Pandemic in Eastern Africa Countries

These effects have been extensive and multifaceted, including problems with health-care, employment, education, and general quality of life. The region's livelihoods were severely damaged by lockdowns, restrictions, and business closures, especially in the informal sectors (Chebby et al. 2021; Mgema and Komba 2020). Many people experienced job loss or income reductions, which resulted in challenges and food insecurity. For instance in Tanzania it was observed that that, the number vof tourists and the associated income and employment reduced drastically (Chebby et al. 2021). For instance, a study conducted in Tanzania found that the COVID-19 pandemic had an impact on the tourism industry, which in turn resulted in a decrease in business revenue and the loss of jobs. Additionally, it was expected that protected areas, particularly the Burunge WMA authority, would see revenue decreases of 32.5 and 76.4% for the fiscal years 2019–20 and 2020–21, respectively, as collection is heavily reliant on the tourism industry. Local people were forced to adapt to a number of alternative forms of income, including livestock raising, Boda-Boda (motorcycle taxis), and fishing, as a result of COVID-19's consequences (Chebby et al. 2021). The socio-economic impact on the livelihoods of people living adjacent to the area is significant. This is due to the multiplier effects associated with the project. For instance, as the project operated, people earned income through employment and small businesses. A study in the coastal zone of mersing showed that, Young people are benefiting significantly, as they are employed and not migrating to other places. For entrepreneurs, the tourism sector has been observed to receive a boost, with increased prices of goods and services (Sapari et al. 2019).

In East Africa, the pandemic-related economic slowdown increased inequality and aggravated poverty as this was associated with dencrease in income as people were not working to earn (Kithiia et al. 2020). The results from Mombasa, Kenya, showed that one of the socioeconomic effects of COVID-19 was movement restriction, which severely interfered with their day-to-day activities and resulted in a loss of income and an increase in living expenses due to price increases for some essential commodities, as well as the ongoing presence of children at home due to school closures (Kithiia et al. 2020). The majority of the crisis' effects fell on vulnerable people, such as those in slums, refugee camps, and rural areas. Production, distribution, and availability to food were impacted by supply chain interruptions, travel restrictions, and income losses, and food scarcity (Bizoza and Sibomana 2020; Kithiia et al. 2020; Lasdun et al. 2023).

Education was affected by school closings and other services like religios institution, particularly for children without access to distance learning resources (Kithiia et al. 2020). The learning gaps and access to high-quality education were both impacted by the further widening of the digital divide. electronic transformation the epidemic hastened the adoption of digital technology across a number of industries, including healthcare, remote labor, and e-commerce (Bizoza and Sibomana 2020; Tumwesige 2020). According to a study conducted in Uganda, the COVID-19 epidemic had an impact on socioeconomic gatherings because of movement restrictions and lockdowns. Affected operations included small and medium-sized businesses of various sizes (Kimuli et al. 2021). Communities had to accept a new way of life, and it was determined that using digital technology was the solution. The systems and tools used by companies to connect and network with many stakeholders in order to support business operations are done through digital infrastructure. In Uganda and other Eastern African nations, the COVID-19 pandemic has led to greater use of social media sites like Facebook, Snapchat, and WhatsApp, as well as digital transportation services like SafeBoda and Uber (Kimuli et al. 2021). While this change offered benefits, it also brought attention to the digital gap, which left some areas behind. Significant numbers of migrant workers reside in East Africa, and the epidemic caused changes in migratory patterns. Sending remittances back home was difficult for many East Africans working overseas, which had an effect on the stability of recipient families' finances. Numerous workers in the area were left without social protection and were at risk of economic vulnerability due to the strong reliance on the informal economy. Particularly impacted were informal traders, street sellers, and workers earning a daily salary. For instance, in Kenya, it was reported that people experienced a loss of income and employment, leading to social implications such as food insecurity (not having proper meals) (Pinchoff et al. 2021).

The pandemic's uncertainty, anxiety, and isolation had a negative impact on East Africa's mental health. Stress, despair, and anxiety all got worse as a result of worries about one's health, future, and way of life. Gender disparities as a result of the pandemic's disproportionately negative socioeconomic effects on women and girls, gender disparities have gotten worse (Bukuluki et al. 2020). A study conducted in Uganda that found that, the lockdown had an impact on refugees' livelihoods and exacerbated income insecurity, sexual and gender-based violence, and anxiety provided evidence of this (Bukuluki et al. 2020). Domestic violence, more demanding caregiving roles, and restricted access to healthcare and education were all issues that women had to deal with. travel restrictions and lockdowns caused a severe fall in the tourism and hospitality sector, which is crucial to many East African economies (Chebby et al. 2021, 2024). This resulted in employment losses and economic downturns in tourist-dependent areas. Challenges to social cohesion particularly physical segregation and movement restrictions hindered communal meetings and social contacts, which had an impact on social cohesion and customs.

Impacts of COVID-19 on Health in Eatern African Countries

Strain on Healthcare Systems is another effect whereby the sudden influx of COVID-19 patients strained healthcare facilities, including hospitals, clinics, and testing centers (Ngongolo 2023). The region reported a significant number of COVID-19-related deaths, highlighting the severity of the pandemic's impact on public health. COVID-19 Infections and Mortality whereby Eastern African countries experienced a surge in COVID-19 cases, leading to an increased burden on healthcare systems.

Personal protective equipment (PPE) shortages rose as a result of the increased demand for PPE, including as masks, gloves, and gowns, which put a pressure on supplies in Eastern African nations and made it difficult for healthcare professionals to fully protect themselves. Disruptions to Routine Healthcare Services (Stöckl et al. 2023). Because of the focus on the COVID-19 response, routine healthcare services have occasionally received less attention and funding. Due to this disturbance, maintaining immunization programs and addressing other health concerns became difficult. Stress, worry, and uncertainty caused by the epidemic had a negative impact on the mental health of the area. Due to the pandemic's effects on daily life and livelihoods, many people experienced anxiety, depression, and other psychological difficulties (Bukuluki et al. 2020; Chaturvedi 2020; Prime et al. 2020; Teng et al. 2020).

Healthcare Services Access as the Access to healthcare services was hampered by movement restrictions and lockdowns, especially for people living in outlying locations. Travel restrictions and virus-related anxiety also discouraged some people from seeking medical attention. Additionally, there have been difficulties with the access and distribution of COVID-19 vaccinations in Eastern African nations. Important problems were vaccine fairness and accessibility, especially for the region's low-income nations (Binagwaho et al. 2022). According to authors from Rwanda, there is still an issue with injustice in the local distribution of vaccines, with some nations blatantly discriminating against minorities and lacking in logistical planning (Binagwaho et al. 2022). The study came to the conclusion that in order to spread the COVID-19 vaccine, pharmaceutical companies should share their technology in order to increase supply and lower prices, governments should give priority to equitable distribution to the most vulnerable populations worldwide, and low-income countries should improve their logistical capabilities in order to support large-scale vaccination campaigns (Binagwaho et al. 2022).The introduction of new and more contagious viral varieties, such the Delta variant, presented new difficulties for Eastern African public health responses, which is another effect.

However, there have been disparities among susceptible populations, and the pandemic disproportionately affected these populations, particularly the poor, refugees, and disadvantaged groups. Existing gaps were made worse by restricted access to resources, information, and healthcare. The impact on healthcare workers has also been observed (Adebisi et al. 2021). During the pandemic, healthcare personnel had to deal with increased workloads, virus exposure, and health hazards. Due to infections and burnout, certain nations have reported a scarcity of healthcare professionals. Last but not least, there have been superspreader events and community Transmission. Community transmission and superspreader events were facilitated by social gatherings, large events, and crowds of people, which resulted in localized outbreaks in specific regions (Nachega et al. 2021). The effects have included health communication and misinformation. During the epidemic, effective health communication was essential, but it was difficult to spread accurate information due to widespread myths and disinformation concerning COVID-19 (Chilongola et al. 2022; Kithiia et al. 2020).

Psychological Impacts of COVID-19 Pandemic in Eatern African Countries

Communities in Eastern African nations have had major psychological effects as a result of the COVID-19 outbreak. There has been fear and anxiety. Many people have reported feeling more anxious as a result of the pandemic's uncertainty and fear. It has been typical to have anxiety about catching the illness, worry for loved ones, and worry about the future. Additionally, mood disorders such as depression The pandemic's social exclusion, financial challenges, and interruptions to daily life

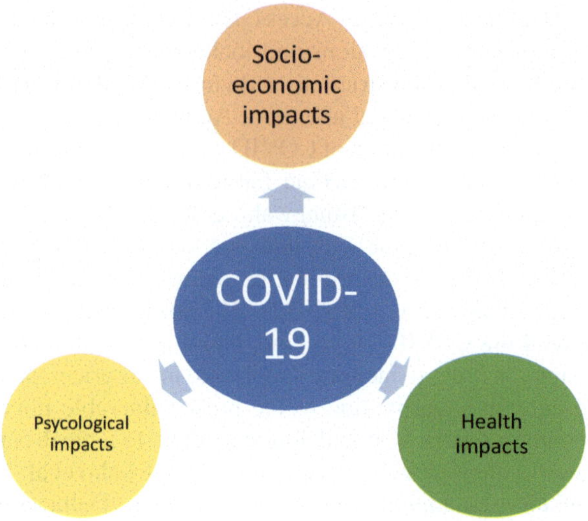

Fig. 17.3 The occurance of COVID-19 imapcted the communities psychologically, health system and socio-economicall

have all led to a surge in depression and other mood disorders in the countries of Eastern Africa. Due to stress and psychological distress The pandemic and its effects, including job losses, financial instability, and health issues, have put a lot of strain on people's mental health, resulting in psychological anguish (Fig. 17.3).

Many individuals in Eastern Africa have reported grieving and feeling a great sense of loss after losing loved ones to COVID-19 (Stöckl et al. 2023). The grieving process has also been made more difficult by restrictions on gatherings and funeral services. It has been noted that COVID-19 patients, medical personnel, and members of the afflicted communities have experienced stigma and discrimination. For people who are impacted, this stigmatization might make their mental health problems worse. According to research conducted in Tanzania, COVID-19 had an impact on mental health both during and after restrictions. This impact was attributed to COVID-19 knowledge, behavior changes, problems with economic livelihood, an increase in arguments and intimate partner violence with partners, as well as stress brought on by childcare-related issues. Tanzania only imposed minimal limitations, yet the COVID-19 epidemic nonetheless caused a rise in the number of women reporting signs of poor mental health (Stöckl et al. 2023). The region has always struggled with a lack of resources and access to care for those with mental illnesses, and the pandemic has made matters worse for those services.

In this instance, frontline workers in the healthcare industry and other occupations have experienced high levels of stress, long workdays, and viral exposure, which has resulted in burnout and emotional weariness as most people are not willing to vaccinate against COVID-19 (Chilongola et al. 2022). The findings in East Africa showed that prenatal and postnatal clinics had been rescheduled, employees had been relocated, and there was less time available for health education and child monitoring. There have been reports of a shortage of necessities like medical supplies, equipment, and immunizations. The evidence of routine use in the three districts, which showed a reduced uptake of prenatal, postnatal, family planning, and immunization services as well as fewer institutional deliveries, partially confirmed such opinions (Shayo et al. 2023). The effect on children and young people. Children and teens in Eastern Africa have had psychological impacts as a result of school closures, restricted social contacts, and disruptions to schooling. Numerous people have felt lonely and worried about the future. Substance addiction and Coping techniques: To cope with the stress and worry brought on by the pandemic, some people have turned to substance addiction or other unhealthy coping techniques (Shayo et al. 2023; Tarimo et al. 2022).

Social support networks have been broken as a result of movement restrictions and physical segregation policies, making it more difficult for people to get emotional support from friends and relatives. Misinformation and the infodemic have also been noted to have an impact. The public has been confused and anxious due to the proliferation of false information and rumors about the virus, which has a negative impact on mental health (Kithiia et al. 2020; Stöckl et al. 2023). The communities have experienced hopelessness and uncertainty. The pandemic's protracted character and its socioeconomic effects have left people feeling despondent and apprehensive about the future.

Limitations of the Study

While the book chapter offers insightful information about the pandemic's effects on the local population, it is important to recognize its limits in order to properly understand the results. The data could be one of the study's weaknesses. Accessibility and Dependability as The study's findings mainly rely on data about COVID-19 cases, healthcare, socioeconomic indices, and people's well-being in East Africa that are readily available and trustworthy. The quality and robustness of the study may be impacted if the data used is insufficient, out-of-date, or prone to reporting errors.

Additionally, a time limit Due to the pandemic's continuing nature and the study's constrained period for data gathering and analysis, it may have certain limitations. The effects of COVID-19 might change over time, and the study might not fully account for their long-term effects. The generalizability, on the other hand, is yet another restriction. The results might only apply to East Africa and not be entirely transferable to other areas or nations with diverse socioeconomic, cultural, and healthcare circumstances. knowledge these limitations can aid readers and decision-makers in properly interpreting the results and motivate additional study to give the subject a more thorough knowledge.

Recommendations

Since East Africa is made up of a variety of nations, each with its own distinct cultural setting, it is also vital to explore regional and cultural variations. Examine the regional and cultural variations in COVID-19's effects on wellbeing to spot regional issues and chances for specialized solutions. On the other hand, it is crucial to take intersectional approaches into account. Analysis of the disparate effects of COVID-19 on disadvantaged and vulnerable populations will be done, taking into account variables including gender, age, income level, and ethnicity. This intersectional strategy will aid in locating and addressing particular needs. Recognizing the possible long-term consequences of the pandemic on people's mental health in the area, more research is required on Assessing Mental Health and Psychological Well-being through conducting thorough assessments of mental health and psychological well-being.

It is necessary to assess the adaptability and capacity of East African healthcare systems to deal with the difficulties of the pandemic and to pinpoint areas in need of development and investment. Additionally, the effectiveness of health education and awareness initiatives during the pandemic must be investigated, as must the effect of correct information transmission on public health behaviors. Create evidence-based policy suggestions for East African governments and organizations based on the research findings to improve public health responses, social safety nets, and overall human wellbeing. The consequences of COVID-19 on human well-being must also be discussed with stakeholders from the government, non-governmental organizations, academia, and local communities. Collaborative efforts will lead to more comprehensive and inclusive solutions.

Conclusions

This book chapter provides a thorough and perceptive examination of the impacts of the COVID-19 pandemic on the local community. The chapter has illuminated the multifaceted issues faced by East Africa during this unprecedented global crisis using a combination of empirical data, qualitative analysis, and literature reviews. Policymakers, researchers, and organizations can benefit greatly from the findings of this study as they work to lessen the pandemic's widespread effects and advance human welfare in the area.

The enormous public health issues brought on by the pandemic in East Africa are one of the chapter's primary conclusions. Healthcare systems in the area were under a great deal of stress, and efforts to contain the virus and provide proper medical care encountered a number of challenges. For targeted remedies to be developed and future health systems to be strengthened, it is essential to comprehend the specific risks and resource limitations in the area's healthcare infrastructure. The chapter also underlines how the pandemic widened socioeconomic gaps already present in East

Africa. The effects of the pandemic have fallen disproportionately on marginalized and vulnerable groups, who now have more difficulty getting access to basic essentials like healthcare and education. Policymakers must prioritize inclusive policies and social safety nets to address these inequalities and promote more equitable well-being outcomes.

Another important conclusion in the chapter was that there were disruptions in education. Education has been disrupted by school closings and a move toward online learning, and the digital divide has made access to high-quality learning possibilities more unequal. Governments and education stakeholders must address these issues in order to create inclusive, resilient educational systems that can endure future crises. The chapter also emphasizes how the epidemic has affected mental health and wellbeing. Health issues, economic uncertainties, and social isolation have all contributed to an increase in anxiety, stress, and depression. For the region to thrive generally, it is crucial to understand the value of mental health assistance and to incorporate mental health services into healthcare systems.

The chapter also emphasizes the value of environmental preservation and sustainable forest management in fostering community resilience. The importance of East Africa's forests and natural resources in sustaining livelihoods and ecosystem functions highlights the demand for ethical and sustainable behavior. In conclusion, the book chapter offers a thorough grasp of the pandemic's numerous effects on the area. In order to address the issues facing East Africa, the findings obtained highlight the importance of focused initiatives, evidence-based policies, and collaborative efforts. East Africa may create resilience and progress toward a more just and sustainable future after COVID-19 by putting a priority on inclusive development, bolstering healthcare systems, tackling socioeconomic inequities, and encouraging sustainable practices. The chapter's findings will undoubtedly serve as a valuable resource for guiding future research, policymaking, and interventions aimed at enhancing human well-being in the region and beyond.

Acknowledgements I am deeply grateful to my wife, Atuhimbye Sigala, for her exceptional contributions to the book chapter. Her academic expertise, support, and encouragement have been invaluable assets in writing this project. I am deeply grateful for Atuhimbye Sigala's insightful suggestions and constructive feedback during the drafting and revision process. Her keen eye for clarity and coherence has played a crucial role in presenting the research in a comprehensive and accessible manner.

References

Aabed K, Lashin MM (2021) An analytical study of the factors that influence COVID-19 spread. Saudi J Biol Sci 28(2):1177–1195

Adebisi YA, Ekpenyong A, Ntacyabukura B, Lowe M, Jimoh ND, Abdulkareem TO, Lucero-Prisno III DE (2021) COVID-19 highlights the need for inclusive responses to public health emergencies in Africa. Am J Trop Med Hygiene 104(2):449

Atzrodt CL, Maknojia I, McCarthy RD, Oldfield T, Po J, Ta KT et al (2020) A guide to COVID-19: a global pandemic caused by the novel coronavirus SARS-CoV-2. FEBS J 287(17):3633–3650

Barman A, Das R, De PK (2021a) Impact of COVID-19 in food supply chain: disruptions and recovery strategy. Curr Res Behav Sci 2(7):100017

Barman A, Das R, De PK (2021b) Logistics and supply chain management of food industry during COVID-19: disruptions and a recovery plan. Environ Syst Decis 42(1):338–349. https://doi.org/10.1007/s10669-021-09836-w

Binagwaho A, Mathewos K, Davis S (2022) Equitable and effective distribution of the COVID-19 vaccines—a scientific and moral obligation. Int J Health Policy Manag 11(2):100

Bizoza A, Sibomana S (2020) Indicative socio-economic impacts of the novel coronavirus (Covid-19) outbreak in Eastern Africa: case of Rwanda. Available at SSRN 3586622. SSRN Elsevier. https://papers.ssrn.com/sol3/papers.cfm?abstract_id=3586622

Bukuluki P, Mwenyango H, Katongole SP, Sidhva D, Palattiyil G (2020) The socio-economic and psychosocial impact of Covid-19 pandemic on urban refugees in Uganda. Social Sci Human Open 2(1):10004

Chaturvedi SK (2020) Covid-19, coronavirus and mental health rehabilitation at times of crisis. J Psychosocial Rehabil Mental Health 7(1):1–2

Chebby F, Mmbaga N, Ngongolo K (2021) Tourism status and sources of income to local communities amidst COVID-19 pandemic and its implications in biodiversity conservation in Burunge Wildlife Management Area in Tanzania. Res Square. https://doi.org/10.21203/rs.3.rs-1175273/v1

Chebby F, Mmbaga N, Ngongolo K (2024) Impact of COVID-19 pandemic on tourism, income of local communities and biodiversity conservation: Evidence from Burunge wildlife management area, Tanzania. Heliyon e24327.

Chilongola JO, Rwegoshola K, Balingumu O, Semvua H, Kwigizile E (2022) COVID-19 knowledge, attitudes, practices and vaccination hesitancy in Moshi, Kilimanjaro Region, Northern Tanzania: COVID-19 vaccination hesitancy in Tanzania. Tanzania J Health Res 23(1):1–12. https://doi.org/10.4314/thrb.v23i1

El Otmani Dehbi Z, Sedrati H, Chaqsare S, Idrissi Azami A, Merzouki M, Raji M, Rhalem W, Al Idrissi N, Nejjari C, Amzazi S, Ghazal H (2021) Moroccan digital health response to the COVID-19 crisis. Front Public Health 9:690462. https://doi.org/10.3389/fpubh.2021.690462

Jahn B, Sroczynski G, Bicher M, Rippinger C, Mühlberger N, Santamaria J, Urach C, Schomaker M, Stojkov I, Schmid D, Weiss G, Wiedermann U, Redlberger-Fritz M, Druml C, Kretzschmar M, Paulke-Korinek M, Ostermann H, Czasch C, Endel G et al (2021) Targeted COVID-19 vaccination (TAV-COVID) considering limited vaccination capacities—an agent-based modeling evaluation. Vaccines 9(5). https://doi.org/10.3390/vaccines9050434

Jindal H, Jain S, Suvvari TK, Kutikuppala L, Rackimuthu S, Rocha ICN, Goyal S, Radha (2021) False-negative RT-PCR findings and double mutant variant as factors of an overwhelming second wave of COVID-19 in India: an emerging global health disaster. SN Comprehen Clin Med 3:2383–2388

Kimani J, Adhiambo J, Kasiba R, Mwangi P, Were V, Mathenge J, Macharia P, Cholette F, Oore S, Shaw S, Icon BM, Musyoki H, Bhattacharjee P, Moses S, Fowke RK, Lyle R, McKinnon RL, Robert Lorwa R (2020) The effects of COVID-19 on the health and socio-economic security of sex workers in Nairobi, Kenya: emerging intersections with HIV. Global Public Health 15(7):1073–1082

Kimuli SNL, Sendawula K, Nagujja S (2021) Digital technologies in micro and small enterprise: evidence from Uganda's informal sector during the COVID-19 pandemic. World J Sci Technol Sustain Develop 18(2):93–108

Kithiia J, Wanyonyi I, Maina J, Jefwa T, Gamoyo M (2020). The socio-economic impacts of Covid-19 restrictions: data from the coastal city of Mombasa, Kenya. Data Brief 33(1):06317

Lasdun V, Harou AP, Magomba C, Aku A (2023) COVID-19, climate shocks, and food security linkages: evidence and perceptions from smallholder farming communities in Tanzania. Environ Dev Econ 28(3):211–229

Louis EF, Eugene D, Ingabire WC, Isano S, Blanc J (2022) Rwanda's resiliency during the coronavirus disease pandemic. Front Psychiat 12(1):589526. https://doi.org/10.3389/fpsyt.2021.589526

Mgema JM, Komba CK (2020) Socio-economic effects of Covid-19 pandemic on the performance of co-operative societies in Tanzania. Moshi C0-Oper Univ 1(1):1

Nachega JB, Atteh R, Ihekweazu C, Sam-Agudu NA, Adejumo P, Nsanzimana S, Rwagasore E, Condo J, Paleker M, Mahomed H, Suleman F, Ario AR, Kiguli-Malwadde E, Omaswa FG, Sewankambo NK, Viboud C, Reid MJA, Zumla A, Kilmarx PH (2021) Contact tracing and the COVID-19 response in Africa: best practices, key challenges, and lessons learned from Nigeria, Rwanda, South Africa, and Uganda. Am J Trop Med Hyg 104(4):1179

Ngongolo K (2023) Impacts, mitigation strategies of Covid-19 on human well-being in Africa: a brief review. Social Sci Human Open 8(1):100709. https://doi.org/10.1016/j.ssaho.2023.100709

Nuwematsiko R, Nabiryo M, Bomboka JB, Nalinya S, Musoke D, Okello D, Wanyenze, K. R. (2022). Unintended socio-economic and health consequences of COVID-19 among slum dwellers in Kampala, Uganda. *BMC Public Health*, *22*(88). https://doi.org/10.1186/s12889-021-12453-6

Petersen E, Wasserman S, Lee S-S, Go U, Holmes AH, Al-Abri S, McLellan S, Blumberg L, Tambyah P (2020) COVID-19-We urgently need to start developing an exit strategy. Int J Infect Dis IJID: Off Publ Int Soc Infect Dis 96:233–239. https://doi.org/10.1016/j.ijid.2020.04.035

Pinchoff J, Austrian K, Rajshekhar N, Abuya T, Kangwana B, Ochako R et al (2021). Gendered economic, social and health effects of the COVID-19 pandemic and mitigation policies in Kenya: evidence from a prospective cohort survey in Nairobi informal settlements. BMJ Open 11(3):e042749

Prime H, Wade M, Browne DT (2020) Risk and resilience in family well-being during the COVID-19 pandemic. Am Psychol 75(5):631

Sapari MS, Shuib A, Ramachandran S, Kunasekaran P (2019) Contributions of tourism micro and small entrepreneurs to local community livelihood. SSRG Int J Econ Manage Stud 6(8):182–189

Shayo E, Nassor NK, Mboera L, Ngadaya E, Mangesho P, Bakari M, Urassa M, Seif M, Tarimo C, Masemo A, Mmbaga TB, O'Sullivan N, McCoy D, Russo G (2023) The impacts of COVID-19 and its policy response on access and utilization of maternal and child health services in Tanzania: a mixed methods study. PLOS Glob Public Health 3(5):e0001549. https://doi.org/10.1371/journal.pgph.0001549

Solymári D, Kairu E, Czirják R, Tarrósy I (2022) The impact of COVID-19 on the livelihoods of Kenyan slum dwellers and the need for an integrated policy approach. PLoS ONE 17(8):e0271196

Stewart R, Madonsela A, Tshabalala N, Etale L, Theunissen N (2022) The importance of social media users' responses in tackling digital COVID-19 misinformation in Africa. Digital Health 8:20552076221085070. https://doi.org/10.1177/20552076221085070

Stöckl H, Mosha N, Dartnall E, Ayieko P, Mtolela G, Mshana G (2023) The short and long-term impact of COVID-19 restrictions on women's mental health in Mwanza, Tanzania: a longitudinal study. PLOS Global Public Health 3(6):e0001781

Tarimo CS, Mmbaga BT, Mboera LEG, Bakari M, Seif M, Mangesho PE, Masemo A, Ngadaya E, Urassa M, Nassor N, McCoy D, Shayo EH (2022) The effects of the COVID-19 pandemic on livelihoods and social support mechanisms in Pemba, Unguja and Ilala, Tanzania: a qualitative study. Tanzan J Health Res 23(1):78

Teng YM, Wu KS, Lin KL (2020) Life or livelihood? mental health concerns for quarantine hotel workers during the COVID-19 pandemic. Front Psychol 11(1):2168

Tumwesige J (2020) COVID-19 Educational disruption and response: rethinking e-Learning in Uganda. University of Cambridge

Weitz JS, Beckett SJ, Coenen AR, Demory D, Dominguez-Mirazo M, Dushoff J, Leung C, Li G, Măgălie A, Park WS, Rodriguez-Gonzalez R, Shashwat Shivam S, Zhao CY (2020) Modeling shield immunity to reduce COVID-19 epidemic spread. Nat Med 26(6):849–854

World Health Organization (2023) WHO coronavirus (COVID-19) dashboard. Situation by region, country, territory & area. WHO. https://covid19.who.int/table

Worldometer (2023) COVID-19 coronavirus pandemic. Coronavirus cases. https://www.worldometers.info/coronavirus/#countries

Chapter 18
Mediating Human Security and Social Protection: A Strategic Scheme for the Social Institution in Indonesia Disaster Cycle

Eko Wahyono, M. Alie Humaedi, Reza Amarta Prayoga, Rudy G. Erwinsyah, and Dian Karinawati Imron

Abstract Striving to overcome national disasters problems and challenges has not been resolved optimally. There is a gap in concept and implementation between the macro level (social protection) and the micro level (human protection). This research discusses the formulation of operational steps in strategic disaster management schemes through institutions in Indonesia. The research used the main methods of literature review, observation, and in-depth interviews. Research data were analyzed using Soft System Methodology. The results of this research show several findings. *First*, Indonesia's geographical conditions as the Ring of Fire area with high disaster frequency require a rapid, organized, and effective response. *Second*, there are pivotal regulations at the ministerial level and the community's potential role and resilience at the root levels to manage disasters. *Third*, this research discovers strategic and operational models at the meso (community) level to deal with disaster threats. *Fourth*, This study provides an overview of the position of social institutions in addressing disaster problems and then formulates strategic and operational schemes for managing social institutions at various macro, meso, and micro levels. *Fifth*, the position and power of social institutions (meso) potentially bridge the concepts of human security and social protection in policy-making and practical

E. Wahyono · M. A. Humaedi · R. A. Prayoga · R. G. Erwinsyah (✉) · D. K. Imron
Research Center for Social Welfare Village and Connectivity, National Research and Innovation Agency, Jakarta, Indonesia
e-mail: rudy006@brin.go.id

E. Wahyono
e-mail: ekow005@brin.go.id

M. A. Humaedi
e-mail: mali002@brin.go.id

R. A. Prayoga
e-mail: reza010@brin.go.id

D. K. Imron
e-mail: dian073@brin.go.id

U. Mukhopadhyay et al. (eds.), *Climate Crisis, Social Responses and Sustainability*, Climate Change Management, https://doi.org/10.1007/978-3-031-58261-5_18

action in Indonesia's national disaster cycle. This research contributes to strategic and operational steps in disaster management in Indonesia, especially at the institutional level. This study provides a new perspective for developing concepts and action plans at the meso-institutional level in dealing with disasters.

Introduction

Indonesia is a country with a high risk of natural disaster and require integrated response to overcome the damage impact. The increased natural disaster potential in Indonesia is caused by geological factors where this country is located in the Pacific Ring Zone (Astawa et al. 2022; Suarjana 2020; Subagia 2015; Suharini et al. 2015). The National Disaster Management Agency documentation (Muhari 2021) shows that throughout 2021, there were 3092 disaster events, namely 1298 floods, 804 extreme weather, 632 landslides, 265 forest and land fires, tidal waves and 45 abrasion, 32 earthquakes, 15 droughts, and a volcanic eruption. Moreover, National Board for Disaster Management (BNPB) released the number of people who suffered and evacuated are 8,426,609 people, 14,116 people were injured, up to 665 people died, and 95 people were missing. The challenges of disaster management response lie in timing, unstructured approach, and neglected local institution influence communities to prone situations (Bronfman et al. 2019; Harijoko et al. 2021; Humaedi et al. 2016; Rachmawati et al. 2018; Shalih 2020; Tsani 2019).

Based on the United Nations (2015) report in the United Nations System Task Team on the Post-2015 UN Development Agenda, social protection is a systematic effort through programs arranged by the government and the private sector that deal with social, environmental, and economic vulnerability and powerlessness (such as education, food security, income security, health insurance, and decent work assurance). Norton et al. (2001, 2002) emphasize social protection as an effort to protect the community from the social vulnerability threat caused by four factors; life cycle, economic, environmental, and social governance; life cycle (e.g., disability, death, and old age), economy (crop failure, pest attack, unemployment, high prices of necessities, and economic crisis), environment (drought, flood, earthquake, and volcanic eruption), social government (crime, domestic violence, and political instability). Budiman et al. (2019) clearly explain social protection as a collective effort from a set of policies and/or government programs to provide immediate responses to various possible risks and vulnerability threats throughout the community's life cycle. Meanwhile, human security is the protection of human security from various threats and disturbances that can harm massively as well as economic and political security, access to food, health, the environment, individuals, cultural systems, and society as a form of response to sudden losses and disturbances within a country (Churruca Muguruza 2017; Jolly and Basu Ray 2006; Simanjuntak 2020; Zervaki 2018).

Social protection and human security approaches are essential in managing disaster risk and social vulnerability (Asfaw et al. 2014; Bayu et al. 2015; Hadis 2014; Sieber et al. 2022). In a disaster cycle, social protection and human security policies is a vital factor of resiliences for a community in a high disaster risk area (Asfaw et al. 2014; Maksum 2021; Tadjbakhsh and Chenoy 2007). Moreover, in mitigating the disaster cycle, it is also necessary to involve local wisdom-based institutions that participate in promoting social protection and human security, including the socio-economic dimensions and considering the typology and character of the local community (Bahransyaf and Probosiwi 2018; Chotimah et al. 2017; Suradi et al. 2022). Disaster mitigation with the decision makers' understanding enhances sustainable social protection and human security for communities affected by disasters (Alkire 2016; Hasegawa 2007; Lowder et al. 2017; Norton et al. 2001; Suradi et al. 2022; Tadjbakhsh and Chenoy 2007). Strategic policy from the government by strengthening social protection and human security at the local institutional level is considered a safety net in dealing with pre, post, prevention, and disaster mitigation.

The empirical evidence shows that the main problem of disaster mitigation. Disaster management and governance are not yet positioned as a strategic scheme for recovery. Next, the implementation of disaster mitigation policies became ineffective because it did not consider the disaster risk reduction investment and post-disaster build back better (Danar 2020). Danar (2020) and Suradi et al. (2022) emphasize disaster governance in disaster risk prevention and reduction: it is necessary to involve a circle of various actors (government, private sector, NGO, local government, and other actors). Various evidence measure the position of human security and social protection as an urgent aspect. Shaw (2006, 2012, 2014) explains that human security is closely intertwined with disaster management. The findings of Asfaw et al. (2014), Patriadi et al. (2015), Hadis (2014), Sieber et al. (2022), Maksum (2021), Bahransyaf and Probosiwi (2018), Wickramasinghe (2014), Holzmann and Jørgensen (2001) and Hasegawa (2007) underlined the concept of social protection and human security as a supplement to a practical approach within the disaster impact reduction policy schemes.

Strategic and operational steps in disaster governance are important actions that should be stated in various levels of policy synergy from central to local. Indonesia is in a complex disaster circle, requiring adaptive disaster management steps limited to physical reconstruction efforts and sustainable social and economic recovery after the disaster. Adopting Nee's (2003) thinking as an initial bridging step in formulating a new institutionalism scheme is compatible and adaptive to understanding the context of Indonesia's disaster. At least Nee (2003) stated five constructive keys in the scheme of community social institutions: the institutional framework risk society, social protection as regulation and collective actions, organizational filed-institutional meso, institutional strengthening, and social group-human security.

Examining the problems above, this chapter chooses to fill them by focusing on the function of the intermediary position between social protection and human security at meso-level institutions during disasters. This research analyzes two important questions: (1) How are the conceptual framework and strategic scheme capable of bringing together social protection and human security mechanisms in the context of a national disaster? (2) How to set the right position role of social institutions to mediate social protection and human security during a disaster? In addition, this study aims to combine the conceptual framework of social protection and human security mechanisms into a strategic disaster mitigation scheme and to initiate social institutions' role in bridging social protection and human security. Social institutions at the meso level are important mediators as a security mechanism for the Indonesian community trying to improve disaster management.

Rationale of the Study

Strategic planning and measurable steps in disaster management are critical considering the high quantity of disasters, extent of distribution and types of disasters in Indonesia. The state has responsibility to prepare the management responses though providing a budget, enact policy and framework and official institutions task. Government institutions has trying diverse approach to included the community. Community in disaster areas have natural abilities and culture to deal with disasters. This research was conducted to provide an empty gap between the role of the state and community needs. Meso institutions bridges the interests of the state and the needs of society. The first rationalization of this research is to offer a scheme at the meso level to deal with national disasters in Indonesia. Second, it provides new information on how local disaster management institutions are formed and implemented. Third, provide direction regarding the steps that local institutions need to take in the community in handling and anticipating unexpected disasters. Fourth, provide a comprehensive study for the state, especially ministries or institutions that handle disaster issues in Indonesia, to involved local institutions. Fifth, this research is expected to be a strategic first step in establishing various local and natural institutions in areas with a high level of disaster risk.

Material and Methods

The approach used in this study is qualitative, using primary and secondary data to present a detailed description of a specific situation, social setting, or social relations (Neuman 2006). This approach accommodates researcher to acknowledge various social phenomena in their natural environment and trying to comprehend the phenomena based on the subject meanings who directly experience the occasions supported by various data and publications ("The SAGE Handbook of Qualitative Research," 2005). We utilize the primary data and secondary data at the same level. The two types of data complement mutually to provide a comprehensive understanding.

The primary data processed are residues from various research experiences that have been carried out previously. This data was collected through participatory observation and interviews. We sorted the data relevant to disaster issues, especially those related to the concept of Human Security and Social Protection, from field notes collected previously. Although mostly in the form of residues, this primary data set can be processed and assembled into a narrative.

Secondary data is collected through structured searches on scientific publication search engines such as Google Scholar, Portal Garuda, and Scopus. The use of various keywords in Indonesian such as "*bencana*" (disaster) and "*perlindungan sosial*" (social protection), and in English, such as "disaster", "human security", and "social protection" is carefully carried out. These two languages are used partly because this chapter's locus is Indonesia. Some articles are found through the Publish or Perish 8 software which extracts structured search results, and others are collected manually through a direct search on Google Scholar. We collected several articles used as secondary data sources in this chapter.

After we collect primary and secondary data, we use Soft System Methodology (SSM) for the analysis process. SSM is a procedure or way to solve a chaotic or disorganized problem through the development of an information system using a structured approach to understand an issue, build a conceptual mode, then formulate the desired change action, and then implement the change (Checkland 2000; Checkland and Scholes 1990; Hardjosoekarto et al. 2014). Initially, SSM was developed from Hard System Engineering to solve dynamic, soft, and unstructured real-world problems. Messy is a complex social condition that requires a more systematic explanation. At the same time, the soft system is an organizational system for human actions, which is the other side of the hard system (technology). Ill-structured is a social condition that is not structured, so there needs to be a more straightforward and honest explanation to explain the phenomenon of person lives. SSM has an action research background as a research effort to solve problems in the real world. Within the framework of this SSM analysis, we take seven steps, starting from understanding the situation on the upstream side and taking action on the downstream one. These steps do not stand alone but are interrelated with each other. These seven steps can be seen in the following chart (Fig. 18.1).

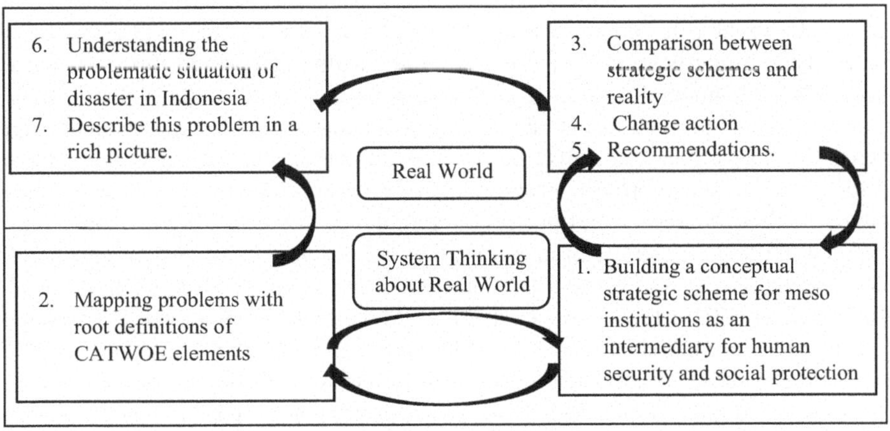

Fig. 18.1 Soft system methodology (Checkland and Poulter 2006)

Result and Discussion

The Role of Social Protection *and* Human Security *to Address the Disaster Issues*

In this section, social protection and human security have an equally substantial role. Social protection as ensures the security of income and social services, especially for communities with high vulnerability. Social protection emerges in the form of basic income security and universal access to social services in national priority programs (International Institute for Labour Studies 2012). Social protection become the first safety net when a disaster occurs. Social protection is obtained from outside the individual, generally provided by the government and tends to be state-centric and formal. Social protection data utilization also support the disaster responses. Through social protection, the safety responses would implement in an immediate time. In the case of volcanic eruptions, tsunamis, and earthquakes in Indonesia, responsive assistance is the result of utilizing the social protection function.

The position of social protection is often confronted with human security, both empirical practice and conceptual theoretical. Both concepts complement and support each other as a new approach. Human security focuses more on individual security. The concept regards individual security as necessary as state security which cannot be explained by realism theory that focuses on security created by the state as the main object of human rescue. As a human being who can think and have the "instinct" to save lives, the position of human security is crucial. Several disaster locations in Indonesia cannot be physically reached by social protection. Human security practices has been shown in the Mount Merapi incident in Yogyakarta (Tessa 2021); and Mount Lokon in Tomohon, Central Sulawesi (Humaedi et al. 2016). Human security has a pattern that tends to be informal and natural (Table 18.1).

Various studies (Churruca Muguruza 2017; Faist et al. 2015; Maksum 2021; Ohemeng and Foli 2023; United Nations 2015; Ville 2019) state similarities and differences in the roles of social protection and human security to address disaster issues. The similarity issue is about community vulnerability and threat of social crisis. The fundamental difference is that there are no study findings that link the urgency of social protection and human security in community protection on disaster issues. Previous studies, describe the two separated issues according to the context of social reality, social protection and human security are mostly studied as a policy regulation that provides social protection guarantees in the reading of labour realities and is often equated with compensation for the protection system and social security guarantees for workers (Chantavanich and Jitpong 2023; Griffiths et al. 2023; Maksum 2021; Teodorescu et al. 2021). Meanwhile, human security (personal security) in the context of Indonesian disasters signalled as a natural entity that arises from a sense of collective concern in providing a feeling of mutual protection from community bond. Moreover, social protection as a regulatory framwork provide social insurance to community, especially related to the social welfare system (Ohemeng and Foli 2023; Ville 2019). In fact, social protection and human security

Table 18.1 The relationship between social protection and human security in dealing with disasters

	Social protection	Human security
Conceptual	People-centred Multidimensional/multisector The long-term objective of human fulfilment Address chronic risk society Inclusive and Participatory Collective	
Different Aim	Providing a regulatory structure Social relations Institutions Structures	Provision of a sub-set of basic capabilities (vital core)
Different time horizon	Long time	Short-time (emergency relief work, peacekeeping) but also, the long-term human and institutional development
Interconnected motivation	Universal and fundamental human rights Activities and strategies in which individual and/or collective agents are engaged	
Different approach	Indivisibility of human rights, the equal priority of all human rights	Open-ended prioritization of human rights
Disaster Context	As a macro and collective protector	Personal security

Source Faist et al. (2015), United Nations (2015), Churruca Muguruza (2017), Ville (2019), Maksum (2021), Ohemeng and Foli (2023), Maksum (2021), Teodorescu et al. (2021), Chantavanich and Jitpong 2023, Griffiths et al. (2023), Ville (2019), Ohemeng and Foli (2023)

approach generate pivotal benefit to strengthen the community's capacity dealing with situational social crises. Both aspect complement mutually as enabler in the pre-, during, and post-disaster periods. The meso institutional system is suitable to typical protection for Indonesian community. It is essential because Indonesia has unique types of social dan cultural community characteristics. One of the unique character is the concept of *gotong royong* or mutual cooperation, which emphasizes the spirit of mutual assistance without formal mobilization (Slikkerveer 2019). This concept is not only at the cultural level and collective conciousness in society, but has become a state ideology for decades (Bowen 1986; Slikkerveer 2019). In the context of disaster, several studies have revealed that *gotong royong* is considered to be one of the bottom-up solutions in handling pre-disaster (Aksa and Afrian 2022; Taylor and Peace 2015), during disaster (Arianti and Koentjoro 2023; Nuryana et al. 2023), and post-disaster (Koopman 2021; Setiadi et al. 2020).

In the context of disasters in Indonesia, for instance earthquakes, tsunamis, landslides, floods, tornadoes, volcanic eruptions, and the Covid-19 pandemic, it is a community learning point to strengthen community resilience (Djalante et al. 2020; Hanifa et al. 2022; Hizbaron et al. 2018; Kurnio et al. 2021; Pribadi et al. 2021;

Wardekker et al. 2023). The resilience of the Indonesian community during the earthquake and Tsunami in Aceh in 2004 is an example of a turning point in the course of a human security that is identical to the uniqueness of Indonesia, namely the collective movement of community from various layers of status, religion, ethnicity, and race naturally reacted and empathised to collect all kinds of humanitarian assistance and was carried out spontaneously without being mobilised by the government (Grayman 2014; Vidyattama et al. 2021). The previous research findings from Faedlulloh et al. (2021) and Muqsith et al. (2021) sees the reality of a solidarity movement from a community of mutual cooperation during the outbreak of the Covid-19 pandemic in Indonesia.

Community needs and government responses are sometimes less syncronize due to late in providing social assistance and overlapping authority to implement mitigation (Coppola 2015). Social protection as collective protection from the government is still not optimal in emergency management, it has not been an accurate step by the government as a regulator for post-disaster recovery (Suppasri et al. 2021). The fast responses from community is coordinated and driven by a sense of solidarity. The community action is critical roles to fill the absence of social protection from the government.

Authentic resilience of Indonesian supported through local knowledge, traditional knowledge, community-based actions, and local practices and experiences from past events in disaster hazard management (Kurnio et al. 2021). The resilience developed from solidarity networks, cultural beliefs, trust in early warning and access to vital resources in a crisis situation (Thouret et al. 2022). Furthermore, flexibility is important to utilize collaborative design. Shmueli et al. (2021) reveal integrated collaborative planning of disaster hazard management risks with community involvement, during and after disaster events occur.

Social Institution on a National Disaster

Social institutions at the meso level, such as informal networks, values, and actors, have an important role in bridging the process between social protection and human security. Social institutions are intermediaries. The function of this social institution is to mediate concepts that are often in a gray area and contradicted. This section offers a perspective of social institutions from the empirical and theoretical context. Social institutions in several locations are reflected in the community that dealing with disasters. This institution is dynamic, where in certain conditions, it is formal, and in other conditions, it can be informal. It depends on the needs of the community and the natural phenomena. Formal and informal institutions have been formed in areas where volcanoes frequently erupt, such as in the Disaster-Prone Areas (KRB) of Mount Merapi. The ecological process of the Merapi eruption has helped shape the behavior and culture of the community (Tessa 2021). Communities on the slopes of Merapi, such as those in Cangkringan (Yogyakarta) and Selo (Boyolali, Central Java), have solid institutions to deal with disaster.

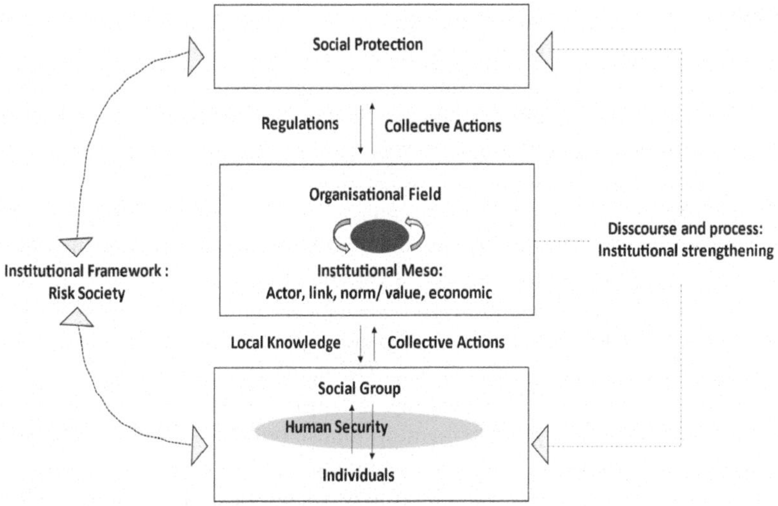

Fig. 18.2 Community social institutional model with disaster risk (adopted from Nee 2005)

Various institutional-forming tools such as local actors, social networks, and values have a crucial role. Informal local actors, such as Mbah Marijan, have a significant role in determining the response and action whenever Mount Merapi eruption. Almost all the slopes of Merapi, both in Central Java and Yogyakarta, have informal local actors who can mobilize and provide services to the community during a disaster. The emergence of these local actors did not happen suddenly but was motivated by cultural values in Java rural communities. The values of *gotong royong* and mutual cooperation is strengthened the community (Fig. 18.2).

Social institutions are interrelated systems between informal and formal elements, habitus, norms, and rules governing social relations where actors have the power to carry out their actions and activities. Nee (2003) also explains that institutions are social structures that provide networks for collective action by facilitating and regulating actors' interests. Social institutions, from the perspective of Nee (2005), explain that institutions interact with social networks and social norms in directing economic action. Local actors cannot be separated from the context and conditions of the community.

Moreover, adopting the framework created by Nee (2005) will mediate the social institutions and human security. This meso-social institution explains the relationship between the social protection function (macro-policy environment) in creating regulations, rules, and policies to address the national disaster problems. Human security is at the micro level and produces local wisdom, knowledge, and distinctive behavior. The principles of social institutions are built on the assumption of rationality that is tied to the context of society, and actors are supported by special interests, norms, networks, and social ties. The actor is an individual who can articulate interests and networks within a community organization in dealing with disasters. Social institutions interpreted as a system interconnected with one another, both formally and informally. This system facilitate, encourage and regulate various quick and appropriate actions. The position of this meso institution is to mediate macro-level mechanisms (social protection) in the form of laws, regulations, and formal rules and micro-level mechanisms (human security) in the form of individual humanitarian actions. Furthermore, the central government leads to institute bottom-up initiatives in short and long-term disaster situations (Shah et al. 2022).

Critical action encourages the abilities and skills in early prevention, psychologist support, and social activists (Indrayani and Wasistiono 2021). Social institutions at the meso-level are part of national and local social networks. This social network is one of the main tools of social capital with the beliefs and norms element. This social network consists of local actors who are capable develop the media of social relations into organizational collaboration. The network is based on the necessity to inform and help each other. This network is formed from interpersonal, personal, and institutional relationships inter-institutional networks. Norms and beliefs also support this meso-social institution as a complete and multidimensional institution. In the case of disasters in Indonesia, the role of social meso institutions is often less visible, whereas, in practical reality, the meso position is noteworthy and crucial.

The meso-social institution is also directed to anticipate disasters not only when disasters occur but also as a preventive and post-disaster action. Economic mitigation has an essential impact on disaster management. Various efforts and strategies can be planned through existing social institutions to overcome economic problems. In the event of a disaster, economic conditions will enhance a significant impact.

Social protection is presented by the government in the Law Number 24 year 2007, stated point (a): "*That the Unitary State of the Republic of Indonesia is responsible for protecting all the people of Indonesia and their entire native land, intending to protect lives and livelihood, including protection against disasters, to realize public welfare based on Pancasila (The Five Basic Principles), as mandated in the 1945 Constitution of the Republic of Indonesia*". The regulation provides protection for community affected by disasters. Indonesian context is unique because the community bonding characters. The community voluntarily involved, participate, and empathize with resolving the problems caused by the disasters immediately.

When Tsunami struck Aceh in 2004 and an earthquake occurred in Yogyakarta 2006, diverse staeholders worked together to manage the disaster impact. In the case of COVID-19 (from 2020 to the present), there is an empirically evidence about the actual support especially economic assistance to those affected. The middle and high-income groups made donations on various manual and digital platforms. Even through charity concerts, public donations amounted to IDR 2 billion (USD 135,000) in hours.

Table 18.2 confirms that various elements have initiated unstructured cooperation. It is reflected in the significant increase in donation trends through digital platforms during the COVID-19 pandemic. The National Disaster Management Agency (BNPB) collected a total COVID-19 grant of IDR 237.3 billion (USD 15,8 million). The total number was divided into foreign grants IDR 104.9 billion (USD 7 million), domestic grants IDR 55.2 billion (USD 3.7 million), and donations IDR 77.2 billion

Table 18.2 Parties involved in social protection and philanthropy in Indonesia during the pandemic

Parties involved		
The state sector	The private sector	The community
IDR 695.2 trillion (USD 46.4 billion USD) for the cost of handling COVID-19. The cost of handling the pandemic is equivalent to 4.2% of Indonesia's GDP; in detail, IDR 87.55 trillion (USD 5.8 billion) for the health sector and IDR 607.65 trillion (USD 40.5 billion) for the Economic Recovery Program	IDR 20 billion (USD 1,3 million) from Adaro, IDR 20 billion (USD 1,3 million) from Bakrie Group, IDR 10 billion (USD 667,000) from Tempo Scan Group, IDR 10 billion (USD 667,000) from XL Axiata, IDR 5.2 billion (USD 347,000) from Yayasan Benih Baik Indonesia, IDR 5 billion (USD 334,000) from Indosat, IDR 2 billion (USD 134,000) from Yayasan Wahana Artha	IDR 15.4 trillion (USD 1 billion) in 2020 (Lembaga Survei KedaiKOPI et al. 2022)

Source References processing results of various information sources

(USD 5.1 million) (Mutia 2022). Table 18.2 illustrates that there were various participations of many parties, whether in social protection or human security. It must be studied and considered for decision-making by all stakeholders, particularly the government.

The findings in this section are related to the improvement of the social resilience of villages in Indonesia to overcome the impact of COVID-19 in a sustainable approach (Kolopaking et al. 2022). The study in this section tries to offer how social institutions can form a shared social resilience through transparent and community empowerment-oriented stakeholder policies, the ability of formal and informal institutions to increase adaptive capacity of disasters management, and local wisdom that is still alive as values, norms and behaviors of community. Moreover, an important emphasis is on collaboration between communities and stakeholders.

Mediating the Potential of Social Protection and Human Security by Using Soft System Methodology Analysis

The Soft system methodology has been equipped with comprehensive tools to analyze and plan strategic schemes for national disasters (Hakami et al. 2013; Akhyar et al. 2023). Data and symptoms of potential disaster can be designed for better activity and system in the future. The disaster cycle and resolving mechanisms are complex issues because of their complicatedness, multi-interpretation, and multi-perspective nature. Problems caused by disasters can be solved either through macro-structure, namely social protection, or through micro-structure, human security. Both of them will be mediated by local social institutions as meso-structure. The meso institution have a role as a strategic scheme to address disaster challenges. The collaboration between field findings (social facts) and system frameworks relevant to disaster conditions in Indonesia will be strongly needed to determine the strategic scheme.

Figure 18.3 describes recommendations for improvement in the disaster cycle using SSM. *Firstly*, the problem in dealing with disasters occurs when there is no conformity between macro (*social protection*) and micro (*human security*) institutions. In some cases, the two elements of the organization can work well together, though it remains unexplainable in detail and structure. *Secondly*, the problem will be described in a rich picture for elaboration. The aims to depict a comprehensive and complete understanding in a large diagram (*helicopter view*). The main objective is to provide a structured, objective, and realistic description of the problems regarding disaster management in Indonesia through a *rich picture.*

Figure 18.3 describes problems and phenomena to provide a complete understanding of meso-social institutions. This picture bring up structural elements, processes, and relations in the context of disaster. *Thirdly*, to map the problems and the main tools of meso institutions in dealing with disasters, it is necessary to define keywords (root definitions) in the CATWOE elements (Client, Actor, Transformation, Worldview, Owner, and Environment) (Suryaatmaja et al. 2020).

The Table 18.3 explains that institutional meso mediates social protection and social security. The root definition of the CATWOE is a source goal known as a human activity system (Checkland and Poulter 2006). The modelling language and strategic scheme are based on verbs. This process consists of assembling and minimum structuring of activities needed to carry out the transformation process for better condition. This research aims to develop a strategic scheme by strengthening

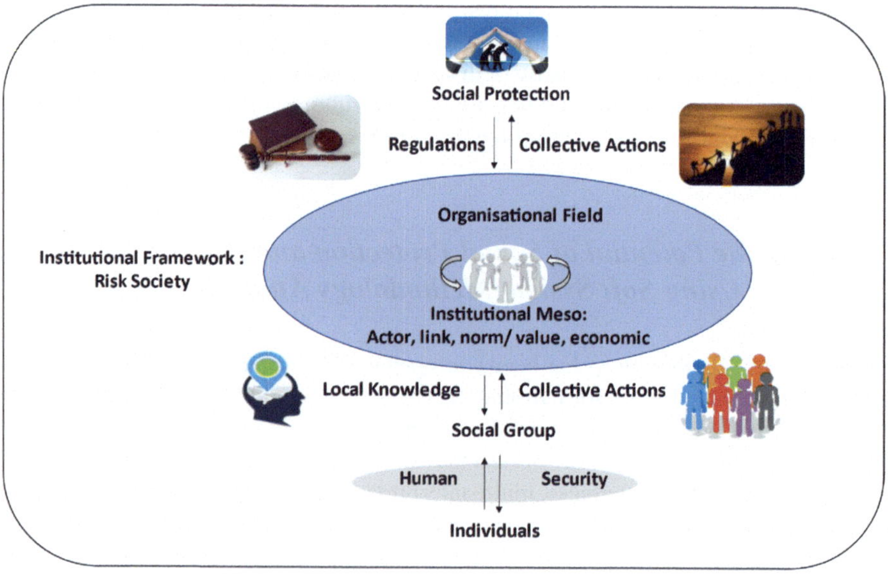

Fig. 18.3 Rich Picture Risk Society in Institutional Meso

meso-social institutions. The comparison between the strategic scheme and reality is carried out at stage 5 of SSM (Reynolds and Holwell 2020) (see Fig. 18.4)

This social modelling scheme viewed in Fig. 18.4 is made on two main bases, which are empirical facts and a working system framework (conceptual). The scheme's implementation of disaster problem-solving in Indonesia is expected to be well-executed. Meso social institutions are the mediators which bridges social protection and human security potentials. The three most essential aspects of a meso institution are actors, networks, and values. The local actors of the community are very important to respond quickly during disasters. They could be natives or those who have lived for a long time in a region and understand its geographical topology, natural resources, and community. Those local actors play a significant role in community mobilization. The rural community is characterized by 5–7 villages around the area. It means that in one sub-district, there are optimally 2–3 local leaders coordinating the community when a disaster occurs. The local figures function as network members for the local community and village, sub-district, and district administration figures.

Informal inter-community networks are as important as formal ones. Local-informal networks usually have information and quickly respond to disasters in the area, especially in locations with volcanic eruption, landslide, and tsunami potential.

Table 18.3 CATWOE analysis

	Social protection macro	Human security micro	Institutional meso
C	*Society*	Individual	*Group community*
A	Government	Individual	*Group community*
T	Regulation, policy	Empowerment	Empowerment
W	Protection for the community	Self-protection	Community protection
O	Formal constitution	*Value*	Local wisdom
E	*Socio-political constraints*	Human capital	Social network

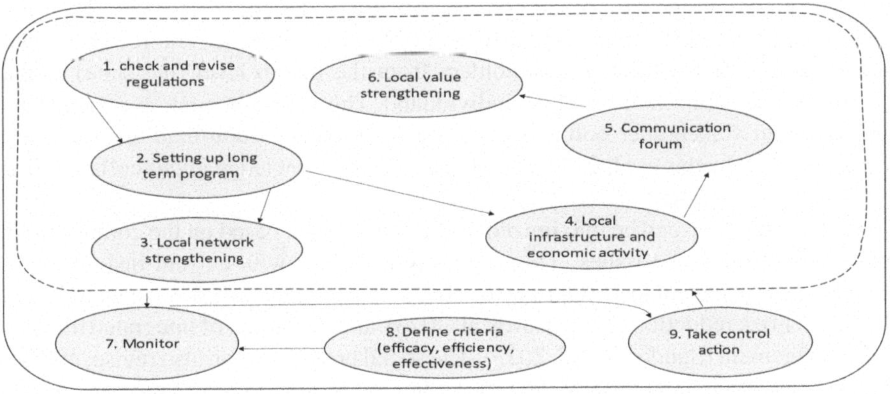

Fig. 18.4 The meso-level social institutions of disaster

The inter-community network can regularly be nurtured and developed both offline and online. It is necessary to hold regular offline meetings to discuss disaster risks and how to manage them.

Next, the social protection implementation based on characteristic. The government have to provide appropriate intervention based on needs and characteristic of target such as an indigenous community, and rural and urban community. If the meso social institutions are successfully formed and sustained, various programs, forums, and activities for pre-, during, and post-disaster management will be well-executed. The diverse background institutions have already existed in many communities in Indonesia. Some are already well-established, while some others are still in the form of embryos. Furthermore, disaster mitigation link to sustainable economy.

Recommendation of the Strategic Scheme as an Action

Indonesia regulation number 24 year 2007 is the manifestation of the Hyogo Framework For Action (HFA). The laws urge disaster risk management as the local priority and strengthen institutional resources. The laws need to be supported by a policy and legal framework that deliberate the aspect of decentralized capacities and responsibilities at all community levels. Disaster management should engage the pre-disaster and post-disaster periods management. Moreover, policy affirmation could play a significant role to hook multi-stakeholder partnerships. This chapter fills the gap about the institutional roles at the meso level. The effort to address the gap would make a vital contibution to the policy and action project.

The main purpose of this article is to mediate a strategic scheme plan for stakeholders at all unit levels. This research recommendation is based on ideal conditions between concepts and research findings. The recommendation was developed systematically from various conceptual frameworks and emic cultural conditions of the community. There are two criteria (Checkland and Scholes 1990) to formulate recommendations and action steps at stage 6 and stage 7 of SSM. They are "systemically desired" and "culturally" possible to implement. The recommendation is to engage multi-level stakeholders from the macro level (the state) to the micro level (the community and the individuals). This recommendation enriches the finding about stakeholder collaboration and inter-agency coordination in disaster management (Quader et al. 2023). Next, the current knowledge about collaborative governance in disaster risk reduction.

Social protection can be maximized at the macro level, based on the community's interests and the latest issues. This necessity is in line with current insight about strategic policy formulation (Ogra et al. 2021), community-driven ecosystem-based for disaster risk reduction (Lansakara et al. 2023), and the notion of integrated disaster risk management (Sandoval et al. 2023). The social protection can also revise, update, develop, and implement the laws that strengthen roles, functions, and responsibilities for disaster management implementation. There should be more flexible and in-depth connections among the President, the Parliament, the Governors, the Regents/

Mayors, and the National Disaster Management Agency (BNPB) in establishing regulatory cooperations and implementation to manage disasters.

At the micro level, a capacity development for human resources as individuals needs to be carried out. Sensitivity and local wisdom in dealing with disasters need to be reinforced as a disaster preparedness mechanism. Empowering and counseling activities, particularly for the younger generation (Generation Z), are essential to drive a disaster responsive generation. This kind of effort can be done through socialization and empowerment for school children who live in disaster-high-risk areas.

Table 18.4 are additional knowledge findings from this section. The findings describe three meso significant institutions roles to fill the todays learning according to the disaster management. First, actor responsibility to broader relationships inter-institution and inter-community. The idea link to todays discussion about engagement of various institution in the frame of actors relation and collaborative governance. It is neccesary for different stakeholders to aware for emergency response and responsibility in critical situation. Second, the finding about network function for cooperation. The effort to broaden network in meso institutional level provide opportunities for multi-sectors holistic support (Alcántara-Ayala et al. 2020). Third, the value and trust is principle to improve institutional development continuously. The institutional roles would enrich cooperation with its community-oriented bond.

The social institutions consisting "intermediary people" and "intermediary institution" that mediate inter-linkage between social protection and human security (Fig. 18.5). The meso institution can be "legal" in its responsibility yet "informal" in its nature as carrying out the responsibility. It means that the meso-social institutions will continue to seek their best form. Often, a local network consisting of local actors preserves significant roles for instance social activists is concerned about people's safety. The institution development encourage through strengthening social and communication networks, and empowerment economic mitigation.

Figure 18.5 shows the research design illustration to develop strategic ways of institutional roles to disasters management in Indonesia. The findings of this research state that there is a need to strengthen local institutions to broader stakeholder network included macro and micro linkage. The case in Indonesia is different from developed

Table 18.4 The Meso Institution Field Actions

Meso institutional tools	Field tasks
Actor	To mediate inter-institution relations To mediate inter-community relations To prepare field strategies pre-, during, and post-disasters To identify disaster and socio-economic potential To increase public awareness and socialization
Networks	To reinforce inter-sectoral cooperation networks
Value dan *trust*	To reinforce values and beliefs

Source Researcher data analysis (2023)

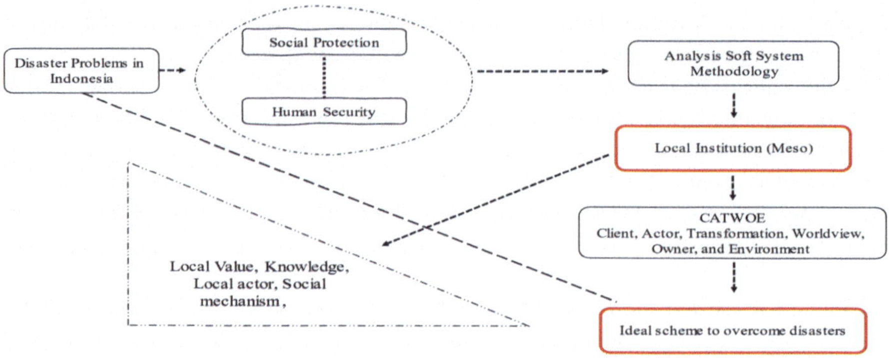

Fig. 18.5 Flowchart research design illustration. *Source* Researcher data analysis (2023)

countries where disaster management is mostly handled by the government and institutions that carry out formal legal responsibilities. The meso institutional approach reveal the local institution critical roles at the sub-district level or in village areas with high level of disaster vulnerability.

Limitations of the Study

There are five limitations to this study. First is the limited locus, where the study primarily focuses on the disaster management context in Indonesia, so the findings may not be directly applicable to other nations with different socio-economic, culture, and ecological conditions. Second is the dynamic nature of disaster management, where the field of disaster management is dynamic, and strategies and regulations may evolve over time. The study may not be able to capture real-time changes or adaptability in disaster management practices beyond the study period. Third is sociopolitical factors, where the study may not fully explore the complex sociopolitical factors that influence disaster management, which can have significant implications for the success of strategic and operational schemes at various levels. Fourth, policy implementation challenges, the study emphasized the existence of good disaster regulations at the national to local levels. However, challenges in the effective implementation of these regulations at the ground level still require more in-depth study. Fifth is the type of study, where broadly speaking this study is more at the conceptual level, so there is a need for further studies that can record empirically and in-depth on this issue.

Conclusion

Various empirical studies and literature show a gap between human security and social protection, that is, the meso institution. The institution's position is essential as a liaison and a driving force for the second mechanism of the human social security system. Disasters often occur quickly and unexpectedly, resulting in material and non-material losses for the community. The reinforcement of the local institution functions as a mediator between social protection and human security. Social institutions which conceive local actors, networks, and local values are capable of bridging two major concepts in national disaster management. Institutional strengthening at the meso (middle) level is important as an intermediate at the macro level (social protection) and the micro level (human security) by adopting a new model of institutionalism (Nee 2003). The macro level of social protection applies various regulations, policies, and rules to deal with disaster problems. At the micro level, human security is the natural ability of humans and society to deal with disasters. The meso level, functioning as the mediator, helps strengthen social institutions as an alternative key in disaster management. This study provides an overview of strengthening the meso level of social institutions in Indonesia's disaster management. Although this research data is based on a literature review, the analysis process aims to be a strategic scheme to deal with disasters effectively and based on local wisdom. Based on the systemic thinking process, a strategic interpretive scheme is formulated for policymakers and practitioners/social activists to deal with disasters. The systemic thinking process must involve various local contexts, such as values and the local wisdom of the community. Further research should explore other aspects, such as social, disaster, and economic mapping in disaster-prone areas as disaster mitigation mechanisms. Economic mitigation as an alternative to social protection and human security of the community will become an important research topic in the future.

References

Akhyar D, Akhyar D, Hardjosoekarto S (2023) Crisis communication in non-tectonic tsunami disaster management policy in Indonesia: the application of soft systems methodology based multi-method. Jurnal Komunikasi Indonesia 12(1). https://doi.org/10.7454/jkmi.v12i1.1054

Aksa FI, Afrian R (2022) Community adaptation strategies toward tidal flood: a case study in Langsa, Indonesia. Jàmbá J Disaster Risk Stud 14(1):1–8. https://doi.org/10.4102/jamba.v14i1.1258

Alcántara-Ayala I, Rodríguez-Velázquez D, Garnica-Peña RJ, Maldonado-Martínez A (2020) Multi-sectoral reflections and efforts in strengthening partnerships to reduce disaster risk in Mexico: the first MuSe-IDRiM conference. Int J Disaster Risk Sci 11:686–691. https://doi.org/10.1007/s13753-020-00302-w

Alkire S (2016) A conceptual framework for human security. Centre for Research on Inequality, Human Security and Ethnicity, CRISE. https://ora.ox.ac.uk/objects/uuid:d2907237-2a9f-4ce5-a403-a6254020052d

Arianti NK, Koentjoro K (2023) Volunteer altruistic behaviour in terms of disaster threat type. Jàmbá J Disast Risk Stud 15(1):1–7. https://doi.org/10.4102/jamba.v15i1.1478

Asfaw S, Tirivayi N, Knowles M, Davis B, Demeke M (2014). Risk management and the productive role of social protection. In: Food Agric Organization (FAO) of the United Nations in Collaboration with The New Partnership for Africa's Development (NEPAD) Planning and Coordination Agency, 1(1), pp 1–7. https://www.fao.org/3/i4240e/i4240e.pdf

Astawa IBM, Budiarta IG, Treman IW, Sarmita IM (2022) Pengembangan Wawasan Warga Sekolah Laboratorium Undiksha Tentang Sekolah Berwawasan Lingkungan Dan Mitigasi Bencana (Swaliba). Widya Laksana 11(1):41–48. https://doi.org/10.23887/jwl.v11i1.41854

Bahransyaf D, Probosiwi R (2018) Membangun Ketahanan Sosial di Desa Serakapi: Sebuah Replikasi Model. Jurnal Penelitian Kesejahteraan Sosial 12(3):225–236. https://doi.org/10.31105/jpks.v12i3.1255

Bayu H, Zaini M, Hamat Z (2015) Human security in local wisdom perspective : Pesantren and its responsibility to protect People. Procedia Environ Sci 28:100–105. https://doi.org/10.1016/j.proenv.2015.07.015

Bowen JR (1986) On the political construction of tradition: Gotong Royong in Indonesia. J Asian Stud 45(3):545–561. https://doi.org/10.2307/2056530

Bronfman NC, Cisternas PC, Repetto PB, Castañeda JV (2019) Natural disaster preparedness in a multi-hazard environment: characterizing the sociodemographic profile of those better (worse) prepared. PLoS ONE 14(4):e0214249. https://doi.org/10.1371/journal.pone.0214249

Budiman MR, Syahbanaf FA, Nurhasanah Y, Widhyharto DS, Pratama AR (2019) Kebijakan, Risiko, dan Perlindungan Sosial Alternatif dalam Transisi Pemuda di Jawa Tengah. Jurnal Studi Pemuda 7(1):46. https://doi.org/10.22146/studipemudaugm.42908

Chantavanich S, Jitpong W (2023) Precarity and social protection of migrant fishers in Thailand: case studies of death and disappearance at sea. Mar Policy 155:105743. https://doi.org/10.1016/j.marpol.2023.105743

Checkland P (2000) Soft systems methodology: a thirty year retrospective. Syst Res Behav Sci 17:S11–S58. https://doi.org/10.1002/1099-1743(200011)17:1+%3C::AID-SRES374%3E3.0.CO;2-O

Checkland P, Poulter J (2006) Learning for action: a short definitive account of soft systems methodology, and its use for practitioners, teachers and students. Wiley, London

Checkland P, Scholes J (1990) Soft systems methodology in action. Wileym London

Chotimah HC, Sande JP, Dwi CE, Yusvarini DR, Hong S (2017) Perkembangan Aspek Keamanan Ekonomi dalam Konsep Human Security. Transformasi Global 4(1):65–76

Churruca Muguruza C (2017) Human security as a policy framework: critics and challenges. Deusto J Human Rights, *4*, 15–35. https://doi.org/10.18543/aahdh-4-2007pp15-35

Coppola DP (2015) Preparedness. In: Introduction to international disaster management. Butterworth-Heinemann, pp 275–320. https://doi.org/10.1016/B978-0-12-801477-6.00005-8

Danar OR (2020) Disaster governance: Sebuah pengantar. Diva Press. https://books.google.co.id/books?hl=id&lr=&id=ewvbDwAAQBAJ&oi=fnd&pg=PA3&dq=Problem+utama+dalam+mitigasi+kebencanaan+secara+empiris+belum+memosisikan+sebagai+sebuah+skema+strategis+dalam+kebencanaan.+&ots=hG7E9jEKVR&sig=KybU1e1v_kkj9zwpTNTSjkZPCvQ&redir_e

Djalante R, Lassa J, Setiamarga D, Sudjatma A, Indrawan M, Haryanto B, Mahfud C, Sinapoy MS, Djalante S, Rafliana I, Gunawan LA, Surtiari GAK, Warsilah H (2020) Review and analysis of current responses to COVID-19 in Indonesia: period of January to March 2020. Progr Disas Sci 6. https://doi.org/10.1016/j.pdisas.2020.100091

Faedlulloh D, Meutia IF, Yulianti D, Karmilasari V (2021) Gotong Royong digital: Praktik Baru Solidaritas Warga di Era Pandemi. Jantra 16(1):1–14. https://doi.org/10.52829/jantra.v16i1.118

Faist T, Bilecen B, Barglowski K, Sienkiewicz JJ (2015) Transnational social protection: migrants' strategies and patterns of inequalities. Popul Space Place 21(3):193–202. https://doi.org/10.1002/psp.1903

Grayman JH (2014) Rapid response: email, immediacy, and medical humanitarianism in Aceh, Indonesia. Soc Sci Med 120:334–343. https://doi.org/10.1016/j.socscimed.2014.04.024

Griffiths D, Di Donato M, Lane TJ, Gray S, Iles R, Smith PM, Berecki-Gisolf J, Collie A (2023) Transition between social protection systems for workers with long term health problems: a controlled retrospective cohort study. SSM—Popul Health 23:101491. https://doi.org/10.1016/j.ssmph.2023.101491

Hadis BF (2014) Risk, social protection and trust amidst cuts in welfare spending. Health Risk Soc 16(5):459–480. https://doi.org/10.1080/13698575.2014.936832

Hakami A, Kumar A, Shim SJ, Nahleh YA (2013) Application of soft systems methodology in solving disaster emergency logistics problems. Int J Ind Manuf Eng 7(12):2470–2477

Hanifa NR, Gunawan E, Firmansyah S, Faizal L, Retnowati DA, Pradipta GC, Imran I, Lassa JA (2022) Unmanned Aerial Vehicles for geospatial mapping of damage assessment: a study case of the 2021 Mw 6.2 Mamuju-Majene, Indonesia, earthquake during the coronavirus disease 2019 (COVID-19) pandemic. Remote Sens Appl Soc Environ 28:100830. https://doi.org/10.1016/j.rsase.2022.100830

Hardjosoekarto S, Yovani N, Santiar L (2014) Institutional strengthening for the role of mass media in disaster risk reduction in Japan and Indonesia: an application of SSM-based action research. Syst Practice Action Res 27(3):227–246. https://doi.org/10.1007/s11213-013-9282-z

Harijoko A, Puspitasari D, Prabaningrum I, Prastika KP, Wijayanti NF (2021) Manajemen Penanggulangan Bencana Dan Pengurangan Risiko Bencana Di Indonesia. UGM Press

Hasegawa Y (2007) Is a human security approach possible? Compatibility between the strategies of protection and empowerment. J Refug Stud 20(1):1–20. https://doi.org/10.1093/jrs/fel027

Hizbaron DR, Hadmoko DS, Mei ETW, Murti SH, Laksani MRT, Tiyansyah AF, Siswanti E, Tampubolon IE (2018) Towards measurable resilience: mapping the vulnerability of at-risk community at Kelud Volcano, Indonesia. Appl Geogr 97(June):212–227. https://doi.org/10.1016/j.apgeog.2018.06.012

Holzmann R, Jørgensen S (2001) Social risk management: a new conceptual framework for social protection, and beyond. Int Tax Public Finance **8**(4), 529–556. https://doi.org/10.1023/a:1011247814590

Humaedi MA, Hakam S, Seftiani S, Propiona, JK (2016) Etnografi Bencana: Menakar Peran Para Pemimpin Lokal dalam Pengurangan Resiko Bencana. LKis

Indrayani E, Wasistiono S (2021) The role of community protection institution in disaster management at West Java, Indonesia. Jàmbá J Disast Risk Stud 13(1). https://doi.org/10.4102/jamba.v13i1.943

International Institute for Labour Studies (2012) World of Work Report 2012. Better Jobs for a Better Economy

Jolly R, Basu Ray D (2006) The human security framework and National Human Development Reports: a review of experiences and current debates. National Human Development Report Series

Kolopaking LM, Wahyono E, Irmayani NR, Habibullah H, Erwinsyah R G (2022) Re-Adaptation of COVID-19 impact for sustainable improvement of Indonesian Villages' social resilience in the digital era. Int J Sustain Develop Plan 17(7). https://doi.org/10.18280/ijsdp.170713

Koopman J (2021) The restoration of gotong royong as a form of post-disaster solidarity in Lombok, Indonesia. South East Asia Res 29(3):279–296. https://doi.org/10.1080/0967828X.2021.1966318

Kurnio H, Fekete A, Naz F, Norf C, Jüpner R (2021) Resilience learning and indigenous knowledge of earthquake risk in Indonesia. Int J Disast Risk Reduc 62:102423. https://doi.org/10.1016/j.ijdrr.2021.102423

Lansakara DC, Le De L, Petterson M, Wickramasinghe D (2023) The potential for community-driven ecosystem-based disaster risk reduction in South Asia: a literature review. Disaster Prev Manag An Int J. https://doi.org/10.1108/DPM-06-2023-0128

Lembaga Survei KedaiKOPI, Wibowo KA, Aryani S, Wibowo YS, Afifah AN (2022) Indonesia Philanthropy Outlook 2022: Perkembangan dan Proyeksi Filantropi di Indonesia Tahun 2022. Filantropi Indonesia. https://filantropi.or.id/download/ipo-2022-id/

Lowder SK, Bertini R, Croppenstedt A (2017) Poverty, social protection and agriculture: levels and trends in data. Global Food Secur 5:94–107. https://doi.org/10.1016/j.gfs.2017.06.001

Maksum A (2021) Social sciences & humanities open indonesian post-migrant workers: a challenging problem for human security. Soc Sci Human Open 4(1):100223. https://doi.org/10.1016/j.ssaho.2021.100223

Muhari A (2021) Catatan Refleksi Akhir Tahun Penanggulangan Bencana 2021. Badan Nasional Penanggulangan Bencana

Muqsith MA, Pratomo RR, Kuswanti A, Muzykant VL (2021) Social solidarity movement to prevent the spread of COVID-19 pandemic in Indonesia. Masyarakat, Kebudayaan Dan Politik 34(2):147. https://doi.org/10.20473/mkp.V34I22021.147-158

Mutia A (2022) Studi: Masyarakat Indonesia Paling Banyak Berdonasi ke Institusi Agama. Kata Data Media Network

Nee V (2003) The new institutionalism in economics and sociology (No. 4; CSES Working Paper Series).

Nee V (2005) The new institutionalisms in economics and sociology. In: Smelser NJ, Swedberg R (eds) The handbook of economic sociology, 2nd edn. Princeton University Press, Princeton, pp 49–74

Neuman WL (2006) Basics of social research: qualitative and quantitative approaches, 2nd edn. Pearson Education, London

Norton A, Conway T, Foster M (2001) Social protection concepts and approaches: implications for policy and practice in international development (No. 43; Working Paper No. 143)

Norton A, Conway T, Foster M (2002) Social protection: defining the field of action and policy. Develop Policy Rev 20(5):541–567. https://doi.org/10.1111/1467-7679.00189

Nuryana M, Erwinsyah RG, Suradi S (2023) Strengthening society preparedness as a strategy for disaster risk reduction in Indonesia. IOP Conf Ser Earth Environ Sci 1180(1):12045. https://doi.org/10.1088/1755-1315/1180/1/012045

Ogra A, Donovan A, Adamson G, et al (2021) Exploring the gap between policy and action in disaster risk reduction: a case study from India. Int J Disaster Risk Reduct 63. https://doi.org/10.1016/j.ijdrr.2021.102428

Ohemeng FLK, Foli R (2023) Towards a paradigm shift in social protection in developing countries? Analysing the emergence of the Ghana national unemployment insurance scheme from a multiple streams perspective. Urban Govern 3(3):219–227. https://doi.org/10.1016/j.ugj.2023.05.002

Pribadi KS, Abduh M, Wirahadikusumah RD, Hanifa NR, Irsyam M, Kusumaningrum P, Puri E (2021) Learning from past earthquake disasters: the need for knowledge management system to enhance infrastructure resilience in Indonesia. Int J Disast Risk Reduc 64:102424. https://doi.org/10.1016/j.ijdrr.2021.102424

Quader MA, Khan AU, Malak MA, Kervyn M (2023) Mainstreaming decentralization and collaboration in disaster risk management: insights from coastal Bangladesh. Int J Disaster Risk Sci 14:382–397. https://doi.org/10.1007/s13753-023-00495-w

Rachmawati TA, Rahmawati D, Susilo A (2018) Pengurangan Risiko Bencana Berbasis Tata Ruang. Universitas Brawijaya Press. https://books.google.co.id/books?hl=id&lr=&id=JfKIDwAAQBAJ&oi=fnd&pg=PA1&dq=Akhir-akhir+ini+timbul+kekhawatiran+bencana+terutama+bencana+alam+yang+tidak+bisa+diprediksi+kapan+terjadinya+.+Indonesia+gempa+bumi+,+tsunami+,+letusan+gunung+berapi+,+Indonesia+

Reynolds M, Holwell S (2020) Systems approaches to making change: a practical guide. In: Systems approaches to making change: a practical guide. https://doi.org/10.1007/978-1-4471-7472-1

Sandoval V, Voss M, Flörchinger V, et al (2023) Integrated disaster risk management (IDRM): elements to advance its study and assessment. Int J Disaster Risk Sci 14:343–356. https://doi.org/10.1007/s13753-023-00490-1

Setiadi A, Andriessen A, Anisa R (2020) Post-occupancy evaluation of Pagerjurang Permanent Housing after the Merapi volcanic eruption. J Archit Urban 44(2):145–151. https://doi.org/10.3846/jau.2020.11265

Shah I, Mahmood T, Khan SA, Elahi N, Shahnawaz M, Dogar AA, Subhan F, Begum K (2022) Inter-agency collaboration and disaster management: A case study of the 2005 earthquake disaster in Pakistan. Jàmbá J Disast Risk Stud 14(1). https://doi.org/10.4102/jamba.v14i1.1088

Shalih O (2020) Strategi Membangun Ketahanan Komunitas (Masyarakat) Terhadap Bencana Tanah Longsor Di Desa Sirna Resmi, Kecamatan Cisolok, Kabupaten Sukabumi [Program Pascasarjana Ilmu Geografi]. https://www.academia.edu/download/61945363/Osmar_Shalih-Tesis-FMIPA-Full_Text-202020200130-22275-t6ry3s.pdf

Shaw R (2006) Community-based climate change adaptation in Vietnam: inter-linkages of environment, disaster, and human security. In: Multiple dimensions of global environmental change. TERI Press, New Delhi, pp 521–547

Shaw R (2012) Community-based disaster risk reduction, vol 10. Emerald Group Publishing Limited. https://doi.org/10.1108/S2040-7262(2012)10

Shaw R (2014) Kobe Earthquake: turning point of community-based risk reduction in Japan. In: Shaw R (ed) Community practices for disaster risk reduction in Japan. Springer, pp. 21–31. https://doi.org/10.1007/978-4-431-54246-9_2

Shmueli DF, Ozawa CP, Kaufman S (2021) Collaborative planning principles for disaster preparedness. Int J Disas Risk Reduc 52:101981. https://doi.org/10.1016/j.ijdrr.2020.101981

Sieber S, Orsholits D, Cheval B, Ihle A, Kelly-irving M, Delpierre C, Burton-jeangros C (2022) Social science & medicine social protection expenditure on health in later life in 20 European countries : spending more to reduce health inequalities. Social Sci Med 292. https://doi.org/10.1016/j.socscimed.2021.114569

Simanjuntak C (2020) Pendekatan Human Security dalam Penanganan Pandemi: Studi Kasus COVID-19 di Jepang. Center for Area Studies Indonesian Institute of Sciences. http://psdr.lipi.go.id/news-and-events/opinions/pendekatan-human-security-dalam-penanganan-pandemi-studi-kasus-covid-19-di-jepang.html

Slikkerveer LJ (2019) Gotong royong: an indigenous institution of communality and mutual assistance in Indonesia. In: Integrated community-managed development: Strategizing indigenous knowledge and institutions for poverty reduction and sustainable community development in Indonesia. Springer, Berlin, pp. 307–320. https://doi.org/10.1007/978-3-030-05423-6_14

Suarjana IGP (2020) *KONDISI Sosial Dan Ekonomi Masyarakat Pengungsi Bencana Erupsi Gunung Agung Desa Ban Kecamatan Kubu Kabupaten Karangasem*. Universitas Pendidikan Ganesha. https://repo.undiksha.ac.id/999/

Subagia IW (2015) Pelatihan Mitigasi Bencana Alam Gempa Bumi Pada Siswa Sekolah Dasar Negeri 1 Pengastulan Kecamatan Seririt Kabupaten Buleleng Bali. JPI (Jurnal Pendidikan Indonesia) 4(1)

Suharini E, Liesnoor SD, Kurniawan E (2015) Pembelajaran kebencanaan bagi masyarakat di daerah rawan bencana banjir DAS Beringin Kota Semarang. Forum Ilmu Sosial 42(2):184–195. https://doi.org/10.15294/fis.v42i2.10143

Suppasri A, Maly E, Kitamura M, Syamsidik, Pescaroli G, Alexander D, Imamura F (2021) Cascading disasters triggered by tsunami hazards: a perspective for critical infrastructure resilience and disaster risk reduction. Int J Disast Risk Reduc 66:102597. https://doi.org/10.1016/j.ijdrr.2021.102597

Suradi, Sumarno S, Sugiyanto, Nainggolan T, Murni R, Erwinsyah R, Andriyani L (2022) Synergy between disaster preparedness area program with local institutions for community disaster preparedness. IOP Conf Ser Earth Environ Sci 1027(1):012018. https://doi.org/10.1088/1755-1315/1027/1/012018

Suryaatmaja K, Wibisono D, Ghazali A, Fitriati R (2020) Uncovering the failure of Agile framework implementation using SSM-based action research. Palgrave Commun 6(1):1–18. https://doi.org/10.1057/s41599-019-0384-9

Tadjbakhsh S, Chenoy A (2007) Human security. Routledge. https://doi.org/10.4324/9780203965955

Taylor H, Peace R (2015) Children and cultural influences in a natural disaster: flood response in Surakarta, Indonesia. Int J Disast Risk Reduc 13:76–84. https://doi.org/10.1016/j.ijdrr.2015.04.001

Teodorescu DI, Muñoz Sastre MT, Kpanake L, Sorum PC, Mullet E (2021) People's positions regarding social security's health insurance programs: The case of Romania. *Revue Europeenne de Psychologie Appliquee* 71(5). https://doi.org/10.1016/j.erap.2021.100716

Tessa A (2021) Pemanfaatan Tiga Worldview dalam Peran Patron Sosial Bagi Penguatan Resiliensi Komunitas Tanggap Bencana Merapi. Jurnal Masyarakat Dan Budaya 23(2). https://doi.org/10.14203/jmb.v23i2.1204

The SAGE Handbook of Qualitative Research (2005) In: Denzin NK, Lincoln YS (eds) Qualitative research in organizations and management: an international journal, 3rd edn. SAGE Publication. https://doi.org/10.1108/17465640610666642

Thouret J-C, Wavelet E, Taillandier M, Tjahjono B, Jenkins SF, Azzaoui N (2022) Defining population socio-economic characteristics, hazard knowledge and risk perception: the adaptive capacity to persistent volcanic threats from Semeru, Indonesia. Int J Disast Risk Reduc 77:103064. https://doi.org/10.1016/j.ijdrr.2022.103064

Tsani RA (2019) Pengorganisasian dalam menciptakan masyarakat tangguh bencana gunung meletus di Desa Puncu Kecamatan Puncu Kabupaten Kediri. UIN Sunan Ampel Surabaya. https://digilib.uinsby.ac.id/32376/

United Nations (2015) Social protection: a development priority in the post-2015 UN development agenda.

Vidyattama Y, Merdikawati N, Tadjoeddin MZ (2021) Aceh tsunami: long-term economic recovery after the disaster. Int J Disast Risk Redu 66:102606. https://doi.org/10.1016/j.ijdrr.2021.102606

Ville I (2019) When social protection and emancipation go hand in hand: towards a collective form of care. Alter—Eur J Disabil Res 13(2):101–112. https://doi.org/10.1016/j.alter.2018.11.006

Wardekker A, Nath S, Handayaningsih TU (2023) The interaction between cultural heritage and community resilience in disaster-affected volcanic regions. Environ Sci Policy 145:116–128. https://doi.org/10.1016/j.envsci.2023.04.008

Wickramasinghe K (2014) Role of social protection in disaster management in Sri Lanka. Sri Lanka J Soc Sci 35(1–2):1–8. https://www.researchgate.net/profile/Kanchana-Wickramasinghe/publication/267736701_Role_of_Social_Protection_in_Disaster_Management_in_Sri_Lanka/links/5458bc4f0cf26d5090aca198/Role-of-Social-Protection-in-Disaster-Management-in-Sri-Lanka.pdf

Zervaki A (2018) Human security and climate change mitigation: the case of ocean governance. *Marine Policy*, September, pp 1–9. https://doi.org/10.1016/j.marpol.2018.09.026

Part IV
Climate Crisis and Sustainable Solution to SDGs

Chapter 19
Resilience and Preparedness Among Coastal Communities on the Natuna Islands

Awani Irewati, Sjarief Widjaja, and Imam Syafi'i

Abstract The occurrence of climate change introduces an element of unpredictability into the fishing activities of local fishers on the Natuna Regency. These seasonal fluctuations in weather conditions lead to drastic changes that have a profound impact on the fishing community. The primary objective of this chapter is to examine the resilience and preparedness of coastal communities (on Lampa Strait/Pulau Tiga) to withstand and prepare for extreme weather conditions, as well as to examine the efforts made by the government to enhance the economic resilience of fishermen. The study employed a purposive sampling strategy, which involved deliberately selecting participants with specific traits in order to obtain a thorough representation of relevant subgroups that are in line with the research question and goals. For technical measures, the research method using the Analytical Hierarchy Process (AHP) with number of samples each group to examine the background of the SKPT development. In conclusion, during the active fishing season, fishermen are able to generate sufficient income to sustain their livelihoods. However, during the off-season, some may seek alternative employment, while others receive support from fellow fishers who gather together to secure their daily earnings. These alternative endeavors contribute to the resilience and preparedness of fishers in facing the challenging weather conditions each year. These efforts are complemented by the government's ongoing initiatives to meet the needs of fishers, such as the construction of roads, fresh water reservoirs, and the establishment of the Integrated Marine and Fisheries in Sclat Lampa (SKPT).

A. Irewati (✉) · I. Syafi'i
PRP-BRIN (National Research and Innovation Agency), Jakarta City, Indonesia
e-mail: awan001@brin.go.id

I. Syafi'i
e-mail: imam016@brin.go.id

S. Widjaja
Faculty of Marine Affairs and Fisheries Technology, Institut Teknologi Sepuluh Nopember (ITS), Surabaya City, Indonesia
e-mail: sjarief@na.its.ac.id

U. Mukhopadhyay et al. (eds.), *Climate Crisis, Social Responses and Sustainability*, Climate Change Management, https://doi.org/10.1007/978-3-031-58261-5_19

Introduction

Overview

Coastal communities depend on conventional beliefs and methods to deal with the yearly changes in weather. They can recognize natural signs that show up the beginning of periodic weather transitions, such as observing the moon's movement and the shift between the north and south monsoon seasons. Several scholars (Ansaar 2019; Hasmawati et al. 2022) have discussed how the traditional knowledge of fishermen in sea fishing is rooted in their understanding of the environment. This includes maritime knowledge encompassing various aspects related to the sea, navigation, weather, and more. Fishermen rely on natural signs in the sea, such as waves, wind direction, and currents, and on land, they recognize features like mountains, capes, birds, and specific landmarks. In the sky, they observe signs like clouds, stars, the moon, and the sun. These natural indicators serve as guides for fishermen in determining the position and direction of their boats while sailing. However, these skills are today somewhat limited because of the uncertain climate shifts. Activities like excessive fishing and illegal fishing further affect the lives of fishers, leading to social repercussions.

In a broader context, a study conducted by Leal Filho et al. (2021) examines the resilience of coastal communities in Small Islands Developing States (SIDS) in the Pacific and the Caribbean Small Island States. The study reveals common characteristics that influence the resilience of these communities, including limited resources and access to technologies, fragile economies dependent on tourism, and vulnerability to climate change and its impacts. The objective of the study is to enhance understanding of policy contexts, appropriate strategy choices, and to present evidence on the limitations of adaptation and adaptive capacities in proposed case studies. Furthermore, the study establishes a framework for comprehending and exploring the inevitable climate-related effects on preventing and reducing future risks. The overall findings of the research provide a significant inspiration for this chapter in understanding the resilience of coastal communities in the face of weather challenges.

Hafsaridewi et al.'s (2020) study examines the impact of extreme weather on fishers' resilience in the South China Sea, specifically focusing on Natuna Island in Indonesia. They discuss how climate change and severe weather affect the productivity of capture fish, which is influenced by shifts in fish distribution due to changes in water temperature, ocean currents, and other marine conditions. Their discussion covers Natuna Island broadly and does not specify the research location. In contrast, the focus of this chapter's study is on how the coastal community in the Lampa Strait and surrounding islands (Pulau Tiga) demonstrates resilience and preparedness, along with the government's corresponding actions. Another article by Wibowo and Satria (2015) explores the adaptation strategies of the community in Pulau Panjang Village on Subi Island, Natuna, to climate change impacts. While other articles discuss the condition of Natuna Island's coral reef ecosystem and its coastal communities, these studies focus on Natuna's marine ecosystem and its natural resources. In contrast, the chapter under review highlights the resilience and preparedness of coastal communities in the Lampa Strait, particularly on Pulau Tiga, South Natuna, where the government constructed the Integrated Marine and Fisheries Center (SKPT) from 2015 to 2019 to support fishing and integrated maritime activities for fishers from nearby areas like Pulau Tiga.

This chapter highlights the primary concerns of coastal communities in the Southern Natuna Islands (Lampa Strait/Selat Lampa, and Pulau Tiga) which revolve around the various natural and social challenges they face. It underscores the significant of resilience and preparedness in addressing these issues. The fishing activity has been significantly affected by climate-related crisis, particularly impacting small and medium scale fishers who mostly operate boats with a capacity of less than 30 GT. The fishing activities have been significantly impacted by clime crisis affecting small and medium-sized fishers with boats average 6 GT capacity. During the famine season of the south wind season (start from July–September, sometime extend to October), the fishers of the southern Natuna experience scarcity and are forced to go further away even approaching to Malaysia sea water (in-depth interviews with key figure in the fishing community of Pulau Tiga, June 2023). These fishers frequently struggle to attain adequate fish catches and rely on support from collectors (*pengepul*) to meet their basic necessities. Moreover, there are also those who pursue additional occupations, such as working in construction, serving as motorcycle taxi drivers, or engaging in manual labor for transporting goods.

Conversely, certain fishers from the northern Natuna area temporarily migrate to the southern Natuna region (Lampa Strait and its surroundings) (as told by the local fisher in Pulau Tiga, August 28, 2023) for fishing. However, **it is important to knowledge that these migrating fishers are not the primary focus of this chapter, as they come from regions beyond the scope of the research area** (as shown in Fig. 19.2) Additionally, fishers also face socio-economic challenges, particularly in relation to conflicts arising from competition over fishing area, involving both foreign and local fishers outside of the Natuna Islands (from Java Island). Following the resurgence of illegal fishing activities post the Susi Pudjiastuti era, the government implemented a policy allowing fishers from the north coast of Java Island to fish in the North Natuna Sea, which has been met with resistance from local fishermen.

This chapter aims to address the underlying question of how fishers' resilience and preparedness have helped them survive these two challenges, as well as the attempts conducted by the state to reinforce the socio-commercial resilience of fishers. The chapter concludes that during the famine time, collectors plays a crucial role in supporting to survival of fishers. Additionally, the government has constructed the SKPT in the Lampa Strait to facilitate easy access for fishers to sell their catches, as well as providing convenience in purchasing Bio-solar for their boats.

This chapter is inspired by the important role of the Natuna Islands, both in terms of politics and economics. From a political standpoint, Natuna's location on a major trade route and the Indonesian Archipelago Sea Lanes 1 (ALKI-1) positions it as Indonesia's northernmost frontier in the South China Sea region. The abundance of marine resources in the area highlights its significance as a crucial fishing hub, known as WPP 711 (Fisheries Management Area-711). The presence of oil and natural gas reserves in the region makes it an area of interest for exploration and exploitation by both domestic and foreign companies, giving Natuna significant economic importance for Indonesia. However, this study focuses on the resilience and preparedness of fishers in the southern part of Natuna, particularly in the Lampa Strait and surrounding islands like Pulau Tiga, in response to unpredictable weather changes. Establishing the Integrated Marine and Fisheries Center (SKPT) in the Lampa Strait in 2019 further emphasizes prizing this subject, as it targets to establish Natuna as a hub for fish and marine production, serving the needs of local fishers in the Lampa Strait. This chapter tries to address this specific research gap in Natuna. The chapter begins with an introduction, followed by explaining the study's reasons, describing the materials and methods used, definitions, presentation of results, discussion, and, a conclusion.

This chapter aims to address the underlying questions of

- How the resilience and preparedness of fishers have enabled them to endure these dual challenges, along with the efforts made by the state to strengthen the socio-economic resilience of fishers. The chapter concludes that during the famine time, collectors play a crucial role in supporting to survival of fishers.
- How the government has established the SKPT in the Lampa Strait to facilitate the easy selling of catches for fishers, along with providing convenience in acquiring Bio-solar for their boats.

The Natuna Geography at the Crossroad of Global Maritime Routes

Natuna belongs to the group of border regions and remote islands within the Riau Archipelago Province. Its administrative center is Ranai, housing a population of 81,495 residents. Geographically, Natuna is situated between 1° 16′ to 7° 19′ North Latitude and 105° 00′ to 110° 00′ East Longitude. The total area of Natuna Regency spans 264,198.37 km^2, comprising 2001.30 km^2 of land and a substantial 262,197.07 km^2 of sea (Natuna Regency in Figure 2023).

As per Law No. 33 of 2008, enacted on July 21, 2008, Natuna was initially structured into 12 districts, and later, the regions of Bunguran Selatan, Bunguran Timur Laut, and Serasan Timur were added. Furthermore, following the enactment of Natuna Regency Regional Regulation No. 14 in 2014, on December 10, 2014, three new districts were established: Bunguran Batubi, Pulau Tiga Barat, and Suak Midai districts (as illustrated in Natuna Regency in Figure 2023). Consequently, by 2016, Natuna Regency was composed of a total of fifteen districts. Detailed information about the number of districts in Natuna can be found in Table 19.2.

The Natuna Islands, a collection of 154 small islets, occupy a strategically significant position near the Sea Lines of Communication (SLOC), connecting the Malacca Strait to the South China Sea. These maritime pathways hold global importance, serving as lifelines for coastal nations and users worldwide. To its north, Natuna

Regency faces the North Natuna Sea, while its southern neighbor is Bintan Regency. To the west, it shares a border with Peninsular Malaysia, and to the east lies the North Natuna Sea. Natuna, situated in the northern part of Republic of Indonesia, is geographically close to countries such as Malaysia, Vietnam, Mainland China. Its location, bordering the South China Sea to the north, offers a valuable advantage in terms of facilitating direct exports of marine resources, especially in the fisheries sector, to neighboring nations. The South China Sea plays a vital role as a trade thoroughfare, connecting the primary trade networks of Southeast Asia. According to estimates from the United Nations Conference on Trade and Development, in 2016, the South China Sea facilitated 21% of global trade ($3.37 trillion worth) (Holt 2022). Natuna Regency strategically positions itself along this vital route, offering the potential for strategic utilization.

The Potential Natural Wealth of Natuna

The marine resources in Natuna Regency are incredibly abundant. These resources are found within Fisheries Management Area 711 (FMA 711) and are spread across the Natuna Sea, Karimata Strait, and North Natuna Sea. FMA 711 encompasses island waters, territorial seas, additional zones, and Indonesia's exclusive economic zone (EEZ). This management area covers a vast expanse of 266,382 Km^2, crossing five countries: Singapore, Malaysia, the Philippines, Vietnam, and Thailand. Among this wealth are various types of fish and marine life, as detailed in Table 19.1. These marine resources include Small Pelagic Fish, Large Pelagic Fish, Demersal Fish, Coral Fish, Penaeid Shrimp, Lobster, Crabs, Soft-Shell Crabs, and Squid (*Sidik Zone*, April 7, 2022). The Decision of the Minister of Maritime Affairs and Fisheries of the Republic of Indonesia No. 19 of 2022 regarding the Estimation of Fishery Resource Potential, Permitted Fish Catch Quantity, and Utilization Rate of Fishery Resources in the Fisheries Management Area. This decision was signed on March 29, 2022, The information in Table 19.1 reveals that lobster (1,1), crab (1,9), and soft-shell crab (1,2) have surpassed the permitted limit, signifying that they are being overharvested. Consequently, measures to decrease fishing activities should be put into action.

Rationale of the Study: Coastal Communities Living in Natuna as Remote Islands

Indonesia is motivated to ensure that the Natuna Islands, being remote in nature, receive a prominent place in the country's Maritime Development Policy. This commitment is to advancing the maritime sector in distant islands. It is clearly articulated in the Nawa Cita Policy, where the first point emphasizes the creation of an integrated Tri Matra based on national interests and the reinforcement of Indonesia's

Table 19.1 The estimation of fishery resource potential, permitted fish catch quantity, and utilization rate of fishery resources in the Fisheries Management Area-711

No.	Type of fishery resources	Estimation of potential (tons)	Permitted fish catch quantity (tons)	Utilization rate (€)
1	Small Pelagic Fish	536.917	375.842	0.9
2	Large Pelagic Fish	163.744	114.621	0.7
3	Demersal Fish	289.300	202.510	0.8
4	Coral Fish	197.580	138.306	0.5
5	Penaeid Shrimp	71.810	50.267	0.6
6	Lobster	1.467	734	1.1
7	Crab	3.388	1.694	1.9
8	Soft-Shell Crab	9.804	4.902	1.2
9	Squid	32.369	22.658	0.5

Notes Utilization Rate (E): $E < 0.5$ = Moderate, catch fish could be increased. $0.5 \leq E < 1$ = Fully-exploited, fishing efforts are maintained under strict monitoring
$E \geq 1$ = over-exploited, fishing activities have to be decreased
Source The Decision of the Minister of Maritime Affairs and Fisheries of the Republic of Indonesia No. 19 of 2022

Table 19.2 Number of capture fisheries households by sub-district in Natuna Regency, 2018–2022

Sub district	2018	2019	2020	2021	2022
Midai	212	163	171	173	173
Suak Midai	70	76	80	82	82
Bunguran Barat	886	736	773	781	781
Bunguran Utara	209	225	272	280	280
Pulau Laut	275	322	338	341	341
Pulau Tiga	388	388	333	336	336
Bunguran Batubi	71	238	250	253	253
Pulau Tiga Barat	173	173	127	130	130
Bunguran Timur	654	654	687	694	694
Bunguran Timur Laut	554	605	635	641	641
Bunguran Tengah	142	193	–	–	–
Bunguran Selatan	671	702	737	744	744
Serasan	572	582	582	590	590
Subi	328	310	340	545	343
SerasanTimur	258	295	265	270	270
Total Natuna	5463	5662	5590	5658	5658

Source BPS. Kabupaten Natuna Dalam Angka 2023, Kabupaten Natuna, BPS

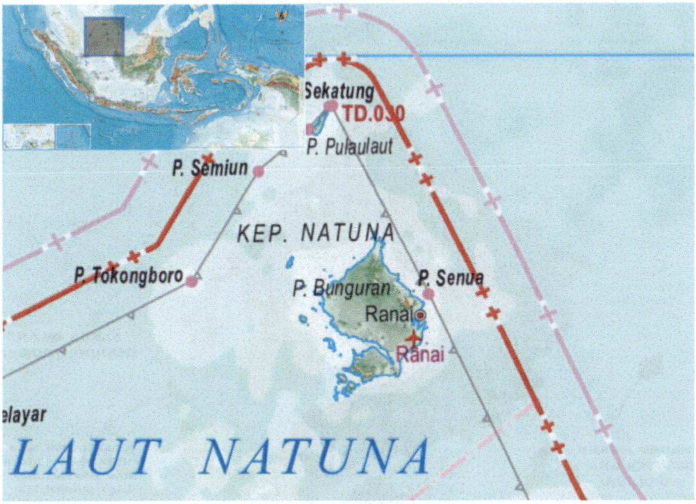

Fig. 19.1 The Natuna Islands. *Source* Retrieved from "Peta Negara Kesatuan Republik Indonesia", https://big.go.id/assets/download/2017/NKRI-2017/Peta_NKRI_Skala_5000000_03102017.jpg. Natuna islands (comprising 154 small islands) are near Sea Lines of Comnication (SLOC) connecting the Malacca strait to the South China Sea

identity as a maritime nation, The Natuna falls squarely within the ambit of this policy and its objectives. In an ideal scenario, the outer islands need to possess comprehensive resilience. This signifies that outermost islands, situated a significant distance from the central government (Fig. 19.1). It should be endowed with complete resilience across all sectors, including social, economic, and political security. The socio-economic well-being of places like Natuna, for instance, emerges as a major priority for the local government due to its direct alignment with the interests of the local population.

Addressing the challenges presented by the annual north wind season is an ongoing struggle for the communities residing in these regions. During this period, fishermen are unable to set sail due to the formidable sea waves, some of which reach heights of up to 6 m. This compels them to seek alternative employment opportunities to meet their daily needs. For instance, they might work as motorcycle taxi drivers, in construction, or in other occupations that can generate daily income. In the case of the southern Natuna region, larger-scale fishermen (who have crews of around 10–20 people) can provide assistance to help them endure the lean season (as told by resource on Pulau Tiga, 2023).

According to a source on Pulau Tiga (in June 2023), it can be inferred that they have significant experience in addressing this annual concern and are already quite knowledgeable in its management. The majority of local fishermen do not possess fishing vessels that exceed a 30 Gross Tonnage (GT) capacity, nor do they have access to advanced communication equipment. They rely on simple fishing gear, which they

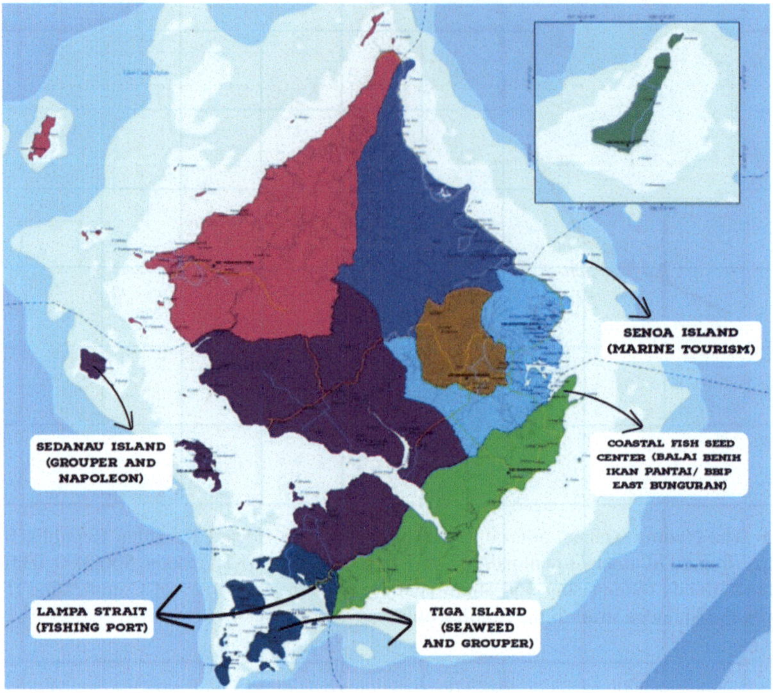

Fig. 19.2 The Location of SKPT Natuna's Activities. *Source* Widjaja (2022, p. 6)

believe is safer for the marine ecosystem. They find themselves unable to reach fishing grounds farther within the Natuna Exclusive Economic Zone (EEZ). Consequently, the North Natuna Sea, which has historically been their primary fishing area, has gradually transformed into an uninhabited region.

Therefore, fishermen in Natuna not only deal with the challenges of seasonal scarcity but also grapple with matters of illegal fishing carried out by neighboring countries within the North Natuna Sea. The island's geographic proximity to the South China Sea could be disadvantageous, as it places Natuna in the midst of global competition within the larger Asia Pacific region. As a result, the government has placed considerable emphasis on Natuna Island by establishing security measures and a military defense facility. However, this specific aspect is not the primary focus of analysis in this chapter.

The authors belief that the strength of a nation, including the significance of Indonesia's numerous islands, is ultimately determined by its people. It is of paramount importance that individuals residing on small and remote islands, such as Natuna, receive increased attention from both governmental and non-governmental entities. The government has formulated plans to bolster and enhance the marine and fishery sector by establishing various Centers for Integrated Marine and Fishery (referred to as SKPT), one of which is located in Natuna and known as SKPT Natuna. At a broader perspective, these SKPTs were envisioned to bolster coastal economies and were identified as a focal point by the Ministry of Marine and Fishery and associated stakeholders. The establishment of an SKPT in the Lampa Strait of Natuna was a governmental initiative. At a more localized level, the primary objective behind the establishment of the SKPT was to provide support and convenience to fishermen during the process of docking and selling their caught fish at the port. This is particularly important in ensuring a fish market for buyers, which in turn aids fishermen in effectively selling their catch. Additionally, the SKPT has the potential to serve as a gateway for activities related to fish processing, thereby adding value to fish production intended for export.

Fluctuating weather conditions have the potential to disrupt the environments of coastal island communities, potentially causing imbalances in the natural ecosystem. Heightened sea waves can erode coastlines and reach further inland. While residents in the southern areas are relatively protected during the high wave season compared to those in the northern Natuna region. It's equally important to pay attention to the fishermen, who constitute an integral part of these isolated Indonesian islands. Understanding the dynamics of coastal communities on remote islands like Natuna can provide significant insights for those residing on the mainland.

Materials and Methods

This study utilizes a combination of qualitative and quantitative analyses, employing a purposive sampling strategy to deliberately select participants with specific traits that align with the research question and goals (conducted in 2023). For technical measurements, the research method involves the use of the Analytical Hierarchy Process (AHP), with a specified number of samples for each group. The AHP was conducted in 2022 to understand the technical measurement background of the SKPT development.

Series of interview have been conducted to the fisherman, local government and fisheries office following the research method using the Analytical Hierarchy Process (AHP) with number of samples each group (fishermen, local government officers and fisheries officers) of 15 peoples. The flowchart below shows the research step conducted by the authors. Some interviews conducted in a Focus Group Discussion (FGD) and other using personally approaches (**Field Research- in-depth interview step**). Analytical Hierarchy Process (AHP) determines the level of influence of an element on a problem through a fundamental comparison scale of an individual's ability to make a comparison in pairs against several elements compared (**Field research—Technical Measurements of AHP step**) For the processing of questionnaire results in this study for each cell, the value is determined (combined assessment) using the average measure (geometry), with the following formula:

$$\sqrt[n]{a_1^x a_2^x \ldots a_n = a_w} \tag{19.1}$$

where

a_w Combined assessment (final assessment)
a_i Rating of respondents to I (on a scale of 1/9 s/d 9)
n Number of respondents.

After obtaining the combined value (a_w) for each cell, a new pairwise comparison matrix is formed. With this matrix, synthesis calculations are carried out by weighting and adding to produce a single number that shows the priority of each element. In pairwise comparisons, it can happen that the choices given are inconsistent so that consistency tests are carried out with steps:

- Calculate the Consistency Index (CI) by using the formula:

$$CI = \frac{\lambda_{maks} - n}{n - 1} \tag{19.2}$$

where

λ_{max} maximum eigen value
n matrix size
CI Consistency Index

- Calculate the Consistency Ratio (CR) with the formula:

$$CR = \frac{CI}{RI} \tag{19.3}$$

The price of CR according to Saaty depends on the matrix formed. From this calculation, if the value of $\leq$ 10% is considered consistent.

Additionally, the authors held a series of discussions and interviews with some key persons (such as the Dean of the Faculty of Technology and Marine Affairs at ITS, who previously served as the Secretary-General of the Ministry of Marine Affairs and Fisheries during President Joko Widodo's first term, Kemenkomarves, KKP, academician etc.) (**Data Collection-Desk Research-In-depth Interview step**). Number of library research have been conducted to gather secondary data relevant to the topic of this chapter, and to determine the topic, the problems, hypothesis (**Choosing a Topic- Identifying the problems-Data Collection step**). (see Scheme 19.1).

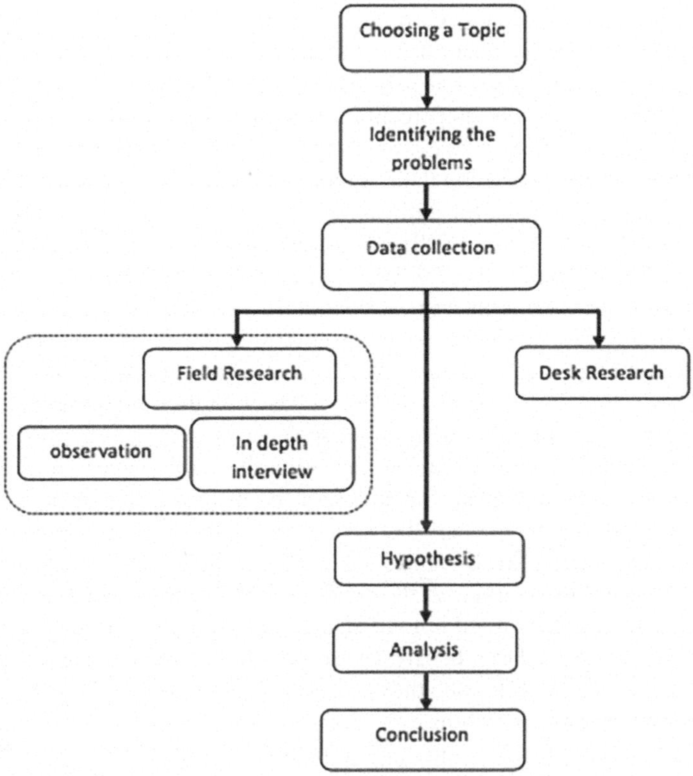

Scheme 19.1 Methodology flowchart of the research process. *Source* The scheme was constructed by the Authors, 2024

The Importance of Coastal Communities, the Impact of Changing Weather, and the Need for Effective Adaptation

Defining and Understanding Coastal Community

Satria (2015) defines coastal communities as people who live in coastal areas with unique cultures and a strong dependence on coastal resources. Coastal areas include land affected by tidal wave, sea winds, and saltwater, as well as sea regions influenced by activities on land, like sedimentation and freshwater runoff (Bengen, 2001). In Indonesia, another definition of coastal communities comes from Law No. 1 of 2014. It identifies three types of communities in coastal and small island areas: customary law communities, local communities, and traditional communities. However, only the traditional community category is associated with fishermen, particularly those who practice traditional fishing with traditional fishing rights in island waters.

In this context, we're focusing on coastal communities, specifically fishermen. According to Bailey (1988), fishermen in Indonesia are mostly small-scale producers who often struggle with poverty due to overfishing in the waters of Java, Sumatra, and Sulawesi. In Indonesia, about 90 percent of those involved in fishing are considered small-scale producers (Loneragan et al. 2018). Kusnadi (2002) divides Indonesian fishers into three groups based on their equipment ownership, investment scale, and technology use. Additionally, the authors refer to two laws for the definition of fishermen. The first, Law Number 7 of 2016 concerning the Protection and Empowerment of Fishers, Fish Cultivators, and Salt Farmers, classifies fishermen into small-scale, traditional, labor, and owner fishermen, distinguishing them from fish cultivators.

We use "small-scale producer" to refer to traditional fishermen, small-scale fishermen, artisanal fishermen, and labor fishermen. Among these groups, except for labor fishermen, each has a distinct definition based on their unique characteristics. In this chapter, when we talk about small-scale producers, it refers to their difficulty accessing distant fishing grounds. This difficulty is often due to limitations like insufficient capital, small boat size, engines, or the fishing gear they use. Generally, there are two types of relationships between owner fishermen and labor fishermen: exploitative and partnership. In exploitative relationships, labor fishermen usually can't directly access the market, so they earn very little. In partnerships, labor fishermen can sell their catch directly in the market and share the earnings as agreed. In some cases, in Indonesia, labor fishermen are essentially workers who provide labor without needing specific skills, and they receive the lowest wages, both in large-scale fishing units and traditional fishing.

Owner fishermen typically come in two forms. First, there are owner fishermen who actively fish at sea. These individuals, known as "skippers" or called various local names like "punggawa," "juragan," or "nahkoda" in different Indonesian regions, often own their production tools or boats and receive higher wages due to greater responsibilities. Secondly, there are owner fishermen who aren't directly involved in fishing. They're often known as land bosses or, in many Sumatran regions, as "tauke." These individuals typically own capital and manage multiple fishing fleets through trusted captains ("nahkoda"). The relationships between land bosses or "tauke" and captains and labor fishermen are based on a patron-client system. This also holds true in Natuna Regency, where a cooperation system based on agreements and trust is shaped by the patron-client relationship.

Changing Weather: The Impact of Seasonal Winds

Indonesia has two main monsoons that influence its seasons. The Asian Summer Monsoon brings the dry season ('musim timur') when winds blow from Australia to Asia. The Australian Winter Monsoon leads to the rainy season ('musim barat') as winds blow from Asia to Australia (Yulihastin 2010). During 'musim timur,' the southern part of Indonesia experiences southeast monsoon winds, while the north encounters southwest monsoon winds. Conversely, during the 'musim barat,' the northern regions have northeast monsoon winds, and the south experiences northwest monsoon winds (Najid et al. 2012). In equatorial regions, during the northwest monsoon, winds shift northward (in Natuna called 'musim utara'), and during the northeast monsoon, they shift southward (called 'musim selatan) (Najid et al. 2012). These seasonal wind patterns have a significant historical impact on maritime and fishing activities in Indonesia. Traditionally, Indonesian communities have harnessed these winds for inter-island voyages and trade. Recent research delves into how variations in monsoon winds affect fishing practices, determining when and where fish are caught and the quantities caught (Suhery et al., 2023; Gustika et al. 2023; Azizi et al. 2017; Kunarso et al. 2016; Ali et al. 2004). This is the seasonal challenge faced by the coastal community in Natuna.

In most of Indonesia's waters, the west monsoon is the most challenging time for fishing, especially for small-scale fishermen. Strong currents, high waves, and powerful winds make it tough for small boats to operate. Additionally, this season affects the movement and habits of certain fish species, leading to changes in where, when, and how fish are caught, which can raise production costs. This difficult period during the west monsoon is often called "musim paceklik" or a season of scarcity. It makes small-scale fishermen more vulnerable and can impact their well-being (Gustika et al. 2023; Sari and Wiyono 2021; Azizi et al. 2017; Kunarso et al. 2016; Fitriani et al. 2023). During uncertain weather, small-scale producers use different strategies. They either diversify their income sources by taking up other jobs, like labor or farming, or they adapt their fishing tools and techniques based on the season and the type of fish they target. Access to various tools isn't limited to large-scale fishermen; even small-scale ones can access them through social connections, like patronage relationships with investors called "tauke", who provide not only capital for fishing but also loans to support them during lean times (Gustika et al. 2023; Yamazaki et al. 2018; Fitriani et al. 2023).

Resilience and Preparedness of Fishermen/Coastal Community

Climate change is affecting coastal ecosystems, and this is especially challenging for coastal communities, including those on small islands (Klein and Nicholls 1999). It's leading to lower fish catches, harming social infrastructure and people's skills, and causing problems with fishing facilities. These issues are making these communities more vulnerable to ecosystem damage, storms, floods, and rising sea temperatures and levels (Ferrol-Schulte et al., 2015; Cazenave and Cozannet 2014; Badjeck et al. 2010).

Certain researchers have pointed out that climate change has consequences in two key areas: the environment and society (Eurich et al. 2023; Mason et al. 2022; Pinsky et al. 2021). They've noted that climate change has brought about physical transformations like coastal erosion, increasing sea temperatures and levels, extreme weather, and changes in the migration behavior and production systems of specific animal species. The 2021 Global Program on Risk Assessment and Management for Adaptation to Climate Change Report mentions that climate change and extreme weather are causing more significant problems for small-scale fisheries in the Caribbean, South Pacific, and West Africa. This is worsened by overfishing, environmental damage, and pollution. However, the effects of climate change, like extreme weather and rising sea levels, differ from one place to another and depend on factors like natural conditions, how local communities respond, and the policies implemented by countries (Cazenave and Cozannet, 2014; Dolan and Walker 2006).

In Indonesia, the impact of climate change on fishing and coastal communities is evident. This is reflected in the declining fish stocks, which pose a threat to the crucial fishing sector that contributes approximately 2.6% to the country's GDP and employs 7 million people (Kaczan et al. 2023). Recognizing the urgency of the situation, the Indonesian government has prioritized the "marine and fisheries sector adaptation" project in various coastal areas (Bappenas 2021). However, government initiatives encounter challenges, including inconsistent implementation, insufficient funding, and weak coordination between agencies (Ferrol-Schulte et al. 2015). How should coastal communities be perceived in terms of their ability to adapt and show resilience in response to climate change? Considering the diverse impacts of climate change across different regions, some experts suggest that successful adaptation begins with a thorough comprehension of local circumstances. This entails valuing and integrating local knowledge and traditions into the decision-making, planning, and management processes, particularly in remote areas highly vulnerable to climate change risks (Kaczan et al. 2023; Dolan and Walker 2006).

In a local context, a study in the Kei Islands, Maluku, found that the success of fishermen largely depends on their social connections. This includes how they interact with others within and outside their community, trust-building, and cooperation. These connections extend to relationships with traditional leaders and government officials. They can also help establish ties with groups beyond fishing, providing opportunities for information and diversifying livelihoods. These networks are crucial for accessing assistance and protection from the government (Kaczan et al. 2023). Natuna fishermen, particularly those in the Lampa Strait and Pulau Tiga, employ a similar approach. They create social connections with both local fishermen and those outside their community to access larger markets.

Results: Empowering for Coastal Communities, Policies and Strategies

Governments take several measures and initiatives to support coastal communities, particularly those engaged in fishing. The enhancement of community-driven economic development in fisheries is delineated in five policies introduced by the Ministry of Marine Affairs and Fisheries, which are: (1) The expansion of marine conservation zones; (2)The implementation of controlled and quota-based fishing practices; (3) The promotion of sustainable aquaculture encompassing marine, coastal, and land-based activities; (4)The improvement of monitoring and governance over coastal regions and small islands; (5) The engagement of fishermen in clearing plastic waste from the sea during the "Ocean Love Month". All these endeavors share a common objective, which is to attain sovereignty, sustainability, and prosperity (retrieved from https://kkp.go.id/AgendaPrioritasKKPdengan5Kebijakan).

One of these policies is executed through the establishment of Center for Integrated Marine and Fisheries (known as Sentra Kelautan dan Perikanan Terpadu or SKPT) situated in Small Islands and Border Areas. This initiative, prioritized by the Ministry of Marine Affairs and Fisheries, seeks to integrate community-based marine and fisheries business operations while optimizing the utilization of marine and fisheries resources in small islands and border regions in an environmentally sustainable manner. One of the crucial aspects in deciding the location of the SKPT facility in Selat Lampa is its strategic position in terms of protection against sea waves and its accessibility from all areas.

The Location of the SKPT: Determinant Aspects

The decision to establish the SKPT development in Natuna Regency is informed by a range of factors, encompassing both technical and non-technical considerations, which include environmental and socio-economic factors. Among these considerations, technical aspects hold significant importance and directly influence the sustainability of the designated area. The technical aspects include factors such as the suitability of the land area for development, the accessibility of the location to the economic center, the presence of essential utilities like water and electricity, as well as transportation availability. Land suitability, in particular, is a critical parameter as it directly relates to how the area earmarked for development is allocated.

In the context of developing the Natuna SKPT area, ensuring a reliable water source is vital, especially for industrial purposes. Access to electricity is equally crucial for sustaining operations. Furthermore, the area's proximity to the economic center is significant. Good road infrastructure within the area enhances mobility, and the availability of transportation facilities further improves accessibility. Lastly, legal considerations and community input play a pivotal role. Evaluating the development in line with existing policies is essential to avoid conflicts and ensure harmonious coexistence with other regions. In the economic evaluation, we consider parameters like production and the availability of market inputs. Input availability is a crucial factor as it directly impacts the Natuna SKPT area's activities. Moreover, the market dimension is essential in location selection because it determines the ease of selling processed products. After analyzing various parameters, scores are assigned, and the total score is calculated for each location. The location with the highest total score is considered the most suitable for developing the SKPT Natuna area. In Natuna Regency, multiple locations, including Lampa Strait, Depeh Bay, Serantas, and Cemaga, were identified as potential SKPT sites. The assessment of these four prospective locations is based on three key aspects: technical, legal, and economic. In the overall assessment, Lampa Strait outperforms the other locations, with a total score of 4.45, while Depeh Bay comes in second with a score of 2.88.

Based on Lundgret's research in Pranoto (1999) and the findings by Nomura and Yamazaki (1997), 11 key characteristics have been identified as criteria for an ideal fishing port location. These characteristics include proximity to the rich fishing grounds of the South China Sea, which boasts the largest potential for marine fisheries in Indonesia (estimated at 1,057,050 tons/year). This estimation does not encompass the potential within the Exclusive Economic Zone (EEZ), which remains underutilized. Presently, the South China Sea yields only about 379,900 tons of fish resources per year. Therefore, Natuna Island holds significant untapped potential for capture fisheries. Additionally, the Lampa Strait's location as a landing port is characterized by safety from various oceanographic influences, including waves, currents, and tides, with ample depth for vessel navigation.

Data from the Ranai Meteorological Station regarding Natuna waters shows that wave heights range from 0.25 to 3.0 m, mainly coming from the Southwest and South. These waves occur in open waters, while the planned port site is sheltered in the Lampa Strait among islands. Regarding sea currents, in 2013, Natuna had currents ranging from 5 to 65 cm/s. However, in the Lampa Strait, the currents predominantly flow from west to east due to its geographical layout. In terms of wind, wind speeds in Natuna sea waters in 2013 ranged from 5 to 15 knots, with various wind directions. Wind speed decreases upon entering the Lampa Strait from the Natuna Sea. As for water depth, it increases as you move from the coast to the sea, ranging from − 0.7 to − 39.8 m. The required -2-meter depth for type C and D fishing ports is found 20–70 m from the coast. The port pool's water flow maintains a relatively significant depth and isn't affected by silt due to the absence of river inflow into the strait.

In terms of land transportation infrastructure, the Lampa Strait offers extensive road access, connecting the port to the fish marketing area. This road covers a distance of 65 km and allows for a travel time of 1.5 h to reach the city of Ranai and Natuna Airport. The authors proved this by having a trip from Ranai City to SKPT in Lampa Strait (Selat Lampa) in June 2023. According to the person in charge of the Management of the Selat Lampa Fishing Port, SKPT Natuna this location is not only convenient for fishermen and those involved in fisheries but also ideal for activities like landing, handling, storage, and marketing (in-depth interview, SKPT Natuna, June 2023). Moreover, it's well-suited for establishing factory facilities and providing repair services for fishing vessels and tools. It offers a comfortable setting for the supply and delivery of essential goods required for fishing operations, including items like oil, ice, bait, food, water, and more.

Challenges for the SKPT and the Government Strategies

Besides the advantages offered by the Lampa Strait, there are several challenges to overcome. These include limited availability of skilled labor in the fishing and fish processing fields, a fishing community culture that is not yet accustomed to meeting the quality standards required by customers or engaging in international trade, and insufficient access to banking services. To address these challenges, the central government has developed strategies to engage various stakeholders at different stages of the development process. For example, fishermen, as primary actors, are encouraged to prepare high-quality raw materials or products. Fisheries cooperatives, numbering 10 in Natuna Regency, receive assistance for fishing facilities and infrastructure. State-Owned Enterprises (SOEs) serve as marketing off-takers, partnering with cooperatives to purchase fish from fishermen. Private business actors, in addition to buying products, also provide training and skill improvement to enhance production quality and increase income for fishermen. These private business actors are further supported by investors who purchase products and contribute to building fisheries infrastructure in SKPT locations. Banking institutions play a vital role by providing convenient access to business financing for fishermen and other players in the fisheries sector, thereby increasing business volume and improving production quality. The acknowledgment of local fishermen from Pulau Tiga regarding the convenience of selling fish around SKPT is evident from a prominent figure's statement (in-depth interview, June 2023). Several fishermen, who previously sold their catch outside Natuna, specifically to Kijang on Bintan Island, now find it more convenient due to the existence of SKPT in Selat Lampa. It translates to increased ease in terms of time, energy, and expenses when utilizing SKPT, including the simplified process of acquiring diesel fuel for their boats.

Discussion: Fishermen's Resilience and Preparedness, and Government Efforts

The Natuna region signifies a vital aspect of our nation's unwavering commitment to national jurisdiction. Its purpose is to guarantee a serene atmosphere in the Natuna Sea, safeguarding the plentiful marine resources while ensuring that they are effectively managed and controlled by the government. The key factor lies in the involvement of the people, many of whom are fishermen.

The Number of Fishermen

Statistical data (of BPS) reveals that a significant portion of the Natuna Regency population is engaged in fishing, and this is represented by number of capture fisheries household (Table 19.2).

The total capture fishery households presented in Table 19.1 demonstrate growth in the number of families involved in capture fisheries in Natuna Regency starting in 2018 until 2019 (5463 and 5662 respectively). However, there was a noticeable decrease in 2020 (5590) and slightly increase in 2021 (5658), likely attributed to the influence of the COVID-19 pandemic on the fishing endeavors of the local fishermen. Yet, if we delve into the data for individual sub-districts, it becomes evident that there are fluctuations in the annual trends, including both rises and falls. To illustrate, take Pulau Tiga as an example, where the figures remained constant from 2018 to 2019 but saw a decline in 2020 (333) before experiencing a modest raise in 2021. Compared to other sub-districts such as Bunguran Barat, Bunguran Timur, Bunguran Timur Laut, and Bunguran Selatan, there are fewer households engaged in capture fisheries on Pulau Tiga and Pulau Tiga Barat. However, fishers living on these islands have a significant opportunity, as they are closer to Lampa Strait (Selat Lampa), where the SKPT is being developed.

Impact of Changing Weather

Fishermen in the Lampa Strait and its nearby areas (include Pulau Tiga) have adjusted to the unpredictable weather patterns over time. However, recent years have seen an increase in erratic weather due to the gradual onset of climate change. Fishers must employ creative strategies to navigate these conditions and maintain their livelihoods. When they are unable to catch fish, some diversify into fields such as construction workers, farming, fish cultivators or various odd jobs. Nonetheless, there are still fishermen who persist in their fishing activities, though they now primarily operate within coastal regions only (Source: Republica.co.id, 30/1/2022). As an example, instead of selling fish based on weight, fishermen choose to sell them individually at higher prices, with rates ranging from 30,000 per fish to 70,000–80,000 per fish (Republica.co.id, 30/1/2022). Furthermore, those who fish near the coast tend to catch specific fish types in limited quantities, such as tuna and bonito. In the lean season, small-scale fishers go on shorter fishing trips closer to shore due to limited fish availability, typically lasting half a day. Upon returning home, they try to catch fish around their residence, even though the catch may not be sufficient. This illustrates their strategic approach to coping with economic challenges, as observed by the author during a visit to Pulau Tiga in June 2023.

Figure 19.2 visually represents the geographic positioning of the Lampa Strait and the nearby islands located to the south of Natuna. These islands (like Pulau Tiga, and Pulau Tiga Barat) are notably distant from the north of Natuna part. In terms of geography, during the period of the northern monsoon season (November–March or "angin utara"), this particular area experiences relatively minor impacts. It seems to enjoy some protection from the ocean waves during this period. Conversely, the eastern monsoon season (April-June), often referred to as the calm season, witnesses a significant number of fishermen venturing out into the sea, resulting in an abundance of fish. Unfortunately, during this season of plenty, fish prices tend to decrease.

According to a report by Yogi Eka Sahputra (Mongabay, August 9, 2023), the price of one kilogram of white skipjack tuna, which was previously 25,000 Indonesian Rupiahs, has now dropped to 14,000 Indonesian Rupiahs.

Subsequently, in the southern wind season (July–September or "musim selatan"), marked by strong winds, it becomes challenging to catch fish, leading to increased fish prices and limited fish stocks. Entering the southwest monsoon season, which occurs in October, arrives abruptly and is feared by fishermen. Presently, due to unpredictable weather changes, the southwest monsoon season can arrive unexpectedly, even while fishermen are out at sea. According to one fisherman's testimony, the southwest monsoon season can strike at any moment. Climate change has rendered this west wind season highly unpredictable, unlike a few years ago (Mongabay, August 9, 2023). Local fishers are facing a challenge with the decreasing availability of fish stocks. This has led them to compete with fishing ships from other parts outside Natuna, as reported by the Fisheries and Marine Office in Ranai, Natuna, in June 2023. In response, local fishers are now searching for fishing grounds in border areas such as Sarawak, where there is still a plentiful supply of fish. This action could be considered a way for local fishers to resist or protest against the problem of illegal fishing by foreign boats in the Natuna Sea, prompting them to explore the waters of neighboring countries.

During January, February, and March, the Lampa Strait, Pulau Tiga, and nearby areas experience minimal impacts, possibly due to some protection against ocean waves. During the fishing season, Pulau Tiga had 18 households engaged in seafish aquaculture, while Pulau Tiga Barat had 16 participating households. In 2022, they respectively produced 265.12 tons and 204.21 tons of seafood. In these conditions, small-scale fishers in Pulau Tiga were able to withstand both the lean season and the season of abundant fish. During the calm season from April to June, many fishermen head out to sea, leading to an increase in fish abundance and a decrease in fish prices. And during the season from July to September fishermen can explore income sources during lean fishing. The arrival of the southwest monsoon season can occur unexpectedly, catching fishermen by surprise while they are at sea. Fishermen must stay vigilant when fishing during this period. While some fishermen choose to refrain from going to sea during these months, others venture out with adequate preparedness to navigate extreme weather conditions. This season typically occurs in October. During November and December, the Lampa Strait, Pulau Tiga, and the surrounding areas experience minimal impacts, likely benefiting from some protection against ocean waves.

Resilience and Preparedness: The Fishermen's Efforts

The preparedness and resilience of the fishermen can be evaluated by considering the following alternatives:

Diversification of Income

Fishermen can explore alternative income sources during lean fishing seasons. In terms of diversifying their sources of income, fishermen have two alternative options available to them. Firstly, they can opt to borrow money from their employer to meet their personal needs. Secondly, some fishermen continue to catch fish, regardless of the fish type, solely for their family's consumption. The latter group typically engages in fishing for just one day, as opposed to several days like traditional deep-sea fishing. However, during the regular fishing periods (lasting around 10 days), they manage to catch an average of 0.5 tons of fish using their 3GT boat capacity, as reported by a fisherman on Pulau Tiga (Table 19.3).

Furthermore, housewives contribute to their household finances by crafting fish crackers known as "amplang," using Tenggiri fish, Banana chip, squid crackers. They market these homemade crackers in Ranai city and its vicinity. As conveyed to the authors by the seller, their primary customers are often government offices in the Ranai municipality (the Author bought some of fish crackers from the seller on *Selat Lampa Port*, July, 2023). In Natuna, various non-traditional production ventures complement the conventional fishing industry. These include activities like Aquaculture of Sea Fish, Aquaculture of Brackishwater Fish, Cultivation of Seaweed, and Rearing of Freshwater Fish. Data sourced from Natuna Regency in Figures illustrates that, on Pulau Tiga, 18 households were involved in sea fish aquaculture, whereas Pulau Tiga Barat had 16 participating households. In 2022, they respectively produced 265.12 tons and 204.21 tons of seafood (Natuna Regency in Figures 2023).

In a broader context, across the entirety of Natuna Regency's fishing sector, the production quantities were 3641.29 tons in 2020, 3554.82 tons in 2021, and 3485.49 tons in 2022. The production trend exhibited a decrease, primarily attributed to the impacts of the COVID-19 pandemic. In the case of Pulau Tiga and Pulau Tiga Barat, these statistics underscore the fishing communities' endeavors to explore alternative business avenues, aimed at supplementing their incomes and providing financial security for their livelihoods. It can help fishermen recover from losses due to extreme weather events or climate-related disruptions.

Environmental Conservation Efforts

To offer a more complete dataset on the number of fishermen, we have opted to provide information about boat ownership and their types. Three motorboat categories are employed, namely non-power boats, inboard motor boats, and outboard motor boats. Table 19.4 reveals that the utilization of non-power boats (traditional) is relatively substantial (1186), even though the highest preference is for inboard motor boats (3069). This suggests that a significant portion of Natuna's fishermen continues to rely on traditional boats, which are inherently eco-friendly. Nevertheless, this trend can also be influenced by the choice of capture equipment utilized.

Their participation is essential for the preservation of the marine ecosystem. According to the 2023 statistics from the BPS of Natuna Regency, the data reveals that

Table 19.3 A calender of fishing cultivation in Pulau Tiga, Southern Natuna

Yearly seasons	January	February	March	April	May	June	July	August	September	October	November	December
The Northern monsoon season	XXXXXX	XXXXXX	XXXXXX								XXXXXXXX	XXXXXXXX
	The Lampa Strait, Pulau Tiga, and the nearby areas experience relatively minimal effects, seemingly benefiting from a degree of protection against ocean waves during this time During the fishing season, Pulau Tiga had 18 households involved in sea fish aquaculture, while Pulau Tiga Barat had 16 participating households. In 2022, they produced 265.12 tons and 204.21 tons of seafood, respectively. Under these circumstances, the small-scale fishers of Pulau Tiga were able to endure both the lean season and the season of abundant fish										The Lampa Strait, Pulau Tiga, and the surrounding areas face relatively minor impacts during this period, possibly due to some protection from ocean waves	
The Eastern monsoon season				XXXXX	XXXXX	XXXXX						

(continued)

Table 19.3 (continued)

Yearly seasons	January	February	March	April	May	June	July	August	September	October	November	December
				The calm season, witnesses a significant number of fishermen venturing out into the sea, resulting in an abundance of fish Fish price tend to decrease								
The Southern wind season							XXXXX	XXXXXX	XXXXX			
							Adverse wind conditions pose a challenge to fishing, resulting in elevated fish prices and restricted fish availability Some fishers rely on support from collectors. Others find out the side jobs					

(continued)

Table 19.3 (continued)

Yearly seasons	January	February	March	April	May	June	July	August	September	October	November	December
The Southwest monsoon season										XXXXXXX		
										The onset of the southwest monsoon season can happen unexpectedly, catching fishermen off guard while they are at sea Fishermen need to remain cautious when fishing during this period. While some fishermen opt to avoid going out to sea in these months, others venture into the waters with sufficient preparedness to tackle extreme weather conditions		

Source Constructed by the Authors, 2024

Table 19.4 The Number of fishing boats by sub-district and type of boat in Natuna Regency for the year 2022

Sub-district	Non-power boat	Inboard motorboat	Outboard motorboat
Midai	90	92	7
Suak Midai	91	36	6
Bunguran Barat	87	266	57
Bunguran Utara	4	172	37
Pulau Laut	238	145	26
Pulau Tiga	66	278	20
Bunguran Batubi	28	75	17
Pulau Tiga Barat	14	113	17
Bunguran Timur	36	591	44
Bunguran Timur Laut	139	288	21
Bunguran Tengah	–	–	–
Bunguran Selatan	110	369	16
Serasan	223	3069	39
Subi	3	292	12
SerasanTimur	57	155	19
Total Natuna	1186	3069	338

Source Kabupaten Natuna Dalam Angka, Kabupaten Natuna, BPS 2023

most Natuna fishermen employ various fishing gear, including fish traps, seines, longlines, purse seines, and beach nets. Moreover, in the regions where fishing communities live, including Selat Lampa Harbor and Pulau Tiga, as well as the nearby areas, the sea conditions are notably pristine, displaying a vibrant deep blue color that reaches the seabed. Residents still have the opportunity to fish directly in front of their seaside homes.

Empowering for Coastal Communities: The Government Efforts

As mandated by Law No. 7 of 2016, which addresses the Protection and Empowerment of Fishermen, Fish Farmers, and Salt Farmers, the government is tasked with the responsibility of taking measures to enhance the welfare of fishermen. Governments can take several measures and initiatives to support coastal communities, particularly those engaged in fishing, to enhance their resilience and well-being. The government's initiatives in this regard are evident through the following actions.

Infrastructure and Access Improvement

Infrastructure development, specifically the building of roads in the fishing village of Pulau Tiga, is reasonably satisfactory. The financing for constructing concrete roads is sourced from the Village Fund Allocation and Village Fund, which began between 2016 and 2018. These concrete roads are intended to improve the ease of movement for the local residents (Fig. 19.3).

The development of a clean water reservoir, referred to as an 'embung,' in the Sabang Mawang District in Pulau Tiga has advanced to a stage where it can store a specific volume of water. Nevertheless, the infrastructure for distributing this water to households through a pipeline network has not been put in place as of now (Tribun-Batam.id, July 10, 2021). Furthermore, in response to the need for clean water, the District Government has erected a water reservoir at a cost of 1.8 billion Rupiah, enabling the distribution of water to the households of the residents (retrieved from https://hariankepri.com, July 2021). There is optimism that the implementation of a clean water supply system for the residents will enable the fishermen to meet their necessities for making ice blocks, for cleaning fish and so fort.

Another noteworthy point is the Natuna Regency builds construction of a road connecting fish market in Ranai to both Lampa Port and the SKPT in Lampa Strait. This coastal road, which traverses Teluk Depeh (Depeh Bay), renowned for its picturesque beach, was reportedly built by excavating through the hills to enhance accessibility people going from and to both Lampa Port and SKPT (refer to Fig. 19.4). The Central Government, under the Ministry of Public Works, built this road with the intention of utilizing the beach's natural beauty to support and promote for tourism development in the future.

Fig. 19.3 The Photos of Construction Progress Concrete Roads in Pulau Tiga. *Source* The photos above are the property of the PRP-IPSH, BRIN Border Team, June 2023. *Notes* This image showcases one of roads on Pulau Tiga that has been built using Village Allocation Funds and Village Funds. The central photo features a plaque that transparently reveals the allocated funds for the road construction

Fig. 19.4 The road from Ranai to Lampa Port and SKPT through Teluk Depeh. *Source* The following image shows the condition of the road along Teluk Depeh Beach, connecting from Kota Ranai to Selat Lampa Harbor and SKPT. The photo was taken in July 2023

Development of Center for Integrated Marine and Fisheries (SKPT) in Lampa Strait

Planning Stages

Planning the Integrated Marine and Fisheries Center Area (SKPT) begins by figuring out how much fish local fishermen will land at SKPT Natuna each year. This target production guides decisions like how many fishing vessels are needed, the size of the dock and fish landing area, and the capacity of cold storage (set at 200 tons). Other factors like fuel, water, electricity, and gas supply also depend on the number and size of fishing vessels and their logistics requirements per trip. These calculations form the basis for the master plan of SKPT Natuna.

The core idea of the SKPT Natuna master plan is to provide an overview of the broader SKPT development, including a map of the development plans. This high-level concept considers the potential and utilization of fish resources, aiming to boost marine and fishery product production, raise incomes for local communities, especially fishermen and fish processors, and establish integrated marine and fisheries economic zones as regional economic growth centers. The sub-centers play a supporting role in building the Natuna SKPT center. Given Natuna Regency's surplus of fishery products, it's vital to connect with other regions. Direct exports may face challenges in terms of economies of scale, so promoting large-scale, two-way trade relationships with neighboring areas, like Hong Kong, is key for efficient import–export activities. For local fishermen in places like Pulau Tiga, Pulau Tiga Barat, and the surrounding islands, SKPT greatly facilitates their activities both before and after fishing trips.

The Lampa Strait has a very ideal position to build a fishing port area, the coastal area is a bay area protected from waves from the South China Sea with a Block Plan design as in Figs. 19.5 and 19.6, as follows:

Fig. 19.5 Lampa Strait—before construction

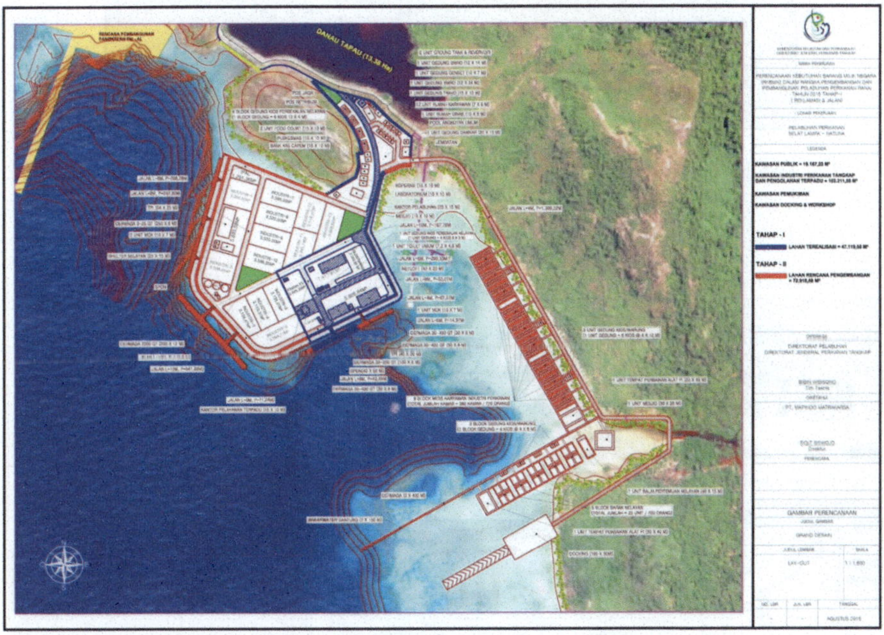

Fig. 19.6 Selat Lampa—Block Plan. *Source* Widjaja (2022, pp. 27–28)

Construction Stages

The development of the Lampa Strait Fishing Port commenced in 2015 and included land reclamation over a 3-hectare area (indicated in blue), the construction of a trestle and a pier, as depicted in Fig. 19.7.

During 2016, significant infrastructure components were established, including the SKPT administrative offices, fish marketing sites, net assembly wards, logistics kiosks, gen-set houses, SWRO installations, fuel filling stations, water towers, and more, as illustrated in Fig. 19.8.

The construction work was finalized in 2017, involving the physical construction of the 200-ton integrated cold storage, causeway to the pier, roads, trestles, area roads, a dock for 5–20 GT boats, employee mess, integrated cold storage yard, fish landing

Fig. 19.7 The development of the SKPT Natuna—Selat Lampa in 2015. *Source* Widjaja (2022, p. 32)

Fig. 19.8 Construction of SKPT Natuna—Lampa Strait in 2016. *Source* Widjaja (2022, pp. 33–34)

Fig. 19.9 ICS construction and Ship Assistance Year 2017. *Source* Widjaja (2022, p. 35)

site yard, a mosque, and various other support facilities. Additionally, assistance was provided to acquire 100 fishing boats for Natuna fishermen, as indicated in Fig. 19.9.

In addition to the main facilities as mentioned above, supporting infrastructure is also prepared, as described in Table 19.5.

Since the official commencement of SKPT Natuna's operations in 2019, the local fishing community in Natuna has adapted to conducting their fishing operations with larger vessels. This transformation has been facilitated by the readily available fuel and fresh water resources, allowing them to engage in fish trading with improved quality and expanded market reach. The government's ongoing efforts to improve and finalize the development of facilities through the SKPT reflect their dedication to enhancing the well-being of both large and small-scale fishermen, particularly those in the vicinity of the Lampa Strait. These initiatives aim to strengthen the preparedness and resilience of fishermen during challenging periods each year.

Table 19.5 Supporting infrastructure

Types	Descriptions	Locations
Electricity	• Construction of Power Plants to support electricity needs in the PP Lampa Strait area	PP Selat Lampa
Clean Water	• Clean water treatment plant by increasing water capacity with BWRO Lampa Strait	PP Selat Lampa
Transportation	• Construction of Road from Port to Ranai City	PP Selat Lampa
	• Create a public transport base/ terminal	PP Selat Lampa
IPAL	• Building WWTP in PP Natuna with a capacity of 10 tons in the short and medium term (1–5 years), and 10 tons per day in the long term (more than 10 years) after PP Natuna developed into the Natuna Ocean Fishing Port	PP Selat Lampa
Telecommunication	• Telecommunication network needs in Bunguran Island and **Pulau Tiga**	Bunguran Island and **Pulau Tiga**
Airport	• The need for access to Natuna Airport as an international airport	East Bunguran District

Source Widjaja (2022, p. 36)

Conclusions

The coastal community in Natuna Regency faces the challenge of adapting to seasonal weather patterns every year, which has become a regular part of their lives. Particularly, small-scale fishermen in areas like Pulau Tiga, Selat Lampa, and the surrounding lead unpredictable lives.

The findings of the research indicate that the coastal community in Natuna Regency has awareness of the significant impact of the seasonal weather pattern which influenced to the livelihood compared with other factors such as fish price, handling and transportation mode, logistic and financial support. The fishermen knew also that within the off season, getting an alternative work is very necessary, such as motorcycle taxi drivers or construction workers, to make a daily income. In the southern Natuna region, larger-scale fishermen with crews of 10–20 people may provide support to help smaller fishermen through the tough times.

Following the awareness of impact of the seasonal weather pattern, the fishermen themselves make efforts to cope with, that include: (1) safeguarding their local environment to prevent pollution of marine life and resources, often using traditional fishing methods, such as non-power boat, various fishing gear, including fish traps, seines, longlines, purse seines, and beach nets. They can catch fish near their homes, where the seawater remains clean and clear; (2) fishermen's wives take advantage of market opportunities by selling products like fish crackers and banana chips, squid crackers in town. This aligns with experts' advice to integrate local knowledge and traditions into managing small businesses, enabling them to connect with a broader social network; (3) larger-scale fishermen provide 'assistance' or loans to laborers during the off-season, with the understanding that these loans will be repaid with the earnings from future catches. These practices are long-standing traditions, even though they may not always have positive implications for the fishermen.

The second finding of the research reveals that the primary focus of the central fisheries office and local government has been on developing a robust infrastructure for the fisheries' industry, rather than direct involvement in areas such as fish trading, pricing policy, enforcement of regulations for traders and fishers, and the socio-economic aspects of affecting the fishers. The government is making efforts in several ways, such as: enhancing and upgrading the main roads, especially with paved roads, on the islands to facilitate the movement of fishers; there has been progress in constructing a clean water reservoir known as an 'embung' in the Sabang Mawang District on Pulau Tiga. However, the infrastructure for distributing this water to households through a pipeline network is not yet in place. This infrastructure is crucial to prevent residents from having to fetch water manually. In response to the need for clean water, the District Government has invested 1.8 billion Rupiah to build a water reservoir, which will enable water distribution to residents' households, and; the establishment of the Center for Integrated Marine and Fishery (SKPT) is a key element in supporting fishermen's activities and needs.

The Center for Integrated Marine and Fishery SKPT has been operational since 2019, its development is ongoing. Key considerations regarding SKPT include:

a. SKPT Natuna is part of the government's strategy to boost the growth of a competitive fisheries industry that can compete in international markets, leveraging Indonesia's abundant fish stocks in the outer islands.
b. The choice of the Lampa Strait in Natuna, with all its pros and cons, results from a thorough study that considers geographical, economic, and social factors to maximize community benefits.
c. SKPT Natuna's development also involves various stakeholders in fishing communities, employing social psychology approaches to engage different fishing community groups, trade businesses, consumer communities, and financial institutions at regional and international levels.

SKPT Natuna not only serves the fishers living near the Lampa Strait but also benefits all fishers in Natuna Regency and beyond. The goal is for both the fishers and the seaside communities near SKPT to experience greater advantages compared to before the SKPT was established. The ongoing process of improving SKPT facilities will lead to a better quality of life for the fishers. Therefore, the resilience and preparedness of the coastal communities and fishers in Natuna, especially those on the southern islands, rely on government support and involvement. **They likewise take their initiatives, driven by a strong sense of local community awareness.** In addition, they engage in activities driven by a strong sense of awareness and involvement in their local community, such as protecting the marine ecosystem in their vicinity and using fishing equipment that is sustainable.

Acknowledgements On this occasion, we would like to express our gratitude to the Border Research Center (BRIN) Political Research Team, especially to the 2023 SKPT research team in Natuna, namely Hayati Nufus, Anta Maulana Nasution, and Sandy Nur Ikfal Raharjo, for their collaborative work. We also appreciate the support from the Sumitomo Foundation for the 2023 SKPT research. Our thanks go to the Dean of the Faculty of Marine Affairs and Fisheries at ITS (Institut Teknologi Sepuluh Nopember) for sharing valuable insights about SKPT Natuna, as well as various relevant government ministries in Jakarta and Natuna. We are also thankful to the fishermen we had the opportunity to meet in person and those we interviewed through online channels.

References

Andries M, Oktavera L (eds) (2023) Kabupaten Natuna Dalam Angka (Natuna Regency in Figure) 2023, Kabupaten Natuna, BPS Natuna

Azizi A, Putri EIK, Fahrudin A (2017) Analisis Faktor-Faktor Yang Mempengaruhi Perubahan Pendapatan Nelayan Akibat Variabilitas Iklim. Jurnal Sosial Ekonomi Kelautan Dan Perikanan 12(2):225–233

Ali SA, Nessa MN, Djawad MI, Omar SBA (2004) Musim dan kelimpahan ikan terbang (Exocoetidae) di Sekitar kabupaten Takalar (Laut Flores) Sulawesi Selatan. Torani 14(3):165–172

Ansaar (2019) Sistem Pengetahuan Pelayaran dan Penangkapan Ikan Pada Masyarakat Nelayan di Kelurahan Rangas, Kabupaten Majene (Shipping and Fish Catching Knowledge System of the Fishermen Community in the Village of Rangas, Majene Regency), Balai Pelestarian Nilai Budaya Sulawesi Selatan, WALASUJI, vol 10, No. 2, Desember, pp 139–154

Badjeck MC, Allison EH, Halls AS, Dulvy NK (2010) Impacts of climate variability and change on fishery-based livelihoods. Mar Policy 34(3):375–383

Bailey C (1988) The political economy of marine fisheries development in Indonesia. Indonesia 46:25–38. https://doi.org/10.2307/3351043

Bappenas (2021) List of priority locations and climate resilient actions. Ministry of National Development Planning (Bappenas), Jakarta

Bengen DG (2001) Ekosistem Sumberdaya Pesisir dan Laut Serta Pengelolaan Secara Terpadu dan Berkelanjutan. (Prosiding Pelatihan Pengelolaan WilayahPesisir Terpadu). Bogor 29 Oktober-3 November 2001. Pusat Kajian Sumberdaya Pesisir dan Lautan IPB

Cazenave A, Cozannet GL (2014) Sea level rise and its coastal impacts. Earth's Fut 2(2):15–34. https://doi.org/10.1002/2013EF000188

Dolan AH, Walker IJ (2006) Understanding vulnerability of coastal communities to climate change related risks. J Coast Res 1316–1323. http://www.jstor.org/stable/25742967

Direktoral Jenderal Perikanan Tangkap, KKP (2019) Master Plan Sentra Perikanan dan Kelautan Terpadu (SKPT), Kabupaten Natuna, Provinsi Kepulauan Riau.Jakarta, KKP

Eurich JG, Friedman WR, Kleisner KM, Zhao LZ, Free CM, Fletcher M, Mason JG, Tokunaga K, Aguion A, Dell'Apa A, Dickey-Collas M, Fujita R, Golden CD, Hollowed AB, Ishimura G, Karr KA, Kasperski S, Kisara Y, Lau JD, Mills KE (2023) Diverse pathways for climate resilience in marine fishery systems. Fish Fish 00:1–22. https://doi.org/10.1111/faf.12790

Ferrol-Schulte D, Gorris P, Baitoningsih W, Adhuri DS, Ferse SC (2015) Coastal livelihood vulnerability to marine resource degradation: a review of the Indonesian national coastal and marine policy framework. Mar Policy 52:163–171

Fitriani F, Kinseng RA, Lubis DP, Adhuri DS (2023) Kemiskinan dan Strategi Penghidupan Nelayan Kecil di Tanjung Kait, Banten. Jurnal Sosial Ekonomi Kelautan Dan Perikanan 18(1):113–126

Global Programme on Risk Assessment and Management for Adaptation to Climate Change (Loss and Damage) (2021) Climate change and small-scale fisheries. A case for a comprehensive climate risk management. GIZ, Bonn

Gustika W, Dharmawan AH, Abdulkadir-sunito MA (2023) Kerentanan Nafkah Rumahtangga Nelayan dalam Tekanan Variabilitas Iklim: Studi Kasus Desan Dendun, Kabupaten Bintan, Kepulauan Riau. Jurnal Ilmu Lingkungan 21(1):43–56. https://doi.org/10.14710/jil.21.1.43-56

Hasmawati, Pamungkas MI et al (2022) Pengetahuan Nelayan Palanra dalam Menangkap Ikan. Department of Anthropology, Faculty of Social and Political Sciences, Hasanuddin University, Indonesia. Jurnal Mahasiswa Antropologi, vol 1 No. 1 Tahun, pp 1–12

Hafsaridewi R, Hikmah H et al (2020) Fisher's resilience towards extreme weather conditions in the South China Sea: a case study of Natuna Island, Indonesia. OMNI-AKUATIKA, J Fish Mar Res 16(3). p-ISSN 18583873, e-ISSN :24769347

Holt SM (2022) Five countries, other than China, most dependent on the South China Sea, Published 17Nov 2022. Retrieved from https://www.cnbc.com/2022/11/18/five-countries-other-than-china-most-dependent-on-the-south-china-sea.html

Kaczan D, Nurhabni F, Cheung W, Frölicher T, Kuswardani A, Lam V, Muawanah U, Puspasari R, Reygondeau G, Sumaila U, Teh L (2023) Hot water rising: the impact of climate change on Indonesia's Fisheries and Coastal Communities. The World Bank, Washington, D.C.

Klein RJT, Nicholls RJ (1999) Assessment of coastal vulnerability to climate change. Ambio 28(2):182–187. http://www.jstor.org/stable/4314873

Kunarso K, Irwani I, Helmi M, Bayu Chandra A (2016) Musim Ikan Di Perairan Laut Jawa Kabupaten Jepara dan Prediksi Lokasi Fishing Ground-nya. In: Prosiding Seminar Nasional Tahunan Ke-V: Hasil-Hasil Penelitian Perikanan dan Kelautan, vol 2, p 05

Kusnadi (2002) Konflik Sosial Nelayan: Kemiskinan dan Perebutan Sumber Daya Perikanan. LKiS, Yogyakarta

KKP. Agenda Prioritas KKP dengan 5 Kebijakan. Retrieve from https://kkp.go.id/

Leal Filho W, Krishnapilla M et al (2021) Climate change adaptation on small island states: an assessment of limits and constraints. This article belongs to the Special Issue Coastal Risk and Adaptation in Small Islands 31 May, 2021. This article belongs to the Special Issue Coastal Risk and Adaptation in Small Islands. J Mar Sci Eng 9(6): 602. https://doi.org/10.3390/jmse9060602

Loneragan N, Stacey N, Wiryawan B, Adhuri DS, Warren C, Fitriana R, Gibson E, Steenbergen DJ, Mustika PLK, Jaiteh V, Fadilah R (2018) Small-scale fisheries in Indonesia: benefits to households, the roles of women, and opportunities for improving livelihoods. Australian Centre for International Agriculture Research, Canberra, Australia

Mason JG, Eurich JG, Lau JD, Battista W, Free CM, Mills KE, Tokunaga K, Zhao LZ, Dickey-Collas M, Valle M, Pecl GT, Cinner JE, McClanahan TR, Allison EH, Friedman WR, Silva C, Yáñez E, Barbieri MÁ, Kleisner KM (2022) Attributes of climate resilience in fisheries: from theory to practice. Fish Fish 23(3):522–544. https://doi.org/10.1111/faf.12630

Nomura M, Yamazaki T (1997) Fishing techniques (1). Japan International Cooperation Agency. Tokyo, pp 1–206

"Nelayan Kepulauan Riau dan Potensi Perikanan WPP 711. Zona Sidik.com, 7 April 2022

Natuna Regency in Figure 2023. Natuna, BPS Kabupaten Natuna, pp 1–409

Najid A, Pariwono JI, Bengen DG, Nurhakim S, Atmadipoera AS (2012) Pola Musiman dan Antar Tahunan Salinitas Permukaan Laut di Perairan Utara Jawa-Madura. Maspari J 4(2):168–177

Peraturan Menteri Kelautan dan Perikanan Republik Indonesia Nomor Per.08/MEN/2012 tentang Kepelabuhan Perikanan

Pinsky ML, Fenichel E, Fogarty M, Levin S, Mccay B, Martin KS et al (2021) Fish and Fisheries in Hot Water: What is Happening and How do We Adapt? Popul Ecol 63(1):17–26

Pranoto (1999) Pengantar Pelabuhan Perikanan. Balai Ketrampilan Penangkapan Ikan, Tegal, pp 1–34

Republika.co.id, Cuaca Buruk sebagian Nelayan Natuna Beralih Profesi, Januari 30, 2022
Sari M, Wiyono ES (2021) Pengaruh Cuaca Terhadap Pola Musim Penangkapan Ikan Pelagis di Perairan Teluk Lampung. ALBACORE Jurnal Penelitian Perikanan Laut 5(3):277–289
Satria A (2015) Pengantar Sosiologi Masyarakat Pesisir. Yayasan Pustaka Obor Indonesia, Jakarta
Suhery N, Jaya MM, Khikmawati LT, Sarasati W, Tanjov YE, Larasati RF et al (2023) Keterkaitan Musim Hujan dan Musim Angin Dengan Musim Penangkapan Ikan Lemuru Yang Berbasis di PPN Pengambengan. Mar Fish J Mar Fish Technol Manage 14(1):77–90
The Decision of the Minister of Maritime Affairs and Fisheries of the Republic of Indonesia No. 19 of 2022
Undang-Undang No. 1 Tahun 2014 tentang Pengelolaan Wilayah Pesisir dan Pulau-Pulau Kecil
Undang-Undang No. 7 Tahun 2016 tentang Perlindungan dan Pemberdayaan Nelayan, Pembudi Daya Ikan, dan Petambak Garam
Widjaja S (2022) Kajian Pembangunan Sentra Kelautan dan Perikanan Terpadu (SKPT) Natuna, Laporan Akhir UG 182105, Surabaya, ITS, pp 1–62
Wibowo A, Satria A (2015) Fisher's adaptation strategies in small islands to the impacts of climate change: a case study in Pulau Panjang Village, Subi District, Natuna Regency, Riau Island. In: Paper proceedings conference of climate change adaptation 2015, Colombo, Sri Lanka. ISBN: 978-955-4543-3
Yamazaki S, Resosudarmo BP, Girsang W, Hoshino E (2018) Productivity, social capital and perceived environmental threats in small-island fisheries: insights from Indonesia. Ecol Econ 152:62–75. https://doi.org/10.1016/j.ecolecon.2018.05.020
Yulihastin E (2010) Mekanisme Interaksi Monsun Asia dan Enso. Berita Dirgantara 11(3)

Chapter 20
Towards a Sustainable Future: Integrating Indigenous Knowledge to Address the Challenges of Climate Change in West Bengal, India

Mainak Sarkar and Biplob Kumar Modak

Abstract The term "indigenous knowledge" (IK) refers to ideas, events, habits, and data that indigenous people have adapted into their culture through stories, folklore, and rituals. It may include oral transmission between generations. Climate change is a major issue, especially for emerging nations' poor. Indigenous societies have learned a lot about environmental changes and developed complicated ways to manage them due to their close interaction with natural ecosystems and reliance on them for livelihood. Indigenous societies have achieved ecological balance and sustainability via reciprocity, guardianship, and respect for nature. West Bengal, with a 2% tribal population, has a tremendous store of traditional knowledge that, if acknowledged, welcomed, and incorporated, might help solve climate change challenges. This article presents ideas grounded in indigenous knowledge to aid in the development of sustainable initiatives aimed at mitigating the effects of climate change. The research methodically selected case study analysis articles. Initial literature searches used PubMed and Google Scholar to find West Bengal, India-specific climate crisis, disaster management, and mitigation measures. Innovation lies in the study's careful assessment and synthesis of indigenous knowledge and climate crisis management literature from West Bengal, India. The study organized this field's expertise by carefully searching for and evaluating relevant publications. The systematic framework for presenting the data promotes awareness of indigenous knowledge's intricacies and involvement in climate change concerns, enriching this essential subject. The findings were presented according to a comprehensive classification framework, enabling a structured analysis and synthesis of the information obtained from the literature review.

M. Sarkar
Department of Zoology, Bankura Christian College, Bankura, West Bengal 722101, India

M. Sarkar · B. K. Modak (✉)
Department of Zoology, Sidho-Kanho-Birsha University, Purulia, West Bengal 723104, India
e-mail: bkmodak09@gmail.com

U. Mukhopadhyay et al. (eds.), *Climate Crisis, Social Responses and Sustainability*, Climate Change Management, https://doi.org/10.1007/978-3-031-58261-5_20

Introduction

Scientists, policymakers, and the public have been concerned about climate change's growing effects for a decade (Nawrotzki and Kadatska 2010). Climate change is accelerating air composition and chemistry, fast urbanization, industrialization, pollution, and the depletion of life-sustaining resources (Rockström et al. 2009; Steffen et al. 2015). These environmental transformations have disproportionately affected poor nations and vulnerable populations (Olsson et al. 2014). It's becoming clear that Western climate solutions may not work in various cultures (Elliott 2012). This global setting makes indigenous knowledge (IK) essential for designing sustainable solutions that respect local cultures and habitats (Zidny et al. 2020). Indigenous knowledge is a bunch of skillful information that is gained from living and engaging with their local environment and applied in daily decisions (Panda et al. 2023). Numerous studies suggest that cultural rituals and oral traditions transmit indigenous knowledge. Health care, agriculture, conservation, culinary arts, education, and other fields have relied on this approach to improve human and environmental well-being (Sarkar and Modak 2023).

India, among the world's most vulnerable nations to climate change (IFRC 2005), faces natural and man-made challenges due to rapid population growth, high densities, poverty, and inadequate infrastructure (Revi 2008). Limited arable land, dense population, and reliance on monsoon-dependent agriculture make India the fifth most vulnerable nation to climate change among 181 nations (Shanabhoga et al. 2020). Consequently, India is significantly affected by climate change and natural hazards, impacting crop productivity due to extreme weather events like droughts and floods (Gupta et al. 2014).

Extreme weather has damaged West Bengal's ecosystem and inhabitants (De et al. 2005). Rising temperatures, erratic precipitation, and coastal vulnerability have modified agricultural systems, necessitating appropriate climate management and economic costs (Ray et al. 2021). Droughts and floods have increased in West Bengal due to unpredictable precipitation (Nandargi and Barman 2018). West Bengal's coastal regions are vulnerable to rising sea levels due to frequent cyclones and storm surges that destroy infrastructure and uproot communities (Mondal et al. 2020). Reforestation, sustainable agriculture, and disaster preparedness are government and stakeholder efforts to address complicated concerns (Srinivas et al. 2020).

The climate challenges in West Bengal highlight the valuable knowledge of indigenous tribes in managing climate change hazards, emphasizing the need for indigenous knowledge-based solutions in developing nations (Panda et al. 2023). The study explored the traditional techniques used by tribal groups in West Bengal, India, to mitigate climate-related hazards and raise awareness of the importance of indigenous knowledge in climate management to develop effective policies and initiatives for social and environmental sustainability.

Materials and Methods

The study's novelty lies in its systematic approach to exploring and synthesizing the available literature on indigenous knowledge and climate crisis management strategies specific to West Bengal, India. By employing a rigorous methodology for identifying and analyzing relevant publications, the study has contributed to consolidating and organizing the existing knowledge in this field. The structured framework for presenting the findings enhances the understanding of the complexities surrounding indigenous knowledge and its role in addressing climate change challenges, thereby enriching the body of knowledge on this crucial subject.

Study Area

West Bengal (Fig. 20.1) is situated in northeastern India. Numerous Indian states and nations border its northeastern location. Bihar, Sikkim, Bhutan, Bangladesh, and the Bay of Bengal border it on the west, north, east, and south. The climate of West Bengal is tropical. The state has a hot, humid summer, a rainy monsoon season, a pleasant autumn, and a cold winter. The monsoon season brings intense rain from June to September. The climate ranges from the southern Bay of Bengal coast to the northern Himalayas.

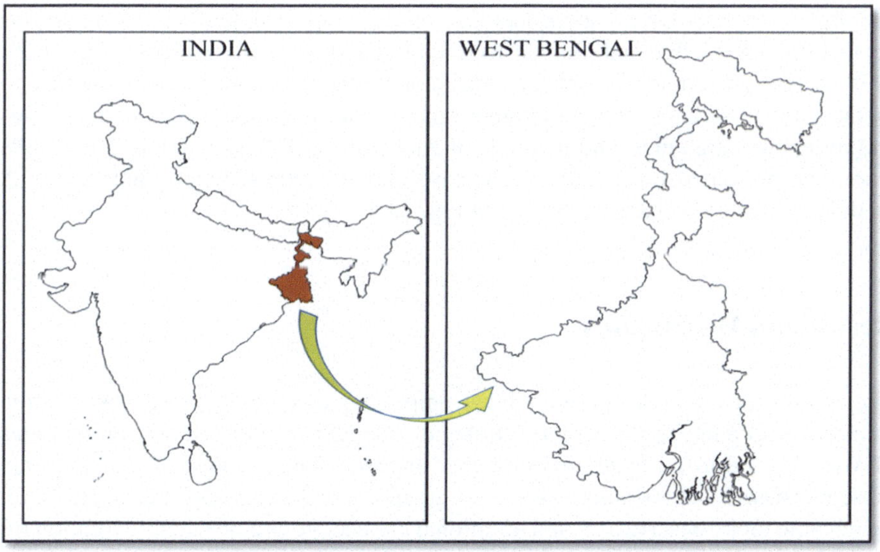

Fig. 20.1 Study area: West Bengal, India. *Source* Author

West Bengal was densely populated and multicultural. It was one of India's most populous states with around 90 million people. West Bengal has hundreds of languages and religions, all of which coexist harmoniously. West Bengal has various indigenous and tribal communities, including the *Santhal, Munda, Oraon, Bihors, Bhumij,* and others. These groups contribute unique languages, customs, and practices to the state's culture.

The northern parts of West Bengal are mountainous and part of the Eastern Himalayas, while the south and center are on the Gangetic plains. Northern Darjeeling and Kalimpong have beautiful hills and tea gardens. Ganges (Hooghly) and Brahmaputra rivers intersect the state and are crucial to agriculture and commerce.

Data Collection

Methodical selection of case study analysis papers was done. PubMed and Google Scholar were used for literature searches. The search phrases included "climate crisis," "climate variability," "disaster management," "disaster risk reduction," and "climate crisis mitigation strategies," coupled with Indian states, with an emphasis on West Bengal. Also included were "local knowledge," "traditional ecological knowledge," "traditional knowledge," "indigenous knowledge," "indigenous strategies," "tribal mitigation strategies," "tribal of West Bengal," "community-based knowledge," "local beliefs," and "traditional beliefs."

After screening documents retrieved by the search engine for their titles, abstracts, and full texts, scholarly papers lacking unique empirical findings or assertions about local knowledge and climate were eliminated in the final phase (Fig. 20.2). This review encompassed several cases and utilized straightforward queries to extract information on climate change recognition, implementation of responses and mitigation strategies, and methods of adaptability skill acquisition. The results were then presented using a comprehensive classification scheme (Table 20.1). A methodological flowchart is depicted in Fig. 20.3.

Result and Discussions

The approach used in the study includes employing a comprehensive classification framework to analyze the various strategies adopted by Indigenous communities of West Bengal to mitigate environmental hazards. The findings of the study are presented based on this framework, covering aspects such as agroecological practices, sustainable food conservation, and traditional knowledge as a sustainable solution for the climate crisis. The study explores how traditional knowledge can be a potential sustainable solution through its analysis of Indigenous communities' strategies.

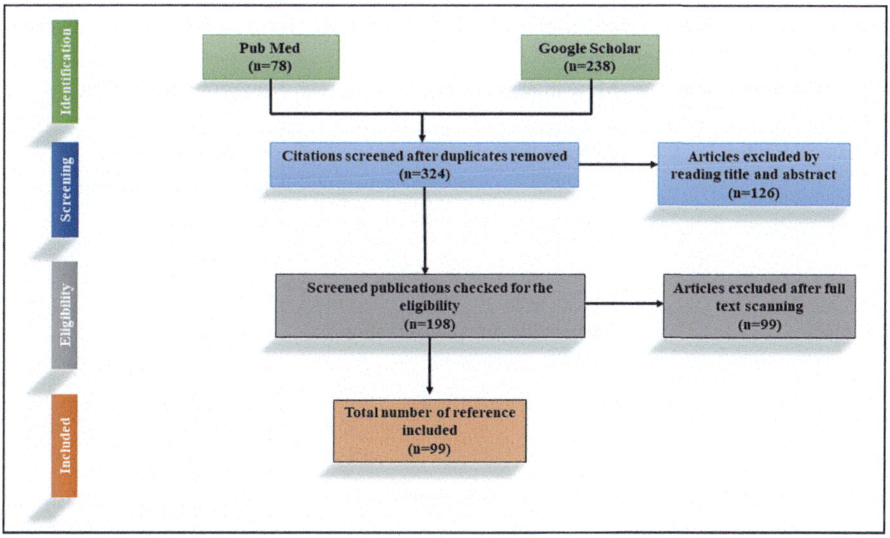

Fig. 20.2 Flow diagram summarizing the approach used in the selection of literature

Indigenous people in West Bengal have devised several methods to address climate change over many centuries. Some of these strategies may vary by indigenous community, however the following are common:

Climate Mitigation Through Agroecological Practices

Indigenous communities in West Bengal, India, have developed agroecological practices that are firmly ingrained in their traditional agricultural practices and knowledge (Datta and Behera 2022). These practices are also environmentally sustainable. These practices (Table 20.2) have been passed down from generation to generation, and their goal is to assure food security while also preserving biodiversity and preserving ecological balance.

Mixed Cropping

Mixed cropping is a common practice among indigenous people, in which multiple types of crops are cultivated concurrently inside the same field (Fig. 20.4a). This serves to reduce the danger of crop failure, promote the fertility of the soil, and reduce the likelihood of outbreaks of pests and diseases. Combinations such as rice with legumes, vegetables with millets, and various fruit trees with understory crops are common examples of common combinations.

Table 20.1 Comprehensive classification framework of various strategies adapted by Indigenous communities of West Bengal, India

Types of environmental hazards	Mitigation strategies	Divisions/categories
Climate variability	Agroecological practices	Mixed cropping
		Traditional seed saving
		Agroforestry
		Terrace farming
		Crop rotation
Drought	Sustainable food conservation	Cereals
		Rice gruel and porridge
		Leafy green vegetables
		Left-over food
		Protein resources
		Fruits and tubers
		Traditional alcoholic beverage
		Food for livestock
	Sustainable water conservation	*Pat Kua* (The Well)
		Pitcher watering system
		Embankment on river
		Hallowed Geyser (*Sita Kund*)
		Roof rain water harvesting system
		Water storages beside the river
		Dari system
		Aquifer
		Naturally created water storages beside the river
		Goira and doba system
		Contour bunds
		Hapa
Flood	Prevention of soil erosion	Bank management
	Food Storage	Crop Banking
	Transport	Buoyant Boats

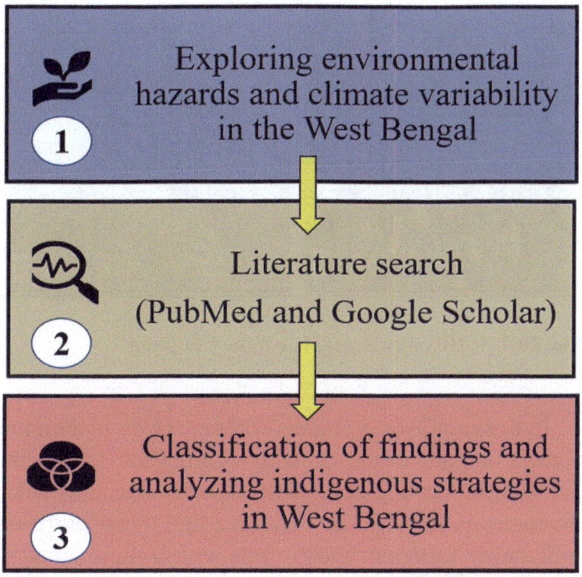

Fig. 20.3 A methodological flowchart indicates a structured and methodical process in the study, encompassing literature selection and classification of mitigation strategies

Table 20.2 Advantages of agroecological practices of Indigenous people

Serial No.	Process	Advantage	Regions
1	Mixed cropping	This lowers crop failure risk. If one crop fails, others may harvest. Mixing crops improves soil fertility and reduces the need for fertilizers and pesticides	Coochbehar, Purulia, Bankura, and Jhargram are predominant
2	Traditional seed saving	Indigenous communities preserve regionally appropriate plant species to diversify and strengthen crops. Some seeds are more pest-, disease-, and climate-resistant. It improves food security and economic independence	Bankura, Purulia, Sundarban, and Dooars are predominant
3	Agroforestry	It is a comprehensive approach to land management, which promotes environmental sustainability	Darjeeling and other hilly regions of West Bengal
4	Terrace farming	Terrace farming help to conserve soil and water by preventing erosion	Darjeeling and Kalimpong are predominant
5	Crop rotation	Rotating crops with differing nutrient needs reduce nutrient depletion and soil exhaustion, maintaining soil fertility	Purulia, Bankura, and Birbhum are predominant

Fig. 20.4 **a** Mixed cropping, **b**, **c** terrace farming

Indigenous West Bengalis combine crops in their fields. Common pairings include cereals like rice or millets with legumes like lentils or chickpeas and vegetables like potatoes or leafy greens with oilseeds like mustard (Mukherjee 2012). This crop diversification maximizes land use and meets nutritional needs (Altieri et al. 2012). Indigenous farmers choose crops with complimentary growth patterns. Legumes fix nitrogen, benefiting nearby grain crops. Crops with varying heights and root architectures maximize sunlight, water, and nutrients (Gebru 2015). Pests and diseases can be reduced by mixed cropping. Interplanting crops makes it harder for pests to discover their host plants (Trenbath 1993). This decreases pesticide use. Mixed cropping optimizes soil nutrient cycling. Leguminous crops add nitrogen to the soil, helping the next crop that needs it (Amossé et al. 2014).

Mixed farming reduces crop failure in rain-dependent West Bengal (Banerjee et al. 2023). Having one crop fail increases the chances of another succeeding, assuring community food security. Mixed cropping reduces soil erosion and nutrient depletion. Plant diversity boosts soil microbes and structure (Shah et al. 2021). Some West Bengali indigenous farming systems use animals in mixed cropping (Sarkar et al. 2020). Indigenous farmers carefully time mixed crops to stagger harvests. This reduces food shortages by providing fresh produce and grains year-round (Altieri et al. 2012). Indigenous knowledge of local ecosystems informs mixed farming practices passed down through generations (Senanayake 2006). Mixed cropping decreases the use of chemical fertilizers and pesticides, making it sustainable. It also protects agricultural biodiversity (Altieri et al. 2017).

Native Rajbanshi farmers of Coochbehar, North Bengal, excel at mixed farming (Das Gupta 2013). Winter-autumn vegetables like sponge gourd (*Luffa aegyptiaca*) and bitter gourd (*Momordica charantia*) can be cultivated alongside mixed cropping. Sponge gourd needs a lattice, although bitter gourd can be grown as a creeper. If kept healthy and provided enough cow dung manure, their productivity and size can increase. Potatoes, radishes, cabbage, cauliflower, carrots, and tomatoes are winter staples. Mixed cropping and inside cropping grow these near together (Das Gupta 2013). Organic manure amends soil. Common bean (*Phaseolus vulgaris*), known as '*Sim*' in Bengali, and Yard-long Bean (*Vigna unguiculate*), known as '*Borboti*', are also mixed-grown as winter crops (Das Gupta 2013).

Indigenous West Bengalis have optimized mixed farming for local conditions for ages. These measures boost food security, environmental sustainability, and climate resilience.

Traditional Seed Saving

Indigenous West Bengalis have traditionally grown and traded seeds (Kingwell-Banham and Fuller 2012). They protect traditional crop types that have been bred to adapt to the local environment over many generations, making them more disease- and pest-resistant (Deb 2017). Traditional rice varieties contain different nutrients. Some may have more vitamins, minerals, and antioxidants than high-yield rice (Deb 2017). Traditional rice cultivars that thrive in certain agroecological situations may be more resistant to droughts and floods (Rao et al. 2016). Traditional rice varieties may provide genetic material for producing climate-resistant crops (Shahzad et al. 2021). Local farmers select and maintain "folk varieties" of rice cultivars for their distinctive features. They are passed down from farmer to farmer and subject to natural and human selects in a region (De 2014). Traditional land races that are extremely resistant to drought, such as '*Kelas*', '*Asnaliya*', and '*Bhutmuri*', are intensively cultivated by indigenous farmers in the districts of Bankura and Purulia in the state of West Bengal. In a similar vein, traditional rice varieties such as '*Matla*', '*Talmugur*', '*Hogla*', '*Kantarangi*', and '*Nona bokra*' of the Sundarban region and '*Himal*' of the Darjeeling district are extremely well-known due to their resilience to high salinity and low temperatures, respectively (Paul 2012). Increasing food security and fostering environmentally responsible agriculture both need the cultivation of traditional rice types (Giller et al. 2021).

Agroforestry

Many indigenous West Bengalis practice agroforestry, planting shrubs and trees in farms (Datta and Behera 2022). This protects biodiversity and soil while selling lumber, fruits, and medicinal plants (Rao et al. 2004). Agroforestry, which uses trees to grow crops and animals, dominates Darjeeling's farming system. Traditional Eastern Himalayan agroforestry systems provide food and livelihood demands while maintaining and improving the environment. Trees, grasses, herbs, bushes, and a large cardamom crop block water flow, reducing soil erosion and pollution. The hamlet plants fodder trees along roadsides, field bunds, and terraces to give animals lean fodder (Mukherjee 2012). Large cardamom-based agroforestry systems promote nutrient cycling, soil fertility and productivity, soil erosion, biodiversity, water and soil preservation, carbon sinking, farm incomes, and mountain community aesthetics (Buckingham 2004). Sustainable large cardamom plantations with shade trees on steep hillsides unsuited for agriculture. The region's varied tree species are preserved by large-scale cardamom agroforestry (Mukherjee 2008).

Terrace Farming

Indigenous West Bengali highlanders practice terrace farming (Fig. 20.4b, c). This involves building level platforms on sloping land to raise crops (Mukherjee 2012). Terrace farming conserves water, prevents soil erosion, and maximizes land utilization (Tefera and Sterk 2010). High-elevation settlements in West Bengal, especially Darjeeling, have fewer irrigated acres and more rainfed terraces. Rainfed terraces are normally outwardly sloping and only suitable for maize and millet. Farmers below 2300 m plant two harvests of maize, millet, barley, wheat, and other cereals and grains on rainfed land each year. Terraces used to cultivate maize are usually dispersed, fractured, have more parcels, and are at least three hours' walk from the farmer's house. Monsoon clouds cover areas above 2500 m with broadleaf evergreen woods (Mukherjee 2012).

Crop Rotation

Indigenous farmers in the lateritic belt of West Bengal widely practice crop rotation, with a prevalence of 94%. This technique aims to preserve soil quality and mitigate nutrient exhaustion by optimizing the growth cycles and nutrient needs of different crops (Adesina et al. 2020). For instance, after cultivating rice, they often plant pulses or oilseeds to restore essential nutrients in the soil (Deka et al. 2019; Borron 2006). Additionally, the use of coal ash in winter paddy fields elevates soil temperature, while its application as a pest control against beetles and the practice of mulching are also prevalent among indigenous farmers in the regions of Purulia and Bankura (Biswas and Biswas 2014).

Nevertheless, it has been observed that certain farmers in these two locations continue to employ the practice of smoke generation by burning paddy husk and dry chilies as a means to address the issue of rodent infestation in their fields (Biswas and Biswas 2014).

Drought Resilience Strategies

Drought poses a significant risk to hydrogeology, the environment, and socio-economic aspects, especially for resource-dependent households in developing countries (Gray and Mueller 2012). Bankura and Purulia in West Bengal are highly prone to drought due to their limited rainfall patterns and geographical characteristics (Goswami 2017; Sarkar and Modak 2023). The indigenous populations in these areas have developed various adaptation strategies to address the impact of drought, with a focus on ensuring food and water security.

Fig. 20.5 Some of the popular indigenous food of West Bengal: **a** Wood Apple/Bel (*Feronia indica*); **b** Coromandel ebony/Kend (*Diospyros melanoxylon*); **c** Yam/Kham alu (*Dioscorea* sp.); **d** Ceylon Oak/Kusum (*Schleichera oleosa*); **e** Mahuwa/Mahul (*Madhuca indica*); **f** Dried Mahuwa (*Madhuca indica*) flower; **g** Traditional food 'bans karoil' made up of *Bambusa* sp.; **h** Taro/Kochu (*Colocasia esculenta*); **i** Water hyssop/Bramhi (*Bacopa monnieri*); **j** Buffalo spinach/Hemcha (*Enhydra fluctuans*), **k** Jute Mallow/Kudrum (*Chorchorus* sp.); and **l** Red water lily/Shaluk (*Nymphaea rubra*). *Source* Author

Drought-Resilient Diet and Feeding Practices

Cereals

In the current period characterized by significant social upheaval, the indigenous population residing in this particular area exhibits a substantial dependence on grain as a fundamental component of their dietary intake (Fig. 20.5). The three crops under consideration are '*Marua*' (*Eleusine coracana*), '*Bajra*' (*Pennisetum typhoideum*), and '*Junar*' (*Sorghum bicolor*). During periods of prolonged drought, individuals engage in the preparation of a distinctive type of sustenance known locally as '*Lapsi Ghata*', which involves the utilization of various cereal crops. Initially, employing the aid of a '*Dheki*' (a farming implement utilized for the purpose of threshing, which involves the separation of grains from their outer husks), the cereals were subjected to forceful impact, resulting in their fragmentation. Additionally, the grains are crushed and separated into three sizes. Sieves isolate the tiniest grains. The largest grains are then dipped in water and boiled in a clay pot. After this, medium-sized grains are added, followed by the smallest. A *Shorea robusta* or '*Sal*' plate is used for the final presentation (Sarkar and Modak 2023).

Rice Gruel and Porridge

'*Marjhola*', a type of sticky rice cooked with a lot of rice gruel and drumstick (*Moringa oleifera*), is one of the most well-known traditional dishes eaten during drought. The rice and vegetables were cooked in a clay oven that was brought to a boil. The final step is to add salt before serving. At night, they eat the same cuisine that was cooked for lunch. Due to prolonged drought, they often have to rely on rice gruel and porridge (known as '*Dhala mar*' in the local language), which is created from broken-up leftover grains. This '*Dhala mar*' plays a crucial role in supplying the necessary water during this dry season.

Green Leafy Vegetables

They chose green leafy vegetables from farms with little water, mixed them with leftover broken rice, and cooked the whole thing with a little water. The oatmeal was seasoned with salt and roasted green chilies. The villagers also use '*Pipre saag*' (*Ficus* sp.), an antipox leafy green vegetable, amid drought (Yarmolinsky et al. 2012). This porridge will be a vital nutrient source during this crucial time. When it's dry, bottle gourd and pumpkin leaves are popular. They are pasted with rice and used. Due to a severe water shortage, they harvest wild leafy greens to augment their meals and cook them briefly to keep as much nutritional content as possible. These greens are rich in calcium, iron, and beta-carotene, according to Ghosh-Jerath et al. (2016). They also like lotus rhizomes, shoots ('*Mulhan*' in local dialect), and bamboo ('*Koril*') of various species.

Leftover Food for Dire Emergencies

Due to the prolonged period of arid conditions, a significant number of indigenous individuals face considerable challenges in fulfilling their daily sustenance requirements. They mostly rely on a distinct type of food often known as '*Potom*' within the region. The aforementioned dish primarily consists of remnants and abandoned edible constituents derived from a ceremonial event that occurred in a proximate town or village. In this time of extreme poverty, this provision will be helpful. This dish is only available in remote indigenous settlements. This meal can be hard to find in grocery stores.

Essential Protein Resources

Protein is an essential part of a balanced diet and vital to human survival. Yet the natives of these areas nonetheless managed to meet their protein requirements by eating a wide range of foods, even during this devastating time. '*Kurkut*' is a popular dish that is made entirely of the eggs of weaver ants (*Oecophylla smaragdina*). The

Santhal people use the word '*Hau*' to describe eggs of this type. As a food source, the eggs are fried in mustard oil with salt, chili spices, and other seasonings after being gathered from the trees. Also, locals in the eastern Himalayas have a strong preference for red ant eggs (Sarkar and Modak 2023; Chowdhury et al. 2015).

Fruits and Tubers

To address the issue of scarcity caused by drought, indigenous populations residing in these areas are compelled to rely on wild edible fruits and tubers as a substitute or unconventional source of sustenance (Sarkar and Modak 2023). *Dioscorea* species tubers, along with other tuber varieties like '*Khesari*' (*Actinoscirpus grossus*), '*Simchiru ara*' (*Cyanotis axillaris*), '*Kana Kanda*' (*Nymphoides indica*), '*Dhai*' (*Tacca pinnatifida*), and fruit like '*Piyal*' (*Buchanania lanzan*), '*Vela*' (*Semecarpus anacardium*), '*Kusum*' (*Schleichera oleifera*), '*Piyal*' (*Buchanania lanzan*) is a plant whose fruit and seed are used for their edible qualities. The edible seed is known as a '*Chironji*'. To prepare a soup called '*Paula*' cooked rice water is combined with a powder made from the dried leaves of '*Tentul*' (*Tamarindus indica*). Green and mature fruits are utilized to create various thorns. The fruit of the '*Kend*' tree (*Diospyros tomentosa*) is delicious and can be eaten. Since the locals consider this fruit to have both a cooling and toning effect (Kumari and Kumar 2021), they regularly eat it during the summer. The indigenous peoples of the area, especially the Bihors, have long termed the region of hills known as '*Kanda*' (*Dioscorea* sp.), also known locally as '*Baula*'. This is often highly bitter, but locals have found several ways to mitigate its unpleasant flavor, making roasted, pelletized, and other cereal preparations conceivable (Gorai et al. 2022; Kumar et al. 2017). During the period of severe scarcity of food resources, a traditional dish known as '*Mohul latha*' gained significant prominence. The preparation of this culinary creation involves the utilization of desiccated '*Mohul*' fruit (*Madhuca indica*) and tamarind seed that has undergone the process of boiling. Then, the substance transforms, resulting in the formation of a circular shape with a somewhat solid consistency, which is then ingested (Sarkar and Modak 2023).

Traditional Alcoholic Beverage

'*Hariya*' represents the predominant traditional alcoholic beverage that is commonly enjoyed by the indigenous communities residing in West Bengal. The aforementioned communities encompass the *Santals, Mundas, Oraons,* and *Lodhas.* During the summer season, indigenous communities residing in this region extensively utilized '*Hariya*', attributing it to a significant source of energy, while enduring the intense heat of the sun (Sarkar and Modak 2023; Dhal et al. 2010).

Feed Provided to Domestic Livestock

When the temperature rises to dangerous levels, people must take care of themselves and their domestic animals at the same time. Locals likely used various plant parts, including fruit and leaves, to supplement their animals' diets. The local population relies on their familiarity with wild flora to survive periods of food scarcity. They travel to the adjacent forests in search of young plants and different sections of those plants (Dangol 2008). The names of several common plants whose byproducts are used as feed are listed below. '*Arjun*' (*Terminalia arjuna*) leaves are used as fodder, and so are the leaves of the '*Assatha/Jore*' tree (*Ficus religiosa*). Fodder can also be made from the delicate branches of the '*Vela*' tree (*Semecarpus anacardium*). The green leaves of the '*Ankra*' plant (*Alangium salvifolium*) are used as livestock feed. The young leaves of the '*Buch*' (*Cordia nyxa*) tree are excellent forage for cattle. Leaves of the '*Kural*' tree (*Bauhinia racemose*) are used as livestock feed (Sarkar and Modak 2023). Table 20.3 sums up the food-related management of indigenous people during drought.

Sustainable Water Resource Management for Drought Resilience

The increasing water demand necessitates careful management planning, incorporating various developmental approaches to achieve optimal results (Sarkar and Modak 2023). Groundwater conservation requires identifying potential zones and optimizing implementation, as it sustains agriculture's water use and quality (Hutti and Nijagunappa 2011; Bera and Das 2021). In drought-prone areas like Bankura and Purulia, micro-watershed development is crucial for regional progress (Deshpande and Narayanmoorthy 2000). Women in remote areas of these regions travel long distances to fetch water in earthen pots, especially in harsh temperatures (Sarkar and Modak 2023). Indigenous communities have developed organic water management methods to address these challenges, while the population growth has led to a doubling of water usage in Purulia (Haldar and Saha 2015). Table 20.4 illustrates how indigenous people use traditional techniques to meet their daily dietary and water needs.

In arid Purulia and Bankura, water management involves collaborative planning among various groups. However, achieving water sustainability in certain locations is challenging due to geographical constraints and changing human–environment interactions (Kar et al. 2023). Indigenous methods are valued for being natural, eco-friendly, and cost-effective (Bauri et al. 2020). The identification of appropriate reservoir and tank locations for water security research, along with the development of water-harvesting systems by diverse scholars, demonstrates promising advancements. Nevertheless, regional underdevelopment can impede the implementation of these solutions, emphasizing the need to integrate traditional knowledge with technological water harvesting in water-scarce areas; Bera and Das 2021; Agarwal et al. 2013).

Table 20.3 List of food consumed by Indigenous people of Indigenous people of West Bengal

Food types	Local name	Scientific name	Consumption type
Cereals	Marua	*Eleusine coracana*	Cooked
	Bajra	*Pennisetum typhoideum*	Cooked
	Junar	*Sorghum bicolor*	Raw/cooked
Rice gruel/Porridge	Marjhola/Dhalamar	*Oryza sativa*	Boiled
Green leafy vegetables (GLV)	Kana Saag	*Comellina benghalensis*	Cooked
	Hemcha Saag	*Enydra fluctuans*	Cooked
	Kural Saag	*Bauhinia* sp.	Cooked
	Notey Saag	*Amarantus viridis*	Cooked
	Pipre Saag	*Ficus* sp.	Cooked
	Lau Pata	*Macaranga harveyana*	Cooked
	Kumro Pata	*Cucurbita* sp.	Cooked
	Marchi Saag	*Catharanthus pusillus*	Cooked
	Hurhuria Saag	*Cleome monophylla*	Cooked
	Kolmi Saag	*Ipomoea aquatica*	Cooked
	Kulekhara Saag	*Leucas mollissima*	Smashed/cooked
	Kedo Saag	*Limnophila indica*	Cooked
	Gara Saag	*Polygonum barbatum*	Cooked
	Barial/Junka	*Sida cordata*	Cooked
	Satgithia	*Spermacoce hispida*	Cooked
	Mulhan (shoot of lotus plant)	*Nelumbo nucifera*	Cooked with spices
	Kural (young shoot tip of bamboo plant)	*Bambusa* sp.	Cooked with spices
Left-over food	Potom	–	Sundried and wrapped in shal leaves (*Shorea robusta*)
Proteinaceous food	Kurkut (Eggs of Weaver ant)	*Oecophylla smaragdina*	Raw/cooked with spices
Fruits and tubers	Bainchi	*Flacourtia indica*	Cooked
	Piyal	*Buchanania lanzan*	Raw

(continued)

Table 20.3 (continued)

Food types	Local name	Scientific name	Consumption type
	Vela	*Semecarpus anacardium*	Cooked
	Kusum	*Schleichera oleosa*	Cooked
	Amla	*Emblica officinalis*	Raw
	Madal	*Annona squamosa*	Cooked
	Amra	*Spondias pinnata*	Raw/cooked
	Dumur	*Ficus hispida*	Cooked
	Kochra	*Madhuca indica*	Cooked
	Kend	*Diospyros tomentosa*	Cooked
	Tentul	*Tamarindus indica*	Raw/cooked
	Khesari	*Actinoscirpus grossus*	Raw/cooked
	Simchiru ara	*Cyanotis axillaris*	Cooked
	Kana kanda	*Nymphoides indica*	Cooked
	Dhai	*Tacca pinnatifida*	Cooked
	Gittha	*Dioscorea bulbifera*	Cooked
	Bhyakur	*Dioscorea pentaphylla*	Cooked
	Tarul	*Dioscorea alata*	Cooked
	Genthi	*Dioscorea hispida*	Cooked
	Mohul (Mohul latha)	*Madhuca indica*	Cooked with tamarind seed
Traditional alcoholic beverage	Hariya	–	Drink as liquid
Fodder (livestock feeding)	Assatha/Jore	*Ficus religiose*	Raw leaves
	Arjun	*Terminalia arjuna*	Raw leaves
	Vela	*Semecarpus anacardium*	Tender branches
	Ankra	*Alangium salvifolium*	Green leaves
	Buch	*Cordia nyxa*	Fresh foliage
	Kural	*Bauhinia racemose*	Raw leaves

(continued)

Table 20.3 (continued)

Food types	Local name	Scientific name	Consumption type
	Bans/bamboo	*Dendrocalamus strictus*	Raw leaves
	Gamhar	*Gomelina arborea*	Raw leaves
	Sisoo/Indian rosewood	*Dalbergia sissoo*	Raw leaves
	Sirish	*Albezia lebbeck*	Raw leaves
	Tun	*Toona ciliate*	Seeds and raw leaves
	Amlaki/Indian gooseberry	*Phyllanthus emblica*	Raw leaves
	Bohera	*Terminalia belerica*	Fruit
	Nishinda	*Vitex negundo*	Raw leaves

Modified from Sarkar and Modak (2023)

Strategies for Flood Risk Reduction

Water overflows borders, flooding territory and generating a deluge. Naturally occurring river and stream flooding. Once every 2.33 years, streams exceed the average annual flood magnitude, according to Jha and Bairagya (2013). Rainfall exceeds soil, river, stream, and coastal area capacity, causing flooding. Overflowing rivers flood surrounding crops.

Some tropical countries experience severe floods from sustained monsoon rains (Mathison et al. 2018). Tropical cyclones can generate massive floods for many reasons. First, the hurricane's front brings 8-m waves. Tropical storms often rain heavily (Hossain et al. 2008). Alpine snow melting in strong heat waves has caused flash floods and deaths (Pomeroy et al. 2016). Marine earthquakes and plate movements cause tsunamis that flood beaches (Narayana et al. 2005).

West Bengal, the lowest-lying Ganga Basin riparian state and the foothills of many Himalayan rivers in the Brahmaputra Basin, has drainage and flood management issues. Most West Bengal rivers are inter-state or international (Annual Flood Report 2019, Government of West Bengal 2020).

Of the state's 88,752 km^2, 22% is prone to annual flooding (Jha and Bairagya 2013). Flooding, waterlogging, bank erosion, embankment breaches, drainage congestion, and cyclonic storms have often worsened the flood disaster. Tidal surges at the Sundarban delta, the world's biggest delta, often erode and flood rural villages. With 42.55% of its land area flood-prone, West Bengal is one of the most flood-prone states in the nation (Annual flood report 2019, Government of West Bengal 2020). Floods threaten 42.3% of the state's land. Susceptibility is spread across 110 blocks in 18 districts. Figure 20.6 depicts flood-prone districts. Some of the indigenous strategies (Table 20.5) of West Bengal to mitigate the effect of flood are as follows.

Table 20.4 Strategies used by indigenous people of this region to manage water resource

Serial number	Indigenous methods	Brief description	Benefit
1	*Pat Kua* (The Well)	Semi-circular underground hole made of brick and cement, depth determined by subterranean water table, provides a reliable source of safe drinking water. *Pat Kuas* are owned by one to several households. *Pat Kua* systems can be open or closed. To prevent water contamination, wealthy farmers built covered *Pat Kuas*	Reliable source of safe drinking water
2	Pitcher watering system	Water filled jars labelled "*Kalsi*" are buried near plants, with bottom holes for soil hydration and plant watering, mitigates desiccation of plants	To mitigate the desiccation of plants
3	Embankment on river	Berms constructed along lower river sections to retain water for agricultural and domestic use during low river flow periods	Plant, grass, and other fodder crop growth
4	Hallowed Geyser (*Sita Kund*)	Geysers extract water from the groundwater table through cracks and fractures. The best-known natural geyser in Purulia is *Sita Kund* in the Baghmundi block of Ayodhya village, protected and used for ceremonies by the *Santals* group	Natural source of Water
5	Roof rain water harvesting system	They excavate a square pit that is fifty feet across next to their house. When it rains a lot, like during *Kal Baisakhi* (a heavy storm with rain) during the rainy season, water runs over the roof and into a pit made of dirt, where it can be used for at least three months	Constant access to water
6	Water storages beside the river	These water reserves, which are extremely useful to the community, appeared as a consequence of the soil being eroded. People would travel great distances from their houses to get to those freshwater depots so that they can use the water for various purposes, including drinking, bathing, and cooking	Availability of water in its natural state
7	Dari system	A tiny, circular *Dari* is dug alongside dry rivers semi-permanently to collect groundwater in summer. The river's upper bed is exposed to air in summer, but water still emerges through these man-made small holes. *Dari* sand filters water naturally. This is why it's used to make potable water	Good source of potable water

(continued)

Table 20.4 (continued)

Serial number	Indigenous methods	Brief description	Benefit
8	Aquifer	Naturally seeping water collects at a specific depth, ensuring minimal evaporation loss. This runoff usually forms a stream, providing water for irrigation or replenishing larger bodies of water	A reliable water supply in the driest of times
9	Naturally created water storages beside the river	The indigenous people of these regions have learned that the water storages generated by river bank soil erosion are a valuable resource. Many people trek far from their homes near the river to collect water for drinking, bathing, and other uses	Access to water during drought conditions
10	Goira and Doba system	In general, *Goira* and *Doba* are long-lasting or nearly permanent rainwater collection systems. *Goiras* are located close to human settlements and serve a variety of purposes, from fishing to bathing. *Doba*, on the other hand, varies in size and shape, and are perfectly suitable for animals	Access to water during drought conditions
11	Contour bunds	Water from runoff is distributed along the contour via a series of bunds. Bunds are built to reduce runoff, promote absorption, and channel any excess water to a centralized drainage system	Reduce runoffs
12	Hapa	Hapa, the huge dimple in the pond's middle. When it rains hard or if it seeps through the ground, this area quickly becomes saturated with water, which then gets caught among the stones and the dirt	Access to water during drought conditions

Adapted from Bauri et al. (2020), Sarkar and Modak (2023), Kar et al. (2023)

Bank Management

The Moribund Delta (Murshidabad and Nadia districts of West Bengal, India) and Mature Delta (Haringhata block of Nadia and the entire Bongaon, Barrackpur subdivision, the northern half of Barasat and the Basirhat subdivision of North 24 Parganas, Kolkata City, Alipore, and Baripur subdivisions) have alluvial soil. Due to its structural and textural soil, the region is prone to erosion, which is linked to flood. This strategy used locally adapted plant species to afforest river and pond banks to reduce erosion. Bamboo or wooden sticks can be used to build a protective barrier along the bank, packed with mud or sandbags. According to Tarafder and Debnath (2021), many indigenous people have lobbied for bamboo plants or bamboo or wooden stick barriers to protect river or pond banks from erosion. Mud or sandbags are sometimes used to fill barriers.

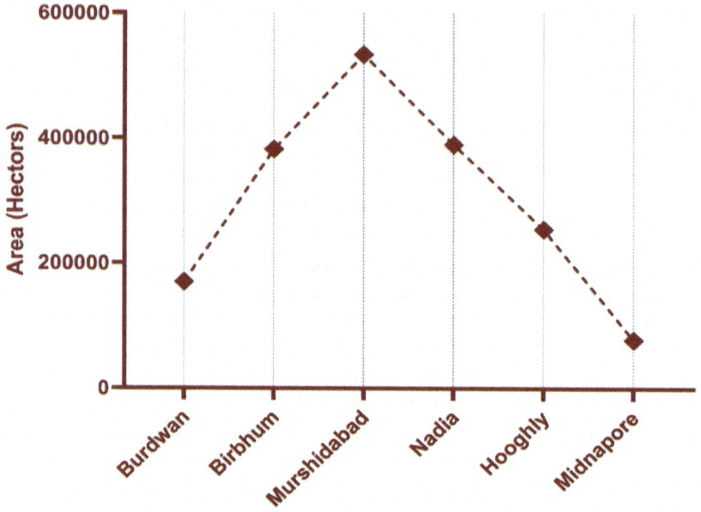

Fig. 20.6 Significant flood prone districts in West Bengal. Modified from Jha and Bairagya (2013)

Table 20.5 Some indigenous strategies of flood mitigation in West Bengal and their advantages

Serial No.	Strategies	Advantage
1	Bank management	1. Reduced soil loss, which is important for maintaining soil fertility and productivity 2. Improve water infiltration, which can help to reduce flooding and improve soil moisture retention 3. Increase soil stability, which can help to reduce landslides and other flood-related hazards 4. Reduce sediment pollution in rivers and streams, which can improve water quality and reduce the risk of flooding
2	Crop banking	1. To preserve the biodiversity of crops, improve food security 2. Reduce the risk of crop failure 3. Increase farmers' resilience to climate change
3	Buoyant Boats	1. Providing a safe and reliable means of transportation and shelter 2. Typically lightweight and easy to maneuver, making them ideal for navigating flooded areas 3. Used for a variety of purposes, including transportation, fishing, and rescue operations

A dynamic delta and a little beach are in Purba Medinipur, North and South 24 Parganas. The Bay of Bengal coastal plain has 24 Parganas (Paul and Chowdhury 2021). Digha is closest to the Bay of Bengal coast, whereas Kanthi is farthest. Maximum dunes are 11–12 m high and 15–16 km from the coast. During monsoon and cyclonic seasons, tides surged. Structure and texture also render soil prone to erosion (Mandal et al. 2019). Residents take precautions because cyclones and flooding are common. They've learned indigenous practices and skills that may lessen the probability of common calamities (Kurnio et al. 2021). They planted native trees and coconuts on river and pond banks to avoid erosion. Bamboo sticks and mud bags safeguard the bank (Tarafder and Debnath 2021). Framework and walls are bamboo, while sheds are Southern cattail leaves. A tent with a mat shelter them in rainy weather. Stalls are built above the ideal high-water mark. Beachgoers occasionally put up stalls on bamboo or wood platforms above high tide. The platforms are above water (Tarafder and Debnath 2021).

Crop Banking

Delta residents traditionally preserve their crops in their homes. A '*Gola*' (huge round-shaped mud ball) is built within the house to achieve this. A cylindrical wall of mud, wood, or concrete is strengthened with matting to form the '*Gola*'. Covering the '*Gola*' with tin or straw and fastening the hut to the base with nylon rope or iron wire protects it. This technique also uses a mud '*Kuthi*' (storehouse) (Tarafder and Debnath 2021). They make a '*Gola*' inside the home boundary to keep their crop on a muddy, wooden, or concrete platform with a cylindrical wall made of mat and tin or straw. Late in the year, they sell surplus grain.

Additional Flood Response Strategies

Banana stem boats are used in flood-prone southern Bengal. The boats transport people and safeguard household animals, boosting their survivability. Additionally, they use small wooden or wood-and-tin watercraft. In flood-prone locations, most households have modest boats. In rural South Bengal, especially the delta, bamboo or wooden bridges connect settlements separated by canals or rivers (Tarafder and Debnath 2021). Delta people wall ponds to preserve resources from monsoon flooding. Always between winter and summer, they empty the ponds, remove a horizontal layer of detritus, and lime the bed to maintain quality. A pump fills the pond, and pipelines remove the water without destroying the bank (Tarafder and Debnath 2021).

Climate change affects rising nations, especially the poor (O'Brien and Leichenko 2000). Climate change worsens droughts and floods in West Bengal, India (Clarke et al. 2022). Governments, international organizations, and NGOs assist in climate change adaptation (Linnerooth-Bayer and Mechler 2006).

The failure of Western technologies to align with growing national cultures has underscored the importance of indigenous wisdom, which encompasses various forms of local and traditional knowledge (Chimhowu et al. 2019). This includes indigenous agroecology, drought resilience, and flood management practices, which have become increasingly crucial in addressing contemporary challenges. Agroecology, practiced by indigenous West Bengalis, involves sustainable farming techniques such as terrace farming, crop rotation, and agroforestry, contributing to improved food security, soil health, and climate change adaptation (Sarker 2017).

Additionally, indigenous cultures in West Bengal provide valuable drought-resistant knowledge and employ traditional methods to enhance drought and flood resistance in the region (Ghosh 2019). Meanwhile, their understanding of local ecosystems and natural processes allows them to mitigate floods using methods like building buoyant boats and implementing early warning systems (Roy et al. 2023). Therefore, leveraging local knowledge can significantly enhance flood management in West Bengal.

Indigenous knowledge in climate change adaptation and mitigation, especially in West Bengal, has far-reaching implications. These findings can guide multi-level vulnerability policies for farm households and society. Here are some comprehensive discussions and potential policy interventions based on the study.

Agroecological Practices and Sustainable Agriculture

Agroecological practices created by West Bengali tribes can help farm households adapt to climate change and preserve agriculture (Rao et al. 2016). These methods emphasize local ecological knowledge, encourage biodiversity, soil health, and sustainable crop production. Indigenous agroecological methods like crop diversification and organic farming could be integrated into mainstream agricultural policies and programmes (Migliorini and Wezel 2017). This could involve supporting indigenous agricultural training and capacity building, promoting traditional seeds and regionally adapted crop types, and incentivizing sustainable farming through agricultural subsidies and extension services.

Drought Resilience and Water Management

Indigenous knowledge has provided effective measures for drought resilience, including traditional water harvesting techniques, water conservation practices, and drought-tolerant crop varieties (Mugambiwa 2018). Policy interventions could focus on integrating indigenous water management strategies into regional water resource management plans. This could involve the restoration of traditional water storage systems, the promotion of community-based water conservation practices, and investments in rainwater harvesting infrastructure (Katusiime and Schütt 2020).

Flood Control and Disaster Preparedness

The indigenous tribes of West Bengal have developed flood management techniques that are rooted in their deep understanding of local ecosystems and environmental patterns. One possible suggestion is to integrate indigenous flood control practices into regional disaster preparedness and management plans. This could involve the restoration of traditional flood-resilient infrastructure, such as embankments and water channels, as well as the implementation of community-based early warning systems and evacuation protocols (Munawar et al. 2021).

Cultural Sensitivity and Community Engagement

Indigenous knowledge must be integrated into climate change adaptation and mitigation strategies in a rigorous, culturally sensitive, and inclusive manner (Lam et al. 2020). Policy interventions could improve indigenous group alliances, include them in decision-making, and honor their ecological knowledge.

Ethical Considerations and Long-Term Sustainability

Integrating indigenous knowledge into climate change activities is practical and ethical (Hossain and Ballardini 2021). To ensure indigenous people receive equitable acknowledgment and reward for their knowledge contributions, policy proposals may include legal frameworks that safeguard and promote Indigenous intellectual property rights (Halewood 1998). This could involve creating culturally appropriate processes for resolving traditional knowledge and resource management issues and equitable sharing of indigenous knowledge advantages.

Policymakers, researchers, and communities can design comprehensive and holistic climate change strategies that address farm households' and society's multilevel vulnerabilities by translating these findings into actionable policy recommendations and interventions, which ultimately promote sustainable development.

Recommendations

a. Indigenous wisdom enables more effective, inclusive, and equitable climate change adaptation and mitigation methods, tapping into traditional skills (Salgotra et al. 2018).
b. Indigenous understanding of ecosystems, particularly climate change, can drive region-specific adaptation and mitigation techniques, enhancing climate impact models (Makondo and Thomas 2018).
c. Indigenous participation in climate projects supports inclusivity and equity, prioritizing their well-being and cultural practices (Black and McBean 2016).
d. Integration of indigenous wisdom promotes environmental justice and a reciprocal relationship with nature (Zimmerman 2005).
e. Collaboration with indigenous people fosters knowledge sharing and co-creation of contextually suitable climate solutions through mutual learning (Zimmerman 2005).

Conclusion

Integration of Indigenous knowledge into climate change adaptation and mitigation in West Bengal is essential for building a sustainable future (Fig. 20.7). Indigenous communities have devised agroecological, drought-resistant, and flood control methods to address climate change, which can benefit policymakers, scholars, and the populace in solving climate-related challenges. This approach acknowledges the wisdom and contributions of millennia-old residents (Al-Mansoori and Hamdan 2023). The inclusion of indigenous knowledge promotes comprehensive, culturally sensitive, and inclusive climate change solutions, benefiting the ecology and its dependent communities (Reid et al. 2013).

Theoretical Implications

This chapter shows how indigenous knowledge must be included into climate change adaptation and mitigation. It promotes indigenous tribes' agroecological, drought-resistant, and flood-control ways to counteract climate change. By showing indigenous societies' long-standing harmonious coexistence with nature, the portion stresses cautious observation and collaboration. This emphasizes the theoretical necessity of adopting indigenous knowledge in supporting sustainable environmental practices and protecting environmental-dependent cultures.

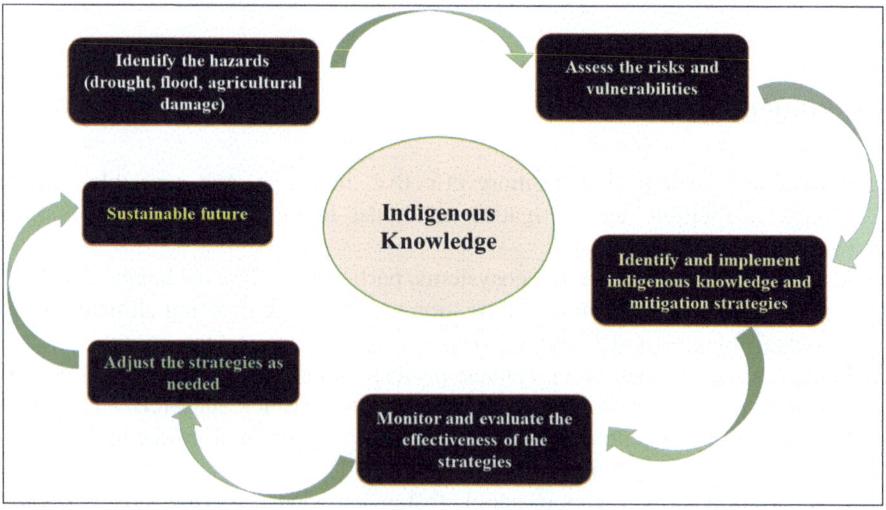

Fig. 20.7 Schematic representation of indigenous knowledge and their mitigation strategies of environmental hazards, that ultimately aimed at sustainable future. *Source* Author

Managerial Contributions

The manuscript highlights the managerial benefits of indigenous knowledge in West Bengal climate change measures. It suggests that indigenous knowledge can improve drought preparedness and agricultural sustainability. Indigenous flood management methods can improve flood resilience, protect vulnerable populations, and reduce economic and environmental consequences. The managerial contributions can increase climate resilience, sustainable agriculture, and catastrophe risk reduction by integrating indigenous knowledge.

Limitations and Future Research Perspectives

This chapter invites conversation and exploration in these areas. Indigenous knowledge integration into official climate change policy and its practical problems and opportunities should be studied in the future. In West Bengal, empirical research are needed to assess indigenous knowledge-based climate change adaptation and mitigation methods.

Further research could address indigenous intellectual property rights and the ethics of using traditional ecological knowledge in climate change programmes.

Acknowledgements We owe a great deal of gratitude to the indigenous populations in the areas where we conducted our research for their help and cooperation.

References

Adesina I, Bhowmik A, Sharma H, Shahbazi A (2020) A review on the current state of knowledge of growing conditions, agronomic soil health practices and utilities of hemp in the United States. Agriculture 10(4):129

Agarwal R, Garg PK, Garg RD (2013) Remote sensing and GIS based approach for identification of artificial recharge sites. Water Resour Manage 27(7):2671–2689

Al-Mansoori F, Hamdan A (2023) Integrating indigenous knowledge systems into environmental education for biodiversity conservation: a study of sociocultural perspectives and ecological outcomes. AI, IoT Fourth Ind Revol Rev 13(7):61–74

Altieri MA, Funes-Monzote FR, Petersen P (2012) Agroecologically efficient agricultural systems for smallholder farmers: contributions to food sovereignty. Agron Sustain Dev 32:1–13

Altieri MA, Nicholls CI, Montalba R (2017) Technological approaches to sustainable agriculture at a crossroads: an agroecological perspective. Sustainability 9(3):349

Amossé C, Jeuffroy MH, Mary B, David C (2014) Contribution of relay intercropping with legume cover crops on nitrogen dynamics in organic grain systems. Nutr Cycl Agroecosyst 98:1–14

Annual Flood Report 2019 (2020) Irrigation & Waterways Directorate, Government of West Bengal

Banerjee D, Chakraborty J, Samanta B, Dutta SB (2023) Determinants and spatio-temporal drivers of agricultural vulnerability to climate change at block level, Darjeeling Himalayan (Hill)

Region, West Bengal, India. In: Ecological footprints of climate change: adaptive approaches and sustainability. Springer International Publishing, Cham, pp 373–397

Bauri K, Gorai P, Modak BK (2020) Indigenous knowledge and practices on water conservation and management in Purulia district, West Bengal. J Environ Sociobiol 17(1):89–98

Bera A, Das S (2021) Water resource management in semi-arid Purulia district of West Bengal, in the context of sustainable development goals. In: Groundwater and society. Springer, Cham, pp 501–519

Biswas MK, Biswas SB (2014) Indigenous knowledge practiced by the farmers of Lateritic Belt of West Bengal in Crop protection. Res J Agric Sci 5(5):930–933

Black K, McBean E (2016) Increased indigenous participation in environmental decision-making: a policy analysis for the improvement of indigenous health. Int Indigen Policy J 7(4)

Borron S (2006) Building resilience for an unpredictable future: how organic agriculture can help farmers adapt to climate change. Food and Agriculture Organization of the United Nations, Rome

Buckingham S (2004) Synthesis report on cardamom cultivation. Fauna and Flora International Community-based Conservation in the Hoang Lien Mountains, Darwin Initiative of the UK Government in Vietnam, p 76

Chimhowu AO, Hulme D, Munro LT (2019) The 'New'national development planning and global development goals: Processes and partnerships. World Dev 120:76–89

Chowdhury S, Sontakke PP, Bhattacharjee J, Bhattacharjee D, Debnath A, Kumar A, Datta J (2015) An overview of edible insects in Eastern Himalayas: indigenous traditional food of Tribal's. Ecol Environ Conserv 21:1941–1946

Clarke B, Otto F, Stuart-Smith R, Harrington L (2022) Extreme weather impacts of climate change: an attribution perspective. Environ Res Climate 1(1):012001

Dangol DR (2008) Traditional uses of plants of common land habitats in Western Chitwan, Nepal. J Inst Agric Animal Sci 29:71

Das Gupta A (2013) Indigenous knowledge of Rajbanshi agriculturists of northern West Bengal, India. Viatcheslav Rudnev and Dorothy Billings, BH04, Indigenous knowledge and sustainable development (IUAES Commission on Indigenous Knowledge and Sustainable Development), University of Manchester, 5–10th Aug, p 65

Datta P, Behera B (2022) Factors influencing the feasibility, effectiveness, and sustainability of farmers' adaptation strategies to climate change in the Indian Eastern Himalayan Foothills. Environ Manage 70(6):911–925

De M (2014) Folk rice varieties and cultural heritage. Beats Nat Sci 1:1–7

De US, Dube RK, Rao GP (2005) Extreme weather events over India in the last 100 years. J Ind Geophys Union 9(3):173–187

Deb D (2017) Folk rice varieties, traditional knowledge and nutritional security in South Asia. In: Agroecology, ecosystems, and sustainability in the tropics, pp 117–134

Deka AM, Kalita H, Borah N, Zaman ASN (2019) Nutrient uptake and nutrient balance as influenced by different rice based cropping patterns in Assam. J Crop Weed 15:72–78

Deshpande RS, Narayanmoorthy A (2000) Appraisal of watershed development programme across regions in India. Gokhale Institute of Politics and Economics (GIPE), Pune (India). https://doi.org/10.21648/arthavij/1999/v41/i4/115909

Dhal NK, Pattanaik C, Sudhakar Reddy C (2010) Bakhar starch fermentation—a common tribal practice in Orissa. Ind J Trad Knowl 9:279–281

Elliott J (2012) An introduction to sustainable development. Routledge, London

Gebru H (2015) A review on the comparative advantages of intercropping to mono-cropping system. J Biol Agric Healthcare 5(9):1–13

Ghosh KG (2019) Spatial and temporal appraisal of drought jeopardy over the Gangetic West Bengal, eastern India. Geoenviron Disast 6:1–21

Ghosh-Jerath S, Singh A, Magsumbol MS, Kamboj P, Goldberg G (2016) Exploring the potential of indigenous foods to address hidden hunger: nutritive value of indigenous foods of Santhal tribal community of Jharkhand, India. J Hunger Environ Nutrit 11(4):548–568. https://doi.org/10.1080/19320248.2016.1157545

Giller KE, Delaune T, Silva JV, Descheemaeker K, van de Ven G, Schut AG, van Wijk M, Hammond J, Hochman Z, Taulya G, Chikowo R (2021) The future of farming: who will produce our food? Food Secur 13(5):1073–1099

Gorai P, Dey A, Modak BK (2022) Birhors of Purulia: marching toward mainstream of the society. In: Indigenous people and nature. Elsevier, Amsterdam, pp 29–60

Goswami A (2017) Analysis of meteorological drought and its trend in Bankura District of West Bengal, India. AFHABEC 1–6

Gray C, Mueller V (2012) Drought and population mobility in rural Ethiopia. World Dev 40(1):134–145

Gupta S, Sen P, Srinivasan S (2014) Impact of climate change on the Indian economy: evidence from food grain yields. Climate Change Econ 5(02):1450001

Haldar S, Saha P (2015) Identifying the causes of water scarcity in Purulia, West Bengal, India—a geographical perspective, IOSR. J Environ Sci Toxicol Food Technol 9:41–51

Halewood M (1998) Indigenous and local knowledge in international law: a preface to sui generis intellectual property protection. McGill LJ 44:953

Hossain K, Ballardini RM (2021) Protecting indigenous traditional knowledge through a holistic principle-based approach. Nordic J Human Rights 39(1):51–72

Hossain MZ, Islam MT, Sakai T, Ishida M (2008) Impact of tropical cyclones on rural infrastructures in Bangladesh. Agric Eng Int: CIGR J

Hutti B, Nijagunappa R (2011) Applications of geoinformatics in water resources management of semiarid region, North Karnataka, India. Int J Geomatics Geosci 2(2):342–373

IFRC (2005) World Disasters Report, Geneva; also, Parasuraman S, Unnikrishnan PV (2000) India Disasters Report. Oxford University Press, New Delhi

Jha CV, Bairagya H (2013) Flood and flood plains of West Bengal, India: a comparative analysis. Revista Geoaraguaia 3(1):1–10

Kar S, Jibanbandhu G, Khan KR, Satpati L, Mukherjee S (2023) Achieving rural water security through traditional knowledge: a comprehensive appraisal of traditional water harvesting techniques used in dry lands of Purulia District, West Bengal, India. In: Panda GK, Chatterjee U, Bandyopadhyay N, Setiawati MD, Banerjee D (eds) Indigenous knowledge and disaster risk reduction. Disaster risk reduction. Springer, Cham. https://doi.org/10.1007/978-3-031-26143-5_16

Katusiime J, Schütt B (2020) Integrated water resources management approaches to improve water resources governance. Water 12(12):3424

Kingwell-Banham E, Fuller DQ (2012) Shifting cultivators in South Asia: expansion, marginalisation and specialisation over the long term. Quatern Int 249:84–95

Kumar S, Das G, Shin HS, Patra JK (2017) *Dioscorea* sp. (a wild edible tuber): a study on its ethnopharmacological potential and traditional use by the local people of Similipal Biosphere Reserve, India. Front Pharmacol 8:52

Kumari A, Kumar A (2021) Assessment of genetic diversity of *Diospyros melanoxylon* Roxb. (kendu) growing in Jharkhand, India using RAPD markers. J Plant Sci Res 37(1):23–31

Kurnio H, Fekete A, Naz F, Norf C, Jüpner R (2021) Resilience learning and indigenous knowledge of earthquake risk in Indonesia. Int J Disaster Risk Reduc 62:102423

Lam DP, Hinz E, Lang D, Tengö M, Wehrden H, Martín-López B (2020) Indigenous and local knowledge in sustainability transformations research: a literature review. Ecol Soc 25(1)

Linnerooth-Bayer J, Mechler R (2006) Insurance for assisting adaptation to climate change in developing countries: a proposed strategy. Climate Policy 6(6):621–636

Makondo CC, Thomas DS (2018) Climate change adaptation: linking indigenous knowledge with western science for effective adaptation. Environ Sci Policy 88:83–91

Mandal UK, Maji B, Mullick S, Nayak DB, Mahanta KK, Raut S (2019) Global climate change and human interferences as risk factors, and their impacts on geomorphological features as well as on farming practices in Sundarbans eco-region. In: The Sundarbans: a disaster-prone eco-region: increasing livelihood security, pp 405–437

Mathison C, Deva C, Falloon P, Challinor AJ (2018) Estimating sowing and harvest dates based on the Asian summer monsoon. Earth Syst Dynam 9(2):563–592

Migliorini P, Wezel A (2017) Converging and diverging principles and practices of organic agriculture regulations and agroecology. A review. Agronomy Sustain Develop 37:1–18

Mondal M, Paul S, Bhattacharya S, Biswas A (2020) Micro-level assessment of rural societal vulnerability of coastal regions: an insight into Sagar Island, West Bengal, India. Asia-Pacific J Rural Develop 30(1–2):55–88

Mugambiwa SS (2018) Adaptation measures to sustain indigenous practices and the use of indigenous knowledge systems to adapt to climate change in Mutoko rural district of Zimbabwe. Jàmbá: J Disaster Risk Stud 10(1):1–9

Mukherjee D (2008) Association of medicinal plants with important tree species in hills of Darjeeling. Environ Ecol 26(4A):1697–1699

Mukherjee D (2012) Resource conservation through indigenous farming system in hills of West Bengal. J Crop Weed 8(1):160–164

Munawar HS, Khan SI, Anum N, Qadir Z, Kouzani AZ, Parvez Mahmud MA (2021) Post-flood risk management and resilience building practices: a case study. Appl Sci 11(11):4823

Nandargi SS, Barman K (2018) Evaluation of climate change impact on rainfall variation in West Bengal. Acta Sci Agric 2(7)

Narayana AC, Tatavarti R, Shakdwipe M (2005) Tsunami of 26 December 2004: observations on Kerala coast. J Geol Soc India 65(2):239–246

Nawrotzki RJ, Kadatska P (2010) Addressing climate change with indigenous knowledge. Int J Clim Change Impacts Responses 2(1):33–47

O'Brien KL, Leichenko RM (2000) Double exposure: assessing the impacts of climate change within the context of economic globalization. Glob Environ Chang 10(3):221–232

Olsson L, Opondo M, Tschakert P, Agrawal A, Eriksen S, Ma S, Perch L, Zakieldeen S (2014) Livelihoods and poverty. In: Climate change 2014 impacts, adaptation and vulnerability: Part A: global and sectoral aspects. Cambridge University Press, Cambridge, pp 793–832

Panda GK, Chatterjee U, Panda S (2023) Indigenous knowledge and disaster risk reduction: insight toward perception, response, adaptation and sustainability. In: indigenous knowledge and disaster risk reduction: insight towards perception, response, adaptation and sustainability. Springer International Publishing, Cham, pp 3–18

Paul, A. (2012). Indigenous rice: ignored in conventional agriculture. In: Crop culture: biotechnology biodiversity. Centre for Advanced Research and Education (CARE), Saha Institute of Nuclear Physics (SINP), Kolkata, pp 29–38

Paul S, Chowdhury S (2021) Investigation of the character and impact of tropical cyclone Yaas: a study over coastal districts of West Bengal, India. Safety Extreme Environ 3:219–235

Pomeroy JW, Stewart RE, Whitfield PH (2016) The 2013 flood event in the South Saskatchewan and Elk River basins: causes, assessment and damages. Canad Water Resour J/revue Canadienne Des Ressources Hydriques 41(1–2):105–117

Rao CS, Gopinath KA, Prasad JVNS, Singh AK (2016) Climate resilient villages for sustainable food security in tropical India: concept, process, technologies, institutions, and impacts. Adv Agron 140:101–214

Rao MR, Palada MC, Becker BN (2004) Medicinal and aromatic plants in agroforestry systems. In: New vistas in agroforestry: a compendium for 1st World Congress of Agroforestry. Springer, Netherlands, pp 107–122

Ray K, Giri RK, Ray SS, Dimri AP, Rajeevan M (2021) An assessment of long-term changes in mortalities due to extreme weather events in India: a study of 50 years' data, 1970–2019. Weather Climate Extremes 32:100315

Reid H, Faulkner L, Weisel A (2013) The role of community-based natural resource management in climate change adaptation in Ethiopia. International Institute for Environment and Development

Revi A (2008) Climate change risk: an adaptation and mitigation agenda for Indian cities. Environ Urban 20(1):207–229

Rockström J, Steffen W, Noone K, Persson Å, Chapin FS, Lambin EF, Lenton TM, Scheffer M, Folke C, Schellnhuber HJ, Nykvist B (2009) A safe operating space for humanity. Nature 461(7263): 472–475. https://doi.org/10.1038/461472a

Roy P, Pal SC, Chakrabortty R, Chowdhuri I, Saha A, Shit M (2023) Effects of climate change and sea-level rise on coastal habitat: vulnerability assessment, adaptation strategies and policy recommendations. J Environ Manage 330:117187

Salgotra RK, Zargar SM, Sharma M, Sood M (2018) Traditional knowledge: a therapeutic potential in the scenario of climate change for sustainable development. Development 61:140–148

Sarkar D, Kar SK, Chattopadhyay A, Rakshit A, Tripathi VK, Dubey PK, Abhilash PC (2020) Low input sustainable agriculture: a viable climate-smart option for boosting food production in a warming world. Ecol Ind 115:106412

Sarkar M, Modak BK (2023) Indigenous knowledge and strategic approaches to combat drought: a study from the Western Rarh Region (Bankura and Purulia Districts) in West Bengal. In: Indigenous knowledge and disaster risk reduction: insight towards perception, response, adaptation and sustainability. Springer International Publishing, Cham, pp 153–185

Sarker MNI (2017) An introduction to agricultural anthropology: pathway to sustainable agriculture. J Sociol Anthropol 1(1):47–52

Senanayake SGJN (2006) Indigenous knowledge as a key to sustainable development

Shah KK, Modi B, Pandey HP, Subedi A, Aryal G, Pandey M, Shrestha J (2021) Diversified crop rotation: an approach for sustainable agriculture production. Adv Agric 2021:1–9

Shahzad A, Ullah S, Dar AA, Sardar MF, Mehmood T, Tufail MA, Shakoor A, Haris M (2021) Nexus on climate change: agriculture and possible solution to cope future climate change stresses. Environ Sci Pollut Res 28:14211–14232

Shanabhoga MB, Krishnamurthy B, Suresha SV, Dechamma S (2020) Adaptation strategies by paddy-growing farmers to mitigate the climate crisis in Hyderabad-Karnataka region of Karnataka state, India. Int J Climate Change Strat Manage 12(5):541–556

Srinivas R, Singh AP, Shankar D (2020) Understanding the threats and challenges concerning Ganges River basin for effective policy recommendations towards sustainable development. Environ Dev Sustain 22:3655–3690

Steffen W, Richardson K, Rockström J, Cornell SE, Fetzer I, Bennett EM, Biggs R, Carpenter SR, De Vries W, De Wit CA, Folke C (2015) Planetary boundaries: guiding human development on a changing planet. Science 347(6223):1259855. https://doi.org/10.1126/science.1259855

Tarafder S, Debnath D (2021) Use of indigenous disaster adaptive practices to reduce risk in the southern part of West Bengal. Res Rev Int J Multidiscip 6(1):256–270

Tefera B, Sterk G (2010) Land management, erosion problems and soil and water conservation in Fincha'a watershed, western Ethiopia. Land Use Policy 27(4):1027–1037

Trenbath BR (1993) Intercropping for the management of pests and diseases. Field Crop Res 34(3–4):381–405

Yarmolinsky L, Huleihel M, Zaccai M, Ben-Shabat S (2012) Potent antiviral flavone glycosides from *Ficus benjamina* leaves. Fitoterapia 83(2):362–367

Zidny R, Sjöström J, Eilks I (2020) A multi-perspective reflection on how indigenous knowledge and related ideas can improve science education for sustainability. Sci Educ 29(1):145–185

Zimmerman EM (2005) Valuing traditional ecological knowledge: incorporating the experience of indigenous people into global climate change policies. NYU Envtl LJ 13:803

Chapter 21
Mobility as a Response to Urban Floods and Its Implications for Risk Mitigation: A Local Area Level Case Study from Guwahati, Assam

Upasana Patgiri, Premjeet Das Gupta, and Ajinkya Kanitkar

Abstract Detrimental effects of climate change and development pressures resulting in hazardous disasters have been a grave challenge of the current Anthropocene. Floods, a common global disaster, are posing a major threat to a large percentage of urban population due to increasing frequency and intensity. Prima facie, the inundations are usually short-term disruptions in case of urban floods. However, research indicates individuals alter their travel behaviour during floods. Mobility is an essential part of human life and when disrupted due to such events can prompt individuals to engage in risk-taking and evacuation behaviours, influenced by various socioeconomic as well as environmental factors. It is vital to comprehend human mobility patterns in both regular and unusual circumstances, from a purview of urban vulnerability, resilience, and sustainability. For this research, the travel behaviour study undertaken at a neighbourhood in Guwahati, Assam, analysed risk taking and evacuation behaviour during floods based on Mobility as a Response (MaaR) framework, comprising built environment, transportation, behavioural control, information-seeking and social response during floods. The primary survey involved collection of data through a structured questionnaire administered to 105 individuals. The analysis techniques include Chi-square and Spearman Rank Correlation tests. The findings suggest that risk taking, and relocation behaviour varies across different demographic and socioeconomic groups and the challenge of

U. Patgiri
School of Planning and Architecture, Bhopal, Neelbad Road, Village Bhauri, Bhopal, Madhya Pradesh 462030, India
e-mail: upasana0002@gmail.com

P. Das Gupta (✉)
Department of Urban and Regional Planning, School of Planning and Architecture, Bhopal, Neelbad Road, Village Bhauri, Bhopal, Madhya Pradesh 462030, India
e-mail: premjeet@spabhopal.ac.in

A. Kanitkar
RV College of Architecture, Banashankari 6th Stage, 4th Block, Bengaluru, Karnataka 560109, India
e-mail: ajinkyak.rvca@rvei.edu.in

U. Mukhopadhyay et al. (eds.), *Climate Crisis, Social Responses and Sustainability*, Climate Change Management, https://doi.org/10.1007/978-3-031-58261-5_21

accessing various transportation modes and a lack of viable alternative routes for vehicles and pedestrians show a significant correlation with both relocation and risk-taking behaviours.

Introduction

This chapter delves into the theme of Climate Crisis, Flood, and Social Responses and Risk Mitigation, using a local area level case study from Guwahati, Assam. It presents an exploration of human mobility related behaviour during urban floods, and context-specific recommendations based on the implications of the analysis regarding the safety of individuals.

Cities are becoming increasingly prone to environmental hazards, particularly those in developing countries, adding uncertainties to urban life, causing fatalities and destruction of assets. The link between society and environmental hazards involves studying the response of individuals to environmental risks, where human ecology plays an important role (Pelling 2003). The scientific study of human mobility in response to environmental hazards assumes great importance in this context (Lu et al. 2012).

Flooding is one of the most frequent natural catastrophes across the world and results in huge loss of lives every year. According to estimates, over 500,000 persons worldwide lost their lives to floods between 1980 and 2009 (World Health Organization 2014). Drowning is one the main causes of fatalities during floods (Doocy et al. 2013; Di Mauro et al. 2012). Extreme weather events due to global climate change are anticipated to cause an increase in the frequency and intensity of floods during the ensuing decades (Doocy et al. 2013; Easterling et al. 2000; Patz et al. 2000). In order to develop effective response strategies to mitigate the human risk during urban floods, it is necessary to comprehend how people behave in and around floodwaters (Hamilton et al. 2016a, b, 2018a, b).

Urban Flooding and Mobility Disruptions

A flood is defined as the presence of water in a place which is typically dry and encompasses both the flash flooding and slow-onset flooding features (Jonkman and Kelman 2005; Peden et al. 2017). Urban flooding differs greatly from rural flooding because urban development creates developed catchments, which raises flood peaks and volumes by up to six times. Due to quicker flow periods, flooding therefore happens rapidly in such cases (NDMA 2010). India has seen an increase in the frequency of urban flood disasters, which have severely impacted the country's major cities. Among these, Delhi in 2002 and 2003, Chennai in 2004, Mumbai in 2005, Surat in 2006, Kolkata in 2007, Jamshedpur in 2008, Delhi in 2009, and Delhi and Guwahati in 2010 are the most significant. The National Disaster Management Authority of India has defined urban floods as a separate category of flood disasters.

Mobility is an intrinsic part of human life, though its connotations may vary depending on the context, e.g., commuting, tourism, migration, etc. (Sheller and Urry 2006; Kwan 2018). The concept of mobility holds significant importance in understanding how severe events like urban flooding affect people and society as a whole. Human behaviour around floods can be affected by socio-demographic, intrapersonal, interpersonal and physical environmental variables (Drabek 2001; Salvati et al. 2018; Bempah and Øyhus 2017; Drobot et al. 2007; Tim et al. 2017; Diakakis and Deligiannakis 2013). There are many studies that have compared normal and flood situations to examine changes in travel behaviour, including travel time, mode of transportation, and route selection, and have shown that a number of factors related to individuals and their surroundings influence their travel behaviour and preferences (Khattak and De Palma 1997; Koetse and Rietveld 2009; Cools et al. 2010; Sakamoto and Fujita 2015).

Evidence suggests that experience of past hazards plays a role in mobility along with individual and household characteristics like gender and income (Gray and Mueller 2012; Hong et al. 2021; Zander and Garnett 2020). Another aspect related to mobility during environmental hazards is the decision to move or remain in the same place and is also subjected to socio-economic factors (Sullivan and Young 2020). It is known that individual travel behaviour and preferences during environmental hazards are influenced by the disruption to transport networks and systems and although studies have been done on this aspect, the decision-making process and coping behaviour in relation to mobility is not fully explored (Xia and Yeh 2022). The latter are manifested through mobility responses to environmental hazards, and the current practices of disaster mitigation can be improved by studying the spatio-temporal pattern of such responses.

Local Area Perspective of Urban Flooding and Mobility

Urban floods often affect certain parts of the city more adversely. As such, a local area level perspective is needed in the study of urban floods and mobility disruptions. The mobility of individuals and groups in a space–time framework have been the subject of interest of urban planners (Song et al. 2010; Simini et al. 2012). It is broadly understood that socio-demographic and socio-economic factors, attitudinal factors, lifestyle, etc., play a role in human mobility (Van Acker et al. 2010). However, Cresswell (2010) conceptualized it in terms of physical movement, meanings, practices and experiences, leading to a paradigm shift in the thinking on mobility.

For the purpose of policy formulation, it is crucial to identify the possible mobility responses as well as the social, economic, and environmental elements that increase their likelihood of happening (Sullivan and Young 2020). Mobility responses encompass decrease in trip distance, change in travel frequency, evacuation, and relocation, and can vary depending on the type of event and on whether the responses occur before, during or after the event (Ahmouda et al. 2019; Yabe and Ukkusuri 2019; Kamol et al. 2019; Wang et al. 2017; Zander and Garnett 2020; Han et al. 2019; Chen et al. 2020b). Xia and Yeh (2022) classified mobility responses to environmental hazards spatially and temporally. Spatially they can be classified into relocation from affected areas to unaffected areas, choosing destinations in unaffected areas over affected areas, and irregular pattern of movement within the affected area. Temporally they can be classified as movement before, during, and after the hazardous event.

Activity-related risk-taking behaviour and evacuation behaviour are two of the important types of human behaviour around floods (Hamilton et al. 2020). Risks are exacerbated when individuals choose to enter floodwaters. Crossing submerged streets, bridges, causeways, culverts on foot can have adverse consequences including injury and death (Coates 1999; Haynes et al. 2018; Ashley and Ashley 2008; Jonkman and Kelman 2005; Few and Pham 2010). Risk-taking behaviour may vary across gender and age (Shabanikiya et al. 2014; Ajibade et al. 2013). Evacuation behaviour before or after hazardous events can vary in different areas (Adeola 2009; Box et al. 2016). In developing nations there may be a high proportion of the population that does not relocate at all which could be due to various social reasons (Ajibade et al. 2015; Alam and Rahman 2014).

The focus of this chapter is on gaining a better understanding of the risk-taking and risk-aversion behaviours of individuals during urban floods to plan appropriate interventions at local area level based on the of concept of community resilience. Risk-taking behaviour in relation to urban floods is studied in terms of individual willingness to venture into floodwaters, while risk-aversion behaviour or evacuation behaviour is studied from the perspective of individual willingness towards long-term and short-term relocation. Income-related travel, travel for essential shopping, and education-related travel have been taken into consideration for this study. The two main objectives of the study are:

1. To study the travel behaviour of individuals with the focus on understanding their travel needs during urban flooding events at local area level.
2. To understand the variations in the risk-taking and risk-aversion behaviour of individuals during urban floods at local area level with respect to demographic, socio-economic factors and built-environment perceptions.

Based on the above two objectives, the chapter presents policy, planning and design interventions to ensure safer human mobility at a local area level within the broader context of disaster risk reduction, in line with the concept of community resilience. Research has shown that community-based disaster management programs have been effective in reducing the impact of disasters (Tariq et al. 2021). The study focuses on the crucial timespan from the onset of inundation in the area till mitigation measures are implemented. Throughout this time, the community plays a significant role in assisting one another.

Study Area

Guwahati is the capital city of Assam and the largest city in the North Eastern Region of India. It serves as an economic, industrial, and educational hub in the region. Situated alongside the Brahmaputra River and a multi hazard prone region with fragile geological base, the city is highly vulnerable to floods, landslides and earthquakes. It is frequently affected by urban flash floods during monsoon season which are primarily caused due to anthropogenic reasons (Sarmah and Das 2017).

Guwahati consists of six drainage basins, out of which the Bharalu basin forms the main drainage channel of the city (Sarmah and Das 2018). The Bharalu channel runs for 17 km with an average width of 5–9 m and an average depth of 3 m and is highly polluted (APCB 2013). Within this basin, the Radha Gobindo Baruah Road (R.G.B. Road) also known as Zoo Road, the Guwahati-Shillong Road (GS Road) and Rajgarh Road are the most flood-affected and therefore the triangular area enclosed by these roads have been chosen as the study area at a local area scale in Guwahati, consisting of the localities Anil Nagar, Nabin Nagar, Tarun Nagar and Rajgarh. The location and spatial characteristics of the sample area is shown in Figs. 21.1 and 21.2. The area for study is 206.19 ha and its total population is 28,048 (projected for 2021 using Census of India data for 1991, 2001 and 2011).

Considering it is one of the first settlement areas of Guwahati as well as the most flood affected, studies have been conducted in the area focussing on different aspects like socioeconomic factors that contribute to vulnerability to urban flooding through an indicator-based approach (Kashyap and Mahanta 2021) and hydrological modelling, land use analysis, and stakeholder consultations to develop the flood mitigation plan for the area, suggesting the implementation of an improved design

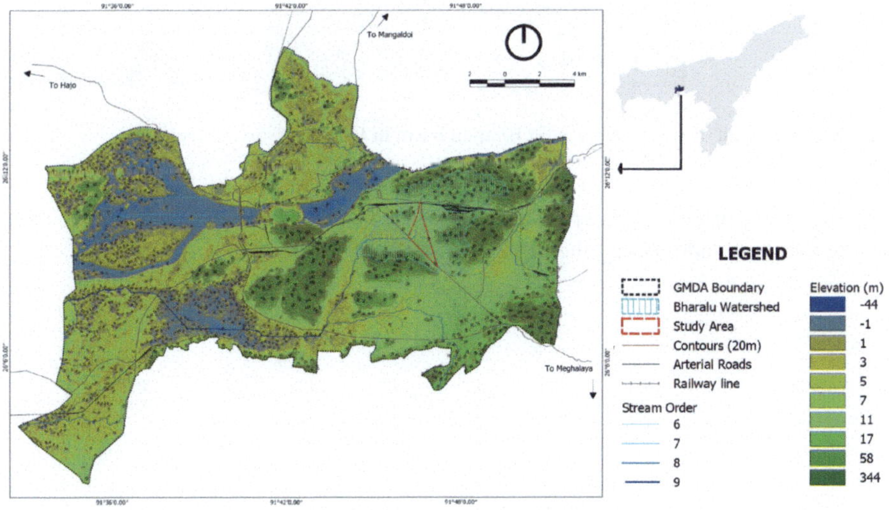

Fig. 21.1 Location of Bharalu basin in Guwahati, Assam. *Source* Author

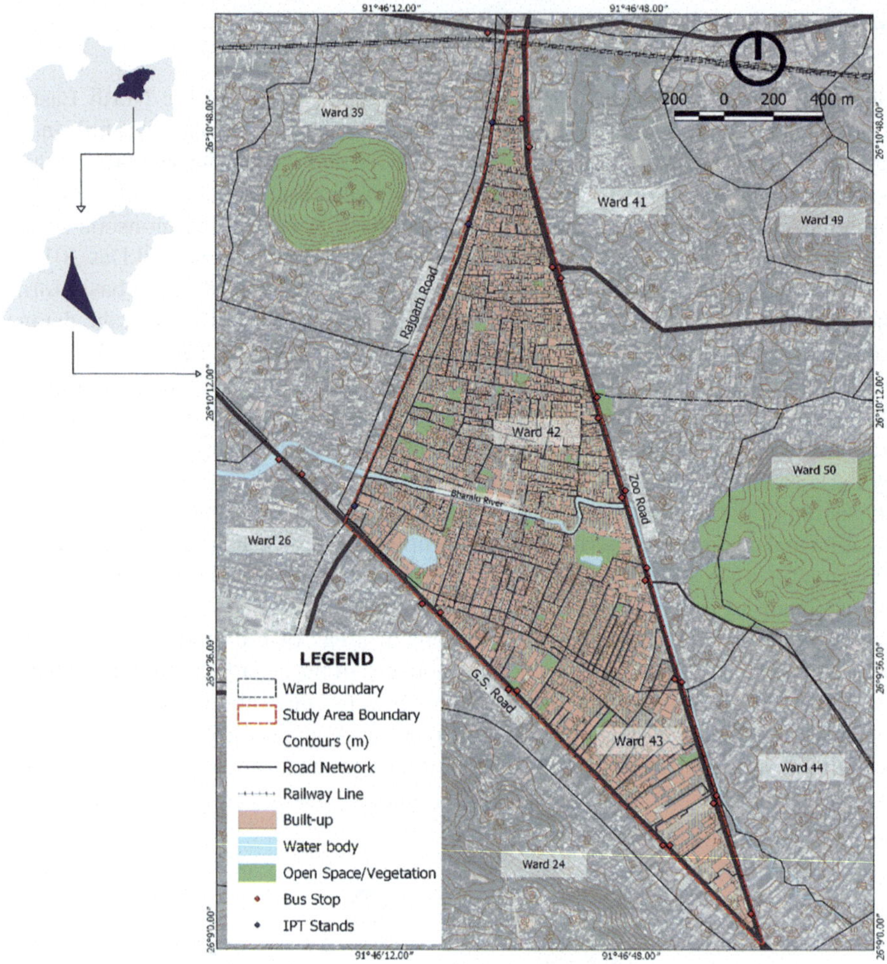

Fig. 21.2 Location of study area within Bharalu basin in Guwahati. *Source* Author

for stormwater drainage system (Sarmah and Das 2017). These have shown the need for interventions addressing the impacts of flooding in the area.

Rationale of the Study

'Mobility as a Response' Framework as the Context

This chapter draws on the 'Mobility as a Response' (MaaR) to environmental hazards framework proposed by Xia and Yeh (2022) in the urban context, encompassing concepts like averting actions from behavioural economics and risk perception from environmental psychology. Examples of averting behaviour include short or long-term behavioural reactions and adaptations to environmental stressors or dangers, and activities performed to limit environmental exposures or balance unwanted results (Cropper 1981; Neidell 2009; Dasgupta 2004; Dickie 2017; Freeman 1979). Environmental psychologists have used constructs such as value orientation, attitude, and intention to study environmental influences on human behaviour (Steg and De Groot 2019). Socio-economic attributes must also be considered in this context (Sun et al. 2017).

Within the MaaR framework, Xia and Yeh (2022) make a distinction between long-term residential migration, short term mobility, and immobility. Risk perception, attitude, residential satisfaction, and environmental considerations in relation to the present dwelling area are important psychological factors that play a part in residential migration, apart from socio-economic and socio-demographic factors, length of residence, etc. (Hunter 2005; Levy and Wadycki 1974; Rossi 1980; Sullivan and Young 2020; Ettema and Nieuwenhuis 2017). However, short-term movements, whether seasonal, weekly, or daily, are better ways to tackle negative environmental consequences than long-term movements like residential migration as it is much more flexible, and the opportunity costs are relatively low (Chen et al. 2020a). As a passive behavioural reaction to negative environmental effects, immobility refers to the continuing engagement of individuals in their typical and intended patterns of movement (Zivin and Neidell 2009). Individuals may also exhibit immobility, characterized by the absence of any behavioural reactions, when they are uninformed of or unconcerned with environmental hazards and risks.

Rationale for Study Area Selection

Building on the preceding discussion, the present study proposes an investigation of risk-taking and evacuation behaviours within the framework of Xia and Yeh's (2022) MaaR model, specifically as they relate to urban floods at a local area level. This research constitutes a pioneering effort in applying this framework to the context of urban flooding at a local area level.

Prior research by Kashyap and Mahanta (2021) established the socioeconomic vulnerability of the chosen study area, alongside three others in Guwahati, to urban floods using the Livelihood Vulnerability Index (LVI). Their work highlighted the impact of floods on households through income generation, education, and transportation disruptions. Transportation is a major area of concern for the area during floods. The roads become waterlogged, resulting in traffic congestion for longer hours or even for a few days. This can cause significant disruptions to the daily routines of residents, including commuting to work or school. A measure for quantifying the transportation disruption in the study area was used in the study, which indicated a high level of vulnerability for the area under consideration. Building on that, this chapter specifically investigates how recurring flood disruptions influence individual travel behaviours and actions within the case study area.

Materials and Methods

Case study method of research is used in this chapter, based on a primary survey of residents of the study area, conducted in the months of January and February, 2023. The area was not experiencing floods at that time and respondents shared experiences of past floods. The data collection tool was a structured questionnaire. Open ended interview was also conducted with selected participants. Given the population size of the area, the primary survey was administered to a relatively small number of 105 individuals, who were chosen for the survey through convenience sampling. As the study is mainly exploratory, and a case study method of research was used, this was deemed sufficient. However, considering the requirements of the study for activity related travel, it was administered only to individuals aged 25–65 who have lived in the area for at least a year. Following the coding and cleaning of the data, and the exclusion of incomplete responses, 101 samples were utilized for the analysis.

The questionnaire used for the primary survey consisted of four sections pertaining to individual and household characteristics, activity-travel characteristics, past experiences of flooding, and perception-based questions related to possible evacuation and relocation and risk-taking behaviours, and perceptions of the built environment and transport related perceptions among others. A six-point Likert scale was used for the perception-based questions related mainly to risk-taking and risk-aversion behaviours and responses collected through multivariate options of 1 = strongly agree, 2 = agree, 3 = somewhat agree, 4 = somewhat disagree, 5 = disagree, 6 = strongly disagree. Due to the small sample size and the six-point Likert scale having no neutral option, the responses were converted to binary, coded as 1 for all levels of agreement and 2 for all levels of disagreement for the analysis.

Both descriptive and inferential statistical methods were used to analyse and draw conclusions. Descriptive analysis was used for analyzing the travel behaviour pattern. To investigate the variation in the risk-taking and risk-aversion behaviour with respect to individual and household characteristics, activity-travel characteristics, etc., Chi-square (χ^2) test was applied using the IBM SPSS v25 software. The Chi-square test is a non-parametric test for examining variations between groups where the scale of the dependent variable is nominal. To analyze the correlation between perception of built environment (and transport-related factors within the built environment) and risk taking and relocation behaviour of individuals, the Spearman Rank Correlation test was performed. The value of Spearman Rank Correlation coefficient (ρ) ranges from -1 to 1. A higher value of the coefficient indicates a stronger correlation between variables, while a lower value suggests a weaker relationship. Additionally, qualitative insights and inferences were drawn from the surveys, from the flood related experiences of the respondents and the factors influencing the likelihood of relocation or risk-taking behaviour.

Results and Discussion

Demographic characteristics in the surveys indicate 54% of the respondents have resided in the area for more than 10 years, 26% for 6–10 years, and 20% for 1–5 years. 56% of the respondents are male and 44% are female. None of the respondents identified themselves other than male or female, therefore no third-gender was included. 37% of the respondents belong to the age group of 25–35 years. 23% belong to the age group of 36–45 years, 28% to 46–55 years and 12% to the 56–65 years. 78% of the respondents have an education level of graduation or above.

As for Household (HH) size, 37% have a HH size of 4, 25% of 3, for 24% it is 5 or more, and 14% have a HH size of 2 or less. A majority 66% of the respondents have a HH income of 50,000 and above. Vehicle ownership varied greatly with majority respondents owning 1 four-wheeler.

The first part of this section focuses on the travel behaviour of individuals during floods, the analysis of the disruptions faced by them in their work trips, essential shopping trips, and educational trips. It also analyses the information-seeking behaviour of individuals during floods and how they practically cope with mobility issues.

The second part focuses on variations in risk-taking and risk-aversion (relocation) behaviour of individuals during urban floods. It discusses correlation between risk-taking and risk-aversion (relocation) behaviour of individuals with built-environment and transport-related perceptions.

Travel Behaviour of Individuals During Urban Flooding Events

This section focuses on the first objective of the study which revolves around the need of individuals to travel for various reasons. Considering that travelling for discretionary activities is most likely to be avoided during floods, obligatory trips undertaken for income related work, education-related and essential shopping were considered. The respondents were therefore asked to provide information, both during normal and flood conditions pertaining to these three types of trips.

Under normal conditions, it was observed that 59% of the respondents travel for income related activity, with the majority of them travelling by personal vehicles primarily within 1–5 km. 94% of the respondents travel for essential shopping, with the majority travelling on foot within a distance of 1 km. For education related purposes, 29% of the respondents reported to travel, majority of whom travel by their personal vehicle within a distance of 5–10 km. Overall it was observed that all of the respondents travel on a regular basis for at least one of the purposes, with the majority of them travelling for essential shopping primarily within the neighbourhood.

To analyse how the travel behaviour of the individuals differ under flooded conditions, the level of disruption faced by them in order to carry out the trips was asked in terms of 'Able to perform most trips', 'Able to perform some trips' and 'Not able to perform any trips'. Cross tabulation analysis was done to see how the level of disruption faced differs with changes in the distance they have to travel for each of the trip purposes as well as the level of inundation they face. This leads into investigating whether they continue to carry out the travel needs even under flooded conditions.

For the work trips, it can be seen that the percentage of respondents unable to perform any trips is highest for those who have to travel within 1 km from their home, mainly because the entire area remains flooded and generally the places of work also are closed during floods (Fig. 21.3). The percentage of people unable to perform any trips increases as the inundation depth increases (Fig. 21.4).

For essential shopping trips, it can be observed that most people are required to perform some trips during flooded times, buying essentials (Fig. 21.5) especially for drinking water due to power outages. When compared with level of inundation, a trend similar to work trips can be observed (Fig. 21.6).

For education related trips, some inconsistency can be observed, which may be attributable to the fact that a smaller number of respondents are engaged in such trips. However, it can be observed from Fig. 21.7 that people choose to not send their children to schools during floods; which also reflects in Fig. 21.8, that the percentage of individuals not able to perform or generally avoid educational related trips is

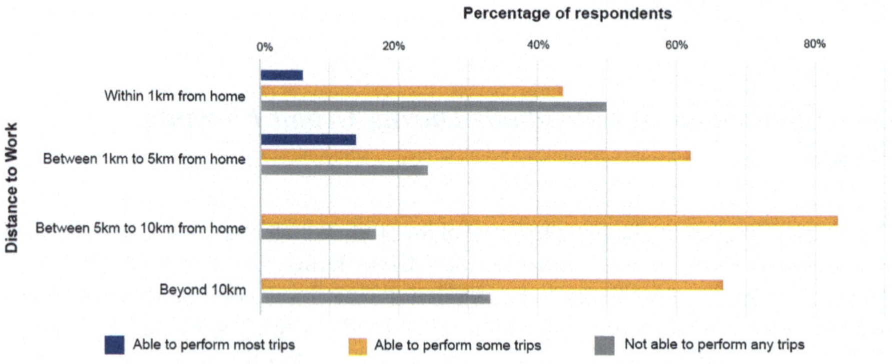

Fig. 21.3 Disruption in work trips—level of disruption versus travel distance. *Source* Primary Survey

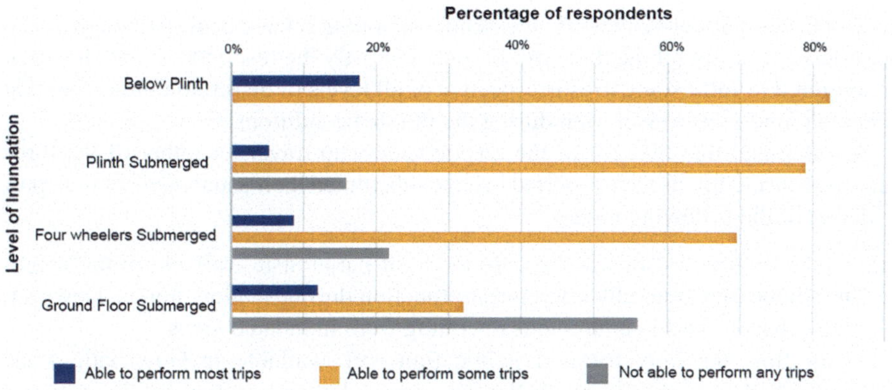

Fig. 21.4 Disruption in work trips—level of disruption versus level of inundation. *Source* Primary Survey

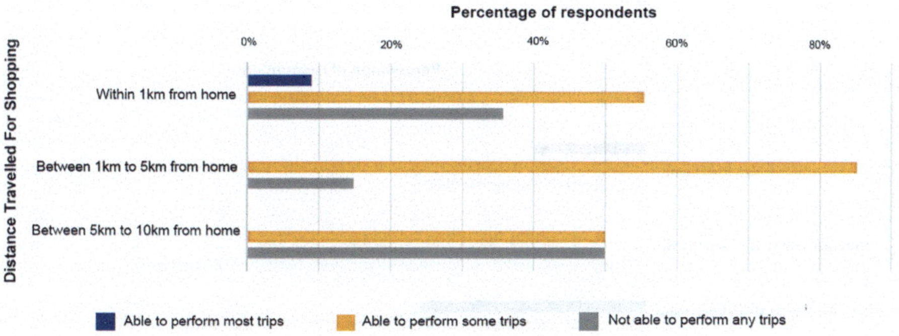

Fig. 21.5 Disruption in shopping trips—level of disruption versus travel distance. *Source* Primary Survey

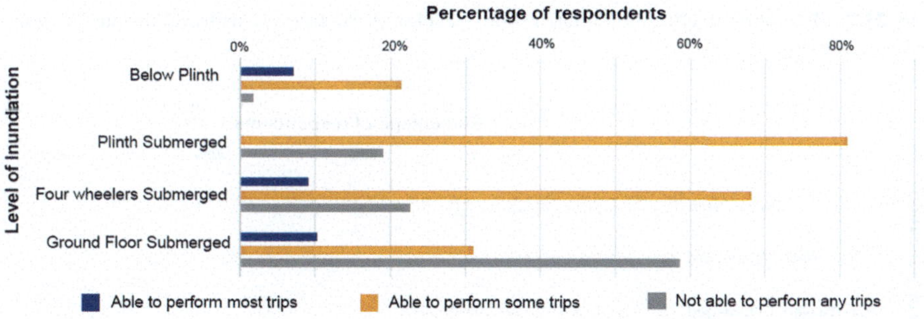

Fig. 21.6 Disruption in shopping trips—level of disruption versus level of inundation. *Source* Primary Survey

highest in comparison with work or essential shopping related trips. Although, it can be observed that for all the three trip purposes, mostly the respondents are able to or are required to fulfil some trips irrespective of all levels of disruption indicating that people do undertake travel even during the flooded conditions.

It was found that 50.54% of the respondents who travel for either of the three activities during floods have reported to have substituted their usual mode of transport with one of the following modes:

i. Cycle Rickshaws: this is the only mode of paratransit available in the neighbourhood and generally continues to function during shallow floods. However, they charge exorbitant amounts operating from inundated areas.
ii. City Bus: the only form of public transport available in Guwahati, if the individuals are able to reach the bus stops, choose to travel by them during floods.
iii. On Foot: individuals also choose to wade through the floodwaters on foot as well in the lack of any other option or to access them nearby.

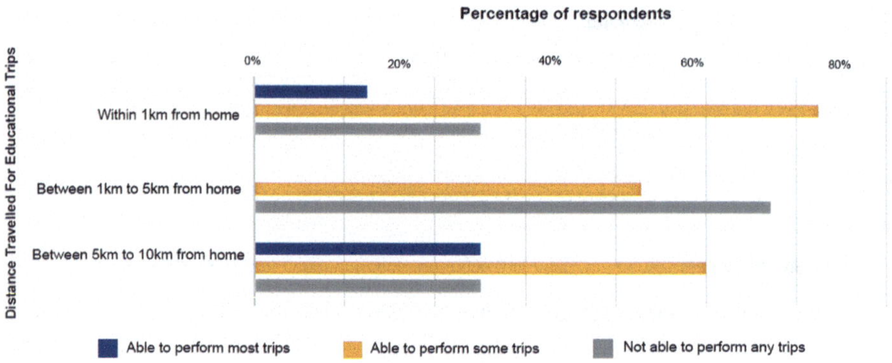

Fig. 21.7 Disruption in education related trips—level of disruption versus travel distance. *Source* Primary Survey

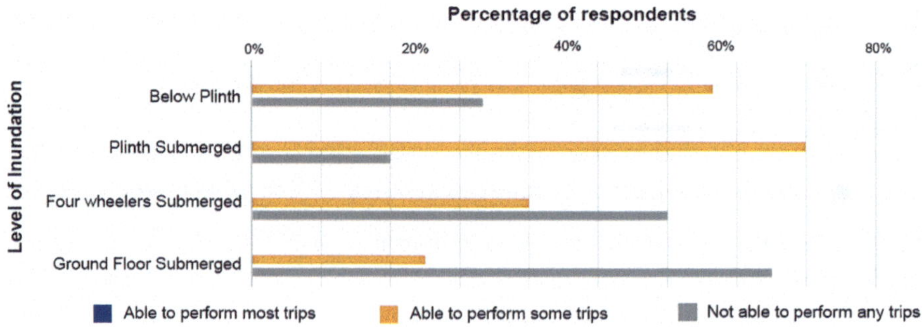

Fig. 21.8 Disruption in education related trips—level of disruption versus level of inundation. *Source* Primary Survey

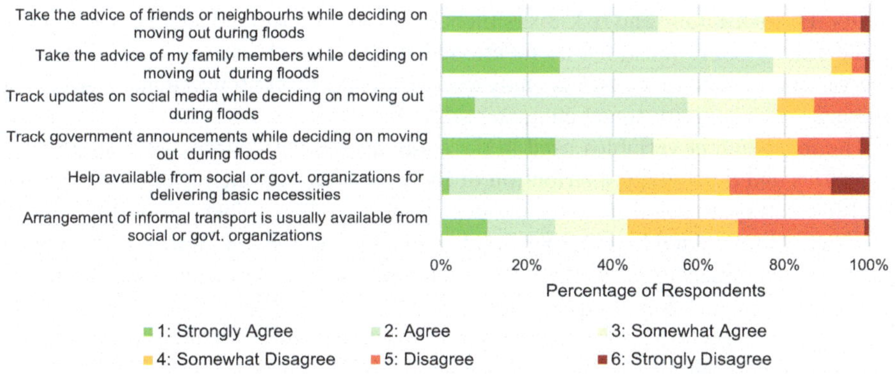

Fig. 21.9 Information seeking and coping with mobility during floods. *Source* Primary Survey

iv. Rescue Boats/Rafts: rescue boats operated by the National Disaster Response Force (NDRF) and State Disaster Response Force (SDRF) along two roads in the highly inundated locality of Anil Nagar are generally operated during evacuation or supply of essential goods at inundated areas.

It also came to light that other than the three trip purposes discussed previously, other travel needs which arise during floods include travelling to get electronic devices charged in case of power cuts in the area, to provide help to others in the neighbourhood and to temporarily stay elsewhere.

Besides the exhibited travel behaviour, it is also necessary to study the information-seeking behaviour of individuals that gets reflected in the actual travel behaviour. Access to correct information during flood events is of prime importance. Through the Likert scale questions, it was aimed to understand the existing norm of information seeking of the residents of the selected area before they decide to enact upon traveling. Also, it was attempted to find the existing coping methods to know whether or not government or social organisations provide help to the residents to fulfil their travel needs during floods.

As observed from Fig. 21.9, majority show some level of agreement on seeking information from different sources including neighbours, friends, family members and social media, regarding mobility related decisions. High disagreement regarding unavailability of help from social or government organizations during floods indicates gaps in providing help for delivering basic necessities or alternate modes of transport.

Variations in Risk-Taking and Risk-Aversion Behaviour of Individuals During Urban Floods

This section focuses on the second objective of the study that is to explore factors impacting risk taking and risk aversion behaviour of individuals during flooding events. The risk-taking and risk-aversion behaviours are assessed using the responses to the Likert scale questions as mentioned in Section "Materials and Methods". The variation in the responses is analyzed with respect to individual and household characteristics, activity-travel characteristics, etc.

The independent variables used in the study are shown in Table 21.1 along with categories and percentage distribution of the sample across the different categories used for testing associations with the dependent variables related to willingness to relocate and take risks by entering floodwaters. Gender, age, monthly HH income, duration of residence, dependent HH members are the variables considered for socio-economic and demographic study. For activity-travel behaviour, categories are income related travel (59.4% of respondents) and non-income related travel (40.6% of respondents). To analyse risk-taking behaviours involving entering floodwaters, the study considered the inundation level and duration of the flood separately. Relocation is considered as part of risk aversion behaviour.

Chi-square test was performed between the independent variables and the Likert scale questions related to their willingness to relocate both temporarily and permanently due to floods as well as their willingness to enter floodwaters (Tables 21.2 and 21.3). Due to the small sample size (n = 101) and the six-point Likert scale having no neutral option, the responses were converted to binary, coded as 1 for all levels of agreement and 2 for all levels of disagreement.

Gender is found to have significant association with the willingness to move to another neighbourhood in the future due to flood related issues ($\chi^2 = 5.119$, $p < 0.05$), with a higher percentage of males showing agreement and higher percentage of women disagreeing. Age shows significant association with some of the temporary relocation questions, with the permanent relocation question, and also with some of the questions on risk-taking by venturing into floodwaters. It has significant association with contemplating temporary relocation to a government shelter due to floods in the neighbourhood ($\chi^2 = 3.451$, $p < 0.10$), and contemplating temporary relocation to a hotel, lodge or other similar paid accommodation due to floods in the neighbourhood ($\chi^2 = 4.723$, $p < 0.05$). It also has significant association with the opinion regarding the effectiveness of temporary relocation before floods ($\chi^2 = 6.076$, $p < 0.05$) and during floods ($\chi^2 = 4.637$, $p < 0.05$), and also the willingness to move to another neighbourhood permanently in the future due to flood related issues ($\chi^2 = 13.040$, $p < 0.01$). As for risk-taking behaviours, age shows significant association with willingness to enter flood waters to ensure the livelihood of the household ($\chi^2 = 3.349$, $p < 0.10$) as well as to ensure the well-being of other household members ($\chi^2 = 5.839$, $p < 0.05$) when the duration of inundation is above the perceived average level.

Household income shows a significant association with wanting to relocate temporarily to a friend's/neighbour's place outside the neighbourhood due to floods ($\chi^2 = 17.419$, $p < 0.001$) and willingness to move to another neighbourhood in the future due to flood related issues in the neighbourhood ($\chi^2 = 6.145$, $p < 0.05$), with the majority in the relatively lower income group showing agreement. For risk-taking, it shows significant associations with willingness to enter floodwaters to ensure both livelihood ($\chi^2 = 2.777$, $p < 0.10$) and well-being of other household members (χ^2

Table 21.1 Sample description (sample size = 101)

Variable	Category	Percentage (%)
(a) Gender	Female	43.6
	Male	56.4
(b) Age (years)	25–45	60.4
	46–65	39.6
(c) Monthly HH income (INR)	Upto 50,000	33.7
	50,000 and above	66.3
(d) Duration of residence in the neighbourhood	1–10 years	45.5
	More than 10 years	54.5
(e) Dependent household member	Present	43.6
	Not present	56.4
(f) Activity-travel behaviour	Engaged in income-related work	59.4
	Not engaged in income-related work	40.6
(g) What is the average level of inundation in the neighbourhood as perceived by you?	Entire ground floor of most buildings was submerged	30.7
	Ground floor of most buildings are not submerged but four-wheelers parked on streets were submerged	24.7
	Four-wheelers parked on streets were not submerged but plinth of most buildings are submerged	22.8
	Water level stayed below plinth level of most buildings	21.8
(h) What is the average duration of inundation in the neighbourhood as perceived by you?	A week or more	4.9
	Two–three days	49.5
	One day	28.7
	Few hours/less than a day	16.9

$= 3.403, p < 0.10$) when the level of inundation is above the perceived average level. In this case there is a higher percentage towards agreement in both income groups.

With respect to duration of residence, significant association is seen with wanting to relocate temporarily to a friend's/neighbour's place outside the neighbourhood ($\chi^2 = 9.102$, $p < 0.01$) and preference to move to another neighbourhood in the future ($\chi^2 = 4.897$, $p < 0.05$), with the majority in the lesser duration group (1–10 years of residence) agreeing. As for presence of a dependent member in the household, no significant association was found with any relocation behaviour. In case of risk-taking behaviour, significant association was found with willingness to enter floodwaters to ensure the livelihood of the household when the duration of inundation is above the perceived average level ($\chi^2 = 4.609, p < 0.05$). Activity-travel behaviour, defined in terms of the requirement to travel for income-related work

Table 21.2 Percentage distribution of responses regarding risk-taking and risk aversion (relocation) behaviour

	I have considered relocating temporarily to a friend's/ neighbour's place in the neighbourhood due to floods (%)		I have considered relocating temporarily to a friend's/ neighbour's place outside the neighbourhood due to floods (%)		I have considered relocating temporarily to a government shelter due to floods in the neighbourhood (%)		I have considered relocating temporarily to a hotel, lodge or other similar paid accommodation due to floods in the neighbourhood (%)		Temporary relocation is an effective option before floods (%)		Temporary relocation is an effective option during floods (%)	
	Yes	No	Yes	No	Yes	No	Yes	No	Yes	No	Yes	No
Gender												
Female	34	66	41	59	32	68	25	75	18	82	70	30
Male	30	70	47	53	44	56	33	67	32	68	81	19
Age (years)												
25–45	31	69	51	49	46	54	38	62	34	66	69	31
46–65	33	68	35	65	28	73	18	83	13	88	88	13
Monthly HH income (INR)												
Upto 50,000	32	68	74	26	35	65	29	71	32	68	82	18
50,000 and above	31	69	43	70	40	60	30	70	22	78	73	27

(continued)

Table 21.2 (continued)

	I have considered relocating temporarily to a friend's/ neighbour's place in the neighbourhood due to floods (%)		I have considered relocating temporarily to a friend's/ neighbour's place outside the neighbourhood due to floods (%)		I have considered relocating temporarily to a government shelter due to floods in the neighbourhood (%)		I have considered relocating temporarily to a hotel, lodge or other similar paid accommodation due to floods in the neighbourhood (%)		Temporary relocation is an effective option before floods (%)		Temporary relocation is an effective option during floods (%)	
	Yes	No	Yes	No	Yes	No	Yes	No	Yes	No	Yes	No
Duration of residence												
1–10 years	35	65	61	39	37	63	28	72	30	70	70	30
More than 10 years	29	71	31	69	40	60	31	69	22	78	82	18
Dependent household member												
Present	34	66	52	48	41	59	34	66	27	73	70	30
No present	30	70	39	61	37	63	26	74	25	75	81	19
Activity-travel												
Engaged in income-related travel	37	63	52	48	35	65	28	72	23	77	85	15
Not engaged in income-related travel	24	76	34	66	44	56	32	68	29	71	63	37

(continued)

Table 21.2 (continued)

	Temporary relocation is an effective option after floods (%)		I would prefer to move to another neighbourhood in the future due to flood related issues in this neighbourhood (%)		I am willing to enter flood waters to ensure the livelihood of the household when the level of inundation is above the perceived average level (%)		I am willing to enter flood waters to ensure the well-being of other household members when the level of inundation is above the perceived average level (%)		I am willing to enter flood waters to ensure the livelihood of the household when the duration of inundation is above the perceived average level (%)		I am willing to enter flood waters to ensure the well-being of other household members when the duration of inundation is above the perceived average level (%)	
	Yes	No	Yes	No	Yes	No	Yes	No	Yes	No	Yes	No
Gender												
Female	14	86	32	68	68	32	61	39	73	27	82	18
Male	26	74	54	46	81	19	72	28	82	18	86	14
Age (years)												
25–45	25	75	59	41	70	30	62	38	72	28	77	23
46–65	15	85	23	78	83	18	75	25	88	13	95	5
Monthly HH income (INR)												
Upto 50,000	29	71	62	38	85	15	79	21	82	18	82	18
50,000 and above	16	84	36	64	70	30	61	39	76	24	40	15

(continued)

Table 21.2 (continued)

	Temporary relocation is an effective option after floods (%)		I would prefer to move to another neighbourhood in the future due to flood related issues in this neighbourhood (%)		I am willing to enter flood waters to ensure the livelihood of the household when the level of inundation is above the perceived average level (%)		I am willing to enter flood waters to ensure the well-being of other household members when the level of inundation is above the perceived average level (%)		I am willing to enter flood waters to ensure the livelihood of the household when the duration of inundation is above the perceived average level (%)		I am willing to enter flood waters to ensure the well-being of other household members when the duration of inundation is above the perceived average level (%)	
	Yes	No	Yes	No	Yes	No	Yes	No	Yes	No	Yes	No
Duration of residence												
1–10 years	26	74	57	43	76	24	72	28	74	26	80	20
More than 10 years	16	84	35	65	75	25	64	36	82	18	87	13
Dependent household member												
Present	23	77	52	48	77	23	70	30	68	32	82	18
No present	19	81	39	61	74	26	65	35	86	14	86	14
Activity-travel												
Engaged in income-related travel	22	78	45	55	88	12	80	20	88	12	88	12
Not engaged in income-related travel	20	80	44	56	56	44	49	51	63	37	78	22

Table 21.3 Chi-square test results

	Considered relocating temporarily to a friend's/neighbour's place in the neighbourhood due to floods		Considered relocating temporarily to a friend's/neighbour's place outside the neighbourhood due to floods		Considered relocating temporarily to a government shelter due to floods in the neighbourhood		Considered relocating temporarily to a hotel, lodge or other similar paid accommodation due to floods in the neighbourhood		Temporary relocation is an effective option before floods		Temporary relocation is an effective option during floods	
	χ^2	p	χ^2	p	χ^2	p	χ^2	p	χ^2	p	χ^2	p
Gender	0.209	0.648	0.419	0.517	1.519	0.218	0.826	0.363	2.332	0.127	1.439	0.230
Age	0.200	0.886	2.447	0.118	3.451*	0.063	4.723**	0.030	6.076**	0.014	4.637**	0.031
Monthly household income	0.011	0.918	17.419***	< 0.001	0.238	0.625	0.002	0.964	1.172	0.279	1.058	0.304
Duration of residence	0.375	0.540	9.102**	0.003	0.980	0.754	0.084	0.772	0.973	0.324	2.076	0.150
Dependent household member	0.209	0.648	1.880	0.170	0.173	0.677	0.719	0.396	0.095	0.757	1.439	0.230
Activity-travel	1.696	0.193	3.026*	0.082	0.814	0.367	0.133	0.716	0.449	0.503	6.264**	0.012

(continued)

Table 21.3 (continued)

	Temporary relocation is an effective option after floods		Would prefer to move to another neighbourhood in the future due to flood related issues in this neighbourhood		Willing to enter flood waters to ensure the livelihood of the household when the level of inundation is above the perceived average level		Willing to enter flood waters to ensure the well-being of other household members when the level of inundation is above the perceived average level		Willing to enter flood waters to ensure the livelihood of the household when the duration of inundation is above the perceived average level		Willing to enter flood waters to ensure the well-being of other household members when the duration of inundation is above the perceived average level	
	χ^2	*p*	χ^2	*p*	χ^2	*p*	χ^2	*p*	χ^2	*p*	χ^2	*p*
Gender	2.424	0.119	5.119**	0.024	2.090	0.148	1.260	0.262	1.380	0.240	0.320	0.571
Age	1.349	0.245	13.040***	< 0.001	1.870	0.171	1.773	0.183	3.349*	0.067	5.839**	0.016
Monthly household income	2.312	0.128	6.145**	0.013	2.777*	0.096	3.403*	0.065	0.514	0.473	0.125	0.723
Duration of residence	1.438	0.230	4.897**	0.027	0.032	0.858	0.748	0.387	0.919	0.338	0.879	0.349
Dependent household member	0.177	0.674	1.880	0.170	0.172	0.679	0.347	0.556	4.609**	0.032	0.32	0.571
Activity-travel	0.069	0.793	0.012	0.913	13.589***	< 0.001	10.792**	0.001	8.877**	0.003	1.932	0.165

Note χ^2 denotes the Chi-Square value and *p* denotes the level of significance
*** Significance: < 0.001; ** Significance: < 0.05; * Significance: < 0.1

or not, shows significant associations with some of the relocation and risk-taking behaviours. Significant association is seen with contemplating temporary relocation to a friend's/neighbour's place outside the neighbourhood ($\chi^2 = 3.026, p < 0.10$) and with the opinion that temporary relocation is an effective option during floods ($\chi^2 = 6.264, p < 0.05$), with more agreement being shown by those engaged in income-related travel in both cases. As for risk taking behaviours, significant associations were found with willingness to enter flood waters for livelihood-related purposes ($\chi^2 = 13.589, p < 0.01$) as well as to ensure the well-being of other household members ($\chi^2 = 10.792, p < 0.01$) when the level of inundation is above the perceived average level. There is a significant association with willingness to enter flood waters for livelihoods also when the duration of inundation is above the perceived average level ($\chi^2 = 8.877$, $p < 0.01$). Here also higher levels of agreement are shown by those engaged in income-related travel.

Spearman Rank correlation test was performed as mentioned in Section "Materials and Methods", to understand which perceptions and thereby characteristics of the built environment correlate to the risk taking and relocation behaviour. Positive correlations signify a consistent alignment of variables, indicating that when one variable increases, the other variable also increases, and when one variable decreases, the other variable follows suit. On the other hand, negative correlations indicate an inverse relationship, where an increase in one variable corresponds to a decrease in the other variable, and vice versa.

Based on the data presented in Fig. 21.10, it can be observed that there are significant correlations ranging from – 0.3 to 0.5. These correlations indicate a weak to a moderate positive relationship. Temporary relocation to a known person's residence, both within and outside the neighbourhood as well as relocating to paid accommodations like hotels or lodges, is found to be positively correlated with facing difficulties in reaching the nearest public transport stops. Additionally, relocating outside the neighbourhood is also correlated with challenges in using various transportation modes and receiving home deliveries. This suggests that individuals opt to move to someone else's house when they are unable to meet their travel requirements from their own place of residence. This finding aligns with the results of the first objective of the study, indicating that people continue to travel and are willing to relocate in order to sustain their travel activities.

Relocating to government shelters exhibits a positive correlation with the presence of alternate walking and driving routes, indicating that the absence of travel options leads to temporary relocation. However, the negative correlation with accessing para-transit at the doorstep and ride-hailing services like Ola and Uber suggests that the unavailability of transportation modes to reach flood relief shelters often deters people from utilizing them, even if they are available. A similar pattern is observed in the case of relocating to paid accommodations, indicating a reluctance to move to such places when transportation modes are scarce. This may be because most hotels within the area are prone to flooding, and those located outside require transportation for access, with an average distance of approximately 3–4 km.

The opinions of the survey respondents regarding the effectiveness of temporary relocation before, during, and after floods mostly demonstrate positive correlations

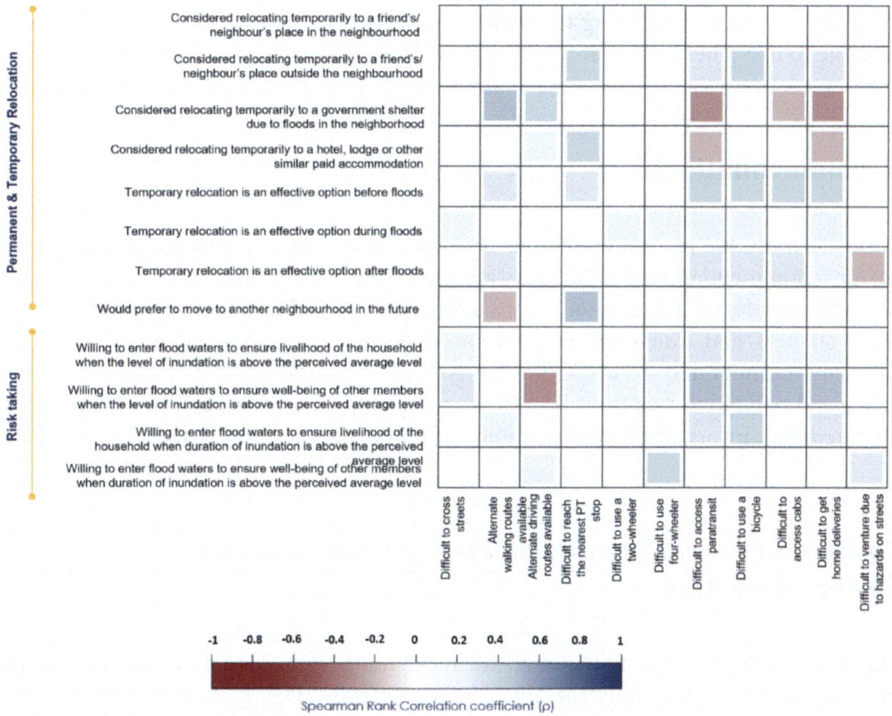

Fig. 21.10 Correlation with built environment and transport-related perceptions. *Source* Primary Survey

with the availability of alternative walking routes, difficulties in crossing streets, reaching bus stops, and using different transportation modes and receiving home deliveries. This indicates that temporary relocation is influenced by the characteristics of the built environment, highlighting the need for appropriate interventions at the built environment level to enable safer mobility.

Permanent relocation to another neighbourhood shows moderate positive correlation with the challenges faced in accessing public transport stops, highlighting the significance of having convenient access to public transportation. Additionally, it displays a minimal negative correlation with the availability of alternate walking routes, suggesting that individuals are more likely to consider relocation when safer alternate routes are not available in the neighbourhood during floods.

As for the risk-taking behaviour, it was observed that difficulty to use transport modes such as cars, two wheelers, bicycles, etc. or even difficulty to access cabs or other public transport shows low to moderate positive correlation to the willingness to enter floodwaters to ensure the wellbeing of other family members as well as to ensure the livelihood of the household when the inundation levels and duration are above the perceived average levels. A lack in availability of alternate routes and difficulty to cross roads also correlate to the same. This shows that people intend to undertake risks by entering floodwaters and wading through them when it is not possible to use other transport modes or safer alternate routes are unavailable.

Recommendations

In this section, the recommendations are given at two levels, namely planning and design recommendations for the study area, and community-based interventions. The planning and design recommendations address locational and operational aspects of relief shelters and transport services based on the correlation analysis involving built environment and floods. Community level recommendations focus on governance and policy interventions for effective organization, accountability for disaster preparedness, information and communication for disaster mitigation.

Local Area Level Planning and Design Interventions for the Study Area

The results clearly indicate the need for more temporary relocation options due to inundation levels, especially for lower income groups. Disagreement in permanent relocation is also observed in most of the cases. These observations highlight the need for local area level strategies to tackle inundation levels and water runoff, alternate temporary shelters and safer means of transport.

This necessitates both active and passive planning and design strategies at the local area level. Active strategies involve identifying existing public infrastructure and amenities that can function as flood relief shelters during times of flooding. Passive strategies involve delineating access routes and preparing access plans for residents who are willing to seek shelter but are unable to due to transport related accessibility issues. Establishments such as schools, colleges, and community halls can serve as multiple purpose relief shelters without requiring the creation of separate shelters which is a challenge in dense urban areas. Figure 21.11 displays several identified locations that can serve as flood shelters. Such shelters can accommodate relief boats/rafts for disaster situations and can easily cater to the basic food, storage and shelter needs during temporary relocation. Necessary amendments are also required in the development control regulations of Guwahati Building Construction (Regulation) Bye-Laws of 2014.

Further, an expansion of routes served by relief boats/rafts has been proposed. It was observed that at present only two roads in the study area are served by SDRF and NDRF boats during floods. The aim is to expand the network of routes served with boats/rafts during floods, covering all the highly vulnerable roads with frequent inundation, connecting the less permeable parts of the neighbourhood to the main roads in order to provide access to the nearest public transport stops and the identified

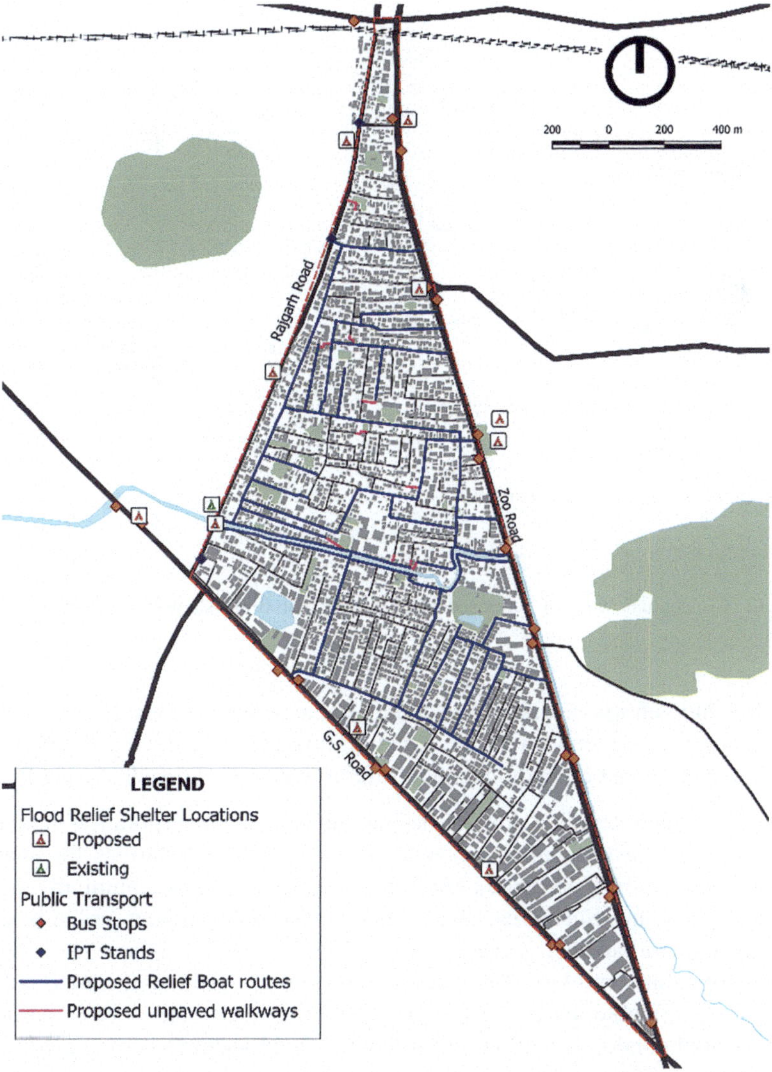

Fig. 21.11 Proposed flood relief shelters and extended network of boat routes. *Source* Authors

relief shelters. This would be extremely helpful to ensure that people get the essentials supplies as well as to ensure their safe evacuation, when necessary. Figure 21.11 shows the proposed network along with location of existing bus and IPT stops as well as the government flood shelters. A detailed proposal as an example for a bridge over the Bharalu river, forming a part of the Rajgarh Road is shown as a suitable evacuation and relief zone. This has been chosen as it remains free from inundation by design and is located very close to Anil Nagar. A proposal for multifunctional placemaking is illustrated in Fig. 21.12.

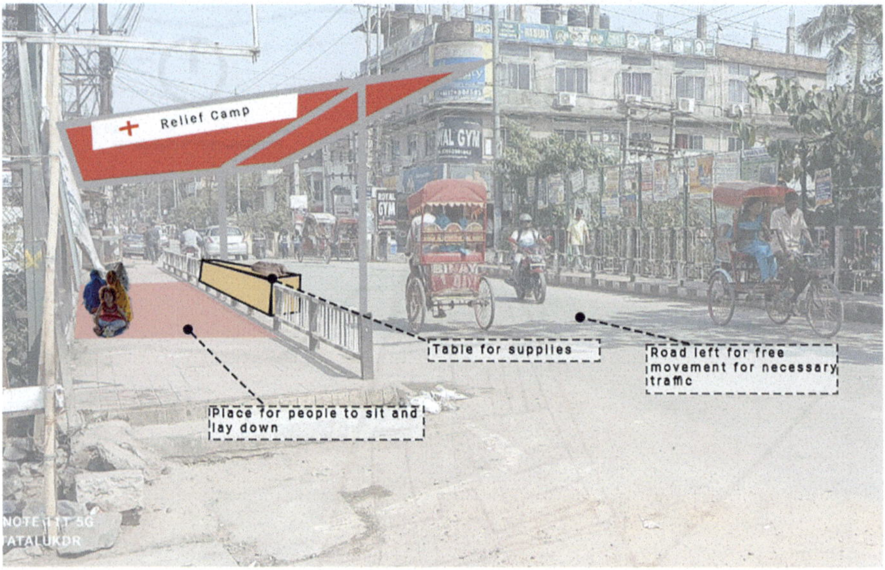

Fig. 21.12 Conceptual view of proposed interventions for flood shelter at the bridge. *Source* Authors

Further interventions are proposed at the local area planning level, which can help in increasing the permeability of the area both in terms of reducing stagnation and quick drainage of water and quicker and safer movement of people during floods.

i. Designing an integrated drainage system through the incorporation of land use, urban street design with storm water drains, design regulations for boundary walls, etc. guiding drainage to recharge locations such as community amenity spaces, maidans. Such areas shall contain permeable surfaces and bioretention areas in the urban environment.
ii. Development control regulations can prescribe construction of compound walls with a permeable surface especially at public edges and porous surfaces at residential edges. This allows accumulated flood water to disperse throughout the area avoiding stagnation, and potentially reducing the level of inundation for safer maneuver.
iii. It was observed that there are various dead ends and roads ending in plots within the study area, which leads to a disconnect for both the road and drainage network, hindering the flow of water and leading to waterlogging during floods. To address this concern, it is proposed to make small pedestrian pathways that can improve pedestrian accessibility of different parts of the neighbourhood till the nearest public transport node.

Community Based Interventions

The observations from Section "Results and Discussion" indicate interdependency of people to stay informed about the flood situation in the locality and to get the essential help during floods especially till the secondary help arrives. Furthermore, instances were noted where the community provided greater assistance in absence of initiatives by local authorities. There are willing individuals to take actions, however perceived failure in betterment of the situation serves challenge. During the survey, the respondents were also asked about their capacity and willingness to provide aid during urban floods, the responses to which are as shown in Fig. 21.13. The following recommendations focus on inclusion of community-led flood disaster management within the broader disaster management framework in an institutionalised manner that can prove to be an effective approach to empower and enhance community resilience at local area level to reduce risks during floods.

The District Disaster Management Plan for the Kamrup Metropolitan District mentions community participation, preparedness and responsibility and volunteer training in their preparedness, capacity building and training measures, yet there is limited evidence of actual on ground implementation. This may be attributed to the absence of a well-structured system for executing these measures. To address the gap, shows a zoning of the area into four different community zones is proposed, such that they cover almost equal area (ranging from about 45 to 55 ha). This is shown in Fig. 21.14, A Community Disaster Management Association (CDMA) can be formed for the study area. This body can serve as a liaison between the community and disaster management authorities, ensuring effective communication, information sharing, and active participation in decision-making processes. Proposed organization structure of the zone wise members is shown in Fig. 21.15. By operating as an elected association, it can work as an organization accountable to the community, thereby fostering a sense of responsibility and motivation to effectively carry out its duties.

Following is a non-exhaustive list of activities that the Community Disaster Management Association can be tasked with:

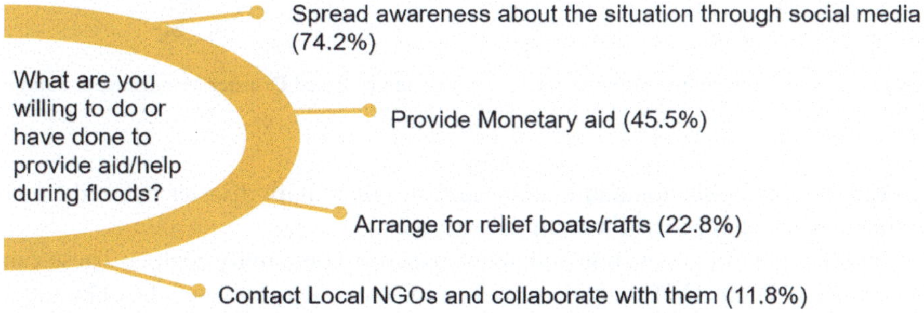

Fig. 21.13 Willingness of the community to contribute during floods. *Source* Primary Survey

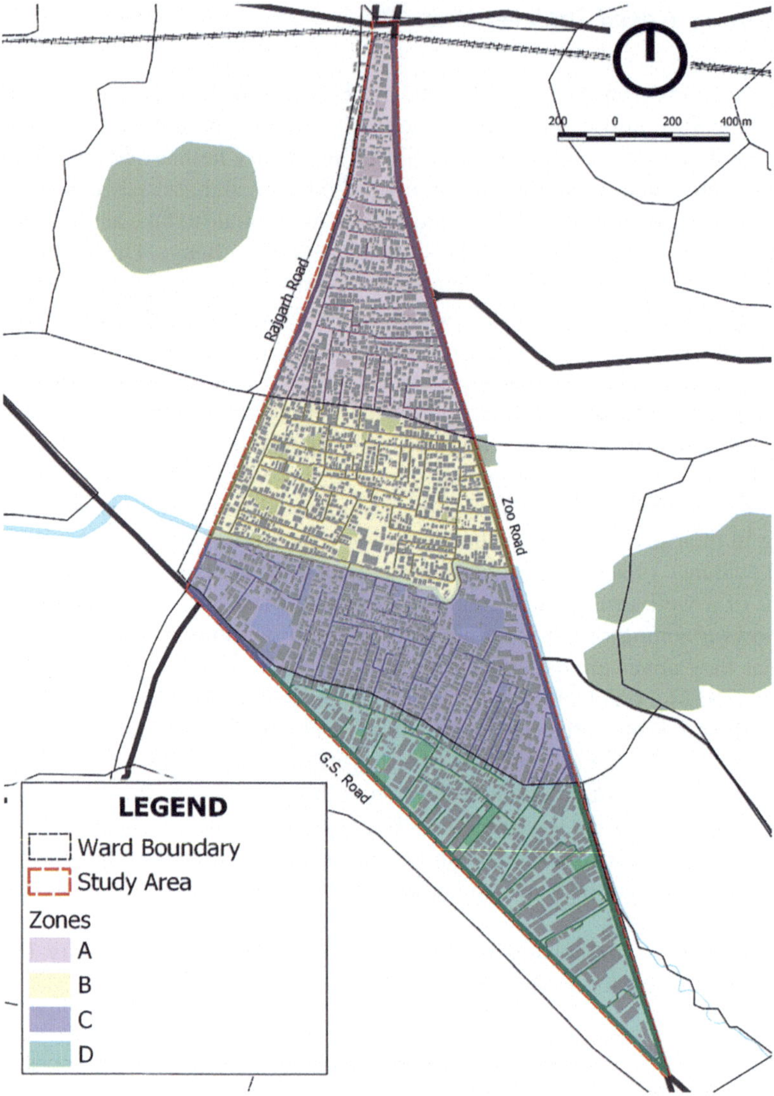

Fig. 21.14 Area divided into different zones for Community Based Disaster Management. *Source* Authors

- Primarily the collection and maintenance of community disaster risk reduction funds, which can be used for various activities.
- Identification or construction and maintenance of community centres. These can be used as flood relief shelters as well as to store emergency supplies like rafts/ boats, life jackets etc. stored in identified shelters.

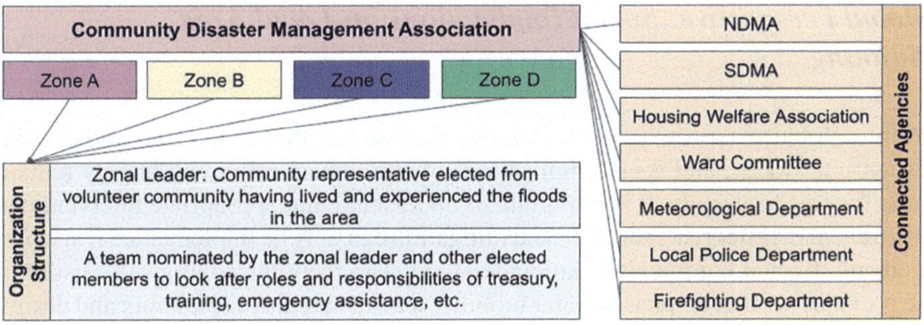

Fig. 21.15 Proposed Community Disaster Management Authority organization structure and connected agencies. *Source* Authors

- Capacity building and training: disasters create impacts on communities, the community members come forward as the first responders. Studies have found that trained volunteers will be a resource and untrained volunteers can exaggerate the impacts on victims.
- Information sharing: signs/information boards put up in public places; e.g. temples, mosques, churches, schools etc. to inform community members on flood risk and risk mitigation methods. Maintenance of evacuation routes delineated.
- Campaign for creating awareness on risks involved in entering floodwaters and disaster preparedness: talks or demonstrations at educational institutions, businesses and public places conducted in Assamese, Hindi and English.
- As the representative of the community, the entity will assume the responsibility of forwarding complaints and concerns to the Disaster Management Authority and other relevant authorities. Its role will be to advocate on behalf of the community, ensuring that their grievances are addressed and advocating for the implementation of improved measures.
- For disaster related communication, utilise the web platform known as Urban Risk Map in the city of Guwahati. This platform uses citizen reporting and social media to quickly map time-sensitive information, without requiring new applications or training. The platform uses real-time sensor information, such as flood gauges, monitoring wells, and pumping stations, and works with Twitter, Facebook, and Telegram. Through the live map, residents can share information about changing situations in the city and assist each other in finding safety. In India it has been used in Chennai piloted in 2017 and 100 Resilient Cities Network. Similarly, through institutional collaboration of DDMA and nodal institutes like IIT Guwahati the project development and implementation can take place.

Global Perspective: Social Implications on Local Area Planning

From a global perspective, methodology discussed in this chapter to study social response to floods; and social implications of subsequent crisis will largely remain same for any case study. It is important to understand that appropriate interventions for efficient disaster risk reduction and mitigation can only be implemented if specific needs of affected communities are understood. The recommendations discussed in the previous section emphasise safer mobility through local area planning and design measures after carefully analyzing correlation between floods affected society and built environment.

To delve deeper into the recommendations as an extension to the MaaR framework, further study needs to be conducted at planning, policy and design levels, some of the evident research areas are mentioned below.

- Mixed use development: Infrastructure suitable for temporary relocation of flood affected communities especially at dense urban cores. Optimum utilisation of the public built infrastructure such as schools, colleges, open grounds, and community halls ensure basic amenities and utilities designed to cater to higher occupancy loads and making them ideal temporary relief shelters.
- Multi-functionality: Even the 'non-building' infrastructure such as bridges, overpasses are designed and constructed considering average and peak inundation levels and make ideal evacuation points via rafts, fire tenders or even aerial modes wherever feasible.
- Integrated urban plans: Nexus of multifunctional social infrastructure through transport-land use planning with a focus on safe access during both regular and flood events.
- E-governance: Smart information and communication tools aid in enhancing this process by connecting the two ends of the spectrum—the government and the society. Ecosystem of smart devices and Internet of Things (IoT) which managed communication even during most critical situations such as COVID-19. Needless to say, development of reliable digital infrastructure is imperative to connect key stakeholders such as community disaster management associations with nodal agencies at district, region and state level.

Lastly, emphasizing on prioritization of bottom-up approach of study for top-down system configuration; former, focusing on resilient society, where community is an actor and a first respondent in disaster risk reduction and mitigation process and latter addressing governance end where local area planning is a result of careful study of people (society) and place (built environment) relationship.

Conclusion

Numerous fast-paced daily activities are essential to the survival of urban dwellers. The ability to be mobile during floods is quite important in this context. The potential of urban floods to disrupt human mobility is increasing with climate change. It is essential to study the mechanisms of risk taking and risk aversion that people adopt in response to urban floods as it can determine whether they choose to alter the built environment or adapt to it. The study conducted in this chapter—possibly the first of its kind in India—uses the lens of human behaviour to identify common threads between the built environment and the local communities that inhabit them. These threads appear to dictate how people react to stressful events like urban floods. The chapter thus looks at the interlinkage between urban floods (as a manifestation of climate crisis), social responses (through analysis of the variations in the behaviour around floods between different socio-demographic groups) and sustainability (in terms of human safety) on ecological perspectives (based on a built environment perspective).

The findings suggest that risk taking, and relocation behaviour varies across different demographic and socioeconomic groups and the challenge of accessing various transportation modes and a lack of viable alternative routes for vehicles and pedestrians show a significant correlation with both relocation and risk-taking behaviours. This chapter draws on the 'Mobility as a Response' (MaaR) to environmental hazards framework proposed by Xia and Yeh (2022) in the urban context. Specifically, it refers to the MaaR framework for using measures of risk taking and risk aversion or relocation behaviours. To the best of the author's knowledge, this is the first such application of the MaaR framework in the context of urban flooding at a local area level in India.

References

Adeola FO (2009) Katrina cataclysm: does duration of residency and prior experience affect impacts, evacuation, and adaptation behaviour among survivors? Environ Dev 41(4):459–489. https://doi.org/10.1177/0013916508316651

Ahmouda A, Hochmair HH, Cvetojevic S (2019) Using twitter to analyze the effect of hurricanes on human mobility patterns. Urban Sci 3(3):87. https://doi.org/10.3390/urbansci3030087

Ajibade I, McBean G, Bezner-Kerr R (2013) Urban flooding in Lagos, Nigeria: patterns of vulnerability and resilience among women. Glob Environ change 23(6):1714–1725

Ajibade I, Armah FA, Kuuire VZ, Luginaah I, McBean G, Tenkorang EY (2015) Assessing the bio-psychosocial correlates of flood impacts in coastal areas of Lagos, Nigeria. J Environ Plan Manag 58(3):445–463. https://doi.org/10.1080/09640568.2013.861811

Alam K, Rahman MH (2014) Women in natural disasters: a case study from southern coastal region of Bangladesh. Int J Disaster Risk Reduct 8:68–82. https://doi.org/10.1016/j.ijdrr.2014.01.003

APCB (2013) Conservation of river Bharalu, preparation of detailed project report, city sanitation plan. The Louis Berger Group, Inc., and DHI (India) Water and Environment Pvt. Ltd., Guwahati

Ashley ST, Ashley WS (2008) Flood fatalities in the United States. J Appl Meteorol Climatol 47(3):805–818. https://doi.org/10.1175/2007jamc1611.1

Bempah SA, Øyhus AO (2017) The role of social perception in disaster risk reduction: beliefs, perception, and attitudes regarding flood disasters in communities along the Volta River, Ghana. Int J Disaster Risk Reduct 23:104–108

Box P, Bird D, Haynes K, King D (2016) Shared responsibility and social vulnerability in the 2011 Brisbane flood. Nat Hazards 81(3):1549–1568. https://doi.org/10.1007/s11069-016-2145-z

Chen S, Chen Y, Lei Z, Tan-Soo JS (2020a) Impact of air pollution on short-term movements: evidence from air travels in China. J Econ Geogr 20(4):939–968. https://doi.org/10.1093/jeg/lbaa005

Chen Z, Gong Z, Yang S, Ma Q, Kan C (2020b) Impact of extreme weather events on urban human flow: a perspective from location-based service data. Comput Environ Urban Syst 83:101520. https://doi.org/10.1016/j.compenvurbsys.2020.101520

Coates L (1999) Flood fatalities in Australia, 1788–1996. Aust Geogr 30(3):391–408. https://doi.org/10.1080/00049189993657

Cools M, Moons E, Creemers L, Wets G (2010) Changes in travel behaviour in response to weather conditions: do type of weather and trip purpose matter? Transp Res Rec 2157(1):22–28

Cresswell T (2010) Towards a politics of mobility. Environ Plan D Soc Space 28(1):17–31. https://doi.org/10.1068/d11407

Cropper ML (1981) Measuring the benefits from reduced morbidity. Am Econ Rev 71(2):235–240

Dasgupta P (2004) Valuing health damages from water pollution in urban Delhi, India: a health production function approach. Environ Dev Econ 9(1):83–106

Di Mauro M, De Bruijn KM, Meloni M (2012) Quantitative methods for estimating flood fatalities: towards the introduction of loss-of-life estimation in the assessment of flood risk. Nat Hazards 63(2):1083–1113. https://doi.org/10.1007/s11069-012-0207-4

Diakakis M, Deligiannakis G (2013) Vehicle-related flood fatalities in Greece. Environ Hazards 12(3–4):278–290. https://doi.org/10.1080/17477891.2013.832651

Dickie M (2017) Averting behaviour methods. In: A primer on nonmarket valuation, pp 293–346

Doocy S, Daniels A, Murray S, Kirsch TD (2013) The human impact of floods: a historical review of events 1980–2009 and systematic literature review. PLoS Curr Disasters (APR 2013). https://doi.org/10.1371/currents.dis.f4deb457904936b07c09daa98ee8171a

Drabek TE (2001) Disaster warning and evacuation responses by private business employees. Disasters 25(1):76–94. https://doi.org/10.1111/1467-7717.00163

Drobot SD, Benight C, Gruntfest EC (2007) Risk factors for driving into flooded roads. Environ Hazards 7(3):227–234. https://doi.org/10.1016/j.envhaz.2007.07.003

Easterling DR, Evans JL, Groisman PY, Karl TR, Kunkel KE, Ambenje P (2000) Observed variability and trends in extreme climate events: a brief review. Bull Am Meteor Soc 81(3):417–426. https://doi.org/10.1175/1520-0477(2000)081%3c0417:ovatie%3e2.3.co;2

Ettema D, Nieuwenhuis R (2017) Residential self-selection and travel behaviour: what are the effects of attitudes, reasons for location choice and the built environment? J Transp Geogr 59:146–155. https://doi.org/10.1016/j.jtrangeo.2017.01.009

Few R, Pham GT (2010) Climatic hazards, health risk and response in Vietnam: case studies on social dimensions of vulnerability. Glob Environ Change Hum Policy Dimens 20(3):529–538. https://doi.org/10.1016/j.gloenvcha.2010.02.004

Freeman AM (1979) The benefits of air and water pollution control: a review and synthesis of recent estimates. National Technical Information Service

Gray C, Mueller V (2012) Drought and population mobility in rural Ethiopia. World Dev 40(1):134–145. https://doi.org/10.1016/j.worlddev.2011.05.023

Hamilton K, Peden AE, Keech JJ, Hagger MS (2016a) Driving through floodwater: exploring driver decisions through the lived experience. Paper presented at the European Health Psychologist, Aberdeen

Hamilton K, Peden AE, Pearson M, Hagger MS (2016b) Stop there's water on the road! Identifying key beliefs guiding people's willingness to drive through flooded waterways. Saf Sci 89:308–314. https://doi.org/10.1016/j.ssci.2016.07.004

Hamilton K, Peden AE, Keech JJ, Hagger MS (2018a) Changing people's attitudes and beliefs toward driving through floodwaters: evaluation of a video infographic. Transp Res Part F 53:50–60

Hamilton K, Price S, Keech JJ, Peden AE, Hagger MS (2018b) Drivers' experiences during floods: investigating the psychological influences underpinning decisions to avoid driving through floodwater. Int J Disaster Risk Reduct 28:507–518. https://doi.org/10.1016/j.ijdrr.2017.12.013

Hamilton K, Demant D, Peden AE, Hagger MS (2020) A systematic review of human behaviour in and around floodwater. Int J Disaster Risk Reduct 47:101561. https://doi.org/10.1016/j.ijdrr.2020.101561

Han SY, Tsou MH, Knaap E, Rey S, Cao G (2019) How do cities flow in an emergency? Tracing human mobility patterns during a natural disaster with big data and geospatial data science. Urban Sci 3(2):51. https://doi.org/10.3390/urbansci3020051

Haynes K, Tofa M, Avci A, van Leeuwen J, Coates L (2018) Motivations and experiences of sheltering in place during floods: implications for policy and practice. Int J Disaster Risk Reduct 31:781–788. https://doi.org/10.1016/j.ijdrr.2018.07.011

Hong B, Bonczak BJ, Gupta A, Kontokosta CE (2021) Measuring inequality in community resilience to natural disasters using large-scale mobility data. Nat Commun 12(1):1870

Hunter LM (2005) Migration and environmental hazards. Popul Environ 26:273–302

Jonkman SN, Kelman I (2005) An analysis of the causes and circumstances of flood disaster deaths. Disasters 29(1):75–97. https://doi.org/10.1111/j.0361-3666.2005.00275.x

Kamol CR, Cebrian M, Hasan S (2019) Quantifying human mobility resilience to extreme events using geo-located social media data. EPJ Data Sci 8(1):18

Kashyap S, Mahanta R (2021) Socioeconomic vulnerability to urban floods in Guwahati, northeast India: an indicator-based approach. In: Economic effects of natural disasters. Academic Press, pp 457–475. https://doi.org/10.1016/B978-0-12-817465-4.00027

Khattak AJ, De Palma A (1997) The impact of adverse weather conditions on the propensity to change travel decisions: a survey of Brussels commuters. Transp Res Part A Policy Pract 31(3):181–203. https://doi.org/10.1016/S0965-8564(96)00025-0

Koetse MJ, Rietveld P (2009) The impact of climate change and weather on transport: an overview of empirical findings. Transp Res Part D Transp Environ 14(3):205–221

Kwan M-P (2018) The limits of the neighborhood effect: contextual uncertainties in geographic, environmental health, and social science research. Ann Am Assoc Geogr 108(6):1482–1490. https://doi.org/10.1080/24694452.2018.1453777

Levy MB, Wadycki WJ (1974) What is the opportunity cost of moving? Reconsideration of the effects of distance on migration. Econ Dev Cult Change 22(2):198–214

Lu X, Bengtsson L, Holme P (2012) Predictability of population displacement after the 2010 Haiti earthquake. Proc Natl Acad Sci USA 109(29):11576–11581. https://doi.org/10.1073/pnas.120388210

National Disaster Management Guidelines: Management of Urban Flooding (2010) A publication of the National Disaster Management Authority, Government of India. ISBN: 978-93-80440-09-5, New Delhi

Nefeslioglu HA, Gokceoglu C, Sonmez H, Gorum T (2011) Medium-scale hazard mapping for shallow landslide initiation: the Buyukkoy catchment area (Cayeli, Rize, Turkey). Landslides. https://doi.org/10.1007/s10346-011-0267

Neidell M (2009) Information, avoidance behaviour, and health: the effect of ozone on asthma hospitalizations. J Hum Resour 44(2):450–478. https://doi.org/10.3368/jhr.44.2.450

Patz JA, Engelberg D, Last J (2000) The effects of changing weather on public health. Annu Rev Public Health 21:271–307. https://doi.org/10.1146/annurev.publhealth.21.1.271

Peden AE, Franklin RC, Leggat PA, Aitken P (2017) Causal pathways of flood related river drowning deaths in Australia. PLoS Curr Disasters

Pelling M (2003) The vulnerability of cities: natural disasters and social resilience. Earthscan, London

Rossi PH (1980) Why families move, vol 2. Sage, Beverly Hills

Sakamoto J, Fujita M (2015) The impact of information on commuters' departure decision on torrential rain day. J East Asia Soc Transp Stud 11(2007):311–325

Salvati P, Petrucci O, Rossi M, Bianchi C, Pasqua AA, Guzzetti F (2018) Gender, age and circumstances analysis of flood and landslide fatalities in Italy. Sci Total Environ 610–611:867–879. https://doi.org/10.1016/j.scitotenv.2017.08.064

Sarmah T, Das S (2017) Urban flood disaster assessment for Guwahati city. In: International symposium on water urbanism and infrastructure

Sarmah T, Das S (2018) Urban flood mitigation planning for Guwahati: a case of Bharalu basin. J Environ Manage 206:1155–1165. https://doi.org/10.1016/j.jenvman.2017.10.079

Shabanikiya H, Seyedin H, Haghani H, Ebrahimian A (2014) Behaviour of crossing flood on foot, associated risk factors and estimating a predictive model. Nat Hazards 73(2):1119–1126. https://doi.org/10.1007/s11069-014-1124-5

Sheller M, Urry J (2006) The new mobilities paradigm. Environ Plan A 38(2):207–226. https://doi.org/10.1068/a37268

Simini F, González MC, Maritan A, Barabási AL (2012) A universal model for mobility and migration patterns. Nature 484(7392):96–100

Song C, Qu Z, Blumm N, Barabási AL (2010) Limits of predictability in human mobility. Science 327(5968):1018–1021. https://doi.org/10.1126/science.1177170

Steg LE, De Groot JI (2019) Environmental psychology: an introduction, vol 2. Wiley, Hoboken, NJ

Sullivan D, Young IF (2020) Place attachment style as a predictor of responses to the environmental threat of water contamination. Environ Dev 52(1):3–32. https://doi.org/10.1177/0013916518786766

Sun C, Kahn ME, Zheng S (2017) Self-protection investment exacerbates air pollution exposure inequality in urban China. Ecol Econ 131:468–474. https://doi.org/10.1016/j.ecolecon.2016.06.030

Tariq H, Pathirage C, Fernando T (2021) Measuring community disaster resilience at local levels: an adaptable resilience framework. Int J Disaster Risk Reduction 62:102358

Tim Y, Pan SL, Ractham P, Kaewkitipong L (2017) Digitally enabled disaster response: the emergence of social media as boundary objects in a flooding disaster. Inf Syst J 27(2):197–232. https://doi.org/10.1111/isj.12114

Van Acker V, Van Wee B, Witlox F (2010) When transport geography meets social psychology: toward a conceptual model of travel behaviour. Transp Rev 30(2):219–240. https://doi.org/10.1080/01441640902943453

Wang Y, Wang Q, Taylor JE, Zhou W-X (2017) Aggregated responses of human mobility to severe winter storms: an empirical study. PLoS ONE 12(12):e0188734. https://doi.org/10.1371/journal.pone.0188734

World Health Organization (2014) Global report on drowning: preventing a leading killer. Retrieved from Geneva, Switzerland: https://www.who.int/violence_injury_prevention/global_report_drowning/en/

Xia C, Yeh AG (2022) Mobility as a response to environmental hazards in the urban context: a new perspective on mobility and inequality. Travel Behav Soc 27:192–203. https://doi.org/10.1016/j.tbs.2022.01.008

Yabe T, Ukkusuri SV (2019) Integrating information from heterogeneous networks on social media to predict post-disaster returning behaviour. J Comput Sci 32:12–20. https://doi.org/10.1016/j.jocs.2019.02.002

Zander KK, Garnett S (2020) Risk and experience drive the importance of natural hazards for peoples' mobility decisions. Clim Change 162(3):1639–1654

Zivin JG, Neidell M (2009) Days of haze: environmental information disclosure and intertemporal avoidance behaviour. J Environ Econ Manage 58(2):119–128. https://doi.org/10.1016/j.jeem.2009.03.001

Chapter 22
Gathering Palm Fruit, Alternative Job for Orang Rimba During Climate Crisis and Poverty in Indonesia

Habibullah, Retnaningdyah Weningtyastuti, Risyanto, and Mijak Tampung

Abstract This research focuses on the Orang Rimba community, an indigenous group in Indonesia known for its nomadic lifestyle and dependence on forest resources. When faced with climate crises, such as long droughts or excessive rainfall, the forest ecosystems in Bukit Dua Belas National Park cannot sustain their livelihoods. As a result, the Orang Rimba struggles to make ends meet and fall into poverty. This study aims to identify climate-related challenges faced by indigenous communities and explore alternative sources of livelihood during such crises. Using a qualitative approach, this research focused on Orang Rimba living in Bukit Dua Belas National Park, Jambi Province, Indonesia. The research revealed that when traditional forest resources are no longer sufficient, Orang Rimba collect oil palm fruits from company oil palm plantations or villagers around the national park. This situation often leads to conflicts between the Orang Rimba, companies, and villagers, so a government role is needed to resolve conflicts and source alternative livelihoods that are sustainable and do not cause a climate crisis. This section recommends future research that discusses the pros and cons of plantation commodities impacting indigenous peoples' livelihoods in the face of climate crises in other countries, especially Africa, South Asia, and Latin America.

Habibullah (✉) · R. Weningtyastuti · Risyanto
Research Center for Social Welfare, Village and Connectivity, National Research and Innovation Agency (BRIN), Jakarta, Indonesia
e-mail: habi003@brin.go.id

R. Weningtyastuti
e-mail: retn026@brin.go.id

Risyanto
e-mail: risy002@brin.go.id

M. Tampung
Makekal United Group, Jambi, Indonesia

U. Mukhopadhyay et al. (eds.), *Climate Crisis, Social Responses and Sustainability*, Climate Change Management, https://doi.org/10.1007/978-3-031-58261-5_22

Introduction

Climate crisis has significant implications for the poverty experienced by indigenous people. Indigenous people face multiple problems, including human rights violations, discrimination, and poverty. Climate change exacerbates these problems, as indigenous people are more vulnerable to their impacts, such as food insecurity, displacement, and threats to critical infrastructure (Townsend et al. 2020).

Climate crises and global poverty are two interconnected issues that have significant impacts on one another. Climate change is a substantial catalyst for the prevalence of global poverty, especially in low-income African and South Asian countries (Leichenko and Silva 2014). Climate change both causes and affects Indonesia, one of the countries in Southeast Asia. Indonesia's causes of climate change include the widespread conversion of forests to palm oil plantations, peatland fires, and the high use of fossil fuels. Meanwhile, the impacts of climate change on Indonesia are higher air temperatures, erratic weather changes, drying up springs, and flooding disasters.

Climate change and poverty are closely linked because climate change can increase poverty and vulnerability. Populations with the lowest socioeconomic status are more vulnerable to the adverse impacts of climate change shocks, mostly because of their high reliance on climate-sensitive industries, such as agriculture, as a means of livelihood. Climate change has the potential to diminish agricultural productivity, heighten food insecurity, and displace populations, all of which collectively contribute to the exacerbation of poverty levels (Dasgupta and Baschieri 2010). Additionally, climate shocks, such as water shortages and extreme weather events, can make agricultural livelihoods more vulnerable, prompting individuals to seek alternative sources of income (Ahmad et al. 2023). The impact of climate change on poverty varies by country. This difference is due to the fact that each country has different environmental conditions and readiness to respond to climate change. Many researchers from both developed and developing countries have studied the impacts of climate change on poverty levels and welfare. These studies recommend a complementary approach to climate change and poverty reduction policies (Letta and Tol 2018). Factors that promote resilience among the poor and explore the nonmonetary dimensions of poverty in the context of climate change are crucial for research (Leichenko and Silva 2014).

Agricultural communities characterised by low socioeconomic status are more vulnerable to the adverse impacts of climate change on household food availability and accessibility, thus increasing poverty levels (Alamgir et al. 2018). The agricultural sector is highly vulnerable to the effects of climate change, owing to its inherent sensitivity to weather and climatic factors. Furthermore, individuals employed in the agricultural sector generally exhibit lower socioeconomic status compared to those working in alternative sectors inside urban areas (Kang et al. 2009).

The climate crisis has had a significant impact on indigenous peoples around the world, even more so than on agricultural communities. In many countries, indigenous people are negatively affected by climate change. Increased frequency and intensity of extreme weather events, such as floods, rainstorms, hurricanes, and hot temperatures. Climate change has a direct impact on indigenous peoples' livelihoods, leading to crop failure, reduced yields, limited water supply, and malnutrition (Leal Filho et al. 2021).

Indigenous peoples have long successfully adapted to climate and environmental change, utilising their traditional knowledge systems and worldviews for climate mitigation and adaptation. The current escalation of colonialism and colonisation presents distinct difficulties for indigenous populations owing to anthropogenic climate issues (Lewis et al. 2020). It is critical to recognise the link between climate change and the historical context of colonialism, which has created economic and social conditions for anthropogenic climate change that limit indigenous peoples' resilience and capacity for resilience (Dent et al. 2023). Despite their knowledge and resilience, indigenous peoples' voices and contributions to climate policy and adaptation are often undervalued and marginalised (Carmona et al. 2023). Although the UN Declaration on the Rights of Indigenous Peoples recognises the political rights of indigenous peoples, they have limited opportunities for active political participation (Tam et al. 2021).

The presence of palm oil plantations in the midst of indigenous peoples is a dilemma in the midst of climate change. On the one hand, palm oil plantations are an alternative source of livelihood for indigenous peoples when natural resources are no longer able to support their livelihoods of indigenous peoples. However, the presence of palm oil plantations worsens the natural and environmental conditions of the indigenous people (Runtuboi et al. 2021). The impact of palm oil plantations on indigenous people is a complex issue involving various social, economic, and environmental factors (Buergin 2016).

In large-scale palm oil plantations, clearing can lead to deforestation and habitat loss, adversely affecting biodiversity and ecosystem services. The negative environmental consequences of extensive palm oil plantations include forest encroachment and land cover loss, which lead to increased flooding and surface runoff (Amalia et al. 2019). Oil palm plantations are associated with deforestation and peatland conversion, which contribute to climate change and severe haze (Paterson 2019). Palm oil plantations drive greenhouse gas (GHG) emissions and biodiversity losses (Beringer et al. 2023). Small-scale palm oil plantations also affect the lives of the indigenous people. Small-scale oil palm and natural rubber plantations do not necessarily help alleviate poverty among indigenous peoples, and even worsen their economic vulnerability (Saifullah et al. 2018). Small-scale plantations do not significantly increase indigenous peoples' incomes (Arthur 2012).

Conflicts between palm oil companies and indigenous communities arise when palm oil plantations are open (Sandbukt 2000; Prasetijo 2017; Najib et al. 2020; Pahmi et al. 2023). The Orang Rimba faces challenges related to livelihoods and land use. The expansion of palm oil plantations has had significant negative impacts on Orang Rimba, reducing access to resources and basic needs (Pahmi et al. 2023). The traditional livelihoods of the Orang Rimba have undergone changes due to industrialisation and the pressure to shift from hunting and gathering practices to rubber and palm oil plantations (Tresno et al. 2022). Due to the inability of the

Orang Rimba to face the pressure of oil palm plantation, expansion and the climate crisis they are experiencing, they are looking for alternative livelihoods by picking up palm fruits (brondol sawit) that fall on plantations that do not belong to them.

Based on the literature search, no author has discussed palm fruit gathering as an alternative source of livelihood for indigenous peoples during the climate crisis and poverty. Previous research focuses more on the impact of climate change on indigenous people (Townsend et al. 2020; Carmona et al. 2023), the expansion of oil palm plantations on climate change iklim (Paterson 2019; Amalia et al. 2019; Beringer et al. 2023) and the negative impact of oil palm plantation expansion on indigenous people (Sandbukt 2000; Prasetijo 2017; Najib et al. 2020; Pahmi et al. 2023) without considering the presence of oil palm plantations as an alternative employment solution for indigenous people to survive. Therefore, the problem of this research is, "How does the collection of oil palm fruits become an alternative job for Orang Rimba during the climate crisis and poverty in Indonesia?"

Therefore, the objectives of writing this section are:

- Knowing the condition of the Orang Rimba during the climate crisis in the tropical forest
- Knowing the condition of the Orang Rimba when facing the expansion of oil palm plantations
- Knowing the condition of the Orang Rimba when experiencing food crisis and poverty
- To know alternative jobs for the Orang Rimba in the midst of climate crisis and poverty
- To understand the challenges faced by palm fruit collectors in the Orang Rimba.

Rationale of Study

The rationale for this section is that a climate crisis is a serious threat to human life, especially for indigenous people (see Fig. 22.1). Indigenous people are highly dependent on natural resources, in the form of forest products. Due to the climate crisis and the expansion of palm oil plantations, natural resources are no longer able to fulfil the needs of indigenous peoples. Indigenous people are looking for alternative jobs to fulfil their needs. Collecting oil palm fruits that fall from oil palm trees that are not their own is an alternative job, although this work creates conflict with other communities. This chapter is important in determining the alternative work of indigenous people when they experience climate crises and poverty.

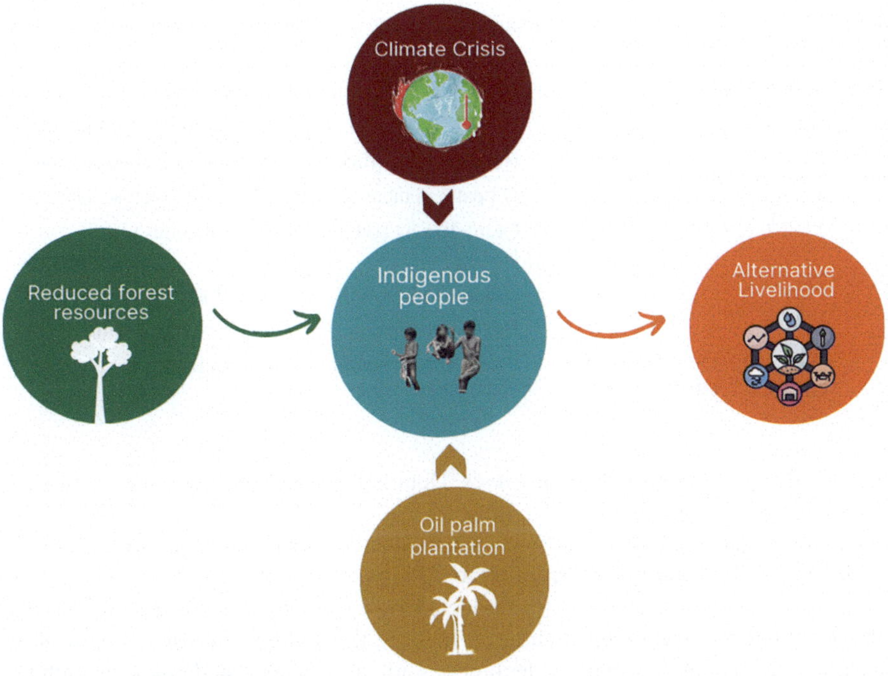

Fig. 22.1 Rationale of the study

Materials and Methods

The focus of our section is to explore how the condition of the Orang Rimba during the climate crisis, environmental changes caused by the expansion of oil palm plantations that are the cause of the climate crisis and a threat to indigenous peoples. On the other hand, the presence of oil palm plantations has become an alternative livelihood for the Orang Rimba. Legally, palm oil plantations do not belong to the Orang Rimba so they do not have the right to officially harvest palm fruit, but they can only collect fallen palm fruit and the remaining harvest. To achieve this goal, we used a qualitative approach using a case study. Case studies are particularly useful for acquiring an in-depth comprehension of a problem, event or phenomenon of interest, in its actual real-life context (Crowe et al. 2011).

The data collection process included interviews, observations, and documentation studies (Denzin and Lincoln 2018). We interviewed 20 participants representing stakeholders and Orang Rimba palm fruit collectors. The interviewees were selected using purposive sampling. The purposive sampling technique was adopted because, in this method, a group of people or specific situations or activities are purposefully chosen (Roy et al. 2022). The stakeholders interviewed were selected for their knowledge and insight into the research focus, their concern and commitment to the

Table 22.1 List of key informants (pseudonyms), affiliation, and category (indigenous people state/non-indigenous people)

No	Informants	Affiliation	Category
1	W1, W2, W3	NGO's	Non-indigenous people
2	B1, B2	National park officer	Non-indigenous people
3	K1, K2	Oil palm plantation owner	Non-indigenous people
4	T1, T2	Orang Rimba community leaders	Indigenous people
5	P1, P2	Orang Rimba youth	Indigenous people
6	S1	Buyer of palm fruit	Non-Orang Rimba
7	OR1, OR2, OR3, OR4, OR5, OR6, OR7, OR8	Palm fruit collectors	Indigenous people

sustainability of the Orang Rimba's lives. Detailed description of the names of each of the interviewees are available in Table 22.1.

The interviews and observations were conducted from June 2 to July 29, 2023. The interview duration ranged from 60 to 90 min.

The documentation study was conducted by exploring various reports, books, scientific journals, and other documents that are the source of information used in this research. Thematic analysis techniques identify potential themes or patterns based on information collected during data collection (Milies et al. 2014). Thematic analysis techniques are particularly beneficial because the flexibility of the process has allowed theoretical freedom that provides a rich, evidence-based, and detailed discussion of all collected data. In summary, the methodological chapter of this book can be seen in Fig. 22.2.

This chapter has various limitations, one of which is the limited time to make observations. In addition, it should be noted that the conditions mentioned above do not provide a comprehensive description of the overall conditions of indigenous peoples in Indonesia, because the Orang Rimba community is only one of several indigenous peoples living in Indonesia, and each indigenous community is unique in climate change adaptation.

Results and Discussion

Who Are Orang Rimba?

Orang Rimba is a community of indigenous people who live in Bukit Dua Belas National Park. Some of the names of Orang Rimba include "Suku Anak Dalam", "Orang Kubu", and "Sanak". The term "Orang Rimba" is not known for certain, but according to the Orang Rimba themselves, the term Orang Rimba is more humane than the term "Orang Kubu". The meaning of the word "Orang Rimba" has more

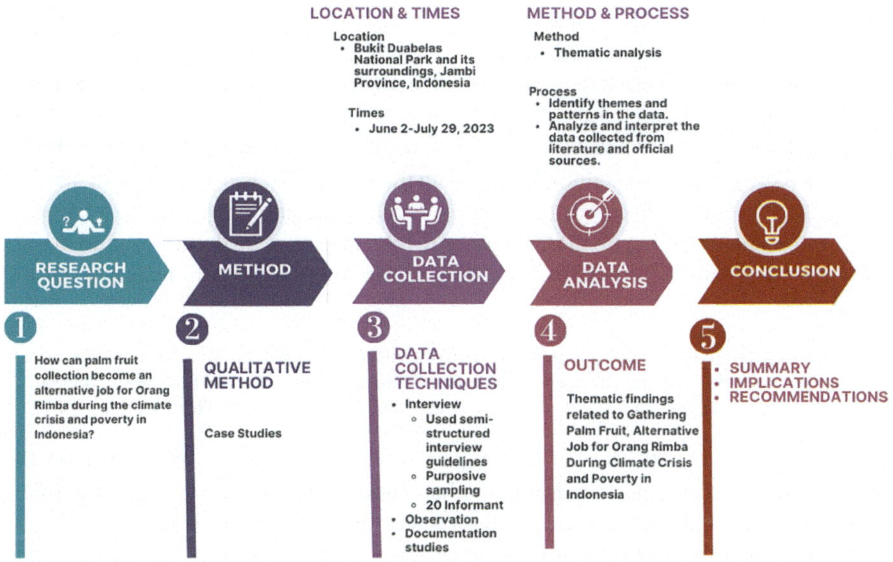

Fig. 22.2 Methodological flowchart

meaning of appreciation and respect. According to Zuhdi (2013), they name themselves jungle people or call themselves "Orang Rimbo" (Zuhdi 2013). However, according to KKI Warsi, an NGOs that assists and empowers Orang Rimba, the term Rimba refers to the place where this community chooses to live in lowland forests.

> Orang Rimbo is a terminology for indigenous people who live in the lowland forests of Sumatra, more humanized than the terms Orang Kubu, Suku Anak Dalam or Sanak. (Interview with W1)

The name Kubu in Melayu has pejorative meanings, such as primitive, stupid, infidel, dirty, and disgusting. The term Kubu was popular in the writings of colonial officials and ethnographers in the early twentieth century.

The Indonesian government defines this community of people as those who live in the forest and do not follow the same norms as Melayu people as Suku Anak Dalam. Suku Anak Dalam refers to people living in the hinterlands. Suku Anak Dalam is a postcolonial categorisation of nomadic groups living in forests (Indrizal and Anwar 2023). The term Suku Anak Dalam makes it easier for policymakers and program managers to identify community groups living in the hinterland of Sumatra. The indigenous people included in Suku Anak Dalam are not only Orang Rimba, but various other communities or ethnic groups found in the provinces of Jambi, Riau, and South Sumatra are also included in the indigenous people of Suku Anak Dalam. The Batin Sembilan and Talang Mamak belong to the Suku Anak Dalam, which has a lifestyle similar to that of Orang Rimba. The Batin Sembilan and Talang Mamak have a relatively settled lifestyle and use the petalangan agricultural system as a source of livelihood, while the Orang Rimba still live nomadically as a source of livelihood hunting and gathering. These tribal people are distinguished by their unique culture, belief system, livelihood, organizational structure, dietary behaviors, traditions, and dialects, which differ from the cultures of the mainstream communities in which they dwell (Barman and Chowdhury 2023).

Orang Rimba collects various fruits such as durian, bedaro/litchi, tampui, forest rambutan, cempedak, and sialang honey from the forest. They also engage in hunting by targeting animals such as deer, antelopes, deer, napu, pangolins, wild boars, porcupines, and kuau (Pahmi et al. 2023). Orang Rimba begins at birth, grows, and breeds in the jungle. Orang Rimba lives according to the rules, norms, and customs of the jungle.

> We usually live in the forest, providing food from hunting animals such as deer, porcupine, pangolin and wild boar and collecting fruits from the forest such as durian, rambutan, longan, sialang honey. But we can't do this anymore because of natural changes. (Interview with T1)

However, along with the various natural and social changes that occur in the surrounding environment, the Orang Rimba cannot fully live according to the rules, norms, and customs that accord with their ancestors' cultural heritage.

Orang Rimba live in groups in the upper reaches of rivers, the forests of lowland Sumatra. There is no accurate data on the population of Orang Rimba in Indonesia, but based on data compiled from various sources, it is estimated that the number is 5630 people. The largest population is in the Bukit Dua Belas National Park, which is the study location, with 3783 people. Bukit Tiga Puluh National Park is the second largest place of habitation with 474 people, and another 1373 people are spread in secondary forests and oil palm plantations, as well as on the edge of the trans-Sumatra road.

The main habitat of Orang Rimba lies within the Bukit Dua Belas National Park zone. Bukit Dua Belas National Park, which is astronomically located between 102° 30′ 00″–102° 55′ 00″ East Longitude and 10° 45′ 00″–20° 00′ 00″ South latitude. The total area of Bukit Dua Belas National Park is 54,760.40 ha. Administratively located in Merangin, Sarolangun, Batanghari and Tebo Regencies, Jambi Province. The map of Bukit Dua Belas National Park (Balai Taman Nasional Bukit Duabelas 2024) can be seen in Fig. 22.3.

The Indonesian Government's recognition of Bukit Dua Belas National Park in 2002 is unique when compared to other National Park definitions. Bukit Dua Belas National Park was officially designated as a protected area with the specific purpose of preserving the Orang Rimba community. This was because of the extensive historical position of the Orang Rimba community in the region, which had given them customary rights to govern and oversee land (Yusuf and Syafrial 2019).

After the designation of the Bukit Dua Belas National Park as a protected area, Orang Rimba communities still face challenges or difficulties. Challenges that often arise are conflicts between the Orang Rimba community and National Park authorities, as well as with surrounding communities. The main purpose of establishing protected areas is to protect the livelihoods and cultural heritage of the Orang Rimba. To achieve this goal, the government implemented zoning restrictions to limit the utilisation of forest resources (Yusuf and Syafrial 2019). This conflict arises as a

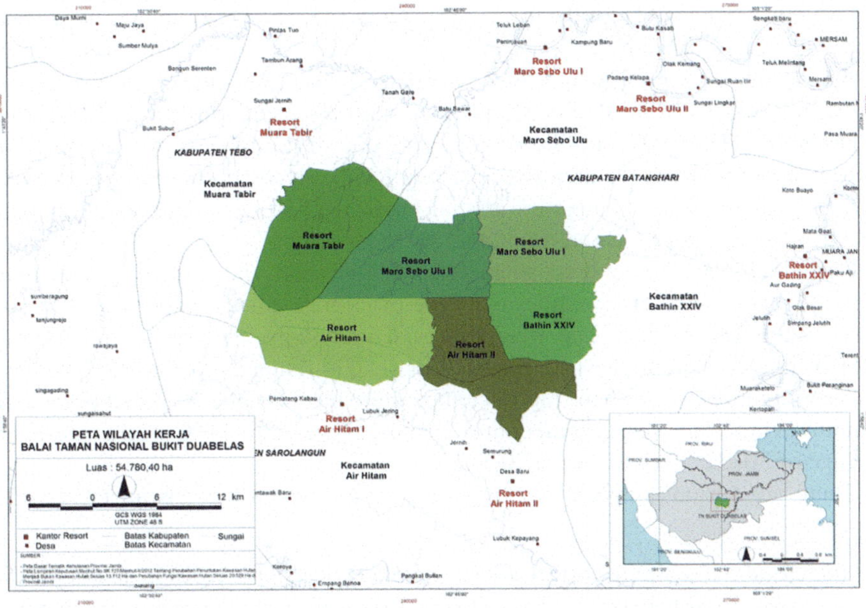

Fig. 22.3 The map of Bukit Dua Belas National Park. Source from Balai Taman Nasional Bukit Duabelas

result of the hegemony and dominance of government authorities, who often do not involve the Orang Rimba community in the development process (Nurfathiyah 2021).

Orang Rimba and the Climate Crisis in Sumatra's Tropical Forests

Climate change has had a significant influence on fruit trees in the Sumatran forests. Climate change in Indonesia, including in Sumatra, has caused an increase in temperature, alterations in precipitation patterns, modifications in hydrographs, and shifts in the Oldeman classification (Ruminta et al. 2018). This negatively affects the growth, development, and yield of fruit trees in the Sumatra forests. Fruit plants in forests are a food source for the Rimba people (Jambi Kita 2019).

> The Deities who regulate this climate are called "petahunon". Petahunon oversees the fruit season; People in Fruit. When this season comes, Orang Rimba call it petahunon. (Interview with T2)

Powerful Gods determine seasons and climates according to the beliefs of the Orang Rimba. When this season arrives, the jungle people call the petahunon. This god was also responsible for the dry and rainy seasons. The Petahunon cycle, associated with the dry and rainy seasons, has been patterned every year with varying levels of intensity from year to year but always at relatively the same time interval (Prasetijo 2017). As part of a tropical climate, the rainy season typically lasts from September to December. The 1997 El Niño drought was one of the longest dry seasons. For more than six months, the area where Orang Rimba lived had no rain. Fires broke out everywhere. Some companies around Orang Rimba have deliberately burned land that has just been cleared (Suprapto 2021).

Current climate change is causing a crisis that has had a huge impact on the lives of the Orang Rimba. Hot weather is a type of climate change experienced by the Orang Rimba community. The Meteorology, Climatology, and Geophysics Agency (BMKG) stated that Indonesia experienced a trend of rising temperatures in western and central Indonesia, with the highest temperature increase of 0.95 °C per decade. In Jambi Province, the average temperature is in the range—26–28 °C with a maximum average of 30–32 °C (KKI Warsi 2019).

There are various impacts felt by Orang Rimba when there is an increase in temperature.

1. Increased temperatures affect a large number of malaria cases in Jambi. There was a relationship between the explosion of mosquito populations and an increase in the temperature. Warming temperatures shorten the life cycle of mosquitoes from eggs to pupae and adults. Malaria is a scary of Orang Rimba. In 2015, the KKI Warsi and Eijkman Bio Molecular Institute (LBH) conducted a health study on the Orang Rimba groups in Tebo, Batanghari, and Sarolangun Regency. The findings indicated that 24.26% of patients in the study population were afflicted with malaria. The prevalence of malaria within the Orang Rimba community exhibited a significantly greater incidence when compared to the overall data on the prevalence of malaria in the Jambi province (Haryanto et al. 2016).
2. The increase in temperature triggers forest fires and disrupts the lives of the forest-dependent Orang Rimba (Purnomo and Anand 2014). Orang Rimba begins at birth, grows, and breeds in the jungle. They live according to the rules, norms, and customs of the Orang Rimba (Zuhdi 2013). Forest fires in Sumatra have been one of the impacts of climate change. Climate change can increase the risk of forest fires by increasing temperatures and decreasing rainfall, making forests drier and more susceptible to fires (Werf et al. 2010). Forest fires in Sumatra and other parts of Indonesia have contributed to global carbon emissions. During 2001–2009, grassland and savanna fires accounted for 44% of the carbon emissions, followed by fires from tropical deforestation and forest degradation at 20%, forest fires in the extratropics at 15%, agricultural waste burning at 3%, and tropical peat fires at 3%. Climate change also causes erratic weather changes, usually in August, the dry season; however, in August, at this time, in several places where Orang Rimba is located, there is still frequent rain.

Orang Rimba and Oil Palm Plantations

Extensive environmental changes have occurred in Sumatra Island since 1980. These changes included government programs for the construction of settlements for transmigration from other islands, construction of highways, expansion of palm oil plantations, and development of industrial timber plantations. These changes have driven changes in Sumatra's environment and landscapes. Alterations in the environment and geography significantly influence the Orang Rimba ethnic group and other communities inhabiting Jambi Province. This is particularly noteworthy, as the livelihoods of the Orang Rimba are intricately tied to forests (Sandbukt 2000).

Changes in the environment and landscape can be recognised through Landsat 8 imagery processed by the KKI Warsi. In 2012, Jambi Province had 1,159,559 ha of forest; in 2019, it was 900,173 ha, and it lost 258,846 ha of forest cover (KKI Warsi 2019). Human activities, dominated by the conversion of natural forests into industrial plantation forest concessions as well as legal and illegal land clearing for plantations and mining that expanded into protected areas, caused this loss of forest cover.

Orang Rimba Are in Food Crisis and Poor

Orang Rimba faces various challenges, including the massive conversion of forest areas to palm oil plantations and climate change. Orang Rimba lost a lot of land for agriculture, reduced forest products as a source of food in the form of: tubers, fruit, and forest animals, because of forest conversion. Makekal Ulu and Air Hitam transmigration programs converted the living areas of Orang Rimba into oil palm plantation plots that were considered capable of multiplying the community's economy in a short time. However, in reality, certain groups of the Orang Rimba obtained oil palm plantation plots, while other Orang Rimba groups still relied on hunting and gathering food from dwindling forests. Orang Rimba, known for its hunting-gathering culture and nomadic life, relies heavily on food sources from the forest even though they are already familiar with farming (Wardani 2022).

The climate crisis made the problems of the Orang Rimba even more complex; in a prolonged dry season, the forest, as a source of food for the Orang Rimba, did not provide optimal results. Tubers cannot thrive as carbohydrate sources during the dry season. Fruit plants also experience drought, so they cannot fulfil the needs of the Orang Rimba. Hunting products as a source of Orang Rimba protein such as pigs, deer, deer and other animals are not maximized because the forest is getting narrower.

Due to the reduction in forest products, Orang Rimba has adapted to the cultivation of various plants. Plants cultivated in Orang Rimba include cassava, jengkol, petai, chili, pineapple, rubber, and palm oils. However, Orang Rimba's efforts to cultivate various commodities have experienced problems owing to the climate crisis. Rice does not thrive during drought, and other crops such as jengkol, petai, chili, pineapple, rubber, and palm oil take a long time to harvest. Cassava is the commodity of choice for planting because it is easy to plant and quick to harvest. However, herds of monkeys and hungry macaques often attack and plunder cassava in Orang Rimba fields.

Forests and land can no longer be the main source of food for the Orang Rimba. Orang Rimba has become highly dependent on external food sources (Prasetijo 2021). The Orang Rimba is forced to buy rice and other food from the surrounding community. The money to buy food for Orang Rimba comes from selling rubber, palm fruits, and honey. The lack of forest products and purchasing power of Orang Rimba have caused it to experience a food crisis. The food crisis had the worst impact on forest loss. The food crisis caused some Orang Rimba to experience hunger and malnutrition (Prasetijo 2021).

Research results from Jambi University provide an overview of maternal and child health in the Orang Rimba Group in the Bukit Suban. Sixty percent of Orang Rimba mothers have experienced child deaths, ranging from miscarriages and newborns to children (Guspianto et al. 2020). The high number of child deaths in Orang Rimba is due to 40% of pregnant women experiencing low nutrition and the stunting of up to 48% of Orang Rimba children. This percentage is very high compared to the WHO standard of 20% or one-fifth of the total toddlers. Orang Rimba, who lives in palm oil plantations, has difficulty obtaining protein and carbohydrate sources, so there are many cases of malnutrition (Suprapto 2021). In cases of malnutrition and unsterile labour, he said, make babies and children of the Orang Rimba very vulnerable, and cases of death are high (KKI Warsi 2020).

The feasibility of a place to live in is an indicator of poverty. The Orang Rimba wander has no fixed place to live there. The government considers that with no houses owned by the Orang Rimba, the Orang Rimba can be categorised as poor. One form of government assistance is to provide housing assistance to Orang Rimba. However, for Orang Rimba, Bukit Dua Belas National Park is home, so there is no need to make a house. The problem faced by the Orang Rimba is not related to housing; however, the source of livelihood is becoming increasingly difficult. Many conflicts outside of Bukit Duabelas arise because their forest resources are depleted (Prasetijo 2021).

For Orang Rimba, accepting resettlement meant choosing to leave the custom. Orang Rimba believes that it would be dangerous if all groups of Orang Rimba accepted being laid off. According to Orang Rimba, the government deliberately wanted to vacate the Bukit Duabelas National Park by building housing outside the customary area and modernising it. The indigenous people must adhere to customs and protect their territories.

> Resettling the Orang Rimba is the same as expelling them from Bukit Dua Belas National Park. Having a place to live, a motorcycle but no job to sustain their lives makes the Orang Rimba even poorer. (Interview with P1)

Some young Rimba people have a critical view that living in housing will actually make them more miserable. Orang Rimba, who lives in settlements, is more miserable than those who live in the Bukit Dua Belas National Park. Settled Orang Rimba can only grow crops and hunts, and is forced to live outside the forest with limited knowledge. As a result, many Rimba were unemployed, despite living in modern times. They own a motorcycle, but have no income. This has led to many cases of palm oil theft, beginning with the Orang Rimba. Orang Rimba, who is critical and knows the conditions that Orang Rimba needs, is not housing but a sustainable livelihood. A sustainable livelihood means that jungle people are able to cultivate commodities that can sustain their lives, and the prices of these commodities are not stable and do not drop during the harvest season.

Alternative Employment for Orang Rimba During Climate Crisis and Poverty

The traditional livelihoods of the Orang Rimba have undergone changes due to industrialisation and the pressure to shift from hunting and gathering practices to rubber and palm oil plantations (Tresno et al. 2022). The livelihood opportunities available to the Orang Rimba community are constrained by progressively diminishing living areas and scarcity of resources (Prasetijo 2017). Today's Orang Rimba does whatever it can make ends meet.

> Orang Rimba currently do anything to fulfill their needs. We can't rely on forest products, we work at whatever we can do as long as we can get money to buy food, including collecting brondol sawit (oil palm fruits). (Interview with OR1)

Historically, the Orang Rimba community relied on a subsistence strategy centred on hunting and gathering within their forest environment. However, owing to changing circumstances, this traditional means of sustenance has become unviable. The Orang Rimba are compelled to adjust their way of life in response to dynamic environmental conditions because of restricted access and the finite availability of forest resources. The Orang Rimba community is faced with the imperative of devising strategies to effectively address several aspects of its existence, particularly pertaining to its means of sustenance. The indigenous community known as Orang Rimba, which resides in the vicinity of oil palm fields, has commenced engaging in the occupation of harvesting oil palm fruits (see Fig. 22.4). The oil palm fruits collected did not come from Orang Rimba's oil palm plantations, but from company-owned oil palm plantations and oil palm plantations owned by the Melayu and Javanese people, who are transmigration communities. Orang Rimba does not own palm oil plantations, although palm oil plantations are customary. On multiple occasions, the Orang Rimba community endeavoured treclaim land within territories previously occupied by businesses, resulting in a state of warfare.

The implementation of the Bukit Duabelas National Park as a designated protected area was primarily intended to preserve the biodiversity and cultural heritage of the Orang Rimba community. However, this initiative has resulted in several conflicts. The interpretation of zoning regulations in the region can be perceived as an endeavour to restrict the spatial availability and accessibility of forest resources

Fig. 22.4 Orang Rimba gathering palm fruit

for the Orang Rimba community, which has historically relied on these resources for their livelihoods (Yusuf and Syafrial 2019). Conflicts have also occurred between the Orang Rimba and Jambi Malay communities because of the increasingly narrow spaces and access to forest resources due to transmigration and deforestation (Harun et al. 2023). Orang Rimba can only make limited use of Bukit Dua Belas National Park, which utilizes non-timber forest products in the form of honey, fruits, game animals that are unable to meet the daily needs of Orang Rimba. Meanwhile, the opening of palm oil plantations, rubber plantations, and fields in Bukit Dua Belas National Park is prohibited because National Park is a conservation area.

Challenges of Oil Palm Fruit Gathering for Orang Rimba

Gathering oil palm fruits (brondol sawit) is an alternative occupation for Orang Rimba when forest resources are unable to fulfil their needs. The Brondol sawit is part of the palm fruit that is detached from the bunch. It falls from a tree when the fruit is ripe or the harvesting process is not properly performed. The Orang Rimba picks palm fruits one by one and collects them into sacks.

Orang Rimba have been working as palm fruit pickers since 2017. Following the Melayu community and the transmigrant community, who do this work in the company's palm oil plantations. The Orang Rimba actively participated in this activity because they considered it "safe," as it involved collecting palm fruits that fell and were not utilized by the palm oil company, rather than being categorized as theft. Collectors who sell palm fruits to small-scale coconut oil-processing companies accept palm fruits collected by Orang Rimba, albeit of lower quality. An overview of the quality of Orang Rimba oil palm fruit is shown in Fig. 22.5.

Almost all members of the Orang Rimba family carry out this oil palm fruit collection activity from their children to the elderly (see Fig. 22.5). In one family, there can be to 2–4 family members carrying out the activity of collecting oil palm fruits. The palm fruits collected by Orang Rimba are sold to collectors, usually local

Fig. 22.5 The oil palm fruits (brondol sawit) and Orang Rimba family activities in oil palm plantations

Melayu people or transmigrant communities around the Orang Rimba. Not all palm fruits of Orang Rimba are suitable for sale. Collectors usually buy palm fruits that are still suitable for resale to small-scale palm oil-processing companies. Orang Rimba's income from gathering palm fruit is uncertain, but Orang Rimba earn an average of Rp. 100,000 (US $6.45) per day from selling palm fruit.

The income of Rp. 100,000 (US $6.45)/day is relatively large compared to the cultivation of crops such as rubber planting rice, planting fruits and vegetables, or opening palm oil plantations. Opening palm oil plantations requires capital and cost until harvest. The cultivation of these crops often leads to crop failure owing to plant pests or droughts caused by weather changes. However, collecting oil palm fruits from other plantations does not require capital to cultivate crops. When viewed from the cultural context of Orang Rimba, the activity of collecting palm fruit is a gathering and hunting activity that does not require capital or technology. Unlike the previous gathering and hunting activities carried out by the Orang Rimba which only concocted and hunted forest products, they could be directly used for daily needs. Orang Rimba actively carry out the palm fruit collection activity at this time. They first sell the collected palm fruits and then use the proceeds to buy necessities. Thus, Orang Rimba developed, namely, buying and selling transactions using money. The Orang Rimba uses the income from collecting oil palm fruits to buy necessities of daily life, especially food, from the village market around the Bukit Dua Belas National Park. The Orang Rimba uses the income from collecting oil palm fruits to buy necessities of daily life, especially food, from the village market around the Bukit Dua Belas National Park.

> We collect oil palm fruits because there is no other work we can do to make a living. At first we were afraid that the owners of the oil palm plantations would not allow us to collect oil palm fruits. (Interview with OR2)

Gathering oil palm fruits is a job in which the Orang Rimba are forced to do in order to fulfil their daily needs. Orang Rimba is also uncomfortable picking up palm fruits in palm oil plantations that do not belong to them. It is not uncommon for plantation owners to not allow Orang Rimba to pick palm oil fruits from their plantations. Oil palm plantation owners often chase away and intimidate the Orang Rimba. They also check the oil palm fruits collected by the Orang Rimba because they worry that the Orang Rimba might be dishonest and take the palm oil fruits directly from the palm trees instead of those that fall from them. If oil palm fruits are taken from an oil palm tree, it is considered that the Orang Rimba stolens it.

Climate change also affects the pick-up of palm oil fruits. In the dry season, the palm fruit harvesting truck can enter the oil palm plantation and park near the plantation to be harvested so that all palm fruits can be transported. This causes the number of oil palm fruits that can be collected by the Orang Rimba to decrease, and certainly affects Orang Rimba's income. Collecting oil palm fruits is one way in which the Orang Rimba survives in times of climate crisis and poverty. To survive in an ever-changing environment, humans constantly adapt their goals. Re-adaptation is an attempt to unify and integrate the re-assembly processes undertaken by individuals or communities in response to social or natural shocks (Kolopaking et al. 2022). Palm oil plantations can lead to deforestation, increased carbon emissions, and climate change, which can disrupt environmental conditions (Obidzinski et al. 2012), whereas palm oil plantations provide alternative jobs for the Orang Rimba under conditions of climate crisis and poverty.

Palm oil plantations are one of the causes of climate crisis and reduce the open space for indigenous peoples' livelihoods. On the other hand, the presence of palm oil plantations provides employment solutions for indigenous peoples, although these solutions are only lifelines. Indigenous peoples have traditional knowledge systems that are very important for climate mitigation and adaptation but are often underestimated and marginalised by companies, government, and environmental activists (Lewis et al. 2020). Environmental activists reject the presence of palm oil plantations because they damage the environment, but are unable to provide livelihood solutions for indigenous peoples in the area. It is important to include indigenous voices and perspectives in regional climate change adaptation efforts to ensure that their unique knowledge and experiences are considered (Dent et al. 2023). It is critical to recognise that indigenous people are not passive victims but active agents in addressing climate change. Their experiences and perspectives should be valued and integrated into the climate change discourse and decision-making processes (Irlbacher-Fox and MacNeill 2020).

The Indonesian government's intervention to the poverty that Orang Rimba faces has long been done but usually fails, such as building housing for Orang Rimba (Prasetijo 2021). The Indonesian government also launched various poverty reduction programs for vulnerable people, including the Orang Rimba. The social entrepreneurship (Yuda et al. 2023; Setiawan et al. 2023) program is a program to expand the livelihoods of the poor, so the Orang Rimba should be considered as a beneficiary of this program. However, it is necessary to adjust the program according to the character, condition, and situation of the Orang Rimba, which cannot be equated with other poor people. Likewise, social protection programs such as the

Family Hope Program (conditional cash transfer) are expected not to preserve the stereotypes (Habibullah et al. 2024) of the Orang Rimba as poor people.

Conclusions

The climate crisis and the expansion of oil palm plantations are considered to be the cause of the Orang Rimba's poverty. However, the Orang Rimba have adapted by changing their livelihoods from collecting forest products and hunting animals to collecting oil palm fruits from oil palm plantations that do not belong to them. Although the collection of oil palm seeds can fulfill their daily needs, the collection of oil palm fruit is not sustainable because it depends on the willingness of the oil palm plantation owner. Oil palm is a climate-resilient plant, but for indigenous peoples, the massive expansion of oil palm plantations has threatened the livelihoods of indigenous peoples who rely heavily on forest resources.

Environmentalists would argue that the expansion of oil palm plantations disrupts ecosystems and is one of the causes of the climate crisis, development economists would argue that the expansion of oil palm plantations will improve people's welfare. Anthropologists, on the other hand, would argue that the presence of oil palm plantations could undermine local norms, values and traditions. The findings of our research on indigenous peoples will always re-adapt to changes in the natural and social environments even though the adaptation is seen to be slow. Indigenous peoples have adapted to the climate crisis, the presence of oil palm plantations although living conditions still lag behind the civilization of the wider community. Adaptive social protection for indigenous peoples is needed so that they are able to survive and adapt to changing natural and social conditions.

This section recommends that similar research be conducted in other countries, particularly in Africa, South Asia, and Latin America that have indigenous peoples and have leading plantation commodities such as oil palm, rubber and acacia but are considered environmentally destructive and violate indigenous peoples' rights. The next section discusses plantation commodities that are considered environmentally damaging, cause climate change, and narrow indigenous peoples' livelihood space. On the other hand, the presence of plantation commodities can provide alternative livelihoods for indigenous peoples.

Acknowledgements The authors would like to thank Prof. Dr. M. Alie Humaedi, M. Ag., M. Hum, Head of the Research Center for Social Welfare, Villages and Connectivity, National Research and Innovation Agency for providing guidance and opportunities to carry out the writing of this chapter. Indonesian Conservation Community KKI-Warsi for providing data and information for the writing of this chapter.

References

Ahmad M, Oxley L, Ma H, Liu R (2023) Does rural livelihood change? Household capital, climate shocks and farm entry-exit decisions in rural Pakistan. Front Environ Sci. https://doi.org/10.3389/fenvs.2022.857082

Alamgir MS, Furuya J, Kobayashi S, Binte MR, Salam A (2018) Farmers' net income distribution and regional vulnerability to climate change: an empirical study of Bangladesh. Climate. https://doi.org/10.3390/cli6030065

Amalia R, Dharmawan AH, Prasetyo LB, Pacheco P (2019) Perubahan Tutupan Lahan Akibat Ekspansi Perkebunan Kelapa Sawit: Dampak Sosial, Ekonomi dan Ekologi. J Ilmu Lingk 17. https://doi.org/10.14710/jil.17.1.130-139

Arthur RD (2012) The effects of Benso Oil Palm Plantation (BOPP) smallholder farmers' scheme on rural poverty reduction in the Mpohor Wassa East District of Ghana

Balai Taman Nasional Bukit Duabelas (2024) Peta Kawasan Tn Bukit Duabelas. http://tnbukitduabelas.id/peta-kawasan-tn-bukit-duabelas

Barman M, Chowdhury IR (2023) Substance uses and associated sociodemographic factors among male tribes and non-tribal communities of eastern and north eastern states in India: an evidence-based study from the national family health survey-5. Glob Soc Welf. https://doi.org/10.1007/s40609-023-00328-0

Beringer T, Müller C, Chatterton J, Kulak M, Schaphoff S, Jans Y (2023) CO_2 fertilization effect may balance climate change impacts on oil palm cultivation. Environ Res Lett 18:54019. https://doi.org/10.1088/1748-9326/accbd5

Buergin R (2016) Ecosystem restoration concessions in Indonesia: conflicts and discourses. Crit Asian Stud 48:278–301. https://doi.org/10.1080/14672715.2016.1164017

Carmona R, Reed G, Thorsell S, Dorough DS, MacDonald JP, Rai TB, Sanago GA (2023) Analysing engagement with indigenous peoples in the intergovernmental panel on climate change's sixth assessment report. npj Clim Action 2:29. https://doi.org/10.1038/s44168-023-00048-3

Crowe S, Cresswell K, Robertson A, Huby G, Avery A, Sheikh A (2011) The case study approach. BMC Med Res Methodol 11:100. https://doi.org/10.1186/1471-2288-11-100

Dasgupta A, Baschieri A (2010) Vulnerability to climate change in rural Ghana: mainstreaming climate change in poverty-reduction strategies. J Int Dev. https://doi.org/10.1002/jid.1666

Dent L, Donatuto J, Campbell L, Boardman M, Hess JJ, Errett NA (2023) Incorporating indigenous voices in regional climate change adaptation: opportunities and challenges in the U.S. Pacific Northwest. Clim Change. https://doi.org/10.1007/s10584-023-03499-z

Denzin NK, Lincoln YS (2018) The SAGE handbook of qualitative research. Sage

Guspianto G, Asparian A, Wisudariani E (2020) Studi Kualitatif : Faktor Predisposisi Sebagai Upaya Pencarian Pengobatan Pada Komunitas Adat Terpencil Di Desa Bukit Suban, Air Hitam Sarolangun, Jambi, Indonesia. J Kesmas Jambi 4:. https://doi.org/10.22437/jkmj.v4i2.10568

Habibullah H, Yuda TK, Setiawan HH, Susantyo B (2024) Moving beyond stereotype: a qualitative study of long-standing recipients of the Indonesian conditional cash transfers (CCT/PKH). Soc Policy Adm 58:108–121. https://doi.org/10.1111/spol.12946

Harun H, Mustiah RH, Najib K (2023) The conflicts of the Orang Rimba and the Jambi Malay community: reality and resolutions. KnE Soc Sci 2023:578–584. https://doi.org/10.18502/kss.v8i4.12946

Haryanto S, Hayati RF, Yohan B et al (2016) The molecular and clinical features of dengue during outbreak in Jambi, Indonesia in 2015. Pathog Glob Health 110:119–129. https://doi.org/10.1080/20477724.2016.1184864

Indrizal E, Anwar H (2023) The indigenous people Suku Anak Dalam Batin Sembilan livelihood: adaptation and socio-cultural dynamics. ETNOSIA J Etnogr Indones 8:24–43

Irlbacher-Fox S, MacNeill R (2020) Indigenous governance is an adaptive climate change strategy. North Rev. https://doi.org/10.22584/nr49.2020.019

Jambi Kita (2019) Perubahan Iklim Ancam Ketersediaan Pangan Orang Rimba. In: Jambi Kita. https://kumparan.com/jambikita/perubahan-iklim-ancam-ketersediaan-pangan-orang-rimba-1s0jtDyYHrQ/full. Accessed 11 Sep 2023

Kang Y, Khan S, Ma X (2009) Climate change impacts on crop yield, crop water productivity and food security—a review. Prog Nat Sci Mater Int. https://doi.org/10.1016/j.pnsc.2009.08.001

KKI Warsi (2019) Kerusakan Ekosistem, Jambi Rugi Lebih Rp 17 T. In: KKI Warsi. https://warsi.or.id/id/kerusakan-ekosistem-jambi-rugi-lebih-rp-17-t/. Accessed 11 Sep 2023

KKI Warsi (2020) Orang Rimba di Luar Hutan Mulai Kesulitan Pangan. KKI Warsi. https://warsi.or.id/id/orang-rimba-di-luar-hutan-mulai-kesulitan-pangan/. Accessed 13 Sept 2023

Kolopaking LM, Wahyono E, Irmayani NR, Habibullah H, Erwinsyah RG (2022) Re-adaptation of COVID-19 impact for sustainable improvement of Indonesian villages' social resilience in the digital era. Int J Sustain Dev Plan 17:2131–2140. https://doi.org/10.18280/ijsdp.170713

Leal Filho W, Matandirotya NR, Lütz JM, Alemu EA, Brearley FQ, Baidoo AA, Kateka A, Ogendi GM, Adane GB, Emiru N, Mbih RA (2021) Impacts of climate change to African indigenous communities and examples of adaptation responses. Nat Commun 12:13–16. https://doi.org/10.1038/s41467-021-26540-0

Leichenko R, Silva JA (2014) Climate change and poverty: vulnerability, impacts, and alleviation strategies. Wiley Interdiscip Rev Clim Change. https://doi.org/10.1002/wcc.287

Letta M, Tol RS (2018) Weather, climate and total factor productivity. Environ Resour Econ. https://doi.org/10.1007/s10640-018-0262-8

Lewis D, Williams L, Jones R (2020) A radical revision of the public health response to environmental crisis in a warming world: contributions of indigenous knowledges and indigenous feminist perspectives. Can J Public Health. https://doi.org/10.17269/s41997-020-00388-1

Milies MB, Huberman AM, Saldaña J (2014) Qualitative data analysis, 3rd edn. Sage

Najib K, Pahmi P, Gunawan B, Iskandar J, Soemarwoto R, Shinta S, Manalu HSP (2020) Government ecology and the indigenous religion of the Suku Anak Dalam: intersubjective relations in forest conservation in Jambi, Indonesia. https://doi.org/10.7226/jtfm.26.3.303

Nurfathiyah P (2021) Ruang Publik dan Komunikasi Dialogis Orang Rimba di Taman Nasional Bukit Dua Belas Provinsi Jambi. J Ilm Sosio-Ekon Bisnis 24:1–11

Obidzinski K, Andriani R, Komarudin H, Andrianto A (2012) Environmental and social impacts of oil palm plantations and their implications for biofuel production in Indonesia. Ecol Soc 17

Pahmi P, Gunawan B, Iskandar J, Soemarwoto R (2023) Reverse social impact of oil palm plantation expansion: a study of three communities in Jambi, Indonesia. For Soc 7:61–75. https://doi.org/10.24259/fs.v7i1.24803

Paterson RRM (2019) *Ganoderma boninense* disease of oil palm to significantly reduce production after 2050 in Sumatra if projected climate change occurs. Microorganisms. https://doi.org/10.3390/microorganisms7010024

Prasetijo A (2017) Livelihood transformations of the Orang Rimba as tacit resistance in the context of deforestation. Endogami J Ilm Kaji Antropol 1:1. https://doi.org/10.14710/endogami.1.1.1-13

Prasetijo A (2021) Displacement as experienced by the Orang Rimba hunter-gatherers. Senri Ethnol Stud 106:235–252

Purnomo EP, Anand PB (2014) The conflict of forest tenure and the emergence of community based forest management in Indonesia

Roy S, Bose A, Chowdhury IR (2022) Darkness under the city lights: a qualitative study on the life of urban homelessness, evidence from Siliguri city of West Bengal, India. GeoJournal. https://doi.org/10.1007/s10708-022-10753-x

Ruminta R, Handoko H, Nurmala T (2018) Indikasi perubahan iklim dan dampaknya terhadap produksi padi di Indonesia (Studi kasus : Sumatera Selatan dan Malang Raya). J Agro 5:48–60. https://doi.org/10.15575/1607

Runtuboi YY, Permadi DB, Sahide MAK, Maryudi A (2021) Oil palm plantations, forest conservation and indigenous peoples in West Papua province: what lies ahead? For Soc 5:23–31. https://doi.org/10.24259/fs.v5i1.11343

Saifullah MK, Kari FB, Othman A (2018) Poverty among the small-scale plantation holders. Int J Soc Econ 45:230–245. https://doi.org/10.1108/IJSE-10-2016-0296

Sandbukt Ø (2000) Deforestation and the people of the forest: the Orang Rimba or Kubu of Sumatra. Indig Aff 39–47

Setiawan HH, Yuda TK, Susantyo B, Sulubere MB, Ganti M, Habibullah H, Sabarisman M, Murni R (2023) Scaling up social entrepreneurship to reduce poverty: exploring the challenges and opportunities through stakeholder engagement. Front Sociol 8. https://doi.org/10.3389/fsoc.2023.1131762

Suprapto Y (2021) Adaptasi orang rimba atasi perubahan Iklim. https://www.mongabay.co.id/2021/09/03/adaptasi-orang-rimba-atasi-perubahan-iklim/. Accessed 11 Sep 2023

Tam C-L, Chew S, Carvalho A, Doyle J (2021) Climate change totems and discursive hegemony over the Arctic. Front Commun. https://doi.org/10.3389/fcomm.2021.518759

Townsend J, Moola F, Craig M-K (2020) Indigenous peoples are critical to the success of nature-based solutions to climate change. Facets. https://doi.org/10.1139/facets-2019-0058

Tresno T, Effendi N, Rizaldi R, Doni H, Harapan TS, Mutashim M (2022) Etnoforestri Orang Rimba Di Taman Nasional Bukit Dua Belas Provinsi Jambi. J Antropol Isu-Isu Sos Budaya 02:195–203. https://doi.org/10.25077/jantro.v24.n2.p195-203.2022

van der Werf GR, Randerson JT, Giglio L, Collatz GJ, Mu M, Kasibhatla PS, Morton DC, DeFries R, Jin Y, van Leeuwen TT (2010) Global fire emissions and the contribution of deforestation, savanna, forest, agricultural, and peat fires (1997–2009). Atmos Chem Phys. https://doi.org/10.5194/acp-10-11707-2010

Wardani EM (2022) The sacred forest of the Orang Rimba hunter-gatherers of Sumatra. Sacred forests of Asia. Routledge, London, pp 292–293

Yuda TK, Setiawan HH, Habibullah, Susantyo B, Suyatna H (2023) Institutionalized social entrepreneurship in Indonesia: a report on the challenges of state-driven social entrepreneurship promotion within the five cities' social assistance program. Asia Pac J Public Adm 1–17. https://doi.org/10.1080/23276665.2023.2270087

Yusuf M, Syafrial S (2019) A big battle: the state vs. indigenous people (case study in Jambi province). J Ilmu Sos Ilmu Polit 23:59–72. https://doi.org/10.22146/JSP.37568

Zuhdi MN (2013) Tuhan, di mana Kau Berada? Kajian Etnografi tentang Sistem Kepercayaan Orang Rimba di Taman Nasional Bukit Dua Belas Jambi. Lap Penelit

Chapter 23
Women's Vulnerability and Adaptive Capacity to Climate Change in Agrarian Settings of Zimbabwe

Douglas Nyathi, Joram Ndlovu, Admire Mare, Munyaradzi A. Dzvimbo, and Mduduzi Ndlovu

Abstract Economic development is impacted by changes in the global environment and how frequently and how intensely climate-related calamities like floods and droughts occur. Even though everyone is impacted by climate change, anecdotal evidence indicates that different social groupings, even within the same polities, have significantly different capacities to adjust to the changing environment. This study examines the factors that make women more vulnerable to the effects of climate change, adaptation tactics, and relative degrees of adaptive capacity in a few Zimbabwean agrarian settings in Matabeleland. The framework for adapting to climate change and the gender and power theory serve as the foundation for this qualitative investigation. The study is based on the knowledge that women face disproportionate workloads, little decision-making authority, and restricted access to and control over productive resources, services, and employment prospects. The ability to control and obtain resources, to use and access information services, and to innovate in response to changing opportunities and challenges are the factors that mediate the capacity to adapt to climate change. The government should support gender-responsive land and climate change policy, the study suggests. A concerted

D. Nyathi (✉)
School of Public Management, Governance and Public Policy, University of Johannesburg, Johannesburg, Gauteng, South Africa
e-mail: douglasnyathi08@gmail.com

J. Ndlovu
School of Social Sciences, Howard College, University of KwaZulu Natal, Durban, South Africa
e-mail: ndlovuj1@ukzn.ac.za

A. Mare
Department of Communication and Media, University of Johannesburg, Johannesburg, Gauteng, South Africa
e-mail: amare@uj.ac.za

M. A. Dzvimbo
Department of Geography, University of Free State, Bloemfontein, South Africa

M. Ndlovu
Department of Development Studies, Lupane State University, Lupane, Zimbabwe

U. Mukhopadhyay et al. (eds.), *Climate Crisis, Social Responses and Sustainability*, Climate Change Management, https://doi.org/10.1007/978-3-031-58261-5_23

effort is required to enhance the ownership and control of productive resources, particularly land and forest products, by women. In the battle against climate change, it is critical to raise women's consciousness and provide them with more knowledge about adaptation and climate change. The empowerment strategy has the potential to greatly increase household resilience to the effects of climate change.

Introduction

In Africa, the anticipated rate of global warming over the twentieth century was 0.26–0.5 °C each decade (IPCC 2007). Livelihoods will be impacted as a result of this trend, which is predicted to persist and possibly even pick up steam. The Intergovernmental Panel on Climate Change (IPCC 2007) projects that by 2080, yearly mean surface air temperatures would rise by 3–4 °C under a medium–high emission scenario. The body of research supporting the claim that climate change poses a serious threat to achieving sustainable development objectives is growing quickly (Davidson et al. 2003; Goodrich et al. 2019). This suggests that the local population, who either directly or indirectly rely on agriculture, particularly rain-fed agriculture, will face challenging times ahead and lack the resources or coping mechanisms to deal with the changes. Additional implications of climate change that have been noted include poor intra-seasonal geographical and temporal distribution of seasonal rainfall, increased frequency of extreme events such protracted dry spells, droughts, and floods, and decreased reliability of rainfall. The above scenario presents humanity with problems of a new kind, especially for the people living in lower-income nations like Zimbabwe (Abedin et al. 2013). The uncertainty surrounding climate change research and the paucity of disaggregated data have contributed to the relative paucity of in-depth analysis of the gender aspects of climate change to date. Maps and analyses of the gendered effects of climate change are therefore challenging. The literature (Mare 2013; Thomas et al. 2019; Yadav and Lal 2018) shows that women are more likely to be found in the poorest segments of society, have fewer resources to help them cope, and are more dependent on resources that are sensitive to climate change due to the gender division of labour, making them disproportionately vulnerable to climate change (Mare 2013). Furthermore, it is widely acknowledged that climate change is not gender-neutral and has substantial long-term detrimental effects on the human community and societal vulnerability. When women's resources, agency, and opportunities are inferior to those of men, they are frequently more susceptible to the effects of climate change (Ambani and Fiona 2014; Dube et al. 2017).

Since women make up the majority of the world's impoverished and depend more on natural resources which are threatened by climate change for their livelihoods, they are more susceptible to the consequences of climate change than males are in many of these circumstances. In addition, they encounter political, social, and economic obstacles that reduce their ability to cope. Firstly, the biggest obstacles are faced by those tasked with securing food, fuel for cooking, and water for heating. Secondly, women in rural regions are disproportionately impacted by climate change due to limited mobility combined with uneven access to resources and decision-making processes. Therefore, it is critical to develop gender-sensitive response plans for the environmental and humanitarian emergencies brought on by climate change. By implementing adaptation and mitigation methods, a 'gender analysis' in the context of climate change fosters knowledge of the ways that men and women are differently impacted by climate-related dangers.

This chapter presents data collected from two provinces in Zimbabwe. Matabeleland North and South provinces are located in a semi-arid area with irregular and scarce rainfall patterns and temperature swings. Significant regional and temporal variability in rainfall can be observed (Muchuru et al. 2016). The variations include changes in rainfall patterns, including an increase in the frequency and intensity of heavy rainfall events, a rise in the percentage of years with low rainfall, a decrease in low-intensity rainfall events, and an increase in the frequency and intensity of mid-season dry spells (Mushore et al. 2021; Unganai 2009). The agro-ecological zones have shifted as a result of climate change, which has made agricultural production more arid in certain areas (Brown et al. 2012). The two provinces' declining water supply is partly due to the protracted drought, which has put enormous strain on the surface and ground water infrastructure. Due to the semi-arid environment and geography, the region does not have large floodplains and swamps. An increasing amount of data indicates that climate change will have an impact on human health through increased frequency of floods, storms, fires, and droughts (Kapuka and Hlásny 2021); altered geographic ranges of infectious disease vectors, such as malaria and other mosquito-borne illnesses like dengue; higher rates of diarrheal illnesses and water-borne pathogens like cholera; and higher rates of cardio-respiratory morbidity and mortality linked to ground level ozone (Brown et al. 2012).

The lack of clear mention of climate change in Zimbabwean urban policy is a result of the increased vulnerability of human settlements. Experts claim that the lack of a national framework for climate change is one of the primary reasons it has not been sufficiently addressed in policy. The great diversity of climate change hazards to ecosystems and associated societal responses are discussed in this chapter, which may serve to direct future research efforts and provide guidance for farmers' adaptation strategies. Thus, the vulnerability of women to climate change and the dynamics surrounding their capability for adaptation are examined in this chapter. The chapter aimed to answer the following research question: What are the contributing variables to the susceptibility of women in rural settings to climate change? What kinds of adaptive techniques do women in the research field employ? What factors influence the study area's women's ability to adapt to climate change? What kind of institutional resources are available to help stressed-out women?

Conceptualisation of Key Concepts and Literature Review

Differential Vulnerability

Vulnerability is the degree to which a system is sensitive to or incapable of adjusting to the negative effects of climate change, according to IPCC (2001). Hazards brought on by climate change can be catastrophic if the community is weak. Exposure, sensitivity, and adaptive ability all influence vulnerability (Smit and Wandel 2006; Abeka et al. 2012). Fraser et al. (2011) argues that the level of vulnerability of a system is determined by its agroecosystem, socio-economic status, and institutional capacity. Vulnerability is a function of factors which determine how much a system may be affected by perturbations. According to Thomas et al. (2019), on the one hand, adaptive ability differential exposure and susceptibility to climatic impacts are driven by access to resources. Some variables impact vulnerability and, consequently, resource access. These include social processes of marginalization and deprivation, governance issues and information access about climate risk (Thomas et al. 2019; Nyathi et al. 2022). The determinants of men's and women's susceptibility and adaptive capacity are gendered and influenced by sociocultural, economic, and political structures and processes (Carr and Thompson 2014: 12). Women's susceptibility is increased and their ability to adjust to the repercussions of catastrophes is impacted by factors including not having control over or access to resources and not having any entitlements (Denton 2002; Sultana 2014).

Women and Climate Change Vulnerability

Human security is seriously threatened by climate change, particularly for women, who account for 70% of the world's impoverished. There is general agreement that the gender dimensions of vulnerability resulting from current socioeconomic inequities and gendered employment divides will be made worse by climate change (Dodman 2010; Atsiaya et al. 2019). The disparities in how exposed and vulnerable men and women are to the hazards associated with climate change can be attributed to several reasons. A person's access to products and services is impacted by gender inequalities. Women's access to credit and employment possibilities is restricted by certain markets and organisations. According to Cuomo (2011), there is a significant gender disparity in productivity and earnings worldwide: women typically make between 30 and 80% of what men do annually. Based on the World Bank's assessment conducted in 141 countries, approximately 103 nations maintain legislative restrictions on women that limit their economic potential (Mishra and Mohapatra 2017). According to data from emerging nations, women are employed in agriculture and basic agricultural production at a higher rate than men. These women, more than their male counterparts, frequently use their money to support their families. Despite making up about half of the agricultural labour force, women only own 10–20% of the land.

According to Yadav and Lal (2018), the main causes of their susceptibility to climate change include poverty, gender inequality, shaky land rights, a strong reliance on agriculture, and limited access to information and education. Brown et al. (2012), concluded that 70% of Zimbabwean women work as smallholder farmers who rely on resources that are sensitive to the environment and rain-fed agriculture. For this reason, women are especially susceptible to the consequences of climate change. The limited pool of assets, social marginalisation, immobility, and exclusion from catastrophic decision-making processes further complicate vulnerability (Yadav and Lal 2018). According to Demetriades and Esplen (2010), women are more likely to be exposed to water-borne illnesses when they use water from unclean ponds as a result of water sources drying up. Dube et al. (2017) state that daily tasks like gathering firewood and water belong to women, whereas seasonal tasks performed by males like hunting and field fencing belong to them. Men sometimes get involved while women do these activities all year round. Droughts brought on by climate change and population growth force women to trek farther to gather firewood and water (Dube et al. 2017). During the agricultural season, planting activities are mainly done by women. Dry spells are one of the major climate related stresses smallholder farmers are facing in Zimbabwe (Jiri et al. 2015). They affect germination rates and replanting will have to be done to try and restore the crops that would have failed to germinate. This doubles the effort by women to produce food for the households.

Adaptation Strategies by Women

In addition to being proactive and agents of hope for adaptation and reduction of the effects of climate change, women are not merely passive victims of this phenomenon. They use their knowledge and skills to create and implement different adaption techniques to lessen the damage. According to Yadav and Lal (2018), women care about environmental issues and strongly support laws that aim to restore the environment. Denton (2010) argues that policies addressing climate change must consider the interests of all parties concerned, particularly women, if they are to provide a sustainable future through the integration of development and environmental challenges. Lawson et al. (2020) report that women farmers reacted differently to the impacts of climate change. These include understanding how to use grains that have been stored, adjusting planting and harvesting schedules, practicing horticulture during the dry season, practicing conservation agriculture, and cultivating new crops (Ndlovu et al. 2023). Most of these adaptable techniques came from personal experience or information shared by farmers.

Conceptual and Theoretical Frameworks

The adaptation framework and the gender and power theory served as the study's pillars. The adaptation framework facilitates comprehension of household reactions to adaptation for academics and practitioners. Planning for climate adaptation and assessing climate risk and vulnerability are the two primary phases of the framework (Lama and Wester 2021). The framework serves as a useful beginning point for planning, notwithstanding the complexity of the process of climate change adaptation (Davoudi et al. 2009). Since climate change adaptation necessitates a reaction to local impacts, there is no one-size-fits-all strategy, and any programme must be contextualised and community-tailored (Dittrich et al. 2016). The adaptation framework places significant emphasis on the analysis of climate risks and their consequences. This makes it possible for people creating adaptation policies and programmes to consider elements like the advantages of mitigation, farmer capacities, and the significance of climate risk (Melisa-Rojas et al. 2017; Dibakoane et al. 2022). This study employed the Adaptation Framework to investigate women's responses to the effects of climate change in agricultural environments. More specifically, this strategy is essential to guaranteeing a broader understanding of the factors that contribute to smallholders' successful climate change adaptation.

According to Eriksen and Lind (2009), the framework lays forth a technique for climate change adaptation that promotes sustainable development as opposed to the other way around. The APF focuses on practice over theory, building on the knowledge that developing nations already have about vulnerable systems like public health, agriculture, water resources, and disaster management. Its goal is to take advantage of existing themes and synergies to facilitate more informed policy-making (Adger 2003). We also embraced the theory of power and gender. Robert Connell's theory of gender and power is a social structural theory that draws on philosophical literature on gender and power imbalance as well as sexual injustice. The sexual division of labour, the sexual division of power, and the cathexis structure are the three main social structures that define the gendered relationships between men and women. Wingood and DiClemente (2000) and Ambani and Fiona (2014), contend that the theory's basic claim is that gender inequality is one of the most enduring patterns in the distribution of power.

Study Materials and Methods

A qualitative case study research methodology was employed in this study. Utilising a case study technique allowed for a deeper exploration of the difficulties and the identification of thought and attitude patterns (Rice and Ezzy 1999; Yin 2006). The research design also allows the researchers to get an in-depth analysis of individuals and groups who are the target population as compared to other models (Creswell 2007). The Matopo and Bubi districts of Zimbabwe were the study's locations. The districts of Matopo and Bubi are in the province of Matabeleland South and North, respectively. Specifically, we purposively selected one farm from each district. In Matopo, the study was done at Fox Farm, while in Bubi we selected Rocksdale Farm. These two farms were selected because of their livelihood patterns. Participants of the study were mainly women above 35 years of age and had resided in the two study farms in the last 10 years. This group is thought to possess an extensive understanding of how women in the research sites can adjust to climate change. Data was collected between the 10th of January and the 8th of March 2020. To establish rapport with the leadership in the study areas and gain a deeper understanding of the two purposefully selected sites, a visit to the two study farms was conducted before data collection. Focus groups, in-depth interviews, key informants, and observations were used to gather data. According to Braun and Clarke (2006), a focus group is a small group of people who have been specifically chosen to participate in an open discussion on a particular topic of interest. In each farm, ten people participated in one focus group discussion.

The traditional leadership in each of the two locations helped with the purposeful selection of these participants. Each focus group deliberation took an average of one hour. Focus group discussions were used because they are capable of providing insights into the creation of opinions which cannot be easily achieved by individual interviews (Punch 2013). Key informants included traditional leadership (4), local authority representatives (2), AGRITEX officers (2), Women's Affairs representatives and NGO field staff (2). The local language *IsiNdebele* was used in our engagements so that participants could easily express their views. To enhance data validity and reliability, digital recorders were used so that all the discussed information is not lost. The participants gave their agreement for all of this to happen. Afterwards, an English transcription of the recorded data was made. Thematic analysis was used as the method for data analysis. In qualitative research, this methodology is the most widely used. According to Braun and Clarke (2006: 143), part of its appeal stems from its lack of reliance on any certain theoretical framework or epistemology persuasion. As part of our data analysis process, we had to become familiar with the data, make some preliminary codes, find, and define themes, and then write the report (Rice and Ezzy 1999). The report writing stage includes the synthesis of the data and the utilisation of first-hand anecdotes from the participants to substantiate study conclusions.

Key Findings

The chapter's findings are divided into four primary sections. The first section examines the factors that make women in rural areas more vulnerable to the effects of climate change. The strategies used by women in rural contexts to adapt to climate change are the subject of the second section. The third section focuses on the variables that affect women's ability to adapt to climate change in these kinds of environments.

The final section examines the two study farms' institutional support for adapting to climate change.

Women's Understanding of Climate Change in the Study Area

In order to gauge women's knowledge of climate change, icebreaker questions were posed to research participants. The conversations revealed that women are aware of and concerned about climate change. It was noted that the participants' education has enhanced their knowledge and comprehension of climate change and inconsistency. The majority of participants referred to the notion that patterns of rainfall and temperatures are changing due to climate change. This runs counter to the findings of Dube and Phiri (2013), who discovered that most farmers were unaware of climate change. Their current understanding of the phenomenon seems to be greatly influenced by their land and forestry-based livelihoods and frequency of drought and high temperatures in the past years. One participant argued that:

> Climate change encompasses changes in rainfall patterns and temperatures. What we have seen here is a reduction is rainfall leading to a recurrent in droughts. We have also witnessed an increase in temperatures as well. These changes have had a negative bearing on our rain-fed agricultural activities. Droughts have affected our crop yields as well as pastures. For those who depend on mopane worms, they have seen a reduction in their number as well. High temperatures kill these precious worms. (Female participant; 10 January 2020)

One participant highlighted that climate change has also brought with it some water stress challenges in the area. She pinpointed that:

> The changing climate is increasing water stress challenges in our communities. A substantial number of boreholes have run dry creating problems in terms of securing water for both domestic and animal consumption. We now travel several kilometres to secure water, something that. (Male participant; 13 January 2020)

When further probed on why they think the climate is changing, participants responded diversely. Some believed that the changing climate was the result of enraged gods. Talks showed that some people thought that the introduction of Christianity had caused many important ceremonies, like those for creating rain, to be abandoned by most societies. However, some did mention that climate change was emanating from the emission of gases that damage the ozone layer.

Drivers of Women's Vulnerability to Climate Change

To investigate the variables that raise women's susceptibility to climate change, study participants were given questions. Participants' awareness of how climate change affects their livelihoods and the gendered disparity in people's susceptibility to this challenge emerged from the conversations. Focus groups and in-depth interviews demonstrated that because of gendered disparities with men, women in rural areas are nonetheless vulnerable to stresses like climate change and variability. It was discovered that women's restricted access to and control over productive resources, particularly land, was significantly feminising poverty and increasing vulnerability to climate change. The unequal power dynamics that govern decision-making involvement and information access further exacerbate the situation. The results are consistent with previous studies (see Fraser et al. 2011; Thomas et al. 2019). Participant discussions also suggested that gendered migration, restricted access to technology, and low levels of education increased women's susceptibility to climate change. One female participant alluded that:

> The situation of women in land reform areas is terrible. Most women you see on this farm and the ones you have visited work on the land that they do not own. Apart from that, women lack information on climate change and adaptation. The other challenge is that as women we cannot diversify our livelihoods portfolios compared to our male counterparts. Livelihood diversification needs capital. Without financial capital, you cannot make it for example in small-scale mining. Given the implications of climate change on rainfed agriculture and induced water stresses in the community, women now have an extra burden in their caring roles. (Female participant; 22 February 2020)

This concurs with Yadav and Lal (2018) in that lack of education and social marginalisation increase women's vulnerability. However, this study found that women are more vulnerable as they are financially constrained to diversify their livelihoods compared to their male counterparts.

Climate Change Adaptation Strategies

Participants were also asked questions designed to help us record the approaches to climate change adaptation that women have taken to address issues brought on by the changing climate. The intention was also to explore what informed the decisions to adopt particular strategies. The interviewees alluded to several adaptation measures they utilise to lessen the effects of climate change concerning agricultural and household food security. Participants' conversations revealed that women are essential to ensuring the food security of households, and as such, they are often obliged to put forth a great deal of effort to obtain the resources and methods necessary to maintain the food supply. The in-depth interviews and focus group discussions revealed that cutting back on meals is a key strategy for addressing food poverty at the family level. Adults ate once a day in the evening while children ate in the morning as well as in the evening. The intention was to reduce household consumption of food. Interviewees posited that they had to diversify their livelihood portfolios to increase household incomes. However, the diversification of livelihoods increased the workload for women who are expected to care for household members. One participant posited that:

> Women in this community are the cornerstone of all households. Providing food for our families is something we do quite often. Given the food insecurity challenge we face, we

> end up being forced to reduce meals a day. Some households in this community have gone to the extent of diversifying their livelihoods to include off and non-farm activities. In this community, most households have ventured into small-scale mining. However, some of us without capital, only diversify into less lucrative livelihood options such as selling our labour while the well-to-do households are venturing into profitable activities such as mining. (Female; 9 March 2020)

Participants also indicated that some women have migrated to nearby urban areas in search of livelihood opportunities. Migration in search of employment rises after some calamities such as droughts. It emanated that women who mainly migrate are those heading households compared to the married. Married women are less mobile because of some culturally embedded roles and responsibilities at the household level. A substantial number of female migrants contribute to the informal labour market in urban areas. The main livelihood adaptation strategies that occupy women in urban settings include vending, domestic work, beer brewing and in extreme cases prostitution (Ndlovu et al. 2023). The findings of this study reveal that most women who migrate become very vulnerable to poverty and other stressors as they continue to lack resources access to and control of resources in urban areas. Remittances were also mentioned as an important adaptation strategy. Some participants indicated that they received remittances from their husbands working in nearby towns, mines, and neighbouring countries such as South Africa and the United Kingdom. However, it emanated that relying on remittances was very much irregular as a survival income.

Discussions with participants also revealed the importance of sending children to school as a long-term adaptation strategy. From the focus group talks, it emerged that participants thought that one strategy to avoid relying on livelihoods based on land and forests, which are damaged by climate change, would be to send children to school. Women desire skilled and educated children who may obtain wage employment and therefore lower the tradition of depending on natural capital. One participant hinted that:

> What we have realised is that those with educated children easily cope with the impact of drought and other challenges such as food insecurity. If they are educated and employed, there is no way they can die of hunger. (Female; 7 March 2020)

Women in the two study farms classified children's education as a critical long-term adaptation strategy in dealing with persistent drought. Simotwo et al. (2018), who contend that there is a high correlation between educational levels and adaptive capacity, corroborate the findings of this study. Education enhances people's ability to grow their abilities and, in the process, increases their potential to hold positions in society that provide them with access to a variety of knowledge on coping with climate change adaptation as well as significant revenue streams (Silvestri et al. 2015).

Adaptive Capacity to Climate Change

To learn more about the variables influencing women's ability to adapt to climate change in the study locations, study participants were also given questions. The goal was to identify the barriers preventing women in these rural, climate-vulnerable communities from responding effectively to change and variability. Interviewees stated that numerous obstacles prevent them from adapting to climate change. The chapter's conclusions demonstrated that several variables, such as a lack of information, limitations on financial resources, a lack of government support, a lack of water supplies, and political difficulties, affect women's ability to adapt to climate change. It emanated from the anecdotes that inadequate knowledge of climate change and adaptive capacity is one crucial factor hindering the success of adaptive strategies. Women spend most of their time on unpaid care and they invest less in educational programs as they have the burden of care. One participant indicated that:

> Most women lack information about climate change adaptation and I think this is one big challenge in this community. It is something that needs urgent attention from our leadership and other stakeholders. (Female participants; 22 February 2020)

Women play an important role in smallholder agriculture and access to agricultural inputs influences productivity. A reduction in agricultural production has a bearing on household food security as well as purchasing power parity. Participants of the study opined that responding to climate change has been made difficult by a lack of farming inputs, especially draught power and seeds. Although some women indicated that at one point, they got small grains from a government and non-state organisation partnership project, they indicated that such support came late into the cropping season. Women were finding it challenging to successfully adjust to climate change and fluctuation due to the government's insufficient help. Discussions also indicated the unaffordability of farming inputs as one challenge affecting women in agrarian settings. Women also emphasised that one of their biggest obstacles to adjusting to climate change was a lack of assets. One participant summarised the main hurdles they face in adapting to climate change as follows:

> You need to understand that land reform areas have their specific challenges that impact on our lives and livelihoods. For your information we still do not have titled deeds for these pieces of land. We are always threated by evictions two decades down the line. How do you invest on your land given what we are experiencing as farmers? We also have the problem of squatters in most resettled farms. Apart from that, we don't have dams for irrigation. Everyone knows that rain-fed agriculture no longer has capacity to sustain our food security. We only get promised dams and irrigation schemes toward elections. (Male participant; 7 March 2020)

Women's ability to adjust to the effects of instability and climate change is impacted by the absence of suitable remedies. The availability and accessibility of technology affects the strategies for adaptation. Agricultural adaptation techniques include direct or indirect use of technology such as plant breeding or drought proofing. Community limited access to technology has limited their adaptability; as technology is expensive. This has made it harder for them to adjust to the consequences of climate change. The resources available are retrograde, for example, women had no clue about greenhouses. The findings of this study are corroborated by Kuhl (2020), who shows that although technology can increase household resilience to climate change, acceptance and resilience are not fully understood in relation to each other. The global south continues to have very low rates of women adopting technology because to the conflict between the rate of adoption and the severity of climate change (Kuhl 2020).

Institutional Support and Capacity Building in Agrarian Settings

The last object of the chapter was to explore the availability of institutional support meant to assist women under stress in agrarian setting. Participants were asked questions around the availability of programmes such as food relief, agro-loans, farming trainings, information dissemination on climate change, smart agriculture, and access to markets. It emanated from the engagements that there is limited access to institutional support for women in land reform areas. The interviewees conjectured that their inability to provide collateral caused difficulties in obtaining credit plans. One participant hinted that:

> We do not have collateral to use to get loans to finance some of our agricultural or entrepreneurship activities. Remember that government has been reluctant to give land reform beneficiaries title deeds. There is no finance house that can give you money without title deeds or collateral in case you fail to pay back their money. At one-point Agribank promised us loans without collateral. However, the scheme never materialised. There is not even a single farmer who got money from that bank through that programme. (Male participant; 6 February 2020)

The only institutional support the participants hinted on was on farming. It emanated that a number of organisations and government were supporting farmers in a number of ways. The participants reported having received training on the effects of climate change and adaptation tactics from several non-governmental organisations (NGOs). However, a few of the respondents worked in non-agricultural fields and supplemented food shortages brought on by crop failures with pension funds. It also came out that NGOs and government were playing a critical role in encouraging the cultivation of small grains such as sorghum and millet. Small grains were seen as drought tolerant compared to maize varieties. However, participants were quick to point out that small grains are labour intensive and needed more attention especially against quelea birds. One participant alluded that:

> In terms of farming, we have received support from both the government and NGOs. Last season we were trained on small grains and climate smart agriculture. We also got trained by the local extension workers on agronomic practices and diversifying cropping patterns. The other support we can mention is access to information on weather conditions. The Department of Meteorological Services updates us on weather related issues. (Female; 15 February 2020)

Participants did not mention much about irrigation development, mechanisation of farming and water harvesting. Due to a lack of resources, cultural beliefs that uphold men's superiority, distorted land ownership patterns, and other factors, female farmers do not receive enough assistance to engage in revenue-generating activities (Monwar et al. 2018). One key informant noted that:

> A lot still needs to be done especially in these newly resettled areas. There is need for government and other development players to consider scaling up irrigation development is such area. A crucial part of enhancing rain-fed agriculture will be irrigation. Farmers must also be introduced to technology, such as water harvesting. It has proved to be an effective adaptation strategy in some countries in the Middle East. (Female; 15 February 2020)

The finding of this chapter is substantiated by Vincent (2007) who argues that strong institutions that are properly governed have capacity to enhance local communities in adapting to climate change. As a result, the majority of respondents said that cutting back on food consumption was a better way to deal with climatic variability. Boko et al.'s (2007) claim lends further credence to the fact that strong local institutions have the 'ability to provide local communities with knowledge on the impact of climate change and how to manage future likely uncertainties and risks'. According to Omolo and Mafongoya (2019), climate data is an important tool that communities, governments, and other service providers can use to manage risks effectively and promptly, make well-informed decisions, and create a variety of diversified and adaptable livelihood options that will increase resilience and decrease vulnerability.

Discussion and Analysis

The adaptation framework enables scholars and practitioners to comprehend household responses to adaptation. The framework is divided into two primary phases: planning for climate adaptation and assessing climate risk and vulnerability. However, the ability of agricultural communities to access various forms of assets has a direct bearing on how vulnerable those societies are to climate variability and change (Lama and Wester 2019). The natural and physical asset bases are the main targets of climate change, which makes farming more challenging and unpredictable. This has been further supported by study results that show lack of knowledge, financial resource limits, low levels of government support, scarcity of water supplies, and political obstacles all negatively impact women's ability to adapt to climate change. Participants emphasised receiving institutional assistance for their farming endeavours. However, many women are prevented from accessing and establishing livelihoods and obtaining control over their lives by a mix of intra-household dynamics, decision-making processes both inside and outside the home, and disparities in responsibilities, employment, and income. According to Atsiaya et al. (2019), between 9 and 12% of Kenyan women had access to secondary and university education. Women are unable to use creative coping mechanisms like irrigation systems since they do not have post-secondary education (Batool et al. 2018). This affects their susceptibility to climate extremes and certainly impairs their capacity to predict and prepare for catastrophic disasters. The situation is further complicated by their limited access to and control of productive resources such as land.

In light of what has been discussed above, the adaptation framework emphasizes on the need to analyse climate risks as well as its impact. This enables those designing the adaptation policies and programmes to ponder factors such as mitigation benefits, farmer capacities as well as climate risk relevance (Abeka et al. 2012; Nyathi et al. 2022). It is imperative to emphasise that initiatives may fall short of improving people's livelihoods and, consequently, their resilience if vulnerability and capability assessments (VCAs) overlook disparities and people's diverse needs. One of the main causes of women and girls' higher susceptibility to climate change is gender inequality. Despite having limited access to post-secondary education, women have implemented agricultural strategies that allow them to adapt to changing climate conditions by using their indigenous knowledge and abilities (Abeka et al. 2012). The results showed that further research is required to illustrate the practical mitigation techniques that are in use, as these can serve as role models for people in disaster-prone locations. Further research is also needed to understand factors determining women's adoption of technology in the context of climate change. Technology adoption has been seen as critical in enhancing climate change adaptation and adaptive capacity (Gonda 2016).

Studies indicate that following the aftermath of disasters, gender disparities may also worsen. In addition, a study conducted by Dibakoane et al. (2022) showed that lack of funds prevented the farmers in the Matibidi community from erecting a fence around their properties to reduce agricultural losses caused by cattle. Many girls may be compelled to drop out of school in order to assist with domestic tasks due to a significant increase in workload (Brown et al. 2012). It has been observed that all governments participating in international climate change discussions should make it possible for gender concerns to be represented in policies and related programming in an effort to increase institutional support (Kuhl 2020; Ndlovu et al. 2023).

Limitations of the Study

Case study research facilitates the investigation and comprehension of intricate problems by means of summaries of previous investigations. It can be regarded as a strong research methodology, especially when a comprehensive, in-depth analysis is needed. However, a case study design offers very little basis for scientific generalisation of the results to other situations. Considering that only two sites were selected in Matabeleland North and South, there is a danger of bias since the study may be skewed by the researchers' preferences and beliefs or prejudiced viewpoints could skew the results and conclusions.

Recommendations

Gender plays a crucial role in climate change adaptation, as the effects of climate change are not gender-neutral. There is need to address climate change which exacerbates inequality between men and women and if not urgently addressed will likely to deepen rural feminisation of poverty and the problem de-agrarianisation. This chapter suggests that programmes and policies should incorporate the voices of women and girls given that climate change affects them differently from man. Specifically, there is need to factor in the peculiar disquiets and complaints of women with respect to how climate-related difficulties could be addressed. Such a stride will go a long way in promoting the notion of "ecological democracy and environmental justice". Efforts to adapt to climate change in different agricultural contexts should be promoted to ensure that women who are traditionally in charge of household farming, food security, and general caregiving have control over productive assets such as land. This can be done through empowering women to have access to and control over assets that support their livelihoods as well as in decision-making processes. Securing women's land rights is cardinal in the fight against the gendered impact of climate change and attempts to increase household resilience to shocks.

There is also need to upscale women's access to financial assets. This can be done through decentralising financing and addressing structural barriers such as lack of collateral that hinders women's access to credit facilities. Enhancing women's access to financial resources could increase their capacity to invest in adaptation strategies towards improved access to water, income, food security and positive health impact. Government and other development partners should also increase women's access to information on climate change. This can be done through the use of music, drama and locally available gender-sensitive information dissemination structures. Improved access to climate change information can play a crucial role in enabling women to allocate household resources in a manner that responds to the unpredicted climate events. Livelihoods diversification seems important in coping with climate change impacts amongst women. It is therefore imperative for development stakeholders to promote livelihood diversification for rural poverty alleviation through reducing asset entry barriers into lucrative activities. Above all, there is need for more sociological studies on climate change adaptive capacity that can help to detect the contestation of power that build and underpin hegemonic gender identities. These hegemonic gender identities determine access to and control over productive resources: a key determinant of successful adaptation to climate change.

Conclusion

Women's involvement in decision-making at all levels, especially in the creation of policies for adapting to climate change, needs to be increased. This will help to ensure that important programmes like these are informed by the objectives and opinions of women. Enhancing women's access to loans, marketing, and climate change adaptation information might significantly increase the resilience of their livelihoods. Enhancing women's access to land in these recently relocated areas is also necessary. Women will own and control more productive resources as a result of this intentional action, particularly land and forestry products. The productivity of smallholder agriculture will probably increase as more women have access to land and other productive resources, which will also increase household food security. We recommend that gender-responsive land use and climate change policies be supported by the government. In the battle against climate change, it is critical to raise women's consciousness and provide them with more knowledge about adaptation and climate change. The empowerment strategy has the potential to greatly increase household resilience to the effects of climate change. Government should also strengthen the extension services in newly resettled areas so as to promote small grains and other importance farming technologies that improve production such as the irrigation approach.

References

Abedin MA, Habiba U, Shaw R (eds) (2013) Water insecurity: a social dilemma. Emerald Publishers, Bingley, UK

Abeka S, Anwer S, Huamaní RB, Bhatt V, Bii S, Muasya BP et al (2012) Women farmers adapting to climate change: four examples from three continents of women's use of local knowledge in climate change adaptation. Dialogue 9(8):8–51

Adger WN (2003) Social capital, collective action, and adaptation to climate change. Ecol Econ 79(4):384–404. https://doi.org/10.1007/978-3-531-92258-4_19

Ambani M, Fiona P (2014) Facing uncertainty: the value of climate information for adaptation, risk reduction and resilience in Africa. Available at: www.careclimatechange.org/files/Facing_Uncertainty_ALP_Climate_Communications_Brief.pdf

Atsiaya GO, Ayuya OI, Nakhone LW, Lagat JK (2019) Drivers and responses to climate variability by agro-pastoralists in Kenya: the case of Laikipia County. SN Appl Sci 1(8):1–12. https://doi.org/10.1007/s42452-019-0849-x

Batool H, Ali W, Manzoor R, Mahmood N (2018) Women's perception of climate change and coping strategies in Pakistan: an empirical evidence. Earth Syst Environ 2(3):609–619. https://doi.org/10.1007/s41748-018-0073-7

Boko M, Niang I, Nyong A, Vogel C, Githeko A, Medany M, Osman-Elasha B, Tabo R, Yanda P (2007) Climate change 2007; impacts, adaptation and vulnerability. In: Parry ML, Canziani OF, Palutikof JP, van der Linden PJ, Hanson CE (eds) Contribution of working group 11 to the fourth assessment report of the intergovernmental panel on climate change. Cambridge University Press, Cambridge, pp 533–647

Braun V, Clarke V (2006) Using thematic analysis in psychology. Qual Res Psychol 3(2):77–101

Brown D, Chanakira RR, Chatiza K, Dhliwayo M, Dodman D, Masiiwa M, Muchadenyika D, Mugabe P, Zvigadza S (2012) Climate change impacts, vulnerability and adaptation in Zimbabwe. IIED climate change working paper no. 3, Oct 2012. International Institute for Environment and Development

Carr ER, Thompson MC (2014) Gender and climate change adaptation in agrarian settings: current thinking, new directions, and research frontiers. Geogr Compass 8:182–197

Creswell J (2007) Research design: qualitative, quantitative and mixed methods approaches, 2nd edn. Sage, London

Cuomo CJ (2011) Climate change, vulnerability, and responsibility. Hypatia. J Feminist Philos 26(4):690–714. https://doi.org/10.1111/j.1527-2001.2011.01220.x

Davidson O, Halsnæs K, Huq S, Kok M, Metz B, Sokona Y (2003) The development and climate nexus: the case of sub-Saharan Africa. Clim Policy 3(supp 1):S97–S113. https://doi.org/10.1016/j.clipol.2003.10.007

Davoudi S, Crawford J, Mehmood A (2009) Planning for climate change: strategies for mitigation and adaptation for spatial planners. Earthscan, London

Demetriades J, Esplen E (2010) The gender dimensions of poverty and climate change adaptation. In: Social dimensions of climate change: equity and vulnerability in a warming world, pp 133–143

Denton F (2002) Climate change vulnerability, impacts, and adaptation: why does gender matter? Gend Dev 10(2):10–20

Denton A (2010) Standards for infusion therapy: Royal College of Nursing, 4th edn. Royal College of Nursing, London. http://www.rcn.org.uk/__data/assets/pdf_file/0005/78593/002179.pdf

Dibakoane S, Siyongwana P, Shabalala AN (2022) Vulnerability, impact and adaptation strategies of female farmers to climate variability. Jàmbá J Disaster Risk Stud 14(1):a1302. https://doi.org/10.4102/jamba.v14i1.1302

Dittrich R, Wreford A, Moran D (2016) A survey of decision-making approaches for climate change adaptation: are robust methods the way forward? Ecol Econ 122:79–89. https://doi.org/10.1016/j.ecolecon.2015.12.006

Dodman D (2010) Urbanization and climate change risk. In: McLean D (ed) World disaster report 2010: focus on urban risk. International Federation of Red Cross and Red Crescent Societies. Imprimerie Chirat, France, pp 114–137

Dube T, Phiri K (2013) Rural livelihoods under stress: the impact of climate change on livelihoods in South Western Zimbabwe. Am Int J Contemp Res 3(5):11–25

Dube T, Intauno S, Moyo P, Phiri K (2017) The gender-differentiated impacts of climate change on rural livelihoods labour requirements in Southern Zimbabwe. J Hum Ecol 58(1–2):48–56

Eriksen S, Lind J (2009) Adaptation as a political process: adjusting to drought and conflict in Kenya's drylands. Environ Manage 43(5):817–835

Fraser ED, Dougill AJ, Hubacek K, Quinn CH, Sendzimir J, Termansen M (2011) Assessing vulnerability to climate change in dryland livelihood systems: conceptual challenges and interdisciplinary solutions. Ecol Soc 16(3):200–213

Gonda N (2016) Climate change, "technology" and gender: "adapting women" to climate change with cooking stoves and water reservoirs. Gend Technol Dev 20(2):149–168. https://doi.org/10.1177/0971852416639786

Goodrich CG, Udas PB, Larrington-Spencer H (2019) Conceptualizing gendered vulnerability to climate change in the Hindu Kush Himalaya: contextual conditions and drivers of change. Environ Dev 31:9–18. https://doi.org/10.1016/j.envdev.2018.11.003

IPCC (2001) Climate change 2001: impacts, adaptation, and vulnerability. The Press Syndicate of The University of Cambridge. https://archive.ipcc.ch/ipccreports/tar/wg2/index.php?idp=642

IPCC (2007) Climate change 2007: the physical science basis. In: Agenda. Durban, p 333

Jiri O, Mafongoya PL, Chivenge P (2015) Indigenous knowledge systems, seasonal 'quality' and climate change adaptation in Zimbabwe. Clim Res 66(2):103–111

Kapuka A, Hlásny T (2021) Climate change impacts on ecosystems and adaptation options in nine countries in southern Africa: what do we know? Ecosphere 12(12):e03860. https://doi.org/10.1002/ecs2.3860

Kuhl L (2020) Evacuation as a climate adaptation strategy for environmental justice communities. 127(3):493–504

Lama P, Wester M (2019) Women as agents of change? Reflections on women in climate adaptation and mitigation in the global north and south. In: Kinnvall C, Rydström H (eds) Climate hazards, disasters, and gender ramifications. Routledge. https://doi.org/10.4324/9780429424861-4

Lama P, Wester M (2021) Gendered dimensions of migration in relation to climate change. Clim Dev 13(4):326–336

Lawson ET, Alare RS, Salifu ARZ, Thompson-Hall M (2020) Dealing with climate change in semi-arid Ghana: understanding intersectional perceptions and adaptation strategies of women farmers. GeoJournal 85(2):439–452

Mare A (2013) Gender and climate change adaptation among agro-pastoral communities: case study of Chivi district in Southern Zimbabwe. In: Mulinge MM, Getu M (eds) Impacts of climate change and variability on pastoralist women in sub-Saharan Africa. Fountain Publishers, Kampala

Melisa-Rojas DM, Nejadhashemi AP, Harrigan T, Woznicki SA (2017) Climate change and livestock: impacts, adaptation, and mitigation. Clim Risk Manag 16:145–163

Mishra AK, Mohapatra U (2017) Int J Humanit Soc Sci 6(5), ISSN (P) 2319–393X

Monwar MM, Hossen M, Kalam AM (2018) Indigenous adaptation practices for the development of climate resilient ecosystems in the Hail Haor, Bangladesh. Glob Soc Welf 5(2):125–136. https://doi.org/10.1007/s40609-014-0014-9

Muchuru S, Botai OJ, Botai MC, Landman AW, Adeola MA (2016) Variability of rainfall over Lake Kariba catchment area in the Zambezi river basin, Zimbabwe. Theor Appl Climatol 124(1–2):325–338

Mushore TD, Mhizha T, Manjowe M, Mashawi L, Matandirotya E, Mashonjowa E, Mutasa C, Gwenzi J, Mushambi GT (2021) Climate change adaptation and mitigation strategies for small holder farmers: a case of Nyanga district in Zimbabwe. Front Clim 3:676495. https://doi.org/10.3389/fclim.2021.676495

Ndlovu J, Ndlovu M, Nyathi D (2023) The role of indigenous climate forecasting systems in building farmers' resilience in Nkayi district, Zimbabwe. In: Chatterjee U, Shaw R, Kumar S, Raj DA, Das S (eds) Climate crisis: adaptive approaches and sustainability, pp 195–210

Nyathi D, Ndlovu J, Phiri K, Muzvaba NE (2022) Climate change and food insecurity: risks and responses in Bulilima district of Zimbabwe. In: Leal Filho W, Djekic I, Smetana S, Kovaleva M (eds) Handbook of climate change across the food supply chain. Climate change management. Springer, Cham. https://doi.org/10.1007/978-3-030-87934-1_24

Omolo N, Mafongoya PL (2019) Gender, social capital and adaptive capacity to climate variability: a case of pastoralists in arid and semi-arid regions in Kenya. Int J Clim Change Strateg Manag 11(5):744–758

Punch KF (2013) Introduction to quantitative and qualitative approaches. Sage, Thousand Oaks

Rice PL, Ezzy D (1999) Qualitative research methods: a health focus, vol 720. Authority of the Minister of Health, Melbourne

Silvestri S, Sabine D, Patti K, Wiebke F, Maren R, Lanetta M (2015) Households and food security: lessons from food security households in East Africa. Agric Food Secur 4(1):23, 210–220

Simotwo HK, Mikalitsa SM, Wambua BN (2018) Climate change adaptive capacity and smallholders in Trans-Mara East sub-Country, Kenya. Geoenviron Disasters 5(5)

Smit B, Wandel J (2006) Adaptation, adaptive capacity and vulnerability. Glob Environ Change 16(3):282–292

Sultana F (2014) Gendering climate change: geographical insights. Prof Geogr 66(3):372–381. https://doi.org/10.1080/00330124.2013.821730

Thomas K, Hardy RD, Lazrus H, Mendez M, Orlove B, Rivera-Collazo I, Roberts JT, Rockman M, Warner BP, Winthrop R (2019) Explaining differential vulnerability to climate change: a social science review. Wiley Interdiscip Rev Clim Change 10(2):557–565

Unganai L (2009) Adaptation to climate change among agro-pastoral systems: case for Zimbabwe IOP Conf Ser: Earth Environ Sci 6, Article ID: 412045. https://dx.doi.org/10.1088/1755-1307/6/1/412045

Vincent AK (2007) Uncertainty in adaptive capacity and the importance of scale. Glob Environ Change 17:12–24. https://doi.org/10.1016/j.gloenvcha.2006.11.009

Wingood GM, DiClemente RJ (2000) Application of the theory of gender and power to examine HIV-related exposures, risk factors, and effective interventions for women. Health Educ Behav 27(5):539–565. https://doi.org/10.1177/109019810002700502

Yadav SS, Lal R (2018) Vulnerability of women to climate change in arid and semi-arid regions: the case of India and South Asia. J Arid Environ 14(9):4–17

Yin RK (2006) Case study methods. In: Green JL, Camilli G, Elmore PB (eds) Handbook of complementary methods in education research. Lawrence Erlbaum Associates Publishers, pp 111–122

Chapter 24
Local Wisdom-Based Food Security in Facing the Climate Crisis in Indonesia

Iwan Setiajie Anugrah and Hari Harjanto Setiawan

Abstract The climate crisis has become an essential agenda for global society, including Indonesia. Climate change causes floods, droughts, and plant diseases, which result in crop failure. This chapter will describe food security in the climate crisis based on local knowledge. This local wisdom is in the form of the Indonesian tradition of storing rice harvests in barns to deal with crises. A qualitative approach was used in this research. The primary data was obtained through interviews and observations. The informants interviewed were taken purposively as many as 15 people representing three areas affected by climate change in the form of floods, drought pests, and diseases. Interviews were also conducted with officials at the Ministry of Agriculture to obtain information regarding handling climate change in the agricultural sector. The research results show that climate change in Indonesia has resulted in droughts, floods, and increased plant pests, resulting in crop failures. The area of rice crops that failed to harvest due to climate change from 2014 to 2019 was 1.5 million ha. This resulted in the loss of 6 million tons of milled dry unhulled grain, or the equivalent of 750 thousand tons of rice. The government's efforts to protect farmers from crises are based on local wisdom, namely by providing insurance for rice farmers, such as the rice barn principle. What is new in this chapter is the development of an insurance program for farmers based on local wisdom values.

Introduction

Climate crisis is a term that describes a very critical situation due to climate change, which hurts the earth and humans. Global climate change is caused, among other things, by increasing greenhouse gas emissions due to various activities that increase the earth's temperature (Huang et al. 2022). The impacts of climate change include

I. S. Anugrah · H. H. Setiawan (✉)
National Research and Innovation Agency (BRIN), Jakarta, Indonesia
e-mail: hari.harjanto.setiawan@brin.go.id

I. S. Anugrah
e-mail: iwan035@brin.go.id

U. Mukhopadhyay et al. (eds.), *Climate Crisis, Social Responses and Sustainability*, Climate Change Management, https://doi.org/10.1007/978-3-031-58261-5_24

severe droughts, air scarcity, severe fires, rising sea levels, flooding, melting polar ice caps, powerful storms, and declining biodiversity. This condition significantly impacts the world of agriculture, which relies on seasons. Rising global temperatures trigger prolonged dry seasons and extreme rainstorms, disrupting agricultural rhythms. Climate change directly affects the changing seasons, so farmers have difficulty deciding when to plant rice. Another impact is increased air temperature and humidity, which causes plant pest organisms to multiply rapidly.

At the national level, climate change is a strategic issue in the agricultural sector because it threatens the national food security of the Indonesian nation (Mashizha and Mashizha 2019). Climate change and food security have become important agendas for Indonesian society. The climate crisis has created problems regarding food shortages and availability. Three things that significantly impact agriculture in Indonesia, namely drought, floods, and an increase in plant pests (OPT), are starting to occur due to the climate crisis (Eka Suranny et al. 2022). Large areas of agricultural land affected by drought, floods, and pests are at risk of causing crop failure and loss of food production.

Rice plants are essential to Indonesia's agricultural strategy because the staple food is rice (Kumar et al. 2022). Rice is the staple food of most Indonesians because 95% of Indonesians consume rice. The majority of Indonesian people view rice as an essential food, so the need for rice will be huge. This makes rice food supplies important for the Indonesian people. During 2014–2019, the rice planting area in Indonesia that experienced drought and floods reached 1.5 million ha. This resulted in a loss of rice production, resulting in 6 million tons of milled dry unhulled grain (GKG), or the equivalent of 750 thousand tons of rice. The area of rice plants at risk of crop failure due to pest attacks reaches 2.1 million ha, which is the equivalent of a loss of grain production of 8.4 million tonnes or 1 million tonnes of rice. Climate change poses a risk of loss for farmers (Pramukti et al. 2021).

Apart from impacting the loss of national food production, climate change also poses a risk of social change in society. The impact of climate change is very influential, especially on community groups whose livelihoods depend on natural sources. National floods, droughts, landslides, and other disasters have impacted the sustainability of local and regional food security. Reduced farmers' income due to crop failure causes farmers' lives to become poor. Poverty creates broader social obstacles in farmers' lives (Osabohien et al. 2020). Farmer poverty is a social problem because it impacts the lives of their families. This poverty will have an impact on children's health and education.

The Indonesian government is trying to overcome the food crisis due to climate change with the National Rice Farming Business Insurance (AUTP) program. This program is based on local community traditions in overcoming the food crisis. The rice barn is one form of local wisdom in Indonesia that we can still find (Sari et al. 2022). The community maintains the availability of rice by storing it in rice barns during the harvest season. Rice barns managed by Indonesian people aim to keep food available, especially rice, the staple food of Indonesian people. This causes farmers to store the rice they have harvested to use as a supply to meet future food needs. By implementing a rice barn system, Indonesian people can create food security.

The Indonesian government created an insurance system based on the rice barn idea to protect farmers from crop failure due to climate change. National Rice Farming Business Insurance (AUTP) is the government's response through the Ministry of Agriculture's program to protect rice farmers from the risk of loss. Droughts, floods, and pests account for these losses so they can continue farming in the next growing season. In Indonesia, the land area of AUTP participants reached 2.5 million ha from 2015 to 2018; self-help premiums amounted to IDR 91.3 billion, and government-subsidized premiums amounted to IDR 365.4 billion. The target area for AUTP participation during 2015–2019 is 4.5 million ha, and only 2.8 million ha (63.29%) have been realized. The AUTP implementation program is part of the community and government's readiness to face the risk of the climate crisis to maintain sustainable food availability (Ani et al. 2022).

Agricultural productivity measures the quantity of agricultural output produced for a certain amount of input (Akano et al. 2023). The decline in agricultural production due to the climate crisis will increase food prices, reducing household purchasing power. This will increase the number of food-insecure areas. This situation will lead to cases of starvation. This chapter will focus on solving food production problems farmers face due to climate change. First, what risks does climate change pose to food production? Second, what social impacts do farmers face due to climate change? Third, how do we protect farmers from climate change based on the local wisdom of the Indonesian people? The novelty of this chapter describes the Indonesian government's efforts to protect farmers through local wisdom-based agricultural insurance so that they can survive sustainably.

The Rationale of the Study

Climate Crisis Global and Food Insecurity

The climate crisis is an urgent global problem and requires immediate and concerted action. The government and private sector are increasing their role in designing policy changes to overcome this problem. Climate change is a top-priority issue that can potentially create long-term change. This issue is considered to be the cause of global warming and increasing extreme events. Seeing this reality, policies in handling climate change are changing from a traditional approach to a more urgent and crisis-based mindset (Hanna et al. 2021).

Realizing food security is a big issue today, especially climate change (Grzelak 2022). Global food production forecasts are constrained by increasing extreme weather events. The accuracy of the planting period on time is difficult to predict. Crop failure must be anticipated because the estimated planting period is the key to ensuring agricultural commodity markets function well so that the food security system can be maintained. Extreme weather events due to changes in geothermal heat cause increased uncertainty in food production (Wu et al. 2021). In facing these challenges, looking for solutions that combine modern technology with traditional knowledge and practices is crucial.

Local Wisdom-Based Food Security

Food security is an important issue that relies heavily on accurate weather forecasts (Radeny et al. 2019). Years of drought often lead to shortages of food and international food aid. Without precise weather forecasts, anticipating and planning for potential deficits becomes extremely difficult, harming agricultural production. Additionally, weather forecasts are vital in helping farmers make informed decisions regarding their farming practices. Farmers can adjust planting and irrigation schedules by providing information about upcoming weather conditions. This can significantly increase crop yields and ultimately contribute to food security.

A rice barn is a barn used to store and dry harvested rice. Rice barns are usually used by indigenous peoples in Indonesia to store their harvests (Mulya 2022). Barns have different shapes in each region. However, barns are generally shaped like houses on stilts with roofs made of palm fiber, and the entire building is made of wood. The barn can be used to store harvests for quite a long time. Barns with unique designs can keep rice for up to a hundred years. Lumbung is a symbol of food security for people in Indonesia. Food security is vital, considering we can fulfill our daily needs independently (Zebua et al. 2019).

Rice barns are a form of local wisdom in food security in Indonesia (Karyadi 2021). Rice barns are used to store rice in the long term. Rice put into the barn is stacked in a specific order and taken out in a particular order to maximize the time the rice can be stored. This traditional storage pattern is said to be able to keep rice for 20 years. Each barn can generally hold 1000 bushels of rice or 2.5–3 tons. Apart from private storage, there are also communal granaries, and residents must donate part of their harvest to these granaries to be used for traditional ceremonies or needy residents.

Agricultural Insurance Based on Local Wisdom

The rice barn transformed into Agricultural Insurance is very useful for farmers to protect their farming business. Agricultural insurance can provide compensation for crops caused by climate change. The sustainability of farming businesses can be guaranteed even if crop failure occurs (Deli et al. 2022). One form of risk management is Agricultural Insurance. Many farmers will participate to generate reserve funds for compensation payments for farmers who experience a decrease in production yields. Agricultural Insurance is a modern approach to food security incorporating local wisdom, as it protects farmers against crop losses caused by certain events.

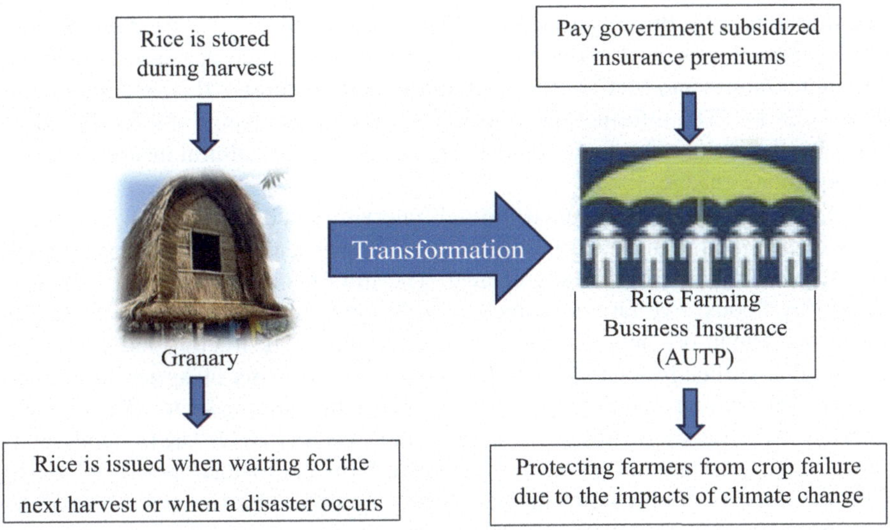

Fig. 24.1 Transformation from traditional food security to modern food security

Apart from traditional rice barns, modern forms of food security based on local wisdom are also emerging in Indonesia. One example is agricultural insurance, which compensates for crop losses caused by climate crisis events (Madaki et al. 2023). This modern food security approach protects farmers and businesses and combines local wisdom with preserving and respecting traditional knowledge and practices. The importance of accurate weather forecasts and contemporary techniques for food security based on local wisdom, such as agricultural insurance, cannot be overstated in ensuring the stability of food security in Indonesia.

In Indonesia, rice barns have traditionally been used as a form of local wisdom in food security. Rice barns are utilized for long-term rice storage, with the rice being carefully arranged in a specific order to maximize its shelf life. The organization and utilization of rice barns in Indonesia is a traditional practice contributing to food security. The need to improve the infrastructure of rice storage and financing schemes in Indonesia is evident to ensure rice quality and durability (Patunru and Ilman 2020). This transformation from local to modern wisdom can be described as shown in Fig. 24.1.

Materials and Methods

This chapter explains local wisdom-based food security in Indonesia in facing the global climate crisis. The impacts felt by communities experiencing the climate crisis are social and economic in the agricultural sector. Furthermore, it describes community adaptation (Pittock and Jones 2000) in facing the climate crisis in the agricultural industry through agricultural insurance based on local wisdom. A qualitative approach was used in this research. Data was collected through documentation studies (Jones and McCoy 2019), observation (Greatorex 2014), and interviews (Prentice 2017). The information collected is: (a) climate change and food production risks, (b) social risks due to climate crisis, and (c) agricultural insurance in the face of climate change.

The primary data was obtained through interviews and observation. The informants interviewed were taken purposively by as many as 20 people representing three regions affected by climate change. The informants consisted of 5 farmers affected by floods, five farmers affected by drought, and five farmers affected by plant pests. The informants also came from 3 agricultural instructors and 2 Ministry of Agriculture officials. Interviews with officials at the Ministry of Agriculture aimed to obtain information on handling climate change in the farming sector. The in-depth interview and observation process took place in December 2022. The in-depth interview lasted approximately 60–90 min. Affected farmers provided their experiences about the impacts of climate change and using agricultural insurance. Meanwhile, agricultural extension workers provide information about policies and programs to protect farmers from the effects of climate change through agricultural insurance (Fig. 24.2).

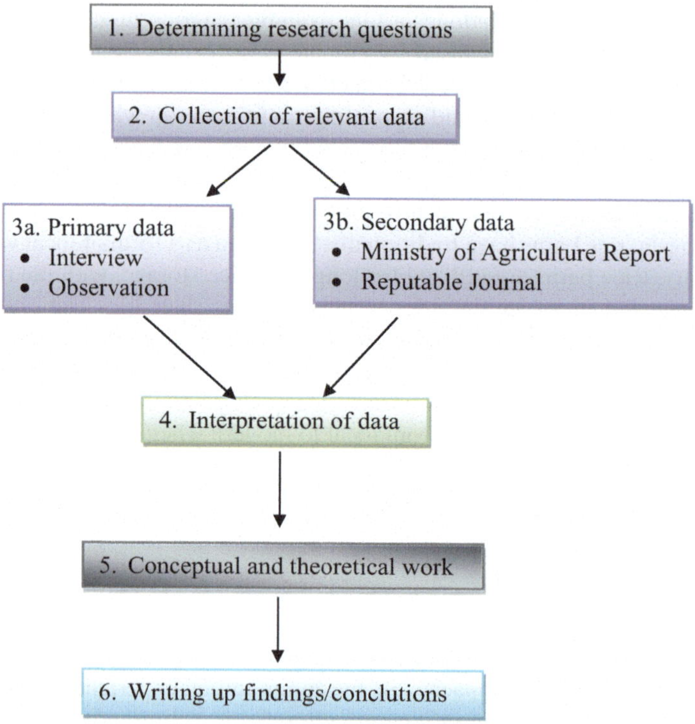

Fig. 24.2 An outline of the primary step of qualitative research

This research was analyzed thematically (Sundler et al. 2019), presenting data according to predetermined themes. The theme chosen was Food Security Based on Local Wisdom in Facing the Climate Crisis in Indonesia. This research was analyzed thematically (Sundler et al. 2019), presenting data according to predetermined themes. The theme chosen was Food Security Based on Local Wisdom in Facing the Climate Crisis in Indonesia. Thematic analysis was chosen to determine the patterns in the data obtained.

Results and Discussion

Climate Crisis and Food Production Risks

Climate change has impacted the risk of loss for business actors in the agricultural sector (Shah et al. 2023). Farmers as food producers, especially rice producers, will face the risk of farm losses due to the impacts of floods, droughts, and high levels of crop pest attacks as a series of threats from climate change. At the national level, the direct risk to rice farming activities is the reduced contribution of the agricultural sector to total rice production as the primary national food source. The sustainability of the agricultural sector's contribution needs attention from various parties, considering that many factors can cause crop failure. The climate crisis is one of the risks inherent in the agricultural sector supply chain (FAO 2011). Climate change causes long dry seasons, floods, and increased plant pest attacks. This condition has an impact on the production system in the agricultural sector, especially rice production, causing various losses for farmers who cultivate rice.

Crop Failure Due to Flooding

Climate change causes continuous rain, causing rice fields to be submerged in water, resulting in crop failure (Islam et al. 2023). This condition will affect grain and rice production, impacting farmers' income. So, it would be good if farmers had insurance protection to deal with crop failure. Figure 24.3 shows data on crop failure due to flooding in Indonesia.

The flood disaster in 2014–2018 caused losses for farmers due to crop failure and puso. Nationally, the most devastating flood disaster during 2014–2018 caused 1,148,793 ha of rice fields to experience crop failure, and another 361,548 ha experienced puso. Under these conditions, the area of rice fields affected by the flood disaster during 2014–2018 reached 1,510,341 ha. If it is assumed that the average production of milled dry unhulled rice per hectare is 8 tons, then rice production losses due to flood disasters during 2014–2018 are estimated to reach 12,082,728 tons. Figure 24.4 shows the condition of agricultural land during a flood.

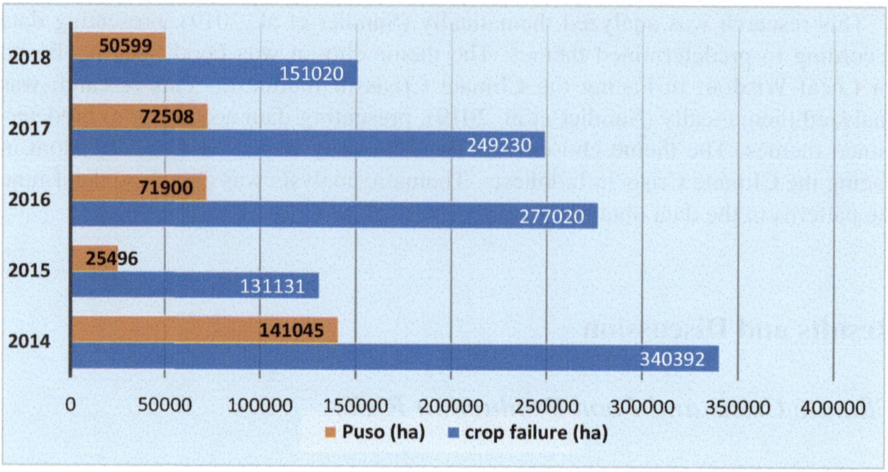

Fig. 24.3 Land area affected by flooding and at risk of failure of rice farming in Indonesia. *Source* Ministry of Agriculture, 2019

Fig. 24.4 Harvest failed due to flooding

When climate change events occur, rice farmers react quickly. Most rice farmers have adopted various climate change adaptation strategies, including planting shade trees, using energy-saving technology, and cultivating organic farming. In addition, farmers are also growing crops other than rice, switching from agriculture to animal husbandry, and looking for alternative food sources (Atube et al. 2021). Nearly 90% of farmers have dynamically implemented this mitigation method to combat the impacts of climate change. So that farmers can more quickly mitigate the event of crop failure due to flooding, an insurance program that can finance it is needed.

Crop Failure Due to Drought

The influence of climate change on agricultural production causes crop failure. Farmers who live in dry areas have a significant impact on the chances of crop failure. Meanwhile, farmers who experience a decrease in yields in arid regions influence farmers' decision opportunities to change planting patterns and shift planting times as a form of adaptation to climate change (Tasnim et al. 2023). The area of land affected by the drought can be seen in Fig. 24.5.

Farmers have to bear losses when a shortage occurs due to the loss of rice production. So far, rice has been a source of livelihood and also a source of household income. Based on the data displayed in Fig. 24.5 it shows that due to the drought in 2014–2018, the impact on crop failure reached 2,076,748 ha. Apart from crop failure, farmers also experience the risk of puso even though their land area is still relatively small. The scope of land threatened by drought due to drought has reached 26,291 ha nationally (Fig. 24.6).

Under these conditions, nationally, the total area of land at risk of loss due to deficiency reached 2,103,039 ha. If it is assumed that rice production per hectare in the form of dry-milled grain runs an average of 8 tons per hectare, then the loss of grain production during 2014–2018 will reach 16,824,312 tons. The climate crisis significantly threatens global food security because it causes drought, so farmers fail to harvest. According to the Food Security Information Network and the Global Network Against Crisis, more than 700 million people are affected by food insecurity

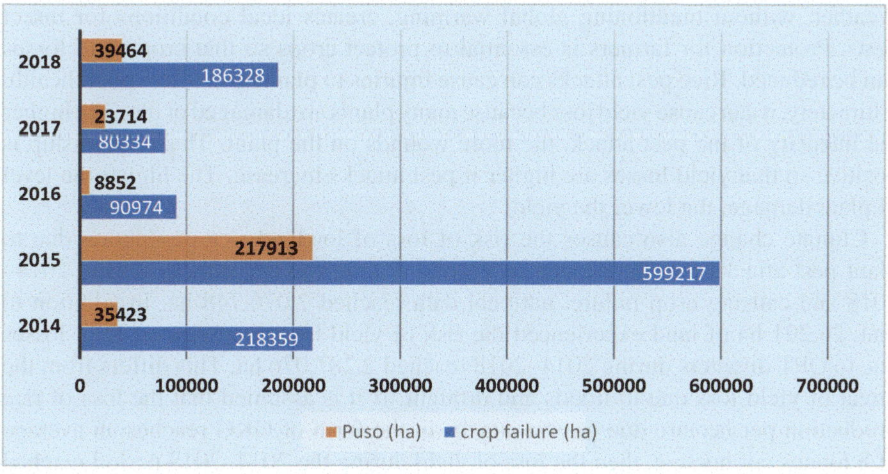

Fig. 24.5 The land area affected by drought and at risk of rice farming failure in Indonesia. *Source* Ministry of Agriculture, 2019

Fig. 24.6 Crop failure caused by drought due to the climate crisis

(Poudel et al. 2021). If the causes of food insecurity are not addressed, this number will increase. Drought conditions resulting from the climate crisis are identified as the primary cause threatening agricultural yields worldwide. Drought causes the highest annual loss of agricultural production compared to other reasons. Lack significantly impacts plant growth and development at various levels, causing a loss of plant productivity (Sun et al. 2020).

Crop Failure Due to Plant Pest Attacks

Weather, without mentioning global warming, creates ideal conditions for insect pests. Protection for farmers is essential to protect crops so that crop yield losses can be reduced. Rice pest attacks can cause injuries to plants and affect plant health. Ultimately, it can cause yield loss because many plants are damaged or die. The higher the intensity of the pest attack, the more wounds on the plant. This relationship is positive so that yield losses are higher if pest attacks increase. The higher the level of plant damage, the lower the yield.

Climate change also causes the risk of loss of lowland rice production due to plant pest attacks (OPT), as seen in Fig. 24.7. Due to OPT attacks during 2014–2018 and causing crop failure, national data reached 2,076,748 ha. In addition to that, 26,291 ha of land experienced the risk of yield loss due to puso. Total losses due to OPT disasters during 2014–2018 reached 2,287,076 ha. This differs from the threat of yield loss due to floods and drought. If it is assumed that the loss of rice production per hectare due to pest attacks in the form of GKG reaches an average of 8 tonnes per hectare, then the loss of yield during the 2014–2018 period reached 18,296,608 tonnes (Fig. 24.8).

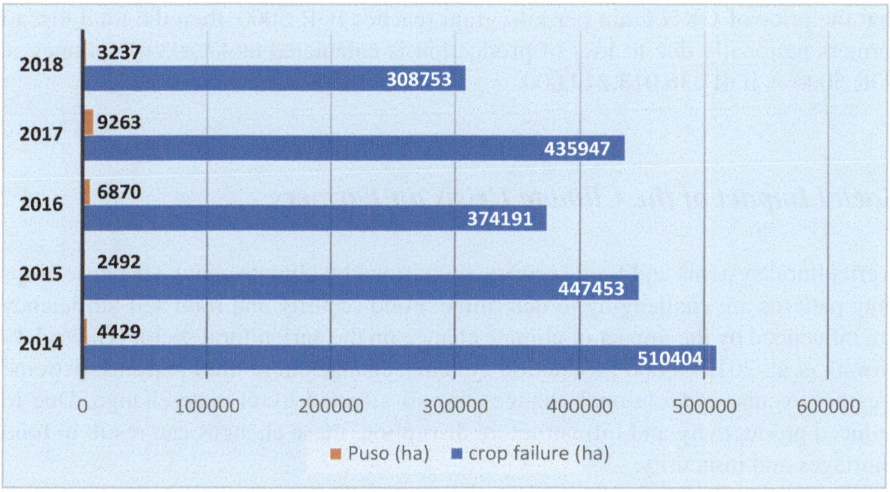

Fig. 24.7 Area of land affected by major OPT attacks and at risk of failure of rice farming in Indonesia. *Source* Ministry of Agriculture, 2019

Fig. 24.8 Crop failure due to OPT

Based on the assumed estimate of the area of land exposed to the risk of drought, flooding, and pest attacks, it can be estimated that the amount or volume of national rice production loss during the 2014–2018 period in the form of GKG will reach: 12,082,728 tonnes (due to flooding), 16,824. Three hundred twelve tonnes (due to drought) and 18,296,608 tonnes (due to pest attacks). From 2014 to 2018, 5,900,456 ha of land were threatened by flooding, drought, and pest attacks.

In detail, 1,510,341 ha were affected by flooding, 2,076,748 ha were affected by drought, and 2,103,039 ha were affected by pests. This land area resulted in a loss of rice production in the form of GKG, amounting to 47,203,648 tons. If it is assumed that the price of GKG grain per kilo gram reaches IDR 5000, then the total loss to farmers nationally due to loss of production is calculated at 47,203,648 tonnes × IDR 5000 = IDR 236,018,249,000.

Social Impact of the Climate Crisis on Farmers

Agricultural systems and food security are harmed by climate change because cropping patterns are challenging to determine. Food security and food self-sufficiency are influenced by the impact of climate change on the agricultural sector (Hawrylak-Nowak et al. 2019). Food production systems, changes in rainfall patterns, extreme weather events, and seasonal changes are all affected by climate change. Due to reduced productivity and infrastructure disruption, these changes can result in food shortages and insecurity.

Food Insecurity

Climate change exacerbates the vulnerability of communities whose livelihoods depend on agriculture. The impact of the climate crisis on farmers is that there are changes in food security and welfare (Kogo et al. 2021). This includes reduced crop yields, increased food prices, reduced access to nutritious food, and increased vulnerability to hunger and malnutrition. The climate crisis poses significant challenges to food production. Climate change also affects agricultural infrastructure, production systems, food security, and community welfare.

Farmers must address the social and economic impacts of climate change (Petersen-Rockney 2022). Farmers believe that there are significant social losses from climate change. They must change their farming management practices in response to the impacts of climate change. Loss of access to membership benefits in formal and informal agricultural groups is one of the social impacts felt. Increasing farmer involvement in reducing climate change's perceived negative social impacts, the Ministry of Agriculture must adapt agricultural policies, programs, and practices to climate conditions. The potential to accelerate and scale up climate change mitigation in agriculture will be more significant if implemented interventions emphasize impacts on livelihoods.

Global concern is being expressed over the issue of climate change. Human activities that ignore the environment and use it carelessly are the cause. Humans, animals, and plants are affected by climate change, so the agricultural sector is particularly affected. One of them is seasonal variations and unpredictable weather, making it difficult to plan when to plant and when to harvest. Additionally, sudden temperature increases can impact plant quality and growth. As a result, there is a food shortage as agricultural production and crop quality decline (Rust and Rust 2013).

Addressing the impacts of climate change on agriculture requires serious and sustained efforts. One way that can be done is to improve the quality of agriculture.

This can be done by reducing the use of chemicals and artificial fertilizers, which can damage the environment. Also, selecting plant varieties resistant to climate change must be done. In conclusion, climate change has a significant impact on the agricultural sector. However, this impact can be reduced with severe and sustainable efforts through agrarian insurance.

Farmer Poverty Due to Climate Change

Farmers are always synonymous with poverty. BPS data shows that of the 27.76 million poor people in Indonesia, 17.28 million are rural residents, most working as farmers. Farmers and the agricultural sector are still considered occupations mired in poverty. This is exacerbated by climate change, which causes crop failure. The impacts of climate change on agriculture have had severe social consequences for farmers in Indonesia, particularly in terms of poverty. Climate change has increased uncertainty and instability in agricultural production, directly affecting farmers' income and livelihoods (Heureux et al. 2022). Farmers in Indonesia rely heavily on agriculture for their income and sustenance. The unpredictable weather patterns and extreme events caused by climate change, such as heavy rainfall and prolonged droughts, have resulted in crop failures, reduced yields, and increased susceptibility to pests and diseases. As a result, farmers' incomes have been greatly affected, pushing many into poverty.

Farmers are people who are vulnerable to the impact of climate change on food security (Kabir and Uddin 2016). They are more susceptible to food shortages and malnutrition because they often lack the resources and resilience to adapt to climate change. In conclusion, climate change has become a major global problem that has profound implications for agriculture and food security. Efforts to reduce the impact of climate change on agriculture and ensure food security are significant. Given the widespread effects of climate change on the agricultural industry, it is clear that urgent action is needed to address this issue.

Climate change significantly impacts agriculture and has far-reaching consequences (Agunbiade and Babalola 2023). It affects food security and agricultural productivity and has social effects, particularly for farmers in developing countries like Indonesia. These farmers heavily rely on agriculture for their income and livelihoods, making them vulnerable to unpredictable weather patterns and extreme events caused by climate change. These impacts lead to crop failures, reduced yields, and increased poverty among farmers. Furthermore, vulnerable communities, particularly the poor and marginalized populations, also bear the brunt of climate change's effects on food security.

They lack the resources and resilience to adapt to the changing climate, leaving them more susceptible to food shortages and malnutrition. In conclusion, the consequences of climate change, including alterations in weather patterns, have a severe impact on agricultural productivity and food security. Efforts to mitigate these impacts and ensure sustainable agriculture are crucial for the well-being of both farmers and vulnerable communities. Moreover, not only does climate change affect food production and food security, but it also has distributional consequences (Deepak 2013). For instance, less developed countries and low-income social groups are more vulnerable to changes in agriculture prices, making them significantly affected by changes in climate (Ciscar et al. 2018). Additionally, developing countries' reliance on rainfall conditions in agriculture makes them particularly susceptible to the adverse effects of climate change on food security (Vigouroux et al. 2011).

Agricultural Insurance Is Based on Local Wisdom to Protect Farmers

The interconnectedness of various climate change threats that the agricultural industry must face impacts production processes that can affect farmers' income stability (Harkness et al. 2021). Since most farmers have less than 0.5 ha of land per person, one of the main problems in the agricultural sector is how to increase their income. To achieve food sovereignty, food independence, and sustainable food security, the state, through the government, provides protection and empowerment to farmers who contribute to the development of the agricultural sector. Agricultural Insurance is a method of protecting income or welfare for farmers.

In the Asia Pacific region, 44 countries pay attention to agricultural insurance programs. The total agricultural insurance premium is around USD 4 billion distributed in this region. China has a portion of 50% of the unlimited premium. Then followed by Japan at 31%, India at 11%, Australia at 4%, and the Republic of Korea at 3%. These five agricultural insurance markets account for 99% of total premiums in Asia Pacific. According to the World Bank (2010), International Agricultural Insurance Practices in the Latin American Region (LAC) cover 32 countries with an area of 205 million ha and inhabited by a population of 561 million people. Of the 75 agricultural insurance companies in the LAC region, half are in Argentina, Brazil, and Mexico. These three countries account for 85% of the total premiums in the LAC region.

Countries have implemented agricultural insurance in 86 countries, which is considered quite successful (FAO 2011). Japan is one of these nations and has had an agriculture insurance program since the 1930s. The Japanese government's backing, which resulted in the passage of laws like the Agricultural Insurance Law, the National Forestry Insurance Law, and the Livestock Insurance Law, indicates this. In contrast, the appropriate legislation in Indonesia was just released in 2013.

Agricultural Insurance Implementation Process

The phrase "agricultural insurance" is still a new thing in Indonesian society which aims to protect and empower farmers. Agricultural Insurance is one of the government's options in this legislation to safeguard farmers against crop failure. Natural disasters, infectious animal disease epidemics, climate change, and other factors can result in crop failure. Following their jurisdiction, the central and regional governments support every farmer's ability to enroll in Insurance under Law 19/2013's Article 39, Paragraphs (1) and (2). Help with premium payments is one of these amenities. Farmers can make their payments, according to regional governments, and this formal help is provided through the APBN and APBD, which the government funds.

Agricultural insurance is the farmer protection solution the government and regional governments created. Farmers are given protection if they (a) grow food crops on land that they do not own and cultivate a maximum area of two hectares, (b) cultivate food crops on land they own on a maximum area of two hectares, or (c) horticultural farmers, planters, or small-scale livestock breeders. Various agricultural insurance products include crop insurance, livestock insurance, greenhouse insurance, forestry/plantation insurance, chicken meat insurance, and aquaculture insurance.

In general, agricultural insurance aims to stabilize farmers' income by reducing losses due to loss of crop yields and encouraging farmers to apply agrarian technology to be more productive (Janowicz-Lomott et al. 2015). The benefit of the farming insurance program is that it makes farmers aware of the risk of crop failure and encourages them to increase their skills and improve agricultural business management. This program can reduce dependence on other parties' capital and help farmers support farming production costs. Agricultural insurance can also increase the awareness and responsibility of local governments in anticipating agricultural business risks and increasing the success of regional food security.

The agricultural insurance program from 2015 to 2019 shows that it can protect rice commodity farmers from losses due to drought, floods, and pest attacks nationally, reaching an average of 1 million ha annually. In 2016, the target was the lowest, namely 500,000 ha, because it was related to the budget. This data also shows that the targets set by the government have not been achieved in each fiscal year. However, in 2016 and 2017, the realization was close to the area of rice fields protected through the insurance program of 99.99% (2016) and 99.80% (2017), respectively (Table 24.1).

In the 2015–2019 trial scale, performance, participation, and claims achievement levels experienced dynamic fluctuations. The implementation of AUTP is still faced with obstacles, problems, and challenges to improve it even better. The biggest challenge is related to the role and participation of rice farmers in implementing the program. AUTP provides tangible benefits for farmers, namely protecting them from disaster risks. Losses in farming can be covered so that rice farmers can also be sustainable (Fig. 24.9).

The system implemented to protect the risk of loss due to floods, drought, and pest attacks has been determined based on the scheme's provisions. Based on the decision

Table 24.1 Data on targets and realization of implementation of agricultural insurance for rice farming

Year	Target (ha)	Realization (ha)	Realization (%)
2015	1,000,000	233,499.55	23.35
2016	500,000*	499,962.25	99.99
2017	1,000,000	997,966.53	99.80
2018	1,000,000	806,199.64	80.62
2019	1,000,000	310,604.08	31.64

Source Ministry of Agriculture, 2019

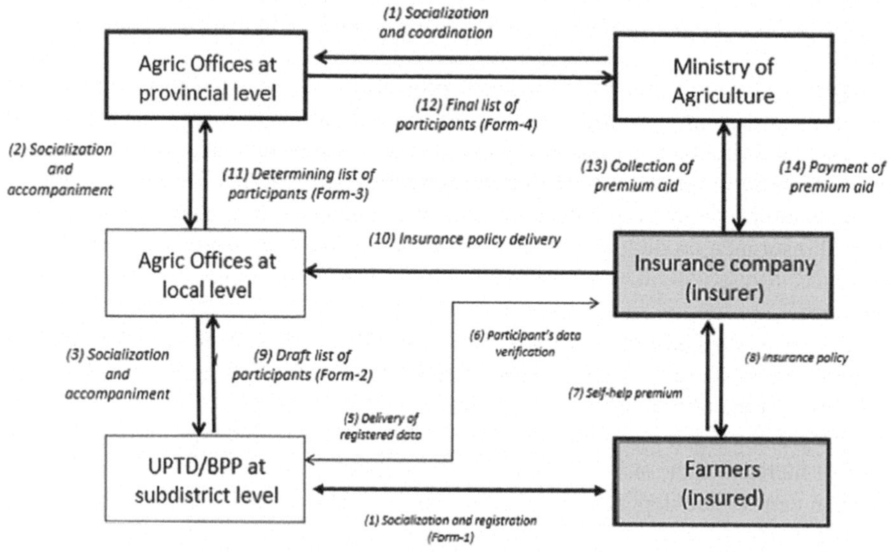

Fig. 24.9 AUTP implementation process. *Source* Ministry of Agriculture, 2019

of the Minister of Agriculture, farmers receive insurance premium assistance from the government. The insurance coverage price for rice farming is IDR 6,000,000/ha/MT. The total rice agricultural insurance premium is IDR 180,000/ha/MT. The government provides a subsidy of IDR 144,000/ha/MT, so with this subsidy scheme, farmers only pay a premium independent of IDR 36,000/ha/MT. Compensation is calculated proportionally, according to the area of land insured.

Agricultural Insurance Effects

Implementing agricultural insurance is a potential solution to the challenges posed by the climate crisis to food security. This insurance can provide farmers financial protection in case of crop loss or damage due to climate crises such as drought, floods, and extreme weather (Nirmal Ravi Kumar and Babu 2021). Agricultural Insurance can help mitigate the economic and social impacts of climate change on food security by providing a safety net for farmers. Implementing agricultural Insurance for farmers can help reduce climate change's economic and social effects on food security.

Furthermore, agricultural Insurance can incentivize farmers to adopt climate-smart practices and invest in resilient farming systems. This can help build rural resilience and reduce farmers' vulnerability to climate-related risks. Overall, implementing agricultural Insurance is crucial in addressing the challenges climate change poses to food security (Atreya and Kaphle 2020). It provides a means to protect farmers from financial losses due to climate-related events and encourages adopting climate-smart practices.

Farmers' responses to the implementation of agricultural insurance in various regions in Indonesia generally state that rice farming insurance is essential (Patt et al. 2009). Especially in rice fields, which often become endemic due to climate change, flooding occurs during the rainy season and drought when entering the dry season. Farmers also feel helped when pests attack their rice farming land. Farmers are vulnerable to disasters due to these three factors, which result in losses due to crop failure. Farmers need capital for further farming activities, especially to catch up with planting time while the rainy season is still occurring.

Capital requirements include production facilities (seeds, fertilizer and medicines, labor, and other costs). The agricultural insurance mechanism still covers some of these conditions. The farmers also said that the administrative process of becoming an insurance participant should be made more accessible. Farmers experience difficulties related to administration. The minimum insurance membership requirement with a land area of one hectare and above is an obstacle for farmers.

Insurance officers limit Agricultural insurance socialization activities because the number of personnel is minimal. Insurance officers are not well coordinated with agricultural assistance officers. This creates confusion for farmers because they deal with different officers. Coordination between agricultural insurance officers is also not well developed. Coordination with regional agricultural departments is also not well developed. Agricultural insurance officers and policy implementers at the central level go directly to the location, so there is often miscommunication between them in understanding the conditions. One policy that has not been well coordinated so far is the budgeting system for operational activities in each region.

Most farmers only have limited land of less than 0.5 ha. Farmers participating in insurance still experience problems in applying for and disbursing it. Reimbursement of funds for claims for losses due to damage to rice plants occurs due to delays in verification by insurance officers. Farmers want to speed up rice planting again because they are catching up to the planting season. There are often differences in determining the value of losses, claims, and payments for the area of rice plants that must be replaced. Even though the current agricultural insurance still has many shortcomings, farmers consider this critical for agrarian sustainability (King and Singh 2020) when affected by disasters due to climate change.

Limitations of the Study

Considering each country has unique characteristics and policies, this study cannot be generalized to all countries. Due to the uniqueness and instability of socioeconomic conditions in communities worldwide, this research can't be used as a benchmark for measurement. However, the findings of this research can be applied to increase community resilience in the agricultural sector when facing climate change. Therefore, local communities from various countries with different characteristics must evaluate the conditions of their communities. This program can be implemented if the conditions are the same as in Indonesia. The small observation time in this research means that the findings cannot be applied to several contexts; for example, this research was not conducted during a flood. However, various situations are strengthened by information from secondary data and social media.

Recommendations

During the evaluation of strategic activities for 2016–2020 in the regions regarding the implementation of agricultural insurance activities, several recommendations were given so that actions could be carried out better. The agricultural insurance program must continue to be socialized massively and continuously to farmers in all locations. The government has increased the budget allocation to support premium cost assistance for farmers who want to become agricultural insurance participants. The selection process for prospective Insurance and participating farmers must be carried out appropriately and transparently to avoid the risk of irregularities.

A breakthrough is needed in assessing land damage, primarily due to floods and drought, using drones to make the results faster and more effective for large areas. The premium amount to increase farmer participation in the agricultural insurance program has been expanded. The government needs to consider an independent insurance model for farmers with specific criteria, such as having a land area of more than 1 ha willing to be registered. The land location criteria requirements for insurable damage (> 75%) need to be reviewed to increase the achievement of the target land area registered for insurance.

Conclusion

One type of agricultural insurance already running in Indonesia is Rice Farming Business Insurance (AUTP). This agricultural insurance is based on the rice barn system that most indigenous communities in Indonesia have implemented. In its implementation, this type of insurance helps rice farmers in overcoming the problem of crop failure. Farmers are guaranteed the continuity of their livestock business with this insurance. However, its implementation is not free from various obstacles, issues, and challenges, so improvements still need to be made. The implementation of this program aims to protect farmers from crop failure due to seasonal changes. This insurance can also provide significant benefits for agricultural activities.

This agricultural insurance program protects farmers from losses due to the risk of crop damage and crop failure. The protection provided also follows the objectives of the issuance of Law Number 19 of 2013, which aims to realize the sovereignty and independence of farmers to improve their welfare. Therefore, there is a need to improve services and implementation to make them more efficient and effective, from registration to claims processing. Utilizing technology (registration, implementation, and settlement of claims) that makes implementation easier is essential. Overall, improvements in various activities are expected to improve the performance of agricultural insurance program implementation.

The continued implementation of this agricultural insurance program will be expanded to include other strategic commodities and will be implemented in various regions in Indonesia. Therefore, the role of policymakers at the central and regional levels is significant to ensure this program's implementation to improve farmers' welfare. The financial assistance allocated by the central government and provincial governments will significantly influence the future success of this program. With government support, farmer protection programs deserve priority in national agricultural development.

The agricultural insurance program will only run well according to targets if there is harmonious communication, cooperation, and coordination between all stakeholders. Two-way communication, outreach, and coordination are the keys to the success of agricultural insurance programs. Communication between service officers in the field and insurance companies is expected to achieve program efficiency. Effective coordination will encourage sustainable improvements in the performance of agricultural insurance programs.

Acknowledgements Both authors were major contributors in completing this book chapter. We thank the Research Center for Social Welfare Villages and Connectivity and the National Research and Innovation Agency (BRIN) for supporting this research.

References

Agunbiade VF, Babalola OO (2023) Endophytic and rhizobacteria functionalities in alleviating drought stress in maize plants. Plant Prot Sci 59(1):1–18. https://doi.org/10.17221/61/2022-PPS

Akano OI, Oderinde FO, Omotayo AO (2023) Agricultural yield, food nutrition and dietary energy supply in Nigeria: evidence from nationally representative data. J Agric Food Res 11:100525. https://doi.org/10.1016/J.JAFR.2023.100525

Ani KJ, Anyika VO, Mutambara E (2022) The impact of climate change on food and human security in Nigeria. Int J Clim Change Strateg Manag 14(2):148–167. https://doi.org/10.1108/IJCCSM-11-2020-0119/FULL/PDF

Atreya PN, Kaphle M (2020) Visible evidence of climate change and its impact on fruit production in Nepal. Int J Agric Environ Food Sci 4(2):200–208. https://doi.org/10.31015/JAEFS.2020.2.10

Atube F, Malinga GM, Nyeko M, Okello DM, Alarakol SP, Okello-Uma I (2021) Determinants of smallholder farmers' adaptation strategies to the effects of climate change: evidence from northern Uganda. Agric Food Secur 10(1):1–14. https://doi.org/10.1186/S40066-020-00279-1/TABLES/11

Ciscar JC, Fisher-Vanden K, Lobell DB (2018) Synthesis and review: an inter-method comparison of climate change impacts on agriculture. Environ Res Lett 13(7):070401. https://doi.org/10.1088/1748-9326/AAC7CB

Deepak AC (2013) A postcolonial feminist social work perspective on global food insecurity. Affilia 29(2):153–164. https://doi.org/10.1177/0886109913516456

Deli A, Pahlevi MR, Rahmaddiansyah R, Afwanudin A, Authar MND, Miswar E, Irham M (2022) Analysis of rice production trends and the effect of harvest failure on insured land area in Aceh Province. E3S Web Conf 339:06004. https://doi.org/10.1051/E3SCONF/202233906004

Eka Suranny L, Gravitiani E, Rahardjo M (2022) Impact of climate change on the agriculture sector and its adaptation strategies. IOP Conf Ser Earth Environ Sci 1016(1):012038. https://doi.org/10.1088/1755-1315/1016/1/012038

FAO (2011) Agricultural insurance in Asia and the Pacific Region, p 225

Greatorex M (2014) Observation. In: Cooper CL (ed) Wiley encyclopedia of management. https://doi.org/10.1002/9781118785317.weom090543

Grzelak A (2022) The relationship between income and assets in farms and context of sustainable development. PLoS ONE 17(3):e0265128. https://doi.org/10.1371/JOURNAL.PONE.0265128

Hanna R, Abdulla A, Xu Y, Victor DG (2021) Emergency deployment of direct air capture as a response to the climate crisis. Nat Commun 12(1):1–13. https://doi.org/10.1038/s41467-020-20437-0

Harkness C, Areal FJ, Semenov MA, Senapati N, Shield IF, Bishop J (2021) Stability of farm income: the role of agricultural diversity and agri-environment scheme payments. Agric Syst 187:103009. https://doi.org/10.1016/J.AGSY.2020.103009

Hawrylak-Nowak B, Hasanuzzaman M, Wójcik M (2019) Biostimulation and biofortification of crop plants—new challenges for modern agriculture. Acta Agrobot 72(2). https://doi.org/10.5586/aa.1777

Heureux AMC, Alvar-Beltrán J, Manzanas R, Ali M, Wahaj R, Dowlatchahi M, Afzaal M, Kazmi D, Ahmed B, Salehnia N, Fujisawa M, Vuolo MR, Kanamaru H, Gutiérrez JM (2022) Climate trends and extremes in the Indus River Basin, Pakistan: implications for agricultural production. Atmosphere 13(3):378. https://doi.org/10.3390/ATMOS13030378

Huang Y, Chen F, Wei H, Xiang J, Xu Z, Akram R (2022) The impacts of FDI inflows on carbon emissions: economic development and regulatory quality as moderators. Front Energy Res 9:820596. https://doi.org/10.3389/FENRG.2021.820596/BIBTEX

Islam A, Ghosh S, Sarkar M, Barman SD, Shit PK, Rahman A (2023) Agrarian and socio-infrastructural vulnerability in the wake of flood: an example from the Mayurakshi River Basin, India. Springer, pp 231–266. https://doi.org/10.1007/978-3-031-21086-0_10/COVER

Janowicz-Lomott M, Łyskawa K, Rozumek P (2015) Farm income insurance as an alternative for traditional crop insurance. Procedia Econ Financ 33:439–449. https://doi.org/10.1016/S2212-5671(15)01727-X

Jones KML, McCoy C (2019) Reconsidering data in learning analytics: opportunities for critical research using a documentation studies framework. Learn Media Technol 44(1):52–63. https://doi.org/10.1080/17439884.2018.1556216

Kabir KMH, Uddin MK (2016) Climatic effects on food security in Bangladesh. J Environ Sci Nat Resour 9(2):135–142. https://doi.org/10.3329/JESNR.V9I2.32184

Karyadi LW (2021) The existence and role of indigenous food institution on strengthening foods security of rural community. Int J Papier Public Rev 2(1):54–66. https://doi.org/10.47667/IJPPR.V2I1.73

King M, Singh AP (2020) Understanding farmers' valuation of agricultural insurance: evidence from Vietnam. Food Policy 94:101861. https://doi.org/10.1016/J.FOODPOL.2020.101861

Kogo BK, Kumar L, Koech R (2021) Climate change and variability in Kenya: a review of impacts on agriculture and food security. Environ Dev Sustain 23(1):23–43. https://doi.org/10.1007/S10668-020-00589-1/TABLES/5

Kumar A, Sengar RS, Pathak RK, Singh AK (2022) Integrated approaches to develop drought-tolerant rice: demand of era for global food security. J Plant Growth Regul 42(1):96–120. https://doi.org/10.1007/S00344-021-10561-6

Madaki MY, Kaechele H, Bavorova M (2023) Agricultural insurance as a climate risk adaptation strategy in developing countries: a case of Nigeria. Clim Policy 23(6):747–762. https://doi.org/10.1080/14693062.2023.2220672

Mashizha TM, Mashizha T (2019) Adapting to climate change: reflections of peasant farmers in Mashonaland West Province of Zimbabwe. Jàmbá J Disaster Risk Stud 11(1):8. https://doi.org/10.4102/JAMBA.V11I1.571

Mulya SP (2022) Determining of peri-urban in rice production centers area: case study in Karawang Regency, Indonesia. IOP Conf Ser Earth Environ Sci 1109(1):012072. https://doi.org/10.1088/1755-1315/1109/1/012072

Nirmal Ravi Kumar K, Babu SC (2021) Can a weather-based crop insurance scheme increase the technical efficiency of smallholders? A case study of groundnut farmers in India. Sustainability 13(16):9327. https://doi.org/10.3390/SU13169327

Osabohien R, Onanuga O, Aderounmu B, Matthew O, Osabuohien E (2020) Social protection and employment in Africa's agricultural sector. Bus Theory Pract 21(2):494–502. https://doi.org/10.3846/BTP.2020.11945

Patt A, Peterson N, Carter M, Velez M, Hess U, Suarez P (2009) Making index insurance attractive to farmers. Mitig Adapt Strateg Glob Change 14(8):737–753. https://doi.org/10.1007/S11027-009-9196-3/FIGURES/2

Patunru A (Arianto), Ilman AS (Assyifa) (2020) Political economy of rice policy in Indonesia: a perspective on the ASEAN economic opportunity. https://doi.org/10.35497/346118

Petersen-Rockney M (2022) Social risk perceptions of climate change: a case study of farmers and agricultural advisors in northern California. Glob Environ Change 75:102557. https://doi.org/10.1016/J.GLOENVCHA.2022.102557

Pittock AB, Jones RN (2000) Adaptation to what and why? Environ Monit Assess 61(1):9–35. https://doi.org/10.1023/A:1006393415542/METRICS

Poudel M, Mendes R, Costa LAS, Bueno CG, Meng Y, Folimonova SY, Garrett KA, Martins SJ (2021) The role of plant-associated bacteria, fungi, and viruses in drought stress mitigation. Front Microbiol 12:743512. https://doi.org/10.3389/FMICB.2021.743512/BIBTEX

Pramukti U, Suryanto, Gravitiani E (2021) Determination of priority locations for the implementation of rice farming insurance: a case study on disaster hazards in Cilacap regency. IOP Conf Ser Earth Environ Sci 819(1):012042. https://doi.org/10.1088/1755-1315/819/1/012042

Prentice CM (2017) Interviews, qualitative. In: The international encyclopedia of intercultural communication, pp 1–8. https://doi.org/10.1002/9781118783665.ieicc0108

Radeny M, Desalegn A, Mubiru D, Kyazze F, Mahoo H, Recha J, Kimeli P, Solomon D (2019) Indigenous knowledge for seasonal weather and climate forecasting across East Africa. Clim Change 156(4):509–526. https://doi.org/10.1007/S10584-019-02476-9/FIGURES/7

Rust JM, Rust T (2013) Climate change and livestock production: a review with emphasis on Africa. S Afr J Anim Sci 43(3):255–267. https://doi.org/10.4314/SAJAS.V43I3.3

Sari NMW, Sjah T, Maryati S, Supartiningsih NLS, Ayu C, Yanuartati BYE (2022) Analysis of the effectiveness of the community food barn program on food security for farmer families during the Covid-19 pandemic in West Lombok Regency, West Nusa Tenggara. IOP Conf Ser Earth Environ Sci 1107(1):012106. https://doi.org/10.1088/1755-1315/1107/1/012106

Shah AA, Khan NA, Gong Z, Ahmad I, Naqvi SAA, Ullah W, Karmaoui A (2023) Farmers' perspective towards climate change vulnerability, risk perceptions, and adaptation measures in

Khyber Pakhtunkhwa, Pakistan. Int J Environ Sci Technol 20(2):1421–1438. https://doi.org/10.1007/S13762-022-04077-Z/TABLES/3

Sun L, Song F, Guo J, Zhu X, Liu S, Liu F, Li X (2020) Nano-ZnO-induced drought tolerance is associated with melatonin synthesis and metabolism in maize. Int J Mol Sci 21(3):782. https://doi.org/10.3390/IJMS21030782

Sundler AJ, Lindberg E, Nilsson C, Palmér L (2019) Qualitative thematic analysis based on descriptive phenomenology. Nurs Open 6(3):733–739

Tasnim Z, Saha SM, Hossain ME, Khan MA (2023) Perception of and adaptation to climate change: the case of wheat farmers in northwest Bangladesh. Environ Sci Pollut Res 30(12):32839–32853. https://doi.org/10.1007/S11356-022-24478-4/FIGURES/7

Vigouroux Y, Mariac C, de Mita S, Pham JL, Gérard B, Kapran I, Sagnard F, Deu M, Chantereau J, Ali A, Ndjeunga J, Luong V, Thuillet AC, Saïdou AA, Bezançon G (2011) Selection for earlier flowering crop associated with climatic variations in the Sahel. PLoS ONE 6(5):e19563. https://doi.org/10.1371/JOURNAL.PONE.0019563

World Bank (2010) Agricultural Insurance in Latin America, Developing the Market. Report No.61963-LAC. Desember 2010. World Bank Insurance for the Poor Program. Washington DC.

Wu F, Wu B, Zhang M, Zeng H, Tian F (2021) Identification of crop type in crowdsourced road view photos with deep convolutional neural network. Sensors 21(4):1165. https://doi.org/10.3390/S21041165

Zebua DDN, Priyanto SH, Sunaryanto LT (2019) An analysis of wheat farming: calculations and perceptions. Caraka Tani J Sustain Agric 34(2):200–212. https://doi.org/10.20961/CARAKATANI.V34I2.28862

Chapter 25
Aging Individuals and Their Capability to Adopt Green Lifestyles

Sheeba Joseph, Tanuja Khan, and Rutwik Gandhe

Abstract Achieving sustainability is the prime objective of this day and age for all societies across nations and cultures, however the world is aging decade by decade and aging as a phenomenon has deep implications on the creation of sustainable societies that have lesser carbon emission. This interconnection of aging and sustainability is poorly understood as of now. Aging societies have a tendency to become more consumption oriented. As individuals age, they naturally desire and require more comforts which leads to more carbon footprints and carbon emission dependent life. The answer to this question is obviously the green lifestyle adoption. However adopting a green lifestyle or any other lifestyle would depend on various psychological as well as circumstantial factors as we understand the human behavior change dynamics. Therefore it behooves us to understand the general and mental wellbeing of older individuals, and special situations associated with them like daily spiritual experiences, depression etc. Studies have shown that depression in the elderly is further linked with cognitive decline and risk of Alzheimer's Disease which indicates how deeply it is linked with individual wellbeing. The current research is divided into three stages in order to understand the interconnections of the lifestyle—how a more or less greener lifestyle, less carbon dependent life interplays with the general wellbeing, mental wellbeing of aged individuals and situations like spiritual experiences and depression among them leading to find solutions towards adoption of green and less carbon dependent life among aged individuals which is the exact antidote to the riddle of achieving sustainability in the globally aging scenario.

S. Joseph (✉)
Department of Social Work, The Bhopal School of Social Sciences, Habibganj, Bhopal, Madhya Pradesh, India
e-mail: sheebajoseph@bsssbhopal.edu.in

T. Khan · R. Gandhe
The Bhopal School of Social Sciences, Habibganj, Bhopal, Madhya Pradesh, India
e-mail: tanujakhan@bsssbhopal.edu.in

R. Gandhe
e-mail: rutwik@bsssbhopal.edu.in

U. Mukhopadhyay et al. (eds.), *Climate Crisis, Social Responses and Sustainability*, Climate Change Management, https://doi.org/10.1007/978-3-031-58261-5_25

Introduction

Aging is a global phenomenon and in the present day and age it is more significant as the majority of the population in developed countries are already approaching it and will be moving towards it in future in developing countries. As societies across the globe become aging societies many social issues and contemporary pressing issues need to be understood in this context. The issue of sustainability and climate change needs to be understood in this direction where at the moment our understanding as a global community is limited (Zheng et al. 2022). We have seen and there are studies to substantiate the point that human behavior is a very significant cause of carbon emission and therefore it impacts the sustainability of the environment (Dietz and Rosa 1997; Dietz et al. 2007). Human attitude, beliefs and norms of societies where people live and age have significant association with the way they are going to adjust the low carbon lifestyle which is quite a significant step moving towards creating a sustainable society. Thus, it is crucial to explore how the lifestyle of an aging population can affect the sustainability aspects and help achieve the goal of creation of low carbon societies. Therefore it is important to address and understand the lifestyle of aged people which has now gained importance across the globe (Fan et al. 2021). Special situations like pandemic etc. also continue to disrupt global life and they leave a profound mark on society, affecting especially aged individuals in multifaceted ways. Even otherwise aged people face a lot of challenges in their life due to age related circumstances which makes it quite difficult for them to adjust to the low carbon lifestyle or for that matter any other lifestyle change. We must recognize that aged ones are a demographic group particularly vulnerable to both the health and socio-psychological consequences of any such situation (van der Velden et al. 2022).

Making the older population change their behavior is quite a complex task and would depend on multiple aspects concerning their life situations, culture context, demography, psychological well-being, general wellbeing, daily spiritual experience and so on (Bardazzi and Pazienza 2017). Considering the daily spiritual experience often rooted in faith, community, and a sense of purpose, it takes renewed significance among many aged individuals who face isolation and loneliness which is again quite likely a social situation among older people. Aged people in these situations resort to spiritual and religious beliefs as a source of comfort and connection. Virtual services and online communities have become a lifeline, bridging the physical gap between maintaining some semblance of spiritual continuity (Zheng et al. 2021). During old age, spirituality also sometimes becomes a coping mechanism, offering hope and solace during these uncertain times. During old age economic repercussions further strains their well-being. This disrupts daily routines quite often, limits access to mental stimulation, and the challenges of maintaining physical health compounds the overall strain on general well-being (López-Sánchez et al. 2022).

Psychological well-being, too, influences the lifestyle of aged individuals. Anxiety and depression surge among older adults due to the uncertainty and isolation normally however it depends on the context of the individual living in. The old age is also quite susceptible to the loss of loved ones, many times spouse, friends etc. which brings additional psychological burden, leading to grief and complicated bereavement that affects the wellbeing of the aged individuals and restricts their possibility to adjust to the new lifestyle. Some, however, exhibited remarkable resilience, adapting to the new lifeways and circumstances through various coping strategies, including mindfulness, virtual connections, and seeking professional support when needed (Khan and Thomas 2022). The physical, economic, social, mental, emotional, cultural, and spiritual factors of health are all included in psychosocial well-being and they all are required to make an old person adapt to the green lifestyle which is the most important step to achieve a sustainable society. To be in good health, a person must be able to manage the many stressors of daily life and reach their full potential as contributing members of society.

So we observe that old aged individuals remain in quite challenging circumstances even if life conditions are normal around them and therefore to make such a vulnerable group adopt the green lifestyle is quite challenging, yet at the same time it is more important these days as society across the globe is aging and hence can not be ignored. Along with other age groups they also must be encouraged to adopt a behavior in their lives that is quite required for achieving the targets of low carbon society.

Research Gap

This section precisely talks about the research gap that is being addressed by the ongoing and proposed research that will be completed in three phases. First let us see the gap. A large amount of research has gone into establishing the fact that climate change issues can effectively be addressed through reducing carbon footprint for which actions have been suggested by experts like using power efficient equipment, imposing restrictions on energy usage of individuals (electricity and fuel both). To achieve this human intervention and behavior has been proven as key shift and has been has been suggested alongside technological intervention, else there is always a danger of getting into the trap of rebound effect (Greening et al. 2000)—which is when society at large end up consuming more energy as only technology based intervention are taken place without any human behavioral restriction on usage of such equipments, instead of low energy usage and hence the objective of achieving of lower carbon emission is defeated. Therefore adopting a green lifestyle and changing human behavior is a must. Here appears the gap as we as a human society do not have enough understanding on how to deal with the complexities of old age while encouraging them to adopt a lifestyle that leads us as a society towards low carbon society (Pais-Magalhães et al. 2020). The challenges of aged individuals in order to meet these objectives are far more serious and call for planned and methodical intervention based on robust understanding of wellbeing issues of old age. We do not have data to design the strategies that help us understand and suggest models and theories that meet the requirements of old age people. Therefore, circumstantial factors pertaining to aging and adoption of green lifestyle by aged individuals in different cultures, traditions and societies are required to be investigated properly. This study precisely intends to fill that gap in the literature. An appropriate method which incorporates a large number of contextual factors concerning circumstances of aged individuals have been discussed in the method section.

Objectives of the Study

- To understand the degree to which circumstantial factors associated with aging and aged individuals are responsible for their adoption of green and sustainable lifestyle
- To study the General, Spiritual and Mental well-being of older individuals with different circumstantial factors
- To understand the intricacies associated with aged individuals and their adaptation for green and sustainable lifestyle
- To understand how different challenging situations characterizing old age e.g., depression that may arise due to loneliness or isolation etc., for better adaptation towards sustainable and green lifestyle.

Methodology

As the study requires us to understand general wellbeing, psychological wellbeing of aged individuals, so that factors associated with aging can be understood properly, which then would help us understand the adoption of green and sustainable behavior for achieving a low carbon lifestyle by older people. First, data general and psychological wellbeing will be obtained at the first stage across gender and marital status. Gender and marital status provides immediate information of circumstances of individuals, i.e. whether an old aged person is male or female? Whether the person is likely to suffer loneliness largely depends on first circle and marital status. In conservative cultures like India, old age loneliness is more determined by marital status than that of a friend circle. So we go by marital status at the first stage. A set of circumstantial factors are to be understood when we gather the same information across three categories of old age population, at second stage by obtaining samples from urban, rural and tribals or from primitive groups mainly following ascetic lifestyle. The third stage of the study would largely be focused on testing multiple models and assumptions with the data collected till the second stage. Since in first stage we are limited with understanding situations of an urban sample of old aged individuals only, we shall collect data from Bhopal city and around 300 respondents participated in this stage. The results obtained from this only are reported in this article. Stages II and III will be presented later.

Stage I of the study is already over and we provide the data from urban samples of old aged people in this manuscript. The Stage II would comprise the data collection from two more groups i.e. (1) rural and old aged and (2) those who are practicing asceticism. These three types of samples would cover all possible kinds of circumstances old aged individuals can possibly have from the point of view of carbon emitting lifestyles. These would cover different styles of aging which would help us understand the green lifestyle adoptions by different groups which are under different conditions. An urban aged person is a normal urbanite who generally follows the comforts and spends on them like room heater, air conditioning etc. and tend to consume more energy with age. In rural areas people live in harmony with nature and hence this group would represent a different level of adoption to a green lifestyle. An ascetic lifestyle practiced by aged ones in the third group representing the third possible way of a green lifestyle. In order to develop a holistic understanding on green lifestyle adoption by old aged people, all these three possible types of behavior patterns are essential to consider as they are three different lifestyles emitting three different levels of carbon emissions.

A quantitative methods design was adopted to gather the quantitative data. Structured surveys were administered to a representative sample of older adults. The surveys included standardized measures to assess daily spiritual experience, psychological well-being, general health, psychological impact, and depression. Furthermore, self-reported questions were also included in the schedule. A stratified random sampling method was used to ensure diversity in the sample, considering variables like age, gender, socioeconomic status, and geographical location. The sample size was determined based on statistical power analysis to ensure the reliability and validity of the self-reported research tool.

The Daily Spiritual Experience Scale (DSES) developed by Underwood (2011) is a 16-item self-report measure designed to assess ordinary experiences of connection with the transcendent in daily life. It includes constructs such as awe, gratitude, mercy, sense of connection with transcendent and compassionate love. It also includes measures of awareness of discernment/inspiration and a sense of deep inner peace. The Warwick-Edinburgh Mental Well-being Scale (WEMWBS) comprises 14 items of mental well-being covering subjective well-being and psychological functioning, in which all items are worded positively and address aspects of positive mental health (Tennant et al. 2007). The Beck Depression Inventory (BDI) is a 21-item, self-report rating inventory that measures characteristic attitudes and symptoms of depression (Beck et al. 1996) and was used to measure the level of depression. Let us know understand the key psychological correlates that we consider essential for determining a positive state for possible green lifestyle adoption by an aged individual.

Mental Well-Being

Mental well-being, often referred to as psychological well-being, is a multifaceted concept that encapsulates the overall state of an individual's mental and emotional health. It encompasses a range of factors, including emotional resilience, positive emotions, optimism, self-esteem, a sense of purpose and meaning in life, positive relationships, autonomy, psychological flexibility, and the ability to cope with stress (Hooper et al. 2023). A person with good mental well-being experiences positive emotions, maintains an optimistic outlook and possesses a positive self-concept. They feel a sense of purpose and meaning in their life, have healthy relationships, and demonstrate psychological flexibility and resilience to stress. Recognizing the importance of mental health and being aware of one's mental state are also integral components. Mental well-being is closely linked to physical health, and it can be influenced by social and cultural factors (Lee et al. 2022). It is a dynamic and evolving aspect of an individual's life that requires active management, support when needed, and an understanding that it can be impacted by various life events. In essence, mental well-being is a critical aspect of an individual's overall health and quality of life, and it is central to their capacity to thrive and adapt in an ever-changing world (Zheng et al. 2021). The pandemic significantly influenced the mental well-being of older adults. The stress, uncertainty, and fear associated with COVID-19, along with social isolation, posed a substantial challenge to their mental health. Many experienced heightened anxiety and depression, impacting their mental well-being negatively (van der Velden et al. 2022).

Daily Spiritual Well-Being

Daily spiritual well-being is a profound and deeply personal aspect of an individual's life, encompassing their connection to their spirituality or religious beliefs on a day-to-day basis. It involves a sense of inner peace, purpose, and a feeling of belonging to a larger spiritual community or something greater than oneself (Hiremath 2020). This daily spiritual experience can manifest in various forms, from the practice of religious rituals to acts of kindness, mindfulness, and contemplation. For many, particularly older adults, daily spiritual well-being can be a source of solace and guidance (Malloy et al. 2013). During any challenging periods that older people are susceptible to, social isolation was prevalent, spiritual practices and beliefs often played a significant role in providing comfort and a sense of continuity. Virtual worship services and online religious communities allowed individuals to maintain their spiritual connections and find support in their faith even in the absence of in-person gatherings (Rivera-Ledesma and Montero 2005).

Daily spiritual well-being goes beyond religious adherence; it also pertains to a sense of inner harmony and balance. Engaging in practices that align with one's spiritual or moral values can contribute to an individual's daily spiritual well-being. These practices might include meditation, acts of kindness, prayer, or simply moments of reflection (Okun and Nimrod 2021). Daily spiritual well-being can serve as a guiding force, offering a moral compass for decision-making and a source of inner strength in the face of life's challenges. It provides a framework for understanding the world and finding meaning in experiences, and it can play a significant role in shaping one's perspective on adversity, death, and the human condition (Okun and Nimrod 2020).

Daily spiritual well-being serves as a coping mechanism, allowing individuals to find hope and resilience in times of crisis. In the situations where fear, uncertainty, and loss were pervasive, many older adults turned to their spirituality to help them navigate these emotional hurdles (Casey and Liang 2014). Faith and spiritual beliefs provided a sense of purpose, comfort, and a sense of belonging to a larger community, even when physical distancing measures isolated them from their usual support networks (Mohammadi et al. 2021). In essence, daily spiritual well-being encompasses the connection to one's innermost beliefs and values, and it can be a vital source of strength, solace, and continuity in the face of life's most challenging moments (Fry 2000). The study found that for many older adults, daily spiritual well-being became a crucial coping mechanism during the pandemic. Isolation and the fear of illness prompted some to rely more on their spirituality, finding comfort, hope, and a sense of purpose in their faith and spiritual practices (Ayoubi-Mahani et al. 2023).

General Health

General health encompasses a comprehensive and multidimensional concept that reflects an individual's overall well-being, encompassing physical, mental, and social dimensions. Physical health is a fundamental component, focusing on aspects like fitness, nutrition, and the absence of chronic diseases. Someone in good physical health is often characterized by vitality, strong immunity, and the ability to perform daily activities with ease. Conversely, mental health, another pivotal aspect of general health, pertains to an individual's emotional and psychological well-being (Attema et al. 2022). A positive state of mental health entails emotional resilience, effective stress management, positive self-esteem, and a stable emotional equilibrium. Mental health conditions, such as anxiety and depression, can significantly impact overall general health (Gadie et al. 2017).

Social well-being is equally essential, highlighting a person's capacity to cultivate and maintain healthy relationships. Having social support, feeling a sense of belonging, and engaging in meaningful social interactions are integral to social well-being (Balachandran and James 2021). These elements foster a positive influence on overall health, as strong support systems can provide vital emotional sustenance. Nutrition and diet form a cornerstone of general health, as they ensure the body receives the essential nutrients it requires for optimal functioning. A balanced diet not only prevents chronic diseases but also supports physical health and contributes to mental well-being. Regular physical activity and exercise are critical for physical health, promoting cardiovascular well-being, muscular strength, flexibility, and reducing stress and anxiety, thereby enhancing mental health (DeSantis et al. 2019).

Adequate sleep is a pivotal aspect, as quality sleep is necessary for physical recovery, cognitive function, and emotional well-being. Sleep deprivation can lead to various health problems, including mental health issues. Preventive care through regular check-ups, screenings, and immunizations aids in identifying health concerns early, allowing for timely intervention and treatment (Hooper et al. 2023). Access to quality healthcare services is integral to general health. Barriers to healthcare access, such as cost or geographical limitations, can hinder an individual's ability to maintain and improve their health. Healthy lifestyle choices, including avoiding smoking, limiting alcohol consumption, and refraining from drug abuse, significantly reduce the risk of chronic diseases and mental health issues (Hadler-Olsen and Jönsson 2021).

Stress management plays a vital role in general health, encompassing techniques like mindfulness, relaxation, and seeking support when needed. High stress levels can adversely affect both physical and mental health. Environmental factors, such as air quality, access to clean water, and exposure to toxins, also influence general health. A safe and clean environment is essential for physical well-being. General health is a dynamic and holistic concept, reflecting the interplay of various factors in an individual's life. It requires a comprehensive approach addressing physical, mental, and social well-being, alongside considerations like nutrition, exercise, sleep, and access to healthcare. Ultimately, a state of general health is essential for individuals to lead fulfilling and productive lives.

Psychological Impact

The psychological impact, often intertwined with emotional and cognitive consequences, plays a central role in the overall well-being of individuals. It represents the emotional and mental repercussions of various life events, which brought profound psychological challenges (Bahl et al. 2017). In the context of the pandemic, the psychological impact was multifaceted. Many individuals, including older adults, grappled with heightened levels of anxiety, fear, and uncertainty, stemming from the pervasive threat of illness and the loss of normalcy (Kieny et al. 2022). The isolation and disruptions to daily routines also contributed to feelings of unease and vulnerability (Norman et al. 2021).

Grief and loss were poignant aspects of the psychological impact, particularly among those who lost loved ones to the virus. This loss extended beyond bereavement, resulting in complicated grief and a deep emotional toll (Candeias et al. 2021). The pandemic tested the emotional resilience of individuals, demanding they navigate a landscape of change and adversity. The psychological impact led to increased rates of depression and anxiety, amplifying the importance of mental health awareness and support (Boro et al. 2023).

However, amidst these challenges, some individuals display remarkable resilience. They found ways to cope with stress, adapt to new circumstances, and seek solace in coping strategies, such as mindfulness, physical activity, and virtual connections (Nagaraj 2017). The psychological impact serves as a testament to the human capacity for adaptation and resilience, but it also underscores the significance of acknowledging the emotional and mental needs of individuals during times of crisis. Psychological impact underscores the emotional and cognitive toll it took on individuals, particularly older adults. While many grappled with increased anxiety, depression, and grief, some demonstrated resilience and adaptive strategies (López et al. 2020). Understanding and addressing the psychological impact is crucial for providing support, promoting mental well-being, and ensuring individuals can navigate such challenges with strength and fortitude. The psychological impact of any adverse life circumstance on older adults is profound. The fear of illness, the loss of friends and family, and social isolation contributed to heightened anxiety and depression. This psychological impact had significant implications for their overall mental well-being (von Humboldt et al. 2023).

Depression

Depression is a complex and pervasive mental health condition that profoundly affects individuals' emotional and psychological well-being. It is characterized by persistent feelings of sadness, hopelessness, and a profound loss of interest or pleasure in daily activities. In the context of other life stressors, depression became a prevalent and concerning issue, impacting individuals of all ages, including older adults (Westerhof and Keyes 2010). Depression can manifest in various ways, including changes in sleep patterns, appetite, and energy levels, as well as difficulty concentrating and making decisions. Feelings of guilt, worthlessness, and thoughts of self-harm or suicide may also be present. The emotional burden of depression is profound, affecting not only an individual's mental health but also their physical health and overall quality of life (López-Sánchez et al. 2022).

Recognizing the signs of depression, seeking timely support, and promoting mental health awareness are crucial in addressing this condition. Professional treatment, such as therapy and medication, can be highly effective in managing depression. Additionally, fostering social connections, maintaining a healthy lifestyle, and practicing stress-reduction techniques are essential steps in mitigating the impact of depression (López-Sánchez et al. 2022). Depression is a treatable condition, and with appropriate care and support, individuals can find hope and relief from its grasp.

As a coping mechanism, many older adults turn to their daily spiritual well-being, relying on their faith and spirituality for solace, connection, and resilience. This link between mental, physical, and spiritual well-being illustrates the intricate relationship between these variables during times of crisis (Sharma et al. 2022). Adverse situations in old age have profound psychological impact, leading to increased rates of depression and anxiety among older adults, further underscoring the interdependency of these factors. Together, these variables shed light on the holistic impact of a global crisis on the well-being of older adults, emphasizing the need for comprehensive support strategies to address their complex needs (Hooper et al. 2023). The research gap in the study of the impact of the COVID-19 pandemic on older adults' well-being lies in the need for a more comprehensive understanding of the nuanced interplay between daily spiritual experience, general health, mental well-being, psychological impact, and depression within this demographic. While existing studies have highlighted individual aspects of this complex relationship, there is a lack of integrated research that explores how these variables intersect and mutually influence each other so that an old aged individual is found or deemed fit for any possible green lifestyle adoption. A more holistic investigation is essential to guide the development of targeted interventions and support systems that can enhance the overall well-being of older adults and encourage them to take positive steps towards possible green lifestyle adoption.

Result and Discussion

Interpretation: The dataset comprises a diverse group of respondents, primarily skewed towards males, though there is representation from females as well. The age groups of the respondents are not explicitly defined, making it challenging to draw specific conclusions about the age distribution. The dataset includes individuals from various castes, encompassing General, OBC, SC, and ST categories, indicating a broad cross-section of society from India. All respondents come from urban areas in the given sample. A significant portion of the respondents receive a pension, but there is a substantial number who do not, highlighting variations in financial support. In terms of annual income, the data demonstrates a range, with some earning between 150,001 and 300,000, and others reporting incomes exceeding 300,000, underlying economic diversity. Marital status and place of stay reveal a mix of life situations, with respondents identifying as Married, Widower, Widow, and Unmarried. The places of stay include Own House, With Son, and Any Other, reflecting diverse living arrangements (Table 25.1).

Ho1: There will be a significant difference between male and female old age adults in terms of Spiritual well-being, Mental Wellbeing, General Health, Psychological Impact and Depression in older adults during the COVID-19 pandemic.

Table 25.1 Summary of demographic data

Respondent category	Number of respondents
Gender	
Male	207
Female	83
Marital status	
Married	177
Widower	23
Widow	16
Unmarried	74
Total respondents	290

The analysis of the data pertaining to various aspects of well-being and mental health in older adults during the COVID-19 pandemic reveals several significant findings.

Firstly, concerning spiritual well-being, a notable gender disparity is observed. Females have a mean score of 113.18, while males have a higher mean score of 119.37. The negative t-value of – 2.848 signifies that females tend to exhibit lower spiritual well-being compared to males, and the *p*-value of 0.005 establishes the statistical significance of this difference. In terms of mental well-being, the data indicates a significant distinction between genders. Females reported a higher mean score of 18.32, reflecting better mental health during the pandemic, in contrast to males with a mean score of 14.23. The t-value of 4.554 and a *p*-value close to zero ($p < 0.001$) affirm the statistical significance of this difference. The analysis demonstrates that general health significantly differs between male and female older adults. Females report a mean score of 25.66, whereas males have a lower mean score of 21.79. The t-value of 3.295 and a *p*-value of 0.001 ($p < 0.01$) validate this discrepancy (Table 25.2).

Conversely, no substantial gender difference is observed in terms of psychological impact. Both males and females report similar mean scores, and the *p*-value of 0.873 ($p > 0.05$) underscores the lack of significant differentiation. Lastly, concerning depression levels, females have a higher mean score of 22.02, indicating a greater level of depression during the pandemic, in contrast to males with a mean score of 17.10. The t-value of 2.761 and a *p*-value of 0.007 ($p < 0.01$) emphasize the statistical significance of this distinction. Analysis highlights pronounced gender differences in spiritual well-being, mental well-being, general health, and depression levels among older adults during the COVID-19 pandemic. These insights can inform tailored interventions and support strategies to address the distinct well-being needs of male and female older adults.

Table 25.2 Measure of spiritual well-being, mental well-being, general health and psychological impact and level of depression on gender

Variables	Education	Mean	S.D.	t-value	*p*-value
Spiritual wellbeing	Female	113.18	19.98	– 2.848**	0.005
	Male	119.37	18.96		
Mental wellbeing	Female	18.32	8.23	4.554**	0
	Male	14.23	7.92		
General health	Female	25.66	10.94	3.295**	0.001
	Male	21.79	9.82		
Psychological impact	Female	25.37	8.31	0.16	0.873
	Male	25.19	9.07		
Depression	Female	22.02	7.14	2.761**	0.007
	Male	17.1	8.91		

**Highly significant

Ho2: There is no significant difference between the marital status of old age adults in terms of Spiritual well-being, Mental well-being, General Health, Psychological Impact, and Depression in older adults during the COVID-19 pandemic.

The data indicates that there is no significant difference between the marital status of old-age adults in terms of their spiritual well-being, general health, and psychological impact. However, there are some noteworthy differences observed in terms of mental well-being (Table 25.3).

Spiritual Well-Being: The analysis shows a significant difference between marital status groups in spiritual well-being, with married individuals reporting lower scores compared to widowed and widower groups.

Mental Well-Being: The F-statistic for mental well-being is not significant ($F = 1.941, p = 0.122$), indicating that marital status does not have a significant effect on mental well-being.

General Health: Marital status groups demonstrate significant differences in general health. Widows report the highest mean scores, followed by widowers, married individuals, and unmarried individuals.

Table 25.3 Measure of spiritual well-being, mental well-being, general health and psychological impact on marital status of older adults

Variables	Marital status	Mean	S.D.	F	*p*
Spiritual wellbeing	Married	19.94	11	7.741**	0
	Unmarried	17.68	12		
	Widow	24.21	12.49		
	Widower	24.58	11.49		
Mental wellbeing	Married	16.91	7.05	1.941	0.122
	Unmarried	17.57	8.13		
	Widow	18.96	9.1		
	Widower	16.2	8.81		
General health	Married	22.91	9.05	6.444**	0
	Unmarried	21.33	12.13		
	Widow	28.28	11.89		
	Widower	25.34	10.05		
Psychological impact	Married	24.65	9.26	2.629*	0.05
	Unmarried	22.34	7.57		
	Widow	23.98	8.39		
	Widower	26.78	9.97		

*Significant
**Highly significant

Psychological Impact: While there is a difference in psychological impact among marital status groups, the statistical significance ($p = 0.05$) is marginal. Widowers seem to experience a slightly higher psychological impact compared to other groups.

The above analysis suggests that marital status does not significantly influence mental well-being and psychological impact in older adults. However, there are significant differences in spiritual well-being and general health among different marital status categories. These findings can be valuable for understanding and addressing the specific needs of older adults in various marital status groups during challenging times to determine whether they are capable of making any shift for adoption of a greener lifestyle to contribute towards achieving a low carbon society.

Challenges/Limitations of the Study

The study has several limitations that should be acknowledged. First, it is just the data to understand the situations of general and psychological wellbeing among aged individuals and the data is subject to multiple interpretations. The core objective of the study is to understand the same general and psychological well-being under different circumstances which are to be measured through different samples which are not presented here as they would be gathered in Stage II of the study. Second, the study focused on understanding gender differences but did not delve into other potentially influencing factors, such as socioeconomic status or pre-existing health conditions. Additionally, the study's scope was limited to a specific region or population, potentially limiting the generalizability of the findings to a broader demographic. Furthermore, the use of self-reported measures can introduce response bias and may not provide a comprehensive view of the participants' well-being. Finally, the cross-sectional nature of the study prevents us from drawing causal conclusions that are compatible with repeated measures. Future research in Stage II will address these limitations by incorporating diverse populations. Additional variables depend upon the context of the sample and longitudinal designs would help in gaining a more comprehensive understanding of the complex relationship between gender, well-being, and mental health in older adults.

Conclusions and Discussion

The study examined the relationship between various aspects of well-being and health in male and female older adults from the urban pocket of the city of Bhopal. This is to understand the mindset and welling status of typical Indian urban middle class aging society. In this kind of society, we have studies that show that with age we consume more energy and comfort hence the carbon footprint of the individuals is more. Such an aging society helps us understand the socio-economic and psycho-social realities of urban areas and the relationship among wellbeing factors therein. The similar needs to be developed in societies that are aging in different conditions with the help of a rural sample and another sample of people living more of an ascetic life. This would also develop insights into how sources of comfort are related with general and psychological well-being of old aged people? Can there be a model which emerges to help us better understand the relationships between general and psychological well-being of aged individuals and their carbon footprint. At the moment in this study we have not gathered any data on carbon footprint of individuals which will be collected in the next stage and then it would be contrasted with each other for different samples taken from different individuals living in different aging conditions and realities i.e. rural or ascetic lifestyle. It would then help us understand the stark realities of relations between psychological measures indicating well-being of people and their carbon footprint indicating their level of carbon emission across gender and marital status and other social realities.

The present data given in this Stage I of the study just reveals significant differences between the two genders. Notably, females reported lower spiritual well-being, while their mental well-being and general health were significantly better. This suggests that older women may be more resilient and employ more effective coping strategies in dealing with challenges of old age and can potentially be more efficient in adopting to green lifestyles. Psychological impact did not significantly differ between genders, indicating that both male and female older adults experienced similar emotional responses to the challenges of old age thus making us conclude that males and females both are equally capable of adopting greener lifestyles. However, the study found that older women experienced significantly higher levels of depression which indicates that women might be insistent in adopting green practices as they grow old. These results emphasize the need for gender-specific support and interventions, particularly for older women who may be more vulnerable to the psychological effects of the old age challenges and need to be perhaps more persuaded to make an effort to adapt to greener lifestyles. Recognizing these gender-related disparities is crucial for tailoring support services to address the unique needs and challenges of older adults to adopt a low carbon emission lifestyle.

Based on the findings of the study, it is recommended that tailored interventions and strategies should be developed to address the unique needs of older women to adopt green practices and lifestyles, who may be more vulnerable to depression during crises in old age. Providing resources and support that enhance spiritual well-being, mental well-being, and general health should be prioritized, with a focus on older women. These findings highlight the importance of a gender-sensitive approach in promoting greener and sustainable lifestyles in urban areas. Later when we blend the information gathered on the same variables from rural and ascetic aging people's sample, there would be deeper insights on what factors are responsible for adopting green lifestyles in an aging society.

Acknowledgements The researchers heartfully acknowledge the time given by the participants for collecting the information.

References

Attema AE, Brouwer WBF, Pinto JL (2022) The role of perceived utility of full health in age weighting. Value Health 25(9). https://doi.org/10.1016/j.jval.2022.04.1733

Ayoubi-Mahani S, Eghbali-Babadi M, Farajzadegan Z, Keshvari M, Farokhzadian J (2023) Active aging needs from the perspectives of older adults and geriatric experts: a qualitative study. Front Public Health 11. https://doi.org/10.3389/fpubh.2023.1121761

Bahl NKH, Nafstad HE, Blakar RM, Geirdal AØ (2017) Responsibility for psychological sense of community and well-being in old age: a qualitative study of urban older adults in Norway. Open J Soc Sci 05(07). https://doi.org/10.4236/jss.2017.57020

Balachandran A, James KS (2021) A multi-dimensional perspective on the gender gap in health among older adults in India and China: application of a new ageing measure. Ageing Soc 41(5). https://doi.org/10.1017/S0144686X19001521

Bardazzi R, Pazienza MG (2017) Switch off the light, please! Energy use, aging population and consumption habits. Energy Econ 65:161–171. ISSN 0140-9883. https://doi.org/10.1016/j.eneco.2017.04.025. https://www.sciencedirect.com/science/article/pii/S0140988317301329

Beck A, Steer R, Brown G (1996) Beck depression inventory. 2 ed, Psychological Corporation, San Antonio, TX

Boro B, Srivastava S, Saikia N (2023) Is there an association between change in old-age living arrangements and older adults' psychological health and subjective well-being in India? Evidence from a cross-sectional study. Ageing Int 48(1). https://doi.org/10.1007/s12126-021-09470-6

Candeias A, Galindo E, Stueck M, Portelada A, Knietzsch J (2021) Psychological adjustment, quality of life and well-being in a German and Portuguese adult population during COVID-19 pandemics crisis. Front Psychol 12. https://doi.org/10.3389/fpsyg.2021.674660

Casey L, Liang RP-T (2014) Stress and wellbeing in Australia survey 2014. Australian Psychological Society, Oct 2014

DeSantis CE, Miller KD, Dale W, Mohile SG, Cohen HJ, Leach CR, Goding Sauer A, Jemal A, Siegel RL (2019) Cancer statistics for adults aged 85 years and older, 2019. CA Cancer J Clin 69(6). https://doi.org/10.3322/caac.21577

Dietz T, Rosa EA (1997) Effects of population and affluence on CO_2 emissions. Proc Natl Acad Sci USA 94:175–179. https://doi.org/10.1073/pnas.94.1.175

Dietz T, Rosa EA, York R (2007) Driving the human ecological footprint. Front Ecol Environ 5(1):13–18. https://doi.org/10.1890/1540-9295(2007)5[13:DTHEF]2.0.CO;2

Fan J, Zhou L, Zhang Y, Shao S, Ma M (2021) How does population aging affect household carbon emissions? Evidence from Chinese urban and rural areas. Energy Econ 100:105356. ISSN 0140-9883. https://doi.org/10.1016/j.eneco.2021.105356. https://www.sciencedirect.com/science/article/pii/S0140988321002620

Fry PS (2000) Religious involvement, spirituality and personal meaning for life: existential predictors of psychological wellbeing in community-residing and institutional care elders. Aging Ment Health 4(4). https://doi.org/10.1080/713649965

Gadie A, Shafto M, Leng Y, Kievit RA, Cam-CAN (2017) How are age-related differences in sleep quality associated with health outcomes? An epidemiological investigation in a UK cohort of 2406 adults. BMJ Open 7(7). https://doi.org/10.1136/bmjopen-2016-014920

Greening LA, Greene DL, Difiglio C (2000) Energy efficiency and consumption—the rebound effect—a survey. Energy Policy 28(6–7):389–401. ISSN 0301-4215. https://doi.org/10.1016/S0301-4215(00)00021-5. https://www.sciencedirect.com/science/article/pii/S0301421500000215

Hadler-Olsen E, Jönsson B (2021) Oral health and use of dental services in different stages of adulthood in Norway: a cross sectional study. BMC Oral Health 21(1). https://doi.org/10.1186/s12903-021-01626-9

Hiremath SB (2020) Stress, depression, coping strategies and spirituality among inmates of old age home in urban areas of north Karnataka. Indian J Psychiatry 62(7)

Hooper SC, Kilpela LS, Ogubuike V, Becker CB (2023) Fat talk, old talk, or both? Association of negative body talk with mental health, body dissatisfaction, and quality of life in men and women. J Eat Disord 11(1). https://doi.org/10.1186/s40337-023-00803-1

Khan T, Thomas S (2022) Promoting positive education through constructivist digital learning heutagogy: an intervention outcome. J Learn Dev 9(2):305–316. https://doi.org/10.56059/jl4d.v9i2.646

Kieny C, Flores G, Ingenhaag M, Maurer J (2022) Healthy, wealthy, wise, and happy? Assessing age differences in evaluative and emotional well-being among mature adults from five low- and middle-income countries. Soc Indic Res 160(2–3). https://doi.org/10.1007/s11205-020-02515-4

Lee J, Kim J, Holden R (2022) Physical and mental health differences reported by three age groups of older adults with diabetes. Gerontol Geriatr Med 8. https://doi.org/10.1177/23337214211055897

López J, Perez-Rojo G, Noriega C, Carretero I, Velasco C, Martinez-Huertas JA, López-Frutos P, Galarraga L (2020) Psychological well-being among older adults during the COVID-19 outbreak: a comparative study of the young-old and the old-old adults. Int Psychogeriatr 32(11). https://doi.org/10.1017/S1041610220000964

López-Sánchez GF, Smith L, López-Bueno R, Pardhan S (2022) Gender differences in the association between physical inactivity and mental-health conditions in people with vision or hearing impairment. Int J Environ Res Public Health 19(6). https://doi.org/10.3390/ijerph19063307

Malloy MC, Rosmarin DH, Forester B (2013) Spiritual struggle and religious coping in older adults with mood disorders. Am J Geriatr Psychiatry 21(3). https://doi.org/10.1016/j.jagp.2012.12.089

Mohammadi F, Enderami SA, Nasiri E, Marzband R (2021) Association between spiritual health and perception of aging in retired teachers in Sari, Iran. J Mazandaran Univ Med Sci 31(197)

Nagaraj M (2017) Anxiety and psychological well-being among adult and old age. Int J Indian Psychol 4(3). https://doi.org/10.25215/0403.099

Norman K, Haß U, Pirlich M (2021) Malnutrition in older adults—recent advances and remaining challenges. Nutrients 13(8). https://doi.org/10.3390/nu13082764

Okun S, Nimrod G (2020) Online religious counseling for older adults. Cyberpsychology 14(4). https://doi.org/10.5817/CP2020-4-7

Okun S, Nimrod G (2021) Lifelong spiritual learning: religious older adults going digital. J Relig Media Digit Cult 10(3). https://doi.org/10.1163/21659214-bja10056

Pais-Magalhães V, Moutinho V, Robaina M (2020) Households' electricity consumption efficiency of an aging population: a DEA analysis for the EU-28. Electr J 33(8):106823. ISSN 1040-6190. https://doi.org/10.1016/j.tej.2020.106823. https://www.sciencedirect.com/science/article/pii/S1040619020301159

Rivera-Ledesma A, Montero M (2005) Spirituality and religiosity in older Mexican adults. Salud Ment 28(6)

Sharma A, Sanghvi K, Churi P (2022) The impact of Instagram on young adult's social comparison, colourism and mental health: Indian perspective. Int J Inf Manag Data Insights 2(1). https://doi.org/10.1016/j.jjimei.2022.100057

Tennant R, Hiller L, Fishwick R et al (2007) The Warwick-Edinburgh Mental Well-being Scale (WEMWBS): development and UK validation. Health Qual Life Outcomes 5:63. https://doi.org/10.1186/1477-7525-5-63

Underwood L (2011) The daily spiritual experience scale: overview and results. Religions 2. https://doi.org/10.3390/rel2010029

van der Velden PG, Marchand M, Das M, Muffels R, Bosmans M (2022) The prevalence, incidence, and risk factors of mental health problems and mental health service use before and 9 months after the COVID-19 outbreak among the general Dutch population. A 3-wave prospective study. PLoS ONE 17(11). https://doi.org/10.1371/journal.pone.0276834

von Humboldt S, Ribeiro-Gonçalves JA, Low G, Leal I (2023) Psychotherapy in old age: older adults' sexual distress concerning their sexual well-being. Educ Gerontol 49(4). https://doi.org/10.1080/03601277.2022.2107842

Westerhof GJ, Keyes CLM (2010) Mental illness and mental health: the two continua model across the lifespan. J Adult Dev 17(2). https://doi.org/10.1007/s10804-009-9082-y

Zheng YB, Shi L, Lu ZA, Que JY, Yuan K, Huang XL, Liu L, Wang YH, Lu QD, Wang Z, Yan W, Han Y, Sun XY, Bao YP, Shi J, Lu L (2021) Mental health status of late-middle-aged adults in China during the coronavirus disease 2019 pandemic. Front Public Health 9. https://doi.org/10.3389/fpubh.2021.643988

Zheng H, Long Y, Wood R et al (2022) Ageing society in developed countries challenges carbon mitigation. Nat Clim Change 12:241–248. https://doi.org/10.1038/s41558-022-01302-y

Chapter 26
Pitfalls of Social Failure to Combat Climate Change Challenges Effectively: A Strategic Approach for a Better Foresight

Ansar Abbas, Dian Ekowati, Fendy Suhariadi, and Rakotoarisoa Maminirina Fenitra

Abstract The world is focusing on the far-reaching consequences of climate change on the Asian peninsula, highlighting the need for increased social awareness and efficacy to prepare for the inevitable shift towards a more sustainable future. We proceeded to do a literature review using the Scopus database, which allows researchers to uncover gaps and areas that require additional investigation by searching for all relevant publications on a particular topic. Through this search, scholars could identify patterns, get a deeper understanding of the corpus of current literature, and generate new lines of inquiry. Individuals and social groups utilize group guidance as a crucial tool for collective problem-solving, which can be made up through social exchange. This approach promotes awareness, cultural aspirations, and responsible behavior while addressing unique challenges. It also aids in identifying suitable individuals for leadership positions within the social sphere. Human

A. Abbas
Faculty of Economics and Business, Jl. Airlangga No. 4–6, Airlangga, Kec. Gubeng, Kota SBY, Jawa Timur, 60115 Surabaya, Indonesia

Hamdard University, 04, Park Link Road, Chak Shahzad, Islamabad, Pakistan

A. Abbas
e-mail: ansar.abbas@hamdard.edu.pk; ansar.abbas-2018@feb.unair.ac.id

D. Ekowati (✉)
Faculty of Economics and Business, and Head of Planning, Jl. Airlangga No. 4–6, Airlangga, Kec. Gubeng, Kota SBY, Jawa Timur, 60115 Surabaya, Indonesia
e-mail: d.ekowati@feb.unair.ac.id

F. Suhariadi
Centre for Postgraduate Program, Jl. Airlangga No. 4–6, Airlangga, Kec. Gubeng, Kota SBY, Jawa Timur, 60115 Surabaya, Indonesia
e-mail: fendy.suhariadi@psikologi.unair.ac.id

R. M. Fenitra
Faculty of Economics and Business, Universitas Indonesia, Depok, Indonesia
e-mail: maminirina.fenitra.r-2018@feb.unair.ac.id; rakotarisoa@ui.ac.id

U. Mukhopadhyay et al. (eds.), *Climate Crisis, Social Responses and Sustainability*, Climate Change Management, https://doi.org/10.1007/978-3-031-58261-5_26

civilization and the future of Earth are threatened by climate change. Effective public awareness and technology are needed to promote sustainable development and environmental conservation in developing nations, especially Asia. The Governments of developed countries should help their less developed counterparts by providing them with robust policy frameworks, adequate funding, and environmental promotion initiatives. Institutions, civic groups, and government agencies must work together to educate the public about alternative energy, construction, and transportation.

Introduction

Climate change is a pressing issue that requires a comprehensive, strategic approach to agriculture (Reilly 1995). Group and individual guidance are essential for addressing social issues and fostering responsible behavior. However, societal failure can hinder effective responses to climate change. Considering both strategic and psychological aspects, a multi-pronged strategy is required to overcome these obstacles (Mochizuki and Bryan 2015). A thorough approach incorporating sustainable land use, ecosystem protection, energy efficiency, and renewable energy sources is required since a lack of strategic planning can impede effective responses. Resources and effort must be poured into the fight, including research, development, and collaboration. Understanding the root causes of climate change, educating the public, and fostering urgency are also crucial. In Asia, a holistic strategy that includes strategic intent, designs, exceptional efforts, and a resolve to analyze the situation more seriously is required to overcome these obstacles and achieve a more potent and effective response to climate change (Vakulchuk et al. 2023).

The responsibility for ensuring societal safety and disaster preparedness in the face of global environmental crises lies predominantly with local and international authorities. These duties are characterized by their high level of demand. For example, Asia comprises the top 10 countries most severely impacted by climate change, including Afghanistan, Bangladesh, India, and Pakistan (Ghazi et al. 2016). Generally, they are not given adequate priority and assistance in promoting societal awareness and preparedness to address the challenges of our planet (e.g., Enander et al. 2015). Language, motivation, and timing influence audiences' interpretation of complicated public health and risk communications.

The interaction between technical language, reasons for processing information, and time available is imperative to understand how such messages are interpreted and received. Therefore, the social elements influencing aspirations and motivation to tackle these challenges can be identified. The interconnections between these factors can be investigated for the betterment of society (Shulman et al. 2021). An advanced theory and a practical guide for public health and communicators address the information processing processes in crisis circumstances. Afterward, with this information, solutions for helping people, groups, and communities deal with disasters can be developed (Enander et al. 2015). There are many challenges confronting society, which have been referred to as a "global crisis". They include an array of additional factors. Disasters that threaten the established social order and public security can result in social crises, which disrupt social mechanisms. The emergence of the environmental crisis has become a prominent issue due to its intricate, interconnected causes and effects. The situation encompasses a variety of challenges, such as the climate crisis, the biodiversity crisis, and the social and community crisis, which are persistent and multifaceted in nature (Xu et al. 2016). Environmental challenges of the twenty-first century include greenhouse gases and global warming, ozone layer depletion, acid rain, and the destruction of tropical forests. Thus, it is clear that the environmental crisis is a complex, global problem that needs to be addressed in the long term.

Combating climate change is difficult, requiring a comprehensive approach incorporating strategic policies, environmental awareness, and education (Jabareen 2013). This conflict is particularly essential in Asia, considering its substantial impact on the global society (Zafarullah and Huque 2018). Good governance, international collaboration, public awareness, and technology are all necessary for addressing climate change effectively. However, attempts encounter challenges, including insufficient policy action, low public awareness, competing interests, and limited resources because of growing social diversity and low income. A multifaceted strategy with important decision-makers is necessary to overcome these obstacles. This approach includes encouraging renewable energy sources, conservation, and sustainable land usage. Thus, This chapter explores the relationship between literature and the risks associated with society's inability to address climate change. It gathers information from many literary sources and constructs persuasive arguments based on the facts discussed in the literature.

Rationale of the Study

All nations, regardless of their level of development or form of government, are at risk from climate variability, such as droughts, floods, and severe storms (Glantz 1990). The research has shown that less developed nations are just as vulnerable to climate change as developed nations, regardless of their natural regimes (Cuervo-Cazurra and Genc 2008; O'brien and Leichenko 2003). In the past two decades, devastating droughts across sub-Saharan Africa have exacerbated food insecurity among broad areas of the African population and sparked considerable discussion about the role of climate change (Coulibaly et al. 2020). Global communities are no longer perceived as isolated entities due to technological and economic interconnectivity (Adger et al. 2009). The recurring increases in sea surface temperatures along the western coast of South America have consistently been linked to severe consequences for South American societies and, to a lesser extent, for communities worldwide (Glantz 2001).

Climate change is causing a rise in temperature, affecting agricultural productivity, transportation, and urban populations. Extreme weather events, precipitation changes, and rising sea levels are causing global societies to experience climate variability (Ummenhofer and Meehl 2017). The increase in temperature is primarily due to escalating carbon dioxide emissions and other trace gases, which trap heat in the atmosphere and cause the temperature to rise. This issue is pressing and needs immediate attention as much as possible to spread knowledge to the masses to be ready to combat these challenges collectively.

Material and Methods

With the climate crisis, social responses, and sustainability from an ecological perspective, this research study seeks to fill a knowledge gap in the existing integrative literature review using guidance for qualitative analysis (Benevolenza and Derigne 2019; Cooper 1984; Flemming and Noyes 2021). We set out to identify the pitfalls of society's inability to adequately address the challenges posed by climate change, even if a societal collapse and climate catastrophe were considered paramount. This book chapter uses social psychology as a strategic approach to better forecasting. The search was carried out using a database that the university provides. This database includes access to several other databases, including Sage, IGI Global, Taylor & Francis, Emerald, and Springer. The search phrases were associated with 615 published works, including journal articles, book chapters, and conference papers, e.g., Fig. 26.1. By selecting the journal option, we narrowed our search to 26 publications with full-text availability; further refining this search limited it to chapters or reviews. A literature review is an invaluable resource for academics, allowing them to search and evaluate previously published works on a subject. Authors can synthesize information, identify trends, and formulate new research questions using keywords like climate change, global societies, and leadership. This comprehensive understanding of the issue and its complexities is crucial for researchers to develop new ideas and contribute to the field (Snyder 2019).

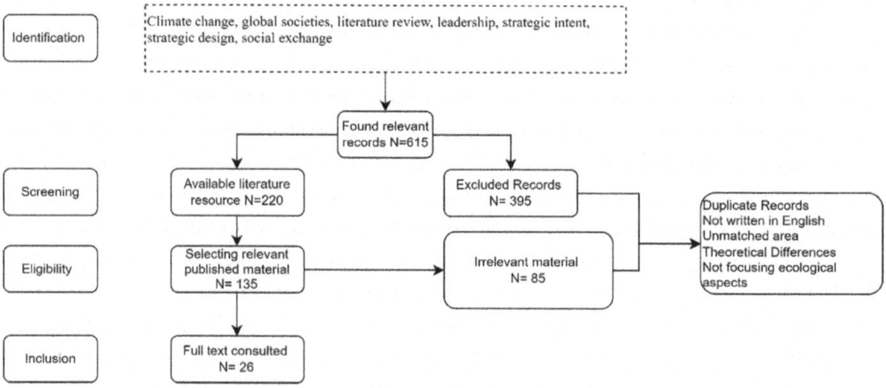

Fig. 26.1 Research protocol (Author's work)

Global Leaders and Climate Change Crisis

For several decades, the study of climate has been somewhat overlooked within social science research. However, a few influential figures have consistently emphasized the importance of addressing climate change to foster intellectual and emotional enlightenment among individuals. Al Gore's organization, The Climate Reality Project, has led this effort, drawing on social science research to develop a comprehensive framework for understanding the effects of climate change on the environment, economy, and society (Abbas et al. 2022b). Researchers vehemently supported increasing public awareness of climate change and informing people and communities worldwide of its potentially disastrous effects if ignored (Fischer 2018). This action has been regarded as a catalyst for a leader's influence in change management. German American social psychologist Kurt Lewin first proposed the concept. A renowned scholar in group dynamics and organizational development, Lewin developed the 3 Stage Model of Change (Burnes 2020). In the first place, he aimed to examine change processes within organizational contexts. The second goal was to explore ways to facilitate effective change by challenging the existing status quo.

Social complexities have been observed deep within the nutshell of change, where disadvantages outweigh advantages. This change model of Lewin has the innate stage of restoring change, which may emerge in resistance to change, as people fear the change will bring about more negatives than positives. Organizations must, therefore, focus on creating an environment that encourages and enables change to be successful. The restitution of the change process is a complex task that should not be overlooked. Understanding and accepting differences may vary when discussing societies in general, as they are not always easily grasped or acknowledged at the same level. (Abbas et al. 2022a).

The 2023 UN Global Climate Action Awards have recognized Climate Leaders recognizing initiatives implementing global climate action. The award category aims to identify leaders who have demonstrated transformative government initiatives and ambition at various levels. The goal is to underscore the urgency for heightened climate action and ambition to mitigate global mean temperature escalation and foster adaptability to climate change. The UN Climate Change Executive Secretary Patricia Espinosa highlights the international community's current climate emergency. This award highlights the need to recognize socially, ethically, and morally responsible individuals for change and the pressing need to ramp up efforts to address climate change effectively (Abbas 2023). Urgent measures are required to foster increased leadership and guidance from national governments and individuals across all sectors of society as the primary means to observe, acknowledge, and cultivate such efforts more responsibly (Fenitra et al. 2022a, b, c; Haryanto et al. 2022). The UN Global Climate Action Awards serve as a platform for recognizing individuals and entities demonstrating proactive measures in addressing climate change while inspiring other governmental bodies, corporations, urban centers, and communities to emulate.[1]

---

[1] https://unfccc.int/climate-action/momentum-for-change/climate-leaders.

Strategic Intent and Strategic Designs for Mitigating Climate Change Crisis

Humans encounter various obstacles and issues that may gradually escalate into crises in daily life. Numerous circumstances can give birth to situations between two parties, including factors beyond their control or controlled by an external moderator. The PESTEL framework represents external factors outside enterprises' control, e.g., political, economic, social, technological, environmental, and legal factors. Businesses must proactively prepare and strategize in response to the emergence and behavior of these factors (Çitilci and Akbalık 2020). COVID-19 has revealed that development concerns about growth and existential challenges related to fulfilling personal aspirations can trigger crises requiring strategic solutions (Fenitra et al. 2022a). Crisis issues conceptually hold significant relevance in efficiently addressing situations, as evidenced by integrating strategic solutions that encompass a third analytical perspective, e.g., (Abbas et al. 2023c). The foundational literature on crisis theory focuses on the individual-level obstacles experienced by each person, as suggested by James and Gilliland (2009). The authors (James and Gilliland) delineate the nine crisis intervention methods: generating awareness, allowing catharsis, offering support, promoting expansion, stressing focus, providing guidance, boosting mobilization, implementing order, and providing protection. These strategies are considered necessary proficiencies for crisis counselors. These techniques are undoubtedly beneficial in dividing the method (or system) into distinct components, including observing, planning, anticipating, designing, and executing strategy.

Being well-prepared to handle and adapt to change is one of the cornerstones of crisis intervention, involving various factors (Fritz et al. 2007). Thus, it is of utmost importance to emphasize prompt action in the aftermath of an incident to stabilize the impacted person and the community to lessen the impact of any perceived responding behavior. Comprehending the events at hand is vital, especially in light of the social movements that oppose climate awareness and reaction measures. The idea of comprehension is crucial in elucidating the reasons for change, such as the social movement in opposition to climate awareness and response actions. It is possible to see it as something that cannot be avoided. Thirdly, it's essential to find solutions that work for the person experiencing the problem and the larger social setting. Kurt Lewin's use of "Calm waters" and "White-water rapids" metaphors to define change is well-established. Two ways of approaching change were depicted by Lewin using these images: one was calm and steady, like a river flowing, while the other was more agitated and violent, like white-water rapids. One possible interpretation of the change metaphor is that people understand change individually, drawing on their experiences, beliefs, and knowledge. The ability to perceive, handle, and adapt to change is contingent upon the management aptitude of persons (O'hare et al. 1998). Ultimately, the purpose is to promote self-reliance to reinstate a sense of independent functioning. Given that change, albeit temporary, necessitates individuals to be adequately prepared, it is crucial for them to carefully consider what they wish to derive from the change and what aspects they should discard to recover their prior state of mind, e.g. (Everly and Lating 2019; Hobbs 1984; Shelby and Tredinnick 1995).

Awareness Strategies

The development of awareness in humans commences throughout childhood, wherein parents play a crucial role in fostering phonemic awareness by reading aloud to their children. The instructional methods encompass discussing a word's orthography, morphology, and phonetics; demonstrating the process of writing a comment while articulating its phonemes; or facilitating interactive activities integrating letter and language manipulation to foster imitation (Rice et al. 2022). In contemporary management and social sciences literature, strategic awareness is infrequent at lower levels but specific to top management only, e.g. (Hambrick 1981). This lack of research may indicate that it has not been recognized as a transformative factor in discussions and decision-making processes at the executive level. This particular subject may receive less scholarly attention when examined in isolation, as the understanding of self-awareness notions is closely intertwined with developing leadership connections rather than being seen as an independent element (e.g., Abbas et al. 2023b). Strategic awareness can be understood as integrating strategic thought with personal attention.

Strategic awareness and leadership are closely intertwined since individuals who possess personal leadership qualities and employ strategic awareness tend to be unaffected by constraining belief systems (Halbesleben et al. 2003). This perspective holds validity as awareness of strategy and understanding play crucial roles in achieving superior individual performance. Individuals in positions of authority have a heightened level of consciousness and possess advanced leadership abilities, enabling them to implement efficient management strategies. Nevertheless, it is essential to note that each approach operates differently. It is worth mentioning that possessing awareness without a well-defined strategy frequently lets them become captivated by the possibilities they have at their disposal (Ghoneim et al. 2011).

Psychological and Emotional Purging Strategy (Catharsis)

Catharsis is a psychological and emotional purging strategy that encourages individuals to purposefully organize, exchange, and implement ideas (Orozco and Harris 2023). It provides a structured framework for individuals to express their feelings, gain clarity, and manage powerful emotions like rage or despair. By adopting this strategy, individuals can better engage with real-world realities, fostering creativity, self-reflection, and exploration. This approach helps maintain psychological equilibrium and manage powerful emotions, ultimately enhancing the effectiveness of the purging process (Pillow 2003). Based on the tenets of catharsis theory, the dynamic release process can potentially reduce aggressive tendencies effectively (Bresin and Gordon 2013). In contemporary times, the concept of catharsis in psychology has undergone adaptations to enhance its relevance and applicability. Catharsis can be a constructive means of emotional release for individuals grappling with stress, rage, and intense emotions. Many individuals have widely utilized it to alleviate these sentiments effectively (Denzler and Förster 2012).

Volunteer behavior is frequently observed at times of crisis, as individuals employ it as a means to elucidate a challenging encounter. This conduct entails either aiding others or reciprocating the aid offered to them. This approach also encompasses community activism, which can serve as a source of catharsis and intrinsic support, enhancing good emotions and self-confidence (Goode 2023). Catharsis can be a good media strategy to call for public sentiment and support for those severely impacted by environmental crises. Invoking a sense of shared suffering and understanding can help bring attention to the plight of those affected and galvanize public support.

Support Strategies

Support strategies must be regularly adjusted to accommodate disruptive or gradual environmental changes. The changes may require quick action or more gradual adaptation, depending on the context. Assessing the situation and deciding which approach is most appropriate is vital. In the dynamic realm of education, change is a constant reality. Strategic management literature supports monitoring, adjusting, and reacting to environmental changes as necessary (Saud et al. 2021; Zhao et al. 2015). It is important to stay attentive to the environment and be prepared for changes. By having a plan in place, organizations can be better equipped to handle any changes that come their way (Davoudi et al. 2012). Educators and policymakers often encounter many changes, such as implementing novel pedagogical or practical approaches, evolving curriculum standards, and incorporating technological advancements (Henriksen et al. 2016). These changes can elicit a diverse variety of emotional responses and are therefore popular among educators for research and development (Van Driel et al. 2001).

Coaching, counseling, and mentoring are all forms of guidance that emphasize students' competence and character growth. Such strategies also help develop self-awareness and motivation, essential for successful learning and career development (Abbas et al. 2022a). Furthermore, these support strategies are also necessary for educating youth about human capital development and environmental awareness (Birasnav et al. 2011; Brooks and Muyia Nafukho 2006). Support strategies literature in times of personal and professional crisis drew attention to mechanisms that restore people's beliefs of having control over their lives despite challenging situations (Abbas et al. 2020). These strategies include problem-focused and emotion-focused coping strategies, self-reflection, and seeking support from family and friends, among other plans. These strategies help individuals to cope with difficult situations and to maintain a sense of self-efficacy. They also provide a sense of meaning and purpose, which can help individuals move forward in their lives (Dada et al. 2019). Burnout and external demands are significant obstacles preventing training and development from enacting new values (Abbas et al. 2021). Because of burnout caused by excessive workloads and emotional strains, one may put learners' needs and district goals ahead of their importance. There is limited room for educators to infuse their basic principles into their teaching when they feel forced to adhere to a preset route due to external pressures such as examinations and administrative instructions (Abidi and Koichi 2019).

To maintain the success of support strategies in garnering mass support for a cause, it is imperative to continually adapt these strategies to accommodate sudden and slow environmental changes that may disrupt their efficacy. Periodic modifications can consider shifts in public sentiment, developing patterns, and novel technology that have the potential to enhance the efficacy of the techniques.

Expansion Strategies

Natural disasters, worldwide pandemics, power and water outages, cyberattacks, operational accidents, violent threats, and supply chain concerns are all examples of potential crises (Barten et al. 2021). However, environmental disasters are a growing crisis that the world must tackle, but the literature on crisis management is scarce (Coombs and Laufer 2018). These disasters can impact the environment, the economy, and human health in the long term. Governments, businesses, and individuals must work together to mitigate and prevent disaster effects. In the manufacturing sector, for instance, a crisis is any unanticipated incident or condition that threatens any company's financial stability, the health and safety of its employees and customers, the continuity of its operations, or the integrity of its brand or reputation (Settembre-Blundo et al. 2021). Organizations will handle and recover from crises differently because of inherent differences in their crisis readiness, maturity level, scope of operations, and financial stability (Mansour et al. 2019). This particular strategy is viable for a few more prominent industries, businesses, and societies that are responsible and can plan better to combat these changes to some extent (He and Harris 2020). However, businesses and communities face widespread impacts from natural climate change, which there is no avoiding in general (Seddon et al. 2021).

Climate change and other environmental matters require leaders to consider the long-term effects and develop strategies to increase organizational and societal resilience (Benevolenza and Derigne 2019; Giudice et al. 2021). According to a 2018 Deloitte poll, 47% of businesses without a crisis management strategy said their finances were severely affected by a recent crisis, while just 31% of those with a plan said the same.[2] Enterprises that weathered a crisis better were more likely to have taken preventative steps (such as having a CMP in place) in 2019, according to a PwC global crisis study of a wide range of enterprises across various industries worldwide.[3] Theoretically, Crisis Management and Business Continuity plan assets would be put forth under NFS's recommendation of Developing an Effective Crisis Management Plan. It is plausible that an organization needs time and strategic efforts to remain guarded against potential new risks while keeping an eye on existing ones.

The COVID-19 pandemic taught us that trouble could arise from anywhere. It can be readily addressed by managing a change cycle, e.g., (Abbas et al. 2022a; He and Harris 2020; Saud et al. 2021; Shulman et al. 2021). The climate change problem, however, is ongoing and will never end. Therefore, these lessons are not learned until the conclusion of the catastrophe. Suppose it can foresee and evaluate potential dangers and has a plan for swiftly responding and recovering. In that case, an organization can emerge from a crisis better, stronger, and more resilient. See further NFS report. The prioritization of climate change education, awareness, and support should be a prominent item on the agenda to garner public interest. It is imperative for governments, corporations, and citizens to collectively recognize the significance of undertaking measures aimed at mitigating the consequences of climate change. Providing educational resources, expansions of such strategies, and support is crucial in facilitating the dissemination of information and fostering a culture of proactive engagement among individuals.

Results and Discussion

Social learning is crucial in promoting a sustainable future, particularly climate change. It helps individuals understand and alter their behavior, enabling them to contribute to climate management and environmental health by considering how their actions will affect others and choosing sustainable methods (Uralovich et al. 2023). The lack of social exchange and educational opportunities would exacerbate social failure to combat climate change challenges effectively. Consequently, it is essential to take a strategic approach to better forecasting to implement a more inclusive approach. Social learning can be facilitated through various means of education, including media, classmates, family, and role models, as more social exchange happens, and more is yet to come. People are more likely to participate in climate action and ecologically beneficial activities when they see others doing so (Tang 2023). Social learning can help with climate action by raising awareness about climate change and its effects, motivating people through prominent leaders and

---

[2] Deloitte. Stronger, fitter, better: Crisis Management for the resilient enterprise. June 2018. (https://www2.deloitte.com/us/en/insights/topics/risk-management/crisis-management-plan-resilient-enterprise.html).

[3] PwC 2019 Global Crisis Survey (https://www.pwc.com/gx/en/services/forensics/global-crisis-survey.html) and PwC 2019 Global Crisis Survey.

figures, and fostering networking and social ties among people with similar interests. Psychological and Emotional Purging Strategy (Catharsis) could benefit from various awareness and support initiatives to increase understanding of this issue.

Motivating people to make changes requires encouraging them to consider how their actions affect the environment. Self-evaluation instruments, reflective writing, and other forms of self-reflection can help pinpoint development areas and promote long-term habits, e.g., independent thinking and inclusivity (Abbas et al. 2023a). However, the challenges faced by our planet cannot be solved by individual acts alone, and there is a lifelong journey of learning and development. Building sustainable communities requires environmental health, which can only be achieved through concerted human effort. Social learning can significantly impact when people, groups, and communities work together; however, the missing link is found by serving others' needs first (Meuser and Smallfield 2023). Businesses, governments, and NGOs can collaborate in partnerships to promote environmental health by sharing information and resources, and collective action should thus take place for inclusive societies. Engaging communities at the local level through community projects, festivals, educational initiatives, and events can also benefit from diverse social learning and more social experiences through exchange programs (Brown et al. 2023).

Moreover, environmental health relies heavily on sustainable policies and laws. A better society for future generations can only be built if we all work together and embrace social learning. Working together can create a more resilient world for future generations (Pooley and Bendle 2023). The process of formulating and putting into action efficient techniques for accomplishing the work can be beneficial to both individuals and organizations. It is essential to have plans since they allow the information to flow in both directions from one set of directions to the next (Cook 2008). Someone less experienced or powerful could try to influence someone they look up to get their way around problems that another (Prahalad and Hamel 1994). Social engagement like this is a critical component of both learning and teaching, and it can potentially foster societal progress. A leader's charisma, charm, and exceptional knowledge can inspire followers to heights (Awamleh and Gardner 1999).

Many people develop strategies for guiding individuals and small groups, which is how the method accomplishes its desired results, e.g. (Northouse 2019). The relational dimension of social interaction can be divided into two overarching approaches (Zaharna 2016). A collectivist standpoint is espoused by various early scholars (Emerson 1962; Homans 1961) and (Baldwin 1978; Blau 1964), who also developed this theory so profoundly that many new spheres in literature are taking benefits from their work (Vasan 2016). In comparison, various prominent researchers match an individualistic perspective, e.g., (Ekeh 1974; Huston and Burgess 1979; Nord 1973). Behavioral or rational-choice motivations characterize the individualistic method, which involves actors within a basic network structure. In this context, micro-orders are inherently fragile due to their close association with individual self-interest. The development and prevalence of relations depend upon the stability of incentives for exchange, actor preferences, and structures of opportunity. Connections, in their essence, arise from individualistic circumstances. On the other hand, the collectivist perspective posits the existence of a broader social entity or affiliation with a group. Social interactions are inherently intertwined with and mirror a cultural or normative structure or system (Lawler et al. 2008).

Regardless of the form of social exchange theory division, researchers agree that social exchange plays a crucial role in guiding societies (Cropanzano et al. 2017). All forms of leadership have been seen as pivotal, as humans seek guidance in times of change, turbulence, stability, or working steadily. Social exchange theory provides a framework for understanding the dynamics of social relationships and the power dynamics that govern them, e.g. (Ruben 1998; Watson and Foster-Fishman 2013). It can also help to explain why people cooperate and engage in social exchange and how they evaluate the rewards and costs that come with it. Social exchange theory can provide a valuable tool for leaders to manage people and guide them through times of change. It can help explain people's motivations for cooperation and social exchange. It can help leaders assess the rewards and costs associated with it (Abbas et al. 2022d; Blanch et al. 2016; Jordan 2016).

Recommendations and Future Directions

Ecological education is driven by social transformation and awareness, which are interdependent processes. When people are socially conscious, they can put themselves in another person's shoes and get insight into their struggles and triumphs, especially those of oppressed groups. This ground plan can inspire the masses to act and make a difference. It can also help them realize how valuable it is to be included in society. Human actions such as pollution and deforestation deteriorate the environment, affecting ecosystems, biodiversity, and human health. It is essential to raise public awareness and effect social change to combat environmental deterioration, leading to consumer habits, public policy, and business practice shifts. Education, advocacy, community engagement, and social media effectively spread awareness. A more ecologically aware way of thinking can inspire people to act, which in turn helps drive conservation and restoration initiatives. Thus, group guidance is a crucial tool in society to address issues faced by any social group. It involves social exchange programs of awareness such as lectures, talks, and role-plays to promote awareness and collective problem-solving. This approach can help organizations develop a unified approach to tackling climate issues, e.g., developing social exchange. Individual guidance, on the other hand, is a personalized approach that helps individuals resolve their unique challenges. This process involves interpersonal communication to investigate thoughts, actions, and emotions. Group guidance can cultivate responsible behavior, foster cultural aspirations, and promote a sense of responsibility to the community (Gancar et al. 2023). It also helps identify suitable individuals for leadership positions within the social sphere.

There are formidable obstacles in the battle against climate change, and viable solutions are needed. However, societal failure can stymie our endeavors and make it harder for us to fight these problems. An all-encompassing strategy that considers both strategic and psychological aspects is required to solve this problem. A failure to plan strategically is a significant obstacle that can hinder an effective response to climate change. It is critical to think about and prepare for the long-term effects of climate change if we want to fight it successfully. Doing so calls for an in-depth familiarity with the issue's dynamics and complexity and the capacity to devise and execute sound tactics.

The psychological and emotional cleansing technique, often called catharsis, is another obstacle that prevents climate change initiatives from succeeding. While it may help people feel better in the short term, catharsis alone might not be enough to inspire lasting social change. People and groups must realize that catharsis is insufficient and embrace more strategic techniques that encourage teamwork. An integrated approach that considers psychological and strategic aspects is necessary to overcome these obstacles and fight climate change. A multi-pronged strategy is required to address climate change effectively, and one component of this is identifying and communicating the strategic goal. For this, you must establish lofty objectives, single out essential players, and include them in every decision-making step. The most effective way to fight climate change is to create comprehensive plans that use a variety of tactics. Sustainable land use, conservation of ecosystems, energy efficiency, and the utilization of renewable energy sources are all part of this. Extra resources and effort must be poured into the fight against climate change in light of the seriousness of the issue and the necessity of extraordinary measures. Spending money on research and development, being open to new technologies, combining forces to fund long-term solutions, reflecting on the situation, and keeping the details in mind is vital for avoiding social failure traps. To accomplish this, we need to determine the root causes of climate change, educate the public about its consequences, and foster a sense of urgency. There is a strong correlation between societal failure and the ineffectiveness of climate change mitigation efforts. Taking a holistic strategy that includes strategic intent, strategic designs, exceptional efforts, and a resolve to remember and analyze the situation more seriously is crucial for overcoming these obstacles. Considering these considerations, A more potent and effective response to climate change is possible.

Conclusion

The climate crisis necessitates a comprehensive approach to address climate change. Sustainable development and social responses are crucial for addressing this issue. Strategic policies, environmental awareness, and education on climate change are essential for promoting sustainable growth and environmental protection in underdeveloped nations, particularly more research in densely populated Asia. However, there must be swift and concerted action on a global scale to promote research on combat climate change issues, including in the Asian region. It is a significant concern because of its effects on the globe and human society, including more frequent and severe weather events, higher sea levels, and dwindling natural resources. Better planning ahead of time is essential if we are to fight climate change successfully. Technology, public awareness, international cooperation, and good governance are necessary for this plan. Due to rising social variety and low income, social efforts to address climate change face insufficient policy action, poor public awareness, competing interests, and insufficient resources and infrastructure. Governments are responsible for providing robust policy frameworks, sufficient funding, and encouraging environmentally friendly behaviors. However, funding is scarce due to COVID-19, the economic collapse, and inflation, and more is needed to address them.

Institutions of higher learning, civic groups, and government agencies should work together to raise public understanding about the dangers of climate change. When countries work together, knowledge, technology, and best practices can be more easily exchanged. Alternative energy sources, more energy-efficient construction, and environmentally friendly modes of transportation are all problems that could benefit through R&D on strategic intent spread. To sum up, the climate problem is a significant danger to human civilization and Earth's future. Cultures that take action to avoid these traps can pave the path to a more sustainable future.

The fight against climate change is brutal, and effective responses are required. But when society fails, it can hinder our efforts and make fighting these problems more difficult. To address this issue, a comprehensive plan is needed to consider both the strategic and psychological components. A failure to plan strategically might impede a successful response to climate change. The key to effectively combating climate change is considering and planning for its long-term consequences. A thorough understanding of the issue's complexity and dynamics and the ability to develop and implement effective strategies are necessary for this task. Another barrier to the success of climate change programs is the psychological and emotional cleaning procedure, often known as catharsis. Catharsis may provide temporary relief, but it may not be sufficient to motivate long-term societal transformation on its own. Groups and individuals need to move past relying on catharsis and adopt more strategic approaches that foster collaboration.

Overcoming these challenges and combating climate change will require an interdisciplinary strategy considering psychological and strategic factors. Effectively addressing climate change requires a multi-pronged approach, one of which is defining and articulating the strategic objective. To achieve this, set ambitious goals, identify key individuals, and involve them in every decision-making process. All-encompassing plans incorporating multiple tactics are the best way to combat climate change. This includes conserving ecosystems, using renewable energy sources, conserving energy, and sustainably using land. Given the gravity of the problem and the need for unprecedented responses, the battle against climate change requires increased funding and human resources. Research and development funding, openness to new technology, and resource pooling to assist long-term solutions are all instances of this. Thinking back on the event and keeping the specifics in mind will help you avoid falling into the pitfalls of social failure. To accomplish this, we need to determine the root causes of climate change, educate the public about its consequences, and foster a sense of urgency. Social failure can significantly impede efforts to mitigate climate change. Overcoming these challenges requires a comprehensive approach that incorporates strategic intent, strategic designs, extraordinary efforts, and a determination to remember and examine the situation more thoroughly. We can respond to climate change more robustly and efficiently by placing these things.

References

Abbas A, Eliyana A, Ekowati D, Saud M, Raza A, Wardani R (2020) Data set on coping strategies in the digital age: the role of psychological well-being and social capital among university students in Java Timor, Surabaya, Indonesia. Data Brief 30:10558. https://doi.org/10.1016/j.dib.2020.105583

Abbas A, Ekowati D, Suhariadi F (2021) Individual psychological distance: a leadership task to assess and cope with invisible change. J Manag Dev 40(3):168–189. https://doi.org/10.1108/jmd-09-2020-0304

Abbas A (2023) The potential of ethically responsible and persistent leadership theory. In: Doğru Ç (ed) Role of human resources for inclusive leadership, workplace diversity, and equity in organizations. IGI Publisher, Hershey, Pennsylvania, pp 255–282

Abbas A, Ekowati D, Shuairidi F, Fenitra RM, Fahlevi M (2022a) Integrating cycle of prochaska and diclemente with ethically responsible behavior theory for social change management: post covid-19 social cognitive perspective for change. In: Pego A (ed) Challenges and emerging strategies for global networking post COVID-19. IGI Global, Hershey, Pennsylvania, pp 130–155

Abbas A, Ekowati D, Suhariadi F, Fenitra RM (2022b) Health implications, leaders societies, and climate change: a global review. In: Chatterjee U, Akanwa AO, Kumar S, Singh SK, Dutta Roy A (eds) Ecological footprints of climate change: adaptive approaches and sustainability. Springer International Publishing, Cham, pp 653–675

Abbas A, Ekowati D, Suhariadi F, Fenitra RM, Fahlevi M (2022c) Human capital development in youth inspires us with a valuable lesson: self-care and wellbeing. In: Clarke KL (ed) Self-care and stress management for academic well-being. IGI Global, Hershey, Pennsylvania, pp 80–101

Abbas A, Saud M, Suhariadi F, Usman I, Ekowati D (2022d) Positive leadership psychology: authentic and servant leadership in higher education in Pakistan. Curr Psychol 41(10):5859–5871. https://doi.org/10.1007/s12144-020-01051-1

Abbas A, Ekowat D, Hussain R, Suhairidi F, Fenitra RM (2023a) The path of an independent thinker to an inclusive leader: the journey of leadership transformation. In: Roache DAM (ed) Transformational leadership styles, management strategies, and communication for global leaders, vol 1. IGI Global, Hershey, Pennsylvania, pp 244–265

Abbas A, Ekowati D, Anwar A (2023b) Authentic leadership journey: an empirical discussion from pakistani higher education employing the lay theory of psychology. Int J Public Leadersh. https://doi.org/10.1108/IJPL-04-2022-0020

Abbas A, Suhariadi F, Ekowati D, Rakotoarisoa FM (2023c) A strategic paradigm for transformational leadership theory in the digital age: scope of the analytical third. In: Roache DAM (ed) Transformational leadership styles, management strategies, and communication for global leaders. IGI Global Publishers, Hershey, Pennsylvania, pp 304–323

Abidi D, Koichi N (2019) Management approach for innovation success in unstable and stable environments: opportunity-based approach vs administrative control approach. Competitiveness Rev: Int Bus J 30(1):22–40. https://doi.org/10.1108/CR-11-2019-0125

Adger WN, Eakin H, Winkels A (2009) Nested and teleconnected vulnerabilities to environmental change. Front Ecol Environ 7(3):150–157. https://doi.org/10.1890/070148

Awamleh R, Gardner WL (1999) Perceptions of leader charisma and effectiveness: the effects of vision content, delivery, and organizational performance. Leadersh Q 10(3):345–373. https://doi.org/10.1016/S1048-9843(99)00022-3

Baldwin DA (1978) Power and social exchange. Am Polit Sci Rev 72(4):1229–1242
Barten DG, Klokman VW, Cleef S, Peters NALR, Tan ECTH, Boin A (2021) When disasters strike the emergency department: a case series and narrative review. Int J Emerg Med 14:1–9. https://doi.org/10.1186/s12245-021-00372-7
Benevolenza MA, DeRigne L (2019) The impact of climate change and natural disasters on vulnerable populations: a systematic review of literature. J Hum Behav Soc Environ 29(2):266–281. https://doi.org/10.1080/10911359.2018.1527739
Brooks K, Muyia Nafukho F (2006) Human resource development, social capital, emotional intelligence: any link to productivity? J Eur Ind Training 30(2):117–128. https://doi.org/10.1108/03090590610651258
Birasnav M, Rangnekar S, Dalpati A (2011) Transformational leadership and human capital benefits: the role of knowledge management. Leadersh Org Dev J 32(2):106–126. https://doi.org/10.1108/01437731111112962
Blau P (1964) Power and exchange in social life. Wiley, New York
Blanch J, Gil F, Antino M, Rodríguez-Muñoz A (2016) Positive leadership models: theoretical framework and research. Psychol Papers 37(3):170–176
Bresin K, Gordon KH (2013) Aggression as affect regulation: extending catharsis theory to evaluate aggression and experiential anger in the laboratory and daily life. J Soc Clin Psychol 32(4):400–423. https://doi.org/10.1521/jscp.2013.32.4.400
Burnes B (2020) The origins of Lewin's three-step model of change. J Appl Behav Sci 56(1):32–59. https://doi.org/10.1177/0021886319892685
Brown C, Luzmore R, Groß Ophoff J (2023) Facilitating the ideas-informed society: a systematic review. Emerald Open Res 1(1). https://doi.org/10.1108/EOR-01-2023-0008
Çitilci T, Akbalık M (2020) The importance of PESTEL analysis for environmental scanning process. In: Olgu Akdeniz Ö, Baykal E, Borda M, Delen D, Ivaldi E, Tatoğlu E (eds) Handbook of research on decision-making techniques in financial marketing. IGI Global, Hershey, Pennsylvania, pp 336–357
Cook S (2008) The essential guide to employee engagement: better business performance through staff satisfaction. Kogan Page Publishers, Philadelphia, PA, USA
Coombs WT, Laufer D (2018) Global crisis management–current research and future directions. J Int Manag 24(3):199–203. https://doi.org/10.1016/j.intman.2017.12.003
Cooper H (1984) The integrative research review: a systematic approach. Sage, Beverly Hills, CA
Coulibaly T, Islam M, Managi S (2020) The impacts of climate change and natural disasters on agriculture in African countries. Econ Disasters Clim Change 4:347–364. https://doi.org/10.1007/s41885-019-00057-9
Cropanzano R, Anthony EL, Daniels SR, Hall AV (2017) Social exchange theory: a critical review with theoretical remedies. Acad Manag Ann 11(1):479–516. https://doi.org/10.5465/annals.2015.0099
Cuervo-Cazurra A, Genc M (2008) Transforming disadvantages into advantages: developing-country MNEs in the least developed countries. J Int Bus Stud 39:957–979. https://doi.org/10.1057/palgrave.jibs.8400390
Dada JO, Babatunde SO, Adeleye RO (2019) Assessment of academic stress and coping strategies among built environment undergraduate students in Nigerian higher education. J Appl Res High Educ 11(3):367–378. https://doi.org/10.1108/JARHE-06-2018-0100
Davoudi S, Shaw K, Haider LJ, Quinlan AE, Peterson GD, Wilkinson C, Fünfgeld H, McEvoy D, Porter L, Davoudi S (2012) Resilience: a bridging concept or a dead end? Reframing resilience: challenges for planning theory and practice interacting traps: resilience assessment of a pasture management system in Northern Afghanistan urban resilience: what does it mean in planning practice? Resilience as a useful concept for climate change adaptation? The politics of resilience for planning: a cautionary note: edited by Simin Davoudi and Libby Porter. Plann Theory Pract 13(2):299–333. https://doi.org/10.1080/14649357.2012.677124
Denzler M, Förster J (2012) A goal model of catharsis. Eur Rev Soc Psychol 23(1):107–142. https://doi.org/10.1080/10463283.2012.699358

Everly GS, Lating JM (2019) Crisis intervention and psychological first aid. In: Everly JGS, Lating JM (eds) A clinical guide to the treatment of the human stress response. Springer, New York, NY, pp 213–225

Ekeh P (1974) Social exchange theory. Harvard University Press, MA, USA

Emerson RM (1962) Power-dependence relations. Am Sociol Rev 27(1):31–41

Enander A, Hede S, Lajksjö Ö (2015) Why worry? motivation for crisis preparedness work among municipal leaders in Sweden. J Contingencies Crisis Manage 23(1):1–10. https://doi.org/10.1111/1468-5973.12067

Fischer AP (2018) Pathways of adaptation to external stressors in coastal natural-resource-dependent communities: Implications for climate change. World Dev 108:235–248. https://doi.org/10.1016/j.worlddev.2017.12.007

Flemming K, Noyes J (2021) Qualitative evidence synthesis: where are we at? Int J Qual Methods 20:1609406921993276. https://doi.org/10.1177/1609406921993276

Fenitra RM, Abbas A, Ekowati D, Suhairidi F (2022a) Strategic intent and strategic leadership: a review perspective for post-covid tourism and hospitality industry recovery. In: Mohanty P, Sharma A, Kennell J, Hassan A (eds) The emerald handbook of destination recovery in tourism and hospitality. Emerald Group Publishing, Bingley, pp 23–44

Fenitra RM, Laila N, Premananto GC, Abbas A, Sedera RMH (2022b) Explaining littering prevention among park visitors using the theory of planned behavior and norm-activation-model. Int J Geoheritage Parks. https://doi.org/10.1016/j.ijgeop.2022.11.002

Fenitra RM, Premananto GC, Sedera RMH, Abbas A, Laila N (2022c) Environmentally responsible behavior and knowledge-belief-norm in the tourism context: the moderating role of types of destinations. Int J Geoheritage Parks 10(2):273–288. https://doi.org/10.1016/j.ijgeop.2022.05.001

Fritz S, Chaitow L, Hymel G (2007) Clinical massage in the healthcare setting-e-book. Elsevier Health Sciences, Amsterdam, Netherlands

Glantz MH (1990) On the interactions between climate and society. Popul Dev Rev 16:179–200. https://doi.org/10.2307/2808069

Glantz MH (2001) Currents of change: impacts of El Niño and La Niña on climate and society, 2nd edn. Cambridge University Press, Cambridge, UK

Gancar CP, Fenitra RM, Masmira K, Sedera RMH, Abbas A (2023) Extended theory of planned behavior and environmentally responsible behavior in the context of beach tourism. Malays J Consum Fam Econ 31(1):364–392. https://doi.org/10.60016/majcafe.v31.14

Ghazi TW, Muniruzzaman ANM, Singh AK (2016) Climate change and security in South Asia. In: Kingham RA (ed) Global military advisory council on climate change (GMACC). Institute for Environmental Security (IES), Hague, Netherlands

Ghoneim AS, Essam DL, Abbass HA (2011, 22–24 Feb) Competency awareness in strategic decision making. Paper presented at the 2011 IEEE international multi-disciplinary conference on cognitive methods in situation awareness and decision support (CogSIMA), Miami Beach, FL, USA

Goode S (2023) Users in crisis response and recovery: catharsis and social learning among social media users during and after a natural disaster. Behav Inf Technol 42(1):108–123. https://doi.org/10.1080/0144929X.2021.2016968

Giudice LC, Llamas-Clark EF, DeNicola N, Pandipati S, Zlatnik MG, Decena DCD, Woodruff TJ, Conry JA, FIGO Committee on Climate Change, Toxic Environmental E (2021) Climate change, women's health, and the role of obstetricians and gynecologists in leadership. Int J GynecolObstet 155(3):345–356. https://doi.org/10.1002/ijgo.13958

Halbesleben JR, Novicevic MM, Harvey MG, Buckley MR (2003) Awareness of temporal complexity in leadership of creativity and innovation: a competency-based model. Leadersh Q 14(4–5):433–454. https://doi.org/10.1016/S1048-9843(03)00046-8

Hambrick DC (1981) Strategic awareness within top management teams. Strateg Manag J 2(3):263–279. https://doi.org/10.1002/smj.4250020305

Haryanto B, Gunawan JAP, Fenitra RM, Abbas A (2022) The role of culture adoption in moderating the influence of country image, corporate image, brand image on brand attitude and purchase intention toward foreign brands. Int J Bus Perform Supply Chain Model 13(1):89–108. https://doi.org/10.1504/IJBPSCM.2022.122437
He H, Harris L (2020) The impact of Covid-19 pandemic on corporate social responsibility and marketing philosophy. J Bus Res 116:176–182. https://doi.org/10.1016/j.jbusres.2020.05.030
Henriksen D, Mishra P, Fisser P (2016) Infusing creativity and technology in 21st century education: a systemic view for change. J Educ Technol Soc 19(3):27–37
Hobbs M (1984) Crisis intervention in theory and practice: a selective review. Br J Med Psychol 57(1):23–34. https://doi.org/10.1111/j.2044-8341.1984.tb01577.x
Homans GC (1961) Social behavior: Its elementary forms. Harcourt, Brace, New York
Huston TL, Burgess RL (1979) Social exchange in developing relationships: an overview. Academic Press, New York, USA
James RK, Gilliland BE (2009) Crisis intervention strategies. Cengage Learning, New York, USA
Jabareen Y (2013) Planning the resilient city: concepts and strategies for coping with climate change and environmental risk 2(31):220–229. https://doi.org/10.1016/j.cities.2012.05.004
Jordan SR (2016) Positive Leadership Behavior. In: Farazmand A (ed) Global encyclopedia of public administration, public policy, and governance. Springer International Publishing, Cham, pp 1–6
Lawler EJ, Thye SR, Yoon J (2008) Social exchange and micro social order. Am Sociol Rev 73(4):519–542. https://doi.org/10.1177/000312240807300401
Mansour HEL, Holmes K, Butler B, Ananthram S (2019) Developing dynamic capabilities to survive a crisis: tourism organizations responses to continued turbulence in Libya. Int J Tour Res 21(4):493–503. https://doi.org/10.1002/jtr.2277
Meuser JD, Smallfield J (2023) Servant leadership: the missing community component. Bus Horiz 66(2):251–264. https://doi.org/10.1016/j.bushor.2022.07.002
Mochizuki Y, Bryan A (2015) Climate change education in the context of education for sustainable development: rationale and principles. J Educ Sustain Dev 9(1):4–26. https://doi.org/10.1177/0973408215569109
Nord W (1973) Adam Smith and contemporary social exchange theory. Am J Econ Sociol 32(4):421–436
Northouse PG (2019) Introduction to leadership: concepts and practice, 5th edn. SAGE Publications Inc., Lodon
O'Brien KL, Leichenko RM (2003) Winners and losers in the context of global change. Ann Assoc Am Geogr 93(1):89–103. https://doi.org/10.1111/1467-8306.93107
Orozco M, Harris S (2023) Psilocybin and the meaning response: exploring the healing process in a retreat setting in Jamaica. Anthropol Conscious 34(1):130–160. https://doi.org/10.1111/anoc.12162
O'Hare D, Wiggins M, Williams A, Wong W (1998) Cognitive task analyses for decision centred design and training. Ergonomics 41(11):1698–1718. https://doi.org/10.1080/001401398186144
Pillow W (2003) Confession, catharsis, or cure? Rethinking the uses of reflexivity as methodological power in qualitative research. Int J Qual Stud Educ 16(2):175–196. https://doi.org/10.1080/0951839032000060635
Pooley AW, Bendle LJ (2023) Long stay professionals and self-reliant leisure sociability in East Asia. World Leisure J 1–14. https://doi.org/10.1080/16078055.2023.2208563
Prahalad CK, Hamel G (1994) Strategy as a field of study: why search for a new paradigm? Strateg Manag J 15(S2):5–16. https://doi.org/10.1002/smj.4250151002
Reilly J (1995) Climate change and global agriculture: recent findings and issues. Am J Agr Econ 77(3):727–733. https://doi.org/10.2307/1243242
Ruben DH (1998) Social exchange theory: dynamics of a system governing the dysfunctional family and guide to assessment. J Contemp Psychother 28(3):307–325. https://doi.org/10.1023/A:1022986025115

Rice M, Erbeli F, Thompson CG, Sallese MR, Fogarty M (2022) Phonemic awareness: a meta-analysis for planning effective instruction. Read Res Q 57(4):1259–1289. https://doi.org/10.1002/rrq.473

Saud M, Ashfaq A, Abbas A, Ariadi S, Mahmood QK (2021) Social support through religion and psychological well-being: COVID-19 and coping strategies in Indonesia. J Relig Health 60(5):3309–3325. https://doi.org/10.1007/s10943-021-01327-1

Seddon N, Smith A, Smith P, Key I, Chausson A, Girardin C, House J, Srivastava S, Turner B (2021) Getting the message right on nature-based solutions to climate change. Global Change Biol 27(8):1518–1546. https://doi.org/10.1111/gcb.15513

Settembre-Blundo D, González-Sánchez R, Medina-Salgado S, García-Muiña FE (2021) Flexibility and resilience in corporate decision making: a new sustainability-based risk management system in uncertain times. Global J Flex Syst Manag 22(Suppl 2):107–132. https://doi.org/10.1007/s40171-021-00277-7

Shelby JS, Tredinnick MG (1995) Crisis intervention with survivors of natural disaster: lessons from Hurricane Andrew. J Couns Dev 73(5):491–497. https://doi.org/10.1002/j.1556-6676.1995.tb01784.x

Shulman HC, Bullock OM, Riggs EE (2021) The interplay of jargon, motivation, and fatigue while processing COVID-19 crisis communication over time. J Lang Soc Psychol 40(5–6):546–573. https://doi.org/10.1177/0261927X211043100

Snyder H (2019) Literature review as a research methodology: an overview and guidelines. J Bus Res 104:333–339. https://doi.org/10.1016/j.jbusres.2019.07.039

Tang KHD (2023) Climate change education in China: a pioneering case of its implementation in tertiary education and its effects on students beliefs and attitudes. Int J Sustain High Educ 24(5):1058–1081. https://doi.org/10.1108/IJSHE-05-2022-0151

Ummenhofer CC, Meehl GA (2017) Extreme weather and climate events with ecological relevance: a review. Philos Trans Royal Soc B: Biol Sci 372(1723):20160135. https://doi.org/10.1098/rstb.2016.0135

Uralovich KS, Toshmamatovich TU, Kubayevich KF, Sapaev IB, Saylaubaevna SS, Beknazarova ZF, Khurramov A (2023) A primary factor in sustainable development and environmental sustainability is environmental education. Caspian J Environ Sci 21(4):965–975. https://doi.org/10.22124/CJES.2023.7155

Van Driel JH, Beijaard D, Verloop N (2001) Professional development and reform in science education: the role of teachers practical knowledge. J Res Sci Teach: Official J Natl Assoc Res Sci Teach 38(2):137–158. https://doi.org/10.1002/1098-2736(200102)38:2%3C137::AID-TEA1001%3E3.0.CO;2-U

Vasan S (2016) Gender and power in the death metal scene: a social exchange perspective. In: Brown AR, Spracklen K, Kahn-Harris K, Scott NWR (eds) Global metal music and culture. Routledge, Lodon, pp 261–276

Vakulchuk R, Daloz AS, Overland I, Sagbakken HF, Standal K (2023) A void in Central Asia research: climate change. Central Asian Survey 42(1):1–20. https://doi.org/10.1080/02634937.2022.2059447

Watson ER, Foster-Fishman PG (2013) The exchange boundary framework: understanding the evolution of power within collaborative decision-making settings. Am J Commun Psychol 51:151–163. https://doi.org/10.1007/s10464-012-9540-8

Xu J, Wang Z, Shen F, Ouyang C, Tu Y (2016) Natural disasters and social conflict: a systematic literature review. Int J Disaster Risk Reduction 17:38–48. https://doi.org/10.1016/j.ijdrr.2016.04.001

Zaharna RS (2016) Beyond the individualism–collectivism divide to relationalism: explicating cultural assumptions in the concept of relationships. Commun Theory 26(2):190–211. https://doi.org/10.1111/comt.12058

Zafarullah H, Huque AS (2018) Climate change, regulatory policies and regional cooperation in South Asia. Public Adm Policy 21(1):22–35. https://doi.org/10.1108/PAP-06-2018-001

Zhao S, Liu H, Ma H, Jiao M, Li Y, Hao Y, Sun Y, Gao L, Hong S, Kang Z (2015) Coping with workplace violence in healthcare settings: social support and strategies. Int J Environ Res Public Health 12(11):14429–14444. https://doi.org/10.3390/ijerph121114429

Chapter 27
Sustainable Livelihood Diversification in The Merapi Volcano Disaster-Prone Area

Lis Purbandini, Eko Wahyono, Reza Amarta Prayoga, Siti Fatimah, Nuzul Solekhah, Fatwa Nurul Hakim, Djoko Puguh Wibowo, and Rachmini Saparita

Abstract The Merapi volcano area is inherently disaster-prone, given the constant risk of eruptions capable of disrupting lives. Mitigating this vulnerability requires diversifying livelihoods, as the choices people make in their means of living significantly impact community resilience both during and after disasters. A significant challenge lies in the fact that many individuals in such areas predominantly rely on agriculture, a livelihood susceptible to the impact of eruptions. The research aims to delineate community livelihood sources and analyze the social, human, financial, natural, and physical capital within these disaster-prone areas. In Mount Merapi's Balerante area, diversifying livelihood sources emerges as a vital survival strategy. Leveraging sustainable agriculture, MSME development, and responsible tourism

L. Purbandini (✉) · E. Wahyono · R. A. Prayoga · S. Fatimah · N. Solekhah · F. N. Hakim · D. P. Wibowo · R. Saparita
Research Center for Social Welfare, Village and Connectivity. National Research and Innovation Agency (BRIN), Jakarta, Indonesia
e-mail: lisp001@brin.go.id

E. Wahyono
e-mail: ekow005@brin.go.id

R. A. Prayoga
e-mail: reza010@brin.go.id

S. Fatimah
e-mail: siti080@brin.go.id

N. Solekhah
e-mail: nuzu001@brin.go.id

F. N. Hakim
e-mail: fatw001@brin.go.id

D. P. Wibowo
e-mail: djok022@brin.go.id

R. Saparita
e-mail: rach001@brin.go.id

U. Mukhopadhyay et al. (eds.), *Climate Crisis, Social Responses and Sustainability*, Climate Change Management, https://doi.org/10.1007/978-3-031-58261-5_27

allows the community to mitigate economic risks and enhance overall welfare. The transformation towards a sustainable livelihood system depends on various resources, each playing a distinct role. Social capital, characterized by trust, networks, and norms, forms the bedrock for cooperation and support during evacuation and recovery in Balerante. Human capital is nurtured through disaster preparedness training and guidance from local authorities, enhancing the community's ability to respond effectively to disasters. Financial capital remains robust, with families diversifying income sources and allocating savings for various purposes. Natural capital, particularly land and water, proves crucial for community livelihoods, with responsible management and conservation being integral for disaster risk reduction. Physical capital, encompassing evacuation routes and early warning systems, facilitates efficient disaster response and preparedness. To anticipate resource trends and foster sustainable recovery, communities should prioritize disaster awareness, responsible utilization of natural resources, and livelihood diversification.

Introduction

In the disaster-prone area of Mount Merapi, volcanic eruptions profoundly influence the livelihoods of the surrounding communities (Sullivan and Sagala 2020). Eruptions have shaped livelihoods by bestowing materials such as volcanic sand and rocks, boosting tourism due to the striking changes in the landscape, and affecting local biodiversity.

A significant consequence of volcanic eruptions is their effect on agriculture. Lands impacted by eruptions often become richer in nutrients post-eruption, as prior research shows (Small and Naumann 2001). While eruptions might momentarily hinder agricultural activities, enhanced soil fertility can bolster agricultural productivity over time (De Bauw et al. 2016; Rahayu et al. 2014; Small and Naumann 2001). Volcanic eruptions can also offer unique opportunities for the tourism industry. Striking transformations in the landscape, such as the Blue Fire phenomenon at the Ijen Crater in Banyuwangi, captivate tourists with their breathtaking natural allure (Bhaskara and Filimonau 2021; Lestari et al. 2019).

Proper utilization of the livelihood potential generated by volcanic eruptions is crucial for sustainability. Hence, the establishment of sustainable livelihood systems becomes imperative to conserve and safeguard available natural resources for future utilization (Abdulai et al. 2022).

The three aspects that must be considered in managing livelihoods to be sustainable are economic, social, and environmental (Shackleton et al. 2019). A sustainable livelihood system can improve people's welfare and reduce poverty by bringing economic and social benefits to the community (Roy et al. 2023). Community involvement, particularly in the decision-making process, is integral to managing such systems (Andreastuti et al. 2023). Communities surrounding Mount Merapi rely primarily on agriculture and livestock, which leaves them especially susceptible to eruptions. Such eruptions can deeply impact the local economy and social sectors, altering people's means of living (Safitri and Fajarwati 2015; Umaya et al. 2020a). Hence, choosing resilient sources of livelihood is pivotal for community resilience during and after disasters (Huhtamaa et al. 2022). Sustainable livelihoods bolster community resilience against disasters, whereas unsustainable ones amplify vulnerability. Goyal's (2019) research corroborates this notion, indicating that sustainable management of natural resources enhances community resilience to disasters and accelerates post-disaster recovery (Goyal 2019).

One method to protect livelihoods is through diversification. This method can encompass the growth of non-agricultural sectors like trade, creative industries, or services that remain operational even during disasters. Expanding these avenues can diminish people's vulnerability to disasters, alleviate poverty, and enhance livelihood sustainability (Uddin et al. 2021). By having a diverse income revenue, residents in disaster-prone areas can ensure continuity of income even if specific sectors are impacted, as they'll have alternative revenue streams to rely on (Irsyad et al. 2020).

The significance of this research stems from the pressing need to address the vulnerabilities of communities around Mount Merapi. These rural communities are heavily dependent on the region's resources, including agriculture and tourism. Yet, the volatile nature of the volcano jeopardizes their primary sources of income. Therefore, it is crucial to comprehend how diversifying livelihood can strengthen resilience and sustainability in such disaster-prone areas. The innovative approaches discussed in this chapter provide original on-the-ground insights into the Balerante case and inform policies and initiatives aimed at building more sustainable and resilient livelihoods in disaster-prone areas.

This chapter attempts to focus on the issue of community livelihoods in disaster-prone areas in Balerante village. The majority of people there make their living from agriculture, making them vulnerable to the impact of the Merapi eruption. To overcome these problems, the purpose of this study is to describe the sources of livelihood of the community in disaster- prone areas and analyze the resources owned, namely social capital, human capital, financial capital, natural capital, and physical capital.

Materials and Methods

With a qualitative approach (Miles and Huberman 1994) with descriptive analysis. Descriptive methods are designed to collect information about the real and present state (ongoing). The use of this method aims to systematically illustrate/describe the nature of a situation that is running at the time it is done and examine the causes of a particular phenomenon. Descriptive analysis is used to answer the research objectives, namely, to explain systematically and in detail the sources of community livelihood in disaster-prone areas and to analyze social, human, financial, natural, and physical capital in the community. This research took place in Balerante Village, situated in the Kemalang Sub-district of Klaten District, Central Java Province. The selection of this location was deliberate, given the village's extensive experience

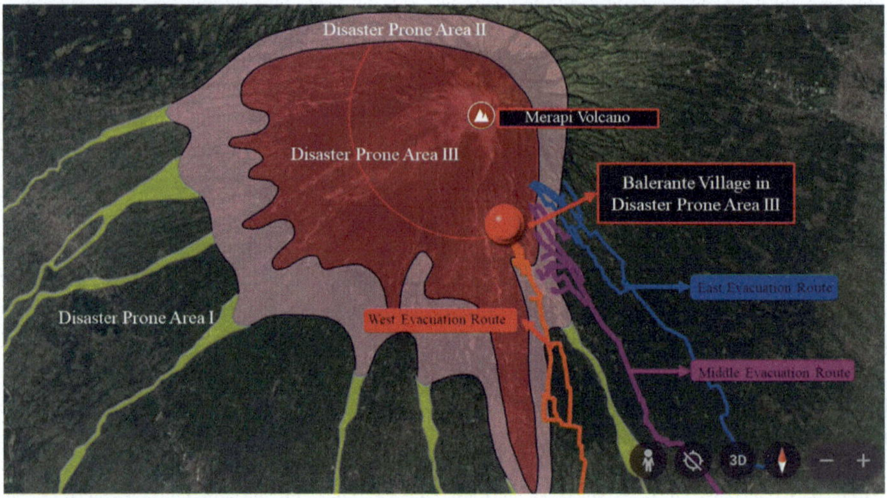

Fig. 27.1 Disaster-Prone areas of mount Merapi. *Source* Google Maps (2020)

with Mount Merapi eruptions. Balerante Village falls within KRB II and III, directly confronting the imminent threat of eruptions and hot clouds from Mount Merapi. Moreover, the village is a mere 5 km from the crater's edge. Figure 27.1 illustrates the eruption's range marked in red, depicting the potential hazard zone during an eruption. It vividly underscores the constant peril faced by Balerante village, where the specter of death and eruption looms ever-present.

The research incorporated both primary and secondary data sources. For primary data, we conducted an observation on aspects tied to community livelihoods, while in-depth interviews were carried out with 10 informants from disaster-prone areas II and III. Informants were determined based on vulnerability and livelihood criteria. In Balerante Village, there are only two vulnerability criteria, namely KRB II and KRB III. The number of informants was ten people, determined by a random sample of 5 people in each KRB. The informants represent people whose livelihoods are in the agriculture/livestock sector and other services (sand mining, trading/shops, stone mining, and massage therapy). Secondary data were also collected from various sources such as reports and research results relevant to the substance of the research. Descriptive analysis techniques were applied to the qualitative data by following the methodology described by Braun and Clarke (2013). This approach explains and analyzes communities and units of analysis in a more in-depth and comprehensive manner. Through in-depth interviews and live-in methods at the research site, researchers can capture and understand the behavior and habits of the community. Through this approach, researchers not only get information from what informants say, but also see the daily life and lifestyle of informants.

This thematic analysis approach illuminated not just the livelihood patterns in Balerante Village but also underscored the significance of harnessing livelihood resources sustainably. The thematic analysis is classified into three categories:

(a) types of livelihood sources in Balerante Village, (b) diversification of livelihood sources in Balerante Village to support community resilience, and (c) sustainable livelihoods in Balerante Disaster-Prone Areas. The underpinning analytical framework drew from the livelihood system theory proposed by Ellis (2000). The community sustains its livelihood based on its assets, namely social capital, human capital, financial and economic capital, natural resources and the environment, and physical infrastructure. Asset utilization activities, as well as efforts to access these assets, are often modified through institutions, organizations, and social relations that are in a context of vulnerability due to disasters, conflicts, and trends such as climate and environmental changes that determine the status of life at both the individual and household levels. Sustainable livelihood strategies include the utilization of natural and non-resources. These strategies not only ensure environmental sustainability (quality of soil, water, forests, and biodiversity) but also ensure more viable livelihoods (improved income stability, level of risk, and seasonal vulnerability).

Results and Discussion

Overview of Potential Resources of Balerante Village

Balerante Village, situated near Mount Merapi, is subject to frequent natural disasters due to its proximity to the active volcano. This unique location presents both challenges and opportunities for livelihood development. While eruptions can obliterate established sources of income, they also carve out avenues for new economic ventures.

Ballantine's strengths lie in its abundant natural resources, infrastructure, institutional backing, and supportive village governance. Notable natural resources include (a) The Tuk Bengbeng water springs in the Merapi mountains, catering to the community's daily water needs and livestock; (b) Farmlands dedicated to agriculture and grass cultivation; (c) sand and rock piles, becoming a new source of livelihood for the community after the eruption of Merapi; (d) Touristic appeal, courtesy of the scenic attractions near Mount Merapi and Kaliworo; and (e) Established evacuation pathways, this evacuation route, in addition to being used for evacuation during the eruption of Mount Merapi, also serves as a means for the community to move to the garden and other activities such as to the market and school.

In addition to the potential for natural resources, Balerante Village also has the potential for human resources. This potential is manifested in the form of community groups that are formed by the community and are informal. The existence of these groups will directly or indirectly affect the livelihood of the community.

Balerante boasts five community-based organizations, prominently featuring farmer groups. There are five such groups, with the Mekar Indah Women Farmers Group (*KWT/Kelompok Wanita Tani*) being one of them. These groups are organized based on regions or hamlets, with two groups each in every hamlet (*Dusun*). For instance, Dusun I hosts the Marsudi Mulya and Giri Mulyo farmer groups, while

Dusun II accommodates the Tani Makmur and Tani Raharjo groups. Typically, each collective comprises 20 to 30 members. Farmer group activities are still limited to regular meetings every 35 days to discuss the problems faced. In the future, the role of farmer groups needs to be encouraged to make a more optimal contribution to supporting business development.

In 2014, Balerante Village spotlighted Kalitalang as a tourist destination, showcasing the majestic beauty of Mount Merapi. With attractions like a selfie zone in the national park and a glamping area, the village paves the way for budding entrepreneurs in the service and trade sectors. Recognizing tourism's potential as a future revenue stream for Balerante, the important role of *Pokdarwis* is emphasized. In addition to tourism management, Pokdarwis provides coffee training, *jatilan* performance art, Talang river tourism, dairy farming, and religious tourism. In Balerante, there are two tombs, namely the tomb of Sri Maulana Akhir and Kyai Ali Jufri, who, according to history, was a student of Sunan Kali Jaga.

Small businesses, operated individually and communally, flourish in Balerante Village. Notable micro, small, and medium enterprises (*UMKM/Usaha Mikro Kecil dan Menengah*) offer products ranging from Javanese palm sugar sticky rice cakes and tempeh to crunchy onde-onde, entik chips, fermented sticky rice, and mini peyek chips. Small business production is still on a household scale, so marketing is limited to the village level and has not yet reached the franchise level because the processed products have not met the standards to be marketed with the conditions determined by the franchise. The constraints faced in small business development include the low skills of entrepreneurs, especially in processing products into other products, as well as limited technology or infrastructure to process products and marketing, especially in terms of promotion and packing of products.

In 2017, Balerante Village initiated *Bumdes*, a venture involved in a plethora of business activities—these range from savings and loans and tourism promotion to *UMKM* product marketing and BRI Link collaboration. During the COVID-19 pandemic, BUMDes had a role in supplying the community's basic needs. The existence of disaster-affected communities with COVID-19 can not carry out their livelihood activities, so their life needs are disrupted. To overcome these problems, BUMDes procures community needs by providing loans through money, basic foodstuffs, and other needs. The repayment period is several months, and if the community has not been able to repay within the given time period, they only need to pay the interest. Post the 2010 eruption, its engagement in CSR initiatives was instrumental in rejuvenating the local economy. In the future, Bumdes will expand its business to include tax services such as land and building tax and motor vehicle tax.

Moreover, the village prides itself on having a Disaster Risk Reduction Organization (*OPRB/Organisasi Pengurangan Resiko Bencana*). This body educates the community about disaster preparedness, from initial response to evacuation procedures and post-disaster recovery. In carrying out its duties, OPRB is divided into several teams, including evacuation and handling teams, data collection teams in refugee camps, medical teams, and logistics teams. If overwhelmed by the number of casualties, OPRB must coordinate and cooperate with Balerante Village volunteers and Search and Rescue (SAR) units to evacuate people and livestock.

The genesis and sustenance of these diverse livelihoods cannot be separated from the support of the village governments. Their support is reflected in regulatory frameworks and financial backing via the Village Revenue and Expenditure Budget (*APBDes/Anggaran Pendapatan dan Belanja Desa*). This is evidenced by the existence of several village regulations on the establishment of OPRB, BUMDes, and other organizations, as well as the allocation of village funds to support these activities.

Vulnerability and Resilience of Balerante Community

Vulnerability is a characteristic determined by physical processes or factors (infrastructure, land use), social factors (poverty, inequality, marginalization, social exclusion, and discrimination based on gender, social status, disability, and age), psychological factors, economic factors (informal sector, vulnerable rural livelihoods, dependence on a single industry, globalization of business and food supply chains), environmental factors (poor environmental management, overconsumption of natural resources, declining risks governing ecosystem services, climate change) that affect the vulnerability of individuals, communities, assets or systems from disaster impacts (UNDRR 2017).

Geographically, Balerante Village is included in Disaster Prone Areas III and II (KRB III and II). KRB III is an area that is often hit by pyroclastic clouds, lava flows, volcanic explosions, toxic gases, and rockfalls (incandescent). In this area, it is not recommended that anyone make permanent residences or utilize the area for commercial purposes. KRB II is an area that has the potential to be hit by hot clouds, possibly lava flows, rockfalls, avalanches, and ashfalls, generally occupying the slopes and foothills of the volcano, and lava flows (https://bnpb.go.id/berita/identifikasi-posisi-terhadap-krb-gunung-merapi-gunakan-cekposisi). In this village, there are five hamlets, namely Sukorejo, Ngipiksari, Gondang, Banjarsari, and Sambung Rejo, in the Directly Affected Area (Area Terdampak Langsung-ATL) of Merapi hot clouds. (Balerante 2021). Given its location in KRB III and II, it is very vulnerable to the possibility of volcanic eruptions that will have an impact on physical factors (infrastructure, land use, social factors (poverty), economic factors (livelihoods), and environmental factors (climate change). Vulnerabilities include immediate threats like eruptions, pyroclastic flows, ashfall, and Lahars flood. Post-disaster vulnerabilities manifest in tendencies like natural resource overexploitation, which provides for activities like sand mining and forest encroachment. These events critically disrupt residents' livelihood, as income sources are crucial for post-disaster recovery.

The 2010 Merapi eruption devastated the region, destroying numerous houses and resulting in livestock losses, especially in hamlets like Sukorejo, Ngipiksari, and Sambungrejo 103 houses were damaged and 155 cattle died (Balerante 2021). Essential agricultural infrastructure was obliterated, including farmlands cultivating edible crops, horticultural commodities, and forage crops/fodder (Depari and Lindell 2023). This situation further strained the lives of Balerante residents, as all their income sources were compromised and rendered unusable. Consequently, many residents found refuge in camps for approximately 14 months, relocating often or staying with relatives outside Balerante. Their sustenance primarily relied on governmental, private, NGO, and organizational aid. Some tried selling snacks in evacuation areas, though this was often short-lived due to frequent relocations. On the other hand, the Balerante community also has community resilience in facing the Merapi volcanic eruption disaster. According to FAO (2016), resilience capacity in the community can be measured from five indicators: community access to basic services, the existence of capital (Asset), Social Safety Net (Jaringan Pengaman Sosial-JPS), sensitivity, and adaptive capacity. Communities with access to basic services can better cope with and quickly recover from the impact of disasters or unexpected economic changes. Capital or assets owned by the community in the form of land, buildings, savings, and insurance are very important in building community resilience because they can be used as resources to recover losses due to disasters or economic changes. With sufficient capital, communities can recover more quickly after experiencing a costly disaster. Adaptive capacity includes knowledge, skills, and resources that enable communities to respond quickly to environmental, social, or economic changes. A community's ability to recognize the signs of a disaster is a form of adaptive capacity, the knowledge of which will prepare the community for a disaster.

Upon deeming Balerante safe, residents would return daily to clear debris and rehabilitate farmlands while nights were spent at evacuation sites. In reconstructing the destroyed houses, the community received material assistance from the government, and the implementation was carried out in cooperation. However, the experts (builders) were paid as daily laborers. For their fresh start, they tapped into various livelihood sources, from savings, residual crops, farming using the "gadoh" system, construction work, massage therapy parlor, and handouts from relatives and other sources.

Notably, the post-eruption period saw the community's remarkable solidarity and collaboration, from taking turns rebuilding homes to cultivating lands. This spirit stemmed from a collective experience of Merapi's eruption, "shared fate," and a unified aspiration for a brighter future.

Types of Livelihood of Balerante Community

The sources of livelihood of the Balerante community are diverse and can be broadly classified into two categories: agriculture (on-farm) and non-agriculture (off-farm). For agriculture-based livelihoods, the community cultivates land and breeds cattle, while non- agricultural activities are carried out as services such as sand miners, trading and manual workers. There are differences in community life before and after the disaster. Before the disaster, people generally made a living from agriculture and livestock, but after the eruption, more people began to work as sand miners who utilize the abundant sand and stones due to the eruption of Merapi, MSMEs and a little agriculture.

Based on the results of field searches, the types of livelihoods of the Balerante community are presented in Table 27.1.

Table 27.1 shows that the livelihoods of the Balerante community are very diverse. Given the threat of the eruption of Mount Merapi, each livelihood option has different sustainability impacts. The agricultural sector is the most vulnerable livelihood to the impacts of the eruption of Mount Merapi. However, this sector plays an important role in providing food and maintaining land productivity, so the solution to this problem is the intercropping system. In contrast to other service industries, such as trade, these livelihoods can support the economy, provide local economic opportunities, and provide a stable source of income for affected communities. The tourism sector is now an important economic alternative in disaster-prone areas. However, tourism development must consider environmental and social impacts to avoid increasing disaster risks.

The various livelihood sources pursued by Balerante residents ensure that the community's economy remains resilient despite ongoing disasters, and even after the eruption, sources of income will increase which will have an impact on increasing people's income.

Balerante Community Livelihood Diversification Supports Community Resilience

Located in a disaster-prone area, Balerante faces major challenges to survival. Livelihood diversification reduces economic and social risks and strengthens overall community resilience. Diversification serves as one of the strategies for ensuring survival amidst Merapi eruptions. Compared to monoculture businesses (Ellis 2000), people can meet daily necessities by broadening current ventures and exploring new business opportunities (Chambers and Conway 1992; Molénat et al. 2023). This approach of diversification not only supports sustainable livelihoods but also bolsters community resilience (Hsu and Peng 2023). Several ways of diversification to support the resilience of the Balerante community are (a) sustainable agriculture through the application of sustainable cropping patterns and the combination of various types of crops can provide stability in agricultural production (Altieri 2009; Peng et al. 2022) (b) encouraging the development of micro, small, and medium enterprises (MSMEs) will provide more diverse economic options, (Alemu 2023). (c) sustainable tourism that utilizes local natural beauty, adhering to sustainability principles to protect the environment and local culture (UNWTO 2015), and (d) increasing community capacity through education and skills training. Capacity building can create new jobs and increase the competitiveness of local communities (Hossain 2024).

Balerante communities usually have multiple livelihoods. After the eruption, their livelihoods increased along with the emergence of various livelihood options due to the impact of the eruption of Mount Merapi.

Table 27.1 Types of livelihood of Balerante community

No.	Source of livelihood	Type of business	Land use	Business/ Work system	Supporting institutions	Remarks
1	*On farm*					
a	Farmers	Plot of land/ Yard	The front yard is more used for medicinal plants, vegetables and ornamental plants, while the back or better known as "kebon" is used for livestock pens (cows, goats, sheep and poultry) and growing grass	Tumpang sari	5 farmer groups have been formed	
		Tegalan/ Farm plot	Planted with vegetables such as chili, bok choy, leeks, and cabbage, intercropping with an system. But there are also people who utilize plotted land by planting horticultural crops intercropped with grass	Tumpangsari		

(continued)

Table 27.1 (continued)

No.	Source of livelihood	Type of business	Land use	Business/ Work system	Supporting institutions	Remarks
		Land owned by TNGM	Planted with grass	Sanggem		Generally, people receive sanggem land from their ancestors for generations, but there are also people who make use of sanggem land by purchasing land from someone who has the right to cultivate national park land in exchange for compensation
b	Ranchers	Dairy cattle Beef cattle	Each household has an average of 2 cows	Monoculture	Cooperative	Cooperatives serve as agents in the purchase of milk and provision of livestock feed
2	*Off-farm*					
a	Sand miners	Sand miner manual workers	Kaliworo river	Daily		Working for a company: sand miner, coker and truck driver Manual: as a sand miner in groups of 5–6 people Stable and large income but high risk because there is no injury or life insurance

(continued)

Table 27.1 (continued)

No.	Source of livelihood	Type of business	Land use	Business/ Work system	Supporting institutions	Remarks
b	Tourism	Kalitalang managers, food and beverage vendors around Kalitalang	Kalitalang tourism destination	Daily		This source of livelihood grew after the 2010 eruption and has the potential to be developed in the future
c	Other livelihood of services	Masseurs, construction workers, grocery stores, and food stalls	Village scope	Daily		As a supporting or part-time source of income

Source researchers' field data findings, 2023

Post-eruption, the community's income sources notably varied, spanning sectors like agriculture, livestock, and other services such as construction, massage therapy, grocery trading, and food stalls. Typically, a family had two to three revenue streams. Dual-income households often combined farming or cattle breeding with another service-oriented job. Those with three income streams juggled farming, cattle breeding, and either sand mining or another service activity. Remarkably, some even had a quartet of income sources: agriculture, cattle breeding, sand mining, and an additional service job. Intriguingly, these income—generating efforts often involved the entire family—husbands, wives, children, and in-laws. They collaborated, pooling their earnings to cater to the community's collective needs.

This spirit of cooperative economics is evident in how the Balerante residents share the proceeds from joint ventures. Based on the analysis, family income is utilized under three main criteria. The primary aim is to meet daily necessities, which often involves relying on dairy farming (specifically for milk), sand mining, and jobs in trading, massage therapy, or construction. Dairy farming, particularly labor-intensive due to the twice-daily milking routines, employs many from the community. Sand mining occurs in two main ways: with heavy machinery and manually. Companies often deploy heavy machinery, with community members taking roles as manual workers, equipment operators, and truck drivers. In contrast, manual mining is typically a community-driven group effort involving teams of 5–6 individuals. Profits are equally distributed among members, with each individual typically earning between 100 and 200 thousand Rupiah daily.

The second approach involves short-term/monthly savings, often dependent on agricultural yields. The community practices intercropping to cultivate various vegetables simultaneously, ensuring continuous harvests throughout the year. This method not only offers higher profits compared to monoculture, as detailed by

Umaya et al. (2020b), but also promotes a healthier environment and conserves local biodiversity (Liebman et al. 2016).

Third, long-term/annual savings, relying on beef cattle farming. Raising beef cattle requires relatively little labor compared to dairy cows, and the trade of products spans a long time (3–4 years), so this business is often run by people with limited labor and the orientation of long-term results/annual savings. Beef cattle are sold when there are urgent needs, such as children's school tuition or family members falling ill.

Members of the community engage in various occupations suited to their skills. Men often take roles in agriculture, milking, and volcanic sand mining, while women tend to be farmers, grass foragers, and merchants. The distribution of their time varies throughout the day. Mornings are dedicated to milking, grass foraging, and sand mining. By midday, following the foraging and milking, many transition to farming. Evenings are again reserved for milking and further sand mining.

Post the 2010 Merapi eruption, the Balerante community's livelihoods diversified further with the emergence of the Kalitalang tourist attraction. This destination unlocked numerous local business opportunities: manning the entrance ticket booth, vending refreshments, and overseeing the parking area. As many visitors flocked to Kalitalang to appreciate the scenic beauty and Mount Merapi, food stalls sprouted in its vicinity. A detailed exploration of the Balerante community's diversified livelihoods can be found in Table 27.2.

New business opportunities are critical in supporting sustainable livelihoods and building community resilience in disaster-prone areas. These fresh avenues for business can serve as a driving force, creating positive impacts and offering solutions to the challenges faced by communities susceptible to disasters. The key to establishing new businesses that adapt and deliver positive impacts is to focus on innovation and sustainability. In this context, some new sustainable business opportunities that can be explored are (a) Renewable energy. Renewable energy is energy obtained from "sustainable natural processes" such as solar, wind, water flow, biological, and geothermal energy. New businesses focusing on renewable energy can reduce the risk of disasters associated with traditional energy infrastructure. Investing in solar panels, micro-hydropower plants, and other green technologies can create reliable and environmentally friendly energy sources. The impact of these ventures is to provide stable access to electricity and reduce disaster-related risks. (b) Adaptive and resilient agriculture: New businesses integrating technology into agriculture, such as vertical farming and hydroponics, can help create more adaptive and disaster-resistant agricultural systems. This contributes to food security and helps reduce the risks associated with natural disasters that can damage agricultural production. Farmers can utilize land efficiently and produce more food using vertical and hydroponic farming systems. In addition, vertical farming also has significant environmental benefits by reducing the use of land, water, and pesticides (c) Surveillance and Early Warning Technologies: businesses that provide surveillance and early warning technologies can help improve disaster preparedness. Sensors, mapping systems, and mobile applications can provide critical real-time information for early evacuation and disaster risk reduction. The impact of these businesses increases community preparedness and enables rapid response to disasters and (d) Resilience Training and

Table 27.2 Livelihood sources, actors and time allocation of Balerante community in carrying out business activities

No	Source of livelihood	Actor	Time allocation	Utilization of business outcomes	Remarks
A	*On farm*				
1	Farmers	Men and women	Towards noon and evening	For monthly/ short term savings	Activities carried out by women in planting and harvesting while men in land work
2	Ranchers				
	a. Grass foraging	Mostly done by women	Morning	Beef cattle for long-term savings	Although there are also men who forage
	b. Milking	Men	Morning and afternoon	Milk to make a daily living	
B	*Off-farm*				
1	Sand miners	Men	Morning and afternoon	To provide for the daily needs	Few women are involved in sand mining and wives commonly assist their husbands
2	Tourism				
	a. Entrance Ticket Counter Attendant	Men and women	Morning until afternoon	To provide for the daily needs	Youths and mostly done on Saturdays and Sundays
	b. Managing the parking area	Men	Morning until afternoon	To provide for the daily needs	Youths and mostly done on Saturdays and Sundays
C	*Other livelihood of services*				
	a. Groceries merchant	Women	All day	To provide for the daily needs	As a side business/supplementary business
	b. Food stalls	Women	Morning	To provide for the daily needs	As a side business/supplementary business
	c. Masseurs	Women	At any time/as needed	To provide for the daily needs	As a side business/supplementary business
	d. Construction workers	Men	At any time/as needed	To provide for the daily needs	As a side business/supplementary business

Source researchers' field data findings, 2023

Education. Efforts focusing on resilience training and education can help communities better cope with disasters. Education programs that cover evacuation, first aid, and emergency planning can improve community preparedness and resilience.

Sustainable Livelihoods in Balerante Disaster Prone Areas

Sustainable livelihood systems are associated with vulnerability in the form of various trends and shocks (Ellis 2000). These shocks and trends will affect livelihood changes (Steenbergen et al. 2017). Consequently, individuals impacted by disasters often adapt or modify their livelihoods (Uddin et al. 2021).

Ellis' (2000) livelihood system theory is used in this study because people living in the disaster-prone area of Mount Merapi rely on various assets such as social, human, financial, natural, and physical capital to survive. Communities living in this area are highly vulnerable to livelihoods and sustainability. Ellis' (2000) theory is very relevant in explaining the disaster phenomenon in Indonesia, especially the livelihood system of communities in the Merapi disaster-prone area.

In analyzing livelihood sources, we refer to the livelihood system theory described by Ellis (2000). Livelihood systems are based on asset capabilities such as social, human, financial, natural, and physical capital. Sustainable livelihood system transformation depends on access to social relations (gender, social class, age, ethnicity), institutions (rules and customs, land ownership and markets), and organizations (associations, NGOs, local government, and state institutions). Sustainable livelihood systems also relate to conditions, contexts, trends, shocks, and vulnerabilities. The sustainable livelihood strategy consists of two activities, one based on natural resource outputs (crops, livestock, non-agricultural natural resources) and the other based on non-natural resource outputs (village trade activities, village industries, remittances). This strategy not only ensures environmental sustainability (soil, water, forest, and biodiversity quality) but also ensures more viable livelihoods (improved income stability, level of risk, and seasonal vulnerability) see Fig. 27.2.

Before, during, and after a disaster, a community's resilience hinges on its available resources or assets. These assets can be categorized into five capitals: social, human, financial, natural resource, and physical (Mbiba et al. 2019). As outlined by Islam and Walkerden (2022), human and physical capitals are pivotal during emergency responses. Conversely, natural and financial capital become more significant during the early and extended recovery phases (Islam and Walkerden 2022).

Social capital encompasses three foundational principles: trust, networks, and norms. These three elements are intricately intertwined, each reinforcing the other. For example, successful networking relies on both trust and prevailing norms. In a disaster context, this social capital manifests during evacuation. The social capital of the Balerante community is very robust; this is very much felt in the relationship of neighborly life. For example, in daily life, they help each other when a neighbor has a celebration, such as a wedding, *sunatan* (circumcision), or death. Strong social

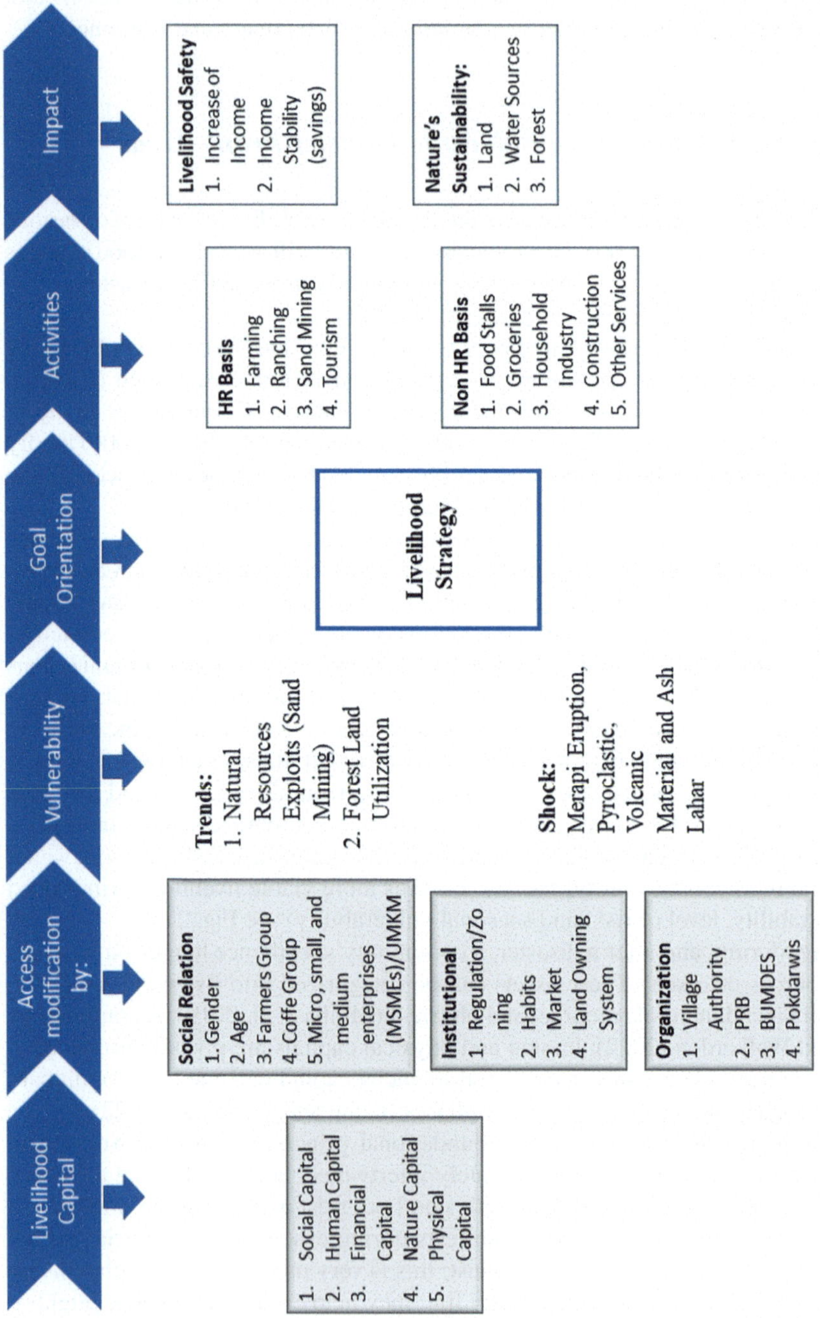

Fig. 27.2 Community livelihoods in disaster-prone areas: Balerante Village. Adopted from Ellis' (2000) livelihood system theory

capital facilitates evacuation, increases community resilience, and is important in disaster risk reduction.

Amid a Merapi eruption, every household and neighbor feels a profound moral duty to ensure everyone's safety, ensuring no one is overlooked. This unwritten yet deeply ingrained commitment is a hallmark of the Balerante community's ethos. Moreover, robust community networks and mutual trust expedite information exchange, coordination, and assistance during post-disaster periods.

Such faith and networks also catalyze inter-party collaborations (Coloman 1990), epitomized by the *"Desa Paseduluran"* initiative—a partnership between Balerante Village and Kebon Dalem Lor Village, Prambanan District, Klaten Regency. Among its provisions are protective measures for the Balerante community during Merapi calamities, offering shelter, parking, and communal livestock enclosures.

Another manifestation of social capital is the creation of informal groups, ranging from farmers and coffee groups, PKK, and Posyandu to disaster risk reduction teams. Such entities collaborate and strengthen Balerante Village's economic vitality. For instance, UMKMs operating near the Kalitalang tourist site demonstrate solidarity, often aiding peers without stalls to vend their products. A similar bond exists among manual sand miners, known to work in tight-knit groups. Each group consists of an average of 6 to 9 people. This working relationship materializes through unwritten rules and becomes a common consensus in the bond of fellow sand miners. In terms of socio-culture, the tradition carried out by the community to establish friendship between residents/neighbors is that the community around Merapi (Balerante) often holds a *kenduri* tradition, with the sole purpose of maintaining good relations with *pepundhen/sesepuh* (the elders). In addition, the community also preserves existing cultures such as jatilan, wayang, and ketoprak for entertainment when there is a celebration.

Human capital in regions prone to volcanic disasters encompasses the community's knowledge, preparedness, and responsiveness. In Balerante Village, facets of human capital include awareness and skills to mitigate disaster risks. Residents can discern impending Merapi eruptions, detecting signs like increasing air temperatures and unusual animal behaviors. This insight stems from past experiences enforced by information from disaster risk reduction volunteers and authorities at the village and regency levels. In addition, technical guidance on disaster prevention was organized to improve the community's knowledge and skills. Apart from that, in order to increase community knowledge and skills, technical guidance on disaster prevention is also provided. Apart from that, in order to increase community knowledge and skills, technical guidance on disaster prevention is also provided.

Moreover, villagers possess the wisdom to act during crises and understand evacuation timings, priorities, routes, and destinations. Such competencies arise from simulations and training orchestrated by the village head in conjunction with Klaten Regency BPBD. Among the practical skills is the community's ability to protect assets, like livestock, during disasters. In Ballantine's KRB III zone, farmers often entrust their cattle to safer locations outside the high-risk area, employing the *"gadoh"* system—a profit-sharing livestock maintenance approach. This system, shaped by mutual agreements, mitigates livestock losses due to sudden Merapi eruptions, including dangers like the *"wedus gembel"* or pyroclastic flows.

The presence of financial capital in the livelihood systems of communities in disaster-prone areas is essential to help communities cope with and recover from disasters (Islam and Walkerden 2022). This capital plays a role in increasing household survival; the better the household's financial condition, the more able it is to survive. The residents of Balerante possess solid financial capital, with many families having multiple income streams. This diversity allows them to allocate their earnings for savings effectively. Typically, they distribute their income into three categories: daily expenses, short-term (monthly), and long- term (annual) savings. These savings can manifest as gold or jewelry, bank deposits, and investments in livestock. Investing in gold or jewelry is the safest strategy for mitigating economic risks. Worn items, such as rings, bracelets, and necklaces, are not only easy to transport during evacuations but can also be swiftly liquidated in case of urgent financial needs.

Natural resource capital encompasses land, water resources, fisheries, forest products, wildlife, biodiversity, and environmental services. In Balerante Village, key natural assets influencing the community's livelihood are land, water resources, and biodiversity.

On average, a Balerante resident owns a yard or garden farmland and makes use of forest land within the national park. This land is a foundational livelihood source, facilitating crucial activities such as agriculture and livestock ranching. Notably, post the 2010 eruption, the agricultural land was enriched with added minerals, leading to enhanced harvest yields (Aini et al. 2022).

Besides land, the Tuk Bebeng water source at Mount Merapi's base is a vital community lifeline. This source meets residents' daily water needs for domestic purposes (like kitchen and sanitation) and livestock sustenance. It benefits the Balerante community and neighboring populations around Mount Merapi—furthermore, the scenic Kaliworo landscape and emerging biodiversity present significant tourism opportunities.

However, this natural capital, while promising, is vulnerable. Eruptions from Mount Merapi can inflict substantial damage. As such, it's imperative for both the local communities and village administrations to safeguard and manage these natural resources judiciously. Efforts in ecological conservation, adept disaster management, and strategic spatial planning can minimize disaster risks and ensure sustainable utilization of the area's natural capit. Furthermore, there's an urgent need to amplify community awareness about the significance of conserving natural resources. This awareness can help mitigate the adverse effects on the livelihood systems in Mount Merapi's disaster-prone areas.

Ballantine's physical capital encompasses infrastructure, transport, and technology. Infrastructure developments, such as evacuation routes, serve dual purposes. They aid in community movement during evacuations and support farmers and livestock keepers in their daily activities, like accessing farmlands or sourcing fresh grass for livestock.

Regarding four-wheeled vehicles, many residents own diverse minivans, including models like APV, Carry, and Avanza, along with trucks and pickup cars. These vehicles are not just transportation means; they become crucial evacuation tools capable of carrying multiple passengers during disasters. Trucks also assist sand miners by transporting material, while pickups prove handy for market runs and crop transportation. Similarly, two-wheelers play an integral role, especially during evacuations. One notable practice is the unique way residents ride motorcycles: sitting at the rear. This method, though unusual, is a tactical decision: it allows the motorbike to carry more passengers (often 3–4 individuals) or a greater volume of farm produce (milk, grass and vegetables). In emergencies, this space can facilitate quick evacuations.

Balerante Village equips itself technologically with an early warning system that utilizes satellite technology and mobile phones. This system is beneficial for providing real-time updates about Mount Merapi's status, ensuring residents can prepare and act promptly should eruption indicators emerge.

Balerante Village equips itself technologically with an early warning system that utilizes satellite technology and mobile phones. This system is beneficial for providing real-time updates about Mount Merapi's status, ensuring residents can prepare and act promptly should eruption indicators emerge. Meanwhile, the function of other technological devices such as mobile phones is to coordinate between volunteers and the community before, during and after a disaster. Meanwhile, other technological devices, such as cell phones, coordinate between volunteers and the community before, during, and after a disaster.

In Balerante Village, various resources are available to anticipate potential over-exploitation of natural resources and aid recovery after a Mount Merapi eruption. Social dynamics can sometimes shape the community's ease of access to this capital. Factors like age, gender, affiliations to community groups (such as farmer or UMKM groups), cultural customs, land and market ownership systems, and village government support can all influence resource utilization.

Individuals can elevate their contributions by enhancing their knowledge and skills regarding human capital. They can be active members in various community entities, from farmers' groups and UMKM collectives to Posyandu and PKK groups. Beyond these, there is an opportunity for community members to engage actively in organizations focused on disaster risk reduction and community resilience, spanning both risk mitigation and tourism development initiatives (Becker 2009).

Similarly, with financial and natural resources, in this case land, communities can utilize their land and capital to develop the village's local economy through cooperatives or BUMDes (North 2012). The existence of cooperatives or BUMDes can maintain production price stability and business continuity because BUMDes can play a role in facilitating the promotion of various community products. In addition, Bumdes has a savings and loan business that the community can utilize to encourage it to develop its business to survive despite the disaster. The community must manage the utilization of natural resources wisely to ensure their longevity and prevent social conflicts (Ostrom 1990).

Regarding social and technological resources, communities can utilize existing cultures or traditions to build strong relationships between communities and create conducive institutions. Advances in information technology must support all these activities to enhance communication access and network expansion (Putnam 2000). Castells also expressed the same opinion (Castells 2009).

To anticipate the shock and trends of Mount Merapi's eruption and exploit resources towards sustainable livelihoods, the community can *first;* increase public understanding and awareness of the Merapi eruption and map disaster-prone areas. These two factors are essential in supporting the development of community livelihoods related to the type and sustainability of the business (Manyena et al. 2019). A community that knows and understands the early signs of the eruption of Mount Merapi will find it easier to take precautions to reduce disaster risk. Mapping disaster-prone areas helps people understand vulnerability levels and decide on the type of business they should pursue to minimize the risk of failure.

Second; optimizing nature-based resources such as agricultural land, livestock, and eruption materials (sand) and utilizing changes in the Kaliworo landscape as a tourist destination. Among these various activities, the agricultural sector is the most widely cultivated sector by the community. On the other hand, the agricultural sector is also the most vulnerable in the event of an eruption, so livelihood diversification is an important strategy for mitigating economic risk (Holzmann and Jørgensen 2001). Business diversification is one of the strategies in post-eruption economic mitigation because diversification is the integration of various businesses whose purpose is, among others, to reduce dependence on certain business resources or financial sectors. Livelihoods that rely on one type of product increase the risks associated with fluctuations in market prices and demand, whereas diversifying into many different industries reduces these risks. Having diverse sources of income makes it easier for individuals and communities to respond and adapt to economic changes that may occur in the future. Some livelihoods and financial sectors may be more vulnerable to certain risks, such as natural disasters or regulatory changes. Diversification can help protect against these vulnerabilities by spreading sustainability risks across different areas (Caballero and Krishnamurthy 2003). *Third;* encourage communities to utilize non-natural resources by developing alternative businesses such as handicrafts or the service sector, including construction manual workers, masseurs, agricultural product collectors, grocery stores and food stalls. Diversifying livelihoods will create conditions where communities have many diverse sources of income, which will impact increasing community income and income stability. This diversification can help communities recover more quickly from disasters and reduce economic uncertainty. Rational and orderly use of land, forest and water resources will create environmental sustainability if utilizing these natural resources does not harm the environment and maintains ecological balance. Some concepts or principles that can be used to preserve the environment and resources are (1) agroforestry and forest restoration. Agroforestry is a profitable agricultural model that integrates annual staple crops with various woody plants and other plants. Implementation of this system will increase soil productivity and conservation, support biodiversity conservation, and improve water quality and the environment. Meanwhile, forest restoration is an effort to restore damaged or degraded forest ecosystems. This includes replanting native species and restoring ecosystem function (Suding et al. 2015). (2) Sustainable agriculture and land management. Good tillage, crop rotation, and appropriate inputs will reduce negative environmental impacts (Pretty 2008). Similarly, good land management, such as the establishment of terraces to prevent erosion, is also important for maintaining ecosystem functions (Lal 2015). (3) Water resources management, the principle of water use based on needs, protection, and conservation of natural water resources, efficient use of technology. Forest protection and restoration to maintain water flow and water quality is an important aspect of sustainable water management (Gleik 2003).

Recommendation

This research provides theoretical and practical recommendations on the livelihood system in the community. Theoretically, this research recommends developing Elis' (2000) theory. The livelihood system is based on the ability of assets such as social, human, financial, natural, and physical capital to diversify livelihood activities. Various livelihood assets owned by the community can be converted into livelihood diversification. Field findings show that most people have multiple livelihood patterns to survive. This research expands on Elis' (2000) theory that capital assets in the community can create livelihood diversification and multiple livelihood patterns in the community. Practically, this research recommends that environmentally friendly livelihood diversification be developed for the sustainability of community livelihoods and nature. This renewable and ecologically friendly livelihood sector needs institutional support and empowerment—for example, the development of arabica coffee commodities. Diversification of livelihood sources and multiple livelihood patterns can be applied at the same time in the case of coffee development by utilizing yard land. This research is also oriented towards a community action plan based on the results. The results of this research also encourage practical action on developing tourism businesses and village-owned enterprises as enablers of community empowerment in disaster-prone areas.

Conclusion

Diversification of livelihood sources is an alternative survival strategy in the Mount Merapi disaster-prone area of Balerante; by utilizing sustainable agriculture, MSME development, and sustainable tourism, the community can reduce economic risk and improve welfare. Sustainable livelihood systems depend on various resources owned, and each of these resources plays a different role before, during, and after a disaster. Social capital, the presence of trust, networks, and norms formed the basis for cooperation and support during evacuation and recovery in Balerante. Human capital was developed through disaster preparedness training and guidance from local authorities, increasing the community's ability to respond effectively to disasters. Financial capital is strong, with families diversifying income sources and allocating savings for various purposes. Natural resources, particularly land and water, are critical to community livelihoods, while responsible management and conservation are essential for disaster risk reduction. Physical capital, evacuation routes, and early warning systems support efficient disaster response and preparedness. Social relations, customs, and support from the village government influence access to such resources, which further enhances community adaptability. To anticipate resource trends and sustain recovery, communities should focus on disaster awareness, responsible natural resource utilization, and livelihood diversification. Cooperative efforts, responsible resource management, and community engagement are essential for long-term sustainability in the face of disaster risk.

Acknowledgements Our gratitude goes to the National Research and Innovation Agency (BRIN), the Education Fund Management Institution (LPDP) of the Ministry of Finance of the Republic of Indonesia and our gratitude goes to all those who have provided advice and input and became sources in writing this book chapter, including Balerante Village Officials, Disaster Risk Reduction Organization Management (*Organisasi Pengurangan Resiko Bencana*), BUMDes chairperson, Pokdarwis chairperson, community leaders and the Balerante Village community who cannot be listed one by one. All authors of this paper are key contributors. Rachmini Saparita is a research instrument designer, data analyst, and manuscript author; Djoko Puguh Wibowo a research designer, data collector, and manuscript author; Eko Wahyono as a theoretical and conceptual designer and manuscript author; Reza Amarta Prayoga is a literature reviewer and research problem formulator and manuscript writer; Fatwa Nurul Hakim and Nuzul Solekha as data collectors and data analysis manuscript writers; Siti Fatimah as a research methodology reviewer and manuscript writer; Lis Purbandini as a field research designer and manuscript writer.

Disclaimer
The views and opinions expressed in this article are those of the authors and do not reflect the official policy or position of any affiliated agency of the authors.

Informed Consent
This article does not contain any studies with human participants performed by any of the authors.

References

Abdulai IA, Ahmed A, Kuusaana ED (2022) Secondary cities under siege: examining peri-urbanisation and farmer households livelihood diversification practices in Ghana. Heliyon 8(9)

Aini LN, Isnawan BH, Hanudin E (2022) Land suitability re-evaluation for fruit crops in the post-eruption Merapi 2010. IOP Publishing, s.l., p 012034

Alemu FM (2023) Measuring the intensity of rural livelihood diversification strategies, and Its impacts on rural households welfare: evidence from South Gondar zone, Amahara regional state, Ethiopia. Methodsx 10:102191

Altieri MA (2009) The ecological role of biodiversity in agroecosystems. Agr Ecosyst Environ 74(1–3):19–31

Andreastuti SD et al (2023) Volcano disaster risk management during crisis: implementation of risk communication in Indonesia. J Appl Volcanol 12(1):3

Balerante D (2021) Profil Desa Balerante. Balerante, Desa Balerante

Becker GS (2009) Human capital: a theoretical and empirical analysis, with special reference to education. University of Chicago press, 3rd edn. University of Chicago Press, s.l.
Bhaskara GI, Filimonau V (2021) The COVID-19 pandemic and organisational learning for disaster planning and management: a perspective of tourism businesses from a destination prone to consecutive disasters. J Hosp Tour Manag 46:364–375
Braun V, Clarke V (2013) Successful qualitative research: a practical guide for beginners. SAGE Publication, London
Caballero RJ, Krishnamurthy A (2003) Excessive dollar debt: financial development and underinsurance. J Finance 58(2):867–894
Castells M (2009) The space of flows. In: The rise of the network society. s.l., Wiley, pp 407–459
Chambers R, Conway G (1992) Sustainable rural livelihoods: practical concepts for the 21st century. s.l., s.n
Coloman JS (1990) Fondation of social theory. BelknapPress, London
De Bauw P, Van Asten P, Jassogne L, Merckx R (2016) Soil fertility gradients and production constraints for coffee and banana on volcanic mountain slopes in the East African rift: a case study of Mt. Elgon. Agr Ecosyst Environ 231:166–175
Depari CD, Lindell MK (2023) Moving or not?: factors affecting community responses to environmental disruption. Int J Disaster Risk Reduction 95:103898
Ellis F (2000) The determinants of rural livelihood diversification in developing countries. J Agric Econ 51:289–302
FAO (2016) Analysing resilience for better targeting and action. Rome
Gleik PH (2003) Water use. Annu Rev Environ Resour 28:275–314
Google Maps (2020) Disaster-Prone area Merapi volcano. Google Maps and emergency operations center national disaster management agency. https://www.google.com/maps/d/edit?mid=158_rRxuyyXdU7wZRW8lNfwTHJtiCXWo&usp=sharing
Goyal N (2019) Disaster governance and community resilience: the law and the role of SDMAs. Int J Disaster Risk Manag 1(2):61–75
Holzmann R, Jørgensen S (2001) Social risk management: a new conceptual framework for social protection, and beyond. Int Tax Public Financ 8:529–556
Hossain MR (2024) Evaluating household's risk management strategies considering heterogeneities in the livelihood diversification: evidence from farm households of Bangladesh. Soc Sci Humanit Open 9:100785 https://bnpb.go.id/berita/identifikasi-posisi-terhadap-krb-gunung-merapi-gunakan-cekposisi, 10 Nov 2020
Hsu K, Peng L-P (2023) Understanding vulnerability and sustainable livelihood factors from coastal residents in Taiwan. Mar Policy 155:105793
Huhtamaa H, Stoffel M, Corona C (2022) Recession or resilience? Long-range socioeconomic consequences of the 17th century volcanic eruptions in northern Fennoscandia. Climate Past 18(9):2077
Irsyad M, Irwan SNR, Budiani SR (2020) Strategi Mencapai Penghidupan Berkelanjutan Pada Sektor Pariwisata Suku Tengger di Taman Nasional Bromo Tengger Semeru. J Kepariwisataan: Destinasi, Hospitalitas Dan Perjalanan 4(1):11–28
Islam R, Walkerden G (2022) Livelihood assets, mutual support and disaster resilience in coastal Bangladesh. Int J Disaster Risk Reduction 78:103148
Lal R (2015) Restoring soil quality to mitigate soil degradation. Sustainability 7(5):5875–5895
Lestari P, Hapsari W, Raharjo B (2019) Potensi wisata alam Gunung Ijen setelah erupsi tahun 2018. J Sains Dan Teknologi Lingkungan 11(2):142–151
Liebman M et al (2016) Ecologically sustainable weed management: how do we get from proof-of-concept to adoption? Ecol Appl 26(5):1352–1369
Manyena B, Machingura F, O'Keefe P (2019) Disaster resilience integrated framework for transformation (DRIFT): a new approach to theorising and operationalising resilience. World Dev 123:104587

Mbiba M, Collinson M, Hunter L, Twine W (2019) Social capital is subordinate to natural capital in buffering rural livelihoods from negative shocks: Insights from rural South Africa. J Rural Stud 65:12–21

Miles MB, Huberman AM (1994) Qualitative data analysis: an expanded sourcebook. Sage, Thousand Oaks, California

Molénat J et al (2023) Diversification from field to landscape to adapt Mediterranean rainfed agriculture to water scarcity in climate change context. Curr Opin Environ Sustain 65:101336

North D (2012) Institutions, institutional change and economic performance. Cambridge University Press, s.l.

Ostrom E (1990) Governing the commons: the evolution of institutions for collective action. Cambridge university press

Peng W, Robinson BE, Zheng H, Li C, Wang F, Li R (2022) The limits of livelihood diversification and sustainable household well-being, evidence from China. Environ Dev 43:100736

Pretty J (2008) Agricultural sustainability: concepts, principles and Eviden. Philos Trans Royal Soc B: Biol Sci 363(1491):447–465

Putnam RD (2000) Bowling alone: the collapse and revival of American community. Touchstone Books/Simon and Schuster, s.l.

Rahayu et al (2014) Dampak erupsi Gunung Merapi terhadap lahan dan upaya-upaya pemulihannya. Caraka Tani: J Sustain Agric 29:61–72

Roy SBA, Basak D, Chowdhury IR (2023) Towards sustainable society: the sustainable livelihood security (SLS) approach for prioritizing development and understanding sustainability: an insight from West Bengal, India. Environ, Dev Sustain

Safitri D, Fajarwati A (2015) Kajian Dampak Erupsi Merapi Terhadap Perubahan Penghidupan Masyarakat Tani Desa Ngargomulyo Untuk Pengembangan Konsep Sister Village Dalam Manajemen Bencana. J Bumi Indonesia 4(4):228315

Shackleton RT, Shackleton CM, Kull CA (2019) The role of invasive alien species in shaping local livelihoods and human well-being: a review. J Environ Manage 229:145–157

Small C, Naumann T (2001) The global distribution of human population and recent volcanism. Global Environ Change Part B: Environ Hazards 3(3):93–109

Steenbergen DJ, Marlessy C, Holle E (2017) Effects of rapid livelihood transitions: examining local co-developed change following a seaweed farming boom. Mar Policy 82:216–223

Suding KN et al (2015) Committing to ecological restoration. Science 348(6235):638640

Sullivan GB, Sagala S (2020) Quality of life and subjective social status after five years of Mount Sinabung eruptions: disaster management and current sources of inequality in displaced, remaining and relocated communities. Int J Disaster Risk Reduction 49:101629

Uddin MS et al (2021) Disasters threaten livelihoods, and people cope, adapt and make transformational changes: community resilience and livelihoods reconstruction Bangladesh. Int J Disaster Risk Reduction 63:102444

Umaya R, Soekmadi R, Sunito S (2020a) Direct economic benefits and human dependence toward Gunung Merapi National Park, Indonesia. Biodiversitas J Biol Divers 21(3):982–993

Umaya R, Soekmadi R, Sunito S (2020b) Livelihood adaptation patterns of sub villages community in the slope of Merapi Volcano. 528(1):012020; IOP Publishing. s.l., s.n., p 012020

UNDRR (2017) Vulnerability. [Online] Available at: https://www.preventionweb.net/understanding-disaster-178risk/component-risk/vulnerability. Accessed 24 May 2023

UNWTO (2015) Global report on adventure tourism. World tourism organization (UNWTO)

Chapter 28
Strengthening Local Food Security for Climate Change Adaptation: Lessons from Merauke-Papua Indonesia

Djoko Puguh Wibowo, Dodi Al Vayed, and Bernadin Dwi Mardiatmi

Abstract Researchers, governments, rural communities, and international institutions are increasingly focusing on local food security as a means of adapting to climate change. The focus has been on the role of formal institutions and donors in supporting these communities to adapt to climate change collectively. This study took place in Salor Indah Village (kampung), Kurik District, Merauke Regency, Papua Province, selected purposively selected for its characteristics. The aim was to deeply understand how local communities proactively and collaboratively build food security in response to climate change. Through qualitative methods such as focus group discussions and in-depth interviews with local and community leaders, as well as selected farmers, this study sought to capture the leadership characteristic in response to climate change, the community's understanding of climate change and its impacts, and their collaborative efforts in establishing local food security. The core findings highlight the importance of community networks, effective information sharing, robust local leadership, active community involvement, trust, and the delegation of authority as fundamental components in crafting community-based local food security.

D. P. Wibowo (✉)
Research Center for Social Welfare, Villages, and Connectivity, National Research and Innovation Agency, Jakarta, Indonesia
e-mail: djok022@brind.go.id

D. Al Vayed
Master of Government and Administrative Affairs, Jusuf Kalla School of Government, Muhammadiyah University Yogyakarta, Yogyakarta, Indonesia

B. D. Mardiatmi
Faculty of Economics and Business, Universitas Pembangunan Nasional Veteran Jakarta, South Jakarta, Indonesia
e-mail: bernadindwim@upnvj.ac.id

U. Mukhopadhyay et al. (eds.), *Climate Crisis, Social Responses and Sustainability*, Climate Change Management, https://doi.org/10.1007/978-3-031-58261-5_28

Introduction

Climate change poses a grave threat to global food security. Its effects can severely impede nations' economic and social development, especially those with lower resilience against extreme weather events (Leisner 2020). Factors like the escalating severity of weather extremes and changes in rainfall patterns can compromise food security, damage infrastructure, and worsen poverty (Muluneh 2021). Those entangled in the cycle of poverty, relying heavily on agriculture and natural resources, are especially at risk from climate-induced disruptions (Ali et al. 2017). Unpredictable growing seasons, declining agricultural yields, and heightened disaster risks can erode their livelihood, thrusting more individuals into entrenched poverty (Karimi et al. 2018). The World Bank underscores the profound developmental implications of climate change, notably its catastrophic effects on the impoverished in various global regions (Castells-Quintana et al. 2018). They advocate for international cooperation to counteract the developmental setbacks climate change imposed (Wellbery et al. 2018). This framework emphasizes the significance of financial and technical assistance to bolster developing nations' struggle for food security (Masipa 2017).

Amid the present global landscape—shaped by climate change, population surges, urbanization, and evolving food consumption patterns (Islam and Wong 2017)—ensuring local food security has become the top priority of many nations' agendas (Kline et al. 2017). Food security refers to a region's capability to sustainably meet its population's food demands in quality and quantity, minimizing reliance on excessive imports (Campos and Ortiz 2019). Strengthening local food security requires multifaceted strategies: agricultural diversification, productivity boosts, sustainable resource management, infrastructural development, and empowering local farmers and communities (Sulaiman et al. 2019). Such initiatives guarantee consistent food access, fortify local economies, alleviate poverty, and achieve environmental conservation through sustainable farming (Mabhaudhi et al. 2019). In the face of global and local challenges, strengthening food security is important in creating a more sustainable and food-secure future for all communities (de Amorim et al. 2019).

Furthermore, it's vital to underscore that adaptation approaches must align with a broader sustainable development framework. This step entails blending local knowledge with modern scientific and technological insights (Xu et al. 2018). Such collaboration can produce resilient and diverse adaptation methods, ensuring decisions are rooted in a comprehensive understanding (Smith and Lawrence 2018). Incorporating local knowledge into climate change adaptation policies benefits the immediate communities and has positive global implications. It strikes a harmony between environmental sustainability and human welfare (Jaakkola et al. 2018). In Indonesia, the Merauke region in Papua exemplifies the unique challenges and prospects surrounding food security (Setiani et al. 2021). It portrays the multifaceted dynamics inherent in strengthening local food resilience (Lasminingrat and Efriza 2020).

Merauke, characterized by its remote frontier setting, presents distinct geographical and societal features (Kause et al. 2013). While blessed with a wealth of natural resources encompassing vast lands and rich biodiversity, the region struggles with adequate food access (Wambrauwa et al. 2022). Constrained infrastructure, a lack of sustainable farming expertise, and a disjointed agricultural-food system have attributed to the challenges to Merauke's food accessibility. However, Merauke's journey provides invaluable insights into strengthening local food security. Pioneering endeavors in the region include sustainable farming practices, crop diversification, and the reinforcement of local market networks (Fitriani and Maturbongs 2021).

The experiences from Merauke offer valuable insights into enhancing local food security, which can be applied in Indonesia and globally. Analyzing Merauke's journey reveals key elements for successful food security initiatives: active community involvement, cross-sectoral collaboration, the application of relevant technology, and the conservation of indigenous knowledge (Holle 2020). The strategies adopted in Merauke serve as an instructive model, emphasizing the significance of ensuring sustainable food sources amidst looming challenges (Kamin and Altamaha 2019).

Several studies have discussed local food security, previous research conducted by (Deininger et al. 2023) which discusses Quantifying war-induced crop losses in Ukraine in near real time to strengthen local and global food security. Then the research conducted by (Guell et al. 2022) which discusses Perspectives on strengthening local food systems in Small Island Developing States. Research related to food security was also researched by (Susanti and Sopia Rukmana 2020) who explained about going local for food security: strengthening local collaborative institutions in the implementation of food security in Indonesia. After that, research was also conducted by (Bouey 2020) which discusses Strengthening Local Food Systems in Times of Concomitant Global Crises: Reflections From Chile.

This study aims to re-narrate the experiences of Salor Indah Village residents in the Kurik District, Merauke Regency, and their endeavors to fortify local food security as a climate change adaptation strategy. Distinct from prior studies that predominantly centered on agronomic adaptation, this research paints a detailed portrait of how local leaders, endowed with intrinsic authority, respond to climate change, and orchestrate community-based food security initiatives. It further explores the collective strategies employed by farmers and institutions in response to climate change threats. In the broader scope of Indonesia's climate challenges, this research underscores the role of village governments in fostering community resilience through native authority, in line with the mandates of Law No. 6/2014 on Villages and its associated regulations.

Material and Methodology

This study employed a qualitative method, engaging local leaders and farming communities through detailed interviews and group discussions. We interviewed village chiefs, officials, heads of village-owned enterprises (BUMDES), and leaders of farmer group associations. Selected farmers also contributed their insights.

Table 28.1 Participants and number of respondents

No.	Participant	Status	Total
1	Village head		1
2	Village apparatus		1
3	Chairman of Bumdes		1
4	Chairman of the farmer group association		1
5	Farmers		10
	Total		14

Source data processing results, 2023

Through these methods, we delved into their understanding of climate change, their responses to its impacts, and their strategies for developing food security (Table 28.1).

This research relies on both primary and secondary data sources. Primary data comprises transcripts from detailed interviews with local leaders, capturing their experiences with climate change impacts and their efforts in building community-based resilience. Additionally, group discussions with farming communities provided insights into their understanding of climate change and their rationale for participating in food security initiatives. Field observations further complement the primary data, offering systematic recordings of phenomena arising from climate change.

Our team reviewed relevant literature for secondary data and organized all collected data into field notes. Field notes became a classification of key postulates of food security and community adaptation to climate change. We extracted these key postulates from informant interviews and field observations. To ensure accuracy and depth, we employed a process of conceptual triangulation. This process required field and conceptual findings comparison, using coding and data reduction based on these key postulates. The field notes are the primary source of data from interviews and observations that contain verbatim transcriptions of interviews and observation notes. As well as the research methodology diagram used as follows (Fig. 28.1).

Figure 28.1 Methodological Flowchart.

Literature Review

Climate Change and Food Security

Climate change refers to long-term shifts in weather patterns and average regional temperatures influenced by human activities and natural phenomena (Prakash 2021). On a local scale, these changes can have profound implications for food security (Gomez-Zavaglia et al. 2020). Local food security is the capacity of a community to ensure consistent access, availability, and nutritionally adequate food through localized efforts (Udmale et al. 2020). The connection between climate change and local food security becomes especially pertinent for communities reliant on natural

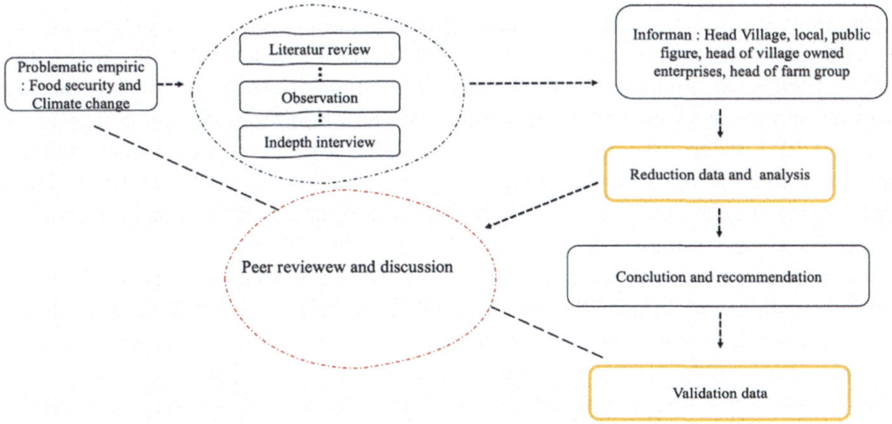

Fig. 28.1 Research methodological flowchart

resources and regional agriculture (Ani et al. 2022). Fluctuations in average temperatures and rainfall can upset traditional farming schedules and impact local agricultural output, leading to potential crop failures and destabilizing food supplies (Wicaksana and Rachman 2018). Additionally, extreme weather events pose risks to agricultural infrastructure and stored food (Praveen and Sharma 2019).

Community-based adaptation emphasizes the local nuances in addressing climate change's effects on food security ((Akanksha) Patnaik 2021). Local knowledge, agricultural traditions, and native wisdom are crucial in formulating strategies for climate-resilient food security (Amoak et al. 2022). Strategies might include introducing farming methods suited to unpredictable weather, cultivating climate-resistant crop varieties, and diversifying food sources (Haque and Khan 2022). However, crafting sustainable local food security in the face of climate change is not without challenges. Factors like limited resources, societal shifts, and local economic dynamics can hinder a community's adaptability (Mbuli et al. 2021). Collaborative efforts among local governments, research institutions, communities, and other stakeholders become urgent. Together, they can shape a supportive framework for evolving adaptive and sustainable local food security (Aggarwal et al. 2018).

Community Food Reserve and Local Leaders

Food reserve or "*lumbung pangan*" is a concept closely related to the provision and distribution of food in a community (Nandi et al. 2021). The idea involves collecting and storing crops or other food resources for equitable distribution to all community members (Osabohien et al. 2020). Food reserves aim to address fluctuations in food supply due to weather changes, natural disasters or other factors affecting food production. (Duchenne-Moutien and Neetoo 2021). Dalam konteks lumbung pangan, peran pemimpin atau tokoh masyarakat lokal sangatlah penting (Béné 2020). In food reserves, local leaders or community figures are crucial (Murrell and Jones 2020). They act as organizers of food distribution, make strategic decisions regarding storage, and ensure sufficient food availability for all community members (Reeve et al. 2020). Leaders also have a role in mobilizing active community participation in maintaining the sustainability of food reserves (Andrew et al. 2022). By building collective awareness about the importance of food reserves, leaders can create strong social ties among community members (Panday et al. 2021).

Local leaders also play a role in balancing food distribution and social justice (Amorim-Maia et al. 2022). They must ensure equitable food distribution, avoiding extreme disparities among community members (Xia et al. 2022). In addition, leaders also play a role in developing an efficient and transparent food storage management system so that people have total trust in the distribution process (Bublitz et al. 2019). The involvement of leaders can also contribute to the empowerment of local communities (Arroyo et al. 2019). Leaders can facilitate collective learning and skills development by involving community members in decision-making related to community food reserve (Raidimi and Kabiti 2019). This involvement can increase the community's capacity to address food challenges and improve long-term food security (Galli et al. 2020).

Overall, community food reserves are a concept that plays an important role in maintaining community food security. The role of local leaders or community leaders in managing food reserves is crucial in distribution arrangements, community empowerment, and building solid social ties. Community food reserves can effectively deal with food fluctuations and sustain community welfare with solid cooperation between leaders and communities.

Result and Discussion

Village's Local Resources Capacity

Salor Indah is a village located in the Kurik District of Merauke Regency, Papua Province. Established in 1980, the village was one of 11 transmigration settlement units (UPT), with some of its settlers originating from Java Island (Andriati et al. 2004). During the development regime, transmigration in Papua served a dual purpose: it was intended to spur regional development through population distribution and capitalize on regional strengths such as food, plantations, forestry, and fisheries. This strategy aimed to expedite progress in Eastern Indonesia. In line with this vision, Salor Indah and similar villages, with their wetland agroecology, were designed to majorly cultivate food crops, particularly rice (Saleh 2005). These areas were earmarked to become significant food reservoirs for the broader Papua region. For irrigation, the village relies on nearby swamplands and river systems. Salor

Indah benefits from its proximity to the Apar and Wapeko Swamps and the Neta River, which provide essential water for agricultural endeavors.

Designed as a food reserve hub, Salor Indah boasts 370.5 hectares (ha) of paddy fields and an additional 500 ha of dry land. However, many villagers have expanded their cultivation into dryland paddy fields outside the village, often through purchases from neighboring communities. Over the years, the village has seen a rise in agricultural initiatives, encompassing 23 farmer groups that represent all farmers in Salor Indah. The Village-Owned Enterprises (BUMDES) play a pivotal role as local economic institutions. Bolstering agricultural ventures in Salor Indah are vital infrastructures like roads and irrigation systems, complemented by a full suite of agricultural equipment, including tractors, harvesting machines, grass choppers, and water pumps.

The introduction of Law No. 6/2014 on Villages marked a significant turning point for village communities in Indonesia. With this Law's adoption, villages gained the autonomy to tailor programs and initiatives based on the specific needs of their communities. The criteria for such authority emphasize service activities and community empowerment. Moreover, the law ensures that initiatives are confined to the village's jurisdiction, impacting the village residents directly and addressing their unique requirements. As a result of this autonomy, villages can undertake vital activities, including adapting to climate change, enhancing food security, and establishing local food reserves, as these align with the community/s needs and fall within the village's purview.

Local Leaders Who Understand Climate Change

To learn the capacity of local leaders—including the Village Chief, Head of Planning Affairs, BUMDES Manager, and Chairman of the Salor Indah Farmer Group Association—we initiated our interview by inquiring about their perceptions and observations of environmental shifts due to climate change. Subsequent questions explored their collective plans and processes for enhancing food security.

The Village Head of Salor Indah, a second-generation transmigrant from Java with a high school education, embodies the philosophies of the Javanese. Holding firm to his principles, he remains committed to steering the community away from the food shortages they experienced during their initial years in Salor Indah.

> As a leader of the Salor community, I don't want to hear little children crying over their parents don't have food supplies at home. (Salor Indah Village Chief)

In this context, the village chief exemplifies leadership and deep concern for the community's fundamental needs, focusing on ensuring an adequate food supply. The statement reflects a genuine worry for the well-being of children, aiming to prevent any potential hardships they might face due to food shortages at home.

This underscores the leader's unwavering commitment to safeguarding the community's welfare, especially for those more vulnerable, such as children. As a second-generation leader, the chief vividly recalls the environmental and social transformations experienced by the Salor Indah community. Despite the community's tangible experiences, the Salor Indah's chief does not recognize these changes as indicative of climate variability. Although these changes are real and impactful, the community lacks official information from institutions that should serve as authoritative sources regarding climate change, its impending consequences, and strategies for adaptation.

> In the last five years, our swamps have dried up, and river water levels have gone down, insufficient for irrigating rice fields. During dry spells, the murky, brown river water signifies increased salinity, rendering it unsuitable for crops. I've observed a significant dip in rice yields compared to our early years here. Recognizing these challenges, as a village chief, I plan to prioritize climate change mitigation and food security as key initiatives. (Salor Indah Village Chief)

Over the past five years, Salor Indah has undergone substantial environmental transformations. The once lush swamps have dried up, and the river's water discharge has significantly diminished, posing challenges in supplying water for rice paddy irrigation. Particularly during the dry season, the river water exhibits a disconcerting murky brown hue, signaling a decline in water quality. This predicament is further exacerbated by an escalation in salt content, rendering the water unsuitable for crop irrigation and posing severe impediments to local agriculture. Numerous community members have expressed grievances about diminished production, attributing it to drought, water scarcity, and pest infestations.

> On various occasions, I've highlighted the climatic changes in Salor Village. Our ancestors referred to this as 'salah mongso'. Previously, such anomalies spanned 1–2 years. However, the persistent challenges we face today, from altered rainfall patterns and frequent pest attacks to declining yields and crop failures, stem from more consistent climate changes. (Chairman of the Farmer Group Association)

Although the changes are real and experienced by the community, there has been no information from official institutions or offices at the district level that should act as a source of information and convey about climate change, the impacts that will occur and strategies to adapt. Instead, the community received information about climate change from researchers who conducted studies in Salor Indah about eight years ago.

> We and the Village-Owned Enterprises (BUMDES) Manager, the Head of the Farmer Group Association, and the Community Leaders received complete information after consulting with researchers from the center (Jakarta) who had visited Salor Indah. In detail, they explained that the phenomenon occurring in Salor Indah results from climate change, namely changes in temperature and rainfall that impact agriculture and disrupt the community's food security. Furthermore, they pointed out that villages can allocate budgets and design activities according to community needs, including adaptation to anticipate climate change, which is feared to have an increasing impact in the future. (Salor Indah Village Chief)

Since the government enacted Law No. 6/2014 on Villages, as independent entities, villages have the authority to develop development plans to implement programs and activities following community needs. Every year, the relevant Ministry issues regulations supporting this policy, empowering village governments to allocate budgets that address their community's specific issues and challenges. Priority programs and activities include climate change, food security, extreme poverty, and stunting elimination.

> Based on the local authority of the village scale, we as village officials will make food security programs and activities a village development priority. However, I realize that this process is not easy and can no longer be handled by the village government but must actively involve community leaders and farmer groups and place BUMDES as a village economic institution to support the needs of business capital, agricultural facilities, and infrastructure. (Village Secretary of Salor Indah Village)

As a form of commitment and authority, the policy taken for adaptation is to mainstream climate change and food security in village development plans. Budget priorities are allocated to support adaptation activities and build food security. Village officials are committed to making food security a priority for village development. While recognizing resource constraints, village officials actively engage community leaders and farmer groups and position BUMDES as the village's economic institution to support capital needs. This approach emphasizes collaboration to ensure the success of food security programs at the village level.

> Due to integrating food security as a village development priority, I had to adjust the planning of the village development budget (APBDes), which was different from previous years. The portion of the budget allocation is no longer for physical development. Still, it is prioritized for strengthening the capacity of farmer groups, building and repairing agricultural infrastructure, increasing BUMDES business capital, empowering women's groups, and repairing buildings for storage facilities for production as food reserves. (Head of Planning Affairs of Salor Indah Village)

As a village economic institution, BUMDES has an important role in supporting various economic activities of the Community, including activities supporting climate change adaptation (Ansori 2022). The existence of BUMDES in the village area opens and gives farmers access to business capital loans to get fertilizers, pesticides, and quality seeds and fertilizers to higher quality (Helwig 2021). Investment opportunities from international investors and donor agencies willing to support climate-smart adaptation efforts (Gcf and Factsheet 2021).

When asked by BUMDES administrators about the impact of climate change on the agricultural business unit, they stated that the business is already experiencing difficulties in providing capital loans and production facilities during the planting season.

> The onset of drought, pest attacks and declining rice crop production have caused losses and difficulties in starting their businesses because the production costs are getting more expensive. This hindrance causes BUMDES to no longer be able to help farmers who lack business capital to pay labor wages and purchase superior seeds, fertilizers, and pesticides whose prices are increasing. To overcome this problem, the Village Government has approved the request of the BUMDES management to increase capital to enlarge the agricultural business to help farmers who lack business capital. (Chairman of BUMDES Salor Indah)

The Head of the Farmer Group Association acknowledged the challenges faced by many members, including difficulties in securing additional business capital and buying production facilities at the store managed by BUMDES. During discussions, he proposed a one-year planting plan detailing planting patterns and fertilizer needs. The document contains a plan for planting patterns fertilizer needs as a basis for providing production facilities and capital loans within one year.

> Considering the impact of climate change, which is feared to increase in the future, we as the Chairman of the Farmer Group Association have invited farmers to do through uniform cropping patterns, the use of certified superior seeds, storing food reserves at the household and group levels as our parents did in the past. (Chairman of the Farmer Group Association)

Evaluating the perspectives of Salor Indah Village's local leaders, it's evident that they are attuned to the environment's gradual changes and potential future impacts. The village head, representing an official local institution, has shown responsiveness towards adapting to climate change. He articulated plans to weave climate change adaptation and food security into village development, implemented through community-based collective actions. Similarly, non-formal local leaders, including the Heads of the Farmer Group Association, Water Management group, and BUMDES, acknowledged the tangible impacts of climate change on agricultural production in the Salor region. They believe that anticipating climate change effects requires collective action, backed by the Village Government and the farmer members of the Farmer Group. These insights align with prior research (Mubaya and Mafongoya 2017; Vignola et al. 2017), emphasizing the importance of local leadership adept at discerning technical and social-managerial challenges. Such studies underscore the important role of community-driven collective actions, facilitated by informed local leaders, in successful climate change adaptation (Adger 2006; Gislason et al. 2021).

Community Knowledge on Climate Change Adaptation

We obtained information about climate change's impact on the Salor Indah community in Kurik District, Merauke Regency, Papua, and its effects on food security through in-depth interviews with 10 participants. These participants included the head of the Farmer's group, the Water Management Group, the Head of RT, and individual farmers. The Salor Indah Village community depends on agriculture with rice commodities. In one year, the community endeavors to plant rice at least three times a year without a break in planting. The first planting season starts in December and harvests in March/April—the second planting season in April and yields in June. The third planting season is in July, and harvests in September. However, changes in climate variability are perceived to be increasing, resulting in drastic decreases in rice production. Rising air temperatures cause swamp water pools around settlements to drop dramatically, so they cannot sustain rice cultivation for three crops yearly. During the dry season, the saltwater content is high when the tide flows into the river.

Meanwhile, pest attacks always occur during the dry and rainy seasons before the harvest season.

Some adaptation models that have been carried out by farmers are no longer able to maintain rice production. Even if they can harvest, the results are not balanced with the production costs incurred. The community is worried that climate change will continue to recur, and its impact will increase if adaptation is not carried out by involving all components of the community.

> Increasing temperature changes, drought, water scarcity, and pest attacks have a huge impact on rice production compared to the period in the 80s at the beginning of transmigrant placement in Salor Indah. In 1980–1990, the rice yield per hectare was approximately 60–70 sacks, equivalent to 6000–7000 dry grain. Currently, the yield per hectare is no more than half of that. When compared to the costs incurred, farmers suffer losses. (Chairman of the Farmer Group Association)

Farmers felt that seawater intrusion into the river caused an increase in salinity and disrupted plant growth during high tides. Meanwhile, temperate changes cause the dry season to lengthen, drastically reducing water reserves in the swamps surrounding the settlements.

> When the tide enters the Neto River as a water source for growing rice, it is brownish white, contains high levels of salt, and cannot irrigate the rice planted. Once I tried to keep the river water flowing into the rice fields; the plants grew stunted, some died, and the harvest failed. (Farmer Group Association Member)
>
> Since three years ago, the swamp water around the settlement, which used to be full of water throughout the year, is drastically reduced in the dry season and is insufficient for cultivating rice three times a year. (Farmer Group Association Member)

Changes in climate variability have also led to increased pest attacks. Additional doses of pesticides and herbicides can no longer withstand the rate of pest and disease attacks, which are growing every year. Likewise, using superior seeds and altering patterns help combat the rate of pest and disease attacks. However, production drops and the outcomes do not justify the costs.

> Currently, pest and disease attacks occur at the end of the dry season and before harvest. Although I have used superior seeds and increased pesticide dosage when spraying. (Farmer Group Association Member)
>
> To anticipate pest attacks, group members have tried to overcome the problem by changing cropping patterns, selecting crop varieties that are claimed to be resistant to the long dry season, and using pesticides by increasing larger doses. The reality is that sometimes these efforts are successful, but the yields are low compared to those of the previous three years and not worth the production costs. Some villagers have even experienced crop failure. (Head of Farmer Group Association)

Adopting altered cropping patterns and intensification is an adaptive strategy for the Salor community amidst the uncertainties of seasonal shifts. Extended dry periods have rendered seasonal transitions unpredictable. To make matters worse, climate change has disrupted the community's ability to forecast the seasonal calendar accurately.

> To anticipate the long dry season and lack of water, I tried to change the planting pattern, use new varieties of seeds and increase the dose of fertilizer. The results were still no better than the previous year and not worth the costs. I don't know who to ask to cope with the current changes. (Farmer)
>
> As the head of the water management group, until now, the community and I have never received counseling or information from either the Field Extension Officer (PPL) or the Agriculture Office regarding these various changes that have hit this village. (Chairman of Salor Indah Farmers Group Association)

The community's adaptation strategy for changing cropping patterns and intensifying crop types is considered an adaptive way to respond to unpredictable seasonal changes. The long dry season has made the farming community in Salor strengthen the basis of farmer group collectivity to respond to contingent seasonal changes. The addition of fertilizers and water management is a surefire way for the Salor Indah farming community to be adaptive to seasonal changes. Increasing the productivity of farming crops no longer provides immense benefits like ten years ago. Finally, changes in farming behavior patterns are the final way to minimize losses from crop yields.

Food Reserve Building Process for Food Security

We gathered information about the community food reserves construction process from a group discussion involving 10 participants. This group included the village head, the BUMDES leader, the head of the farmer group association, and seven farmers actively engaged in the food reserve's development. The village head emphasized that the decision to build the reservoir was collective, not determined solely by the village government. The decision-making process was comprehensive, involving community consensus achieved in village meetings and subsequent approval from the district government at district-level gatherings,

> In the village deliberation forum, the community has reached an agreement to establish a food reserve to overcome food vulnerability caused by climate change. The management of the community food reserves is fully handed over to the community. Through the local food security program, the village government will provide support through business capital assistance for Village-Owned Enterprises (BUMDES), budget allocations for training, repairing irrigation channels, providing agricultural equipment, and various empowerment activities. (Salor Indah Village Chief)
>
> I informed the community during village deliberations that the district-level meeting almost canceled the Salor Indah Village Food Security Program funded by the Village Fund due to skepticism about climate change. Despite feeling its impact on food security, the forum believed Salor always had a yearly harvest. Emphasizing the village's authority in program development, I clarified that the current focus is on food security. Eventually, the forum approved the proposal, suggesting BUMDES business capital support specifically for aiding business and providing production facilities for farmers. (Salor Indah Village Chief)

During the village deliberation forum, the community reached a consensus to establish a food barn as a strategic response to food vulnerability stemming from climate change. The responsibility for managing the food barn was entrusted to the community, supported by the village government through capital assistance for BUMDES, allocations for training, repair of irrigation systems, provision of agricultural equipment, and overall empowerment initiatives. However, at the district level, there was initial hesitation to proceed with the proposed Food Security Program, largely due to the forum's skepticism regarding climate change. To address this concern, the village head clarified that the village possesses the authority to spearhead a food security program. Eventually, the forum embraced the proposal, underscoring the importance of BUMDES capital support to assist farmers, with a specific focus on business capital and production facilities. In essence, the community unanimously agreed to adapt to climate change by implementing a food storage initiative, utilizing harvests as the primary source, and entrusting the management to the community, as suggested by the farmer group.

> I don't mind if every harvest, I have to store about two quintals of the harvest, some of which is stored at home, and the rest is stored in the community food reserves and released during the next harvest. (Farmer Group Association Member)
>
> To avoid fraud, the management of community food reserves should be handed over to people trusted and appointed by the community. I also have no objection to storing part of the harvest at the household level and part at a designated place. (Farmer Group Association Member)

Farmer Group members agree to keep a portion of the harvest (about 2 quintals) each harvest. Part of it is held in the household, and the rest is stored in the food granary to be released at the next harvest. The importance of avoiding fraud was recognized, and they suggested that the management of the food granaries be left to trusted individuals appointed by the community. Established procedures can also store Some of the harvests in the household. The farming community agrees that the community manages the food granaries openly, and the village government, through BUMDES, will provide business capital loans and superior seeds, fertilizers, and medicines on time.

> With the storage model at home and group food reserves, the certainty of getting capital assistance and production facilities from BUMDES, I feel safe. My family's food availability is guaranteed throughout the year. (Farmer Group Association Member)
>
> I am more at ease with the food reserve and the BUMDES plan to provide business capital and serve the needs of agricultural facilities every planting season. Given that so far, we have had difficulty getting good seeds and fertilizer at the beginning of the planting season. Even if there is a limited amount and the price is expensive. (Farmer Group Association Member)
>
> To ensure the continuity of agriculture throughout the year, we will ask field extension officers to assist farmers and farmer groups to develop group work plans. I remember in the past, farmers and farmer groups have been asked to develop a work plan that includes the needs of production facilities (seeds, fertilizers, pesticides) at planting time. The work plan makes it easier for BUMDES to provide production facilities throughout the year. (Head of Farmer Group Association)
>
> With the launch of the Da Salor Indah Food Security Program, the task of the agricultural business unit that is a BUMDES is getting harder. In order to serve the needs of farmers in the right amount and on time, we have asked the community through their farmer groups to

prepare a plan for capital needs and spares for each planting season and an estimate of the remaining production that will be sold through BUMDES. (Chairman of BUMDES).

This study resulted in learning that climate change has been felt to have a detrimental impact and must be addressed together. This awareness cannot be separated from the role of local leaders who continue to communicate about the realities in their areas that endanger community food security. As a formal leader, the village head also realizes that maintaining food security cannot be imposed on the community or borne by the village government. Therefore, the policy is to integrate resilience and climate change into the village development plan. This commitment is reflected in allocating village development funds for various supporting activities such as community capacity building, improvement of agricultural facilities and infrastructure and financial assistance to BUMDES. Delegation of management to the community is one of the strategies to build trust and give responsibility to the community to make rules together. BUMDES, as a community-based economic institution, is one of the essential pillars that plays a role in supporting agricultural activities and the shelter and marketing of production products.

The construction of the Salor Indah Village Community Food Reserves in Kurik District, Merauke Regency, is evidence that climate change has complex impacts and affects the agricultural sector, encouraging communities to adapt to build local food security. Drought, changes in rainfall patterns, and seawater intrusion have led to decreased rice production and crop failure. Communities have undertaken various adaptation efforts but have struggled to mitigate the escalating impacts of climate change. This struggle exists alongside a growing awareness and a collective resolve to counter food insecurity by developing food reserves. The growth in understanding owes much to local leaders who consistently document the effects of climate change over time and actively communicate the realities of these changes and the potential consequences of not addressing them collectively.

The results of our research show that local leaders, especially the village head of Salor Indah in Kurik District, are a mirror of local leaders who are responsive to climate change and can predict its impact on food security in their area. Through a participatory and tiered process, based on its authority, the village government, as a formal institutional representation, sets a policy that integrates the strengthening of local food through food reserves as a form of climate change adaptation. With budget support, programs and activities are designed inclusively, involving farmers and farmer groups—communities and BUMDES as economic institutions as actors by their functions and capacities. Management is designed transparently to build trust, not burdensome and thoroughly carried out and controlled by the community. These principles are considerations, and the main components of the granary serve as a platform for the parties involved on how they should work and the benefits to be gained from building food security collectively (See Table 28.2).

Establishing a community food reserve to store collective agricultural produce serves as a resilient livelihood instrument. For subsistence farmers in Salor Indah, strengthening food security resilience through such collective storage presents a strategic alternative. This approach encourages farmers to economically evaluate

Table 28.2 Components of the Salor Indah food reserve Kurik district Merauke regency

No	Component	Description
1	Goal	Ensure food availability and food accessibility throughout the year
2	Principle	Independent, collective and transparent
3	Actor	
	a. Village government	Integration of food security in village development – BUMDES capital assistance – Increased farmer capacity – Agricultural infrastructure improvement – Provision of storage area (Lumbung Pangan) – Women group empowerment
	b. Farmer group association	– Organizing climate change adaptation training – Develop a cropping pattern plan and group needs
	c. Farming community	– Intensification with adapted agricultural capital – Sustainable water management – Develop cropping plans and capital requirements and production facilities – Storing food in households and reservoir
	d. BUMDES	– Business capital loan – Provide production facilities on time and in the right amount – Storage and marketing of surplus production
4	Source of food reserves	Farmers' community
5	Needs	400 kg dry grain for For 4 months
6	Storage	– 200 kg stored in households – 200 kg stored in community food reserves
7	Rotation	Storage turnover is done at the end of each harvest season
8	Manager	Households, and designated personnel

Source data processing results, 2023

their produce, considering both immediate needs and future livelihoods. It is crucial that local leaders, especially through BUMDES, lead this food security initiative, ensuring that stored harvests act as a buffer during times when cultivation is unfeasible. By consciously setting aside a portion of their yield post-harvest, and through shifts in cropping patterns and land intensification, households can further fortify their food security.

Recommendations

The impact of climate change, marked by rising temperatures and altered rainfall patterns, has reverberated through the agricultural sector, posing a threat to the food security of local communities. This challenge is becoming increasingly prevalent on both a global and local scale, exacerbated by a lack of comprehensive knowledge on effective adaptation models due to insufficient information. Drawing insights from the successful experience of the Salor Village community in Merauke, where they implemented a food granary model as a proactive measure for climate change adaptation, several key determinants of success emerged. These include the transformative power of information, the presence of adaptive local leaders, and the implementation of collective actions designed and executed through participatory approaches. Despite these successes, critical and pressing policy agendas that warrant immediate attention for effective climate change adaptation within the framework of fortifying food security remain outlined as follows.

1. Local Government

   - Map areas impacted by climate change, establish accessible information systems for all stakeholders and implement community-based monitoring and evaluation.
   - Facilitate comprehensive capacity-building initiatives for all stakeholders, enabling them to apply climate change adaptation models tailored to the region's specific characteristics.
   - Formulate regulations as a foundational framework, prioritizing climate change and local food security as key focal points on the agenda for all stakeholders.

2. Village Government

   - As local authorities, integrate climate change adaptation and food security into village planning and development, aligning with available financial resources.
   - Formulate a participatory work agenda to serve as the foundation for implementing collective actions related to climate change adaptation and food security development.
   - Entrust the execution of climate change adaptation initiatives to community groups.
   - Establish farmer group-based alert teams as pivotal actors in climate change adaptation.

3. Farmer Groups

   - Develop plans addressing production facility requirements, annual cropping patterns, and implementing sustainable agriculture practices.
   - Monitor the identification of local, sustainable agriculture practices and observe climate change changes to provide feedback for refining climate change adaptation models.

4. Village-owned Enterprises (Badan Usaha Milik Desa - BUMDES)

As a village economic institution, BUMDES is an economic actor recommended as a provider of production facilities and business capital and as a collector of products when the production of agricultural products exceeds the needs of households and the needs of food granaries.

Conclusion

The analysis reveals that even with limited knowledge about climate change, the Salor Indah community in Kurik District, Merauke Regency, Papua Province, has been collaboratively strengthening their food security against the escalating effects of climate change. Collaborations between local formal leaders and external researchers have resulted from exchanging comprehensive information about climate change and its implications for agriculture and food security. Armed with this information, formal and informal local leaders, communities, and local economic entities have rallied together, establishing food security as a primary measure against climate change.

Local leaders and institutions are instrumental in promoting climate change adaptation. Their commitment manifests in participatory strategies incorporating climate change concerns into village development plans, laying the foundation for collective efforts to reinforce local food security. For the farming community, entrusting authority over community food reserve management elevates trust and promotes transparency, independence, and a collective sense of responsibility.

References

(Akanksha) Patnaik H (2021) Gender and participation in community based adaptation: evidence from the decentralized climate funds project in Senegal. World Dev 142:105448. https://doi.org/10.1016/j.worlddev.2021.105448

Adger WN (2006) Institutional adaptation to environmental risk under the transition in Vietnam. Ann Assoc Am Geogr 90(4):738–758. https://doi.org/10.1111/0004-5608.00220

Aggarwal PK, Jarvis A, Campbell BM, Zougmoré RB, Khatri-Chhetri A, Vermeulen SJ, Loboguerrero AM, Sebastian LS, Kinyangi J, Bonilla-Findji O, Radeny M, Recha J, Martinez-Baron D, Ramirez-Villegas J, Huyer S, Thornton P, Wollenberg E, Hansen J, Alvarez-Toro P, Aguilar-Ariza A, Arango-Londoño D, Patiño-Bravo V, Rivera O, Ouedraogo M, Tan Yen B (2018) The climate-smart village approach: Framework of an integrative strategy for scaling up adaptation options in agriculture. Ecol Soc 23(1). https://doi.org/10.5751/ES-09844-230114

Ali S, Liu Y, Ishaq M, Shah T, Abdullah IA, Din IU (2017) Climate change and its impact on the yield of major food crops: evidence from Pakistan. Foods 6(6):1–19. https://doi.org/10.3390/foods6060039

Amoak D, Luginaah I, McBean G (2022) Climate change, food security, and health: harnessing agroecology to build climate-resilient communities. Sustain (Switz) 14(21). https://doi.org/10.3390/su142113954

Amorim-Maia AT, Anguelovski I, Chu E, Connolly J (2022) Intersectional climate justice: a conceptual pathway for bridging adaptation planning, transformative action, and social equity. Urban Climate 41:101053. https://doi.org/10.1016/j.uclim.2021.101053

Andrew NL, Allison EH, Brewer T, Connell J, Eriksson H, Eurich JG, Farmery A, Gephart JA, Golden CD, Herrero M, Mapusua K, Seto KL, Sharp MK, Thornton P, Thow AM, Tutuo J (2022) Continuity and change in the contemporary Pacific food system. Glob Food Sec 32:100608. https://doi.org/10.1016/j.gfs.2021.100608

Andriati RD et al (2004) Membangun Daerah Bersama Transmigrasi. Departemen Tenaga Kerja dan Transmigrasi

Ani KJ, Anyika VO, Mutambara E (2022) The impact of climate change on food and human security in Nigeria. Int J Clim Change Strat Manage 14(2):148–167. https://doi.org/10.1108/IJCCSM-11-2020-0119

Ansori M (2022) Peningkatan Ekonomi Lokal dan Pemberdayaan Masyarakat melalui Bumdes. J Sains Komunikasi Dan Pengembangan Masyarakat [JSKPM] 6(4):480–493. https://doi.org/10.29244/jskpm.v6i4.1016

Arroyo CG, Barbieri C, Sotomayor S, Knollenberg W (2019) Cultivating women's empowerment through agritourism: evidence from Andean communities. Sustain (switz) 11(11):1–14. https://doi.org/10.3390/su11113058

Béné C (2020) Resilience of local food systems and links to food security—a review of some important concepts in the context of COVID-19 and other shocks. Food Secur 12(4):805–822

Bouey J (2020) Strengthening China's public health response system: from SARS to COVID-19. Am J Public Health 110(7):939–940. https://doi.org/10.2105/AJPH.2020.305711

Bublitz MG, Peracchio LA, Dadzie CA, Escalas JE, Hansen J, Hutton M, Nardini G, Absher C, Tangari AH (2019) Food access for all: empowering innovative local infrastructure. J Bus Res 100(October 2018):354–365. https://doi.org/10.1016/j.jbusres.2018.12.027

Campos H, Ortiz O (2019) The potato crop: its agricultural, nutritional and social contribution to humankind. In: The potato crop: its agricultural, nutritional and social contribution to humankind. https://doi.org/10.1007/978-3-030-28683-5

Castells-Quintana D, del Lopez-Uribe MP, McDermott TKJ (2018) Adaptation to climate change: a review through a development economics lens. World Dev 104:183–196. https://doi.org/10.1016/j.worlddev.2017.11.016

de Amorim WS, Borchardt Deggau A, do Livramento Gonçalves G, da Silva Neiva S, Prasath AR, Salgueirinho Osório de Andrade Guerra JB (2019) Urban challenges and opportunities to promote sustainable food security through smart cities and the 4th industrial revolution. Land Use Policy 87:104065. https://doi.org/10.1016/j.landusepol.2019.104065

Deininger K, Ali DA, Kussul N, Shelestov A, Lemoine G, Yailimova H (2023) Quantifying war-induced crop losses in Ukraine in near real time to strengthen local and global food security. Food Policy 115(July). https://doi.org/10.1016/j.foodpol.2023.102418

Duchenne-Moutien RA, Neetoo H (2021) Climate change and emerging food safety issues: a review. J Food Prot 84(11):1884–1897. https://doi.org/10.4315/JFP-21-141

Fitriani F, Maturbongs EE (2021) Model Keterlibatan Perguruan Tinggi Dalam Pengembangan Pangan Lokal. Societas: J Ilmu Administrasi Dan Sosial 10(1):27–39. https://doi.org/10.35724/sjias.v10i1.3394

Galli A, Moreno Pires S, Iha K, Alves AA, Lin D, Mancini MS, Teles F (2020) Sustainable food transition in Portugal: assessing the footprint of dietary choices and gaps in national and local food policies. Sci Total Environ 749:141307. https://doi.org/10.1016/j.scitotenv.2020.141307

Gcf NDA, Factsheet I (2021) Role of microfinance in advancing adaptation finance facility, pp 1–5

Gislason MK, Galway L, Buse C, Parkes M, Rees E (2021) Place-based climate change communication and engagement in Canada's provincial north: lessons learned from climate champions. Environ Commun 15(4):530–545. https://doi.org/10.1080/17524032.2020.1869576

Gomez-Zavaglia A, Mejuto JC, Simal-Gandara J (2020) Mitigation of emerging implications of climate change on food production systems. Food Res Int 134:109256. https://doi.org/10.1016/j.foodres.2020.109256

Guell C, Brown CR, Navunicagi OW, Iese V, Badrie N, Wairiu M, Saint Ville A, Unwin N, Kiran S, Samuels TA, Hambleton I, Tukuitonga C, Donato-Hunt C, Kroll F, Nugent R, Forouhi NG, Benjamin-Neelon S (2022) Perspectives on strengthening local food systems in small Island developing states. Food Security 14(5):1227–1240. https://doi.org/10.1007/s12571-022-01281-0

Haque MI, Khan MR (2022) Impact of climate change on food security in Saudi Arabia: a roadmap to agriculture-water sustainability. J Agribusiness Dev Emerg Econ 12(1):1–18. https://doi.org/10.1108/JADEE-06-2020-0127

Helwig K (2021) The role of microfinance in climate change adaptation: evidence from rural Rwanda Akamaro k'inguzanyo ntoya n'iziciriritse mu guhangana n'imihandagurikire y'ikirere: ubuhamy. https://edshare.gcu.ac.uk/id/eprint/5179

Holle Y (2020) The Marind tribe's behavior in cultivating paddy field in Merauke regency, Papua Province. J Penyuluhan 16(2):185–198. https://doi.org/10.25015/16202033564

Islam MS, Wong AT (2017) Climate change and food in/security: a critical nexus. Environ—MDPI 4(2):1–15. https://doi.org/10.3390/environments4020038

Jaakkola JJK, Juntunen S, Näkkäläjärvi K (2018) The holistic effects of climate change on the culture, well-being, and health of the Saami, the only indigenous people in the European Union. Curr Environ Health Rep 5(4):401–417. https://doi.org/10.1007/s40572-018-0211-2

Kamin ABM, Altamaha R (2019) Modernisasi Tanpa Pembangunan Dalam Proyek Food Estate Di Bulungan Dan Merauke. BHUMI: J Agraria Dan Pertanahan 5(2):163–179. https://doi.org/10.31292/jb.v5i2.368

Karimi V, Karami E, Keshavarz M (2018) Climate change and agriculture: impacts and adaptive responses in Iran. J Integr Agric 17(1):1–15. https://doi.org/10.1016/S2095-3119(17)61794-5

Kause WL, Helfiarne MR, Komba YT, Salim A, Djesse ST (2013) Kajian Status Provinsi Nusa Tenggara Timur Sebagai Provinsi Kepulauan Ditinjau Dari Perspektif Hukum. J Borneo Administrator 9(2):137–161. https://doi.org/10.24258/jba.v9i2.102

Kline KL, Msangi S, Dale VH, Woods J, Souza GM, Osseweijer P, Clancy JS, Hilbert JA, Johnson FX, McDonnell PC, Mugera HK (2017) Reconciling food security and bioenergy: priorities for action. GCB Bioenergy 9(3):557–576. https://doi.org/10.1111/gcbb.12366

Lasminingrat L, Efriza E (2020) Pembangunan Lumbung Pangan Nasional: Strategi Antisipasi Krisis Pangan Indonesia. J Pertahanan Bela Negara 10(3):243. https://doi.org/10.33172/jpbh.v10i3.1053

Leisner CP (2020) Review: climate change impacts on food security—focus on perennial cropping systems and nutritional value. Plant Sci 293(December 2019):110412. https://doi.org/10.1016/j.plantsci.2020.110412

Mabhaudhi T, Chibarabada TP, Chimonyo VGP, Murugani VG, Pereira LM, Sobratee N, Govender L, Slotow R, Modi AT (2019) Mainstreaming underutilized indigenous and traditional crops into food systems: a South African perspective. Sustain (Switz) 11(1). https://doi.org/10.3390/su11010172

Masipa TS (2017) The impact of climate change on food security in South Africa: current realities and challenges ahead. Jamba: J Disaster Risk Stud 9(1):1–7. https://doi.org/10.4102/jamba.v9i1.411

Mbuli CS, Fonjong LN, Fletcher AJ (2021) Climate change and small farmers vulnerability to food insecurity in Cameroon. Sustain (switz) 13(3):1–17. https://doi.org/10.3390/su13031523

Mubaya CP, Mafongoya P (2017) The role of institutions in managing local level climate change adaptation in semi-arid Zimbabwe. Clim Risk Manag 16:93–105. https://doi.org/10.1016/j.crm.2017.03.003

Muluneh MG (2021) Impact of climate change on biodiversity and food security: a global perspective—a review article. Agric Food Secur 10(1):1–25. https://doi.org/10.1186/s40066-021-00318-5

Murrell A, Jones R (2020) Measuring food insecurity using the food abundance index: Implications for economic, health and social well-being. Int J Environ Res Public Health 17(7). https://doi.org/10.3390/ijerph17072434

Nandi S, Sarkis J, Hervani AA, Helms MM (2021) Redesigning supply chains using blockchain-enabled circular economy and COVID-19 experiences. Sustain Prod Consumption 27:10–22. https://doi.org/10.1016/j.spc.2020.10.019
Osabohien R, Ufua D, Moses CL, Osabuohien E (2020) Accountability in agricultural governance and food security in Nigeria. Braz J Food Technol 23:1–12. https://doi.org/10.1590/1981-6723.08919
Panday S, Rushton S, Karki J, Balen J, Barnes A (2021) The role of social capital in disaster resilience in remote communities after the 2015 Nepal earthquake. Int J Disaster Risk Reduction 55:102112. https://doi.org/10.1016/j.ijdrr.2021.102112
Prakash S (2021) Impact of climate change on aquatic ecosystem and its biodiversity: an overview. Int J Biol Innovations 03(02). https://doi.org/10.46505/ijbi.2021.3210
Praveen B, Sharma P (2019) A review of literature on climate change and its impacts on agriculture productivity. J Public Aff 19(4). https://doi.org/10.1002/pa.1960
Raidimi EN, Kabiti HM (2019) A review of the role of agricultural extension and training in achieving sustainable food security: a case of South Africa. S Afr J Agric Extension (SAJAE) 47(3). https://doi.org/10.17159/2413-3221/2019/v47n3a520
Reeve B, Thow AM, Baker P, Hresc J, May S (2020) The role of Australian local governments in creating a healthy food environment: an analysis of policy documents from six Sydney local governments. Aust N Z J Public Health 44(2):137–144. https://doi.org/10.1111/1753-6405.12968
Saleh HH (2005) Transmigrasi: Antara Kebutuhan Masyarakat dan Pemerintah (Cetakan pe). Pustaka Sinar Harapan
Setiani SY, Pratiwi T, Fitrianto AR (2021) Tenaga Muda Pertanian dan Ketahanan Pangan di Indonesia. Cakrawala 15(2):95–108. https://doi.org/10.32781/cakrawala.v15i2.386
Smith K, Lawrence G (2018) From disaster management to adaptive governance? Governance challenges to achieving resilient food systems in Australia. J Environ Plann Policy Manage 20(3):387–401. https://doi.org/10.1080/1523908X.2018.1432344
Sulaiman AI, Chusmeru C, Kuncoro B (2019) The educational tourism (Edutourism) development through community empowerment based on local wisdom and food security. Int Educ Res 2(3):1. https://doi.org/10.30560/ier.v2n3p1
Susanti G, Sopia Rukmana N (2020) Going local for food security: strengthening local collaborative institution in implementation of food security in Indonesia. Int J Manage (IJM) 11(8):1998–2009. https://doi.org/10.34218/IJM.11.8.2020.176
Udmale P, Pal I, Szabo S, Pramanik M, Large A (2020) Global food security in the context of COVID-19: a scenario-based exploratory analysis. Prog Disaster Sci 7:100120. https://doi.org/10.1016/j.pdisas.2020.100120
Vignola R, Leclerc G, Morales M, Gonzalez J (2017) Leadership for moving the climate change adaptation agenda from planning to action. Curr Opin Environ Sustain 26–27:84–89. https://doi.org/10.1016/j.cosust.2017.03.005
Wambrauwa MSF, Ohee K, Anastasia A (2022) Analysis the impact of the forest crisis on Merauke women in the perspective of ecofeminism. J Perempuan dan Anak 5(2):104–130. https://doi.org/10.22219/jpa.v5i2.22152
Wellbery C, Sheffield P, Timmireddy K, Sarfaty M, Teherani A, Fallar R (2018) It's time for medical schools to introduce climate change into their curricula. Acad Med 93(12):1774–1777. https://doi.org/10.1097/ACM.0000000000002368
Wicaksana A, Rachman T (2018) Pembangunan bidang pertanian dalam rangka meningkatkan ketahanan pangan nasional. Angew Chem Int Ed 6(11):951–952; 3(1)
Xia L, Robock A, Scherrer K, Harrison CS, Bodirsky BL, Weindl I, Jägermeyr J, Bardeen CG, Toon OB, Heneghan R (2022) Global food insecurity and famine from reduced crop, marine fishery and livestock production due to climate disruption from nuclear war soot injection. Nat Food 3(8):586–596. https://doi.org/10.1038/s43016-022-00573-0
Xu L, Gober P, Wheater HS, Kajikawa Y (2018) Reframing socio-hydrological research to include a social science perspective. J Hydrol 563:76–83. https://doi.org/10.1016/j.jhydrol.2018.05.061

Chapter 29
Ecosystem-Based Adaptation with Bamboo-Based Livelihood Diversification: A Case Study in India's Eastern Himalayas

Lumgailu Panmei and Thiru Selvan

Abstract Climate Change is a pressing issue globally. It affects all the earth's entities, threatening its own continuance. One's ability to absorb climate shocks is often linked to the socio-economic status of the individual. People in developed nations are less sensitive, and the sensitivity increases with poverty and the absence of livelihood diversification options. The current chapter is a case study in two hill districts of Manipur, India's North East, Eastern Himalayas. The state is moderately vulnerable to climate change, but alarmingly, the studied area shows relatively high vulnerability and is identified among India's least developed and economically deprived districts. Extreme events directly impacting livelihood include cyclonic storms, landslides, slips, flash floods, hailstorms, etc. This chapter elaborates on the various livelihood diversification options from bamboo-based economic activities as safety nets for enhanced resilience. The abundance of bamboo in the natural forests and the conversion of Jhum (slash-and-burn agriculture) lands to commercial bamboo plantations have opened up new horizons to fight impoverishment. The bamboo-based activities include growers, shoot collectors, artisans, fermenters, and vendors; the observed values are 27.92%, 20%, 15.8%, 5.42%, and 1.25%, respectively.

Introduction

Nature-based solutions (NbS) are a soft measure that can maintain the integrity of the communities and the infrastructures through restoration and protection from the consequences of climate change (Seddon et al. 2020). Climate change and development fields of study have brought forth the urgent necessity for financial upliftment of rural poor through 'Community-Based-Adaptation' using community experiences of adaptation to local vulnerabilities (Huq and Reid 2007). Huq et al. (2013) noted

L. Panmei · T. Selvan (✉)
Department of Forestry and Biodiversity, Tripura University, Suryamaninagar, Agartala 799022, India
e-mail: selvanthirudr@gmail.com

U. Mukhopadhyay et al. (eds.), *Climate Crisis, Social Responses and Sustainability*, Climate Change Management, https://doi.org/10.1007/978-3-031-58261-5_29

that despite being much talked about for about two decades now, critical adaptation interventions such as livelihood are yet to be mainstreamed, albeit the diversity and sustainability for materializing in the achievement of SDGs through imbibition with local ethnic communities' knowledge and involvement. A community's capacity to deal with climate-related stresses is enhanced through resilient livelihoods, which depend on their access to capital and natural resources (Pouliotte et al. 2009). These local traditional knowledge systems are inherited and practiced for generations, abetting building informed decisions on means to retort the changes in their local environment. Although, spatio-temporally-culture specific, these practices arm the individual and community to adapt and enhance their resilience capacity towards climate extremes, variability, and change (Heller and Zavaleta 2009, Rahman and Alam 2016). A nuanced comprehension of climatic variables and vulnerabilities and their various factors, including food and livelihood security, can significantly contribute to increased resilience (Warner et al. 2012). Rural agrarian-based communities could overcome shortcomings in their prevailing practices by adopting innovative ways to find livelihood options that can be suited to the changes around them and need to be documented to serve as models for livelihood development programmes (Rodima-Taylor et al. 2012; Schroth et al. 2009). Several regions especially the poorer nations have been adopting bamboo as a nature based solution to climate change impacts either through Community-Based-Adaptation or with government and institutional assisted adaptation programmes (FAO and INBAR 2018; INBAR 2014; INBAR 2023a, b; Ling et al. 2016; Lobovikov et al. 2012; Marsh and Smith 2007; UNESCO and INBAR 2023). This chapter will address and elaborate on the various climate change impacts and adaptation requirements; the concepts, prospects and challenges of adopting EbA; livelihood diversifications; and also the scope of bamboo for EbA/NbS to various challenges in the Eastern Himalayas.

Climate Change Impacts and Need for Adaptation

Climate change and its associated impacts as an environmental issue are archaic. It has been discussed as a socio-economic quandary that must be tackled through concerted efforts from all spheres (Popkova and Shi 2022; Salzmann et al. 2016). Climate change affects the human and biological systems, mainly due to climate variability (Thornton et al. 2014). There are changes in climate variability as well as the frequency, distribution, duration, and time of occurrence of extreme weather and climatic events, especially surge in the frequency and ferocity of heat waves and precipitation (Geneletti and Zardo 2016; IPCC 2012; Meehl et al. 2007; UNFCCC 2007). Ecosystems over the world are vulnerable to climate change-related phenomena (Mukherjee and Shaw 2021; Seddon et al. 2020) as well as the communities that dwell around such vulnerable ecosystems (IPCC, 2012; Rahman and Alam 2016; Seddon et al. 2020) as they are highly dependent on the ecosystem services (Seddon et al., 2020). Socio-economic and human livelihood threats are in sectors like agriculture, forestry, fisheries, food security, and health (Chew and Sarabia

2016; IPCC 2007; MEA 2005; UNFCCC 2007; Wale et al. 2022). The prevalence of poverty and poor infrastructure among remote ethnic communities limits their adaptive capacity as they cannot quickly and effectively respond to natural disasters and hydro-meteorological emergencies such as flash floods, storms, and droughts (UNDP 2010). Extreme weather events exacerbated by climate change (NASEM 2016) are the most likely risk that can disintegrate the economy (Seddon et al. 2020; UNFCCC 2007). Global Risks Report of the World Economic Forum (2020) reported that the inability to control and cope with climate change poses the most prominent peril to the global economy. Least developed countries are very much reliant on natural resources and agriculture (FAO 2007) and are observed to be tormented more by the negative consequences related to climate change. Climate extremes and irregularities caused detrimental impacts on the natural ecosystems; biodiversity and human population are further aggravated due to non-climate inconveniences, for instance, rural destitution, below-par infrastructure, and low employment rate (Marin et al. 2020; McLean et al. 2016; Wale et al. 2022). The authors further emphasize that families that will be less unshielded to climate change and other societal stressors are those who practice livelihood activities from the forests and natural ecosystems (Wale et al. 2022). Islam and Winkel (2017) add that the risk for lower-income countries is likely to be higher primarily due to their geographical location and also the dynamics of the economy. Households living in poverty must experience more disproportionate suffering from the pernicious consequences of climate change as compared to financially affluent households. Islam & Winkel (2017) have further presented evidence of disproportionality in sufferings by disadvantaged social groups owing to their gender, age, race, and ethnicity. It should also be noted that the so-called developed countries themselves are susceptible to climate change impacts, so climate change adaptation is essential (Huq et al. 2013; Neef and Pauli 2020). IPCC (2007) elucidated climate change adaptation as "*adjustment in natural or human systems in response to actual or expected climatic stimuli or their effects, which moderates harm or exploits beneficial opportunities.*" It can be realized through "Hard measures" and "Soft measures" where the first mentioned measure are generally pursued with futuristic technologies, design, and engineering of climate resilient infrastructures and machineries that can prove somewhat maladaptive bearing in mind the factuality that economically incapacitated nations cannot take on such measures to adapt (Colls et al. 2009; Huq et al. 2013; IPCC 2012; Jones et al. 2012; Klein et al. 2007).

A careful and detailed look at the report by Colls et al. (2009)—'The Atoll Ecosystem Conservation Project' in the Maldives (Cattermoul et al. 2008; IUCN 2008), which scrutinized the worth of the physical safeguarding systems being bestowed by littoral and pelagic ecosystems divulged that a whopping estimated sum in the range of USD 1.6 TO 2.7 billion USD would be required for building seawalls, breakwaters and other structures that would be built instead of the natural barriers customarily provided by coral reefs. Moreover, expenses were expected to be higher should the so-called complex measure project cause harm to existing towns, villages, resorts, and other infrastructure and establishments (IUCN 2008). It must also be taken into consideration that not just the costs accompanied, but the loss would be irreversible, supposing that the physical safeguarding provided by coral reefs degenerated or vanished by such intervention (Colls et al. 2009). It can further lead to a very grave situation threatening the natural ecosystem and biodiversity as well as impeding the access of rights of indigenous remote communities to make use of ecosystem services that can make overall adaptation goals futile (CBD 2009; de Block 2018; Pérez et al. 2010; Jones et al. 2012).

On the contrary, natural ecosystems serve as bulwarks and are customarily inexpensive to maintain, usually more effective than built physical engineering edifices (Colls et al. 2009). Soft measures emphasized on pre-informed preventive and preparedness through forecasting and warning systems for extreme events (CARE International 2013; Enríquez-de-Salamanca et al. 2017) while also acquiring knowledge of detrimental effects of climate change to the forests and forest-dependent populations and their vulnerabilities that are paramount for amplifying adaptation and resilience (Rahman and Alam 2016; Salzmann et al. 2016; Thornton et al. 2014). Khanal et al. (2018) conveyed that people with a high education possessed an increased potential of acquiring formal employment and adapting to climate change as they tend to be better-informed and aware with higher perception of climate change along with adaptation (Luís et al. 2018; Wale et al. 2022). However, a good proportion of the rural community households in forest fringe villages lack educational proficiency, with a poor likelihood of picking up other livelihood activities, and therefore continue being dependent upon natural resources to sustain livelihood (Wale et al. 2022).

EbA Concepts, Prospects and Challenges in Adoption

IUCN (2016) defines nature-based solutions as "*actions to protect, sustainably manage and restore natural or modified ecosystems, which address societal challenges effectively and adaptively, while simultaneously providing human well-being and biodiversity benefits*". The endowment provided by the natural ecosystem is exploited to improve the capacity of the inhabitants or the dependent people in the fringes of such areas to meet considerable environmental, social, and economic goals that curtail risks resulting from climate change-related extreme events and variability that (CBD 2009; Hashida and Lewis 2019; IPCC 2007; IUCN and NERI 2011). Climate change adaptive provisional services of natural eco-structural features like lush green vegetation, woodlands, meadows and grasslands, and aquatic water bodies essentially provide regulatory services of the local climatic, aquatic, and edaphic situations and moderates and gives protection against hydro-meteorological and geological hazards (Bolund and Hunhammar 1999; Niemela et al. 2010). Other adaptation germane services include but are not limited to, provisioning services like food, fodder, fiber, protection from pests and diseases, and conserving endemic gene pool and genetic diversity (UNFCCC 2012). Demographic characteristics, typically ethnicity and race, are reciprocal to assets and income and opportunities to public resources, including healthcare, education, sanitation, housing, etc. (Islam and Winkel 2017). The societal or socio-economic challenges associated with climate change include food and water soundness, loss of livelihood, and damage and destruction of lives, assets, and properties due to natural disasters (FEBA 2017; UNFCCC 2007).

Ecosystem-based adaptation is a comparatively novel concept defined as the "*use of biodiversity and ecosystem services as part of an overall adaptation strategy to help people adapt to the adverse effects of climate change*" (CBD 2009; Kabisch et al. 2017). In layman perceptions, it can be approaches taken up for planning and implementing climate change adaptation through utilization/ harnessing of ecosystem services from the natural landscape and biodiversity for the well-being of the society (Girot et al. 2012; IUCN 2016; MEA 2005) to cushion communities from climate extremes and variability (de Block 2018; Munang et al. 2013). Ecosystem-based Adaptation is acknowledged as a worthwhile, approachable measure of alleviating poverty and climate change-associated risks by many Least Developed Countries like Bangladesh, Ethiopia, Haiti, Lesotho, etc. (Colls et al. 2009). A thorough understanding of the dynamics of the socio-ecological landscapes is a pre-requisite for practitioners of EbA need to exploit the social, environmental, and economic possibilities that are inherent to the existing scenarios, including indigenous traditional expertise of utilization of the local natural biodiversity and local institutions (CBD 2014, 2016; FEBA 2017). UNFCCC formulated the National Adaptation Programme of Action for Least Developed Countries to single out pre-eminent and prompt requirements for adaptation through community-level efforts, endorsing that grassroots communities are the primary stakeholders (Colls et al. 2009). That being said, EbA's notion is people-centric and hails that humankind's resilience is contingent on the integrity of ecosystems. While more authors exhort that EbA facilitates low-risk, cost-effective defense and provides other ecosystem services (Reid et al. 2018, 2019; Seddon et al. 2020).

Pérez et al. (2010), Vignola et al. (2009) along with Mercer et al. (2012) expressed that EbA promotes the implementation of both vernacular and external comprehension about the existing ecosystems to recognize and connect climate change adaptation measures, acknowledging the diversity in local environmental and societal situations and fabricate a facilitating environment for adequate localized adaptation and ecosystem stewardship. EbA also identifies natural resource conservation as an essential practice that promotes the efficacy of adaptation (Girot et al. 2012). Agroforestry practices, though underappreciated, can be seen as human intervention and adaptation, which is a close-to-nature approach to seeking solutions to climate change impacts (Hart et al. 2023; Selvan and Kumar 2017). Huq et al. (2013) summarized the major principles and indicators of EbA, of which five are relevant in the present context which includes *Knowledge based adaptation—Best available science and local knowledge, culturally appropriate* (Colls et al. 2009; Mercer et al. 2012), *Maximum stakeholder involvement—Maximum stakeholders, entailing local communities* (Colls et al. 2009; Renaud et al. 2016; Watson et al. 2012), *Ensuring governance—Gender balancing and equity* (Jones et al. 2012), *Resilience building—Reducing the vulnerability and shortcomings in Disaster Risk Reduction* (Renaud et al. 2016; Watson et al. 2012) and *Integrating development—Providing social, economic and environmental benefit* (Colls et al. 2009; Pérez et al. 2010).

Of late, anxiousness over forest resources is propelled, as it has been seen as a front burner for global mitigation and adaptation measures in tackling and confronting multifarious climate change issues (Ravindranath et al. 2006; Renaud et al. 2016). Adaptation to climate change should embody the founding and efficient oversight of protected-area complexes for warranting perpetual conveyance of ecosystem services (Colls et al. 2009). Over the past five decades or so, climate change-associated events have strained several facets of forest ecosystems, such as the development and die-down of trees, the rapid expansion of area under invasive species, species distribution, and migration, alterations in natural phonological cycles of trees and even species extinctions (Abalo et al. 2017; Locatelli et al. 2008; Rahman and Alam 2016; Seppälä et al. 2009).

The legitimacy of calling a practice as EbA is often debatable as we are often unsure of how sustainable the knowledge and practices are and how well the practices are implemented (FEBA, 2017). There is widespread acceptance that healthy ecosystems buffer communities against risks. Still, there remains an obligation to investigate how best a blend of hard and soft adaptation measures can be implemented in accordance with specific local set-ups (FEBA 2017; IUCN 2014, 2016). Colls et al. (2009) elaborated that ecosystem-based adaptation entails a broad scale of ecosystem management ventures that mitigate the vulnerability to climate change to extend the resilience of social as well as natural environments. These ventures include but are not limited to the formation of diverse agricultural practices, in which utilization of indigenous proficiency of local crops and livestock, all the while conserving their genetic diversity, and conserving the agricultural landscapes to ensure food security despite the changes climatic conditions (Selvan et al. 2023). The effectiveness of any EbA measures could be downplayed if they are taken up as a stand-alone effort. So, there is a necessity to integrate into mainstream adaptation strategy involving decision-making, planning, and implementation in a bottom-up manner from local to national to regional or to global level (FEBA 2017; Girot et al. 2012; IUCN 2011, 2016). Therefore, the FEBA (2017) established that every element mentioned in the elucidation of EbA as per CBD (2009) must be fulfilled for any practice, approach, project, etc., to be entitled to EbA. The three elements of EbA in the definition are (a) assisting humans to adapt to climate change, (b) through active utilization of biodiversity and ecosystem services, and (c) obliging to the all-inclusive adaptive strategies (CBD 2009).

Consequently, ecosystem-based adaptation presents methods that are effortlessly accessible for the rural population that can be incorporated into community-based adaptation and oversees several issues and precedent problems recognized by the highly vulnerable nations and communities. Moreover, many healthy natural ecosystems, including terrestrial systems like forests and grasslands, pelagic and estuarine systems like wetlands, mangroves along with coral reefs, are better armed to wrestle with climate change themselves and have a high potential of recovery after exposure to extreme weather events (Colls et al. 2009). The distinction of EbA or "business unusual" from "business as usual" i.e. conventional adaptation approaches, is that it connects traditional conservation approaches for biodiversity and ecosystem with feasible socio-economic development as an element of comprehensive designs for assisting people cope with the perilous situations linked with climate change (FEBA 2017).

Livelihood Diversifications to Face Climate Change

Discussions on climate change impacts impact were initially focused on its physical element basis, but over time, the societal impacts attracted recognition, and evidence was conferred concerning the association between climate change, poverty, and livelihood (IPCC 2007; Islam and Winkel 2017; UNFCCC 2007). In compliance with the sustainable livelihoods framework (Scoones 1998; Wale et al. 2022), the adaptation designs in rural communities are supposedly determined by benefaction with veritable facades of assets (Wale et al. 2022). Here, the term assets refers to "*natural, human, physical, financial, social and cultural capital,*" as part of the "*ensemble or opportunity set*" including capabilities and activities to build their livelihoods (IPCC 2014). Injustice, inequality, and unsustainability in the extraction and utilization of natural resources sabotage efforts for conservation, economic improvement, and human well-being. At the same time, threats to the rights and access of indigenous and rural forest fringe village communities call for equitable governance and implementation of environmental obligations (IUCN 2020). Nature-based solutions (NbS) are a fortified league to approach social trials like climate change, biodiversity loss, and food security. While the world is striving to overcome the COVID-19 pandemic and dash forward to realize the UN Sustainable Development Goals, ensuring that ensuing investments attain their prospects that confer the vigor and well-being of people and the planet (IUCN 2020). Agriculture in Southeast Asian Nations like Vietnam and the Philippines are facing several impediments with the increase in the frequency of landslide and flash floods, causing roadblocks and bridge wash that hamper the communication systems and transportation of agroforestry products to the markets (CARE International 2013; Cruz 2015). Moreover, forest-dependent farmers are witnessing crop damage, yield reduction, and irregularities in the fruiting and harvesting cycle, accompanied by the prevalence of pests and diseases due to rising temperatures and changes in precipitation patterns (Cruz 2015). To overcome the adversities in their agriculture, farmers modify their usual practice by multiple cropping or intercropping, modifying the planting cycle, and seeking alternative livelihoods (Cruz 2015).

There is an expanding acceptance of the capacity of robust ecosystems to stimulate and aid communities in tackling climate change as they give various provisional services like drinking water, habitat, non-timber forest products that can serve as food as well as raw materials for several forest-based products, providing livelihood diversification opportunities to overcome impacts of climate change to one's primary occupation (Colls et al. 2009). Forests are intrinsic to the poverty alleviation endeavors for rural households and communities all over developing nations (Cheng et al. 2019; Rasmussen et al. 2017; Wale et al. 2022). The involvement of indigenous communities and rural populace in NbS activities overrides other adaptation measures and efforts as they hold 37.9 million sq km of terrestrial land, especially in Earth's prime regions of biodiversity importance (IUCN 2020). The mantle of the forest ecosystem to support livelihoods with sustainable utilization and conservation of forest resources depends on several governmental, corporate, financial, social, and ecosystem factors (Ferrara et al. 2017; Wale et al. 2022). Thus, there is an upshot in research attention on the dynamics of climate change with forest ecosystems and rural livelihood (Wale et al. 2022; Weiss et al. 2019). It is to be noted that forest-based livelihoods of ethnic rural communities are essentially associated with forest resource conservation. They collect wild edible fruits and vegetables, hunt wild animals, indulge in fishing from creeks and brooks, and acquire various ecosystem services from these forests to sustain their livelihoods (Panmei et al. 2019; Singh et al. 2018; Wale et al. 2022). This is true in the case of traditional mountain communities of India's Eastern Himalayas (Panmei et al. 2019; Ramakrishnan 2007; Singh et al. 2018).

Indigenous communities have been recognized for their vulnerability to climate change due to the proximate connection between forests to their culture, spirituality, and socio-economic systems (Colls et al. 2009; Islam and Winkel 2017; IUCN 2020). Traditional livelihoods are markedly reduced as a negative consequence of climate change, prompting people of affected communities to explore and take up different income generation activities commonly termed as IGAs (Rahman and Alam 2016; Singh et al. 2018). The IGAs vary with communities' needs and knowledge, like animal husbandry and pisciculture-based activities like poultry, cattle, and fish rearing; weaving of traditional tribal attire; crafting of bamboo and rattan-based value-added products; inclusion of fuel wood species in agroforestry and cultivation of medicinal plants (Rahman and Alam 2016). The utilization of natural resources by traditional communities is tailored to the local ethos and is exploited and maintained (Turner 2005) to sustain their livelihoods (Singh et al. 2018). Native populations tend to have various traditional measures to manage the locally available resources (Agrawal and Gibson 1999) to obtain maximum benefit (Heubach et al. 2011) and equip themselves to face the worst situations by being self-conscious about their remoteness and accessibility.

The most crucial factor that can serve as effective retaliation to climate change is ingress to the market as it dramatically influences the opportunities for rural populations' livelihood adaptation strategies (Hoque et al. 2018; Wale et al. 2022). Communities with access to markets for forest-based products tend to pursue forest-dependent adaptation means. In contrast, communities with access to agriculture-based products will turn to agricultural-based local entrepreneurship (Wale et al. 2022). The assets decide the types of IGAs households and communities adopt, and hence, assets influence livelihood modification designs as retaliation to climate change (Jezeer et al. 2019). Formulation and promotion of Pro-poor entrepreneurship serving as the cornerstone for sustainable forest resource management resulted in curtailing of vulnerability and social frictions, upliftment in the rural economy, and skill enhancement at the local level through diversification in income generation activities of locals (Charmakar and Chivhenge 2018). Rural forest-dependent households with fecund assets are presumed to be more likely to adapt and adopt strategies from the forests near their villages. Moreover, there is anticipation that the land holding of a household plays a remarkable part in reducing households' dependency on community forests (Wale et al. 2022).

Farmers in the least developing and developing nations are often small-holder farmers, and they find difficulties in overcoming challenges concerned with sustaining food production in the present scenario of unforeseeable climatic conditions that can exacerbate droughts, pests, and diseases (CARE International 2013; Colls et al. 2009; Vermeulen et al. 2018). Such farmers have less land and often practice intense, diversified crop and livestock-based production with little to no input of external fertilizer and pesticides (Selvan et al. 2023). Small holder farms in Roslagen, Sweden, were under threat due to exposure to mild winters in the 1990s that considerably magnified the intensity and severity of pest outbreaks, chiefly damping off due to fungal infestation of crops (Tengo and Belfrage 2004). This elicited the farmers to conduct trials of pest resistance with both new and old crop varieties through which they learned that farming systems are more resilient to veritable climatic conditions by modifying old cropping systems with multiple-species cropping systems used in olden days. Diversification and modifications of existing ecosystem management practices can enhance progressive and proactive farmers, such as those in the Roslagen area, to improve their resilience to climate uncertainties, simultaneously conserving local and regional biodiversity (Tengo and Belfrage 2004). Indigenous communities have cultivated multiple crops in organic systems following a mix of synergistic crops to maintain a harmonic production of all the species from the same plot (Cruz 2015; Selvan et al. 2023). Readopting of traditional natural farming system termed '*Quesungual*' Slash and Mulch Agroforestry System (QSMAS) by resource-poor smallholder farmers, which involves planting small crops as understory among sparsely distributed local trees, accompanied by making terraces and other soil conservation methods (Colls et al. 2009; FAO 2009). This activity significantly upgraded their livelihoods and indicated that they acquired the capacity to thrive in climate extremes like grievous droughts and heavy rainfall during el Niño and la Niña phenomena, including Hurricane Mitch in 1998 (FAO 2009).

An example of a successful community-level EbA is the Kudawa rural fishing village in Sri Lanka, which heavily depended on the Bar Reef Marine Sanctuary. The sanctuary is a zone of prime pelagic diversity but suffers from overfishing and illegal exploitation. Unfortunately, the reef system is highly vulnerable to climate change and sustained a mass coral bleaching event in 1998 (Colls et al. 2009). IUCN formulated the SLED, IMM 2008, "*a tool that can facilitate adaptation among communities dependent on climate change sensitive resources that work by helping communities making informed choices about their livelihood options.*" The tool as mentioned above, was implemented in the village using the help of NGO Community Help Foundation, emphasizing the utilization of locally available resources, withholding from indiscriminate exploitation of natural resources, and conforming to the social environment. This was done by promoting sustainable livelihood diversification activities, including improving home gardens and culturing seaweed and seabass, which were earlier exploited from the reef. Implementing this tool has dramatically helped diversify livelihoods while realizing adaptation to existing pressures and enhancing their capacity to face changes in the future (Colls et al. 2009; IUCN 2008). The authors reiterate that livelihood enhancement and diversification stimulate community members to retreat from unsustainable extraction and deterioration of local natural resources, which improves the resilience of both the social and environmental dimensions to the impacts of climate change. Meanwhile, elsewhere in tsunami-disturbed regions of India, Indonesia, Malaysia, Sri Lanka, and Thailand, the Green Coast Programme is pursuing local livelihood improvements of affected communities through the adoption of a community-led approach by building ecologically sound livelihood or income generation activities by engagement in restoration of mangroves along with other native flora. Over and above the income generated from the restoration labour works, the community-based organizations involved in the mentioned programme were also awarded collateral-free and interest-free loans for capital to create alternative livelihoods (Colls et al. 2009).

Bamboo Resource Advantages and Opportunities in Eastern Himalayas

Bamboo ecosystems have been considered a powerful tool for NbS (especially as EbA approach) projects by FAO, INBAR, and other relevant organizations working in landscape restoration, livelihood and food security, climate adaptation and resilience (FAO and INBAR 2018; Ling et al. 2016; Lobovikov et al. 2012). A reported 1642 species in 88 bamboo genera grow in different regions of the world, primarily in the tropical and subtropical zones (Vorontsova et al. 2016). China has the richest bamboo diversity in the world, followed by India, where bamboo is more diverse in the Indian section of the Eastern Himalayas. There are 136 species under 23 bamboo genera in India; 125 are indigenous, and the remaining 11 are exotic species (FSI 2021; Selvan 2018; Varuvel et al 2021). The North Eastern Region of India is a merger of two biodiversity hotspots, viz., the Eastern Himalayas and the Indo-Burma, making the region rich in biodiversity, boasting about 90 bamboo species, of which 41 are endemic (Loushambam et al. 2017). As per Environment Information System-ENVIS, the North Eastern state of Manipur is home to 51 species of bamboo under 12 genera. A new restoration project in the Tamenglong District of Manipur has utilized *Bambusa blumeana* Schult. & Schult. f., the total species in Manipur stands at 52 under 12 genera. Bamboos are entwined with the region's art, culture, tradition, literature, livelihood, and culinary heritage. Fifteen edible species have been listed in Manipur (Premlata et al. 2020), most available in the wild and cultivated in the valleys and hills. Bamboo is a common feature in agricultural fields and home gardens across the Eastern Himalayas (Sharma and Borthakur 2017), where such land use has become an essential area for the ex-situ conservation of bamboo (Tangjang and Arunachalam 2009, 2017). The total bamboo-bearing area of the country is about 15 million hectares, of which the eight states of North East India bear a total of 5.35 million hectares (FSI 2021). India State of Forest Report-2019 (FSI 2019) reported that a total of 23,740 tons of bamboo was collected by populations in the forest fringe villages of the eight states of Northeast India.

Marsh and Smith (2007) opined that the profitability of bamboo in climate change mitigation efforts does not lie solely in carbon payments and concerted efforts by government, industries, research institutions, and ethnic communities by establishing a highly versatile and viable economy based on bamboo remarkably uplift rural population from poverty through livelihood diversification. Thus, any agroforestry land use system that incorporates bamboo as a component exhibits three wholesome attributes viz. Productivity, Sustainability, and Adaptability (Selvan and Tripathi 2017), which further guarantees it as an essential species for the Ecosystem-Based-Adaptation approach to climate change and enhancing resilience of farmers and communities through livelihood diversification options (Lobovikov et al. 2012; Selvan and Kumar 2017).

Objectives of the Study

India's Eastern Himalaya, or the Indian Himalayan Region consisting of 12 states, is very prone to climate-related disasters (DST 2021). Climate change is a very grave threat to rural livelihood, food security, natural resources, and agriculture (Hashida and Lewis 2019), and adaptation is absolutely and immediately required to tackle existing situations and also enhance the resilience of vulnerable communities, mainly concentrated in the fragile mountain ecosystems (DST 2021; Marin et al. 2020). Community-based adaptation using natural resources for livelihood diversification can be acknowledged as an effective EbA. The study intends to bring to light the various traditional bamboo-based livelihood activities that are adopted by the Rongmei ethnic community of India's Eastern Himalayas, who have taken advantage of the vast bamboo resources in the region. This work attempts to serve as a guide for the relevant organizations to identify avenues for sustainable community development programmes. When community members are informed of the existing and future challenges accompanied by financial and technical aid from relevant authorities, the two groups can work together to bring sustainable bamboo-based livelihood opportunities. This can also help in the conservation of local bamboo species.

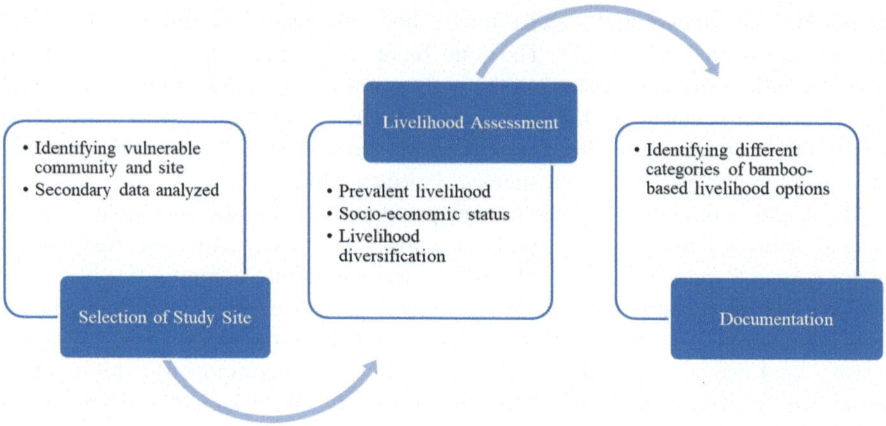

Fig. 29.1 Methodology adopted for identifying the vulnerability of the site and ethnic community concerned, assessment of livelihood and documentation of the bamboo based livelihood options

Methodological Approach

The field research was conducted from November 2021 to December 2022. During this period, a household survey (Semi-structured questionnaires) was conducted at six Rongmei-inhabited villages. Chaengdai, Duiluan, Khumji, Makhuam, Puiluan and Luangrang. Socio-economy status determination is done by the Modified Kuppuswamy Scale, 2021 and 2022, and Udai Pareek Scale, 2019. This document represents a complete report on the SES, prevalent livelihood activities and alternative livelihood options, perceptions of climate change, and impacts on the livelihood activities of Rongmei people of all study sites/villages. The village-level data will be provided to understand the outlook of each village on the climate change scenario of the socioeconomic standings. The method being adopted is given in Fig. 29.1.

Description of Study Sites

The study was conducted in two hill districts of Manipur in the NE section of the Indian Eastern Himalayas. The state's total geographical area is 22,327 sq km, constituting 0.7% of India's total area. The average elevation of the lands ranges from 760 to 3000 m above the mean sea level. There are 16 districts in the state; however, official documents are mainly available for nine districts, as seven were created recently and have yet to be included in the Census of India. For this chapter, despite the study being carried out in two Rongmei inhabited hill districts of Manipur, the land area is taken for the erstwhile pre-bifurcation Tamenglong as the official land area. The total land area of the districts is 4391 sq km. The hilly study area has tropical evergreen forests, subtropical forests, and bamboo brakes. As per the Koppen-Geiger climate classification systems, the climate of Tamenglong is humid subtropical (Cwa). The district has witnessed several landslides, river bank breaching, and inundation of agricultural fields in the last few years. The map of the study areas is arranged in Fig. 29.2a–d depicting the Manipur state map, the Tamenglong District, the Noney District, and the six studied villages, respectively. The distance from the state capital ranges from 65 to 150 km.

Socio-Economic Survey (Including Livelihood)

Six Rongmei inhabited villages in the study area were selected for the study. Forty households were randomly surveyed in each of the six villages. The SES and livelihood survey was done based on the Modified Kuppuswamy Scale 2021 and 2022 and Udai Pareek Scale 2019, along with semi-structured questionnaires. The modified Kuppuswamy Scale for 2021 and 2022 was updated based on the modified Kuppuswamy Scale 2019 (Wani 2019). The update was required as the CPI (Consumer Price Index) value varies occasionally. While the Udai Pareek Scale 2019 (Wani 2019) was used as it is since the questions/indicators don't involve any monetary value. These two scales are used to determine SES (Socio-Economic Status), a vital source for measuring the household's capacity. The Kuppuswamy Scale determines the SES based on Profession, Education, and Income. In contrast, the Udai Pareek Scale is based broadly on the social class of the studied population, the primary source of income, landholding, farm power, assets, social participation, etc. Both these scales are used as the villages are remote and have less access to market facilities for robust business-based activities. Monetary income may be meager, but they may have large land holding for pastoral, silvicultural, or horticultural-based agriculture.

The attached questionnaire sets for different farm and household material assets or access to community forests to collect non-timber forest products, which may give them livelihood diversification options. The SES tools give mark weightage to each question, and the accumulated score is used as the standard scoring range for identifying the socio-economy of the surveyed community members. The standard scoring for the Kuppuswamy SES Grading Tool and Udai Pareek SES Grading Tool are given in Table 29.1.

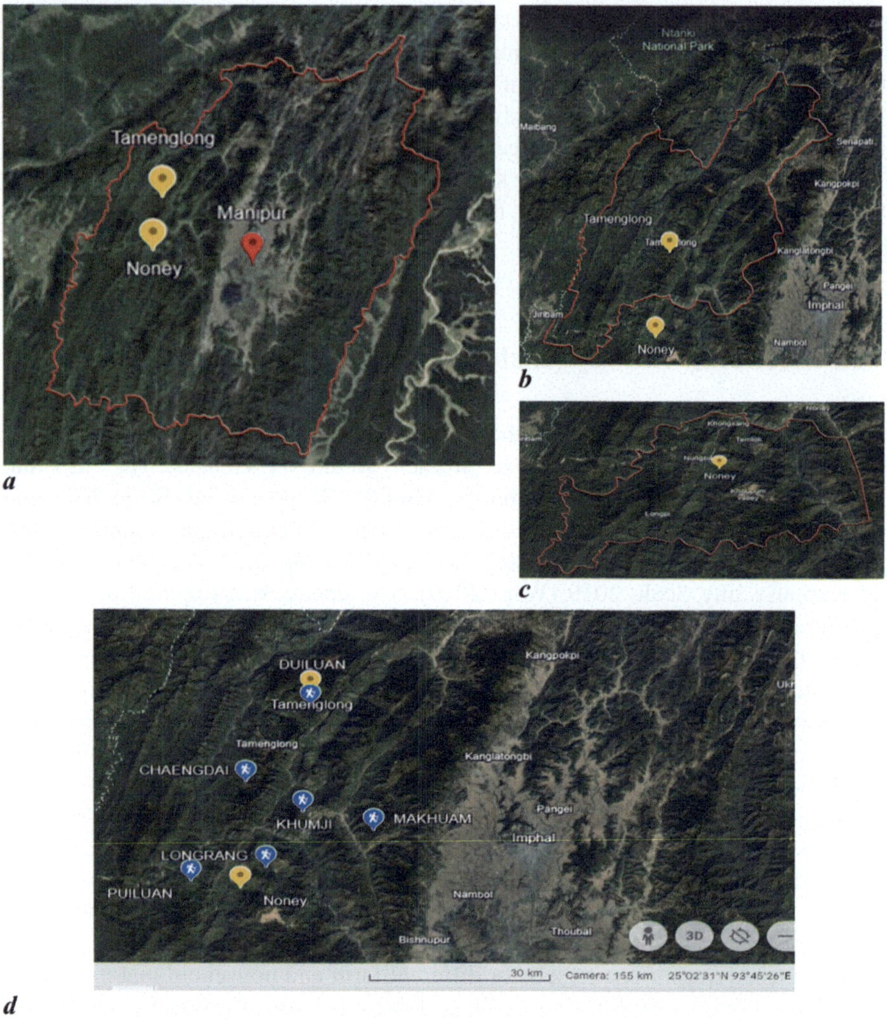

Fig. 29.2 **a** Study area in Manipur, NE India part of Eastern Himalayas; **b** Tamenglong district and **c** Noney district in Manipur and **d** Six Rongmei villages viz. Chaengdai, Duiluan, Khumji, Longrang, Makhuam and Puiluan which falls in Tamenglong and Noney districts in Manipur where the bamboo based livelihood activities were studied

Results and Discussions

Climate change issues are not alien to any people or communities. Still, the worst hit are the marginalized minorities and remote ethnic communities who are deprived of infrastructure and aptitude to cope with changes in the environment. The states in

Table 29.1 Standard scoring for Kuppuswamy SES grading tool and Udai Pareek SES grading tool (modified from Wani 2019)

Kuppuswamy scale 2021 and 2022		
Score	Category	Grade
26–29	Upper class	I
16–25	Upper middle class	II
11–15	Lower middle class	III
05–10	Upper lower class	IV
< 5	Lower class	V
Udai Pareek scale, 2019		
> 43	Upper class	I
33–42	Upper middle class	II
24–32	Middle class	III
13–23	Lower middle class	IV
< 13	Lower class	V

the Indian Eastern Himalayas are vulnerable to climate change and climate extremes and climate variations, the primary drivers being remoteness from mainland India, poor connectivity and infrastructure, rain-fed agriculture, variability in agricultural productivity, poor implementation of social development schemes and projects, a high number of marginal and ethnic rural/ tribal communities. The lands are degraded primarily because of over-exploitation of forests for fuel, wood, timber, and fodder, Jhum cultivation, poor land-use strategies, infrastructure development, etc. The studied state of Manipur is home to more than 33 indigenous tribal groups with agriculture and natural resource-intensive livelihoods, making them vulnerable to climate change. The state's overall climate vulnerability index (VI) is 0.52, which is within the range of moderate vulnerability as per Climate Vulnerability Assessment for Adaptation Planning in India Using a Common Framework (DST 2021). The forest is the primary land use in the NE states of Eastern Himalayas, including Manipur, more so in the remote hill districts of Tamenglong and Noney districts, which is almost exclusively inhabited by the Zeliangrong group of Naga tribes, of which Rongmei is a significant ethnic community. The studied districts have agrarian economies, with a substantial population that depends upon agriculture as their primary livelihood. The VI for Tamenglong district (inclusive of Noney district) is 0.69, which is relatively highly vulnerable (DST 2021), where district ranking for the state is based on the agricultural sector. The drivers for agricultural vulnerability in the state include lack of crop insurance, landless, marginal and small farmers, low number of livestock, lack of irrigation, etc. One of the significant aspects of the mountain-based rural life of the Rongmei populace is gathering plants and fungi, fodder, and medicinal purposes (Panmei et al. 2019). Improved income can increase adaptive capacity in climate emergencies and better prepare them for future uncertainties (UNDESA 2013; UN 2019).

Land degradation is rampant as several tracts of forests are claimed for Jhum cultivation, and bamboo forests, which is the prevalent forest type in the studied districts itself, are also under pressure from unsustainable use, rapid clearance of lands for NF Railways project and Asian Development Bank funded Act East policy road extension. A large number of the respondents have perceived climate change happening, while some have claimed that climate change-related extremes, especially hydro-meteorological influenced events such as torrential rains, hailstorms, cyclonic storms, landslides, slips, droughts, and forest fires, have impacted their livelihood activities. Moreover, such climate extremes have caused crop failures affecting farmers for many years.

Altogether 240 households and three organisations were surveyed in the Rongmei villages, 40 families each from Chaengdai, Duiluan, Khumji, Makhuam, Puiluan, and Lonrang in the Tamenglong and Noney Districts. The results from the analysis for each village using the tools mentioned above are displayed in Table 29.2 for Kuppuswamy SES Grading Tool and Table 29.3 for Udai Pareek SES Grading Tool respectively.

The above compiled data indicates that the villages do not have a high income group of families in SES Grade I-Upper Class (who have a high level of education and income) as per the Kuppuswamy SES Grading Tool. Here, we see one household each in the Lower Class in the villages of Puiluan and Longrang. It should be borne in mind that these two villages are the remotest among the studied villages. Puiluan village is about 140 km from Imphal, the capital of Manipur, and relatively far from the Tamenglong and Noney towns; hence, it has a minimal scope of livelihood diversification options for income generation. We see that in all the villages, the number of households in the mid-SES range i.e., Grade III is the highest.

Table 29.2 SES grades of Rongmei households in Tamenglong and Noney districts of Manipur using Kuppuswamy SES grading tool

Grade	Chaengdai		Duiluan		Khumji	
	Count	Percentage	Count	Percentage	Count	Percentage
I	0	0	0	0	0	0
II	8	20	2	5	3	7.5
III	20	50	23	57.5	31	77.5
IV	11	27.5	15	37.5	6	15
V	1	2.5	0	0	0	0
Total	40	100	40	100	40	100
Grade	**Makhuam**		**Puiluan**		**Longrang**	
	Count	**Percentage**	**Count**	**Percentage**	**Count**	**Percentage**
I	0	0	1	2.5	0	0
II	7	17.5	1	2.5	0	0
III	25	62.5	16	40	9	22.5
IV	8	20	21	52.5	30	75
V	0	0	1	2.5	1	2.5
Total	40	100	40	100	40	100

Table 29.3 SES grades of Rongmei households in Tamenglong and Noney districts of Manipur using Udai Pareek SES grading tool

Grade	Chaengdai		Duiluan		Khumji	
	Count	Percentage	Count	Percentage	Count	Percentage
I	1	2.5	1	2.5	0	0
II	10	25	4	24	13	32.5
III	19	47.5	24	60	22	55
IV	9	22.5	11	27.5	5	12.5
V	1	2.5	0	0	0	0
Total	40	100	40	100	40	100
Grade	**Makhuam**		**Puiluan**		**Longrang**	
	Count	**Percentage**	**Count**	**Percentage**	**Count**	**Percentage**
I	5	12.5	1	2.5	0	0
II	14	35	13	32.5	5	12.5
III	18	45	14	35	26	65
IV	3	7.5	12	30	9	22.5
V	0	0	0	0	0	0
Total	40	100	40	100	40	100

With the Udai Pareek SES Grading Tool, we see that the households in all the studied villages are widely distributed into different SES Grades as there are more parameters consisting of primary occupation, land holding, assets, household member numbers, social participation, etc. Here, the households are distributed in comparative proportions in Grade III and Grade IV.

The cumulative analyzed results using both the SES grading tools are shown above in Table 29.4, and the data show that the proportion of households in grade III is highest in both SES grading systems. The percentage of households in the Grade III Lower Middle Class on the Kuppuswamy SES Scale is 51.66%. In contrast, the percentage of households in the Grade III Middle Class of Udai Pareek Scale is 51.25%. Households' income range as reflected in the SES Grades using the Kuppuswamy SES Grading Tool is used as an indicator of vulnerability as household with less income possessing the least capacity to adapt to the impacts of climate risks (DST 2021; O'Brien et al. 2008; Wani 2019). The lowest percentage for SES grading with the Kuppuswamy SES Grading Tool is seen in Grade I Upper Class. The weakest Udai Pareek SES Grading Tool is seen in Grade V Lower Class. The best differentiating feature between the Kuppuswamy SES Grading Tool and the Udai Pareek SES Grading Tool could include landholding in the latter, which is very applicable for assessing climate vulnerability. Community members with small landholdings, like marginal and smallholder farmers, are subjected to climate hazard risks as they do not have adequate land for sowing and other livelihood operations, leading them to have low adaptive capacity (Sathyan et al. 2018). For better visual understanding, the following chart Fig. 29.3, depicts the overall percentage shares of SES w.r.t Kuppuswamy SES Grading Tool, and Fig. 29.4 illustrates the broad scale w.r.t Udai Pareek SES Grading Tool.

The primary occupation of the people is agriculture and livelihood activities from the exploitation of forest bioresources; this is evident from the survey results

Table 29.4 Comparison of SES grades of Rongmei households (240 families in 6 villages) in Tamenglong and Noney districts of Manipur using two SES grading tools

Kuppuswamy SES			Udai Pareek SES		
Grade	Count	Percentage	Grade	Count	Percentage
I upper class	1	0.42	I upper class	8	3.33
II upper middle class	21	8.75	II upper middle class	59	24.58
III lower middle class	124	51.66	III middle class	123	51.25
IV upper lower class	91	37.92	IV lower middle class	49	20.42
V lower class	3	1.25	V lower class	1	0.42
Total	240	100	Total	240	100

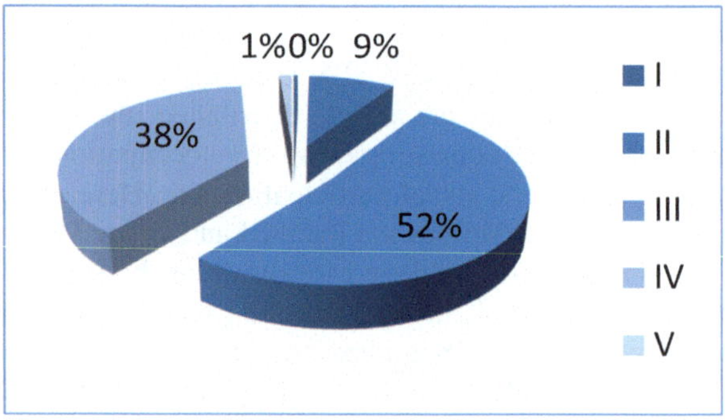

Fig. 29.3 Overall percentage shares of SES of Rongmei community in Tamenglong and Noney districts of Manipur using Kuppuswamy SES grading tool

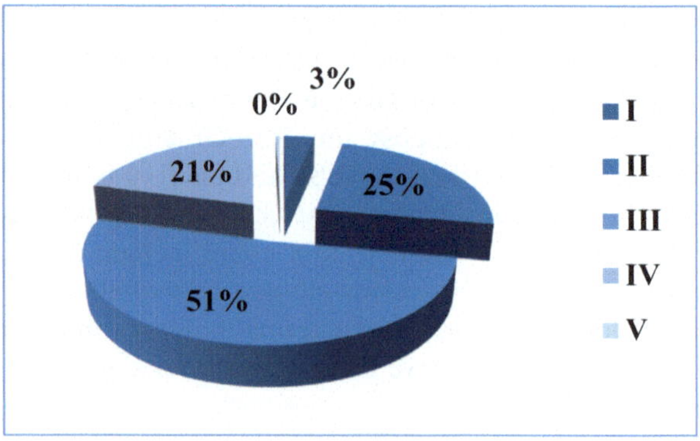

Fig. 29.4 Overall percentage shares of SES of Rongmei community in Tamenglong and Noney districts of Manipur using Udai Pareek SES grading tool

Table 29.5 Primary occupation/livelihood classifications based on Q2 of Udai Pareek SES grading tools

Primary Occupation	Count	Percentage
None (Unemployed)	1	0.42
Laborer	17	7.08
Caste occupation	0	0
Business	18	7.5
Independent profession	58	24.2
Cultivation	118	49.2
Service	28	11.6
Total	240	100

conducted in the rural Rongmei villages. Differences in primary occupation are shown in Table 29.5.

The studied Rongmei community household has diverse livelihoods or primary income generation activities. The percentage of households in Caste occupation is non-existent in the villages as the Rongmei community does not have caste and class distinction and discrimination. Unemployment rates are meager as they are industrious and have rich traditional knowledge of exploiting the environment and existing situations to sustain their livelihood. The study revealed that the number of households practicing agriculture as the primary source of income is highest at 118 out of 240, i.e., 49.2%. The number of families involved in independent professions and self-employment is 24.2%. A mere 11.6% are employed as government servants or in formal institutions like schools. Creating a new district has also given scope for working in masonry and labour works for 7.08% of the studied households.

The survey has revealed that individuals or households, as well as village-level community organizations, are utilizing bamboo for deriving income from various bamboo-based livelihood activities, including the collection of fresh shoots from individually owned, wild or community-owned lands, shoot fermenters, commercial growers, artisan, sellers/vendors of NTFP and Jhum produce whose major items is bamboo shoots in its season or fermented and dried shoots during and off-season. Diffrent uses of bamboo by the community members are shown in (Annexure I) Plates 29.1, 29.2, and 29.3. Communities that depend on healthy forests, for instance, bamboo forests, for alternative livelihood by extracting NTFPs and other important ecosystem services have better rural economies and higher adaptive capacity (IHCAP 2018, 2019). Others are involved in more than one bamboo-based economic activity, such as a grower, who can also be an artisan, a collector, having a family-owned bamboo plantation, etc. The involvement of rural community members and households in bamboo-based sectors like growing, management, and marketing provides various livelihood options that can help lift them out of poverty (INBAR 2014; Panmei and Selvan 2022; Phimmachanh et al. 2015). Vulnerable communities can adapt to climate change and make themselves resilient to climate shocks through economic development through bamboo-based livelihood diversifications (INBAR 2014).

Community-Based-Adaptation taken up at the village level with assistance from government agency is seen in the study site. It is learnt through interactions with community/village members that in Duiluan Village, the Youth Society invested 40,000 INR in June and July of 2022 buying *Dendrocalamus hamiltonii* Gamble shoots from collectors and made a profit of 20,000 INR in November 2022. Moreover, with funds from NEC and GOI, the village authority has started *Bambusa blumeana* Schult. & Schult. f. plantation in about 5 Hectares of community lands within the village lands. It is to be noted that the land for the plantation was reclaimed/restored land that had been left fallow from Jhum cultivation in the preceding years. Petrosillo et al. (2023) implores that rehabilitation of degraded lands can be successfully achieved through careful selection of crop species that can withstand the existing edaphic and climatic challenges. Reclamation of Jhum lands with bamboo is seen as a lucrative option for rural communities as forest lands are restored, carbon offsetting and other ecosystem services derived from such bamboo forests can be sustainably exploited that can reduce anthropogenic disturbance to the fragile forests ecosystem (FAO and INBAR 2018; Guleria et al. 2020; INBAR 2023b; Pan et al. 2023; Petrosillo et al. 2023). The '*Utthan*' (uplift) successful case of land restoration with bamboo in Allahabad, Uttar Pradesh, India (INBAR 2014) can be taken as a model for further development and assimilated into the current study area's abandoned Jhum lands. The project rehabilitated agricultural lands in 96 villages in Allahabad that had been stripped of its topsoil because of the brickfields. As evident from the project mentioned above, the Jhum lands can be restored by adding humus from bamboo leaf litter and providing raw materials for local artisans, which can also significantly help pressure the natural forests.

The various livelihood activities explored and adopted by the Rongmei community are documented during the field surveys. The livelihood diversification activities are shown in Table 29.6 and Fig. 29.5, respectively.

Table 29.6 Livelihood diversity/classifications of Rongmei community in Tamenglong and Noney districts of Manipur based on involvement in Bamboo-based activities

Classifications	Count	Percentage (in 240 samples)	Percentage (in the 169 bamboo-based livelihood diversity samples)	
Growers	67	27.92	Growers	39.64
Collectors	48	20	Collectors	28.40
Artisan	38	15.83	Artisan	7.69
Sellers	3	1.25	Sellers	22.49
Fermenters	13	5.42	Fermenters	1.78
Sub total including Mixed category (29 households involved in two or more of the above activities)	169	70.42	Mixed	17.6
None	71	29.58		

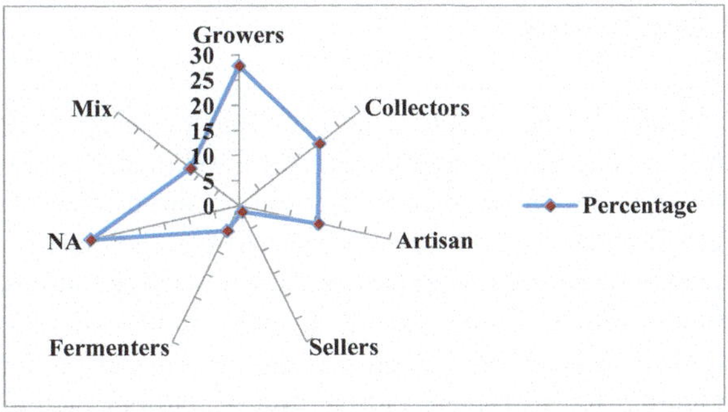

Fig. 29.5 Percentage shares of Rongmei community (in Tamenglong and Noney districts of Manipur) members' involvement in Bamboo-based livelihood activities

Bamboo is the world's most important NTFP, and most rural communities view bamboo as a shared resource that can be conveniently exploited for a part-time or seasonal supplement to other livelihood pursuits (FAO and INBAR 2018; INBAR 2014, 2023b). This statement is echoed in the current chapter as the study has revealed that the number of households dependent on bamboo for self-consumption and utilization and derivation of monetary income is very high. The activities are divided into different classes viz. Growers, Collectors, Artisan, Sellers, Fermenters, and Mixed. Growers form the largest group of the population i.e., 27.92% of the whole studied population of 240 or 39.64% of the 169 households who have claimed to be utilizing bamboo for livelihood or income-generating activities. Growers are those individuals or households that have purposely planted bamboos for extracting shoots and fulfilling requirements for domestic use and sale of poles. Incorporation of bamboo to existing agroforestry land use such as ginger (*Zingiber officinale*) which is a common crop in the Jhum of the study sites could yield substantial benefit (Sharma et al. 2022). Collectors who form 20% of the sample collect edible fresh shoots from the wild in large volumes for self-consumption, sale to fermenters and vegetable vendors, or for processing and preserving for future use. A small yet significant portion (15.83%) of the studied population is found to be skilled bamboo artisans. Sellers or vendors are those individuals who sell shoots brought to them by collectors, making up 1.25% of the surveyed population. Fermenters, which comprise 5.42% of the whole surveyed population, have large fermenting units for fermenting the bamboo shoots, especially of *Dendrocalamus hamiltonii* Gamble. Bamboo shoot fermenting is one of the oldest, most profitable, and most sustainable traditional industries for the Rongmei ethnic community. In the Mixed category are those households or individuals involved in two or more of the above-mentioned diverse classifications in bamboo-based livelihood options. The proportion of households and individuals in this class stands at a fair 17.16% of the total 169 households involved.

Conclusions

Climate change dispenses unforeseen challenges to the vulnerable populations, and they are encouraged to pursue livelihood diversification to adapt to and transform in this dynamic environment. It could be summarized that villages that are located in remote locations and far away from towns and markets have poorer populations. Vast bamboo resources, rich traditional knowledge, and market demands have led many households to explore various livelihood diversification activities. Lasting solutions to climate issues will require more than local communities' efforts. There is vast potential for sustainable utilization of bamboo resources, and with encouragement from policymakers, community leaders, and NGOs, the Rongmei community can continue to pursue bamboo-based livelihoods sustainably. Increase in the number of growers and reclamation of jhum fallows can significantly reduce pressure on the forests. This will also mitigate vulnerabilities and risks associated with womenfolk who venture into the fragile topographic situation in the monsoons, the bamboo shoot season. Over time, the population will increase, and education and modern lifestyle may soon lead to abandoning traditional bamboo-related heritage. If community artisans impart the rich art of crafting traditional sustainable products like baskets and other bamboo crafts, then the indigenous traditional knowledge can be preserved. Diversification in livelihood activities with bamboo can be seen as a solution to enhance their adaptive capacity and make them resilient to climate change extremes and variability. Thus, pursuing bamboo-based EbA will lead the vulnerable society to sustainable development through the achievement of SDG 1 (No Poverty), SDG 2 (Zero Hunger), SDG 8 (Decent Work and Economic Growth), SDG 11 (Sustainable Cities and Communities), SDG 12 (Responsible Production and Consumption), SDG 13 (Climate Action) and SDG 14 (Life on Land).

Communities vulnerable to climate change impacts in different regions of the world that has access to rich natural forests can adopt various sustainable forest-based livelihood options to curb themselves against poverty and climate related vulnerabilities. Case studies such as the current work on Rongmei community of India's Eastern Himalayas with attention and assistance from relevant institutions, government and non-government organizations can help in achieving sustainable development.

Acknowledgements The authors are indebted to the Rongmei community leaders and experts, household members, artisans, and guides from all the 6 selected Rongmei villages viz. Chaengdai, Duiluan, Khumji, Makhuam, Puiluan and Longrang of Tamenglong and Noney Districts of Manipur for assisting in the field work and providing valuable information, without which the study would have been impossible. Due gratitude also goes to the Range Officer of Tamenglong Forest Range, Manipur, India. The authors also express sincere gratitude to the Department of Forestry and Biodiversity, Tripura University, for providing research facilities.

Annexure I

Field Photographs.

See Plates 29.1, 29.2 and 29.3.

Plate 29.1 Bamboo fermentation unit in Khumji village, Tamenglong, Manipur, India

Plate 29.2 Bamboo Artisan with his crafts in Longrang village, Noney, Manipur, India

Plate 29.3 Rongmei women in Chaengdai village, Tamenglong, Manipur, India preparing for traditional dance holding bamboo culm of Melocanna clarkei (Gamble)

References

Abalo EM, Divine PP, Appiah DO, Sarpong BSA, Amankwaa G, Nakoja Y (2017) Perceived synergy between deforestation and/or forest degradation and climate variability and change in the Ejisu-Juaben Municipality, Ghana. Environ Socio-Econ Stud, Sciendo 5(4):40–56

Agrawal A, Gibson CC (1999) Enchantment and disenchantment: the role of community in natural resource conservation. World Devel 27(4):629–649

Bolund P, Hunhammar S (1999) Ecosystem services in urban areas. Ecol Econ 29(2):293–301. https://doi.org/10.1016/S0921-8009(99)00013-0

CARE International (2013) Annual report 2013: fighting poverty and empowerment, women and girls in the poorest communities in the world. CARE Internationall Secretariat, Geneva, Switzerland. CARE-International-Annual-Report-2013.pdf

Cattermoul B, Townsley P, Campbell J (2008) Sustainable livelihoods enhancement and diversification (SLED) 2008 a manual for practitioners. IUCN, Gland, Switzerland http://www.iucn.org/about/work/programmes/marine/marine_resources/marine_publications/?2653/SLED

CBD (2009) Connecting biodiversity and climate change mitigation and adaptation: report of the second Ad Hoc technical expert group on biodiversity and climate change. Technical Series No. 41. CBD, Montreal, Quebec, Canada. https://www.cbd.int/doc/publications/cbd-ts-41-en.pdf

CBD (2014) Promoting synergies in addressing biodiversity and climate change adaptation issues: linking national adaptation plans and national Biodiversity strategies and action plans. https://www.cbd.int/doc/meetings/cop/cop-12/information/cop-12-inf-29-en.pdf. Accessed 17 Aug 2023

CBD (2016) Ecosystem approach. https://www.cbd.int/ecosystem/ Accessed 17 Aug 2023

CBD (United Nations Convention on Biological Diversity) (2017) 12 Principles of ecosystem-based approaches. https://www.cbd.int/ecosystem/principles.shtm Accessed 17 Aug 2023

Charmakar S, Chivhenge E (2018) Community forest enterprises in developing countries: a potential strategy for climate change mitigation and adaptation. In: Forest in climate change research and policy: the role of forest management and conservation in a complex international setting. At: Pietermaritzburg and Durban, South Africa

Cheng SH, MacLeod K, Ahlroth S, Onder S, Perge E, Shyamsundar P, Rana P, Garside R, Kristjanson P, McKinnon MC, Miller DC (2019) A systematic map of evidence on the contribution of forests to poverty alleviation. Environ. Evid 8(1):1–22

Chew SC, Sarabia D (2016) Nature–culture relations: early globalization, climate changes, and system crisis. Sustainability 8(78):1–29 https://www.mdpi.com/2071-1050/8/1/78

Colls A, Ash N, Ikkala N (2009) Ecosystem-based adaptation: a natural response to climate change. IUCN, Gland, Switzerland

Cruz A (2015) Holistic adaptation needed for smallholders in the Philippines. Agroforestry world blog, 2015 http://blog.worldagroforestry.org/index.php/2015/12/17/

De Block D (2018) Entrepreneurship in ecosystem-based adaptation to climate change. Ph.D thesis, Wageningen University, Wageningen, the Netherlands (2018). 210 pp. https://doi.org/10.18174/442949

DST (2021) Climate vulnerability assessment for adaptation planning in india using a common framework. DST, GOI and SDC

Enríquez-de-Salamanca A, Díaz RD, Martín-Aranda RM, Santos MJ (2017) Environmental impacts of climate change adaptation. Environ Impact Assess Rev 64:87–96

FAO (2007) Adaptation to climate change in agriculture, forestry and fisheries: perspective, framework and priorities; FAO inter-departmental working group on climate change, Rome, Italy

FAO (2009) Luiz Alvarez Welchez agroforestry expert at the FAO lempira extension system project (SEL), Lempira, Honduras. In: Sudmeier-Rieux K, Ash N (eds) Environmental guidance note for disaster risk reduction: healthy ecosystems for human security, IUCN, Gland, Switzerland

FAO, INBAR (2018) Bamboo for land restoration. INBAR policy synthesis report 4. INBAR, Beijing, China

FEBA (Friends of Ecosystem-based Adaptation) (2017) Making ecosystem-based adaptation effective: a framework for defining qualification criteria and quality standards (FEBA technical paper developed for UNFCCC-SBSTA 46). In: Bertram M, Barrow E, Blackwood K, Rizvi AR, Reid H, von Scheliha-Dawid S (eds) GIZ, Bonn, Germany, IIED, London, UK, and IUCN, Gland, Switzerland. 14 pp

Ferrara C, Carlucci M, Grigoriadis E, Corona P, Salvati L (2017) A comprehensive insight into the geography of forest cover in Italy: exploring the importance of socioeconomic local contexts. For Policy Econ 75:12–22

FSI (2019) India state of forest report—2019. Forest Survey of India. Dehradun, Uttarakhand, India

FSI (2021) India state of forest report—2021. Forest Survey of India. Dehradun, Uttarakhand, India

Geneletti D, Zardo L (2016) Ecosystem-based adaptation in cities: an analysis of European urban climate adaptation plans. Land Use Policy 50:38–47. https://doi.org/10.1016/j.landusepol.2015.09.003

Girot P, Ehrhart C, Oglethorpe J (2012) Integrating community and ecosystem-based approaches in climate change adaptation, global wetlands portal bio. http://global.wetlands.org/Portals/0/Major%20Projects/WLP/Lessons%20Learnt%20in%20Aceh%20(English)-GC%202.pdf

Guleria V, Choudhary P, Vashisth GA, Selvan T (2020) Studies on the physical characteristics, properties and productivity potential of exotic and indigenous bamboos in rainfed subtropical environment. J Pharmacognosy Phytochem 9(3):1552–1558

Hart DET, Yeo S, Almaraz N, Beillouin D, Cardinael R, Garcia E, Kay S, Lovell ST, Rosenstock TS, Sprenkle-Hyppolite S, Stolle F, Suber M, Thapa B, Wood S, Cook-Patton SC (2023) Priority science can accelerate agroforestry as a natural climate solution. Nature Climate Change. https://doi.org/10.1038/s41558-023-01810-5

Hashida Y, Lewis DJ (2019) The intersection between climate adaptation, mitigation, and natural resources: an empirical analysis of forest management. J Assoc Environ Resour Economists 6(5):893–926

Heller NE, Zavaleta ES (2009) Biodiversity management in the face of climate change: a review of 22 years of recommendations. Biol Conserv 142(1):14–32. http://linkinghub.elsevier.com/retrieve/pii/S000632070800387X

Heubach K, Wittig R, Nuppenau EA, Hahn K (2011) The economic importance of non-timber forest products (NTFPs) for livelihood maintenance of rural West African communities: a case study from northern Benin. Ecol Econ 70(11):1991–2001

Hoque SF, Quinn C, Sallu S (2018) Differential livelihood adaptation to social-ecological change in coastal Bangladesh. Reg Environ Change 18(2):451–463

Huq S, Reid H (2007) Community-based adaptation: a vital approach to the threat climate change poses to the poor. International Institute for Environment and Development, London, UK

Huq N, Renaud F, Sebesvari Z (2013) Ecosystem based adaptation (EbA) to climate change—integrating actions to sustainable adaptation. United Nations University, Institute for Environment and Human Security (UNU-EHS)

IHCAP (2018) Climate vulnerability assessment for the Indian Himalayan region using a common framework. Indian Himalayas Climate Adaptation Programme, Department of Science and Technology, Ministry of Science & Technology Government of India NMSHE. http://www.ihcap.in/resources.html

IHCAP (2019) Climate vulnerability and risk assessment: framework, methods and guidelines for the indian himalayan region. New Delhi, NMSHE, DST, GOI and SDC

INBAR (2014) INBAR policy synthesis report-bamboo: a strategic resource for countries to reduce the effects of climate change. INBAR, Beijing, China

INBAR (2023a) Combating climate change with Bamboo. INBAR, Bamboo Rattan Update 4(1):1–24. https://www.inbar.int/resources/inbar_publications/bru-4-1/

INBAR (2023b) Rehabilitating land with Bamboo. INBAR, Bamboo Rattan Update 4(2):1–24. https://www.inbar.int/resources/inbar_publications/bru-4-2/

IPCC (2007) Climate change 2007: impacts, adaptation and vulnerability. In: Parry ML, Canziani OF, Palutikof JP, van der Linden PJ, Hanson CE (eds) Working group II contribution to the fourth assessment report of the intergovernmental panel on climate change. Cambridge University Press, Cambridge, UK. https://www.ipcc.ch/site/assets/uploads/2018/03/ar4_wg2_full_report.pdf Accessed 17 Aug 2023

IPCC (2012) Summary for policymakers. In: Field CB et al (eds) Managing the risks of extreme events and disasters to advance climate change adaptation. Cambridge University Press, Cambridge. pp 1–19. http://ebooks.cambridge.org/ref/id/CBO9781139177245

IPCC (2014) Climate Change 2014: impacts, adaptation, and vulnerability, contributions of WGII AR5. In: RK Pachauri, LA Meyer (Eds). Climate Change 2014: Synthesis Report. Contribution of Working Groups I, II and III to the Fifth Assessment Report of the IPCC. IPCC, Geneva, Switzerland p 151

Islam SN, Winkel J (2017) Climate change and social inequality. UNDESA. DESA Working Paper No. 152ST/ESA/2017/DWP/152. https://www.un.org/esa/desa/papers/2017/wp152_2017.pdf

IUCN (2008) The economic value of marine and coastal biodiversity to the maldives economy. http://cmsdata.iucn.org/downloads/policy_briefs_the_economic_value_of_marine_and_coastal.pdf

IUCN (2011) Draft principles and guidelines for integrating ecosystem-based approaches to adaptation in project and policy design. https://portals.iucn.org/library/efiles/documents/2011-064.pdf

IUCN (2014) Nature based solutions for human resilience: a mapping analysis of IUCN's ecosystem-based adaptation projects. https://www.iucn.org/sites/dev/files/content/documents/eba_in_iucn_mapping_analysis.pdf

IUCN (2016) Nature-based solutions to address global societal challenges. In: Cohen-Shacham E, Walters G, Janzen C, Maginnis S (eds) https://portals.iucn.org/library/sites/library/files/documents/2016-036.pdf

IUCN (2020) IUCN global standard for nature-based solutions. A user friendly framework for the verificatiob, design and scaling up of NbS, 1st edn. https://portals.iucn.org/library/sites/library/files/documents/2020-020-En.pdf

Jezeer RE, Verweij PA, Boot RG, Junginger M, Santos MJ (2019) Influence of livelihood assets, experienced shocks and perceived risks on smallholder coffee farming practices in Peru. J Environ Manage 242:496–506

Jones HP, Hole DG, Zavaleta ES (2012) Harnessing nature to help people adapt to climate change. Nat Clim Chang 2(7):504–509. https://doi.org/10.1038/nclimate1463

Kabisch N, Korn H, Stadler J, Bonn A (2017) Nature-based solutions to climate change adaptation in urban areas-linkages between science, policy and practice. In: Kabisch N, Korn H, Stadler J, Bonn A (eds) Nature-based solutions to climate change adaptation in urban areas. Theory and practice of urban sustainability transitions. Springer, Cham. https://doi.org/10.1007/978-3-319-56091-5_1

Khanal U, Wilson C, Hoang VN, Lee B (2018) Farmers' adaptation to climate change. Its determinants and impacts on rice yield in Nepal. Ecol Econ 144:139–147. https://doi.org/10.1016/j.ecolecon.2017.08.006

Klein RJT, Eriksen SEH, Naess LO, Hammill, A, Tanner, TM, Robledo C, O'Brien KL (2007) Portfolio screening to support the mainstreaming of adaptation to climate change into development assistance. Climatic Change 84(1):23–44. http://www.springerlink.com/index/10.1007/s10584-007-9268-x

Ling M, Christensen M, Donnison A, Belmonte KD, Brown C (2016) Scoping study to inform the global assessment of Bamboo and Rattan (GABAR). Prepared for the international network of Bamboo and Rattan (INBAR). UN Environment World Conservation Monitoring Centre (UNEP-WCMC). UNEP, 219 Hurlington Road, Cambridge, UK

Lobovikov M, Schoene D, Yiping L (2012) Bamboo in climate change and rural livelihoods. Mitig Adapt Strateg Glob Change 17:261–276. https://link.springer.com/article/10.1007/s11027-011-9324-8

Locatelli B, Kanninen M, Brockhaus M, Colfer CJP, Murdiyarso D, Santoso H (2008) Facing an uncertain future: how forests and people can adapt to climate change; Center for International Forestry Research (CIFOR), Bogor, Indonesia

Loushambam RS, Singh NR, Taloh A, Mayanglambam S (2017) Bamboos in North East India. Indian J Hill Farming 30(2):181–185

Luís S, Vauclair CM, Lima ML (2018) Raising awareness of climate change causes? Cross-national evidence for the normalization of societal risk perception of climate change. Environ Sci Policy 80:74–81. https://doi.org/10.1016/j.envsci.2017.11.015

Marin M, Clinciu I, Tudose NC, Ungurean C, Adorjani A, Mihalache AL, Davidescu AA, Davidescu ŞO, Dinca L, Cacovean H (2020) Assessing the vulnerability of water resources in the context of climate changes in a small forested watershed using SWAT: a review. Environ Res 184:109330

Marsh J, Smith N (2007) New bamboo industries and pro-poor impacts: lessons from China and potential for Mekong countries. International conference on managing forests for poverty reduction: capturing opportunities in forest harvesting and wood processing for the benefit of the poor, Ho-Chi Minh City, Vietnam, FAO, RECOFTC and SNV

McLean N, Lawson CR, Leech DI, van de Pol M (2016) Predicting when climate-driven phenotypic change affects population dynamics. Ecol Lett 19(6):595–608

MEA (2005) Millennium ecosystem assessment—ecosystems and human well-being: biodiversity synthesis. World Resources Institute, Washington

Mercer J, Kelman I, Alfthan B, Kurvits T (2012) Ecosystem-based adaptation to climate change in Caribbean small island developing states: integrating local and external knowledge. Sustainability 4(8):1908–1932. http://www.mdpi.com/2071-1050/4/8/1908/

Mukherjee M, Shaw R (2021) Ecosystem-based disaster and climate resilience-integration of blue-green infrastructure in sustainable development. Springer Nature Singapore Pte Ltd. 2021. eBook ISBN: 978-981-16-4815-1 Edition Number 1. XVII. https://doi.org/10.1007/978-981-16-4815-1

Munang R, Andrews J, Alverson K, Mebratu D (2013) Harnessing ecosystem-based adaptation to address the social dimensions of climate change. Environ: Sci Policy Sustain Dev 56(1):18–24. https://doi.org/10.1080/00139157.2014.861676

NASEM (2016) Accounting for social risk factors in medicare payment: identifying social risk factors. National academies of sciences, engineering and medicine, Washington DC. The National Academies Press, Washington DC. p 110. http://www.nap.edu/21858

Neef A, Pauli N (2020) Climate-induced disasters in the Asia-Pacific region-from response and recovery to adaptation (Community, environment and disaster risk management), vol. 22. Emerald Publishing Limited, Bingley, pp 1–9. https://doi.org/10.1108/S2040-726220200000022001

Niemela J, Saarela SR, Sodermann T, Kopperoinen L, Yli-Pelkonene V, Vare S, Kotze DJ (2010) Using the ecosystem services approaches for the better planning and conservation of urban green spaces: a Finland case study. Biodivers Conserv 19(11):3225–3243. https://doi.org/10/1007/s10531-010-9888-8

O'Brien K, Sygna L, Leichenko R, Adger WN, Barnett J, Mitchell T, Schipper L, Tanner T, Vogel C, Mortreux C (2008) Disaster risk reduction, climate change adaptation and human security. Report prepared for the Royal Norwegian ministry of Foreign affairs by the global environmental change and human security (GECHS) project, GECHS report, p 3

Pan C, Zhou G, Shrestha A, Chen J, Kozak R, Li N, Li J, He Y, Sheng C, Wang G (2023) Bamboo as a Nature-Based Solution (NbS) for Climate Change Mitigation: Biomass, Products, and Carbon Credits. Climate 11(175):1–22. https://doi.org/10.3390/cli11090175

Panmei R, Gajurel PR, Singh B (2019) Ethnobotany of medicinal plants used by the Zeliangrong ethnic group of Manipur, northeast India. J Ethnopharmacol 235:164–182

Panmei L, Selvan T (2022) Bamboo-based green economic development of Rongmei community In: Collection of abstracts published for the proceedings of the second global bamboo and rattan congress (BARC 2022) thematic area 2. Green industry and economic recovery. INBAR, p 62 https://www.inbar.int/resources/inbar_publications/barc-2022-abstracts/

Pérez ÁA, Fernández BH, Gatti RC (2010) Building resilience to climate change building: resilience to climate change ecosystem-based adaptation and lessons from the field. IUCN, Gland, Switzerland

Petrosillo I, Valente D, Scavuzzo CM, Selvan T (2023) Editorial: land degradation pattern and ecosystem services. Front Environ Sci 11:1137768. https://doi.org/10.3389/fenvs.2023.1137768

Phimmachanh S, Ying Z, Beckline M (2015) Bamboo resources utilization: a potential source of income to support rural livelihoods. Appl Ecol Env Sci 3(6):178–183

Popkova EG, Shi X (2022) Economics of climate change: global trends, country specifics and digital perspectives of climate action. Front Environ Econ 1(935368):1–4. https://doi.org/10.3389/frevc.2022.935368

Pouliotte J, Smit B, Westerhoff L (2009) Adaptation and development: livelihoods and climate change in Sudarnabad, Bangladesh. Clim Dev 1(2009):31–46

Premlata T, Sharma V, Bisht MS, Nirmala C (2020) Edible bamboo resources of Manipur: consumption pattern of young shoots, processing techniques and their commercial status in the local market. Indian J Tradit Knowl 19(1):73–82

Rahman HMD, Alam K (2016) Forest dependent indigenous communities' perception and adaptation to climate change through local knowledge in the protected area—a Bangladesh case study. Climate 4(12):1–25. https://doi.org/10.3390/cli4010012

Ramakrishnan PS (2007) Sustainable mountain development: the Himalayan tragedy. Curr Sci 92(3):308–316

Rasmussen LV, Watkins C, Agrawal A (2017) Forest contributions to livelihoods in changing agriculture-forest landscapes. For Policy Econ 84:1–8

Ravindranath NH, Joshi NV, Sukumar R, Saxena A (2006) Impact of climate change on forest in India. Curr Sci 90:354–361

Reid H, Argumedo A, Swiderska K (2018) Ecosystem-based approaches to adaptation: strengthening the evidence and informing policy-Research Results from the Potato Park and the Indigenous Peoples Biocultural Climate Change Assessment, Peru. IIED, London, UK. http://pubs.iied.org/17619IIEED

Reid H, Jones XH, Porras I, Hicks C, Wicander S, Seddon N, Kapos V, Rizvi AR, Roe D (2019) Is ecosystem-based adaptation effective? Perceptions and lessons learned from 13 project sites. IIED, London, UK. http://pubs.iied.org/17651IIED

Renaud FG, Sudmeir-Rieux K, Estrella M, Nehren UM (2016) Ecosystem-based disaster risk reduction and adaptation in practice. Springer Science and Business Media LLC, 2016. ISBN: 978-3-319-43631-9

Rodima-Taylor D, Olwig MF, Chhetri N (2012) Adaptation as innovation, innovation as adaptation: an institutional approach to climate change. Appl Geogr 33(2012):107–111

Salzmann N, Huggel C, Nussbaumer SU, Ziervogel G (2016) Climate change adaptation strategies—an upstream-downstream perspective, 1st edn. 292 p. Springer International Publishing Switzerland 2016. eBook ISBN: 978-3-319-40773-9 https://doi.org/10.1007/978-3-319-40773-9

Sathyan AR, Funk C, Aenis T, Winker P, Breur L (2018) Sensitivity analysis of a climate vulnerability index—a case study from Indian watershed development programmes. Clim Change Responses 5(1)

Schroth G, Laderach P, Dempewolf J, Philpott S, Haggar J, Eakin H, Castillejos T, Moreno JG, Pinto LS, Hernandez R, Eitzinger A, Ramirez-Villegas J (2009) Towards a climate change adaptation strategy for coffee communities and ecosystems in the Sierra Madre de Chiapas, Mexico. Mitig Adapt Strateg Glob Change 14:605–625

Scoones I (1998) Sustainable rural livelihoods: a framework for analysis. IDS Working Paper 72. http://opendocs.ids.ac.uk/opendocs/bitstream/handle/123456789/3390/Wp72.pdf?sequence=1

Seddon N, Daniels E, Davis R, Chausson A, Harris R, Hou-Jones X, Huq S, Kapos V, Mace GM, Rizvi AR, Reid H, Roe D, Turner B, Wicander S (2020) Global recognition of the importance of nature-based solutions to the impacts of climate change. Glob Sustain 3(e15):1–12. https://doi.org/10.1017/sus.2020.8

Selvan T (2018) Bamboo resources, their status, conservation and strategies for improvement. In: Sood KK, Mahajan V (eds) Forest, climate change and biodiversity. Kalyani Publishers, New Delhi, India, pp 263–286

Selvan T, Kumar S (2017) Agroforestry in the North-Eastern Himalayas. In: Gupta SK, Panwar P, Kaushal R (eds) Agroforestry for increased production and livelihood security. New India Publishing Agency, New Delhi, India, pp 23–40

Selvan T, Tripathi KM (2017) Economic potential of Bamboo for rural livelihood. Agrobios XV(09):42–44

Selvan T, Panmei L, Murasing KK, Guleria V, Ramesh KR, Bhardwaj DR, Thakur CL, Kumar D, Sharma P, Umedsingh RD, Kalyavizhi D, Deshmukh HK (2023) Circular economy in agriculture: unleashing the potential of integrated organic farming for food security and sustainable development. Front Sustain Food Syst, Sec Agroecology Ecosyst Serv 7(2023):1–17. https://doi.org/10.3389/fsufs.2023.1170380

Seppälä R, Buck A, Katila P (2009) Adaptation of forests and people to climate change—a global assessment report; International Union of Forest Research Organizations (IUFRO), Helsinki, Finland

Sharma TP, Borthakur SK (2017) Bamboos of North-East India. EBH Publishers. Guwahati, India. ISBN: 978-93-86302-47-2

Sharma U, Bhardwaj DR, Sharma S, Sankhyan N, Thakur CL, Rana N, Sharma S (2022) Assessment of the efficacy of various mulch materials on improving the growth and yield of ginger (Zingiber officinale) under bamboo-based agroforestry system in NW-Himalaya. Agroforest Syst 96:925–940. https://doi.org/10.1007/s10457-022-00753-8

Singh RK, Shah MH, Riba T, Singh A, Padung E, Rallen O, Lego YJ, Bhardwaj AK (2018) Classification and management of community forests in Indian Eastern Himalayas: implications on ecosystem services, conservation and livelihoods. Ecol Process 7:27. https://doi.org/10.1186/s13717-018-0137-5

Tangjang S, Arunachalam A (2009) Role of traditional home garden systems in Northeast India. Indian J Tradit Knowl 8(1):47–50

Tangjang S, Arunachalam A (2017) Traditional homegardens in rural landscapes of Northeast India—A model for conserving plant species for sustainability. Indian J Hill Farm 30(2):313–325

Tengo M, Belfrage K (2004) Local management practices for dealing with change and uncertainty: a cross-scale comparison of cases in Sweden and Tanzania. Ecol Soc 9(3):4. http://www.ecologyandsociety.org/vol9/iss3/art4

Thornton PK, Ericksen PJ, Herrero M, Challinor AJ (2014) Climate variability and vulnerability to climate change: a review. Glob Change Biol 20(11):3313–3328. https://doi.org/10.1111/gcb.12581

Turner NJ (2005) The earth's blanket: traditional teachings for sustainable living. University of Washington Press, Washington

UN (2019) Climate action and support trends 2019. UN climate change secretariat, Bonn, Germany. Climate_Action_Support_Trends_2019.pdf_(unfccc.int)

UNDESA (2013) Report on world social situation 2013: inequality matters. UNDESA, UN, NY, USA. RWSS_2013-20_Jan.indd_(un.org)

UNDP (2010) Human development report 2010, 20th anniversary edition. The real wealth of nations: pathways to human development. UNDP, 1 UN Plaza, NY 10017, USA. Palgrave Macmillan, NY, USA. HDR2010ENCompletereprint-1.pdf

UNFCCC (2007) Climate Change: impacts, vulnerabilities and adaptation in developing countries. https://unfccc.int/resource/docs/publications/impacts.pdf

UNFCCC (2012) Slow onset events. Technical paper. FCCC/ TP/2012/7. UNFCCC, Bonn, Germany. https://unfccc.int/resource/docs/2012/tp/07.pdf

UNESCO, INBAR (2023) Bamboo for carbon neutrality in rural areas. Collection of selected case studies. UNESCO and INBAR, 2023, p 16. https://www.inbar.int/resources/inbar_publications/bamboo-for-carbon-neutrality-in-rural-areas-collection-of-selected-case-studies/

Varuvel GJ, Selvan T, Palaniyappan K (2021) Preferential use of bamboos for industrial production of incense sticks. Environ Sci Proc 13(1):7. https://doi.org/10.3390/IECF2021-10821

Vermeulen R, Schets E, Lohuis M, Kolbl B, Jansen DJ, Heeringa W (2018) An energy transition risk stress test for the financial system of the Netherlands, DNB Ocassional Studies 1607. Netherlands Central Bank, Research Department

Vignola R, Locatelli B, Martinez C, Imbach P (2009) Ecosystem-based adaptation to climate change: what role for policy-makers, society and scientists? Mitig Adapt Strat Global Change 14(8):691–696. http://www.springerlink.com/index/10.1007/s11027-009-9193-6

Vorontsova MS, Clark LG, Dransfield J, Govearts R, Baker WJ (2016) World atlas of Bamboos and Rattans. INBAR—international network for bamboo and rattan & the board of trustees of the Royal botanic gardens, Kew. ISBN: 978-92-990082-9-4

Wale E, Nkoana MA, Mkuna E (2022) Climate change-induced livelihood adaptive strategies and perceptions of forest-dependent communities: the case of Inanda, KwaZulu-Natal, South Africa. Trees, Forest People 8:100250. https://doi.org/10.1016/j.tfp.2022.100250

Wani RT (2019) Socioeconomic status scales-modified Kuppuswamy and Udai Pareekh's scale updated for 2019. J Family Med Prim Care 8(6):1846–1849. https://doi.org/10.4103/jfmpc.jfmpc_288_19

Warner K, Afifi T, Henry K, Rawe T, Smith C, de Sherbinin A (2012) Where the rain falls: climate change, food and livelihood security, and migration. Global policy report of the where the rain falls project. UNU and CARE, Bonn, Germany. https://www.un.org/en/development/desa/population/events/pdf/11/warner_climate_change_food_security_migration.pdf

Watson JEM, Rao M, Li KI, Yan X (2012) Climate change adaptation planning for biodiversity conservation: a review. Adv Clim Chang Res 3(1):1–11. https://doi.org/10.3724/SP.J.1248.2012.00001

Weiss G, Lawrence A, Lidestav G, Feliciano D, Hujala T, Sarvašová Z, Dobšinská Z, Živojinović I (2019) Research trends: forest ownership in multiple perspectives. For Policy Econ 99:1–8

World Economic Forum (2020) The global risk report 2020. WEF, Geneva, Switzerland. WEF_Global_Risk_Report_2020.pdf_(weforum.org)

Chapter 30
Exploring the Spatial Fabric of Urban Well-Being Through Geographically Weighted Techniques and PPGIS in Darjeeling, India

Subham Roy, Suranjan Majumder, Arghadeep Bose, Prosenjit Kayal, and Indrajit Roy Chowdhury

Abstract The purpose of this present study is to investigate the spatial distribution of urban well-being (UWB) in the town of Darjeeling, situated in the Himalayas and often referred to as the "Queen of Himalayas." However, due to its multifaceted nature, quantifying UWB proves to be a highly intricate undertaking. Hence, this study aims to empirically measure the extent of UWB and explore its geographical pattern, association, and heterogeneity. Geographically Weighted Principal Component Analysis (GWPCA) was utilized in order to establish a dimensional spatial index for developing an Urban Well-Being Index (UWBI). The research evaluates UWB by employing a composite set of criteria that encompasses socio-economic factors, natural surroundings, public and infrastructure services, recreational-cultural-spiritual amenities, built-up characteristics, and emergency services. The findings reveal that overall UWBI levels in Darjeeling are suboptimal with several domains displaying deficiencies that negatively impact resident well-being. To validate these assessments and gain insights into local perspectives on UWB, a survey involving 238 respondents was conducted. Public Participation in GIS (PPGIS) technique were utilized for spatial mapping and to validate the final UWB index. By promoting a comprehensive understanding of resident well-being through this research endeavor contributes towards enhancing well-being and sustainability in hilly regions.

S. Roy (✉) · S. Majumder · A. Bose · P. Kayal · I. R. Chowdhury
Department of Geography and Applied Geography, University of North Bengal, Siliguri, West Bengal 734013, India
e-mail: subhammn2@gmail.com; rs_subhamroy@nbu.ac.in

S. Majumder
e-mail: rs_suranjan@nbu.ac.in

A. Bose
e-mail: deeparghageo@nbu.ac.in

I. R. Chowdhury
e-mail: irchowdhury_geo@nbu.ac.in

U. Mukhopadhyay et al. (eds.), *Climate Crisis, Social Responses and Sustainability*, Climate Change Management, https://doi.org/10.1007/978-3-031-58261-5_30

Introduction

Over half of the global population currently resides in urban areas, projected to increase to more than 70% by 2050 (UN 2014). Such rapid urbanization introduces multifaceted challenges stemming from heightened environmental pressures and widening socioeconomic divides, which are intimately intertwined with urban health and well-being (Ripp et al. 2020). This swift urban expansion has intensified issues such as air and water pollution, social disparities, and other threats to human health (Chen 2023; Lafortezza et al. 2019). The intricate relationship between urban environments and health or well-being has been a focal point of research across various disciplines, including public health, urban planning, geography, and epidemiology. Prominent institutions, including the World Health Organisation (WHO), The London School of Economics and Political Science (LSE), and the International Council for Science (ICSU), have spearheaded global projects that pivot around well-being within the dynamic urban milieu.

The United Nations' Sustainable Development Goals (SDGs) represent a pivotal blueprint to reshape global trajectories. They epitomize a clarion call to eliminate poverty and inequality, protect our environment, and ensure a life of peace, justice, and prosperity for all. The SDGs provide an imperative framework for addressing urbanization and the well-being of urban inhabitants. As Vaidya and Chatterji (2020) delineate, the nexus between urbanization and well-being profoundly resonates with several SDGs, including SDG 1 (No Poverty), SDG 4 (Quality Education), SDG 5 (Gender Equality), SDG 6 (Clean Water and Sanitation), SDG 7 (Affordable and Clean Energy), SDG 8 (Decent Work and Economic Growth), and SDG 13 (Climate Action). Notably, India's urbanization trajectory has witnessed a marked surge. While the 1901 census reported 11.4% of the population in urban settings, this proportion burgeoned to 28.53% by 2001 and reached approximately 34% in 2017, according to the World Bank. Projections by the United Nations indicate that by 2030, urban inhabitants will constitute 40.76% of India's population. Hence, there is an imperative to meticulously assess urban spatial growth, the quality of urban life, the well-being of its residents, and pertinent management strategies. The concept of "well-being" remains nuanced, encompassing various dimensions and interpretations (Dodge et al. 2012; Whulanza and Kusrini 2023). Despite its prominence in academic and popular discourse, achieving a singular definition proves challenging, as terms such as "happiness," "life satisfaction," and "quality of life" are often used synonymously (Alatartseva and Barysheva 2015). Notably, subjective well-being (SWB) has garnered attention across diverse academic disciplines, including psychology, economics, sociology, and geography (Clark et al. 2008; Diener et al. 2009; Schwanen and Atkinson 2015), and has emerged as a critical objective in policy formulation (Sanna Ala-Mantila et al. 2018). Recently, the focus has evolved from broader scopes to more localized dimensions, such as the well-being of specific cities (Florida et al. 2013) or states (Oswald and Wu 2010). Urban well-being (UWB) is a comprehensive term encapsulating the myriad facets contributing to the quality of life within cities and urban settings (Sanna Ala-Mantila et al. 2018). At its core, UWB intertwines socio-economic dynamics, wherein opportunities for employment, education, and income stability play pivotal roles in determining urban dwellers' overall satisfaction and security (Roy et al. 2023a, 2024). Beyond economic dimensions, the physical environment is paramount: clean air, access to green spaces, and sustainable environmental practices directly influence residents' health and comfort (Roy et al. 2023b). Equally integral to UWB are the infrastructural underpinnings of a city, including the efficiency of public services like transportation, healthcare, and sanitation (Douglas 2012). The cultural vibrancy of an urban space, marked by recreational avenues, cultural events, and social hubs, augments a sense of belonging and community among its inhabitants (Roy et al., 2023a, b). However, the essence of UWB is not solely anchored in tangible amenities; it profoundly resonates with intangibles like safety, community cohesion, and mental well-being (Whulanza and Kusrini 2023). Cities that offer a secure environment, foster community interactions, and prioritize mental and physical health invariably elevate their urban well-being quotient (Guite et al. 2006). Also, effective governance, marked by transparency and avenues for public participation, serves as the linchpin, ensuring that the urban fabric remains responsive and adaptable to the evolving needs of its residents (Guite et al. 2006; Douglas 2012). The study on urban well-being in Darjeeling, utilizing Geographically Weighted Techniques and Public Participation in GIS, is inherently connected to the core theme of Social Responses and Sustainability. It delves into the dynamics of socio-economic elements, environment, infrastructure, and services within urban settings, highlighting how these aspects are pivotal in addressing climate change and enhancing urban resilience. Emphasizing the significance of community involvement and localized insights, this research sheds light on effective social strategies for sustainability. This participatory and localized research methodology serves as a vital link in crafting socio-ecological approaches for climate mitigation and sustainable urban planning, embodying the book's mission to blend socio-ecological views for addressing the multifaceted nature of climate change and sustainability.

Central to this discourse is the notion of "urban well-being" (UWB), which encapsulates urban populations' holistic health and contentment, influenced by myriad factors ranging from socio-economic conditions to the quality of local amenities and services. The current study introduces a novel approach to assess UWB in Darjeeling through the integration of Geographically Weighted Techniques and PPGIS. It also studies the spatial distribution of UWB in the picturesque Himalayan town of Darjeeling, often revered as the "Queen of the Himalayas." Given the multifaceted nature of UWB, devising a quantifiable measure presents inherent complexities. Thus, this research endeavours to empirically evaluate the magnitude of UWB, delving into its geographical distribution, correlations, and spatial variations. To this end, an Urban Well-Being Index (UWBI) was conceptualized and operationalized through Geographically Weighted Principal Component Analysis (GWPCA), resulting in a multi-dimensional spatial index. The study employs a comprehensive framework to appraise UWB, factoring in socio-economic dynamics, environmental considerations, public infrastructure, recreational amenities, cultural and religious facets, urban morphology, and emergency services. Further enriching this analysis, the PPGIS (Public Participation in GIS (PPGIS) technique was employed, harnessing the perceptions of 238 respondents. The objectives are twofold: first, to map and analyze the spatial patterns of UWB, revealing the nuanced interplay between environmental, socio-economic, and infrastructural factors; and second, to engage local communities directly in the research process, ensuring that their perceptions and priorities are reflected in the findings. This methodology not only provides a comprehensive understanding of UWB in the context of climate change but also pioneers a participatory framework for urban sustainability research, aiming to foster more resilient and adaptable urban environments.

Rationale of the Study Area

Darjeeling, often termed the "Queen of the Himalayas," is the primary urban hub of its district and doubles as its administrative nucleus. Its designation arises from its breathtaking landscapes, ideal climate, and cultural relevance (Fig. 30.1). Framed by the majestic Himalayas and verdant valleys, the town's renowned tea estates further augment its scenic allure. Serving as a magnet for international tourists, scholars, and migrants, Darjeeling epitomizes urban diversity (Basak et al. 2021). Established in 1850 by the British, the Darjeeling Municipality, one of India's pioneering municipalities, spans 32 yards, integrating two cantonments: Lebong and Jalapahar. Darjeeling seamlessly merges colonial architectural relics, indigenous Himalayan traditions, and contemporary urban facets. Key landmarks, including the UNESCO-listed Darjeeling Himalayan Railway and the Tiger Hill viewpoint, further underscore its distinctiveness (Sharma et al. 2020).

Understanding the "urban well-being" of high-altitude Himalayan towns like Darjeeling is of paramount importance for a multitude of reasons. Firstly, these towns, owing to their unique topography and ecology, face specific challenges related to sustainable urban development, environmental preservation, and infrastructure provision. Ensuring the well-being of residents in these settings requires a delicate balance between conserving natural landscapes and facilitating urban growth. Secondly, the cultural and historical significance of such towns demands that urban development endeavors respect and incorporate local traditions, values, and heritage. Moreover, high-altitude regions often have unique health and socio-economic considerations. The thin atmosphere and colder climate can lead to specific health challenges, while the isolation can bring about distinct socio-economic dynamics, which directly impact the well-being of the inhabitants. Additionally, as these towns often rely heavily on tourism, understanding urban well-being is crucial for ensuring sustainable tourism practices that benefit both the local community and the environment. A comprehensive grasp of urban well-being in such towns can guide policies and interventions that foster holistic and sustainable growth, benefiting both the residents and the larger ecosystem.

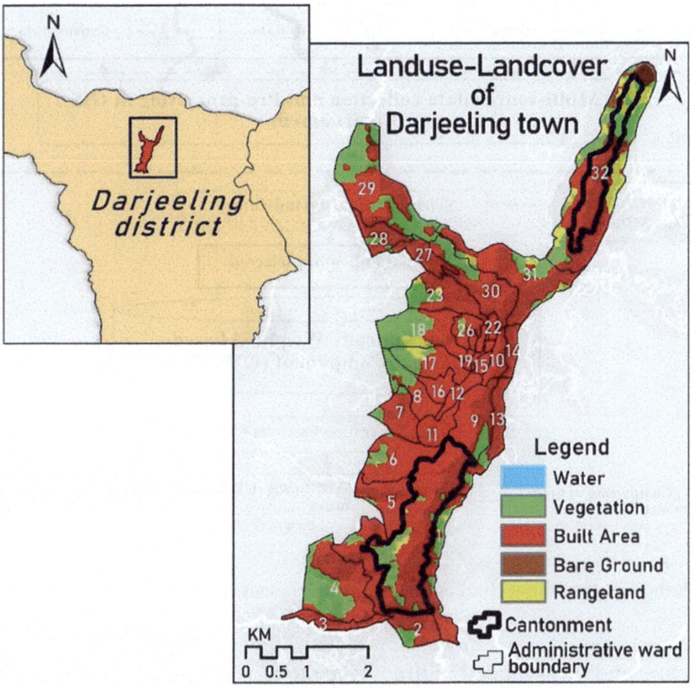

Fig. 30.1 Location map of the study area

Materials and Methodology

Data Retrieval and Processing

Figure 30.2 introduces an innovative assessment framework for UWB in Darjeeling, incorporating 12 indicators spanning nine domains. These evaluative markers derive from the previously discussed UWB conceptual framework, data accessibility, literature insights, and the distinctive physical, cultural, and historical attributes of the Darjeeling Himalaya region. The evaluation employs 12 unique variables sourced from diverse origins. The study integrates a rich array of both non-spatial and spatial datasets, segmented into six categories: GIS datasets, Landsat satellite imagery, Digital Elevation Model (DEM), Census records, Municipal archives, and primary datasets from perception surveys. GIS databases, capturing features such as roads, parks, hospitals, and police stations, comprise point and polygon data, initially extracted in shapefile formats, subsequently processed within a GIS framework. The Normalized Difference Vegetation Index (NDVI) was derived from Landsat imagery, while the Shuttle Radar Topography Mission (SRTM) DEM of 30 m resolution, sourced from NASA Earth Data, facilitated slope preparation. Given Darjeeling's

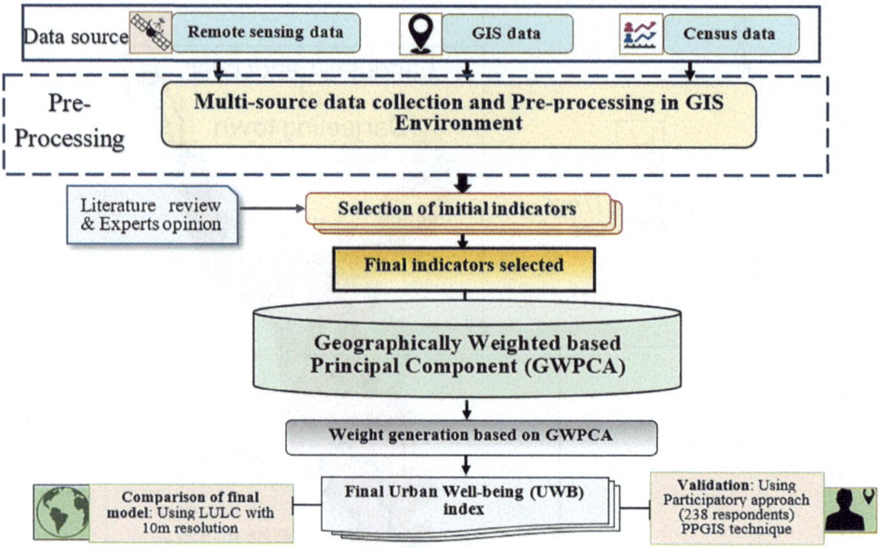

Fig. 30.2 Methodological flow chart adopted for the present study

hilly landscape and pronounced inclines, the slope is deemed a vital determinant for UWB in this study, influencing various facets of the region's holistic UWB.

Other data detailing habitable household conditions, potable water access, female literacy rates, and the working populace were acquired from the Office of the Registrar General & Census Commissioner, India. Municipal archives supplied data regarding educational institutions' locales. Evidently, this research weaves in diverse datasets of varied resolutions to meet its objectives. Notably, a substantial analytical portion relies on 30 m resolution Landsat imagery. To maintain spatial accuracy, all UWB indicators were adjusted to UTM Projection with a WGS-84 Datum in the Zone 45°N format, standardized to a uniform 30 m cell size.

Indicator Normalization

Before combining indicators into a single synthetic composite index, it is necessary to perform standardization due to the presence of variability among indicators in a data set that often exhibits varying measurement units and intervals (OECD 2008; Pareto 2021). Standardization is a technique employed in composite indicator construction to transform a group of independent indicators into a format that allows for coherent comparisons (Pareto 2021). In the study we used 'Z-score normalization' technique to transform the twelve selected indicators for the assessment of urban wellbeing into a standard dimensionless distribution with a mean of zero and a standard deviation of one. This procedure is attained by subtracting the mean of the dataset from each data point and then dividing the result by the standard deviation.

So, consider $[X_{ij}]$ as the matrix containing the raw data values for the indicators of the ith ward and the jth indicator. In this context, the element "i" represents the numbers 1, 2, 3... up to "n", which represents the total number of urban units (ward) in Darjeeling town (here, n = 32). Similarly, the element "j" represents the numbers 1, 2... up to "k", which represents the total number of indicators (here, k = 12). $[Z_{ij}]$ is the matrix of standardized indicators as follows:

$$Z_{ij} = \frac{S_{ij} - \overline{X}}{\sigma} \tag{30.1}$$

In this manner,

$$\overline{X}_j = \frac{\sum_{i=1}^{N} X_{ij}}{N} \quad \text{and} \quad \sigma_j = \sqrt{\frac{\sum_{i=1}^{N} (X_{ij} - \overline{X}_j)}{N}} \tag{30.2}$$

Here, Z_{ij} is the standardised value of 'i' indicator for 'j' urban unit, S_i is the original value of 'i' indicator for 'j' urban unit, $\overline{X}$ and σ represents the mean, and standard deviation of the dataset.

Application of Geographically Weighted PCA (GWPCA) and Construction of Urban Wellbeing Index (UWBI)

Principle Component Analysis (PCA) is an extensively employed technique for reducing dimensionality. It achieves this by maximizing variance using eigenvalues from a normalized correlation matrix and data rotation. In global PCA, the principal components (PC) are produced by executing an orthogonal transformation that minimizes collinearity across variables and capture maximum variance across the entire dataset. Nonetheless, global PCA fails to exploit the spatial non-stationarity inherent in the input data, resulting in principal component loading vectors and cumulative variances that do not encompass the geographical variations present in the dataset (Harris et al. 2011; Wang et al. 2022).

To address the limitation of the global PCA, we utilized robust GWPCA to develop the Urban Wellbeing Index (UWI) for Darjeeling town. This cutting-edge GIS-based approach takes into account the distinct geographical characteristics encompassed in the dataset. By adopting a localized approach, it allows for the creation of region-specific Principal Component Analyses, thus providing a means to account for variations that may exist across different geographic locations (Benita et al. 2020; De Silva et al. 2022; Harris et al. 2011). Further, the generation of the weighted score at the local level is the most significant aspect of GWPCA-based analysis while it is applied to generate the composite map. Various studies (e.g., Benita et al. 2020; Mishra 2018; Roy et al. 2023b) have employed the GWPCA to create a comprehensive scenario encompassing different dimensions of urban sustainability and deprivation. They propose the GWPCA as a highly sensitive technique that effectively captures the spatial heterogeneity at the local scale (Harris et al. 2015; Roy et al. 2023b; Tsutsumida et al. 2019). Additionally, it takes into account localised component scores and variances.

Moreover, before considering GWPCA, it is necessary to check that data suitability for PCA by using two distinct techniques: Kaiser–Meyer–Olkin (KMO) test and Bartlett's Test of Sphericity (Budaev 2010; Sinesilassie et al. 2018). The KMO assessment evaluates the suitability of the dataset for PCA analysis by examining the adequacy of the data's correlation pattern (Roy et al. 2022). Its value ranges from 0 to 1, with higher values indicating more suitability for factor analysis. Studies typically use the KMO test in conjunction with Bartlett's test of sphericity, which checks whether the correlation matrix is significantly different from an identity matrix. Whenever the p-value from Bartlett's test of sphericity is small (usually < 0.05), it indicates the presence of substantial correlations in the data, which makes it suitable for performing PCA (Roy et al. 2022).

The urban wellbeing map is generated using the ward-level weighted index outcome (as per Eq. 30.2) from robust GWPCA, employing the variance and scores of each observation, as demonstrated by Sekhar et al. (1991) and Roy et al. (2023a, b). The study applied a robust GWPCA incorporating an adaptive bandwidth with automatic selection based on AICc (Akaike Information Criterion corrected) and a Gaussian kernel spatial weightage scheme. The weighted value of urban wellbeing index for Darjeeling town for each ward is mathematically expressed as:

$$UWI_{WIVi} = \left\{\left(Var_{PC1_i} \times S_{PC1_i}\right) + \left(Var_{PC2_i} \times S_{PC2_i}\right) + \left(Var_{PC3_i} \times S_{PC3_i}\right)\right\} \tag{30.3}$$

Here, UWI_{WIVi} refers to the urban wellbeing index obtained from robust GWPCA based weighted index value; *Var* is the localized variance for '*i*' ward, and *S* is the localized Scores for '*i*' ward, and *PC1, PC2*, and *PC3* refers to principle components generated as the outcome of GWPCA.

Finally, as proposed by Roy et al. (2023a, b) and Sekhar et al. (1991), the weighted value of urban wellbeing is standardized using Eq. (30.3), resulting in a standardized urban wellbeing map. The resultant Urban Wellbeing Index (UWI) extends from 0 (indicating low wellbeing) to 100 (indicating optimal wellbeing).

$$\mathrm{UWBI_{STND_i}} = \frac{\mathrm{UWI_{WIV_i}} - \mathrm{Least_{UWI_{WIV}}}}{\mathrm{Highest_{UWI_{WIV}}} - \mathrm{Least_{UWI_{WIV}}}} \tag{30.4}$$

Derivation of Winning Variables Using GWPCA

GWPCA seeks to uncover the most influential variables within a dataset, assigning weights to each variable based on their contribution to the principal components. The process of deriving winning variables involves identifying those with the highest weightings or scores in these principal components. The spatial heterogeneity within the data of wellbeing indictors can serve as a means to gauge the influence of respective indicator on urban wellbeing. These winning variables are pivotal in understanding and interpreting the underlying structure of the data, enabling researchers to focus on the most relevant factors driving the variation in their dataset. By pinpointing the winning variables in GWPCA, analysts can make more informed decisions and gain valuable insights for various applications, including dimensionality reduction, feature selection, and data-driven decision-making.

Mapping Public Perception Through GIS (PPGIS) for Validation of Urban Wellbeing Assessment

In recent time, Public Participation GIS (PPGIS) is a popular approach that integrate community supports on planning and management. PPGIS is an approach that integrates GIS system with public participation approaches to involve local community members or inhabitants in spatial decision-making processes and generate place-based information through maps (Brown and Kyttä 2014; Engen et al. 2018; Waleghwa and Heldt 2022). Currently, PPGIS arise with great popularity and effectiveness in spatial information visualization that directly involved the stakeholders who shares their knowledge, and experience about the place and phenomenon for better management, and policy making (Brown and Weber 2012; Waleghwa and Heldt 2022). PPGIS is an effective application that widely used in landscape planning, urban governance, and management, urban planning, transportation planning, and environment management (Cao et al. 2015; Engen et al. 2018; Kotzebue 2022; Kahila-Tani 2015; Lehto et al. 2022; Waleghwa and Heldt 2022).

In this research we have developed a novel approach, where the urban resident perception were mapped using the GIS techniques. Mapping Public Perception through GIS (PPGIS) is an innovative technique that extends the capabilities of traditional PPGIS practices. This integrated approach provides insights regarding how surroundings influence and alter individual perceptions and preferences, generating crucial insights into the spatial dimensions of human perception. Here, all the respondents were geo-tagged to capture the spatial locational information that further add in GIS. In this research, the innovative PPGIS approach provides a comprehensive framework for analysing and visualizing the multifaceted relationship between urban residents' perception and geographical space by integrating qualitative data regarding individual perceptions and experiences with spatially explicit information. Essentially, PPGIS identifies spatial relationships, trends, and individual perceptions.

Survey Design for PPGIS

In this research, PPGIS applied to identify the perceived response and feelings of the urban residents of Darjeeling town about their wellbeing scenario and living experiences. The response used to detect the community participation on sustainable urban wellbeing condition, stakeholders knowledge on the existing urban service, and potential issues in Darjeeling. Additionally, the knowledge of the stakeholders used to map the pockets with potential bottlenecks and validate the results of UWI computed using GWPCA. The data for community participation is obtained through the use of questionnaires consisted 14 question linked with the basic components for sustainable urban living. The survey instruments includes questions related with natural environment, scenic beauty, drinking water availability, transportation viability, educational facilities, healthcare, safety, public convenience, and service. The sample consisted of 238 permanent residents of Darjeeling town as participants (Table 30.1). The participants was quite diverse in terms of age, residential span time, and educational level which made it an excellent choice for studying the community participation on urban liveability assessment that used to mapped in GIS environment.

Table 30.1 Demographic characteristics of the respondents (N = 238)

Characteristics	Frequency	Percentage
Sex		
Male	185	77.13
Female	53	22.27
Age (years)		
$\leq$ 18	13	5.46
19–30	53	22.27
31–45	37	15.55
46–60	78	32.77
61 and above	57	23.95
Educational level		
Illiterate	26	10.92
Middle school	24	10.08
< Higher secondary	128	53.79
Higher secondary and above	60	25.21
Residential spans (Years)		
< 5	43	18.07
5–10	83	34.87
11–15	52	21.85
> 15	60	25.21

Results and Discussion

Spatial Variations of Selected Variables

Figure 30.3 illustrates the distribution of the twelve variables chosen for this research. Notably, four key indicators—livable housing conditions, on-premises water availability, female literacy rates, and the working population—predominantly converge in the central sectors of Darjeeling. In stark contrast, the peripheral regions display reduced concentrations of these metrics, underscoring superior socio-economic conditions in the town's heart. There is marked variability across areas when assessing natural attributes such as slope, NDVI (Normalized Difference Vegetation Index), proximity to green spaces, and distance from primary viewpoints. The NDVI, for instance, reveals a diminished presence of natural vegetation in the town's central region, suggesting higher population densities in these parts. Conversely, the core town areas show higher concentrations for attributes like parks and gardens and major viewpoints. As for public infrastructure components—comprising roads, hospitals, educational establishments, and police facilities—the northern and central portions of Darjeeling exhibit elevated concentrations. Contrastingly, the town's southern outskirts are characterized by sparser distributions of these amenities, hinting at potential disparities in infrastructure accessibility.

Spatial Variability of Urban Well-Being and Its Winning Variables

In order to have a holistic perspective of urban well-being in high-altitude Himalayan locales, it is necessary to consider a combination of twelve distinct criteria at an integrative scale. These act as vital indicators assimilation, each offering distinct perspectives on urban wellbeing in these unique settings. By employing a weighted urban wellbeing index score based map, we can analyze the geographical distribution of urban wellbeing across the 32 wards in Darjeeling. Further, this analysis unveils a multifaceted and diverse urban wellbeing landscape within the town, with many wards exhibiting lower levels of urban wellbeing. The UWI map of Darjeeling Town, based on the GWPCA approach, classifies UWI scores into five levels using a quartile system. In this representation, areas colored in red signify very low urban wellbeing, while green areas indicate very high urban wellbeing. This visual representation offers valuable insights into the disparities and variations in urban wellbeing across the town.

Figure 30.4 reveal a noteworthy polycentric spatial pattern in the urban wellbeing of Darjeeling town. The analysis highlights that a cluster of five wards situated in the extreme northern fringes of the town, specifically wards 27, 28, 29, 30, and 31, constituting 16% of the total wards, stands out with a significantly elevated level of urban wellbeing. This exceptional urban wellbeing experience extends to

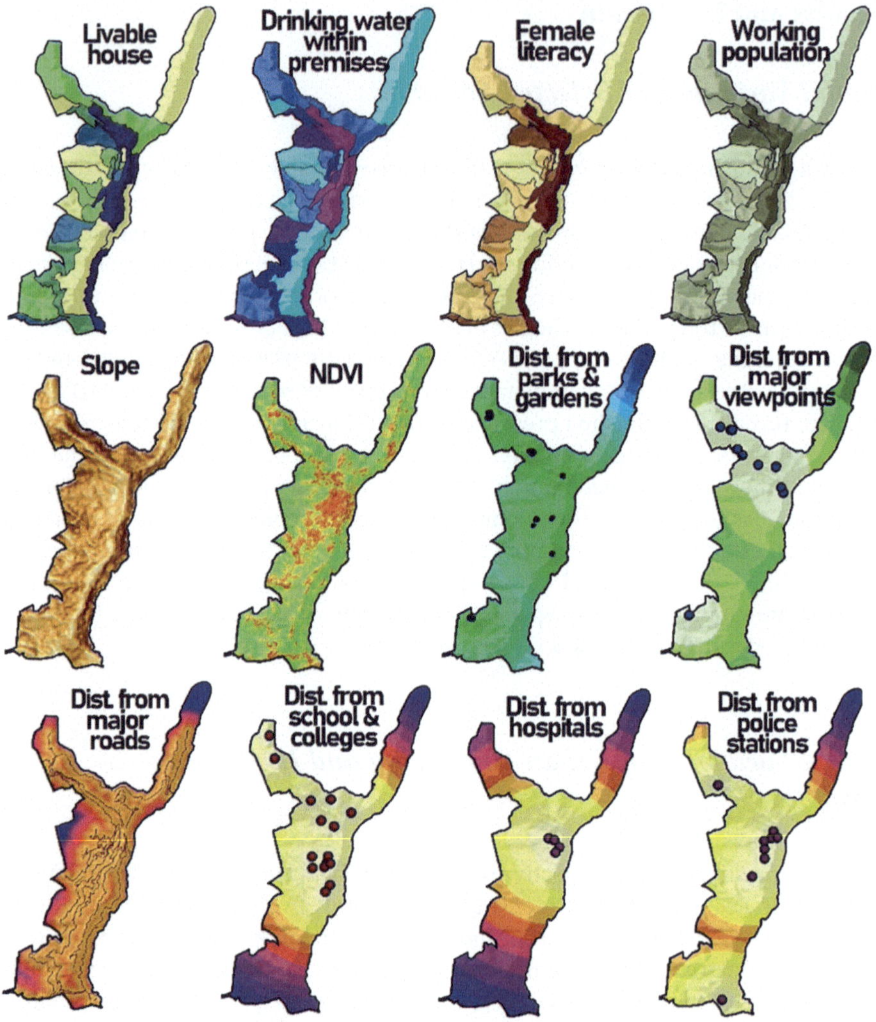

Fig. 30.3 Spatial distribution of selected variables for the present study

the region between the north and central parts of the town, where approximately 19% of the wards also exhibit a high urban wellbeing index score. The findings is particularly interesting considering over 35% of the Darjeeling town area experiences a high to very high degree of urban wellbeing. This phenomenon can be attributed to several factors, such as a higher proportion of habitable housing conditions, widespread access to drinking water within residential premises (with the exception of a few wards like ward 32 and 28), a significant female literate population, a thriving workforce, relatively gentle terrain, and close proximity to essential amenities like schools, medical facilities, and parks. These combined attributes contribute to the overall urban wellbeing of the town, creating a multifaceted landscape of well-being. The existence of low and extremely low urban wellbeing clusters in Darjeeling has significant social and economic consequences. Citizens residing in these regions may encounter difficulties pertaining to the accessibility of healthcare services, educational resources, and employment prospects. Consequently, this could have a significant influence on their general standard of living and financial possibilities.

Furthermore, the analysis revealed that seven wards in the study area exhibit a moderate level of urban wellbeing. These wards are primarily located in the central part of the city, surrounded by both high and low urban wellbeing index (UWI) wards. In contrast, another seven wards were found to have low urban wellbeing scores. Notably, this low wellbeing cluster is closely associated with factors such

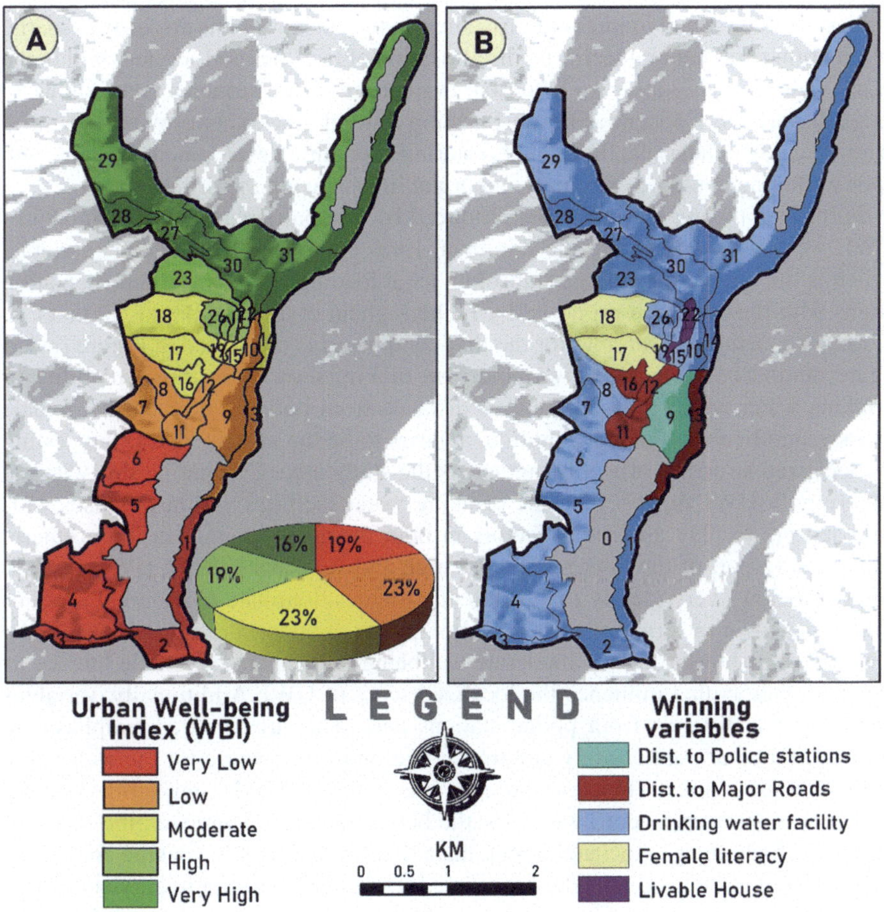

Fig. 30.4 Spatial variability of urban well-being of Darjeeling town. **a** UWBI & **b** Winning variables

as a limited presence of green spaces (as indicated by NDVI values), a comparatively smaller working population, greater distances to schools and hospitals, and reduced access to essential services. A substantial portion of Darjeeling town, approximately 42% of its wards, falls within the category of experiencing low to very low composite urban wellbeing. This points to a significant well-being disparity within the city. Particularly, the outskirts of Darjeeling, mainly in the southern regions, show a noticeable lack of standard urban wellbeing. These areas are distant from the city's central core and face various challenges and hardships in daily life, indicating a need for targeted interventions to improve the living conditions of residents in these locations (Roy et al. 2023b). Conversely, the moderate UWI class is concentrated within the inner periphery of Darjeeling town, forming a cluster between the city center, characterized by high vitality, and the northern areas, where citizens experience very high levels of wellbeing. This distribution suggests a gradient of wellbeing across the city, with a more favorable urban living experience in the central and northern regions, but significant room for improvement in the outskirts. This disparity in spatial distribution stimulates concerns on the allocation of facilities and services within the municipality, necessitating an extensive evaluation of the variables contributing to these disparities. Concentrating on boosting natural spaces, improving access to schools and healthcare facilities, and fostering economic development in areas with poor wellbeing could be extremely important in narrowing the difference.

Further analyzing the outcomes produced by this particular GWPCA model of UWI, we can also illustrate the dominant local component's 'winning' variable, which is the one with the largest absolute local loading. This visual representation aligns with the approach proposed by Fotheringham et al. (2002) and Harris et al. (2002), the 'winning' variable for the first component is illustrated in Figure [insert figure number]. The winning variable map of Fig. shows that the availability of drinking water within house premises, distance from police station, roads, female literacy, and livable housing condition, appears to be the critical factors suggesting the localized structure of wellbeing across the study area. Comparing this with the pattern with UWI suggest that, the drinking water facilities within premises variable appears to play an critical factors in both the areas (mainly in the northern, and southern part of the town) with very high and very low level of UWI. Further, GWPCA shows very clearly that different variables dominate in different areas of Darjeeling town in terms of citizen wellbeing. In contrast, the central part of the town (mostly showcasing the moderate wellbeing) appears to be female literacy as a distinct factors that influence the localized score of UWI. Additionally, variables related to the distance from police stations and roads are featured, emphasizing the importance of public safety and transportation infrastructure for the well-being of these communities. The surrounding wards with low UWI, suggesting different wellbeing factors as a contribution for the better loadings. Moreover, the repeated presence of variables associated with female literacy and livable housing conditions indicates a clear focus on gender-specific education and housing quality standards, highlighting a commitment to promoting education and enhancing the overall quality of life for residents.

Comparing UWBI with Land Use and Land Cover

Furthermore, in order to gain a comprehensive understanding of the prevailing conditions on the ground, the calculated UWBI was juxtaposed with a Land Use and Land Cover (LULC) analysis derived from high-resolution 10-m Sentinel satellite imagery. This comparative analysis yielded intriguing insights, leading to the delineation of three distinct zones within the study area. Zone 1, situated in the northern part of the region, emerged as a notable focal point characterized by very high to high levels of UWB. This area exhibits a landscape that combines moderate to high levels of human settlement with a blend of natural vegetation and rangelands. Notably, Zone 1 encompasses locales such as Lebong, Nayabazar, and Mull Dhara within the Darjeeling town. These regions are geographically distant from the central urban area, experiencing comparatively lower population pressures and benefitting from a significant vegetative cover, both of which contribute to the higher UWB observed in this zone (Fig. 30.5).

Conversely, Zone 2 encapsulates the central business district (CBD) of Darjeeling town, encompassing areas like Chauk Bazar, Mall, and Manabari. In sharp contrast

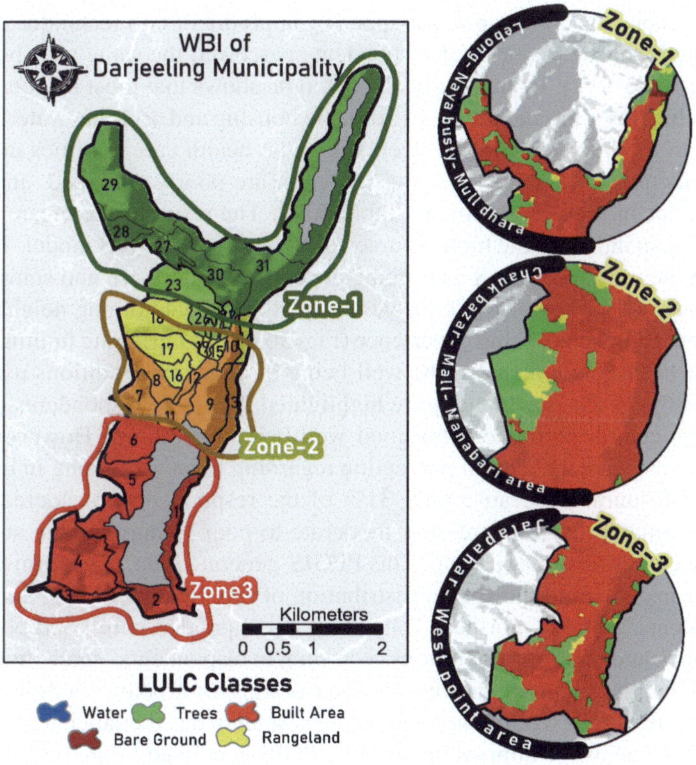

Fig. 30.5 Comparing UWBI with LULC prepared using Sentinel-10

to Zone 1, this region grapples with maximum population pressures, dense built-up infrastructure, and severe congestion. Consequently, the UWB in Zone 2 registers at moderate to low levels, reflecting the challenges posed by these urban characteristics on the overall well-being of its inhabitants. Zone 3, located in the southern part of the study area, is characterized by very low UWB. This zone includes localities such as Jalapahar and West Point, which share similarities with Zone 2 in terms of facing substantial population pressures and congestion. Moreover, Zone 3 exhibits low socio-economic indicators and a lack of essential amenities, all contributing factors to the notably poor UWB experienced in this area. In summary, the comparative assessment of UWBI and LULC has facilitated the demarcation of these three zones, each bearing distinct socio-economic, environmental, and urban development characteristics that significantly influence the well-being of the respective populations residing within them.

Validating Using PPGIS and Community Voices

In this research, the validation was done using urban residents' perceptions. The output of perceived well-being was mapped by applying PPGIS technique. The findings revealed that the majority of respondents perceived the same well-being level (Fig. 30.6, Table 30.2). Interestingly, the outcome shows that most respondents feel that Darjeeling town has the issue of affordable housing and drinking water facilities. Also, due to the hilly terrain and steep slope, the health care facilities and infrastructural amenities like roads and transportation are poorly equipped, and there is high scope for further development (Table 30.3). The result of the mean perceived well-being map indicates the high accuracy of the computed UWB model. The figure shows that the mean perceived well-being is poor in the northern and some extreme southern parts of the town, while the central part and the adjoining neighbourhood experienced a good well-being experience (Fig. 30.6). An intriguing finding from the analysis of PPGIS data revealed the well-being of zonation conditions using interpolation methods. The resultant figure highlighted that some respondents perceived Darjeeling's central part to exhibit good well-being conditions. However, despite this positive assessment, overall perception regarding town well-being in Darjeeling was not as favourable because only 31% of the respondents considered it good, with approximately 69% expressing moderate-to-poor ratings for the sustainable well-being condition of Darjeeling. This PPGIS-generated Fig. 30.6 furnished valuable visual insights into the spatial distribution of reported well-being experienced by respondents. Notably, northern Darjeeling corresponded with good perceptions surrounding standard living conditions according to respondents' perceptions. While the central and the southern part reveals deteriorating well-being conditions—highlighting specific areas within the town where residents faced challenges and identifying potential areas for improvement. The PPGIS-generated figure revealed spatial distribution and aligned well with the UWB model prepared using GWPCA analysis. The zones based on PPGIS delineated through the interpolation technique exhibited

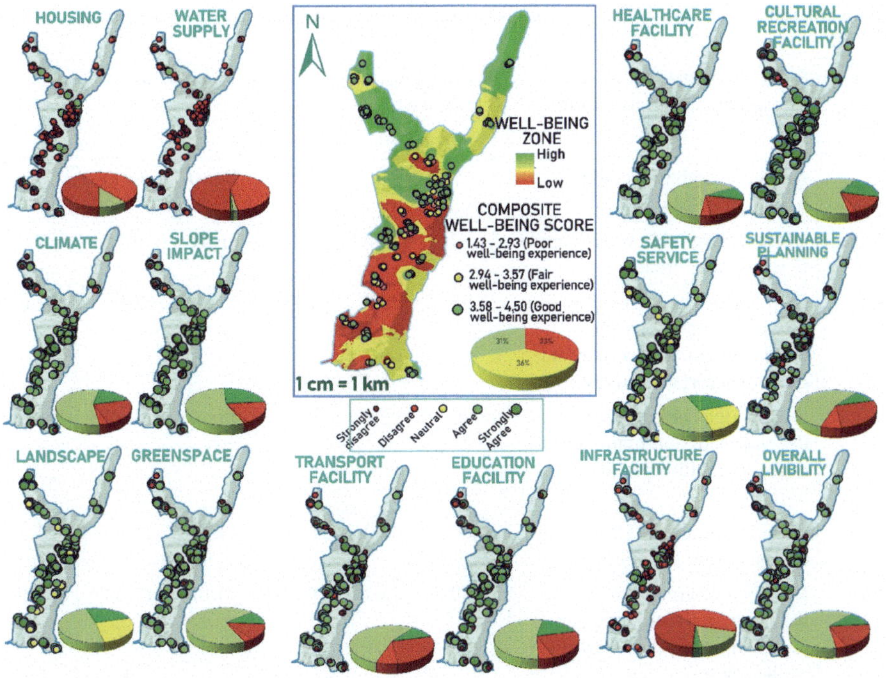

Fig. 30.6 Public participation through GIS (PPGIS) approach for UWB assessment. This map indicates variability of perceptions (from 1 = strongly disagree to 5 = strongly agree) related to likert-scale questions. The final map represents the composite well-being zone based on perception survey of residents

striking correspondence with three categories of well-being conditions (poor/fair/good). This match between subjective perceptions and objective measurements reinforces robustness & validity while lending credibility to findings & strengthening overall assessment.

All the 14-question related to perceived well-being experience corresponds with a significant chi-squared value, and Cronbach's alpha value of 0.797 reflects the high consistency in the response that established that our computed model has satisfactory explanatory power to the neighbourhood sustainable well-being condition in the study area (Table 30.2). The present study endeavours to capture the narratives of inhabitants of Darjeeling town from their experiential realities and perspectives through unstructured dialogues that took place during perception surveys. A few narratives are given in Table 30.3. The primary objective was to interact with local denizens to obtain invaluable insights into extant well-being predicaments within the region. These interactions presented a distinctive avenue for residents to articulate their views, apprehensions, and encounters pertaining to the standard of living in Darjeeling. The narratives from these exchanges illuminated various well-being challenges confronted by those residing within this locale.

Table 30.2 Likert based assessment of resident perceived urban well-being and goodness of fit (N = 238)

No.	Subjective question related to sustainable urban liveability	Residents perceptions on sustainable urban liveability (%)					χ^2
		1	2	3	4	5	
Q1	Adequate and affordable housing options	41.18	48.32	0.00	10.50	0.00	57.6
Q2	Clean and reliable water supply in residential areas	54.62	42.86	0.00	1.68	0.84	221.0
Q3	Hilly terrain of Darjeeling provides suitable living condition due to a cooler and more comfortable climate	23.95	2.94	0.00	55.04	18.07	137.0
Q4	Steep slopes affects the development and maintenance of infrastructure (roads, buildings)	18.49	1.26	0.00	55.46	24.79	146.0
Q5	Hilly landscape of Darjeeling offers natural beauty	0.00	0.00	26.05	48.32	25.63	24.1
Q6	Presence of green spaces, botanical gardens, zoos and parks enhances the liveability of the town	13.87	2.94	0.00	70.17	13.03	266.0
Q7	Transportation infrastructure in Darjeeling (e.g., roads, public transportation) is well-maintained and convenient to access essential amenities	27.31	7.14	0.00	54.62	10.92	133.0
Q8	Educational institutions in Darjeeling provide adequate opportunities for personal and professional development	23.53	5.04	0.00	53.36	18.07	119.0
Q9	Healthcare facilities in Darjeeling meet the needs of the residents	18.91	3.78	0.00	64.29	13.03	207.0
Q10	Cultural and recreational establishments adds to the vibrancy and liveliness	15.13	2.52	0.00	60.92	21.43	181.0
Q11	Safety and security measures contribute to a positive liveability	0.00	0.00	26.05	48.32	25.63	24.1
Q12	Urban planning and development prioritize environmental sustainability and preservation of natural resources	27.31	7.14	0.00	54.62	10.92	133.0

(continued)

Table 30.2 (continued)

No.	Subjective question related to sustainable urban liveability	Residents perceptions on sustainable urban liveability (%)					χ^2
		1	2	3	4	5	
Q13	Darjeeling town has reliable infrastructure facilities	38.24	40.76	0.00	18.07	2.94	91.2
Q14	I feel that Darjeeling Himalaya provides a sustainable and liveable urban environment	20.59	3.36	0.00	63.45	12.61	202.0

Scale Reliability Statistics: Mean = 3.17, SD = 0.635, Cronbach's $\alpha = 0.797$

Table 30.3 Few narratives of prevailing well-being challenges-insights from open-ended conversations in perception surveys with residents of Darjeeling town

Well-being challenges	Respondents narratives
Water availability	*Mr. A (72 years retired male)*—"The dire need for clean and readily available water is a persistent issue that we encounter on a daily basis. It's an uphill battle to ensure that the water we use is not only potable but also safe for consumption,….However, with every passing day, urban development continues to expand relentlessly, exacerbating the scarcity and making it even more challenging for us"
Lack of easy transportation	*Ms. B (22 years female student)*—"The accessibility of public transportation has emerged as a major issue in Darjeeling, one that puts undue pressure on our wallets due to the escalating costs of travel. This problem is especially pronounced in areas with varying altitudes where geographical barriers present additional challenges"
Fragility	*Mr. C (39 years male lawyer)*—"The scourge of illegal constructions in Darjeeling poses a grave threat to the delicate balance of our environment….the situation is further exacerbated by an incessant influx of hotels and tourist pressure, which only serves to worsen the issue at hand"
Pollution	*Mrs. D (45 years female hotel manager)*—"The beauty of our beloved town is besieged by a relentless onslaught of pollution, with some areas even surpassing safety limits for air quality. The root cause of this distressing situation can be attributed to the ceaseless urbanization and tourist pressure that plagues our community. A concerted effort must be made to prioritize regular cleaning and upkeep in order to restore the pristine beauty of our town"
Traffic congestion	*Mr. E (31 years male police)*—"Bustling traffic in Darjeeling is particularly pronounced along the mall road and its adjacent areas—Chauk Bazaar, Chowrasta Market, Lebong Road, to name a few. Unfortunately, this has resulted in an overwhelming sense of congestion and overcrowding that impedes the tranquility of our town. The relentless commotion amplifies noise levels and augments visual pollution to a great extent"

Conclusion

The study identifies a polycentric pattern in urban well-being (UWB) in Darjeeling. Northern and central areas show notably high UWB due to habitable housing, water access, education, and amenities. In contrast, the southern outskirts face low UWB, affecting healthcare and education. Targeted interventions are vital for improvement. The winning variable analysis underscores the role of water access, education, housing, and infrastructure in influencing well-being. Tailored interventions and comprehensive planning are needed to address disparities and enhance UWB across the town. The research identified three distinct zones within Darjeeling, each exhibiting varying levels of well-being. In Darjeeling, the study identifies three distinct zones with varying levels of urban well-being (UWB). Zone 1 in the north enjoys notably high UWB, striking a balance between human settlement and nature. Conversely, Zone 2, encompassing the central business district, faces congestion and moderate to low UWB, necessitating sustainable urban planning and infrastructure improvement. Zone 3 in the south exhibits very low UWB, attributed to population pressure and inadequate amenities, demanding urgent, targeted interventions to enhance residents' well-being. The validation process, incorporating Public Participation in Geographic Information Systems (PPGIS), has enriched our insights by integrating urban residents' perceptions. The alignment between subjective perceptions and objective measurements bolsters the robustness of our findings.

In light of the multifaceted assessment of urban well-being (UWB) in Darjeeling, several policy recommendations emerge to enhance the quality of life for its residents. Firstly, addressing the pressing issues of affordable housing and access to clean drinking water is paramount. Policy measures should be implemented to ensure housing affordability for diverse income groups and to guarantee equitable access to safe drinking water across the town. Secondly, recognizing the unique geographical challenges posed by Darjeeling's hilly terrain, targeted investments in healthcare infrastructure and improved road networks and transportation systems are essential to provide better access to healthcare services and enhance overall connectivity. In the central business district, characterized by congestion and high population density, prioritizing sustainable urbanization is crucial. Policies should encourage the creation of green spaces, pedestrian-friendly initiatives, and efficient public transportation systems to mitigate congestion and elevate the overall urban living experience.

Furthermore, fostering community engagement and participation in urban development initiatives is essential, as residents' perceptions and local knowledge are invaluable resources for informed decision-making. Lastly, a zone-specific approach is recommended to tailor interventions based on the well-being characteristics of each distinct zone. By implementing these policy recommendations, Darjeeling can enhance the well-being of its residents and pave the way for a more sustainable and inclusive urban future. The findings of this study are intrinsically linked with the assessment of urban well-being in Darjeeling with broader societal implications, aligning closely with the focal theme of the book. This research elucidates how urban planning and policy-making, informed by community-engaged methodologies, can significantly contribute to mitigating climate crisis effects, enhancing urban resilience, and promoting sustainability. It underscores the vital importance of incorporating local voices and experiences in creating adaptive strategies that address both immediate and long-term socio-ecological challenges. Thus, this chapter not only provides a nuanced understanding of urban well-being in the context of climate change but also advocates for a socio-ecological approach that prioritizes societal well-being and sustainable development, reflecting the book's overarching aim to explore and address the complex interrelations between climate crisis, social responses, and sustainability.

References

Ala-Mantila S, Heinonen J, Junnila S, Saarsalmi P (2018) Spatialnature of urban well-being. Reg Stud 52(7):959–973. https://doi.org/10.1080/00343404.2017.1360485

Alatartseva E, Barysheva G (2015) Well-being: subjective and objective aspects. Procedia-Soc Behav Sci 166:36–42

Balk D, Montgomery MR, Engin H, Lin N, Major E, Jones B (2019) Urbanization in India: population and urban classification grids for 2011. Data 4(1):35. https://doi.org/10.3390/data4010035

Benita F, Kalashnikov VV, Tunçer B (2020) A spatial livability index for dense urban centers. Environ Plann b: Urban Analytics City Sci 48(7):2006–2022. https://doi.org/10.1177/2399808320960151

Basak D, Bose A, Roy S, Chowdhury IR, Sarkar BC (2021) Understanding sustainable homestay tourism as a driving factor of tourist's satisfaction through structural equation modelling: a case of Darjeeling Himalayan region, India. Curr Res Environ Sustain 3:100098

Brown G, Kyttä M (2014) Key issues and research priorities for public participation GIS (PPGIS): a synthesis based on empirical research. Appl Geogr 46:122–136

Brown G, Weber D (2012) Measuring change in place values using public participation GIS (PPGIS). Appl Geogr 34:316–324

Budaev SV (2010) Using principal components and factor analysis in animal behaviour research: caveats and guidelines. Ethology 116(5):472–480

Cao H, Weiler L, Liu Z (2015) Sharing knowledge on PPGIS: engaging urban indigenous minorities in Western China; final technical report

Chen CW (2023) Can smart cities bring happiness to promote sustainable development? Contexts and clues of subjective well-being and urban livability. Dev Built Environ 13:100108. https://doi.org/10.1016/j.dibe.2022.100108

Clark AE, Frijters P, Shields MA (2008) Relative income, happiness, and utility: an explanation for the Easterlin paradox and other puzzles. J Econ Lit 46(1):95–144. https://doi.org/10.1257/jel.46.1.95

Dasgupta M, Das Gupta SN (2020) Urbanization in India in the globalized era: revisiting the energy aspect and policy issues. Energy, Environ Globalization: Recent Trends Opportunities Challenges India 59–94. https://doi.org/10.1007/978-981-13-9310-5_4

De Silva SSA, Meeker MA, Ryan V, Algur Y, Long DL, Carson AP, Black N, McClure LA (2022) Comparing global and spatial composite measures of neighborhood socioeconomic status across US counties. J Urban Health 99(3):457–468

Diener E, Wirtz D, Biswas-Diener R, Tov W, Kim-Prieto C, Choi DW, Oishi S (2009) New measures of well-being. In: Diener E (ed) Assessing well-being: the collected works of Ed Diener (Social Indicators Research Series No. 39). Springer, Amsterdam, pp 247–266. https://doi.org/10.1007/978-90-481-2354-412

Dodge R, Daly A, Huyton J, Sanders L (2012) The challenge of defining wellbeing. Int J Wellbeing 2:222–235
Douglas I (2012) Peri-urban ecosystems and societies: transitional zones and contrasting values. In: The periurban interface (pp 41–52). Routledge
Engen S, Runge C, Brown G, Fauchald P, Nilsen L, Hausner V (2018) Assessing local acceptance of protected area management using public participation GIS (PPGIS). J Nat Conserv 43:27–34
Florida R, Mellander C, Rentfrow PJ (2013) The happiness of cities. Reg Stud 47(4):613–627. https://doi.org/10.1080/00343404.2011.589830
Fotheringham AS, Brunsdon C, Charlton M (2002) Geographically weighted regression: the analysis of spatially varying relationships. http://www.dlfiles24.com/pdf60/168092213-geographically_weighted_regression_the_analysis_of_spatially_varying_relationships_chris_brunsdon.pdf
Guite HF, Clark C, Ackrill G (2006) The impact of the physical and urban environment on mental well-being. Public Health 120(12):1117–1126
Harris P, Brunsdon C, Charlton M (2011) Geographically weighted principal components analysis. Int J Geogr Inf Sci 25(10):1717–1736. https://doi.org/10.1080/13658816.2011.554838
Harris P, Clarke A, Juggins S, Brunsdon C, Charlton M (2015) Enhancements to a geographically weighted principal component analysis in the context of an application to an environmental data set. Geogr Anal 47(2):146–172
Kahila-Tani M (2015) Reshaping the planning process using local experiences: utilising PPGIS in participatory urban planning
Kotzebue JR (2022) Integrated urban transport infrastructure development: the role of digital social geocommunication in Hamburg's TEN-T improvement. J Transp Geogr 99:103280
Krefis AC, Augustin M, Schlünzen KH, Oßenbrügge J, Augustin J (2018) How does the urban environment affect health and well-being? A systematic review. Urban Sci 2(1):21. https://doi.org/10.3390/urbansci2010021
Lafortezza R, Sanesi G (2019) Nature-based solutions: settling the issue of sustainable urbanization. Environ Res 172:394–398. https://doi.org/10.1016/j.envres.2018.12.063
Lehto C, Hedblom M, Öckinger E, Ranius T (2022) Landscape usage by recreationists is shaped by availability: insights from a national PPGIS survey in Sweden. Landscape Urban Plann 227:104519
Mishra SV (2018) Urban deprivation in a global south city-a neighborhood scale study of Kolkata, India. Habitat Int 80:1–10. https://doi.org/10.1016/j.habitatint.2018.08.006
OECD (2008) Handbook on constructing composite indicators. In: Methodology and user guide. OECD Publications, Paris
Oswald AJ, Wu S (2010) Objective confirmation of subjective measures of human well-being: evidence from the USA. Science 327(5965):576–579. https://doi.org/10.1126/science.1180606
Pareto A (2021) Normalization (of an Indicator) in composite indicator construction. In: Maggino F (eds) Encyclopedia of quality of life and well-being research. Springer, Cham. https://doi.org/10.1007/978-3-319-69909-7_104641-1
Ripp J, Peccoralo L, Charney D (2020) Attending to the emotional well-being of the health care workforce in a New York City health system during the COVID-19 pandemic. Acad Med. https://doi.org/10.1097/ACM.0000000000003414
Roy S, Bose A, Majumadar S, Chowdhury IR, Abdo HG, Almohamad H, Dughairi AAA (2022) Evaluating urban environment quality (UEQ) for class-I Indian city: an integrated RS-GIS based exploratory spatial analysis. Geocarto Int. https://doi.org/10.1080/10106049.2022.2153932
Roy S, Majumder S, Bose A, Chowdhury IR (2023a) GWPCA-based spatial analysis of urban vitality: a comparative assessment of three high-altitude Himalayan towns in India. J Spat Sci. https://doi.org/10.1080/14498596.2023.2267011
Roy S, Majumder S, Bose A, Roy Chowdhury I (2023b) Does geographical heterogeneity influence urban quality of life? A case of densely populated Indian city. Papers Appl Geogr 1–29
Roy S, Majumder S, Bose A, Chowdhury IR (2024) Spatial heterogeneity in the urban household living conditions: A-GIS-based spatial analysis. Ann GIS 1–24

Schwanen T, Atkinson S (2015) Geographies of wellbeing: an introduction. Geogr J 181(2):98–101. https://doi.org/10.1111/geoj.12132

Sekhar C, Indrayan A, Gupta SM (1991) Development of an index of need for health resources for Indian states using factor analysis. Int J Epidemiol 20(1):246–250

Sharma SK, Choudhary N, Kotnala G, Das D, Mukherjee S, Ghosh A, ... Mandal TK (2020) Wintertime carbonaceous species and trace metals in PM10 in Darjeeling: a high altitude town in the eastern Himalayas. Urban Clim 34:100668

Sinesilassie EG, Tabish SZS, Jha KN (2018) Critical factors affecting cost performance: a case of Ethiopian public construction projects. Int J Constr Manag 18(2):108–119

Smyth R, Mishra V, Qian X (2008) The environment and well-being in urban China. Ecol Econ 68(1–2):547–555. https://doi.org/10.1016/j.ecolecon.2008.05.017

Tsutsumida N, Murakami D, Yoshida T, Nakaya T, Lu B, Harris P (2019) Geographically weighted non-negative principal components analysis for exploring spatial variation in multidimensional composite index. GeoComputation

United Nations (2014) Department of economic and social affairs, population division. World urbanization prospects: the 2014 revision, CD-ROM edition; United Nations, Department of Economic and Social Affairs, Population Division: New York, NY, USA

Vaidya H, Chatterji T (2020) SDG 11 sustainable cities and communities. In: Franco IB, Chatterji T, Derbyshire E, Tracey J (eds) Actioning the global goals for local impact: towards sustainability science, policy, education and practice. Springer Singapore, pp 173–185. https://doi.org/10.1007/978-981-32-9927-6_12

Wang Q, Jiang D, Gao Y, Zhang Z, Chang Q (2022) Examining the driving factors of SOM using a multi-scale GWR model augmented by geo-detector and GWPCA analysis. Agronomy 12(7):1697. https://doi.org/10.3390/agronomy12071697

Waleghwa B, Heldt T (2022) Exploring the use of public participation GIS in transportation planning for tourism at a Nordic destination. Scand J Hospitality Tourism 22(3):210–234

Whulanza Y, Kusrini E (2023) Defining healthy city and its influence on urban well-being. Int J Technol 14(5):948–953. https://doi.org/10.14716/ijtech.v14i5.6578

Yadav P (2023) Urbanization in India: growth and its challenges. J Econ, Tourism Serv 2(8):9–22. http://jets.innovascience.uz/index.php/jets/article/view/80

The manufacturer's authorised representative in the EU is Springer
Nature Customer Service Centre GmbH, Europaplatz 3, 69115 Heidelberg,
Germany. If you have any concerns regarding our products, please
contact ProductSafety@springernature.com

Printed and bound by CPI Group (UK) Ltd, Croydon, CR0 4YY
01/10/2026
02225864-0020